To forget how to dig the earth and to tend the soil
is to forget ourselves.
~ Mahatma Gandhi

This garden book is our personal attempt to help our planet's beautiful landscape survive. It's our belief that when you create a home garden you reconnect with our surrounding valuable and fragile nature.

Now, do your part.
Ralph Trout & Dr. Claire Hu

When tillage begins, other arts follow. The farmers, therefore, are the founders of human civilization. ~ Daniel Webster

Agriculture is the most healthful, most useful and most noble employment of man. ~ George Washington

The home garden can be your sanctuary, tension reliever, and the best location for exercise in the fresh air and sunshine. Your garden will provide a direct connection with the elements of our planet's nature, and can help to reconnect with your family and friends. Gardening isn't easy: it requires discipline and scheduling. It is the best learning experience available to comprehend the cycle of life. Consider your garden a continual experiment. Make gardening a hobby to pleasantly and productively fill the hours and your dinner plate during retirement. Whether the garden is big or a few pots on your patio, it will satisfy all your senses. Try it; you'll enjoy time well spent. This book provides quality, tried and true information to begin a new cycle of your life.

The ultimate goal of farming is not the growing of crops, but the cultivation and perfection of human beings. ~ Masanobu Fukuoka

THE HAWAIIAN HOME GARDEN GUIDE

BY

RALPH TROUT

and

Dr. CLAIRE HU

LIBERTAD PUBLISHING LLC

 All photos and drawings were provided by the authors, 912crf.com, and R. Douglas

Published by
Libertad Publishing, LLC
Libertadpublishingllc@gmail.com

ISBN: 978-0-9992239-9-4

ADVISEMENT & DISCLAIMER

Readers are urged to take all necessary precautions before undertaking any how-to task. Always read and follow instructions and safety warnings for all tools and materials. Call in a professional if the task is beyond your abilities.

Neither the publisher nor the author are responsible for accidents, injuries, or damage incurred as a result of tasks undertaken by readers. This book is not a substitute for professional services.

All material in this book has been researched and compiled from reliable sourc-es and is provided for your information only. No action or inaction should be taken based solely on the contents of the information. Readers should consult appropriate health professionals on any matter relating to their health and well-being. DO NOT consider any of this information to be medical advice or instruction. No action or inac-tion should be taken based solely on the information in this book. Readers, please consult health profes-sionals on any matter pertaining to your health and well-being.

THE HAWAIIAN HOME GARDEN GUIDE will not take any responsibility for any adverse effects from the use of plants. Always seek advice from a professional before using a plant medicinally. The use and dosage of herbs and foods vary with cir-cumstances such as climate, the person's health and age, and the manner the dose is administered. Herbs and foodstuffs should be used for medical purposes only with proper medical advice. The medicinal plants must be grown with-out chemical fertilizers or pesticides.

This book is dedicated to my parents who taught me to garden before I knew I loved it.

THE HAWAIIAN HOME GARDEN GUIDE

ACKNOWLEDGMENTS

I began this book several years ago after realizing there was no comprehensive tropical gardening handbook. Initially, I gathered information came from acquaintances, neighbors, and friends. Usually, useful information, proven remedies, only came when a garden disaster struck. When luck smiled and I had good, healthy crops, I never bothered to ask for suggestions or advice.

There are many free courses available through various government agricultural departments. I advanced my knowledge with several free courses, including: safe use of garden chemicals, grow box construction, and every available course on specific fruits and vegetables. Many locations have government sponsored agricultural laboratories to perform soil tests and analyze plant and tree diseases. Government agriculture field officers have a wealth of information.

Many friends advised with common sense gardening techniques to gain the most production from my garden labor. Most important, I learned to love the garden whether in heat or mud. The process of growing my own food led me to other questions and to discover more answers.

The Internet and especially Mr. Google has been the best information source for most of my crops. If you can correctly phrase the question, somewhere the Internet has the answer. Trying to ascertain the correct name for a local fruit, vegetable, or root will always provide work for an amateur garden detective. Local districts, villages, and even families have their own gardening vocabulary and terms, and especially crop names. When inquiring via the Internet, first find the scientific-botanical reference name for that particular fruit or vegetable. This may be difficult as a few have various scientific names, but you will usually succeed.

Most of my Internet gardening questions were answered by the Purdue University Center For New Crop and Plant Products, *http://www.hort.purdue.edu/newcrop/*. If you can ask the food crop question with reasonable skill, there is a good probability that this site has the existing research information. Several international universities are accessible for regional agricultural info. Since areas with similar climates, latitude, and elevations produce similar crops, Universities throughout the Caribbean, Florida, and Hawaii are valuable sources of proven agri-knowledge. There are several web sites dedicated to tropical gardening. Please refer to Sources at the end of this book.

I have lived and gardened in several tropical areas: Caribbean, Florida, Hawaii, Sri Lan-ka, and Southeast Asia. All my tropical gardens provided tasty, fresh foods, but also provided valuable gardening information that I share with you, my readers.

Many friends in Hawaii, Trinidad, the USVI, and Cuba helped and contributed information for this book.

Mike Sine, Dan Boyle, and Caribbean Sally were the critical proofreaders. Tom and Sandy assisted. Dr. Claire Hu helped with all the graphics, photography, and inspiration.

GOOD LUCK AND GOOD GARDENING.

INTRODUCTION
TO THE
HAWAIIAN HOME GARDEN GUIDE

The Hawaiian Home Garden Guide does not promise gardening success. To achieve with agriculture, even in the home garden, you must invest whole-hearted enthusiasm. A garden is a work of love: the love of nature, love of family, and love of good healthy food, but the garden must also feel loved. Gardening then becomes an art form, and true love of the garden will bring knowledge and success.

Garden love grows as the garden matures and vegetables succeed. The garden inhales your perspiration from labor and exhales positive energy with beautiful crops. Take the time to slow your life, enjoy being outdoors, and get healthier through exercise working in the home plot with family and friends, completely knowing the food you consume, and learning the cycles of nature.

We are adamant about the resurgence of the home garden. Families and countries need to become more self-sufficient, provide for themselves, and reduce dependency. The mod-ern world is separating the person from nature, from natural food, and separating the family structure. Time seems to be the main concern of everyone. Everywhere is a clock, on our phones, TVs, and microwaves; and time is quickly passing. Everyone speaks about 'quality time,' spending time together, or wasting time. The home garden provides all the best for the time well spent.

Whether your garden space is large, or just a few containers, watching plants ma-ture into food that feeds the human engine is a miracle of life. With gardens there are trials, and sometimes no great success, but never errors. With each attempt, a lesson is learned and that is the profit. Now it appears our world is fragile, and we must as people, nations, and a species reconnect with nature and conserve what we have for future generations. The home garden is not only an investment for you and your family, but also an investment for the family and future of our world. Invest quality time in your world, your nation, your island, and nature.

When the sun rises, I go to work. When the sun goes down, I take my rest,
I dig the well from which I drink, I farm the soil which yields my food,
I share creation, Kings can do no more. ~ Chinese Proverb, 2500 B.C.

FOOD SECURITY BEGINS IN THE HOME GARDEN

Since ancient times, the Hawaiian Islands have been rich with food and agriculture. Both have been imperative for survival economically and socially. The Hawaiian Islands are blessed with a splendid climate, enough freshwater, and elevations that could produce various edible crops. Food security in the 21st century is a crucial part of independence and personal freedom. Every citizen and visitor must respect the environment. The home garden will increase personal knowledge of the effort necessary to grow your own. Whether it is only fresh herbs, or leafy veggies, roots, and fruits, you grew it with patience and love.

Every country, province, community, and person should become garden-smart and strive to be self-sufficient, reduce waste through composting, and clean our precious and fragile personal environment. As a country of individuals, the home garden builds respect for the ag-riculture systems that supply our food. To become food secure is to become more independent with more food available without destroying more natural forests. It is necessary to utilize ev-ery possible existing cleared space to grow food. These public and commercial garden spaces will be centers for education on living better with nature. Annually, together, the Hawaiian Islands produces more than a million tons each of fruits and vegetables.

The ancient tribal chiefs realized the importance of food security and availability to keep the population content with a healthy national spirit. Napoleon remarked that 'To be effective, an army relies on good and plentiful food.' Today that army is every citizen. To be healthy and happy, every island/state/country must grow adequate supplies for a nutritious, inexpensive diet. What is grown at home is not only food, but free lessons on how every family unit can be more compatible with nature and help not harm the fragile eco-systems. Producing locally, whether in the backyard or community gardens, means fresher foods. The Hawaiian's agricultural perspective is transitioning from the sugar cane and pineapple fields to specific cash crop gardens designed for the export market.

If food is grown locally, the transport costs are reduced, meaning it will be less expen-sive. Less storage will be necessary, less refrigeration, fewer preservatives, less money going to other countries for importing foodstuffs. Less food exchange is crucial for better foreign exchange. Home and locally grown food is more accessible and affordable.

Everyone has personal food preferences. When you grow your own, it's easier to eat what you like. If you produce an excess, you can barter, or trade, with friends and neighbors. A family can experiment with various crops they might not easily locate in the markets.

Utilize what you grow to provide for a healthier diet. Educate yourself and family members about the nutritional value of each vegetable, fruit, and root. Eating locally grown will also produce new friendships and bonds. Get the expensive middleman out of the agricul-ture picture. The local food should be safer with fewer chemicals during production, with less processing, and packaging.

Everywhere in the Hawaiian Islands is a garden waiting to be planted. Each person who plants will quickly learn immense respect for the traditional farmer. With urbanization, the love of nature is becoming lost. Whether it's chili peppers growing with pandan in a clay pot on the balcony or patio, a series of grow boxes filled with greens, herbs, and beans, or yam vines twining around moringa trees, grow something! Enjoy the exercise of becoming more food secure. Food safety and security can be achieved when we all start a home garden.

Great Hawaiian chiefs, the ali'i, knew that great gardens were equal to armaments, necessary to protect their territories. We are all chiefs of our homes and can have a magnificent home garden.

CONTENTS

THE GARDEN

Page:

STATEMENT FROM THE AUTHOR

If this book influences only one reader to grow some healthy, nutritious food, my work has been a success. If this book helps a family enjoy better health through wiser nutritional choices, the book's been a triumph. If a few people, friends, relatives, or strangers work together on a garden project, I am overjoyed. The garden, large or small, not the supermarket, is where our food begins. Farmers are crucial to every society, yet a heavily criticized and disrespected workforce. There's a saying, *'A few times during your life you'll need a doctor, lawyer, or policeman, but three times each day you need a farmer.'* As a society, nations, and world, we need to relearn how to grow good chemical-free, nutritious food in home gardens and be more self-sufficient rather than dependent.

Today's media-packed world has separated people from their food sources, nature, and even their neighbors. Watch a vegetable seed sprout and grow in a ceramic pot on your porch, or enjoy a small garden, and you'll witness one of nature's wonders at its best. Our natural, native environment means everything, and it's slipping away. A home garden will help people, families, and neighbors reconnect with the basic issues of life, growing food and working together. In the 80s, I had the opportunity to work in the Hawaii. This book is meant to honor all my friends on many islands. Also, it is my effort to help readers create a new bond with Mother Earth.

Most islands looked to tourism, which is importing people to pay them for warm weather, sunshine, and beautiful views. While importing business, local people forgot to build a strong base for a constant food supply, and decided to also import their food. Islands forgot the need to be self-sufficient. The islands now want to go green and save our natural beautiful environment. Creating a love for the home garden is the best, least expensive, most self-satisfying method available to save Hawaii's environment. Get involved with nature while you grow your own food.

From your rural and residential front and back yards to condo balconies and window ledges, there's a space to grow something. Plant a tree for fruit and you also can use it to tie your ham-mock to relax in its shade. You can transfer the gardening love that grows inside you, to your friends and family. The love of the garden will grow everyone closer to nature and our fragile environment.

Yes, the print and the photos are small. That's because this book rapidly expanded. Between these covers are many fruits, vegetables and nuts because everything grows in the rich fertile soil and excellent climate of the Hawaiian Islands. There's plenty of basic, common sense information, but it remains to the readers' efforts to make it useful within their own lives and gardens. Every part of your existence is enriched as your garden grows.

Each person must do their part and share the responsibility and labor to their island homeland food secure. Work together, pull in one positive direction. Communities can create compost areas for wet refuse. Recycle everything. Teach the imprisoned to garden and it will help recycle people. Villages can compete for the best, and biggest harvests. Agriculture and community garden fairs will increase tourism and national pride. Enjoy your home garden.

As a person, community, and nation – work through the home garden to learn how to become more self-sufficient and food secure.

The single greatest lesson the garden teaches is that our relationship to the planet need not be zero-sum, and that as long as the sun still shines and people still can plan and plant, think and do, we can, if we bother to try, find ways to provide for ourselves without diminishing the world. ~ Michael Pollan

THE HAWAIIAN HOME GARDEN GUIDE

I'd love to see a new form of social security ... everyone taught how to grow their own; fruit and nut trees planted along every street, parks planted out to edibles, every high rise with a roof garden, every school with at least one fruit tree for every kid enrolled. ~ Jackie French

CHAPTER ONE - REQUIREMENTS & COMMITMENTS

At one time or another most residents and tourists have come in close contact with a garden on one of the beautiful Hawaiian Islands. That may sound odd, but as a culture we have been moving away from agriculture. In just the past few years, many beautiful plots of tended crops have been replaced by roads, large homes, and buildings. It wasn't many years ago, pineapple was the king crop and sprawling fields were visible everywhere. Now, it's possible to exist isolated from the island environment in window-tinted, climate-controlled condos, offices, stores, and automobiles. Perhaps, we believe we have evolved beyond the manual labor necessary to grow our own food. Yes, gardening involves some hard work, but if performed correctly, gardening can be relatively easy, belly filling, and soul fulfilling. Hawaii farmers produce about $18 million in fruit and vegetables. Oahu produces over 55%, The Big Island 20 %, Maui 20% Kauai 3%.

This type of home plot is termed, 'the Hawaiian Home Garden' because so much has changed so quickly in today's world. In a flash, food shortages are in every corner of the globe, inflating costs and emptying our pockets. Middlemen in the market, not the farmers, are making the real profits from agriculture. Bandits steal garden produce where there are no watchmen. Food prices are increasing weekly. This is the new age of self-sufficiency, make your garden at home. The Hawaiian Islands are blessed with an excellent climate to grow a multitude of different crops. The Hawaiian Islands have plenty of fertile land, water, and energy to be a food basket for ourselves and the world. All it takes is a firm commitment to grow our own.

We've all seen photos of fields with straight rows, or neatly staked crops. Gardening is also an art form. As with blooming flowers or trimmed shrubbery in a landscaped yard, a home vegetable garden can be a rewarding and satisfying hobby. There may be some toil and expense in the initial phases, but by the end the self-satisfaction and the produce will be well worth every drop of sweat.

COMMITMENT

You must be dedicated to a Home Garden. Don't begin if you can't finish.

First, before starting, is the commitment and dedication to create a garden.

PHYSICAL CONDITION: Are you physically able to do the work? Almost anyone young and old can tend some type of garden. Be honest about what you can do easily without stressing yourself. Work only in the cooler mornings and evenings. Don't overdo it and drink plenty of water. In a short few weeks, you'll notice as your garden begins to grow, your stami-na will also increase.

MONEY: Can you afford the initial investment for tools, materials, seeds or seedlings, fertilizers, etc.? A garden is an investment in both time and money. Investments take budgeting and maybe some sacrifice of fun money. With good investments there are profits. Your better health and fresh home-grown food will be some of those profits.

TIME: Can you afford, or make the time necessary to daily tend your garden? Depending on the size of your garden, the time can be minutes or hours. Even if you are financially strapped, working two jobs, you can still plant a few seasonings, peppers, or tomatoes in pots and water as you come and go. As your passion for agriculture develops, (and it will) pride of your gardening achievement will make more time for the satisfying, stress relieving gardening.

TRANSPORT: Can you find adequate and reasonable transport to get the necessary materials? Surely someone you know or can hire has an appropriate vehicle and is willing to help you. This is an immense part of the gardening equation: getting the materials you need reasonably delivered to your site.

You need to answer these questions honestly. Do you really want to cut your home food costs, grow safer vegetables with less hazardous chemicals, get physical exercise, and share this garden project spending quality time with other family members?

SPACE: As part of your commitment, consider the size garden you and your helpers can handle. More help means less individual work, but the produce must be shared. More help also means the commitment of every individual involved. I suggest a garden's labors be shared among family, or housemates. Sometimes the more people involved tend to cause unnecessary confusion. If you have committed to a small, perhaps eight by eight-foot garden, then little extra help is needed. You also need to find adequate space for the gardening tools and supplies. Agri-chemicals must be stored in a secure place out of reach of children and pets.

Think of every way you can save money and time in this garden project. Simplify it as much as possible or it may beat you down before you begin. Can you borrow or get tools from someone or a relative who is either elderly, or no longer in the mode or mood to garden. Check house-garage sales in the newspapers for those who might have garden tools. You will need to budget some money weekly, and that's difficult in these times of inflating prices, until you have enough to begin. You may have to barter some of your future vegetable produce for tools or transport now. You will also need to budget your time, trading relaxing for more exercise in the garden. Every day something gets accomplished in the garden.

SPACE

How much room you have for a garden dictates the expense of time and money. There is always someplace to grow a few veggies. Start at your immediate home. Even if the entire lot is concrete, you can build a partition to make a box or find suitable containers that can be filled with dirt. Barricade a corner, or an entire wall used so materials will be less. Don't overestimate your energy, time, or money. Start with a small space and progress as your love for for gardening grows. It doesn't have to begin pretty or built of new materials.

Historically, Hawaiian gardens have been away from the home separated because it was considered a workplace and usually unattractive or smelly due to chemicals and fertilizers. Somehow farming in Hawaii became considered a 'lesser'- almost demeaning occupation to keep hidden. However. it was farming that paid for many educations of lawyers, doctors, and engineers. Today's home plot can be productive and appealing. Modern landscape architects are building vegetable gardens as visible parts of posh estates.

You will need some basic things for your plot. Sun is ultimately necessary for all crops. Select a space that collects a minimum of six hours of direct sun every day, but ten hours is the best. The amount of sunshine will dictate what crops will grow best. Leafy veggies like lettuce, bok choy, or broccoli will survive with less sun.

Soil, or a growing medium, is the next crucial element. Soil - good old dirt - is being replaced with other materials like bagasse (crushed sugar cane, rice husks, or compost. *(More on growing mediums in Chapter 12.)* Great soil is not necessary, but it must be fertile. Sand and gravel can grow some things, yet isn't the best. Hard clay or stony soil won't make a great garden. Your growing medium (soil must be able to be softened with a fork, and/or a hoe. Most important – it must drain away excess water, especially during the rainy season. If you have transport seek out good soil, but some manure, limestone, and a bit of sweat will improve almost any backyard plot.

Grow boxes, framed plots, or raised beds can be developed almost anywhere. *(More on grow box farming in Chapter 2.)* Raised boxes were developed to drain off excess water especially in the rainy season, and to provide better, easier soil to work. If you are on a hillside, try terracing by staking boards across it and filling behind them with dirt or pulling a bit of the hill down. Make your rows cut across the slope like stairs.

Water is an absolute necessity, whether it is from a tap, hose, barrel, or river. A nearby water supply reduces your labor. It must also be constant. During the hot dry season, a small garden can drain a fifty-gallon barrel in a few days. Other methods of watering are a watering can, bucket, sprinkler, or soaker hose. It must be adequate, and gentle so it doesn't erode the roots, knock off blossoms, or over soak the soil. Water should be applied just before the heat of the day and not later than three in the afternoon. This timing gives the plants a good drink when they need it, during the heat of the day, and doesn't leave the soil damp for fungus to develop. Keeping the garden watered properly is a problem we day workers have with the home garden and this means watering early in the morning before pursuing the almighty paycheck. *(More on water in Chapter 5.)*

The closer the garden is to your home the better. Consider driving to your garden, with today's traffic; you'd get there late and stressed. Build your garden close to your home and you will find it an easy satisfying hobby. Garden watching, watering, and weeding will become a healthy pastime. For cooking closeness is a big plus; just walk and pick dinner.

Most house lots are fifty feet by one hundred feet. If you use the length for a garden three or four feet wide you will have a three to four-hundred-square-foot garden just outside your door. You want to make it wide enough to permit easy access to the plants at the rear without stepping into the bed. You also want to be able to pass easily along it. If you do the width of your lot, it will amount to half the size. A garden the length of your lot will feed your entire family and a lot of friends. If you don't have friends already, your garden will attract them.

Again, don't go overboard with the garden at the beginning. Years ago, I took a grow box course at an agricultural center and built a small bed – about ten by ten. That small beginner garden still feeds two families and many friends. It took two afternoons to build from old concrete blocks in the rear corner of our lot. Chayote, sweet, hot, and seasoning peppers, lettuce, culantro, chives, and several spices - plus blooming flowers are being harvested.

Good drainage is necessary for water otherwise your garden will be soggy, and the surrounding area will be muddy. This is not what you want. Damp soil will cause serious fungus problems. Stones or aggregate placed at the very bottom beneath the soil of your garden will permit the water to drain. The garden should have good air circulation to keep fungus and disease to a minimum.

Security. Will yard fowl ruin your efforts? Does your dog like to sleep on the cool, tilled soil where you plant? Is your proposed plot possibly going to feed strangers who pick the fruits of your efforts in your absence? Again, keep your garden where you can see it.

PLAN FIRST TO SAVE LABOR LATER

There are few 'perfect' places for gardens, but make do. Once your space has been chosen, the next decision is what to grow. Plan first to save on labor later. Try to grow at least half of what your family needs. A kitchen garden usually contains cilantro, culantro, celery, lettuce, bok choy, okra, chives, beans, tomatoes, and both sweet and hot, peppers. All of these are relatively easy to grow. Make a drawing of your plot and place your veggies so they are not crowded. This will keep you from buying more plants or seeds than needed. Planting haphazardly will waste time, money, energy, and deflate a positive attitude. It is important tall veggies like eggplant don't block the sun from smaller plants like bok choy or tomatoes. Tomatoes and peppers will waste energy growing to the sun rather than making the desired fruit. Also, separate crops that produce longer than others. Cilantro leaves can be snipped with scissors and almost grow forever. Quick-bearing crops, like bok choy or beans, should be the most accessible as they will need to be replanted. Allow a small area for a seed-bed where you sprout the plants from seeds.

You can bury any number of headaches in the garden. ~ Charles Barnard (1889)

To be the most productive, your garden should always be fully planted. Change the veggies you plant to confuse the insects and diseases. An eight-foot by eight-foot garden, or slightly bigger, is a perfect size 'starter' plot for a small family. If this goes well, you can expand to a larger, or an additional plot for larger vegetables that need considerable space like pumpkin, squash, corn, or melons. Don't over plant or crowd your vegetables. This is counterproductive. It might seem like you should get more, but you will get less. The real trick is to have your seedbed working so you have new plants at the ready when a set is getting old. Good garden timing comes with practice.

You'll need some garden clothes and maybe a widebrimmed hat not only for the sun, but to get into the 'farmer look.' Depending on your career, you may require gloves for 'soft' hands. If you go into gardening in a bigger way, boots are a requirement. Not many tools are needed for a small plot. Some tools, like a fork (big or small), shovel, and wheelbarrow might be borrowed for the original construction of your plot. A small hand fork, trowel, cutlass, or a hoe, a hose, water barrel, watering can, and a spray can are necessities. Don't buy cheap tools, as they won't last. Spend for quality and it will be dependable for several years. Keep your tools clean and sharpened; it will reduce your work. You will need some light rope or string and some sticks or bamboo to tie up your pepper and tomato trees to keep them off the dirt and conserve space.

As your garden expands, the need for water will increase. Irrigation is a convenience for larger plots. Piped water through a spray nozzle requires a further investment, but again drastically reduces the labor of carrying heavy water buckets. Soaker hoses are now available at various agri shops and hardware stores. Carefully place the hose between your rows of veggies. Connected to a tap, it constantly provides a light spray through many tiny holes if it is turned up. Turned down it soaks where it lies. It is recommended to fill your water barrels or buckets at least eight hours before your water your garden so the chlorine will evaporate. Remember, mosquitoes lay their eggs in pools of water so keep your containers covered. Don't over water as this causes fungus and rots the roots of your vegetables.

BE CREATIVE

Part of the fun of new age gardening is scavenging or locating the materials you need to build your plot. Broken blocks, rocks, chunks of concrete, old galvanize roofing, boards, buckets, and even old auto tires can be used and tastefully arranged into a proud statement of your energy. Leaky plastic buckets, old wash tubs, or oil kegs can be filled with soil and plant-ed. These are perfect for chives, celery, lettuce, or bok choy. With a bit of paint, they'll look great. Add a few bright flowers like marigolds or zinnias and it will look fantastic. *(More on container growing in Chapter 2.)*

Involve the family in the search and gaining of materials, and the actual building of the family plot. Perhaps it will assist in bonding the family together in these tough times. Children and elderly relatives can assist in almost every facet of the garden from planting to harvest and learning to cook with fresh vegetables. Once the plants begin to bear, everyone learns the success and failures full-time farmers face every day. As much as your garden feeds your belly, it will also teach lessons valuable for an entire lifetime. Everyone will learn about the four vital elements of growth: sun, wind, water, and earth. (*More on the family garden in Chapter 22.)*

Gardening teaches responsibility and scheduling. Plot chores like watering or weeding can be scheduled by the week to various family members. Certain plants can be assigned to individuals and judged to who grows the best. Each involved learns that the plot's success de-pends on everyone doing their part on time and responsibly. A home garden will demonstrate where our food comes from, how dependent man as a species is to his planet, and how fragile nature has become.

THE GARDEN JOURNAL

Before starting your home garden plot acquire a notebook and begin your garden journal / diary. This seems like schoolwork, but it will take only minutes a day and save you so much time and expense in the future. Mark when you plant so you know when to expect the produce. Write the phase of the moon when you planted a specific vegetable. Many farmers swear by planting by a waning or waxing Moon.

(More on the Moon and farming in Chapter 14.)

Enter everything that works and doesn't. Include all expenses to later tally and see if your efforts really saved money. Keep all addresses and phone numbers of transport and material sources. Write down the garden schedule for watering, weeding, fertilizing, etc. The world is embarking on a new era of the haves and have-nots.

According to the news, the world has many more 'nots.' With some energy and slight expense, grow a home garden and become one of the haves.

Nature does not hurry yet everything is accomplished. ~ Unknown

BASIC GARDEN TOOLS

Below: big fork and hand fork, hoe, cutlass, bucket, watering can, hose with spray nozzle. Spray can and sacks.

Once in your lifetime, you need a doctor, lawyer, policeman, or priest, but every day, three times a day, you need a farmer. ~ Anonymous.

CHAPTER TWO
CONTAINER GARDENS

Deciding to grow at least some of your vegetables requires organizing several items. Before you start locating tools and getting recipes, you need to decide what you can grow, and where you can grow it. Shop at home. Look around your home for what you can improvise into a garden.

Container gardens are raised gardens of various sizes with improved soils. I've seen beautiful tomatoes grown in paint buckets, thriving celery in over-used cooking pots, house spouting/gutters filled with parsley, and amazing peppers grown in plastic bags. A lot of people fill old auto tires to grow a single vegetable. Styrotex grape/fish boxes are easy to find and work well. Cut a leaky blue plastic barrel in half. Small to large, there's something you can fill with dirt, add seeds, water, leave in the sun, and grow something good to eat.

GROWING MEDIUMS

Potting soil, what most people call 'peat moss' is probably the best growing medium for small containers. A bag fills a lot of pots. Some agri stores sell bags of a type of compost. If you cannot locate rich soil, the garden's growing medium will be one of the biggest ex-penses. The good side is, if you take good care of the soil, feed it, and balance the acidity, it will feed you.

Dirt, soil, good old earth is the basic garden medium. Medium translates into a type of agent for growth. Different mediums for gardens are used because of unique features. A truckload of good topsoil doesn't go far and is awfully expensive. Rotted, mashed sugar cane is a good choice if you can find it. Rice husks are priced about the same. All growing mediums have the same issue of controlling moisture, insects, and weeds. Treat the medium with a pre-emergent herbicide, and an insecticide when you are first turning it into your container or plot. It will save time and money later. In a few weeks, the effects of the chemicals will be gone, and the fruits of your garden shouldn't be affected by the chemicals.

SOIL PH

Measuring if a soil medium is acid or alkaline is called soil pH (potential Hydrogen. The soil scale ranges from 1 to 14 with 7 being neutral. Soil below 7 is acidic and above is alkaline. Before you invest time, energy, and money in a big attempt at gardening it's wise do a soil test. An agricultural station can usually perform a soil test, or they can recommend somebody. First ask how to obtain a sample. The best pH for a home vegetable garden is 6 to 6.5. If the soil is too acid, add limestone. Where soil is often wet, as in a flood plain, it tends to be acidic. Sandy soils become acidic easier than clay. Purchase a pH tester online for a reasonable price once you've made the commitment to be a gardener.

I believe soils should be changed or rejuvenated every six months or between crops. Powdered limestone is an inexpensive soil additive. Let your fingers do the shopping as prices vary on a sack. At the producer it will be cheaper than at upscale agri-shops. There are two types hydrated and unhydrated. The former is chemically calcium hydroxide and made by mixing regular or quick limestone with water. The hydrated version, calcium oxide, is more expensive, is pure, white, and acts on soil problems quickly. Usually, effects can be seen within a week. Regular lime is gray and is takes about three months or more to change the soil pH. *(More on the benefits of limestone in Chapter 11.)*

If you have a fungus problem try mixing white, hydrated lime with water at two TBS (tablespoons per gallon and watering the base of the affected plants. For severe problems lightly dust the bases. Fungus and bacteria thrive on acidic soil. There are many chemical treatments, but white lime is successful in fighting fungus and strengthening plants. Every time we prepare ground to plant, we lightly spread regular lime so months later the soil will be strong and balanced.

Soils become acidic because the natural limestone washes away either from floods or over watering/irrigation. Over fertilizing also causes a soil's pH to drop. Decay of organic ma-terials such as chicken manure will also drop the pH. Acid soil will cause the vegetable plants' roots to be stunted, underdeveloped. This will cause the plants to need more watering and not fully accept fertilizer nutrients.

Beans, corn, cucumbers, pumpkin, tomatoes, peppers, and eggplant will withstand soils to 5.6 pH. Lettuce, bok choy, cauliflower, spinach, okra, and beetroot want a soil that is at least 6 or better to thrive.

Each medium needs to be fortified by adding well-rotted manure and limestone. The proportions depend on the size of the garden or container. Chicken manure is either 'bag it yourself' free, or for small money per feed sack. Horse, goat, and cow or bison manure are more expensive. If you see a horse or dairy farm, stop, and ask if they'll give you manure. Sometimes it is adverted in the newspapers under 'Flowers and Plants.' Manure increases the acid content of the soil, so too much can harm young plants. Limestone lowers the acidity and fights bacteria, but use it sparingly. My soil recipe, for most vegetables grown in containers, put soil in a five-gallon bucket, add about two cups of manure and one-half cup of limestone.

FERTILIZER

Fertilizer usually is represented in three numbers like Starter – reddish brown 12-24-12 works good until blossoms appear. Use a pinch in a hole before setting young plants. Special Green – 15 –5 –20 –2 – great for green leafy veggies like lettuce and bok choy. Blue – 12-12-17-2 after blossoms begin to make the fruit come. *(More on fertilizer in Chapter 7)*

The first number in a fertilizer formula is the nitrogen content. Nitrogen is used by plants to promote leaf growth. Too much nitrogen can burn the plant. The second number is the phosphorus content. Phosphorus is used by plants to increase fruit development and to produce a strong root system. The third number is the potassium (potash content. Potassium is important to the size and strength of the plant. For example, a 10 lb. bag of 12-24-12 converted to weight equates to 1.2 lbs. nitrogen, 2.4 lbs. phosphate, and 1.2 lbs. potash.

Bagasse (chopped sugar cane), potting soil, and other mediums contain no nutrients. Manure isn't recommended for grow boxes. They require a starter mix of various nutrients or using a stock starter fertilizer.

CONTAINERS

Red clay pots in all shapes and sizes are reasonable in most plant shops. Expensive, lightweight, designer boxes cost more at various upscale hardware stores, but they're pretty.

First - plant your garden in the correct place that gets morning sun and afternoon shade. Each plant needs at least six hours of sun daily. But too much sun will dry out the soil. You can create some shade for your plants if you build a fence. *(More on fences in Chapter 3.)*

Plants can be started from seed in plastic or clay containers instead of directly in the garden. The containers need to be at least five inches wide with drain holes in the bottom. Moisten the soil before you plant the seeds. If you water after placing the seeds, the seeds may wash away. Once the seeds have sprouted, they will need at least six hours of sunlight a day, but not the heat of the afternoon sun at the start. Your plants will grow without a lot of labor. This is a great way to grow if you're pressed for time and space. Container gardening isn't best method to grow every type of vegetables. Most root crops, beans, corn, cucumbers, melons, pumpkins, and squash grow better if you plant them directly in the garden.

Containers can be placed almost anywhere. Put them on the porch, along the side of your driveway or parking area. Small containers can be hung by wires from trees, or porch railings. Make certain the containers won't fall if you put them on window ledges. Be careful to protect the surfaces underneath the containers from stains. The containers must have drainage holes to let out excess water. That drainage will be acid or alkavline, and that is what makes whitish stains.

A bit of paint or aluminum foil can dress up almost any container. Again, be a creative scavenger. The main requisites for container growing are that the container is durable- won't quickly rust away and has drainage holes to permit excess water to drain. All contain ers should have at least an inch of some rocks, broken brick, or gravel at the bottom to further permit draining and give the roots something to grip. Fill with the soil medium to about an inch below the top. As the medium sinks and compacts with watering, add more. Always make certain the top roots are covered. I prefer watering the potting soil before putting it in the pots and planting seedlings.

There are many sites on the Internet for various types of grow systems. Most use space age materials and claim huge yields. I like dirt or crushed sugar cane because it feels good and is usually cool to the touch. Rice husks is really hot in the beginning (until it has soaked for about a week and can shock the plants.

The most well-known container and certainly well-organized is Earthbox. Its website reads: The US patented EarthBox was developed by commercial farmers and proven in the lab and on the farm. Our maintenance-free, award-winning, high-tech growing system controls soil conditions, eliminates guesswork and more than doubles the yield of a conventional garden-with less fertilizer, less water and virtually no effort.

I saw a TV advert on the Earthbox and Googled to discover American container farming is now a state of the art, expensive commodity making big profits, but hopefully helping feed people. The Earthbox technique uses a perforated partition about an inch above the container's bottom to separate the water and soil. The plants in the soil send their roots down through the partition into the water mixed with nutrients. The water is filled through a tube extending above the soil. The water tube is kept full, and you can't over fill. Earthbox provides only the plastic box and a snug box cover. They recommend potting mix, not potting soil, and recommend the type of fertilizer to purchase depending on what you intend to plant, and how many plants per container. It is necessary only to fertilize at the set up The cover is cut to accept the plants and will drastically conserve water, fertilizer, and end weeding forever. I suspect as with all container gardens the soil must be rejuvenated at periodic intervals.

The Earthbox should function well for several years. Earthboxes are 29" long, 13.5" wide, and x 11" tall, reusable, a variety of colors and cost about $60 USD. In one box they recommend sixteen beans, or six sweet pepper plants, or two tomato plants, or six cabbages, etc. (see more online). Ace Hardware and Amazon can probably supply Earthbox in Hawaii. The manufacturer claims this version of the container garden can double yields, with less water, fertilizer, and labor with a satisfaction money back guarantee. This is truly maintenance-free – container gardening for beginners.

Plant and your spouse plants with you; weed and you weed alone.
~ Jean-Jacque Rousseau

CHAPTER THREE – FENCE GARDENS

Hawaiians love to fence their property to stop encroachment and deter crime. Many modern style homes decided concrete is the best and easiest landscape technique. With less visible dirt, no mess, and no grass to mow, the surrounding fence may be the only area available for a garden. Whether wrought iron, chain link, or BRC, all can support a hanging garden. Make certain the fence can take the weight of a heavy vine with fruits. If it is chain link it must have a heavy pull wire near the top for support or the vine will bend it down.

Many vegetables and fruits need the support of a fence or trellis (an up right frame) to develop. A variety of plants can be easily adapted to grow on a fence. Why? Fence vegetables take less space, away from the soil with fresh air there is less chance for diseases, and ripe produce is easy to spot. A good vine such as passion fruit, chayote, or bitter gourd, offers shade and privacy from the luscious green leaves and beautiful blossoms bear tasty fruit. Tomatoes are usually staked for support as the trees be come heavy as they mature, but they will climb a fence. Long beans need to be away from the ground, or the vines would get trampled during pick ings. It is the same for chayote and lablab beans. I've been successful with climbing cucumbers, pole beans, bottle neck squash, and small pumpkins. Using your fence as part of your garden will benefit other plants. The leafy fence will also provide shade for your regular, flat soil plot or containers. The shade will fight sunburn from the wilting afternoon sun of the dry season and help conserve water at the plants' roots. Staking or fencing keeps the fruit away from the damp soil and reduces rot.

Plant at the bottom of the fence if there is any soil available. If not, place a planted container there. Either way, it only takes a few minutes a day to train the plant to grow up the fence. Once you've decided what and where to plant and have acquired the plants or the seeds, fork a spot two foot in diameter and about a foot deep. Add about a cup well-rotted manure and a half cup of limestone to each hole you are going to plant. Mark your seed spot so you can water as it grows.

Six long bean plants are enough to feed most families. Seeds can be obtained at agri shops, from friends, or go to the market and ask vendors for overripe beans. The seeds will be very visible when long beans have been on the vine too long. Allow the beans to fully dry before shelling. Long beans' thin yet strong flexible vines are easy to lace through any type of fence. The long beans dangle like Christmas ornaments. Let a few beans get full so you'll have some seeds to plant back. Long beans need little water and little fertilizer.

The lablab bean is another quick climber that needs little work and is very tasty and nutritious. Find the seeds in the market or through friends. Two vines will produce plenty. I re-cently planted some purple vines on my back fence to provide shade for christophene. Winged beans, add a nice, broad pointed, deep green leaf with white flowers as a backdrop to my vegetable garden. Getting the tasty, winged beans was a second benefit. Unless you take good care, the long winged bean vines will overgrow and crowd your garden.

Passion fruit will take over and seems to grow with little attention once it is estab-lished. Plant it where it will not interfere with other vines. Occasionally toss a bit of blue fertil-izer (12-12-17-2 at its roots and hit the leaves with hose spray and it will continually produce the delicious juicy yellow or mauve fruits.

Bitter melon/caralli is another of the low-maintenance fence vines that offers a nice light green leaf and yellow blossoms. The seeds may be difficult to locate, again go to the market and look for a fruit that has turned yellowish orange. Permit the seeds to dry before planting. Plant about four seeds per hole and space the holes at least ten feet apart. Water daily until the sprouts are about a foot long. Lay a stick or a string for the new plants to use as a ladder to the fence. Once bitter melon/ carailli has attached there is no stopping it. Two bitter gourd plants are plenty.

The bumpy green-skinned fruit is the strangest and bitterest member of the melon family. A mature fruit should first be sliced and salted, then squeezed to remove the bitter juice before cooking. However, smaller young fruit are almost sweet. Bitter gourd is nutritional, and the bitter juice can be helpful to diabetics.

Tomatoes need to be started against the fence. I have a raised bed about a foot wide. Once the tomato tree is a foot tall, I start weaving the branches carefully into the chain link. I grow tomatoes and bitter gourd together on the fence and that permits room for sweet pepper trees in the front of the same small box. Bitter gourd will try to take over so attention is needed so the tomato can thrive. Six tomato trees will feed most families.

Chayote needs an area to itself. The easiest method to grow this vegetable is to locate a farmer and beg a plant. Failing that, select two chayote at the market. Ask the vendor if they have any that are overripe and budding. If not, set the chayote in a warm window, but not in the direct sun. In a few days it will start to shrivel and wrinkle and soon sprout a bud. Plant the seed bud upwards in a clay pot with sandy soil. As I began growing chayote, I realized that some fruit, no matter the size or age, sprout while growing on the vine. Two chayote plants are all that you will need.

After seeing loaded trellises of chayote in highland-mountainous areas, I incorrectly believed this vine needed coolness to develop. This vine loves the sun, but also needs plenty of water, yet can be grown almost anywhere. Where I lived, almost beachside, I had a small corner of fence as a backdrop to a grow box with a sprinkler. Two chayote vines produced for over six months. In the intense heat of the dry season, I watered it at least three times a day.

Cucumber vines will support the weight of large fruits, but this vine is fragile so be careful when you lace the young vine onto a fence. Cukes will take almost the same care as chayote as they like water and special green fertilizer (15 –5 –20 –2 in small doses. Plant three cucumber vines about six feet apart and that should produce an adequate supply.

Most vines need little attention, water when necessary, except for chayote. A weekly or biweekly spraying with a light insecticide like Pestac or Fastac mixed with a water-soluble fertilizer like Nutrex 20-20-20 should prevent any infestation and give plenty of nourishment. It is best to spray late in the day after the heat has diminished. With chayote, bitter gourd, and cucumber, try the spray on a small part and watch for a few days to determine if it has any burning –yellowing side effects. Birds will present the biggest problem to the fruits of your fence labors.

If you don't have a fence, make a trellis of any old PVC pipes, boards, or bamboo. Stake four posts in the earth and lace four together at the top. Make it sturdy enough and string your hammock underneath the soon-to-be lush vines. If this is too much work, find or buy a piece of 6-inch BRC (wire mat for concrete, and stand or lean it upright for the plants to climb on. The main reason to grow on a fence is to save space, hopefully, to plant more veggies on the ground. You'll quickly see its appearance is tasteful and the vegetable or fruits are tasty. Your neighbors will surely see the value of your green thumb.

If you have a garden and a library, you have everything you need.

~ Marcus Tullius Cicero

CHAPTER FOUR – THE TRADITIONAL GARDEN PLOT

Gardening is a great hobby, relaxing, and a diversion from a stressful world. For some people growing their food is quite an accomplishment. The home garden provides family togetherness while saving money on fresh vegetables.

GROW YOUR OWN

Home gardens are visibly transforming landscapes from lushtropical plants and hedges to at a partially edible garden. Vegetable gardens are no longer hidden behind the home. A vegetable garden, tastefully planned, is a living art form. The produce has the great taste of extremely fresh homegrown. A garden doesn't need to be unsightly and can be inter-mixed with flowers. Some gardeners grow hard-to-find, difficult-to-grow, or unique varieties of veggies, while the majority of us are serious growers for the table, doing it as easily as possible. Our gardens, large or small, are functional and attractive. At least to us!

With proper planning, the vegetable garden can be both functional and attractive. Land-scape designers today often plant flowering annuals in vegetable gardens. This gardening philosophy, coupled with our favorable climate, can offer gardening opportunities all year.

Constructing your garden is not something that needs to be done instantly, unless you want to hire a crew. Make your garden a place to relax, be patient. Plan what you want to plant and make a drawing to scale if possible, so you don't waste time, energy, and money. Think of the garden as split into parts, one for short crops (close to the ground) like bok choy, lettuce, spinach, bush (green) beans, celery, seasonings, and lady fingers/okra arranged by height from shortest (lettuce) to the tallest (lady fingers/okra). The tallest shouldn't block the sun from the shortest. Prepare another section with plants like peppers, tomatoes, and cu-cumbers. Other sections can be developed for root crops such as long white radish/daikon and beetroots, a section for seasonings like chives, celery, spices, and cilantro, or extremely tall crops like, brinjal/ eggplant, cassava/manioc, peas, and corn. I prefer to have the tallest in the eastern part of the garden, so they get the morning sun and don't block the afternoon sun from the shorter plants. Vines need plenty of space or they will overgrow other vegetables.

SOIL

You need reasonably good soil that receives preferably eight hours of direct sun, and close to a source of water. If you can pull the soil higher into a 'raised bed' it will drain better. Drainage is crucial. Once you select the area, to reduce the labor of forking, soak the soil at least overnight. Fork the soil to eight or ten inches deep. Then let it sit in the sun for at least two full days. This is termed 'solarization' and is a method of killing some diseases. Good soil will help plants grow strong, healthy, and produce excellent fruits.

After a few days in the sun, cover the clumps of dirt with well-rotted manure. Chicken manure is the most available and least expensive, but I have seen horse and bison manure for sale. Then dust with powdered limestone. If possible, wet the soil to ease breaking the clumps with a hoe. Break the soil as soft as possible into a nice mix. Pull the soil into beds bordered by drains that will lead excess water away from the plants.

Mark the rows with a line or a board and cut a furrow, a two to three-inch indentation into the loose soil. Space rows about eighteen inches apart so you have room to walk after the plants mature. Space your plantings properly. It might seem more plants per square foot will produce more, but crowding vegetable plants is counterproductive. They need fresh air to circulate; you need to see the plants' conditions, pull weeds, and fertilize differently. Plants need regular watering, so hopefully, your plot is near a water tap and you have a hose. Dry season tip: Make a furrow four inches wide and four inches deep. Spread a one-inch-thick layer of rotted manure on the bottom. Plant seeds in the furrow and cover with soil.

Make certain the furrow isn't filled in to the top and will hold some water. Water the furrow daily for the first few weeks. As the plants grow mold the soil around the stems to protect the roots from the sun. Allow a bit of original furrow to remain to continue to retain the water.

TRANSPLANTS

Leave a small area, maybe a two-foot square, for a seedbed to start your plants, like tomatoes and all types of peppers, from seeds. Most seed packets and seedling dealers will tell you the mature size of the plants you are selecting. Check the date on the seed packets to assure the seeds are fresh. Wet the seedbed to four inches deep, plant the seeds, and cover with dry soil. Keep the soil wet to help germination. Care for the seeds until the seedlings are four to six inches tall. Then for about two weeks gradually reduce the water while increasing the exposure to sunlight to 'harden' the plants for transplanting. When pulling the seedlings, totally wet the bed to make removal easy without any damage to the seedlings' fine roots.

Transplanting is taking a plant from one soil or medium (the seedbed) to another (the garden) where plants will have more space to mature. Most plants suffer a bit of shock being transplanted. Carefully dig, remove, and replant. When buying vegetable transplants, look for stocky, full plants. Their roots should be white or light tan, not brown, and they should not be root bound in their container. Transplants that have already started to fruit are not a bonus! They are stressed from being confined too long. It is best to transplant seedlings in the evening. If the transplants already have flowers or small fruits remove them to ensure adaptation to your garden before fruit production begins. Keep the garden damp when transplanting.

I make a small hole for each transplant and put in just a pinch of starter (like five to ten tiny pellets) of 12-24-12 fertilizer. I pull over a bit of soil and then put in the plant. This keeps the roots away from the fertilizer until it adapts to its new setting. Then the roots hit the fertilizer and it is like getting a vitamin boost!

Direct garden planting can be done in properly wet shallow furrows. This is termed 'drilling'. This is good for lettuce, beans, etc. Place the seeds and cover them with soil, but not too deep. Seeds that are covered with too much soil may never come up. After germination, excess plants are thinned to the correct spacing. 'Hilling' or 'mounding' places several seeds in one mound with fixed spacing. Squash, pumpkins, and melons are often planted in this manner. This is also a good technique for the wet season.

Thinning seedlings to get the best spacing reduces competition for nutrients and water. It provides better conditions for growing healthier vegetables to produce higher yields. Thin the plants directly seeded to the soil when they have one or two pairs of true leaves, about 3 inches tall. In the cool evening, when the soil is damp and soft, is the best time for thinning. This is a must for leaf lettuce, beetroot, white radish, and spinach.

The entire garden plot should be kept fully planted, but not over-planted. Once a set of plants stops producing, pull them and replant. Try to switch plant families to confuse insects and repel some diseases. **Remember, vegetables subject to the same diseases and insects should not follow each other.** Tomatoes are cousins to all types of peppers and eggplants. Cabbage, cauliflower, and broccoli are related. Cucumbers, pumpkin, squash, and melons are in the same family. Remove diseased plants as soon as you notice. It will be more problems and ex-pense to cure than to replant. Do not place sick plants in a compost pile. The biggest enemies to transplants are a combination of too much sun and a lack of water.

WATERING

Plants need about an inch of water every week. I recommend watering in the morning so the soil is not overly damp at night, which may promote fungus growth. I try to soak the plants heavily once a week. Water the base of the plants, slow and evenly, and try not to wet the leaves or erode the roots. Try to dampen the soil to about four inches deep. It is the roots that must receive the water. Deep watering helps strong roots develop that can find water hidden deep within the soil. (*More on watering in the next chapter.*)

FERTILIZE

Once the plants catch (become adapted and start to grow), throw just a pinch of 12-24-12 starter fertilizer around the base. As you water, it will seep down to the roots. I do not recommend water-soluble fertilizers as they can burn the plants unless you only spray it at the roots. The leaves absorb little water; that is the job of the roots. Soluble fertilizers often are high in nitrogen and that can delay fruit production.

As you water, the soil erodes at the base of the plant. This can cause the roots to get burned by the hot sun or the fertilizer. Weekly, I carefully pull soil around the stem of the plant. This is called molding. Many gardeners over-fertilize. When I mention just a pinch, it is all you can grab between two fingertips. Use the reddish-brown starter fertilizer until blossoms appear, then use the blue 12-12-17-2. It is best to apply fertilizer just before a rain or watering the garden. Never permit fertilizer pellets to remain on the leaves, as it will 'burn' them.

WEEDS

Weeds are undesirable in any garden since they compete for nutrients and water. Some weeds spread diseases. Start controlling weeds when you plant by pulling or hoeing. Watering just the base of the plants discourages weed growth. Work the soil properly after each deep watering to kill sprouted weeds and to leave the soil's surface loose to absorb more water.

HARVEST

Harvest is what you have worked and waited for. Remember to harvest when the fruit is ripe. Do not permit it to sit on the vine attracting damage from birds and insects. Once your garden begins to produce, try to pick at least every other day. Now is the time to try new recipes and enjoy the fresh veggies. Remove the old plants immediately after harvest, cultivate, and replant.

Save seeds from one of the best fruits you harvest for your seedbed. Always let the seeds completely dry before planting.

THINGS TO REMEMBER:

If you have a serious gardening problem, contact the professionals at a government agri-station. They are garden specialists and are very busy, so don't waste their time. Don't take advice from just anyone. Other gardeners may be envious and give erroneous and spiteful advice. Agri-shops want to sell you something.

Remember, dry plants are stressed plants. Too much water can damage plants by blocking oxygen to the roots or causing a fungus attack. Don't over water or over-fertilize. Excess ni-trogen fertilizer can cause some root-rotting fungi and bacterial growth. Stress from too much nutrient can make plants more susceptible to diseases and insect damage.

Each hour in your garden will benefit you, your family, and your friends. Every year your garden skills will improve, and your plants will bear more. You may never have 'the perfect garden,' but who does? Disorganized, beginner gardening can be a lot of work, yet the accomplishment of growing your own will outweigh the labor. Your garden is a work of art always in progress.

Everything that slows us down and forces patience, everything that sets us back into the slow circles of nature, is a help. Gardening is an instrument of grace.

~ May Sarton

My passion for gardening may strike some as selfish, or merely an act of resignation in the face of overwhelming problems that beset the world. It is neither. I have found that each garden is just what Voltaire proposed in Candide: a micro-cosm of a just and beautiful society– Andrew Weil

I grow plants for many reasons: to please my eye or to please my soul, to challenge the elements or to challenge my patience, for novelty or for nostalgia, but mostly for the joy in seeing them grow. ~ David Hobson

CHAPTER FIVE – WATER / IRRIGATION

Water is life, and life is mostly water. All living things need water to stay alive, but plants use much more water than animals do. Plants are about 90% water, while our bodies are about 60% water. Plants absorb water, and with the sun's energy, transform it into food. Gardens are like cities: too much, or too little water makes living difficult.

Even though water seems so plentiful, useable freshwater is extremely limited. Our planet is 70% water, but 96 % is saltwater. More than two-thirds of the Earth's fresh water is frozen in ice caps or glaciers; 30% is under the ground. Only about two-thirds of one percent, (.67%), is available for use as surface water from lakes and rivers. Too many people in our world do not have access to safe drinking water.

Hawaii is blessed it has so much freshwater resources. It is estimated that all the is-lands combined get more than 20 million gallons of rain a day. *(USGS Ground Water)* Agriculture uses about 50% of the available fresh water, Industry 30%, and domestic use is 20%. The average person on the island uses 2,500 cubic meters of water a year. That's 40 large shipping containers (40 ft. long) in a year or ¾ of a 40 ft. shipping container every week per person!.

The USA uses over three hundred billion gallons of water a day just from surface water and eighty-five billion gallons from underground resources! 40% of the USA's water is used in agriculture. It's using more water than is being naturally replenished.

The Central African countries as Congo and Cameroon have the most renewable sur-face and underground freshwater per national citizen. Brazil has the most freshwater with 12% percent. Russia has 10%, China has 8%, Canada with 300,000 lakes has 6.5%, and the US has 5%. With the future's demand for agriculture, these water-rich nations will be the important growers.

Plants absorb water at their roots and by a process called 'transpiration' and trade it for carbon dioxide. It is a very complicated process, but simply, leaves have tiny pores that let the water evaporate out and absorb carbon dioxide. Transpiration pulls water from the roots up the stem to the leaves. Every day a full-grown vegetable plant 'transpires' three-quarters of its weight in water. If humans did that, we'd have to drink more than twenty-five gallons of liquids a day. Less than one percent of water pulled to the leaves is used to produce fruit growth. Most of it just evaporates.

Farmers are the biggest consumers of freshwater using about half of the available water. Correctly used, almost all agricultural water is recycled into either surface or ground reserves. Developing countries devote most of their water supplies to agriculture. India, for instance, uses 90% of all water for agricultural purposes, with just 7% for industry, and 3% for domestic use. As more people move away from the countryside into cities, and agriculture depends more and more on irrigation, it will be difficult for cities to meet the increased demand for freshwater. In developing countries, rapid urban growth often puts tremendous pressure on antiquated, inadequate water supply systems.

Plants use various amounts of water depending on age, size, type, amount of sun, tem-perature, etc. Correct watering is vital to growing good vegetables; enough applied properly, yet not too much. Garden vegetable plants need varying amounts of water. All watering tech-niques, from sprinkling can to sprinklers, are considered irrigation. It is important to get the water to the roots. The leaves absorb very little water. The soil at the roots should be moist to a depth of at least four inches. Water slowly and evenly, preferably through a sprinkler spout. Place the sprinkler spout close to the plant's stem, below the lowest leaves, and gently tilt the can. With vines like cucumber, melon, or pumpkin, mark the roots when you plant. Try not to get water on the leaves. This helps fight fungus. If you pour water directly, you may erode the soil at the root, which could potentially harm the plant either by sun burning the visible root,

Different soils are another water factor. Clay-type soils hold water better than sandy ones. With clay, it is best to hoe and roughen the soil's surface and mold around your plants weekly to keep the soil from becoming glazed. Glazed soil will cause the water to run off rather than soak in. Organic material, like compost or well-rotted manure, worked into the garden helps hold water. Fertilizers make plants thirsty, so use as little as possible just before watering. Rough watering will knock off blossoms, or break branches bearing fruit. This happens with sweet pepper. To check how well you are watering, dig a small hole at a row's end and see how many inches down it is moist.

It is hard to calculate the amount of rain that hits your garden unless you have a rain gauge. Just put a jar or a can in among your plants and check how much accumulates. I like to water in the early morning. I make time before I begin my day. It gives me a peaceful, cool period to check out my plants and give them a nice long drink. Watering in the heat of the day loses a lot to evaporation. Shallow or quick watering will only wet the surface and the roots may stay shallow. Deep roots that seek moisture are better and create stability.

You can make a simple sprinkling can from a gallon bleach bottle. First, make certain it is clean. Then use a drill if available with a quarter-inch bit or smaller and put several holes in the plastic cap. (This can also be done with a hammer and a nail.) At the top of the neck where the handle begins, drill or punch one hole into the body to let out the air. Once a week during the dry season, I like to give them a good soaking with a hose. This is especially import-ant for eggplant and vine crops. Tomatoes don't need much water.

FRUIT-VEGETABLE PERCENT WATER

Bananas	74%	Broccoli	91%
Cabbage	92	Cantaloupe	90
Cauliflower	92	Carrots	87
Cucumber	96	Eggplant	92
Lettuce	96	Peppers – sweet	92
Pineapple	87	Ceylon spinach	92

The easiest method I've found is the soaker hose. I've found two types. One is a round black material that literally lets water soak through. The black type is fragile because it seems the sun's UV rays dry it out and breaks easily if moved. The other is a usual green and white vinyl hose that has sprinkler holes. Turned up it sprinkles, faced down it soaks. Both types are placed among the plants, and water seeps to the roots. The seeper holes in the green vinyl type clog and must be cleared with pressure once a week. It can be used on just pipe pressure.

Most sprinkler systems waste water and require a pump. Drip irrigation is another method that brings water directly to the roots. This style of irrigation wastes less water, but is costly and requires regular maintenance. You can use water from the pipe, river, pond, well, or roof. I recommend collecting water in a plastic barrel so the chlorine can evaporate before the plants get it. Keep the barrel covered as a protection for children and mosquitoes.

Planting in furrows that hold water rather than in mounds where it runs off is another dry season tip. As the rains begin, mold the plants and make certain they can drain. A few crops like manioc/cassava and onions can do with minimal water. Modern agricultural science has created different strains of seeds for some vegetables that are more drought tolerant. When I plant, I soak my seeds overnight, especially corn and long beans, and get a good rate of germination.

Over-watering can lead to serious garden problems as fungus, diluting, or washing out fertilizers. Make sure your garden will drain quickly and adequately after a good rain. If you are planting in a raised bed or grow box first lay a layer of stones to let the excess water pass. Keep sloped drains between the beds and rows that lead surplus water out of your garden.

Knowing when and how much water to apply at various stages of the plant is a true garden skill. Every variety of vegetable requires more water at different stages of development. In some vegetables, excess water brings out better fruit, while too much water can harm many plants. Lablab, long bean, passion fruit, papaya, pineapple, and manioc/ cassava need little wa-tering. Tomatoes and peppers need regular water, but not much, and preferably in the morning. Beans need regular water, chow chow/chayote, lettuce, and leafy veggies like bok choy, take a lot several times a day. Try watering your fruit trees in the dry season at least once a month. Fill a five-gallon bucket and add some soluble 20-20-20 fertilizer for spectacular citrus. For early succulent peas (red gram), water those trees in the dry season. Good melons need plenty of water.

Some countries are blessed with water. At times, we get too much too fast. The world seems to have only two seasons: fire, and flood. As your garden experience grows you will learn to grow great vegetables in both the heat and the rain. Even though water is a naturally replenished resource, it is always wise to conserve by accurately and adequately watering our gardens.

Water is an essential resource to sustain life. Governments must make it a priority to deliver adequate supplies of quality water to people. Individuals must conserve and protect this precious resource in their daily lives. According to the World Water Institute, a mere 2.5 per-cent of the earth's ground and surface water is accessible for human use. This finite resource, maintained by the earth's natural water cycle, is used for everything from drinking water to sanitation, agriculture, and industrial processes. Overuse, pollution, and inefficient infrastructure as well as natural occurrences like drought, have pushed humankind's water supply nearly to its limits. Scientists project that by 2030, almost half of the world's population will be living in areas with severe water shortages. This situation represents one of the greatest human development challenges of the new millennium. How water is conserved, used, and distributed in communities, and the quality of the water available will determine if there is enough to meet the demands of households, farms, industry, and the environment. More than 10% of people worldwide consume foods irrigated by wastewater that can contain chemicals or disease-causing organisms.

DID YOU KNOW?

Water *(H2O)* is the most abundant molecule on our planet, comprising almost 70% of the Earth's surface. In nature, it exists in liquid, solid (ice), and gaseous (water vapor) states. It is in a natural balance between thecliquid and gaseous states at standard temperature and pressure, it is a nearly colorless with atnge of blue, tasteless and odorless liquid. Drinking water is polluted by pesticides, fertilizers, sewage, and saltwater. Hawaii uses about 200 million gallons of fresh water daily. Yearly, municipals and industries each use almost one billion cubic meters a year. Agriculture uses about 12 billion cubic meters of water. Every day, each resident uses 150 gallons.

You can appreciate the flowers in another's garden while watering your own – Jennae Cecelia

Water methods—sprinkler, watering can, striped sprinkler hose, black soaker hose.

Watering can made from a bleach bottle. Cut a notch at the base of the handle with a knife. Use a hammer and nail to punch holes in the cap.

Remember that children, marriages, and gardens reflect the kind of care they get. ~H. Jackson Brown, Jr.

CHAPTER SIX - TRANSPLANTS

One of the secrets to a great home garden is starting plants from seeds and knowing how and when to transplant. This is the process of changing plants' locations. This is tricky because the original location, where the plants sprout from seeds, either containers or a seedbed, usually is the best of all environments in. The young plants are destined for either the usually hot, Hawaiian weather or serious wet conditions of the rainy season.

Relocating plants that you have raised can save considerable expense. By acquiring and sprouting, seeds you can grow plants from the seeds of anyone's vegetables. Seeds from fruits of those beautiful hybrid vegetables usually won't sprout, though. Making your own plants lets you coordinate the timing of your garden plot. If you sprout the seeds before you start to fork and prepare the soil, the plants should be ready when your garden is organized. Sprouting is a chance to have well-developed plants that should be able to fight off diseases. This is much better than trying to grow the plants of vegetables like tomatoes, peppers, or eggplant directly in the garden row from seeds.

To get seeds from peppers or tomatoes, let the fruit fully ripen almost to the rotting stage, remove the seeds and let them dry in the air. A window ledge is one of the best places. Once the seeds are fully dried, plant them in a small, shallow container. There are plenty of appropriate containers like styrotex cups, pie tins, etc. Specialized starting trays that can sprout one hundred and twenty seeds can be purchased. These are good for starting several different types of vegetables together at the same time. Potting soil or compost are the best mediums to start the seeds, but good quality dirt can also be used.

One of the best things I've discovered for sprouting seeds is to save eggshells cracked in half and stored in the egg tray. Put a bit of potting soil in each half shell and set the seeds. I dampen the potting soil with a light dilution of a foliar fertilizer like Powergizer 45. Put the seed into the shell, or another container, a quarter-inch deep. A pencil or pen makes a good tool to place the seeds. Keep the sprouting soil damp throughout the nursery period. As the plants grow the shell can easily be cracked to let the roots out as they are put into either a larger container or the soil. The eggshell gives the baby plant some needed calcium. Keep them in the small containers until they are two to three inches tall, perhaps three weeks or longer.

I usually go from the eggshell size to the small coffee cup, and then into the garden. While the small seedlings are in the egg crate or sprouting trays, they should be kept well-watered and out of the heated direct sun. In the coffee cup or larger container, the plants are growing to about 4-6 inches tall. This usually takes 2-3 additional weeks. During this period, the plants should be hardened by slowly increasing the time in the direct sun until they can handle a full day of dry season sun.

Healthy transplant roots should be white and developed, but not packed tight and brown. This is important if you choose to purchase young seedlings from a plant nursery.

Another way of starting seeds is directly in the soil. This is called a 'seed' or 'nursery' bed. To make a bed, fork the soil powdering all the clumps. Mix a small amount of well-rotted manure. A good seedbed size is two-feet square. For that size use about a cup of poultry manure. Make the bed in a location that gets the morning sun, but is shaded in the heat of the afternoon. Wet the soil until it is damp, but not soaked. Make slight indentations or rows, an inch deep with a trowel or your thumb, about six inches apart. Plant the seeds carefully about an inch apart and a half-inch deep. This is good for lettuce, bok choy and other vegetables. Carefully dig out the plants with a trowel or a knife. Be careful not to damage the roots.

The best time to put the young plants in the soil is in the late afternoon. Check the phases of the moon. *(More on this in Chapter 14.)* First soak where you are putting the plants. Pull all the weeds, as they might hide insects or diseases, and compete for moisture and sun-light. I usually make a hole about four inches deep - put in a literal pinch (what you can hold between your index finger and your thumb) of starter fertilizer 12-24-12 and then cover with an inch of dirt. Then place the plant and pull the soil around it to the stem level as it was in its last starting container. Make certain the ball of roots is completely covered or it will permit moisture to easily evaporate, and stunt or kill the plant.

In the garden row cabbage, cauliflower, broccoli, tomato, and pepper plants should be spaced at least a foot apart. Eggplant and lady fingers/okra should be spaced every two feet, while lettuce and bok choy six inches.

During the heat of the dry season, it is very important to water the transplants every morning. It usually takes about two weeks for them to 'catch' in the garden soil. A soaker hose is useful to keep the garden soil moist till the roots adapt to the garden. Mulching with either dried leaves or shredded newspaper will also hold moisture and fight weeds.

Shock, either from too much sun, lack of water, or chemicals, or a combination during the first two weeks in the garden soil, can critically stress and kill the plant.

If you choose, after two weeks in the ground is the best time to spray the young plants with systemic chemical treatments for both insects and fungus.

Timing is everything with transplants. Sprout the seeds to have plants when your gar-den is ready. Keep sprouting plants to have a constant supply. Develop strong healthy plants that can withstand disease. Don't put them into the soil until they can take the sun. Sprouted plants, especially hard-to-get peppers or a special tomato, make excellent gifts for other gardeners.

The love of gardening is a seed once sown that never dies. ~ Anonymous.

NOTES:

CHAPTER SEVEN
GARDEN FERTILIZERS

Fertilizers are another garden chemical mystery to be solved. They simply are food for plants, like vitamins for humans, but must be applied properly at the correct time to be most effective. It is possible to harm your garden with an overdose of fertilizer.

Garden vegetables need various nutrients to grow and produce. Nitrogen (N), phosphorus (P), and potassium (K) are the key elements for healthy plant growth. Others like calcium (Ca) and magnesium (Mg) are supplied by limestone, while more nutrients come from the soil, sun, water, and air. Packaged, 'man-made' fertilizers usually contain two or three of the prime garden chemical elements with numbers. The numbers refer to the ratio or percentage of each element, like N (nitrogen), P2O5 available phosphate, and K2O available potash. Remembered their garden duty as 'N'-leaves and shoots, 'P'-stems and roots, and 'K'-flowers and fruits.

Fertilizers are created from either inorganic or organic materials and exist in many different forms. Organic or natural composts and well-rotted manure are better substitutes than synthetic inorganic pellet mixes because they contain the same nutrients released to the vegeta-ble, but over a longer time. Organic fertilizers include horse, cattle, swine, or poultry manure, and fish emulsion. Compost is a mixture of decayed organic matter (plants, vegetables, weeds, etc.) used to fertilize and condition the garden soil. The organic material of compost or rotted manure will improve the soil structure, help it drain better, and hold nutrients longer. Man-made fertilizers come in different types, a hard pellet meant to slowly dissolve and a water-soluble, granular or liquid, foliar form sprinkled on the leaves or roots of the plant.

Both organic and inorganic fertilizers provide the same essential nutrients that plants need to grow properly. Plants use nitrogen for leaf growth. Timing of the application is partic-ularly important. Leaf crops like lettuce, salad greens, spinach, broccoli, cauliflower, cabbage, thrive on nitrogen. Plants lacking nitrogen are stunted with yellow-green leaves. Over-fertiliz-ing with nitrogen causes excessive growth of leaves at the expense of the fruit. Appling at the wrong time causes too many leaves and can delay blossoms. Too much nitrogen fertilizer can kill the plant by burning the leaves and roots. More than half of the nitrogen from synthetic fertilizers is wasted, usually washed away. All types of beans and legumes like long bean and lablab build soil by injecting the soil with nitrogen they take from the air. Other heavy feeding crops like corn, 'clean' the soil of nutrients. This is one of the reasons for crop rotation.

Phosphorous is necessary for blossoms and fruits. Pumpkins, squash, cucumbers, mel-ons, tomatoes, all types of peppers, and eggplants flourish with phosphorous. Plants lacking phosphorus usually have a purple tint to the leaves. Potassium develops healthy roots and stems important to the strength of the plant and helps to combat the weather elements and diseases. Larger roots with many root hairs have a greater capacity to absorb water and fertilizers. On-ions, garlic, carrots, beets, radishes, leeks, and scallions need potassium.

Fertilizers, whether inorganic or organic, usually have three numbers on the container. The numbers refer to the percentage of main nutrients nitrogen, phosphorus, and potassium in the mixture. A hundred-pound sack of starter growth mix 12 –24-12 has 12 percent nitrogen, 24 percent phosphate, and 12 percent potash, or converted to weight, 12 lbs. nitrogen, 24 lbs. phosphate, and 12 lbs. potash. There are also micronutrients like boron and magnesium, but you and I both ask, what's in the other 52 pounds that I'm carrying? Fillers!

The most important ingredient in a garden is soil. Garden dirt is comprised of mineral matter, sand, silt, and clay or rocks with microorganisms of bacteria, and fungus, that house and feed organisms like insects and worms. Decaying compost from plants and animals provide open spaces for air and water, so roots can easily spread.

Generations of use change the character of the soil. What was once fertile can be worked until it is dead. Fertilizing without balancing the pH can render soil toxic. Gardeners must respect soil as a heritage from our past to our future. Good soil conservation will assure our continued existence. We garden out of the necessity for food. The amount of fertilizer your garden needs depends on the fertility of your plot's soil. The condition of your garden's soil will determine the amount of organic matter, type of fertilizer, and the crops you intend to grow. **First, get a soil test to determine your garden's nutrient deficiencies.**

SOIL TEST

Soil tests can be done at the various extensions of the agriculture divisions of the government. Mr. Google can help. They are not free, but money well spent that will save many dollars later. Soil tests examine the levels of nutrients in your plots soil and the pH of the soil. (pH means potential hydrogen.) Nutrients become unavailable to plants if the pH is above or below a certain range. Soil is based on a scale of 1 to 14 with 7 being neutral. Below 7 is acid and above alkaline. Most vegetables grow best between pH 6.0 to 6.5. This range permits the maximum intake of nutrients and particularly essential micronutrients. If the pH is too low, limestone may be necessary. Constant use of fertilizers will raise the soil's acid level. Flood, rain, and constant irrigation runoff of limestone and alkaline sources, combined with the decaying of some organic matter will also raise the acidity.

There are several types of limestone. The most common is unhydrated lime, gray, and looks like cement. Unhydrated must be added in the initial garden planning stages after your soil test. It takes about three months to work into the soil. Hydrated lime is white and resembles flour. It is fast-acting and can be used by broadcasting around the plants or can be mixed with water. Hydrated lime acts almost immediately.

Acid soil causes poor root growth. Poor roots mean the plant has difficulty taking in water or nutrients. Limestone is the most effective and least expensive method to improve your soil. To increase your soil's pH by one point more alkaline, add four ounces of lime per square meter to sandy soils, a half-pound per square meter in loamy soil, and twelve ounces to clay per square meter. Plan on adding limestone every other year. *(See Chapter 11 on the benefits of limestone.)*

If available, rock sulfur may be used to reduce your soil's acidity and fight potential fungus. Combine 1.2 ounces of powdered sulfur per square meter for sandy soils, or a quarter pound per square meter for all other types of soil. The sulfur should be thoroughly combined with the soil at least a week before planting. Sawdust from untreated wood, or compost will somewhat lower the pH.

Have the technician fully explain your soil test. There are trace elements like magnesium that can help or hinder your garden, depending on the existing amount. If low in magnesium, dolomite (a combination of dolomite and calcium), or calcitic (calcium) limestone can be added. Calcium builds plant tissues. A lack of calcium makes plants struggle to grow in very acidic soil. Most vegetables like slightly acid soil and can be hurt by adding too much lime. Follow the soil technician's recommendations. It is necessary to determine the square footage of your plot because lime and fertilizer recommendations are based on 1,000 square feet. Measure the length and width and multiply together to get the size of your plot. Less than a thousand square feet, then divide your plot's square footage by a thousand and multiply the decimal result by the rates recommended for lime or fertilizer application.

SOIL TYPES

Crops grown on sandy soils usually require good doses of potassium, but crops grown on clay soils do not. Heavy clay soils can be fertilized considerably heavier at planting than sandy. Clay soils and those high in organic matter can safely absorb and store fertilizer at three times the rate of sandy soils. Poor sandy soils, which need fertilizer the most, unfortunately, cannot be fed as heavily and still maintain plant safety. Feed poor thin sandy soils more often in smaller doses.

The main plant nutrients are nitrogen, phosphorus, and potassium. Secondary level plant nutrients are calcium, magnesium, and sulfur. Micronutrients like boron, chlorine, manganese, iron, zinc, copper, and molybdenum are all necessary for development in some veggies.

Animal manure gives the best results of all the fertilizers for it improves the quality and composition of the dirt and is usually the least expensive. A backyard garden needs about one-quarter sack of well-rotted horse, cattle, or chicken manure per hundred (10ft. by 10 ft.) square fThe main plant nutrients are nitrogen, phosphorus, and potassium. Secondary level plant nutrients are calcium, magnesium, and sulfur. Micronutrients like boron, chlorine, manganese, iron, zinc, copper, and molybdenum are all necessary for development in some veggies. eet. First, soak the ground overnight and then fork the soil to a depth of about eight to ten inches apply the necessary manure and lime (if required). The following day, hoe the plot breaking the clumps to as fine a powder as possible. This technique will loosen the ground and provide good drainage. This will improve your soil's ability to transfer nutrients to the roots. Adding manure will not cause fast results, but will have long-lasting benefits. If stinky manure isn't you, try inorganic 5-10-10 or 12-24-12 broadcast at a rate of 2 pounds per ten by ten area. With the proper attention, as the years go by your garden's soil will become better and better.

There are many types of fertilizers as reddish-brown starter 12-24-12, green for foliage and vines 15-5-20--2, fruit and vegetable bearing 12-12-17-2 soluble or dry. Almost any combination of main and secondary plus trace elements can be purchased.

Starter fertilizer at planting time is recommended to help the plants overcome trans-plant shock and to ensure proper nutrients during the initial growth period. Dig the hole for the plant about two inches deeper than necessary and throw in just a pinch of 12-24-12 and with an inch of soil before putting in the plant. You do not want the transplant's roots to touch the fertilizer. As the plant adjusts to the garden soil it will send down a root and hit the food. (A pinch is exactly that – what you can hold between your thumb and index finger.)

As the veggies grow, more fertilizer will be necessary, but not much. The roots of most vegetables spread to considerable distances. A fertilizer applied to the base of the plant is not always the best method although it seems better because it is closer to the plant. If too near, it will burn young plants' roots or stems. Fertilizer should be carefully placed alongside the rows and worked with a hoe into the soil.

A regimen working for me is, first while forking the soil add about a quarter feed sack of extremely rotted chicken manure and two cups of unhydrated lime per ten-by-ten area. I use the pinch of starter (12-24-12) at planting and another pinch about three weeks after sprouting. If it is corn or any type of peppers a mix a half-cup of white calcium nitrate to three cups of starter to get the pinch. I never use this on tomatoes. My experience with tomatoes is to only use starter throughout their life and just a little bit every three weeks. Tomatoes need little nitrogen, but plenty of phosphorus and potassium. As soon as the plants begin to blossom, I hit each plant with a pinch of a mixture of two cups blue bearing salt (12-12-17-2), a half-cup calcium nitrate, and a half-cup of special green (15-20-2). Again, I toss just a pinch at the tips of the branches away from the stem. Good roots should almost reach that far. This is dry fertilizer that should be leached to the root tips by watering, rain, or heavy morning dew. As the plants continue to grow, I water twice a month with a teaspoon of calcium and a teaspoon of magnesium to two gallons of water. Every 3 weeks I hit them with another pinch of blue bearing salt and mixed with calcium nitrate.

The special green fertilizer (15-5-10-2) is great for leafy vegetables like bok choy and lettuce and vining vegetables like cucumbers, pumpkin, and squash. Carefully place it because one pellet on a leaf means a burn spot. Try your own mixes or use the factory mix straight from the sack. Corn especially needs a lot of nitrogen once it goes into tassel, but again be careful not to get it in the stalks.

Any fertilizer applied as a liquid to plant leaves is termed 'foliar.' The most basic foliar is animal manure soaked in water. Modern high-tech foliar fertilizers are very concentrated mixtures of nitrogen, phosphorus, and potassium with trace elements, even vitamins, chlorophyll, and growth hormones. The recommended dilution drastically reduces fertilizer burn. Dry fertilizer must be dissolved, by watering, rain, or dew. I find it easier to overdose with liquid fertilizer than granular.

Today's modern agri-store has many fertilizer mixtures in various forms. There are singles, or with trace elements like boron or magnesium, and combinations, inorganic or organic, mineral fertilizers, composts, and animal manure, to list a few.

Producing great veggies depends on available plant food in the soil. However, too much fertilizer, especially inorganic, can damage the plants causing them to not bear fruit or even die. Years of fertilizing without balancing the pH with limestone can make the soil toxic for all plants. Such small amounts of fertilizers are necessary for small home gardens. New gardeners usually apply too much - too close, believing 'more is better.' Burning happens when fertilizers are too close, too concentrated, or watered too soon after applying. I find it best to water about an hour before fertilizing.

Proper fertilizer use can enhance plant growth without polluting the environment. There are environmental dangers to fertilizer overuse. Fertilizer nitrogen is changed by soil bacteria into nitrates. Nitrates trickle into the groundwater or to streams and rivers. High nitrate levels in drinking water are dangerous to human health. Phosphorus can be washed into rivers and stimulate the development of algae in slow-moving water. Look at the rivers. Algae plants clog many and will eventually decay, reducing the oxygen from the water, killing fish, and increasing the chance of floods. A good rule is to use as little fertilizer as possible. That will save you money and keep the environment safer.

REMEMBER: No fertilizer will compensate for bad soil. Be sure to give your garden plenty of compost to maintain organic matter in the soil, promote good plant growth, and retain moisture.

Drawing: R. Douglas

Gardens are not made by singing 'Oh, how beautiful,' and sitting in the shade. ~ Rudyard Kipling

CHAPTER EIGHT
GARDEN CHEMICALS

The use of chemicals in agriculture is a highly debated topic. Gardening at home gives you a chance to keep your food safe from dangerous chemicals. There are basically two perspectives to gardening, use chemicals safely, or grow organically. (*More on organic gardening in Chapter 13.)* Hawaii's climate and soil make large-scale farming difficult without chemicals. A temperate climate, with freezing winters, kills a lot of insects and diseases, so there are fewer garden enemies to fight.

Everything in our entire world is composed of chemicals. In agriculture, there are two main types, pesticides, and fertilizers. The term 'pesticide' encompasses all chemicals that kill specific things, insects, fungus, or vegetation. There seems to be a 'cide' for everything. Insecticides control insects, miticides for mites, fungicides battle fungus, and bactericides fight bacteria. Herbicides kill vegetation. It may be necessary to control pests and diseases in commercial crops for better harvests. The use of chemicals is a decision that must be made concerning your garden.

In a perfect world, nature's balance would protect gardens. Gardens have good in-sects and bad insects. If the good breed too slowly, the good are outnumbered to combat the bad ones. Some bad have no natural enemies. Gardeners resort to their own chemical control program and all the bugs, good and bad, are exterminated. Diseased plants die or develop abnormally. Bacteria, fungus, insects carrying viruses or parasites can infect them. Bad soil, an extreme dry season, or lack of nutrients can also cause plant problems. Often two different causes have similar appearances in the abnormal plants. Nature's system balances pests and controls, yet gardeners may not want to lose a quantity of their produce, so they fight nature with chemicals.

One easy control is to be certain the seeds are fresh. The 'fresh year' should be printed on the packet or can. There are disease-resistant varieties of seeds. It is less expensive to pur-chase better seeds than to try and cure problems with chemicals.

An example of disease-resistant plants is the 'Gem Pride' tomato type. It is resistant to the Gemini virus transmitted by the whitefly. Once a regular tomato tree is bitten with the virus, the leaves curl up and mutate limiting fruit production. The Gem Pride resists the virus and bears best during the hot weather of the dry season when white fly is more prevalent.

Disease prevention rather than cure is the best garden technique. Keeping weeds out of your garden is a good start. Weeds hide many insects that carry diseases. Weeds compete for water, sunlight, and nutrients. Gardening is always an education mostly learning by too many trials accompanied by twice as many errors. I put in a few extra plants in case some get sick. It is less costly to let the plant die rather than try to cure the disease. Agricultural medicines are better used as preventatives than curative. Garden chemicals cost a lot of money!

Be certain you have correctly identified the plant's problems and decided it is serious, and all non-chemical measures have been tried. Many vegetable diseases have similar appearances, and the real trick is to know from experience what will be the cure. Beware of advice from other gardeners, as you may be their joke! But the real question is the cure worth the cost? If you have sx tomato, peppers, and eggplant; is it worth it to buy chemicals and a

A person who is growing a garden, if he is growing it organically, is improving a piece of the world. ~ Wendell Berry sprayer? Each type of plant has its usual enemies. Try to keep your garden balanced. Too much nitrogen fertilizer can support fungus. Water is necessary, yet too much water can rot the roots. Remove sick appearing plants immediately. Harvest and then pull the old plants.

Any chemical, natural or synthetic, can be poisonous in a certain dosage. Too much table salt can kill you. Pesticides are poisonous to specific insects or diseases, and in the prop-er use aren't harmful to the consumer. However, there is considerably more danger to the individual applying the poison. Used correctly, (that is the key), human-safe chemicals, in the correct application and dilution, should dissipate by the time a consumer makes dinner. It is the applier who inhales the spray without a respirator or gets it on their skin who must really worry.

Chemicals are why I decided to grow my own. I saw local farmers spraying virtually every day and immediately bring their crop to market. Some of them spray serious poisons like lanate on bok choy and citrus. A lot of these gardeners feel chemical use is a more modern tech-nique; yet pay little attention to the label warnings or directions. More is not necessarily better with garden chemicals; in fact, it usually increases the danger to you and your garden. The three 'Cs" of successful garden chemicals are 1-Correct problem identification, 2-Correct selection of the chemical, 3-Correct timing and application. All pesticides have specific instructions that reduce injury to both you and your garden plot.

Getting the correct medicine is the real trick. There are so many: synthetics, or botan-icals, insect growth regulators, broad or narrow spectrum, contact or systemic, short term vs. residual, pheromones, and many more. They are sprays, dusts, lacquers, gels, baits, smokes, and powders. Make your choice, carefully read the label and use only as directed.

If your garden has a serious problem, take a few sample plants to one of the govern-ment's agri-stations. They are extremely competent and extremely busy, so don't waste their time. With agri-shops it is best to know what you need walking in otherwise you must trust the experience behind the counter. Remember, now your loss not only includes crop damage, but the cost of the insecticides.

After gardening for several years, I believe a balance of preventative chemicals and good garden practices is necessary to grow commercially in a tropical climate. The garden chemical downsides are that they are not selective in their targets and regular use can develop resistant varieties of pests. Even in a small garden probems can be avoided by giving plants adequate distance so enough sun light will penetrate, provide good air circulation, and give careful fertilizing and watering. Plants with those benefits are more tolerant to disease than stressed plants.

When you've decided to become a spray-man, it is import-ant to read the chemical label for instructions and warnings and follow them explicitly. Wear rubber gloves and a respirator when using them, especially during the mixing. Wear long-sleeve shirts, pants, and boots. Never spray on a windy day. and always spray late in the day after the bees have visited your veggies.

Insecticides should kill the pest, but not other 'good' insects, dissipate quickly and are nonpoisonous to humans. Bees are 'good' insects. Almost all plants need bees or other insects to pollinate to reproduce seeds and continue the species.

Wear protective clothing and glasses, and don't smoke when spraying. Take chemi-cals seriously. The lethal dosage rate on the label is for ingestion and does not consider what lands on your skin. Bathe after all applications. **Remember, every stage of a pesticide's ex-istence from production, transport, storage, use, and disposal – is dangerous to humans and our environment.**

Chemical residues in food are dangerous. Check the label for the lethal dose rating of a chemical. If the number is small, the chemical is extremely poisonous. Gardeners and farmers need to be concerned with exposure limits. Many gardeners are questioning a chemical's potency from one batch to another or if the pests are becoming resilient to it.

There are botanicals, garden chemicals made from natural plants, which either kill or repel garden pests. Neem or neem oil, from the seeds of the neem tree, fights insects and fungus. Nicotine from tobacco fights aphids, whiteflies, and mites. Pyrethrum is obtained from the tropical chrysanthemum and is an excellent insecticide. Rotenone, derived from the derris tree, is a contact insecticide. Botanical repellants usually cost more than synthetic varieties.

Another non-synthetic approach is insecticide soaps produced from plant oils or animal fat. These soaps must directly contact insects such as aphids, whiteflies, mealybugs, thrips, and spider mites, but no residues remain on your plants.

Sulfur is the oldest known pesticide. It can be used as a dust, or a liquid, for fungus control, and fights some insects, especially on beans, tomatoes, and peas. There are several inexpensive natural ways to fight fungus and insects in your garden. A line of wood ashes or chalk (or limestone) around your garden will repel ants and especially leaf-cutter ants. The aroma of flowers, especially African marigolds, fends off some insects. A spray made from garlic skins with a few chopped garlic cloves and hot peppers mixed with two tablespoons of dishwashing soap in a gallon of water works against pests. (Try on only one plant first.)

To fight fungus, mix one kilo of cornmeal into a ten-by-ten area once a year. A spray can be made by combining one cup of cornmeal in a gallon of water and let sit a day before straining and using. Baking soda can prevent powdery mildew fungus, use one-tablespoon baking soda mixed with half a teaspoon dishwashing soap, in one gallon of water. Water your plants before using the mixture late in the day. (Try these remedies on one plant at first to see the effects.)

The nasturtium flower protects beans, broccoli, cabbage, cucumber, corn, and tomatoes by trapping whiteflies, aphids, and red spider mites. The French marigold will fight nematodes by trapping them if placed throughout the garden.

Soap can be used as an insecticide by dissolving one pound of common laundry deter-gent in six gallons of water. This is good for fighting aphids. Make tobacco water by soaking a pack of broken cigarettes in a gallon of water for a day. Mix a cup of this to a gallon of clean water and use as a spray. Nicotine is a lethal poison.

Unless you are farming on a large scale as a livelihood, I say use chemicals as a last resort. First, try only gardening with healthy seedlings and fresh, treated seeds. Water the roots in the morning and keep the leaves dry. Watering in the evening promotes fungus growth. Use minimal fertilizer, more nitrogen will not help but hurt the crop. Visually check your plants regularly. Occasionally use a magnifying glass to inspect stems and leaves. Pull sick plants immediately and keep the garden weed free. Rotate your vegetables. Pull the old plants quickly after the last harvest and place them downwind of your garden plot.

Every time I visit an agri-shop, I'm amazed at the variety and cost of garden chemicals, and the varied types of applications. Some chemicals work on contact, some are systemic, and are natural or synthetic, etc. The list continues with various chemical companies. Applied wrong they do harm. If you choose to use chemicals, it's best to learn about them first.

If the bee disappears from the surface of the Earth, man would have no more than four years left to live. ~ Albert Einstein

CHAPTER NINE
PESTICIDES IN FOOD

Eating fresh vegetables and fruit, safe from most chemical contaminants is one reason to have your home garden. Do you have any idea of the amount of pesticide remaining on the produce you buy at the market? It may look and taste great, but do chemicals cause the nice appearance? Serious and dangerous health effects result from continuous exposure to low lev-els of pesticides. Continual exposure may cause cancers; damage to the immune, endocrine, reproductive, and nervous systems. 6000 cases of cancers per year in the USA are suspected to be pesticide-related. Recent data links pesticide exposure and brain cancers and attention deficit syndrome in children. Parkinson's disease is linked to rotenone, the active ingredient in many pesticide products.

Some harmful things pesticides can do to humans are to cause low birth weight, birth defects, and interfere with child development and learning abilities. Pesticides can cause neurological problems and disrupt hormone functions. They cause a variety of cancers, includ-ing leukemia, kidney cancer, brain cancer, and non-Hodgkin's lymphoma. Children and fetuses suffer more from pesticides than adults because children's bodily systems are still developing. Children's systems are not developed enough to detoxify most pesticides. Pesticide residue in the unborn and infants can have lifelong effects.

In the U.S., several major organizations regulate the use of pesticides including the Environmental Protection Agency, the Food & Drug Administration, and the U.S. Department of Agriculture. More than 14 separate regulations govern the use of pesticides. All of these regulations are in place to help protect human health. Pesticides must be toxic to kill pests, but a pesticide can be useful only if it kills pests at a small enough dose that causes little or no harm to people, domestic animals, and wildlife. It is not easy to establish what dosage of a pesticide is 'safe' for people. Despite many studies done on the health effects of each pesticide, there is still uncertainty in the long-term health effects of many pesticides. It is essential pesticide ex-posure be minimized and the presence of pesticide residues in food be regulated and monitored. One of the current regulations in the US requires pesticide manufacturers to conduct toxicity testing before it can be permitted for use on products either directly or indirectly destined for human consumption. This includes feed for livestock. This toxicity testing determines the health effects of pesticides, and the level at which there are no toxic effects on children and the elderly, which are the most sensitive parts of the population. This **'No Toxic Effect Level' (NOEL** is the basis of the permitted residue limit. Regulations set the permitted residue level at a level from 10 to 100 times lower than the NOEL.

If a pesticide is tested and a NOEL cannot be determined, then it is unlikely to be permitted for use on food crops. This helps ensure that if a person, child or adult, eats a larger than normal amount of a particular food, or several different foods with the same or similar pesticide residue, they will still not reach the level of exposure required for a toxic effect to occur, even if they are more sensitive than the general population.

All of these safety precautions are used in the US, but the hazards of pesticides to human health are not distributed equally worldwide. Latin American farmworkers are more than ten times more likely to suffer pesticide poisoning than US farm workers. This is because products condemned/banned by the developed countries are often sold to undeveloped nations. Farmers in developing countries often handle extremely toxic pesticides while wearing little or no protective equipment. During the decade of the '80s more than 80% of all pesticide related deaths occurred in developing countries, but they only used 20% of the chemical pesticides. This is because of the ignorance of both workers and management. Garden chemicals are a huge business in the Caribbean. Herbicides and pesticides are the most used.

Check garden chemicals before you buy to be certain they are not banned. These are not the garden chemical names, but what makes up the toxin. Some banned pesticides include: aldrin, benzene hexachloride, cadmium compounds, chlordane, kepone, copper arsenate, DBCP, DDT, fluoroacetamide, mirex, silves, and thallium sulfate. There are many more banned chemicals and several severely restricted in usage as carbofuran, daminozide, lindane, and sodium arsenate. Everyone should be concerned about eating poison in the form of pesticides used on vegetables and fruit. Careless farmers use pesticides on their produce without allowing proper time for it to dissolve. Usually, pesticides, which include insect repellents, fungicides, and weed killers take a minimum of a week to dissipate to a safe level after spraying. Organic food is a great way to reduce or eliminate exposure to pesticides. Organic growing may not be possible for everyone, or in every food choice.

Sweet bell peppers, celery, lettuce, bok choy, and carrots are the vegetables most likely to expose consumers to pesticides. In several samplings at various markets, celery had the highest of percentage of samples test positive for pesticides, 94% followed by sweet bell peppers at 81%, and carrots at 82%. 80% of the celery tested had multiple pesticides on a single vegetable, followed by sweet bell peppers with 62% of the samples.

Sweet bell peppers had the most pesticides detected at 11, detected on one sample. Lettuce and celery samples had nine different pesticides identified. Sweet bell peppers were the vegetable with the most pesticides overall, with 64, followed by lettuce with 57, and carrots with 40.

Vegetables least likely to be contaminated with pesticides were onions, sweet corn, as-paragus, peas, eggplant, broccoli, and sweet potatoes. 51% of the tomatoes tested showed no pesticide residue, while 65% of broccoli, and 75% of eggplant / lady fingers had no detectable pesticides. Tomatoes had the highest likelihood of multiple pes-ticide residues with a 14% chance of more than one pesticide when ready to eat. Onions and corn both had the lowest chance with zero samples containing more than one pesticide. This is because both corn and onions have a protective disposable skin.

BE CAREFUL OF AGRICULTURAL CHEMICAL RESIDUE.

Fruit and vegetables to be wary of, always wash and peel if possible before consuming: peaches, apples, sweet bell peppers, imported and local celery, local cauliflower, cabbage, and broccoli, imported carrots, strawberries, and pears.

Consume with caution: spinach, lettuce, grapes, potatoes, hot peppers, cucumbers, oranges. Better choices: grapefruit, Portugal citrus, tomatoes, sweet potatoes, imported cauli-flower, and watermelon.

Best to eat: cabbage, bananas, mangos, pineapple, sweet corn, onions, and avocados.

HOW TO REDUCE THE AMOUNT OF PESTICIDE RESIDUE IN FOOD:

Grow as much of your food as possible. Wash your food with clean water (but no soap!) before it is cooked or eaten. Peeling helps reduce the levels of pesticides that may be on the surface, but some residues are absorbed into the food. Trimming excess fat from meats helps to reduce the amount of such pesticides that would be eaten. Pesticides have been found to accu-mulate in animal fatty tissue. Cooking helps reduce some of the pesticide residues in food that are not removable by washing or peeling. Specific pesticides are used for specific food crops. Eat different fruits, vegetables, and grains, not just one kind from one source. This prevents eating a particular food with the pesticide residues that it may carry concealed.

GROW YOUR OWN FOOD!

CHAPTER TEN - GARDEN SPRAYING

A spray can seems to be an integral part of today's garden equipment, whether it is to water, fertilize, or medicate vegetable plants. Spray cans may be either small and handheld in the liter to two-gallon size or worn as a backpack with three to five-gallon capacities. Although a spray can saves time and water, especially in a drought, they are dangerous to the health and welfare of the user unless safety precautions are strictly followed.

First, if spraying anything but water, always dress in long pants and a long sleeve shirt, boots, and rubber gloves, wear a respirator certified to protect against toxic chemicals. Most garden chemicals are poisonous on contact with your skin. The skin will accept the chemical into the human body just like a plant takes it in. However, instead of attacking an insect, bacteria, or a fungus; it attacks the liver. Most serious farmers who tend large gardens wrap a cloth over the respirator to avoid and contact with the spray due to a breeze. Sunglasses or safety goggles are also recommended.

Wear the respirator and gloves, especially when filling the sprayer because your face is closest to the chemical when it is filling. The chemical will either give off a cloud if it is a powder such as a fungicide, or stench if it is an insecticide.

Select a sprayer that can be handled easily and the weight, when filled with water, isn't too heavy. You can calculate the weight you will be carrying as a liter weighs 2.2 pounds. The weight of one US gallon of water is approximately 8.35 pounds. An imperial gallon (UK mea-surement) of water weighs 10 pounds.

Second, learn everything you can about the chemical you will use. Garden chemicals are expensive and dangerous. Done wrong, it's a dangerous waste of time and money.

Teach yourself how to repair your sprayer. They are simple pumps with few parts. Read the instruction manual and be certain before buying that repair parts are available. A good tip is to fill the sprayer with water to the correct level before adding the chemical. Pump it and spray. If it doesn't work or is clogged, you haven't wasted any chemical and do not have to work on the sprayer while it is filled with a toxic liquid. Always clean out your sprayer when done and flush with fresh water to prevent the chemicals drying and clogging tiny orifices.

Vegetables may contain very minute remainders of the chemicals after it has grown through its life cycle. The real danger with chemicals is to the person who applies them. Never smoke, or drink while spraying. Clean yourself immediately after spraying. Wash your hands, face, and all exposed areas with soap and water. Shower as soon as possible, washing your hair. Gargle with clean water and clean your nose, neck, and ears where residue may remain.

Some basic guidelines for safe garden spraying are: Never spray during a windy day. The wind may carry the chemical to vegetables that don't need it or may be harmed by it. The wind will swirl the fine spray back onto you. Use any type of spray in the early morning or the cool of the evening. Try to spray after the bees have pollinated. You do not want to kill bees, as they are necessary to make fruit from your vegetable plants. Target just the area you need to treat. Be careful... try not to harm the good bugs! You don't want to run off your allies. Do not spray when temps are above 80 degrees Fahrenheit! Your plants may 'burn' or react to what you are using in excessive heat. Always spray only a small portion of the plant material first. Wait 24 hours to observe any negative reaction. Proceed if there is no damage. Using more of a chemical spray is not better. If you are not getting good results don't increase the strength of these remedies without testing first. There are sprays made from combinations of relatively safe items usually found in your household. Relatively safe means anything taken over the limit will be toxic to the human body. **Warning:** As a precaution, always test on one plant first to check for any negative reactions. Do not proceed if there is any visible damage such as burning or discoloration.

APPLE CIDER VINEGAR FUNGICIDE is good for leaf spot, or powdery mildew. Mix three tablespoons of apple cider vinegar (five percent acidity with one-gallon water and spray in the morning on infested plants. Good for black spots on roses

BAKING SODA SPRAY is good for anthracnose, early tomato blight, leaf blight, powdery mildew, and as a general fungicide. Sodium bicarbonate commonly known as baking soda has been found to possess fungicidal properties. It is recommended for plants that already have powdery mildew, to use clean water to wash all infected leaves before to spraying. This helps to dislodge as many of the spores as possible to help you get better results. Use as preven-tion or treatment at the first sign of any disease. Mix 1 TBS baking soda, 2 ½ TBS vegetable oil with one gallon of water. Shake very thoroughly. To this mix add half teaspoon of pure Castile soap and spray. Be sure to agitate your sprayer while you work to keep the ingredients from separating. Cover upper and lower leaf surfaces and spray some on the soil. Repeat every week as needed.

CHIVE SPRAY is good for preventing apple scab or downy mildew on cucumber, pumpkin and zucchini. Put a bunch of chopped chives in a heatproof glass container, cover with boiling water. Let this sit until cool, strain and spray as often as two to three times a week.

COMPOST OR MANURE TEAS are an inexpensive foliar spray. Many people have success with manure tea keeping blight and other pathogens away from plants. Soak the area around plants and use it as a foliar spray. Do not use on seedlings as it may encourage damping-off disease. Fill a thirty-gallon barrel with water. Let sit for a day to evaporate chlorine and other additives. Add about 4 shovels worth of manure to this and cover. Let it sit for two weeks, stirring once a day. Strain and apply as needed.

NOTE: Different manures supply various nutrients when used as above. Chicken manure is rich in nitrogen and good for use for heavy feeders such as corn, tomatoes, and squash. Cow manure has potash and better for root crops. Rabbit manure will promote strong leaves and stems. Horse manure will aid in leaf development.

COMPOST TEA - Make and use just the same as you would the manure tea. This is another terrific reason to compost all those prunings, grass clippings, and kitchen wastes.

No. 1 GARLIC FUNGICIDE SPRAY is good for leaf spot and mildews. Combine three ounces of minced garlic cloves with one ounce of mineral oil. Let soak for a day or longer. Strain. Mix one teaspoon of fish emulsion with sixteen ounces of water. Add one tablespoon of castile soap to this. Slowly add the fish emulsion water with the garlic oil. Kept in a sealed glass container this mixture will stay viable for several months. To use: Mix two tablespoons of garlic oil with one pint of water and spray.

No. 2 GARLIC FUNGICIDE SPRAY is good for fungicide and as an insect repellent.

In a blender combine one whole head of garlic, three cups water, two tablespoons canola oil, four hot peppers, and a whole lemon. Blend until finely chopped. Steep mixture overnight. Strain through fine cheesecloth. Use at a rate of four tablespoons per gallon of water. Store the unused portion in the refrigerator.

HORSERADISH TEA FUNGICIDE - The cleansing properties of horseradish have been known for more than a decade. This method has proved to be effective, and inexpensive.

Process one cup of roots in a food processor till finely chopped. Combine this with 16 ounces of water in a glass container and let soak for a day. Strain liquid, discard the solids. Now mix the liquid with 2 quarts of water and spray.

HYDROGEN PEROXIDE TREATMENT is used to prevent bacterial and fungal problems. Simple hydrogen peroxide that you can buy most anywhere will prevent the disease spores from adhering to the plant tissue. It causes no harm to plants or soil, however, don't use on young plants. Spray plants with undiluted three percent hydrogen peroxide. Be sure to coat the tops and bottoms of leaves. Do this once a week during dry weather and twice a week in wet weather. This works as a preventative. If you already have problems, use this as a direct treatment.

MILK FIGHT MILDEW - Milk's natural enzymes and simple sugar structures can be used to combat various mildews on cucumber, tomato, squash, and zinnia foliage. This works by changing the pH on the surface of the leaves, so they are less susceptible to powdery mildew. Use an equal mixture of milk and water. Thoroughly spray plants every four days at the first sign of mildews or use weekly as a preventative measure. Milk can also be mixed at a rate of one ounce of milk to ten ounces of water and used as a spray every week to treat mosaic disease on cucumber, tomato, and lettuce.

TOMATO VIRUS PROTECTIVE SPRAY— Skim milk will prevent many viruses that attack tomato plants. This protects the plant surface against disease spores. The skim milk provides the tomato plant with calcium. A calcium deficiency is common in tomato plants. An-ti-transpirants are used on Christmas trees, cut flowers, newly transplanted shrubs, and in other applications to preserve and protect plants from drying out too quickly and can be purchased at most garden and cut flower shops.

Mix a half teaspoon of antitranspirant (like Cloudcover, Wiltpruf, etc.) with eight ounces of skim milk, and one gallon of water. Spray plants. NOTE: an equivalent of prepared powdered milk may be substituted for the skim milk.

DAMPENING OFF DISEASE - Always use a sterile growing medium like potting soil for your seed starting as these should not contain the fungi that cause damping-off. If using regular dirt or compost, put it in a suitable container such as old baking dishes in an oven heated to three-fifty and shut off. Let stand overnight. This will kill all bacteria, fungus, and most unwanted seeds. Water your seedlings with warm water that has been left to sit for an hour or more to dissipate most of the chemicals that are present in tap water. Using cold water stresses the seedlings leaving them vulnerable to harmful organisms.

CHAMOMILE SPRAY is an excellent preventative for damping-off. Use on seed starting soil, seedlings, and in any humid planting area. Chamomile is a concentrated source of calcium, potash, and sulfur. The sulfur is a fungus fighter. This can also be used as a seed soak before planting. To make: pour two cups boiling water over a quarter cup of chamomile blossoms or packaged tea bags. Let steep until cool and strain into a spray bottle. Use as needed. This keeps for about a week before going rancid. Spray to prevent damping off and anytime you see any fuzzy white growth on the soil. Chamomile blossoms can be purchased at larger grocery stores.

REMEMBER

Learn how and when to apply the specific mixture you want to spray. Know the amount to apply - such as drench the plant, just the roots, spray under the leaves, or give the entire plant a light spray. The amount of the mixture that hits the plant is a ratio of the dilution of the solution, the pressure you have pumped the spray tank, and the opening of the nozzle. Try to keep your spray uniform and keep it off of your skin. Wear a respirator, boots, gloves and glasses. Most chemicals are toxic.

If you want to have a balanced garden, feed the soil and build the ecosystem.
~ Cindy Conner

CHAPTER ELEVEN – WEEDS

Weeds are undesirable vegetation that grow naturally, but are not intentionally planted among your crops. Weeds are like unwanted relatives who seem to know you've just bought a pizza and a bucket of chicken. Weeds are competitive and persistent plants that interfere negatively with human activity. Weeds suck nutrients and moisture that your intentional garden plants require. In nature, without the human intervention of transforming a plot of land into a productive garden weeds have purposes. But to the gardener and farmer, weeds are plants that must be controlled, in an economical and practical way, in order to produce food.

No plant is a weed until it grows in an area that's designed for a purpose. While we intentionally plant certain cultured vegetation for our own purposes, mainly food for survival, nature evolves 'weeds' for their own survival. In nature's grand plan, weeds hold soil, shade small saplings so they grow into tall trees. Weeds and their seeds provide nests and also feed a multitude of insects, birds, and small animals, including garden villains such as slugs and snails.

Weeds can also be good for bees to gather pollen for honey. Some weeds are nec-essary for traditional medicinal treatments. Weeds die and make organic compost, but that compost is filled with more seeds of undesirable weeds. Some weeds are edible. None of the preceding natural elements are good for your planned garden.

Because of natural selection, weeds usually produce an overabundance of seeds that rapidly germinate, spreading across your home garden plot. Weed seeds can lie dormant for months, seasons, even years, withstand floods and severe droughts, readily adapting to new conditions. Weeds thrive when they get the regular moisture of a fresh tilled, well-maintained home garden. Indirectly, weeds negatively affect all with less food production leading to higher prices.

If a home garden is not regularly weeded, usually by hand on weekly intervals, weeds will win and strangle your crops. Weeds reduce garden yields by competing for wa-ter, light, soil nutrients, and space. These unwanted plants provide homes to diseases and insects that will infest your garden and waste your hard work. I find the best time to weed is a day after a rain or a heavy watering. Weeds' natural time clock senses moisture and more will sprout. The soil is soft and pulling the weeds out including the roots is easier.

Weeds will easily outnumber your garden plants if left untended. If you toss weeds into your compost pile, you are recycling your problems. Weeds love the warm moist compost and that preserves their seeds' viability. In all cases, it's better to pull the weeds before they flower and make seeds.

If only vegetables grew as fast and as hardy as weeds! Weeds are super-plants that develop difficult-to-destroy root systems. New weed plants sprout from the roots, mature, and go to seed quicker than most vegetables. Many weeds have evolved natural systems to spread seeds over a large area.

Herbicides, chemicals that kill vegetation, work slowly, might require several applications, and aren't selective. If, while applying, the wind changes direction and blows on your garden plants, the chemical will harm or kill vegetable plants. Chemical weed killers can be harmful to the user. (See the previous Chapter 10.) I use a pre-emergent weed killer shortly after I have my home garden ready to plant. First, after tilling the soil, I let it sit in the hot sun for a few days. The sun and heat can kill some seeds and root systems of weeds. Then I spray the pre-emergent, which keeps weed seeds from germinating. This can be liquid or granular.

Both seep into the soil, and create a barrier around the weed seeds. As long as there's not a hard rain that washes these chemicals away, your home garden plot can be mostly weed free for up to 3 months.

If you let weeds tower over your tomatoes or intermix with your lettuce or bok choy you'll have a tough time getting them out. Weeds' roots can intertwine with your veggies. Pull a big weed and you can damage your good plant. When weeds are small and the soil is moist and loose, their roots have a weaker grip and are easier to pull out. Do a walk-through of your garden at least every other day; it'll take only a few minutes to yank out the young weeds. You may be tempted to reach down and snatch, but go slow, grab the weeds by the base and pull out the roots.

Removing weeds by hand depends on the type and size of weeds and the size of the area that requires weeding. Consider your daily-weekly schedule; how much time can you allot to weeding? Hoeing chops weeds just below the surface of the soil. A good long handled hoe kept sharpened with a bastard file, carefully used, standing, can clean a good-sized home garden in a few cool evenings. Remember, the emphasis is on cool. If you choose to get up-close-and personal with the weeds, try kneeling or sitting on a bucket and use a short handled claw/rake.

Another easy method to fight weeds is to cover the soil with a material that will block the sun and elements so weeds can't sprout. Materials like cardboard and newspapers will work. More than one news page layered, because thin paper will dissolve with the elements. Organic mulch is another weed blocker, especially wide leaves – bananas are great. A few wide leaves will cover a lot of area. I use white plastic more than black because black grabs and holds the sun's heat. I use white plastic because it reflects the sun. In some cases, like pineapples, I roll out a sheet of white plastic and poke holes where I want the plants. There is also a black or a green loose woven cloth named weeds preventer or landscape fabric. This is relatively inexpensive and lasts for a few years. Landscape fabric has the best results combined with a surface mulch.

All of the previously mentioned methods of covering the soil will stop weeds from growing by creating a barrier that doesn't let them reach sunlight, but unfortunately also discourages garden-friendly earthworms from freshening the soil because they can't reach the surface.

If you stay constant with your home garden and don't permit your efforts to lapse for a week or longer, then weeds shouldn't become a problem. Remember, your garden will always be a work in progress and only must measure up to your personal standards. Kneeling or standing and hoeing wearing your wide-brimmed hat with a chilled beverage close, makes weeding more palatable. Share the labor of cleaning your garden patch with a relative, friend or neighbor and chat. If alone, add headphones and listen to sports, news, or music. Gardening is fun.

If you fail to sow good thoughts in your mind, the weeds of evil thoughts will take root there and eventually they will take up all the space and not let good thoughts flourish.

~Awdhesh Singh: 31 Ways to Happiness

A weed is a plant that has mastered every survival skill except for learning how to grow in rows. – Doug Larson

CHAPTER TWELVE
GARDEN PESTS

A home garden is a pride, meant to be enjoyed. The backyard garden should be fun, productive, and save money while providing fresh vegetables for your family. Without proper precautions, your garden may also be feeding, and providing a home for unwanted, very unpopular, and certainly counter-productive guests.

Household pet dogs are a joy, and almost a necessary early warning system in today's world. Your garden's soft, cool, tilled soil can be an attractive bed no matter what you have planted. Our dog seems to wait until the soil is prepared and planted before he chooses that exact place to take a snooze. Meeting crushed pepper and tomato seedlings, breeds harsh shouts and reprimands.

The stakes we later use to support the plants vertically are laid horizontal across the garden to make it uncomfortable as a doggie bed. If it is a grow box try making a fence barrier by staking pipes or sticks into the block cells. The best and safest way to keep a dog from your garden is to surround chicken wire. Hopefully, it won't persist by jumping or digging.

Birds, especially squawking crows, are a constant threat to gardens from planting till the picking. Every wild creature's daily goal is an easy meal. Ripe tomatoes and peppers are at the top of the menus. The seeds you plant are vulnerable from birds scratching them out of the soft soil. Usually the odor of insecticide used to ward off the even more damaging mole crickets will keep birds away. Try tying string between stakes above your plants. Wrap aluminum foil, and especially pie tins on the strings. The breeze will flip the pieces and tins causing reflections and sounds that should keep birds away. The downside of staked strings are they provide a closer perch for the more adventurous winged garden predators.

A scarecrow – something that looks like a person is a time-proven simple method. Stuff an old long sleeve shirt with grass. Fix it to a staked cross, and set a ball on top as a head with an old cap. Make it simple, and move it around the garden every few days. Try to do any repellent tricks in the evening so the bird's or dog's curiosity isn't aroused.

Rats and mice do damage, especially to pumpkins, squash, and peppers. Rats, like birds, want seeds and will eat a shaft into a pumpkin and clean out the seeds overnight. If your garden is small, try to keep everything tidy. Rats and mice will take anything as string, old sacks, and coconut husk to construct a nest for breeding. Then you'll have more to deal with. Traps are recommended as pets can be harmed by misdirected poison.

Snakes are creatures that may frighten you from your garden. Always be careful and look before you put your hand anywhere dark or shaded. Nonpoisonous snakes might be frightening, but the little coral snakes are lethal. It is best to consider all are extremely dangerous and lethal. Snakes like the same places as rats; don't leave sacks, boxes, or wood around where they can use as shelter. Mosquitoes swarm at dusk. Either leave the garden, or have repellent that works. If you have chores that can only be done as the sun sets; I recommend the old trick of burning coconut husk. Horse or deer flies are a blight in the morning. These are the big flies that can land, and you don't feel a thing until they take a bite. The best thing is a thick shirt.

There are more than twelve thousand types of ants that have colonized almost every landmass on Earth. Three types especially hinder gardening. The atta variety is a prolific leaf-cutter. This type of ant is big enough to easily see, especially when toting a bigger piece of leaf. Leaf cutters can decimate a garden. The best time to view is either in the evening or use a torchlight at night. They follow a trail carrying your vegetables' tender leaves to their nest to grow the fungus they feed on.

The second type of ant enemy is the fire ant, also called scorpion ants. They are the size of the atta, but their bite feels like a wasp sting. The sharp pain is acute, and lingers. Fire ants nest in the soil, often near moist areas, such as river banks, the edge of drains, or in old, rotting banana stools. Usually the nest will not be visible. All ants love to make a home in the blocks used for grow boxes. Fire ant nests can be as deep as five feet. These biters makes me believe ants are really descendants of wasps.

The last type is the small biting ants, and they are the most common unwelcome garden visitor. These little ants are usually about a sixteenth of an inch, just little black dots that can inflict a bite that will immediately itch. We call them 'biting ants,' or several other names not fit to print. They will swarm on you if you disturb their nest. Both the big and small biting ant attacks will swell into painful bumps, especially when stung repeatedly by several at once. The bump usually forms a white pustule, which can become infected if scratched. If left alone, nearly impossible, it will go down within a few days. If the bites become infected, they can turn into scars. Topical steroid creams as hydrocortisone, or one containing aloe vera are good to rub on the afflicted areas. Regular toothpaste can offer quick and simple relief. A simple solution of half chlorine bleach and half water applied immediately to the area can reduce the pain, itching, and, perhaps, pustule formation.

The first step is to identify the ant enemy. Once you know what you're dealing with, your tactics will vary. A spray of a recommended chemical will kill them on the spot. Ants will not cross a border around your plot of powdery wood ash. Spread a few packets of aspartame, 'Equal or Nutra Sweet', around your garden and puncture each. It's kind of scary that this sweetener kills ants. Sprinkle cornmeal around the plants the ants are feeding on. Supposedly cornmeal impedes fungus, and most ants feed on fungus. The cornmeal upsets their entire feeding structure.

Make a simple inexpensive ant poison by mixing an eighth of a teaspoon of boric acid with a teaspoon of jelly and place on an ant path. Don't add more as they either won't eat it, or won't bring it back to the nest. Some ants prefer fat over sugar, use some vegetable oil in the recipe, or peanut butter. Move your bait stations every two to four days to keep it fresh, occa-sionally mixing up the materials you use. It may take a couple of weeks to completely kill all of the nests near your home and garden, but you should see an immediate reduction in the number of ants after only a few days. With consistency and perseverance, you can get rid of your ants permanently for only pennies compared to commercial ant bait.

Commercial baits take longer to work, but it kills the nest. Cut the neck from a plastic soda bottle and stuff half a sweet orange inside and cover with the ant bait. A worker ant comes across it and carries it back to the nest to share. When many ants do the same, the colony dies. Apply baits wherever ants are foraging and at a time when they are actively looking for food. A mature leafcutter colony can contain more than eight million ants, mostly sterile female work-ers. That's a lot of mouths for a garden to feed! Nests are founded by small groups of queens, or a single queen. Even if only one queen survives, within a month or so the colony can expand to thousands.

Other garden predators are the slug and snail. Both are members of the mollusk group and are similar except slugs lack the external shell. Moisture is critical to their survival and is why they are active only at night or during cloudy days. On sunny days they are hiding in moist, shady places. During hot, dry weather they seal themselves off with a membrane while attached to tree trunks, fences, or walls. An infestation of these slimy critters is easy as adult snails and slugs each lay about 80-100 eggs and they may do this up to six times a year.

Land is not merely soil, it is a fountain of energy flowing through a circuit of soils, plants and animals. ~ Aldo Leopold

Slugs completely ate rows of sweet pepper trees. We discovered what was eating the plants by searching at night with a torchlight. We used table salt, sprinkling it directly on them. The African snail is another garden villain. One trick is to place bowls of beer in the garden. The slugs and snails are attracted and drink themselves literally to their death. A barrier of oat bran will kill slugs but it's only useful in a small plot.

The last garden villains are scorpions and centipedes. Once the plot is cleared, scorpions are seldom seen. Centipedes like rotting wood, so that should be removed. The main deterrent to most of the garden bad guys is just to keep it clean of debris and waste. Be careful around the compost pile as that's where you want insects to do their job.

Enjoy your home garden, don't let a few little pests take all the fun and veggies from it.

NOTES:

If all mankind were to disappear, the world would regenerate back to the rich state of equilibrium that existed ten thousand years ago. If insects were to vanish, the environment would collapse into chaos. ~ E. O. Wilson

CHAPTER THIRTEEN
THE BENEFITS OF LIMESTONE

If your soil is too acidic, then nutrients will not be available to the plants even if they are present. The letters pH stands for Potential of Hydrogen and measures the molar concentration of hydrogen ions in the solution and as such is a measure of acidity. That may seem a bit difficult to understand for us, non-chemist gardeners. The pH scale runs from 4.00, highly acidic soil, to 8.00 which is alkaline. 7 is neutral.

Various types of pH test kits are available. Simple ones require mixing a soil sample with water and comparing it to a color chart. More expensive are electronic meters, which read the soil through a metal stake. Whichever kit you use will come with instructions and will give you a reading. Never make a judgment on just one test. You may have hit a spot with particularly high or low pH. Take samples from several spots, and this will give you a much better general view of your soil's acidity or pH level.

The government will come and take soil samples. This is free but it will take some time, like a few months – to get the results. You can take your own samples per their instructions, and take it to an agricultural center. This is quicker and usually free. Without an accurate soil pH test, there is no way to know if lime is a friend or foe.

To **LOWER** soil acidity, you must **RAISE** the pH value. That always confused me. Reducing soil acidity will help deter some weeds because they are evolved for acid soils, unlike our garden plants. Opposite to what you expect, adding manure year after year will reduce soil fertility by making it too acidic so the plants cannot access the nutrients.

Different plants require different levels of acidity. Most vegetables thrive when the soil is slightly acidic at a pH level between 6.5 and 7. Potatoes tend to prefer a lower pH, a more acid soil. Members of the Brassica family – cabbage, cauliflower, and broccoli prefer slightly alkaline soil, with a pH slightly higher than 7.0.

Agricultural lime or garden lime is gray and made from pulverized limestone. As well as raising the pH, it will provide calcium for the crops and trace nutrients. Dolomite lime is similar to garden lime, but contains a higher percentage of magnesium.

White hydrated lime is produced by a two-stage process. First rock limestone is burned in kilns. This produces quick lime, which is highly caustic and cannot be applied directly to the soil. Quicklime reacts with water to produce slaked, or hydrated, lime. Quicklime is spread in heaps to absorb rain and form slaked lime, which is then spread on the soil. Their use is prohibited by the organic standards and while fast-acting, the effect is short-lived in comparison to garden lime.

Garden soil needs a total of 17 individual nutrients in the right amounts to grow beautiful plants. Carbon, oxygen, and hydrogen are taken in as carbon dioxide and water.

Although air contains vast quantities of nitrogen, most plants cannot use it. Instead, plants take up nitrogen from the soil together with other nutrients such as phosphorus, potassi-um, calcium, magnesium, etc. These nutrients enter the roots dissolved in water.

All 17 nutrients are needed for proper plant growth. No single nutrient is less important than others, but the individual nutrients are needed in different quantities. Soil typically contains an ample supply of micronutrients such as boron, aluminum, manganese, etc. Nutrients required in larger quantities: nitrogen, phosphorus, potassium, etc. must be resupplied because a healthy garden depletes these resources faster than nature can recreate them.

A too low pH level will make heavy metals such as aluminum and iron very mobile in the soil. That can cause the plants to get too much of these essential nutrients and poison themselves. A too high pH will tie up nutrients such as iron and phosphorus and starve the plants. The goal is to balance pH around 6.0-7.0. This can be done by liming once every year or every 2-3 years. The timing for liming depends on soil tests. It neutralizes soil acidity making the fertilizer more accessible to plants and is a natural source of calcium and magnesium.

Horticultural lime can be used as a soil amendment and is a blend of calcium carbonate and magnesium carbonate. The white 'crusty stuff' on showerheads and faucets is largely comprised of calcium and magnesium carbonates. Calcium is necessary for cell division and healthy plant growth. Often, when soil is depleted of calcium, plant leaves may curl, yellow, and fade. When magnesium is deficient in soil, leaves turn yellow, vegetables lack flavor with a poor yield. When your garden is missing both calcium and magnesium, other nutrients are more quickly depleted.

Old plasterboard, or gypsum, can sometimes be found for free at home remodeling sites. Gypsum is also called hydrated calcium sulfate, and contains calcium and sulfur, two elements vital to plant growth. If you get this free and all you have to do is smash it up, it is a bonus, but I would not search it out. Gypsum will loosen clay soil and will provide better soil drainage.

The best time to apply limestone is just at the end of the dry season, after the first rains when the soil is easy to fork or plow. Lime usually takes three months to combine with the soil to lower acidity. The soil will be ready when the rainy season has passed. Spread the limestone evenly. Applying too much can damage your plants. When you are finished, make sure to wash your hands thoroughly. Your garden should show great improvement.

Amount of Lime to Raise Soil pH from one point (from 5.5 to 6.5)

SOIL TYPE	POUNDS PER SQUARE YARD/METER
Clay	1.6
Sand	1.3
Organic-cleared forest	2

Limestone in heavy sacks.

A garden is a grand teacher. It teaches patience and careful watchfulness; it teaches industry and thrift; above all it teaches entire trust. ~ Gertrude Jekyll

CHAPTER FOURTEEN – COMPOST

Compost is decayed organic matter that naturally occurs. Compost can be made from a mixture of garden, home, and kitchen waste to fertilize and condition the soil. Gardeners have been making compost since crop cultivation began. Compost is well-rotted plant matter that improves fertility, structure, and water absorption. It's a natural process; all plants die and decompose.

In 2018, the Hawaiian Legislature passed a bill to assist schools with compost projects to produce nutrient rich soil. The Big Island and others have very organized organic refuse collection. Most islands publish free compost guides.

The purpose of composting plant materials is so that it decays, then it can be easily worked into the soil. By adding substance to the soil, compost improves water absorption. This is a real benefit for sandy soil. Adding compost prevents soil from packing tightly together and that aspect makes the growth and spreading of plant roots easier. It adds some nutrients and better retains fertilizers. All that translates to healthier plants. Every garden can be improved by adding compost, and it is good potting soil for houseplants or seedling trays.

Plant (organic) material decomposes at a rate depending on the size of the pieces, and exposure to elements. Nature decays plant material using microorganisms living naturally in soil. As these organisms eat, their wastes becomes compost. This process creates enough heat to keep the feeders alive, kill off nearly all harmful bacteria, and unwanted seeds from weeds. The internal heat is benefited by hot weather. In the best conditions, it'll take about 6 months to make compost.

Everyone has to clear the bush, weeds, and remove old plants. Locate your compost pile in a place convenient to your garden plot so it's not a long haul. It should be located where it gets a lot of sun. Keep it somewhat out of the way as it will tend to be unappealing and may attract some insects and perhaps rodents. An easy way to pile compost is to first get old concrete blocks, and stack them without mortar about four feet square. Start with three or four rows high and you can easily increase if it is necessary. Keep about an inch space between the blocks. These spaces permit air to circulate so the organisms can keep feeding comfortably. Fresh air can be further increased to your pile by weekly stirring it with a fork and keeping it moist with water. About a week after a big clean up, stick your hand into the pile and feel the generated heat.

Your pile is the place to dump your garden and kitchen waste such as cut bush, old plants, weeds, leaves, fruit and vegetable remains and peels, eggshells, shredded newspapers and cardboard, sawdust, wood ashes, and manure. Your sick garden plants should not be included in this pile as they may transmit the disease. To speed up the decay, get a couple of buckets of rich rotted manure and dump on your pile as the height increases every foot. This will increase the feeder organisms. Earthworms, if you can find them, also increase decay. The ideal pile height is three to four feet. The compost will be ready when it is dark brown and crumbles easily. Remove the concrete blocks, load it into a wheelbarrow, and work the compost into your home garden plot.

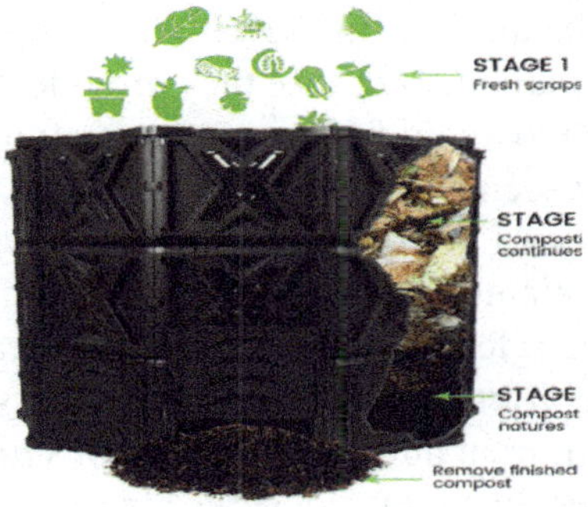

Compost is the end and another beginning for gardens. All gardens need to be constantly weeded and cleaned between plant ings. Keep a covered bucket in the kitchen for rotted or overripe fruits and vegetables, peelings, eggshells; and your compost will grow fast. A neat compost pile obscured from public view, will benefit your gardening efforts. Compost is another of nature's good cycles.

If you build up the soil with organic material, the plants will do just fine.
~ John Harrison

CHAPTER FIFTEEN – ORGANIC GARDENING

GO NATURAL - GROW ORGANIC - STAY HEALTHY

To most people, organic gardening means simply not using synthetic fertilizers or pesticides. Yet to 'go organic' you need to adopt a philosophy that the course of nature doesn't need any changes, especially chemical-wise. An organic gardener comprehends the cycles of nature, in soil, water, sun, air, and weather, work in agreement, and tries to replenish and preserve all resources. 'Organics' try not to use synthetic fertilizers or pesticides whose residues could harm the natural process. Since most home gardens are small, organic gardening can be a relatively easy and inexpensive alternative to chemical medicines and fertilizers. Most farmers and gardeners, who use synthetic fertilizers and pesticides, don't understand the chemical 'fix' is quick, but doesn't last. Synthetic fertilizers are never fully absorbed by the vegetable plants and the remainder stays in the soil or is drained off with rain or irrigation.

Organic gardeners believe all living things exist for one another in a natural system. Everything in nature is related. The treatment of our earth is what we leave for future generations. Organics use natural materials and methods, and avoid using chemical synthetics that may undermine our surroundings.

> ***It is vitally important that we can continue to say, with absolute conviction, that organic farming delivers the highest quality, best-tasting food, produced without artificial chemicals or genetic modification, and with respect for animal welfare and the environment, while helping to maintain the landscape and rural communities.*** ~ King Charles III

Soil maintenance is the most important aspect of organic gardening. The majority of garden problems are caused by poor soil. Increasing your soil's fertility is the first step. Synthetic fertilizers do not improve the soil. Soil doesn't quickly rejuvenate. It will take planning and work to change from chemicals to organic. Diseases, insects, and weeds will always need attention, but there are natural alternatives.

Compost is the best and least expensive natural fertilizer. Composting is a great habit to acquire as it takes healthy vegetative waste from the toxic landfill. Great soil has a 'fluffy' texture and easily absorbs water. Also, good bacteria in the organic matter supply nutrients via loose soil at the plant's roots. Household refuse as rotted fruits, vegetables, weeds, bush and non-treated sawdust, coffee grounds, shredded paper, etc. can be composted. Every time a section of your garden is replanted, the soil can be recharged by forking a generous amount of compost (at least an inch) into it. Good compost will help balance the pH level.

Controlling weeds is very important to the organic gardener because they provide shelter for insects and diseases. Sick plants should be destroyed to avoid further contamination.

Growing the same crop in the same place at first seems like a good idea since you should be able to refine your gardening techniques and have a better harvest with less work. Not so, definitely not so! As you have been feeding your plants the same nutrient blend, insects and diseases have also been feeding. Your soil will be infested with insects and diseases that flourish with a particular vegetable. Always rotate the different vegetable sections in your garden. This will confuse the insects and diseases, bring different nutrient combinations to the section, and discourage certain weed types. Some vegetables, like corn, consume almost all the nutrients in a section of your garden. Beans replenish nutrients, especially nitrogen. So, beans should be planted in the section following corn. I recommend moving rows of beans, especial-ly green "string' beans, to different row locations every three or four months. Green beans are tasty and they certainly revive the soil. Peanuts are legumes or beans, and add nutrients to soil especially when the entire bushes are forked back into the soil.

Mulching is covering the spaces between your plants with materials like bamboo leaves or dead grasses. This technique conserves water, nutrients, and fights weeds. The soil around a transplant should be covered with mulch to reduce root shock from the sun and hot weather. Leaves or shredded newspapers will work to retain moisture and protect roots. Wet the garden section before and after covering it with mulch.

The location you choose for your garden can influence the health of your plants. It should get at least six to eight hours of direct sun, drain well, and have a good flow of fresh air. Choose plants that have been adapted to fight certain diseases.

Animal manures are probably the best organic fertilizer. Use manure that has rotted for several months. It should also be mixed into the compost pile. Manures vary greatly in their content of fertilizing nutrients. To recharge a ten by ten plot at least twenty-five pounds of cattle manure are needed to be forked in, but only twelve pounds of chicken or sheep. This is one typ-ical farmer complaint, so much more manure is needed versus 'synthetic' chemical fertilizer. Actually, the average nutrient value of manure is ten pounds of manure provides one pound of nutrients. Your soil benefits more from the decayed plant matter. Synthetic fertilizers are never fully absorbed by the vegetable plants and the remainder stays in the soil or is drained off by rain or irrigation, or sinks in and makes the soil even more acidic.

Another manure complaint is that it works slowly while synthetic fertilizers are faster acting. Yes, that's true, but the vegetable plants produce quicker, yet have a shorter life span. Garden soil building can be compared to bodybuilding. It takes a lot of exercise to get a person's body in shape, or you can use chemicals like steroids. You get a well-toned body, but it doesn't last. Your soil needs organic material so your plants will have a long healthy life.

Micronutrients are in most organic materials like compost and manure. Some are con-centrated in natural materials like gypsum, which provides calcium and sulfur. Dolomite has calcium and magnesium. Limestone has calcium and reduces acidity in the soil.

Pesticide is a term that includes insecticides (bug killers), fungicides (fungus killers), and herbicides (vegetation killers). One problem with using synthetic pesticides is they are usually not very selective, meaning they kill more than the pests that are hurting your veggies. There are plenty of good bugs that actually help your garden thrive, like bees. Most of the living things have a natural balance between good and bad. Sometimes the bad outnumbers the good and a crime wave takes place in your garden.

A very small percentage of insects are harmful. Beneficial insects like spiders and wasps fight harmful insects. Bees pollinate your plants. Other good bugs assist in decaying organic matter. Bees and butterflies are needed to fertilize plants. Some ant types are nature's cleaners, while others like fire ants, leaf cutters, and stinging ants can literally be a pain.

When garden pests are high, control by natural means may be very difficult. Disease and pest resistant varieties should be planted if possible. Use ten-inch cardboard circles at the base of plants to fight mole cricket damage. Spade garden early so vegetation has time to rot before planting. Remove sick or infested plants immediately. Pull old plants as soon as the harvest is over. Remove weeds as they provide a habitat for insects and diseases and compete for moisture and nutrients. Water in the morning. Your plants should not be wet at night. Damp leaves cause fungus problems. Plant a few extra to allow for some damage from pests.

Another problem with pesticides is that insects may acquire resistance to that particu-lar chemical. Modern organic gardeners use the insect's own natural instincts against it. Ants can carry a poisonous, natural bait back to the nest and after a few weeks, the queen and the entire nest will be dead. Borax is a good naturally occurring chemical for this. Spray aphid infested plants first with a mixture of one teaspoon dishwashing soap to a gallon of water then rinse with a spray of clean water. Another spray is a combination of one cup of soy cooking oil with a tablespoon of dishwashing soap. Combine one tablespoon of this mixture with two cups of water and spray on the plants. Keep shaking the bottle to continually combine the oil with the water. For a more potent spray, combine garlic skins with a few minced garlic cloves and one hot pepper in a quart of water. This will repel most insects. Be careful, and try these remedies on just one plant in your garden to test the outcome before risking the entire plot.

Cornmeal worked into the soil at the rate of a kilo per hundred square feet will fight fungus. Put at the base of plants, it will fight mole crickets. A shallow bowl filled with beer, yes good old Budweiser Light or Carona will kill slugs and snails. They die with a smile, though.

Milk also helps a garden. A weekly spray of milk at 1-part milk to 9 parts water will reduce the severity of powdery mildew fungus infection on the plants.

To repel birds from eating peppers and tomatoes, put a stake slightly taller than the plants on each end of the row. Tie a cord between the stakes and attach pieces of used aluminum foil. As the pieces flap in the breeze, the sounds and reflections keep the birds away.

If your plants are battling for space with pesky weeds or grass, wait till a bright sunny day and carefully spray the weeds with full-strength household vinegar (acetic acid). It is a nontoxic, environment-friendly weed killer. Plant insect repulsing flowers throughout your garden. French marigolds will trap nematodes.

Organic gardening is not new and there are a multitude of suppliers of 'necessary' organic gardening materials. In fact, there are catalogs full of every item from predator insects to biological soaps, etc. They are all labeled as 'non-toxic.' But in reality, too much of anything can be poisonous.

These are a few natural controls for pests. A garden with healthy soil will produce strong resilient plants. If an infestation occurs be certain you are killing the correct insects. If you become a full 'organic' or just partial, your produce will be more nutritional than a chemically 'boosted' sample. Your garden will extend all of nature's cycles to con-tinue human life on planet Earth. Even slightly organic gardeners realize that to work soil like a human body, must be constantly restored. Everything is part of the great life cycle.

Courtesy: Natural Growers

Cultivators of the earth are the most valuable citizens. They are the most vigorous, the most independent, the most virtuous and they are tied to their country and wedded to its liberty and interests by the most lasting bands.

~ Thomas Jefferson

CHAPTER SIXTEEN – GARDENING WITH THE MOON

Gardening by the moon is an ancient practice that may give you, the modern gardener, an edge while working with nature. Over 4,000 years ago, the Sumerians planted by the phases of the moon as many gardeners do today. Amazingly, vastly different cultures around the world from the ancient Chinese, Mayans, Amerindian Cherokees, Hawaiians, and Scottish Highlanders all followed the same Moon planting traditions, but each garden culture developed separately. This gives credence that there must be some substance to this natural practice of gardening.

Many great prehistoric monuments such as Stonehenge in Great Britain, the Great Pyramid of Egypt, Machu Picchu in Peru, and many others, are oriented to certain stars. Those civilizations' essential activities were coordinated with the phases of the moon. Even though we are living in a futuristic, microwave, Internet world, we can try these ancient techniques. What do we have to lose? Science has demonstrated that the Moon controls the water tables and tides.

Our ancestors were not ignorant as they aligned their gardening with the cycles of nature. People across the globe attempt to predict when the last frost, the dry season, or the rains will occur. I keep a plant-to-harvest diary to record the productiveness of the Moon phase garden.

Everything seems to have a life cycle and the gardener's goal is to plant in harmony so their crops will flourish. The moon revolves around the Earth while the Earth revolves around the Sun. Our planet's position in the great universe is constantly changing in relation to other planets. The moon's orbit or cycle lasts 29 days and is divided into quarters. Each quarter is designated as the part of the moon we can see. The new or dark of the moon is when the moon is directly between the Sun and the Earth and is not able to reflect sunlight. The first quarter shows the crescent bulging to the right like the letter 'D'. The Sun's reflection on the moon increases until it reaches full moon that looks like a big 'O'. Traveling to the last quarter the bulge is to the left and looks like the letter 'C'. The days from the dark until the moon increases to full is called the 'waxing' phase. Between the full moon and the dark of the moon as light decreases is the 'waning' period.

Vegetables are affected differently when planted at varying moon phases. Vegetables that produce fruit above ground are best planted when the moon is filling (waxing), while root bearing vegetables are better planted when the moon is decreasing in size (waning). It is reported that it is best to repot houseplants, sow seeds for above-ground plants, fertilize and graft trees during the waxing phase, by the light of the moon.

The first quarter, beginning at the dark of the moon, pulls the water table up towards the surface providing more sustenance and moisture. This supposedly makes seeds sprout better. This is the period that is best to plant leaf veggies like bok choy, lettuce, spinach, cabbage, celery, broccoli, or cauliflower.

During the second quarter, the moon supposedly pulls less on our planet, but there is more moonlight that makes leaves develop. Tomatoes, beans, peas, and vines like cucumbers, squash and melons are best planted in this phase. This is also reported to be the best time to harvest as the fruit will have the most moisture.

In the third quarter, the waning phase, the moon's size decreases. It's pulling the least on the water table and plants are supposedly growing the least. This is the best quarter to plant and prune fruit trees as less sap will seep from the cuts. The plants are supposedly more compact in this phase and more oriented to their roots. It's prime time to harvest crops like cassava/manioc, taro/eddoes, lehua maoli /yautia, taro/dasheen, and sweet potatoes. The roots keep longer.

The fourth quarter, with the least moonlight, is best for clearing ground, cutting grass, and weeding your garden, but little planting. It is felt that the lack of moonlight keeps the bush or grass from returning quickly as their seeds don't germinate. Before the dark of the Moon, it is best to plant below-ground crops, especially yams.

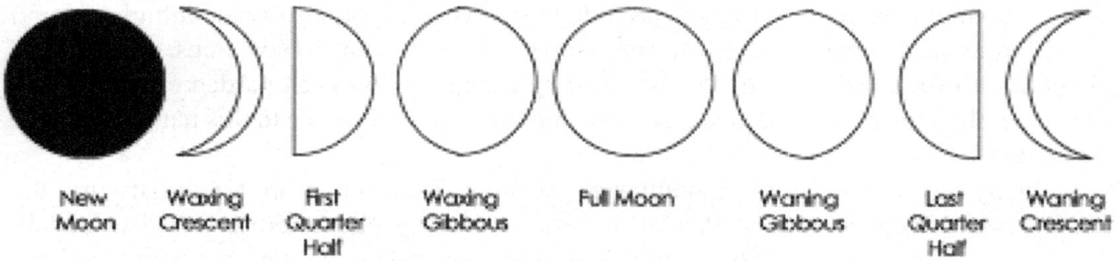

Courtesy: R. Douglas

To add more information to this complex universe planting guide; as our planet travels throughout its orbit, the moon rises in various Zodiac signs that further increase or decrease its growing power. It's most fertile to plant when the moon's in Pisces, Scorpio, or Cancer. In Taurus, Libra or Capricorn it's slightly less fertile. Still less is in Aries, Sagittarius, or Aquarius and don't bother to plant at all in Gemini, Leo, or Virgo. My advice is to buy *McDonalds Almanac* early each year to get the timetables. That little yellow book is easy to read and info's available online.

Again, this is where your garden diary comes in handy. Try a few experiments, if you have enough garden space, by planting the same vegetables during different moon phases. Give them the same care and see if there's a difference in the harvest. I do give some credence to cutting bush during the dark of the moon as we cleared a new piece of land and it took longer before it even started to grow back. That may have been hotter sun, less rain, or the phase of the moon.

People have been planting and harvesting by the moon for millenniums. Building a garden isn't easy; perhaps working with the Moon can make it more productive. Try it; nothing to lose!

There seem to be but three ways for a nation to acquire wealth. The first is by war, as the Romans did, in plundering their conquered neighbors. This is robbery. The second by commerce, which is generally cheating. The third by agriculture, the only honest way, wherein man receives a real increase of the seed thrown into the ground, in a kind of continual miracle, wrought by the hand of God in his favor, as a reward for his innocent life and his virtuous industry. ~ Benjamin Franklin

Courtesy Wikipedia

The moon looks upon many night flowers; the night flowers see but one moon.
~ Jean Ingelow

CHAPTER SEVENTEEN – HOW TO GROW

It's easy to have a productive, self-satisfying garden and still have an eye-catching backyard landscaped with attractive vegetables. It can be low budget, but adequate self-subsistence elegance. A raised box garden built with staked bamboo, or planted paint buckets can look good. It all depends on effort and continual maintenance. Make a small seedling area to start your plants. A good rule is to replant when you have consumed half of that crop or when you see the first blossoms.

Every vegetable should be planted regarding available sun, the height of the mature plant, and how long to harvest. Shallow straight drains will lead the water from your garden and keep your yard from being soggy. Balance and symmetry are important so a garden doesn't become cluttered. Plant with a plan, not at random. One border fence line might be attractive wide leaf edible roots, while another can be cassava. Papaya and banana spaced properly can blend in beautifully. The important aspect is not to shade your main garden plot and keep ev-erything maintained, but keep the hammock shaded and fireside nearby.

DECORATIVE ROOTS

Lehua maoli/malagna, colocasia/eddo/taro, sweet corn root/toppee tambu, ginger, turmeric (Indian saffron), and white turmeric: All are attractive wide leaf plants that produce tasty roots in less than a year. They are virtually effortless to grow. The only danger is damaging them when mowing or trimming your yard. They do not have to be located in a traditional garden plot. For all roots, fork the soil to a foot deep and at least a foot wide.

Regular small colocasia/hairy taro/eddo/*Colocasia antiquorum* - Plant about a foot apart and they will grow into a clump. As they grow in two or three months and every following month, pull dirt up (mold) around the base of the stem. No fertilizer necessary except maybe a little starter 12-24-12. Water daily until they catch and then maybe twice a week. Doesn't matter if in full or partial sun. Taro/eddo will take at least six months to grow and sometimes a year. The bulbs will protrude with their identifying brown flaky skin and the upper leaves will wither as if they are not getting enough water. Do not water when they start to wither as that may harden them and they might not soften when boiled. Only harvest in the mid to late dry season, because those harvested in the rainy season won't soften when boiled. Carefully dig them using a trowel. Harvest separate bulbs. Save some for replanting; what you are going to eat dry in the sun for at least a few days. Plant at the beginning of the rainy season and it should be ready to harvest near the end of the following dry season or just let them keep growing.

Taro/lehua maoli/tannia/yautia/malanga/*Xanthosoma sagittifolium* is a tasty root with a large leaf shaped like a spear point, which makes a great backyard landscaping plant. They can grow to three feet with at least the same distance apart. To plant, fork a hole soft and free of clumps of dirt. Blend in some well-rotted chicken manure, a quarter-cup of crushed limestone, and then form mounds. If you are using a root-head, split it and put it in the mound split side up and cover with about five inches of loose soil. If you are planting a shoot attached to the seed, slide it in so the attached root or eye is facing up. Cover with about five inches of loose soil. If you plant shallower it will produce many small side shoots rather than one big root. Wait at least a year for this enjoyable, crunchy root.

Big taro/green taro/taro/cocoyam/*Colocasia esculenta* - Preferably plant in a damp spot of your yard or the drain edge, about 18 inches to two feet apart. If you cut the leaf for soup, the bulb head won't develop. Use the same growing procedure as with small hairy taro/

eddo. Both make nice landscaping plants when intermixed with some flowers especially as a type of 'arrow leaf drain hedge.'

White turmeric (*Curcuma zedoaria*) also known as zedoary will grow almost anywhere with proper attention. Zedoary likes wet lowland forests, close to drains, streams, and rivers. This root's foliage produces a beautiful yellow and purple flower that is excellent as a hedge or in rock gardens. Its use as a spice and in Ayurveda/traditional medicine is becoming better known. The root is white inside and smells like a mango.

East Indian arrow root/Sweet corn root/*Calathea allouia* – This root is very easy to grow and makes a nice landscaping border for gardens, a sort of edible hedge. The broad green leaves can reach five feet tall and a few will bear a white flower. The almost round root can be up to two inches in diameter and resemble 'new', first harvested small potatoes, or water chestnuts. To locate some seed roots (rhizomes) find someone selling them along a roadside. They either have some for planting or will direct you to whom they got the sweet corn root. To prepare for planting find a well-drained area, fork, and mix in well-rotted chicken manure. Harvests will be less in clay soil than sandy. Plant the 'seeds' about a foot apart. This plant will even grow in shade, which makes it perfect for interspersing between fruit trees, cassava, or plantain. It is a long crop taking at least nine months before harvest.

Ginger has a very different leaf and looks good in the rear of a flower or spice bed. Wrap a root in damp paper until it sprouts, plant the root and wait.

Turmeric (Indian saffron) is another tall green leafed root that can look good in a corner of your garden or backyard. Just find some roots and plant. Turmeric is extremely good for you and has many uses. A square foot of turmeric is a lot.

Giant taro, elephant ear, Egyptian lily *(Alocasia macrorrhizo)* Elephant ear taro is a giant plant growing to 4 meters with distinctive leaves, valued as an ornamental. The roots are used for food. Both the roots and leaves are used medicinally in some countries.

SALADS, GREENS, AND SPICES

These are best planted in a well-prepared area or box that could be as small as two feet by four feet. Fork the soil eight to ten inches deep. While breaking the clumps, work in well-rotted manure. All can be fertilized with either blue or green salt or a combination of the two. If you choose, a light insecticide like Pestac or Fastac could be sprayed every ten days. A good rule is to replant when you have consumed half of the vegetable. I replant as soon as blossoms come. Grow what you like to eat, and experiment. Using limestone between plantings will cut down on many green vegetable problems.

Celery needs a minimum of six hours of direct sun a day and plenty of water regularly. Use green salt, again just a pinch every week. Plant at least two dozen stalks and replant as you pick to always have a supply. Plant at least six. Keep plants three to six inches apart and harvest every other one, so the remaining will continue to grow and spread.

Bok Choy – same as celery. If the leaves get wilted and have many holes, it is probably web lace fungus and they need a weekly spray with Roval. You will need to grow at least 20 plants if you are feeding a family Plant eight inches apart. Carefully pull out and then trim the roots off.

Lettuce- same as bok choy. Plant eight inches apart and plant at least six.

Parsley – same as bok choy. Plant six to eight inches apart. Depending on your usage in the kitchen plant either six or twelve at a time. When only half remain, plant again.

Spices as basil, thyme, and sage take a bit of room and grow tall. You only need one or two of each spice, but plant them where they won't shade the shorter garden veggies.

Culantro, saw tooth cilantro– Plant at least 6. Same as the celery, but six inches apart. Doesn't need much water, but grows well when it is regularly maintained. Also grows wild. Nothing kills it. Trim with scissors and it will keep regenerating tasty leaves.

Chives – Same as celery, but doesn't need any spray. Plant at least a dozen. Plant chives everywhere you have space. This is an effortless and tasty seasoning.

Garlic- same as chives

Daikon, long white radish- grow like chives. They need plenty of sun and extremely well-drained soil that's worked loose to a half-meter deep.

TREE VEGETABLES

Peppers, tomatoes, okra, eggplant, seasoning peppers, chilies, spinach, and bush beans need well-drained soil in a sunny spot. Each vegetable could be grown in a box, container or the traditional plot. Fork the soil until it is loose either in a plot or hole by hole. Work in some rotted manure and limestone. I would put a stake of some kind, wood, bamboo, BRC, or PVC pipe just as I plant so later when they are about to bear and need support, staking won't damage the mature roots. These should be planted according to height.

Eggplant and okra should be the tallest and fullest, which means it will make shade, so it should get the morning light first and be on the east side of your garden. You do not want to shade either the tomatoes or peppers as they will spend their energy and nutrients to get taller reaching for the sun rather than producing fruits. Tomato trees get full and tall, but you can pinch the top of the main stalk at about a month and it will spread rather than get tall. Make certain either to stake or cage your tomato trees. Seasoning peppers, sweet peppers, and hot chilies need individual space. All can grow to two feet and bear for five months with proper care.

Tomatoes – when planting, put a pinch of starter fertilizer in the hole add some dirt and then plant the tomato tree. Plant at least a foot apart with a stake two foot tall. Stake when planting so the roots aren't later damaged. After a week, spray with a systemic insecticide. Three days later, drench (heavy spray) with a systemic fungicide. The three biggest enemies of tomatoes are 1) mites and the whitefly – when the leaves start to curl up, 2) fungus – when they start to wilt, and 3) over fertilizing – the leaves turn brown. Use only starter fertilizer 12-24-12 sparingly. Spray weekly with Pestac and make sure to get under the leaves where the mites live. Water daily, but never let the ground stay damp at night. Replant every two months for a constant supply. If birds peck the fruit run a string along the top of the stakes and tie pieces of aluminum foil to it. If the end of the tomato is rotten add calcium - Calmax- 1 teaspoon to a gallon of water. Just add to the watering on the roots and do not touch the leaves. If you have Calmax, use it before they bear fruit, in the third week for great tomatoes. I'd plant at least four and wait about a month to six weeks and plant another four to have a constant supply. My favorite types are Heatmaster for the dry season and Gem Pride for the wet months. Healthy trees can bear for six months.

Sweet Peppers- Plant with a pinch of starter fertilizer, place about a foot apart. Install a stake at planting. Use the starter fertilizer weekly until you see the blossoms. Then use blue salt. If you are planting on a large scale for income, after a week drench with systemic insecti-cide and then three or four days later a systemic fungicide. Use Pestac insecticide once a week mixed with a foliar spray like Powergizer 45. Mites are the biggest problems for peppers. Water regularly, but never soak. Always try to keep the water from the leaves. The type I like to plant is King Henry. **Pimento** – treat exactly as sweet peppers. Plant about two feet apart. Stakes are not necessary. **Hot pepper** (chilies)–bird, Congo-Scotch bonnet, cayenne, jalapeno, ancho. Needs the least water of all peppers. Use less water when the plants are bearing to make the fruit even hotter.

Eggplant: Same as sweet pepper, but at least two feet apart. Stake at planting with a stake at least a meter or better long. Water regularly but once they are bearing soak them one morning a week for excellent fruit. Spray under the leaves with Pestac mixed with a miticide. Watch for whiteflies. You'll see tiny white dots on the underside of the leaves.

Okra is perfect for a hedge around a backyard garden, because the three to six-foot-tall plants produce beautiful blossoms that rival its cousin the hibiscus. Plant the seeds one inch deep and a foot apart. Okra usually grows well in any good garden soil. Four or five plants produce enough lady fingers for most families.

Spinach – This is another very easy to grow vegetable. You might think it should be classified a leafy, but it can be grown and the leaves trimmed as needed. Red stalk spinach grows like a weed and perhaps the hardest part will be getting rid of unwanted plants. It grows to three feet or more but the main stem can be pinched so it grows out not up. Two or three good plants will provide enough.

Beans – green, snap, string, and contender types are great to rejuvenate soil. The bushes take up little space. An eight-foot row should provide a lot of beans. As soon as one row blossoms, plant again so you will always have the delightful beans. Beans need little attention except for well-drained soil, a little water, and plenty of sun.

GROUND VINES

Cucumber, melon, squash, pumpkin. All vines need a bit of uninterrupted space and you need easy access to pick the fruit.

Pumpkin and melon will take a lot of space. Pumpkins are hearty, but need to remain undisturbed. A specialty melon like cantaloupe can be a nice treat.

Squash is separated into summer and winter types. Hawaiians are accustomed to winter squash such as the butternut. The long season, odd-shaped, hard-skinned squash that store well are usually referred to as winter squash. Smaller, short season types, which are eaten before the skin and seeds begin to thicken are the summer squashes. They have a mild somewhat nutty taste that resembles corn. Summer squash, like zucchini, take up little garden space compared to butternut types. Zucchini is very tasty in a stir fry or grilled.

Cucumber – start as with the other veggies, but plant at least two feet apart. Put a stake where you plant so you will know where the roots are for water. Try not to get the water on the leaves. Use special green fertilizer weekly, but just a pinch. Spray as with other plants. Two vines are plenty, but they need about a four by four-foot space each.

FENCE VINES

Long bean, lablab bean, chayote, winged bean, and bitter gourd need little to produce well. Work the soil adding some rotted manure. Water daily till the sprouts reach up. Be patient when starting a vine to climb.

Long bean: This needs to be planted close to a fence or trellis so it can climb. You can use old eggplant or hot pepper trees. Water regularly. Spray weekly with Pestac. Doesn't need much fertilizer. Replant every two months. A family only needs about 4-6 vines. There are several varieties, red, long, and short green types.

Lablab bean: This bean is now grown in almost every part of the world. It is an integral part of most Chinese back yard gardens. The purple variety type most common is called Ruby Moon. It is very easy to grow, resistant to most diseases, and needs little water, but does need a fence, trellis, or sturdy pole to climb up. Hyacinth bean has purplish stems and alternate, white, pink, or purple flowers in a long bunch that produce maroon bean pods containing 3-5 green beans. When the beans dry, they become black with a white streak. Pick the bean pods while the skin is smooth and the beans pushing out yet.

Chayote – christophene: This is an attractive vine, but it takes a lot of attention to grow. This vine loves the sun, but also needs plenty of water and humidity, and a fence or a trellis. The easiest method to grow this vegetable is to locate a farmer and beg a plant. Failing that, select two chayote at the market. Ask the vendor if they have any that are overripe and budding. If not, set the chayote in a warm window, but not in direct sun. In a few days, it will start to shrivel and wrinkle, and soon sprout a bud. Plant the seed, bud upwards in a clay pot with sandy soil. Lightly fertilize with 12 – 24 –12. Once the plant catches, move it outdoors where the vine can climb. Provide it with some shade using a banana leaf or a board. Do not fully cover it. Water regularly and use 12 –12 – 17 –2 mix when it begins to blossom.

Bitter melon is easy to grow from seeds. It is best planted along a fence or anywhere the vine can climb. Because bitter melon seeds are scarce, first visit the market and search for an overripe fruit. Set the fruit out until it softens and then remove the seeds. Dig several small holes along a fence. Plant about four seeds per hole. Water regularly and in a few days, bright green sprouts will appear. As the vine grows, carefully start it onto the fence. Bitter melon is a natural climber. Spray occasionally with a mild pesticide and water-soluble fertilizer. In a few weeks, yellow blossoms will appear. Water every other day. Birds will be the biggest pests to your bitter melon.

EASY FRUIT

Papaya – Keep trees at least eight feet apart. Use starter fertilizer in the hole at planting. Use all the sprays, but if it gets bunchy top pull it out and replant. Over-watering is the biggest enemy. Once they start to blossom, use blue fertilizer weekly. I plant three trees every six months. Papaya is great for you and a healthy, refreshing dessert fruit from an attractive tree. Keep away from any chlorine.

TALL VEGETABLES

Pigeon peas, cassava, corn. These need to be on the east side of your garden so they don't block the sun from the other ground-level veggies. Four **pigeon pea** trees will provide plenty of peas, but a mature pea tree takes up a good bit of your yard. They can be attractive nicely spaced about four feet from any property line or fence and about six feet apart. This should permit easily picking from all sides. With proper care you might get two pickings from each tree. Keep the soil well drained and fertilize every month with red (12-24-12) salt. This is a long crop that takes six months to bear.

Cassava trees could be planted along your fence line behind the peas as both grow about the same height. Cassava is another virtually effortless foodstuff. Plant it where it gets plenty of sun, in loose well-drained soil. Since it is a root the softer the soil the better. Cassava can take more than a year to mature. The butterstick variety has a nice red colored tree, and takes up more space because it branches. MX is drab gray, usually a single stalk, and makes a better hedge. After forking an area, stick two pieces of the stalk every two feet. After six months carefully brush away the soil and see how the roots are growing. Whenever you feel they are big enough, dig a root. Keep growing the others. Replant as you harvest. Occasionally apply red salt.

Corn must be planted with at least eight rows of corn for it to properly pollinate. That can be a major portion of a yard. I have seen a productive two rows along a driveway. Expect corn to grow six to seven feet tall, so it could dominate your garden. `

Soil mix - add about half a gallon bucket – a paint pail – of manure and two cups of limestone to two five-gallon buckets of dirt or fork into a small garden of eight by four. First fork soil and let sit for two days in full sun. Add manure and lime and hoe till soft. Then soak with water. Let sit for a day and plant.

Water-Irrigation - It is best to water in the morning and no later than three in the af-ternoon, so the ground can dry and not get fungus.

Fertilizers – use just a pinch. Over-fertilizing is a major problem for beginning gar-deners. Don't push your plants and they will bear better and longer. **Red -Starter** – Reddish brown 12-24-12 good until blossoms appear. Use a pinch in the hole before setting young plants. **Green** – 15 –5 –20 –2 – great for green leafy veggies like lettuce and bok choy. **Blue** – 12-12 17-2 after blossoms begin to make the fruit.

Courtesy R. Douglas

NOTES:

The first farmer was the first man.
All historic nobility rests on the possession and use of land. ~
Ralph Waldo Emerson

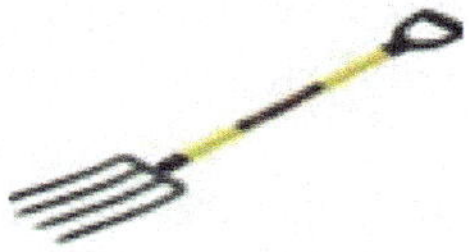

CHAPTER EIGHTEEN
GARDEN TOOLS

To break apart soil, to begin a basic garden, you will need a few essential tools; a hoe, a fork, a bastard file, and a cutlass (long knife). These four are truly the necessities, while others like a wheelbarrow will make life easier. If you are committed to growing your own, then you can get by with just these. Purchase good quality tools that will service you for many years. A saying to remember is, "Cheap things not good, and good things not cheap!"

The garden fork or digging fork has a 'T' or 'D' handle connected by a short sturdy shaft to usually four points called tines. Garden forks are different from pitchforks used for moving compost or manure. The emphasis is on sturdy, if not, the tines will break or bend, or the shaft will break. Many garden forks have thick flat or square tines rather than round. It is used for loosening and turning hard soil. Shop around if you have time and transport because sturdy doesn't have to mean heavy. A few pounds less can save a lot of sweat. You may think you can get by with a garden spade or a shovel, but either of these tools will increase your labor. The fork's tines are pushed into the ground, and not easily stopped by stones. Later, when you are harvesting roots like cassava and eddoes, the fork is a true labor saver. The majority are now made of steel. Ours has been working hard for six years and still going strong with help from a few repair welds.

The best method of using the fork is to stick it into the soil with the help of gravity. Don't try to thrust it. Rest your foot on the top of the tines and apply your body weight as you wiggle the shaft. This will ease the sharp tines into the soil. You decide how deep you want to go before levering out a clod of dirt. Start small to save your back from later aches. If you have had any back problems, do stretches before you start and consider wearing a back-support belt to share the weight with your arms and shoulders. Be careful of the tines. The more you use a garden fork the sharper the tines. Always be wary not to fork one of your feet.

Gardeners use hoes to break the clods made with the fork. Once the hoe breaks down the dirt it can be used to mix limestone and manure into the soil. After the garden is growing it's used to scrape weeds without having to bend down and pull them. Hoes haven't changed much in centuries, but they have changed. There is the basic steel hoe head cast with a ring to insert the handle. The handle must be wedged to hold the head secure. The head must be pitched, angled slightly, towards the operator. This will permit easier cutting. The next tip is to use the file, a bastard file, to keep the blade sharp. Nothing is worse than trying to work with a dull hoe.

A more modern version of the hoe has the blade attached to the handle with a piece of steel called a gooseneck. This is a shock absorber to save your back and shoulders. Again keep the blade sharp. I have seen hoes that started with a six or eight-inch blade decades ago, sharpened down to three and four inches.

The cutlass/long knife/machete is an absolute garden necessity for cutting the tropical bush before you start forking. Get a good one with some weight. I recommend a true three canal made with British steel and having a riveted wooden handle. I also recommended painting the blade a color you that makes it easy to find in the bush, like brilliant red. More sooner than later you will think you stuck it one place, and it is not there. The dull steel and wooden handle won't jump out at you from the green or dead grass.

As with everything these days there are Chinese versions of the cutlass that will not keep a sharp edge. Spend the few extra dollars and get good steel. The blade should be engraved with the maker. Again, the bastard file is needed to keep it sharp, which makes your work easier. Be very careful when drawing the file across the blade. You will not be the first or the ten thousandth person to cut themselves. The best method is to place the cutlass so the blade is pointed up and run the file upwards keeping the sharp blade edge away from your tender fingers. I also recommend wrapping the cutlass handle with black vinyl electrical tape to give it less slippage in your sweaty hand.

There is also a version of the cutlass where the blade is bent and inserted into a wooden handle. This is called a brush cutter/swiper and will clear serious bush with rhythmic swinging. The blade is adhered to the handle with wraps of wire. The length of the handle depends on the height of the bush and that of the operator. A file is a hand tool that cuts fine amounts of material from another piece of steel. It is a hardened steel bar with a series of sharp, parallel ridges, called teeth. Most files have a narrow point at one end called a tang where you can slip on a wooden handle. It is called a bastard file because it is not classified as a 'coarse file' or as a 'second cut file', but one cut finer than coarse. Again, buy wise and your file will last until you misplace it.

Buy decent gloves to save your hands, a hat to save your head, and either safety or sunglasses to save your eyes.

You will need a watering can. Get sturdy steel or make one from a plastic gallon jug by punching holes in the cap with a nail. You need to also cut a breather hole at the neck. Between the sun's effects and the weight of the water, plastic water cans don't last. Get one that has a sprinkler nozzle that can be easily removed when it clogs. If you get a leak, whether steel or plastic, seal it with a piece of roof repair Flashband. Hoses; buy at least 5/8 inch diameter reinforced otherwise you will spend too much time looking for water stopping kinks. Buy a simple, cheap plastic sprayer nozzle because they all break. There is a spring inside the nozzle that is supposed to force back the handle to shut off. They all rust and fail. Buy two at one time so you are not caught short.

Mechanical evolution has hit the garden. If you have the money, transport and strength, rent a gas-powered rotavator or rototiller. I recommend renting this tool because it is an absolute workout. It can be a labor saver, but see if you can handle it. These are motorized cultivators that work the soil by means of rotating tines or blades. Rotary tillers are either self-propelled or you must push. Again, be careful. If you have your plot or grow box ready, a day rental should be enough time.

I feel the best recent garden innovation is the motorized string trimmer, also known as a weedeater, weedwhacker, or bushwacker. It is a powered handheld device that uses a flexible monofilament line. Some have a blade attachment for cutting thick grass and other plants. It consists of a cutting head at the opposite end from the motor separated by a long shaft with a handle or handles. Get a shoulder strap. String trimmers are usually powered by a two cycle gas engine. Some have electric motors, but don't consider them for the garden. Again good models are not inexpensive, but they do save a lot of time and labor when clearing land. Ours gets plenty of duty from initial clearing, to weeding between rows, and keeping drains open and flowing. Buy the appropriate size as there are lightweight, light-duty models for trimming lawns that will not stand up to serious garden chores. Get a brand name for which you can easily get parts. Ours has lasted five years and is still going strong.

That's the garden tool list. Take care of them and they will take care of you

Courtesy R. Douglas

CHAPTER NINETEEN - GARDEN DISEASE PREVENTION

Vegetable gardening can be a hobby that can become a side business, and blossom into a livelihood. Successful gardening requires attention to the environment - soil, water, sunlight, and air circulation. Your garden's environment will determine its susceptibility to diseases.

Just as if you get wet in the rain and sit in damp clothes; you might catch a virus if your immune system is low. Diseases attack vegetable plants when conditions are favorable. Garden damage from disease can be reduced through a combination of proven disease-prevention methods. First, select adapted, disease-resistant varieties. When you buy seeds, check if they are fresh, packaged during this year. Seedlings should look healthy and not be root bound in the growing tray. Use transplants that are supposed to be resilient to disease.

Whether you grow transplants or buy them, every transplant should be inspected for anything abnormal above and below ground, insect damage on the leaves, or insects on the lower leaf surface (especially whitefly). If growing your own transplants, purchase steam-sterilized growth medium/potting soil or sterilize your own in your oven. Always disinfect seed trays with chlorine bleach, or use new containers.

Separate vegetables from the same vegetable family to various parts of the garden. Try not to plant them together because the same diseases and pests will attack all. Tomatoes, all types of peppers, and eggplants belong to the same family. A good rule is not to replant the same veggies in the same area of the garden for at least three years. However, this will not prevent diseases with long-lived resting spores, such as Pythium, Fusarium, and Rhizoctonia.

Keep weeds under control as they will compete with your veggies for nutrients and may have diseases. Weeds may hide pests that can carry specific diseases. Some weeds attract insects that transmit diseases, especially viruses.

Once a plant is sick, quickly remove and destroy it before the disease spreads to more plants. Remove all plants soon after harvest. Disinfect garden tools and shears that have been used on sick plants by washing in a weak solution of chlorine bleach.

Removal of the sick plants will reduce the chance of certain diseases increasing over the years. It also reduces the chance that healthy plants will become infected or infested early. Some plant diseases occur naturally with time late in the season and should not be a problem for healthy mature plants. However, these same diseases can ruin your garden if young plants catch the disease or pest from plants remaining from the last gardening season. This is a good reason not to add young plants among older ones.

Diseased plants should not be put in the compost pile. Instead, it is better to put them in a specific location that can later be safely burned. I recommend burning rather than burying because some diseases as the bacterial wilt fungi can survive in the soil for many years. Prevention, rather than cure, is the best way to manage these diseases. It may be necessary to destroy or disinfect support structures such as wooden stakes and poles used in the garden.

Keep your garden soil well fertilized and the pH properly balanced. Extremes in temperature, rainfall, nutrients, and misapplied herbicide may present similar appearances to diseased plants. However, these conditions won't respond to chemicals and will make conditions more favorable for disease development. The garden should be well-drained. Wet soil encourages the development of root rotting fungi. Good drainage promotes good growth of plant roots. Once a healthy root system develops the entire plant should be in good condition. Raised beds may solve the drainage problems.

Fork the soil and leave it to bake in the sun at least a week. This is solarization. This either kills or reduces plant pathogens, and weed seed. It's believed beneficial organisms are harmed less by solarization than by chemical treatment. Solarization also stimulates the release of nutrients from organic matter present in the soil.

A disadvantage of solarization is the area treated must be out of production for at least a few weeks. Solarization requires the soil is free of clods and plant debris. This prevents pockets of contaminated soil. It is best to add fertilizer before beginning solarization. Dry soil should be moistened to a level that is ideal for planting. Wet soil conducts heat better than dry soil and will allow the heat to move deeper into the soil to remove diseases present.

Plants with the proper nutrients can withstand environmental stresses and diseases better than plants growing in poor, unbalanced soil. Powdered limestone, either hydrated or unhydrated, can help fight a lot of fungi and bacteria. Do not over-fertilize. Excess nitrogen promotes root-rotting fungi. Nutrient stress makes plants susceptible to diseases and insect damage.

Water when the plants are dry to avoid drought stress. Excess water can lead to plant death from lack of oxygen to the roots or because of pathogen attack. Harvest produce at peak maturity. Overripe vegetables will attract insects and other pests.

No matter how hard you work to keep your garden's soil balanced, remove all weeds and diseased plants; sometimes disease still attacks. Many leaf diseases can be managed by spraying or dusting plants with an effective fungicide. Apply fungicides appropriately and in a timely manner when resistant seed or transplant varieties are not available. Most fungicides are designed to protect, not cure. They work on the plant's surface and protect against infection. They do not eliminate established infections. If the disease is not detected early, the plant may die and the disease may spread despite fungicide treatment. Some fungicides are systemic and will move into the plant. Some of these have curative properties and will kill infections already established in the plant, but they will not remove the spots already present on the leaves.

Plant Disease Name	Plants That Are Susceptible
Bacteria Wilt	Cantaloupe, Pumpkin, Squash, Cucumber, Watermelon
Gummy Stem Blight	Cantaloupe, Pumpkin, Squash, Cucumber, Watermelon
Fusarium Wilt	Cantaloupe, Pumpkin, Squash, Cucumber, Watermelon, Tomatoes, Peppers, Eggplant, Potatoes
Cercospora Leaf Spot	Cantaloupe, Pumpkin, Squash, Cucumber, Watermelon
Anthracnose	Cantaloupe, Pumpkin, Squash, Cucumber, Watermelon, Tomatoes, Peppers, Eggplant, Potatoes
Powdery Mildew	Cantaloupe, Pumpkin, Squash, Cucumber, Watermelon
Black Rot	Broccoli, Mustard, Collards, Cabbage, Brussels Sprouts, Kale, Rutabaga
Black Stem Disease	Broccoli, Mustard, Collards, Cabbage, Brussels Sprouts, Kale, Rutabaga
Downy Mildew	Broccoli, Mustard, Collards, Cabbage, Brussels Sprouts, Kale, Rutabaga
Verticillium Wilt	Tomatoes, Peppers, Eggplant, Potatoes
Early Blight	Tomatoes, Peppers, Eggplant, Potatoes
Late Blight	Tomatoes, Peppers, Eggplant, Potatoes
Septoria Leaf Spot	Tomatoes, Peppers, Eggplant, Potatoes
Mosaic Virus	Tomatoes, Peppers, Eggplant, Potatoes

Chart courtesy UK College of Agriculture

The farther we get away from the land, the greater our insecurity. ~ Henry Ford

CHAPTER TWENTY
GARDENING IN THE HEAT

The price of fresh garden vegetables increases during the dry season due to lack of water and the constant heat. There are some simple ways the home gardener can combat most of the perils of the dry season.

First, ask yourself, can you take the heat? Be honest, laboring in daily temperatures of above 33 C can be dangerous. Carry and drink plenty of fluids so you can rehydrate. Your body loses moisture through perspiration and it must be replenished. If you are going to be out in the sun for extended periods, consume some salt to help your body retain fluids. Salt will also help to curb muscle cramps from exertion in the heat. Dress to make your own shade in long sleeve shirts, wide-brimmed hats, and sunglasses. Use sunblock of SPF 30 or more. I tie a large handkerchief around my neck to keep it covered.

The main thing about the home garden is that it is supposed to be enjoyable. Don't over-exert yourself and have a heat stroke to save the price of a few pounds of veggies. Most of us home gardeners work a regular day job, so it isn't too hard to keep our limited gardening to the cool early mornings and early evenings. Limit your labor in the heated afternoons to getting in and out of the hammock. It is said, 'Only mad dogs and Englishmen go out in the midday sun.'

When the rain has ceased, our bush turns to tinder. People lacking common sense and decency for our environment set it afire without a thought to homes, gardens, or wildlife. For your protection, keep a firebreak by clear-cutting at least 6 feet around your garden and home. Birds and animals, which depend on the bush for food will turn toward gardens. Trapping these hungry displaced creatures is another problem of the heat wave.

Think of your garden as an adrift lifeboat with only a certain amount of water to survive. Don't waste water on weeds. When watering, use a sprinkler can, and douse the plant so that the roots are dampened. Be careful not to pour the water hard, which will erode the soil around your plant's roots. These delicate roots will burn in the sun and stunt your plant. A soaker hose is a good thing to have. This type of hose is usually flat and has a multitude of holes on one side. Connect it to a tap and place the holes up and it is a miniature sprinkling system depending on the pressure. I recommend turning the holes down as it will soak the soil around the plants' roots and zero water will be wasted. If you are being severely pinched by your water cost, collect drain water from your kitchen sink and lavatory basins. Push the drain pipe through the wall and catch in a 5-gallon bucket.

I try to thoroughly soak my garden every morning. It invigorates my psyche to check my work with nature in the garden while I'm watering. A good strategy is to surround all plants and fruit trees with a circular dam about two inches high. For plants like peppers, cucumbers, and tomatoes, you can make it about a foot in diameter or include several plants into one dam. This will hold all the water and it will soak down to the roots not wasting a drop. Long rows of corn, long bean, or eggplant can be dammed at the end, and water is poured in between, watering two rows at once. At least every other week, pull some soil from within the dam back to recover the roots to protect from sunburn and mold the plant for strength.

Fruit trees need dams at least four feet in diameter around the roots. I have a five-gallon bucket that I have pounded four equally spaced nail holes at the bottom edge. I put the nails in to fill it, and then take them out when I have the bucket at the base of the tree.

This permits the water to seep out slowly and thoroughly soak the tree's root system. A weak solution of soluble fertilizer can be added for nutrient.

Again, think of the garden as a lifeboat and imagine how often you'd be thirsty, all the time. How often do you really need a drink? At least once a day, twice is best, but not during the hot afternoon. I measure out a Vienna sausage can of water to every small plant, less than eighteen inches tall, like tomatoes or peppers. The bigger plants, especially eggplants, get 2 full cans. As plants start to produce fruit, it is wise to give them more water. Pineapples and orange are 87% water, grapefruit, broccoli, cabbage, eggplant, cauliflower, spinach, sweet potatoes, watermelon, and sweet peppers are 92%, tomatoes are 94%, and cucumbers, lettuce, celery 96%. Good, juicy vegetables take water!

Go back to the garden lifeboat again. What would make you even thirstier? Eating anything salty. Don't over-fertilize (salt) your garden anytime but especially during the heated season. Use the bare minimum, and I find it is best to use it mixed with water, but in a weaker than recommended solution. I do not recommend spraying soluble fertilizer directly onto the leaves in hot weather. The salt residue may burn or otherwise damage the plant. Leaves absorb very little water or nutrients. It is best sprayed or gently poured around the roots. Keep your plants molded or staked so it won't waste energy fighting to stand straight.

What would you want in your garden lifeboat? Shade. The next strategy is to shade your garden. I use shade cloth on a light PVC pipe frame. 'Shade' or 'garden' cloth can be either black or green. This lets some light pass but protects the plants from the scorching afternoon sun. Price varies so shop around. Make a simple frame with four tees' and four elbows. The tees are for the legs, so place them so the frame will balance. Tie the cloth to the frame with very light wire or fishing line. A half an hour and about thirty dollars and you will cool your plants. This is especially good protection for bok choy and lettuce, but it can help sweet peppers and tomatoes withstand the dry season. The shade will permit more of the water to reach the roots and actually cool the plant. Mulching around the plants also keeps them cooler and helps the soil retain water.

The rainy season is a few months away, and then gardeners will be complaining about too much water and floods. Now we need to conservatively water our home gardens in the mornings and evenings, and test our hammocks during the heat of the day.

GARDEN SHADES

The nation that destroys its soil, destroys itself. - Franklin Delano Roosevelt

CHAPTER TWENTY-ONE
GARDENING AFTER THE FLOOD

Seems like everything changes so fast in the tropics. Months and months of scorching heat and drought, and then a rain instantly transforms dry river beds into roaring floods. Gardeners praying for rain get answered to the extreme, more water than they want or need. Gardens that were cared for by lugging heavy sprinkling cans must now be saved from the effects of a water overdose. The garden's enemies have switched from whitefly and beetles to fungus and bacteria.

After the flood, market prices skyrocket, so try a few simple steps to keep yours alive. No one thing will guarantee to revive a flooded garden, but a few simple steps will make a dif-ference. Try to save the vegetables that take the longest to mature as peppers, sweet, bitter, and seasoning. Tomatoes don't like a lot of water, so do your best. Papayas, well on this one you better pray because this fruit's biggest enemy is excess water. Most roots like hairy taro/eddoes and big taro/ tannia can handle the water, but a flood can make cassava rot. Okra and eggplant need a little effort but usually pull through after a serious rain. Don't worry about short-term leafy veggies like lettuce and bok choy. With any luck, you can replant and have it fresh again in three weeks.

Forethought from prior wet season garden experiences is always a help. The best plan is to have the garden beds raised as much as possible so moisture will seep out and the bed will dry even if thoroughly saturated. Higher grow boxes are the best solution. If you have a traditional plot garden, you must have adequate drains at least a foot wide and deep between each garden bed. Keep those drains clean from grass and debris. Run a bushwhacker over each drain and wait a few days and spray the just appearing grass with a systemic herbicide. That should keep it clear for a few months.

Once the soil is wet, it is as slippery as ice. A walking stick is recommended for support while maneuvering around the garden. Imagine what would happen to you if caught in a flood. You'd be tumbled in filthy water, and probably drink enough to make you very sick. Soaked to the bone, your body temperature drops and you probably catch a cold or virus. Scrapes get infected. It would take weeks to regain your strength, if you did make a recovery.

That's what happens to garden plants. The first thing you would need is a good wash with a disinfectant and a few vitamins. That's why plants need some white hydrated powdered limestone and some magnesium. This is the least expensive and most beneficial means to re-vive plants after a flood. White lime is expensive for an eighty-pound sack and magnesium is small money. You should have your soil pH tested. Ours is acidic and once the rain comes bacteria and fungus thrive. One method I use is to wait a few days till the water receded, rain isn't threatening, and the garden drains have no standing water.

If your yard was flooded, you would remove the debris, before a good cleaning can be-gin. Likewise, the first step is to weed and loosen the soil at the plants' base without damaging the roots. Fungus is the big enemy and it comes in many types, but they all love damp conditions with stagnant airflow. Weeding clears some of the path so air can better flow between the plants. Rough up the soil at the base of each plant – carefully giving a slight rotavating with a hand fork to make it porous to dry quicker.

Once the storm is over take off all your dirty flood clothes or you'll get a worse sick-ness. The same precautions apply, pull off any dead branches and leaves from tomato and pepper plants. Dispose of any dead or seriously ill plants. Trying to save a seriously sick plant is a waste of effort. Do not plant back while the soil is soaked and when you do, work a little white lime into each hole.

Even though the place is damp and muddy, I mix half a Vienna sausage can of white lime with a gallon of water. Add a pinch of magnesium. I mix it well and sprinkle it, drenching each plant, especially the roots. The earlier weeding and slight rotavating will permit the mag-lime water to access the roots. Bacteria is another enemy after a deluge. Wash all the produce you pick to protect yourself from contamination. Don't use any fertilizer salts as this will feed bad bacteria instead of feeding your plants. The bacteria will vigorously attack your weakened plants and that outcome is not good. Magnesium will help ward off bacteria and is an essential nutrient that is often lacking in the soil.

Hopefully, there have been only a few slips and no falls. After all the cleaning and vitamins, it's time to treat your illnesses and infections. With your garden plants, that's accomplished by waiting a day and spraying with an appropriate fungicide like Alliette or Acrobat, and / or Ban-rot and Rizolex. I like to use a chemical with a copper base like Cuprasan. This is an excellent chemical to get your plants healthy. This sounds like a testimonial for chemicals, but remember you have been bruised, and cut, tumbling in dirty bacteria-infested water that has gone down your gullet. Of course, you'd take a few pills! Try to spray early and hope that it's sunny as the chemical(s) should work better. Make sure you have the nozzle adjusted for a fine mist and coat the leaves. The results of the lime and the fungicides should be visible in three or four days. Your plants will either die or spring back, but you did your best to save them.

Even though I do not recommend using fertilizer salts while the soil is moist, I do use a foliar chemical spray the fourth evening after the lime application. There are several like Powergizer, GreenStim, and Cytotkin. Use them at half the dosage in the evening several days after spraying with fungicides. By this point, you should either see some recovery or some may have died. This combination will give a booster shot to the plants.

It doesn't matter if you're a seasoned farmer, you can't fight the weather. You can pre-pare and give your garden your best effort at revitalization. If the flood takes it all, get out the fork and start preparing now for the next season. Dig better drains!

Courtesy R. Douglas

The way I see it, if you want the rainbow, you gotta put up with the rain.

Dolly Parton

CHAPTER TWENTY-TWO
GARDENING AND GOOD HEALTH

Do you really want to shape up, have a healthy good looking body, and eat better? A home garden may be the best, least expensive means available with plenty of wholesome benefits.

Gardening is a great whole-body workout that leads to better physical conditioning, mental stamina, and weight loss. Digging and lifting will build muscles. Aerobic activities such as raking, cutting, and hoeing will burn calories while building a stronger heart and pulmonary system. A half an hour of digging, forking, weeding, or clearing bush will burn about two hundred calories. This is absolutely free – no gym membership necessary, no embarrassment about your body's weight or shape, and the perks are eating and enjoying better homegrown foods.

It is essential to stretch before beginning garden work. Change from forking to weeding or hoeing, to molding plants to avoid overusing specific muscles. To prevent back injuries, bend from the knees when you shovel or lift heavy objects. Use a wheelbarrow to save your back. Repetitive strain injury can occur after doing one task for too long. Muscle strain, back injuries, and blisters can result from moving a greater weight than your body can handle, bending, and improper use of garden tools.

Knee muscles are the quadriceps (front of thighs and the hamstrings (back of the thighs. To ease the strain on the knees, practice strengthening exercises regularly, and stretch before starting gardening. Ask your doctor to recommend specific exercises and stretches that are appropriate. Incorrect posture while squatting can put an unnecessary strain on the knees. It is best to keep feet flat with weight evenly distributed. Squatting with heels off the ground can potentially damage knee ligaments. The best posture would be having one knee on the ground, working on hands and knees using a kneeling pad, or sitting on a chair or stool.

To avoid aggravating a back injury, it is important to know how to move, sit, stand, and work in ways that will reduce strain. Use correct postures when doing garden chores. When walking, keep a slight arch in the lower back, slightly tensing the abdominal muscles, and don't slouch, or bend forward. Sit with feet supported and knees level or higher than hips. Always bend from the knees, never from the waist. Follow these suggestions when lifting a large or heavy object; stand the object upright, position feet shoulder-width apart, close to the object, squat or bend at the knees, tighten stomach muscles, roll the object onto bent knees and then up into arms. H the object close to your body so that the thigh muscles are doing most of the work, and slowly lift by straightening knees. Lower loads by reversing this process.

Long-handled tools can make garden work easier by extending your reach and reducing body movement necessary to complete a task. Lightweight and small-bladed tools can reduce the amount of load and resistance. Stand as close to the work area as possible, and use arms and legs to do work instead of the back.

Protect from excess sun exposure by always wearing a hat and using a sunscreen of at least SPF 15. Avoid gardening between 11 AM and 4 PM when the sun is the strongest or garden in the shade during those hours. Wear protective clothing such as a large-brimmed hat, long sleeves, and long pants. Over-exposure to the sun can cause sunstroke, sunburn, and over the long term, skin cancer.

Some good health gardening tips: warm-up with five minutes of slow stretches. Find a comfortable posture, keep your work in front, and close to you. This will avoid reaching and twisting. Use pads or a padded kneeling stool for work at ground level, and periodically change work tasks to avoid injuries from repetition. Use the right tools for the job. Use a wheelbarrow to transport earth and equipment. Work within your strength and endurance, pace yourself, and take a break when you're tired. Occupational Health Standards list safe lifting loads as 64 pounds for men and 28 pounds for women. Always wear proper equipment in the garden including proper shoes or rubber work boots, gloves, and sunglasses or safety glasses.

Always safely store all equipment out of reach of young children. Keep a close eye on children when they are in the garden. Keep them away from equipment such as brush cutters and cutlasses, fertilizers, and other chemicals, bulbs and seeds. Also, water barrels, which may be a drowning hazard.

Keep safety first and remove anything that you don't need or want such as broken tools and furniture, bush, or dead plants. Hoe around plants to break up the compacted soil. Once this is completed, decide what new plants you can add. Look for those that offer long growing season, perhaps some color and fragrance. Perhaps add in some herbs such as thyme, rosemary, chives or oregano, which will be well used in recipes.

Play your favorite music while you are gardening. Nature provides its own music, but after living in today's noisy environment it may take some time to get accustomed to the birds and other natural sounds. Adding the music, especially with comfortable headphones, is a good way to adjust to comfortable gardening.

SO MANY SPEND THEIR HEALTH TO GAIN WEALTH.
THEN MUST SPEND THEIR WEALTH TO REGAIN THEIR HEALTH.

Perspiration, or sweat, is the body's way of cooling itself, when muscles or nerves are overworked. As sweat evaporates from the skin's surface, it removes excess heat and cools you. All sweat does not evaporate, but rather runs off your skin. This is especially true in humid weather as the air has enough moisture and cannot acquire the body's perspiration. Each person has about two and a half million sweat glands distributed over the entire body. High sweat production occurs from exercise and or hot temperatures. Sweat cells that usually reabsorb send perspiration to the skin's surface. The chemical makeup of this hard work sweat is a concentration of sodium and chloride (salt) and potassium.

The body's largest organ is the skin, and perspiration helps the body detox and renew itself. Perspiration expels toxins and even disease from the body. Some viruses and bacteria can't live in temperatures above 98.6, so sweating can burn away illness. Garden work promotes perspiration, which may harden your hands while softening your skin and helping to keep pores clear and clean. After sweating through a chore, you should shower and scrub off all the toxins you have excreted.

A person who is not well acclimated to hard work in hot weather can produce is about two to three liters of perspiration per hour. Drink adequate fluids to avoid becoming dehy-drated, and retreat to a cooler place if you feel yourself getting overheated. The loss of excessive amounts of the body's salt and water can quickly dehydrate you. This can cause circulatory problems, kidney failure, and heatstroke. Sports drinks contain some salts to replace those lost in sweat. A nice homemade thirst quencher is a combination of a liter of water with one teaspoon salt, one teaspoon brown sugar or honey, and the juice of one citrus, a lime or an orange. Also, eat a banana every other hour to replace potassium lost through perspiration. The banana will curb muscle cramps. Dress appropriately for the garden wearing loose clothes that will contain the evaporating perspiration helping to cool you off.

Gardening is the best method to prevent osteoporosis (bones become porous from a loss of calcium) in women age 50 and older. Researchers compared gardening to bicycling, aer-obics, dancing, and weight training. Gardening and weight training were the only two activities shown to be significant for maintaining healthy bone mass. The best part of gardening may be the edible rewards. Home-grown fruits and vegetables contain fiber that may reduce the risk for colon cancer, as well as antioxidants that reduce the risk of heart disease and some cancers. Low in fat, fruits and vegetables can help with weight loss. It is recommended adults get three to five servings of vegetables, and two to four servings of fruits each day.

Home gardening could also be termed 'bio-philia', which means the love of living things. A love of nature is beneficial to the human system because we are part of nature and would prefer to look at flowers and grass rather than concrete. As part of the natural world, we are connected to and regenerated by it. This natural love can lower blood pressure, boost the immune system, and reduce stress. Peace and quiet in today's hectic world can be found weeding or watering the garden. In just a few minutes you will leave the day's stress and find yourself focused on the task at hand. It is a healthy delight to watch a tiny seed grow into a mature, fruitful plant.

HEALTH NOTE: HOW TO BURN CALORIES

Active for one hour

Pushups and situps	550	Take care of a baby	250
House cleaning	250	Play basketball	420
Playing cricket	350	***Farming with shovel or a hoe***	**400**
Running	500	Scrubbing floors	400
Swimming	400	Walking	350
Weight lifting	300		

Most people consume between 1500 – 2000 calories every day.

Watching your garden grow might keep you healthy. Simple research has discovered patients recovering from surgeries who looked out at a view of trees had significantly shorter hospital stays, fewer complaints, and took less pain medication, than those who looked out at a brick wall. Other studies have found looking at scenes of nature lowers systolic blood pres-sure in five minutes or less, even if the person is only looking at a photo of nature. Humans positively respond to caring human faces, nature scenes, and certain types of music. Looking at nature scenes produces positive changes in brain electrical activity, muscle tension, respi-ration, and shifts in emotional states, all of which may be linked to a better immune system. This will protect people from disease and help recovery if they are sick.

As you tinker in your garden listening to some pleasant music on your portable radio you can achieve a self-styled type of meditation. You can stop thinking, obsessing, or worrying. Your senses are awakened, which brings you into dealing only with the present, not the fu-ture, or the past. This simple technique is very effective at reducing stress. The basic question is should we spend money on drugs to medicinally calm us, or should we spend time in the garden? Having problems sleeping; work in the garden will tire you enough to get a full eight hours of delicious, sound sleep.

Don't let physical challenges keep you from gardening. Gardening will help maintain joint flexibility, range of motion, and your quality of life. Everyone, and especially those suffering from arthritis, will appreciate these benefits. Small garden stools or five-gallon buckets with lids provide both seating and tool storage. They provide relief for the knees and reduce trips to get forgotten hand tools. If kneeling or squatting is too painful, try raising your garden to a comfortable height. The width of raised beds should be narrow enough to allow the gardener to work without straining or reaching. Containers, fence gardens, and raised beds look good while allowing those with stiff joints, back problems, and other physical limitations to keep gardening. This is your garden, build it to be comfortable to you.

Pace yourself and enjoy gardening with the peppers and tomatoes. Your garden can become your relaxation area. Take time to enjoy the smell of flowers and the unique beauty of nature in your home garden. A few well-placed fragrant flowers will create a soothing atmo-sphere in your garden and attract colorful birds and butterflies.

If you discover any work is too big or your time is limited, don't be afraid to ask for help. Gardening can also build great friendships. Working with a friend or neighbor can fill the day with work, conversation, and laughter. An overwhelming task suddenly becomes a chance to spend time with new or old friends, enjoy the garden, and create new memories. After helping or being helped, end the day with a refreshment or two. Definitely take the time to admire your combined efforts. Sharing your knowledge, plants, or other talents like cooking or sewing may be the perfect trade for your friends' time and energy.

Planting a garden is a way of showing you believe in tomorrow. Keep yourself healthy enough to enjoy it. If you feel tired, it's time to take a break and rest for a while. The garden should never be stressful. You are not in a race or a competition. Your garden will always be a work in progress and will be better next year.

The best six doctors anywhere,
and no one can deny it, are:
sunshine, water, rest,
fresh air, exercise, and diet.
These six will gladly you attend,
if only you are willing
Your mind they'll ease,
Your will they'll mend
And charge you not a shilling.

~Nursery rhyme quoted by Wayne Fields,
What the River Knows, 1990

Better joint flexiblity
Sunshine provides vitamin D
Burn calories
Lowers blood pressure
Increases respect for nature
Provides healthy fresh food
Creates friendships

Gardening is the greatest tonic and therapy a human being can have. Even if you have only a tiny piece of earth, you can create something beautiful, which we all have a great need for. If we begin by respecting plants, it's inevitable we'll respect people. ~ Audrey Hepburn

$ € ¥ £ **CHAPTER TWENTY-THREE** $ € ¥ £

GARDEN ACCOUNTING

With gardening, you either love it or quickly learn to love it. The learning part is directly related to cost. The price tag for your home garden should be hours of satisfaction enjoying quality time with family and friends while growing quality vegetables. With all ventures that takes time and money, it's wise to keep an accounting of everything to determine what it costs.

This is another reason to keep an accurate garden diary and collect all the costs of growing your own. Before you plant anything or turn a shovel of soil, list in the diary the vegetables you buy, amounts and costs every week for a month, including travel expenses. This will assist you in deciding what to grow. Also, look at the veggies you'd like to buy, but seemingly don't want to pay the price, like cauliflower or broccoli. Both are expensive and are usually grown by using a lot of chemicals. Garden chemicals /pesticides create another long-range expense, for health care.

During the same time frame, keep track of the hours you spend together with family and friends, quality time, not just hi and bye. List the projects you do with children, parents, or siblings.

When your garden project begins, list all tools, chemicals, clothing, etc. you purchase and the associated travel costs. On a separate page, keep track of your hours laboring to prepare the soil. This will reduce after the first tilling. Add in limestone and manure. Keep telling yourself, it will get easier. Once ready, add in the costs of seeds and/or seedlings.

Once you've planted, every day record the time spent in the garden, like punching a time clock. Again, this first garden event will become better organized; meaning it'll become efficient and take considerably less time. Also, note what time is spent in the garden with your close relatives and housemates. Even if they are not initially drawn to your garden plot, it will eventually become almost a magnetic force where people will gather to look, ponder, and chat.

About three months later, on a separate page in your garden diary, it's time to start recording the positive outcomes, how many peppers, tomatoes, etc. Calculate what they would have cost you at the present market price including car fare and time spent. Note if the family seems any closer. Are people communicating over and about the garden?

You'll undoubtedly discover when you add in all the costs and hours, home garden produce isn't inexpensive for the novice, unless you're very lucky. But you grew it yourself and that counts for a lot. You didn't have to buy a membership for a gym to get in shape, that's a plus.

Once a home garden is up and running for a few seasons and you have some experience, you'll find it almost therapeutic. You didn't have to pay a psychiatrist to reduce workplace stress or road rage. That's another plus. In the final tally, produce from your garden should save you almost two-thirds of your original market bill.

I knew a full-time farmer who would never calculate how much he made an hour in wages. He spent every day from shortly after sunrise to sunset toiling, but his crops were good. His rationale was: if after harvesting a crop money remained in his pocket, it was a success. It was his rendition of the frequent credit card commercial, this and that cost so much, yet some things are priceless; like enjoyable time in the peaceful tranquility of the home garden.

The farmer is the only man in our economy who buys everything at retail, sells everything at wholesale, and pays the freight both ways. ~ John F. Kennedy

CHAPTER TWENTY-FOUR
THE FAMILY PROJECT

Gardening is an excellent opportunity to group the family together and share quality time working on a project that can satisfy every member, young or old. Today, whether it is the movies, a beach trip, a sporting event, or a meal in a restaurant, it seems to cost a lot of money just to have the family sitting side by side. That certainly doesn't mean members are communicating. Working in the home garden provides a unique occasion in today's hectic world for a free period for one-on-one communication among family members working towards a common goal, fresh food for all.

Starting a garden as a family group teaches valuable lessons in becoming dependable to others and seeing a project to its completion and reaping the benefits. Good life skills gained from family gardening are following instructions, performing step-by-step procedures, adhering to a schedule, and most important being responsible. Participants may learn important lessons, like the ability to communicate and work with others. Love for each other, our environ-ment, and an understanding of the cycles of nature are important ideals offered by sharing the simple tasks of the home garden.

Every child should be encouraged to be involved in gardening. Invite children to join you among the vegetables and try to explain what you are doing, planting, weeding, watering, and why each is so important. Try not to make gardening *work*, but instead an enjoyable past time. The first perception of a young mind towards agriculture, drudgery or fun, will steer a course either among rows of tranquil plants, or far from your garden. Demonstrate your love for them and the garden by sharing quality time together among the veggies every day. Give them great garden memories.

Beginning gardening for children should be educational and involve small, easy jobs like discerning weeds from plants, or watering each plant with a sausage can, or digging in the dirt to make new rows. Children have a powerhouse of energy, but a short attention span. Have many simple activity options ready to fight possible boredom.

My first memorable gardening experience was picking green beans when I was five. My mother gave me my own straw basket and straw hat. I remember it being hot, but I didn't get sunburned. I was ready every day to pick the lowest branches clean. Had I met with biting ants under one of the string bean bushes, I probably would've never again returned to a garden!

Planting flowers can be a great lead into gardening. Have a young preschooler mix the soil, fill the pot, and plant the seeds. Marigolds and zinnias are very pretty and usually grow. A flower with a delightful scent would get better attention. This will be a great demonstration of raising something living from the soil, then take seeds from the mature plant to grow more. It'll be more personal if you give the plant a human or pet name. With several children try not to make gardening competitive, but instead cooperative.

As children mature and understand a few garden practices, try to find or make some kid-sized tools. This will create a more personal experience. My parents equipped me with a small trowel and hand fork. They explained my tools were to grow veggies and gardens aren't play areas. Plants can't be, harmed, stepped on, or picked until ripe.

Witnessing garden progress, nature will become a reality rather than an abstract to a youngster. The importance of adequate sun and water, and seasonal weather trends, teaches the cycles of life and regeneration of our valuable environment. Insects, like bees and butterflies, get attention when their garden role is explained. Respect for the actual fruit of their labor coupled with respect for our planet often is difficult to comprehend.

The child must be shown that gardening can be fun. At eight or nine, depending on the child, they should be ready to start their own small garden. Keep it small and easy, and make their private plot extremely fertile. Try not to be a perfectionist. Saving all the garden and kitchen refuse for the compost pile can become another kids' garden project. The outcome will be an equally pleasant experience and truly quality shared time. Always keep in mind children have a short attention span. They can't wait weeks for seeds to sprout without forgetting what they planted. Try giving their small garden prime transplants. Beans sprout quickly and long bean climbs rapidly once it catches. Okra and red spinach are hearty, attractive plants that will produce for many months.

Gardening provides a special chance to inject wholesome homegrown, vitamin and fiber-rich vegetables into childhood diets. Eating well-balanced fresh vegetables is a great habit for a lifetime. Spinach or okra can be picked almost every day and used in many dishes. Everything that is associated as a result of a fun time in the garden should be welcomed to the dinner table.

REMEMBER, never have dangerous chemicals or fertilizers within reach of a child. Keep all sharp tools like cutlasses or shears out of sight while children are in the garden. Large, full buckets of water can be extremely dangerous to small children. Be responsive if a child shows an allergic reaction to anything in the garden. Keep a protective eye for stinging insects like wasps, bees, or centipedes.

Teenagers with some gardening experience should participate almost equally with adults in vegetable choices, and schedule responsibilities that do not necessarily interfere with educational or social activities. Within the garden may be one of the few quality times to communicate directly with your teenager. Gardening, including exercise, outdoor time, and the fresh veggie diet may make them better students.

The family garden plot must be planned to be user-friendly. No slippery paths with nothing to trip over, are necessities for the youngest and eldest gardeners. Make it comfortable with stools and maybe a hammock. If this garden experiment is a first try in a newly purchased house, plan the garden with some grafted fruit trees that will eventually shade a home comfort area.

No matter the age, family gardening provides great exercise to increase each individu-al's strength and flexibility. Slowly working into kneeling, carrying, weeding, forking, hoeing, are good exercises for elderly members. For those overweight, an hour of earnest gardening will burn over three hundred calories, not to mention the positive diet change to wholesome garden produce.

Don't push yourself or anyone else, or you'll be turned off by your family plot. If you're basically inactive (behind a desk) at your job, have been retired, or been ill for a while, stretch while looking through your garden. If you have problems with arthritis, sit on a stool or a bucket to keep above damp earth. If your knees tend to ache, make a kneeling pad. Consciously try to keep your back straight while kneeling, or using garden tools like forks, hoes, and shovels. Always consider your posture and don't overdo it. If necessary, get a waist support belt. Don't stoop.

Break up the activities, like weeding or forking, so your muscles don't become strained. Simple cotton gloves will save hands. Always remember, the garden is supposed to be fun. Don't obsess on what needs to get done. There's always more time. Work during the early morning or evening. If you must be out in the heat wear a hat and sunscreen and drink plenty of liquids.

The family garden may present a unique chance for the older family members to work alongside younger members, doing similar chores. The elders get a chance to share experiences, tell stories, and build a family friendship that's rare today.

Gardening may be an excellent way to recuperate from an illness or injury. Again start slow so you don't re-injure or regress. Know your personal time, weather, and effort limits. My good friend had a bypass over seven years ago. Exercise was recommended so he choose to walk along the roads in the early morning until passing traffic became too dangerous. He saw my garden and decided that his backyard wouldn't need the grass cut ever again. He slowly, a ten by ten plot at a time, created a garden that feeds his family and friends while healing his health and spirit.

Like so many, my older friend's garden became his exercise yard, social arena, and his focal point. In his mid-seventies, he awakes before the sun and is totally revitalized. His quiet garden now has two hammocks and a nice area to do a fireside cook. My friend's energy level and optimism from his garden are excellent for a man of his age. In fact, since he started gardening, he seems to be twenty years younger.

My mother is eighty-eight and she gardens every day, hoeing weeds, watering; and our garden is not small. She is still very flexible, with great posture. She was born to a farm and never stopped loving to grow her dinner from the dirt. Her gardening desire is a major common interest.

If something is stressing you, the family garden may be the best therapy. Creating a beautiful garden, with your family will lead to inner contentment. Perhaps a corporate garden could help subordinates and upper-level managers work jointly to create a beautiful space and relax.

A communal garden might be the path to rejuvenating the village spirit by sharing produce, saving money, while giving time. Neighbors might learn to depend on and communicate with neighbors again outside our homes now sealed with bars and separated by high walls. Food is now a national and international priority. Isn't it time we realize we are a global family?

A simple backyard garden is a good start to enhance your entire family's physical condition, your table, and combined attitudes.

A farmer travelling with his load picked up a horseshoe on the road,
And nailed it fast to his barn door, so that luck might down upon him pour;
That every blessing known in life might crown his homestead and his wife,
And never any kind of harm descend upon his growing farm. ~ James T. Fields

Courtesy R. Douglas

THE HAWAIIAN HOME GARDEN GUIDE
HERBS and SPICES

Spices and herbs are common in most foods today. Herbs and spices have been prominent throughout human history to flavor foods and for medicinal purposes. The ancient Egyptian, Chinese, and Indian cultures have the earliest written records of herbs and spices. More than 800 herbal medicinal remedies and numerous medicinal procedures were translated from the Egyptian Ebers Papyrus dating from 1550 B.C.E. Herbs usually come from the leafy part of a plant and can be used fresh or dried. Spices are obtained from seeds, fruits, roots, bark, or some other vegetative substance. Spices are not necessarily as fresh. Herbs are native to many places and climates of the world, while spices are more commonly found in the Far East and tropical countries.

Traditionally, herbs are the leafy parts or flowers of low-growing shrubs as parsley, oregano, coriander, rosemary, marjoram, etc. used to flavor or as a garnish on food. Herbs are consumed for their micronutrients and used for homeopathic/traditional medicinal purposes, or for fragrances. Herbs were used as far back as 5000 BCE in prehistoric medicine. Cuneiform tablets record the Sumerians used medicinal herbs. In 162 CE, the physician Galen was known for concocting complicated herbal remedies that contained up to 100 ingredients.

Spices are usually dried rather than fresh, and from the fruits, buds, seeds, or flowers of plants as cloves, cardamom, saffron, paprika, the bark of trees as cinnamon, or even the roots like ginger are valuable spices. The American Spice Trade Association notes the terms have become largely interchangeable with a spice being defined as any dried plant product used to season and enhance the taste of food.

Some argue that there is no distinction between herbs and spices, considering both have similar uses. Most botanists define an herb as a plant that doesn't produce a woody stem. One of the important differences between herbs and spices is in how to use them when cooking. Herbs are usually used in larger portions than spices for flavoring foods. Spices are generally stronger tasting and used in much smaller amounts.

Some plants are both herbs and spices. Cilantro is an herb while coriander, from the plant's seeds, is considered a spice. Dill seeds are a spice while dill weed is an herb obtained from the plant's stems and leaves.

PAGE:

But in truth, should I meet with gold or spices in great quantity,
I shall remain till I collect as much as possible,
and for this purpose I am proceeding solely in quest of them.

~ Christopher Columbus

ALLSPICE
Pimenta dioica

Allspice is the grown commercially worldwide. Until the 1970s, it was cultivated only in the New World/Western Hemisphere. Its aroma smells like a combination of several spices, especially cinnamon, cloves, ginger, and nutmeg. That is how it acquired its name. Allspice's botanical name is *pimenta dioica*. It is also referred to as English spice, Jamaica pepper, clove pepper, myrtle pepper, pimenta, and pimento. (Allspice is called pimento because the early Spanish explorers thought it was black epper, which they called *pimiento*.)

The allspice tree has undoubtedly been spread by birds and is now recognized as naturalized on the islands of Hawaii, Kauai, Oahu, and Maui. This spice is still wild in the rainforests of Central and South America. The world's best variety is cultivated in Jamaica. Historically, allspice was used as an embalming agent by the Mayans. The Arawaks used allspice to preserve meats. Europeans attempted to transplant it to Asia without any profitable results. Despite a delightful and distinctive aroma, it never became as valuable as nutmeg, cinnamon, or pepper. The main purveyors were English, so it became known as English spice. Today allspice is used in men's shaving scents like Old Spice. Allspice is great to flavor your favorite alcohol.

HOW TO GROW: These trees are available in nurseries throughout the island chain and would be unique and useful in your home garden. To buy, check the social networks. This spice would make an excellent shade tree for your hammock. It is an evergreen tree, with glossy, dark green leaves and gives off a refreshing scent even in a slight breeze. All it needs is well-drained soil and water during the dry season. Initially feed it 12-24-12 at a rate of a cup every 2 months. An allspice tree reaches 30 feet tall and wide. Once it begins to blossom switch to a high nitrogen fertilizer as 12-12-17-2 at a rate of one cup every month. Small white blossoms precede green berries, which ripen to purple. Trees start to develop fruit after about five years and will bear for decades. Allspice trees are either male or female, so you need both for berries to develop, and one at each end of your hammock. Handpicked, allspice berries are harvested green and dried in the sun until the two ¼ inch seeds inside rattle.

MEDICINAL: In the past, allspice was used to treat indigestion and gas. It was also eaten as a treatment for stomachaches, vomiting, diarrhea, fever, flu, and colds. It has been used to flavor toothpastes. Allspice is a digestive aid and carminative similar to cloves. Its oil ex-pands blood vessels; increases blood circulation, and makes the skin feel warmer. During the Napoleonic wars Russian soldiers kept allspice in their boots as a foot warmer. A poultice or steeped in a hot bath, it is a 'bush' remedy for muscles aches and arthritis.

Eugenol, the compound that gives allspice the spicy kick, treats nausea and is used in over the counter toothache remedies. Research has found eugenol is a topical pain reliever if correctly applied. Allspice tea may help settle an upset stomach.

NUTRITION: A teaspoon of ground allspice has only 6 calories with some vitamin B-5 and C with calcium, manganese, copper, and iron.

FOOD & USES: Allspice is a main contributor to the essence of Jamaican Jerk. Created by the Arawak Indians, the spices and peppers preserved meats just like drying them over a fire. The term 'jerk' is derived from the Spanish *'charqui'*, or dried meat. It is often used in European cuisine especially for pickling or marinades.

RECIPES: JERK SEASONING MIX

Ingredients: 2 TBS dried minced onion, 1 TBS dried thyme, 2 TS ground allspice, 2 TS ground black pepper, 2 TS ground cinnamon, 1 TS dried cayenne/red pepper, ½ TS ground nutmeg, 1 TS salt, and 2 TBS cooking oil

Method: Combine ingredients except oil in a small bowl. Coat meat with oil, and then rub seasoning onto meat. Let sit chilled for two hours. Cook meat slowly over coal, stovetop, or oven.

MANGO CHUTNEY

Ingredients: 20 almost ripe mangos peeled, 2 large onions chopped small, 6 garlic cloves - minced, 2 cups raisins (golden preferred), ½ cup fresh ginger root peeled and minced, 3 cups distilled white vinegar, 6 cups white sugar, 6 cups brown sugar, 1 TS ground cinnamon, 2 TS ground ginger, 4 TS ground allspice, 1 TS fresh ground cloves, two TS fresh ground nutmeg, three hot peppers - seeded and minced (more to your taste), one TS salt

Method: In a large pot, mix all the spice ingredients with salt, hot peppers, and vinegar. Boil for 30 minutes before adding onions, raisins, and ginger. Simmer over low stirring frequently for 30 minutes. Add mangos, and simmer for another half hour. Keep stirring. Refrigerate after cooling.

WATERFRONT SPICY SHRIMP PASTA

Ingredients: 1 kg/ 2lbs. medium shrimp - peeled and deveined, 250 g ½ lb. pasta cooked al dente' - (I like bow ties, but any will work), 6 garlic cloves, minced, 3 hot chilies seeded and minced, ¼ cup chives chopped, 1 medium onion chopped, 1 TS fresh ginger root peeled and minced, ¼ cup culantro chopped small, 3TBS oil, 1 green bell pepper, seeded and chopped, 3 large tomatoes chopped, 2 TS curry powder, ½ TS whole or ground allspice, ½ cup chicken stock, 2 TBS soy sauce, 1 TBS brown sugar, 1 TS cornstarch

Method: Combine garlic, onion, ginger, and oil in a small bowl. In another bowl, combine green pepper, tomato, culantro, curry powder, allspice, chicken stock, soy sauce, brown sugar, cornstarch, and hot peppers. In a large skillet, heat the garlic-oil mixture before add tomato mixture, cook for 3 minutes. Stir in the shrimp and cook for 2 minutes. Add pasta and stir until pasta is warmed to taste. Serve immediately.

SWEET POTATO MUFFINS

Ingredients: 4 cups shredded peeled sweet potatoes, 1 cup brown sugar, ½ cup oil preferably canola, 1 TS vanilla extract, 2 eggs, 2 cups flour (bakers), 2 TS baking powder, 1 TS cinnamon powder (fresh grated if possible), 1 TS fresh grated nutmeg, 1 TS allspice powder, 1 TS salt, ½ cup raisins (optional golden preferred), ½ cup almonds or cashews (optional), ½ cup water

Method: In a bowl, blend wet ingredients brown sugar, oil, vanilla, and eggs. In another bowl blend together the flour, baking powder, spices, salt, and sweet potatoes. Then combine both and add raisins. Add water if too lumpy. Place mixture in a muffin greased tin or baking dish. Bake at 350 for half an hour. If muffins stick, carefully ring each with a knife and gently tap out of the tins.

DID YOU KNOW?

The Mayan Indians used allspice to embalm the bodies of their kings. It became known as allspice because its flavor is similar to a combination of many spices, cloves, cinnamon, and nutmeg.

ALO'E - ALOE VERA
Aloe barbadensis

Since I was a small child, my family used the aloe plant to coat cuts and especially burns. It is known in Hawaii as Alo'e. Aloe vera's botanical name is *Aloe barbadensis*. The word 'aloe' in Sanskrit means 'goddess. The name aloe is from the Arabic word, 'alloeh.' That translates as bitter, probably referring to the taste of the leaf juice.

Aloe vera is probably native to the area of Northern Africa, the Canary Islands, and the Cape Verde Islands where it gained another name - Cape Aloe. It is a short-stemmed plant with thick leaves that looks like a cactus. It is considered a succulent because it retains water in its leaves, fat with gel, and adapts to almost any climate or soil, even extremely dry, heated conditions. Due to its reputed herbal medicinal qualities, aloe now grows worldwide in warm climates. There are more than 400 species of aloe but fantastic medicinal healing qualities are only attributed Aloe babbadenis variety.

HOW TO GROW: Aloe is considered a tropical plant, but will withstand severe heat and droughts. It will also survive floods if the soil eventually dries within a week. It cannot live in standing water. It resists most insect pests. Once you put it in the soil it only needs slight attention. A mature aloe plant can reach a meter tall and will spread naturally with sprouts or offshoots. The greenish-gray leaves are thick and fleshy, with rough almost serrated edges. Yellow flowers are produced on a tall center spike.

I have a small cactus garden where I have planted aloe. Small aloes sprout around the mother plant similar to the pineapple. Aloe is a perfect kitchen counter container plant. Sepa-rate the shoots and plant in well-drained sandy potting soil. Set where it will get bright sun and will drain. Water regularly, but do not keep the potting medium wet. Potted aloe's soil medium should completely dry before watering. As they grow, transplant to larger containers.

Due to easy growth, many countries have large-scale aloe farms supplying the cosmet-ics industry Bangladesh, Cuba, the Dominican Republic, China, Mexico, India, Jamaica, Kenya, and South Africa, along with the USA and Australia. Every Hawaiian Island has some large aloe farms in the dry areas. Aloe cultivation is a great, east, low investment home garden business. 15,000 suckers per acre, good drainage with 8.5 pH soil and harvest in 2 years.

MEDICINAL: A cuneiform clay tablet from Sumeria circa 2,000 BCE records aloe being used as a treatment for constipation. The ancient Egyptians had twelve formulas that combined aloe with other herbs and spices to treat both internal and external health issues.

When you slice open an aloe leaf, it has two fluids. A bright green liquid - the sap - seeps out as soon as you scrape the surface, and this fluid is an irritant. When you cut the leaf open or crush it, you'll see the inner gel that soothes burns and helps healing. The latter is what is termed - aloe juice. Drinking aloe vera juice relieves heartburn and irritable bowel syndrome. It is common practice for cosmetic companies to add gel or other derivatives from aloe vera to products such as makeup, tissues, moisturizers, soaps, incense, and shampoos.

For centuries aloe vera has been used for medicinal purposes. The secret is its concen-tration of nutrients and vital substances that include water, vitamins A, B, C, and E, more than 20 minerals, and fatty and amino acids. It seems scientists didn't believe aloe actually cured burns, so it was tested. Aloe is now proven to heal first and second-degree burns, but it has not been proven to protect from sunburn. Drinking aloe juice improves glucose levels for diabetics and patients with liver disease. Aloe juice may reduce the inflammation of stomach ulcers. It reduces gum disease and dental plaque. Aloe vera extracts are antibiotic and fight fungus.

Aloe vera is used to treat skin ailments as insect bites, acne, sunburns, rashes, scars, blemishes, sores, eczema, and psoriasis. It is drunk to treat blood pressure, hypoglycemia, arthritis, ulcers, constipation, poor appetite, digestive disorders, diarrhea, and hemorrhoids. Aloe should be simply patted on the skin just once a day is sufficient to get the desired results.

NUTRITION: Per cup, aloe has 36 calories. Aloe vera contains a good amount of vitamin C - 9.1 g for every 1 cup of aloe vera juice. Aloe vera also contains essential vitamins like vitamin A - beta carotene, E, B12, B - folic acid, and choline. The plant also contains calcium, chromium, copper, iron, selenium, magnesium, manganese, potassium, sodium, and zinc.

FOOD & USES: Drinking aloe vera helps losing weight and boosts the immune system. Many people are disciples of the aloe vera plant and have found taking the juice daily helps to maintain overall good health and provide needed energy.

RALPH'S ALOE CUT AND BURN CURE- fast healer salve

Combine the gel from 1 aloe leaf, the juice from 4 vitamin E capsules, and 1 TS benedine, iodine, or mercurochrome. Apply to cut or burn and allow to dry. Keep area from getting wet and reapply as necessary.

ALOE ENERGY PUNCH 1

Carefully peel 1 long aloe leaf and combine with milk, a banana, and flesh from 1 mango. Blend and add honey if the taste is too tart.

SICK MAN'S PUNCH

Combine the following and blend as smooth as possible; the juice of half a lemon or sour orange, 1 raw tomato sliced, 1 cup raw broccoli, 1 raw carrot sliced, 1 cup raw spinach, 2 cloves raw garlic, ½ cup raw almonds, half seeded hot pepper (optional if discomfort is digestive), the jell from 2 large aloe leaves.

RECIPES: ALOE CHICKEN SOUP – cure for a cold or cough. Raw aloe is *very* slimy, more than the slimiest okra. Cooking it reduces the slime considerably, but it does still have a slippery taste. Cooking the aloe will give off a ton of liquid and the cubes will shrink and soften without losing their shape.

Ingredients: 5 aloe leaves, 1 chicken chunked, 1 cup fresh grated coconut, ½ seeded hot pepper, 2 ripe tomatoes chopped small, 2 garlic cloves minced, 2 onion chopped small, 2 bunch parsley chopped, salt to taste, 2 quarts water.

Method: Wash aloe, slice the skin and chop into chunks. Combine chicken with all ingredients, except aloe, in a suitable pot. Bring to a boil, reduce heat and simmer for 2 hours. Add aloe chunks. Return to boil and then simmer for half an hour.

ALOE BEAUTY MASK:

Ingredients: 1 TS red clay, 1 TS aloe gel, 1 TS witch hazel (from a pharmacy), 2 drops pepper-mint oil or 10 mint leaves minced and crushed, enough water to make a paste.

Method: Mix with a fork or whisk and apply to clean skin and let sit for 15 minutes. Rinse with warm water and follow with cool to close the pores. Air dry or gently blot dry.

DID YOU KNOW?

A papyrus dating from Pharaoh Amen-Hotep's reign in 1552 BC (and found between the knees of a mummy excavated in 1858!) gave twelve different formulas for aloe vera preparations used during the pre-ceding two thousand years.

ASPARAGUS ROOT – SHATAVARI – THE HERBAL QUEEN
Asparagus Racemosus

Asparagus root an herb and not the root of the vegetable asparagus. Also called Shatavari, it is known as the Herbal Queen because it is associated with love, devotion, and fertility. The plant is native to India, Sri Lanka, and Nepal. This is a tropical herb, and with the proper attention it will grow anywhere throughout the Hawaiian Islands. Its botanical name is *Asparagus racemosus.* Seeds are available from the Internet. Grow it for the health benefits.

HOW TO GROW: This is a slender, very adaptable, low maintenance, perennial, climbing shrub with many branches and some curved spines (thorns). There are three variet-ies: large leaves, small leaves, and without thorns. From seed, it grows like a bush and prefers well-drained fertile sandy-loam soil, with a pH of 6–8 up to 1400m elevation. This herb can withstand droughts and only requires weekly watering. With the right conditions, it can grow up to 4 meters tall and usually requires staking. The leaves are thin, like pine needles. The fra-grant tiny blossoms are white. A green, round 6mm (¼ in.) berry forms that will mature to red, with one seed. Save the seeds because the germination rate is low. This herb has a root system with multiple long, tapering tubers that can grow to one meter. The roots are desired, so loose soil with rotted manure produces more. This is a well-known medicinal plant, but because of destructive harvesting and deforestation, this plant is endangered. 'Save the Herbal Queen!' Grow a few plants in your yard.

MEDICINAL: Asparagus root has been used for centuries by women to help balance hormones and increase fertility, and as a general female reproductive tonic. This herb aids the female reproductive system of every age group, through life's natural changes, including menopause. It nourishes female reproductive organs and helps regulate ovulation. It is also used to increase breast milk production in nursing mothers. The male reproductive system can also benefit from asparagus root. Most of the Herbal Queen's medicinal properties are asso-ciated with phytonutrients in its roots, especially saponins. Plants with saponins have anti-in-flammatory and anti-cancer properties, improve blood pressure, and help regulate blood sugar and fat levels.

Scientists found that asparagus root blossoms contain the flavonoids quercetin and rutin. Its leaves contain the photosteroid diosgenin and the berries have rutin and glycosides. An infusion of this root can help to rejuvenate the body and improve mental functions. India is producing a root powder, herbal tablets, and syrups from this valuable herb for local and overseas markets.

NUTRITION: Asparagus root is rich in calcium, iron, and zinc. The Herbal Queen contains a lot of vitamins C and E, K, and B3 niacin, plus nutritious starches.

FOOD: Asparagus root is eaten in a rice porridge to treat body ailments as urinary tract problems, and it can make a green herbal soup named kola kanda to repair digestive difficulties.

RECIPES: KOLA KANDA - treats the reproductive organs

Ingredients: 2 cups of asparagus root 2 cloves of garlic chopped, 2-4 spring onions cleaned, 1 cup freshly grated coconut, 1 cup water, 1 cup brown rice, and salt to taste

Method: In a suitable pot with a cover, steam the rice with the garlic until the rice is soft. Stir in the coconut. Put the asparagus root in a blender with 1 cup of water and blend for 2 minutes. Strain through a sieve and add that leaf-juice to the cooked rice. Add salt and enjoy.

NAMES: The English name is asparagus root. (*Asparagus racemosus* is known in Sanskrit as Shatavari, which means: who possesses a hundred husbands or acceptable to many. Shatavari also means the cure for a hundred diseases.

BASIL – KĪ ‘A ‘ALA
Ocimum basilicum

Basil is one of the world’s most widely grown and used herbs. It is recognized with a warm, clove-like flavor and smell. I have a few plants in my herb garden, and use it in many dishes, but especially when I experiment with Italian recipes. This herb is a relative of mint. There are over sixty varieties with various tastes and colors: with names like sweet, mammoth, dark opal, cinnamon, and licorice. The leaves used in cooking can be green, reddish, or purple. Basil is an important ingredient of Italian and Asian cuisines. Egypt is the main cultivator of basil for the world’s market. Basil is in local Hawaiian is kī ‘a‘ala. It’s also known by the Indian name, *tulsi,* which usually refers to the holy basil variety.

Botanists believe it originated in northeastern India. Basil is from the ancient Greek word meaning ‘royal’ because its medicinal properties made the herb noble and sacred. Ancient Egyptians used basil for preserving mummies. East Indians respect basil as sign of generosity, while the Italians see it as a symbol of love. Superstition is basil plants around a house make a happy home. Basil is commercially grown throughout Hawaii on a limited scale of about 60 acres producing $1.5 million at harvest. Most is consumed locally.

HOW TO GROW: Basil rapidly loses its taste when picked and is not well preserved by drying or refrigeration. It is easy to grow from seed, and a great addition to any garden. Basil needs loose soil and occasional water. Once it has about eight leaves, literally pinch the top of the plant between your thumb and forefinger. This will stunt the plant’s growth upward and force it to grow more leaves. The pinching will also keep the plant alive longer as it will delay the flowers. It is easily grown in pots, and if placed near a window it will deter houseflies and mosquitoes. The heat may present a problem if you are growing on a large scale in a tra-ditional plot garden. If you find that big leaf, traditional Italian basil burns or blooms quickly in the heat, try other basils. Spicy holy basil (*Ocimum tenuiflorum*) has green leaves and hairy stems with a taste of a mix of black pepper and cloves. Sweet Thai basil is the basic botanical *Ocimum basilicum*, but may be more adapted to the heat. Thailand is the world’s largest basil producer. This variety has an aftertaste of licorice/anise. Its leaves are red to purple and the stems are purple.

MEDICINAL: In ancient times, basil was used as an antidote for poison. It is an herbal remedy for diseases related to the brain, heart, lungs, bladder, and kidneys. Mixed with another herb, borage, makes a revitalizing tea tonic. An infusion of lemon-scented basil was used by Hindus to ease the symptoms of diabetes. Basil oil massaged into the skin will enhance the luster of dull looking skin as well as hair. Basil is used cosmetically as a toning body rub mixed with coarse sea salt and vegetable oil. It is also used for acne and skin infections. Basil tea is used to fight coughs, and relieve asthma, bronchitis, and sinus infections. Basil is used to suppress nervous tension, mental fatigue, melancholy, migraine headaches, and to fight de-pression. Basil is a good source of heart healthy magnesium, which relaxes blood vessels im-proving blood flow and lessening the risk of irregular heart rhythms or heart spasms. This herb improves blood circulation by fighting bad cholesterol, while reducing the chance of irregular heartbeats. A basil leaf will relieve the pain of a mouth ulcer. Basil tea will soothe sore gums, and a good soothing remedy for arthritis or rheumatism.

NUTRITION: Basil has very few calories and is a good source of vitamins A, C, and K, magnesium, and potassium.

FOOD & USES: Fresh basil is a requirement for most chefs, especially in Italian cuisine. Its taste accents tomatoes, onions, garlic, and oregano; the basic Italian seasonings.

To get the best flavor from basil add the fresh leaves towards the end of cooking. For a good sauce for fish just mix minced basil leaves with mayonnaise. It is a great addition to a stir-fry, and vegetable dishes with eggplant, cabbage, or peppers. Basil ground with garlic, mixed with olive oil and cheese grated into a paste is what the Italians call 'pesto', which is usually served with pasta.

Note: ½ ounce of fresh basil leaves equals 1 cup chopped fresh basil. When substituting dried for fresh, triple the amount. Basil, or Sweet Basil, is a common name for the culinary herb *Ocimum basilicum,* sometimes known as Saint Joseph's Wort.

RECIPES: BASIC PESTO

Ingredients: 1/3 cup fresh basil leaves, 2 garlic cloves pounded, ½ cup pine nuts, (Unsalted peanuts, almonds, or cashews can be substituted.) ¾ cup grated parmesan cheese, ½ cup olive oil, slight salt (One leaf of culantro can be added to vary the taste.)

Method: Put basil leaves in blender and chop, while adding oil and garlic. Add nuts slowly until everything is a thick cream. Use it as a pasta topping with two tablespoons of pesto per every person. Pesto can be made in volume and frozen, but only add the cheese when you are ready to prepare a meal.

MUSHROOM AND CABBAGE PIE

Ingredients: 2 premade pie crusts, 3 cups cabbage sliced thin, 2 garlic cloves minced, 1 large onion chopped fine, ½ pound mushrooms, 1 cup basil leaves chopped, ½ TS marjoram, ½ TS tarragon, ¼ cup cream cheese softened, 4 eggs hard-boiled, 1 egg beaten, ½ TS fresh dill, salt and spice to taste

Method: Coat a large frying pan with oil and sauté onions and garlic. Add cabbage and mushrooms, and simmer for about 20 minutes. Stir in basil and seasonings. Spread the softened cream cheese in the bottom of pie and arrange a layer of sliced hard-boiled eggs. Cover with cooled cabbage, onion, and mushrooms. Sprinkle with dill and cover with second piecrust. Make slices to let steam escape, and brush with beaten egg. Bake at 350 degrees for half an hour. Serve cool.

TREMENDOUS TOMATOES

Ingredients: 6 medium to large tomatoes, 1 avocado chopped, 1 medium onion chopped fine, 1 cup cheddar cheese grated, ½ cup fresh basil leaves chopped fine (or 4 TBS dried basil), ¼ TS oregano

Method: Halve tomatoes and place cut side up in a baking dish. Cover with the mixture of avocado, basil, onion, and cheese. Sprinkle with oregano and broil in oven or on barbecue grill (covered) for 5 minutes. Serve hot.

DID YOU KNOW? It was believed that Salome hid John the Baptist's head in a pot of basil to cover up the rotting odor. Basil/tulsi symbolizes Goddess Lakshmi, the consort of Lord Vishnu. Those who wish to be righteous and have a happy family worship the tulsi. In Italy, basil has always been a token of love. In Romania, when a boy accepts a sprig of basil from his girl, he is engaged.

A man taking basil from a woman will love her always.
~ Sir Thomas Moore

BAY LEAVES
Laurus nobilis

Bay trees aren't common in Hawaii, but are available on several islands through nurseries. As a 'wannabe' chef, I use all types of spices and find those I use fresh usually have the best flavor. That's not the case with the bay leaf. Its flavor intensifies when properly dried. The bay leaf is a kitchen staple the world over. Although there are a few different types of trees, each having its specific flavor, the sweet bay trees are descendants of the Mediterranean or Turkish variety. Bay belongs to the same family as cinnamon, laurel, and avocado. East Indians call their variety 'Tej Patta,' and it has more of the cinnamon flavor. Historically bay leaves adorned victorious warriors and intellectuals of both the Greek and Roman empires. This special leaf supposedly protected them against lightning and plagues. The English be-lieved bay brought good fortune and a leaf tucked behind your ear would prevent you from getting drunk. Botanically sweet bay is *Laurus nobilis,* while West Indian bay (bay rum) is *Pimenta racemosa.*

HOW TO GROW: Bay trees can be shaped and are a great addition to a landscaped home garden. The bay tree can also be trimmed and kept growing in a big pot of 15 gallons or larger. These trees can be grown from seed if you are very, very patient as it can take 6 months to germinate. Keep the potting soil damp, but not wet enough to rot the seeds. Try to grow several plants as few survive the first year or buy a seedling. Bay likes sun, not too much wind or water, but don't let it dry out. Plant where it will drain and fertilize with well-rotted chicken manure twice a year. The tree can reach 60 feet, but if you trim and top it at 20 ft., a bay tree will complement any back yard. Trimming is best done during the late year rainy season and then you can make holiday gifts of the leaves. In a shaded area (so the leaves dry slowly and retain their flavor) arrange the fresh leaves between flat boards as plywood to prevent curling.

MEDICINAL: Put bay leaves in a damp washcloth to alleviate the pain of headaches, especially migraines. Bathing in a bath with bay leaves can treat skin rashes and pain from over sore muscles or arthritis. Bay oil (if you can find it) increases blood circulation and is reported to prevent baldness. Bay leaf helps the body process insulin more efficiently, which lowers blood sugar levels. It has also been used to treat stomach ulcers. Bay leaf has anti-inflammatory, antioxidant properties, anti-fungal, and anti-bacterial. Bay also treats rheumatism and colic.

NUTRITION: Since a recipe might only call for 1-2 bay leaves, their nutrition value is negligible. 100 grams of bay leaves – probably enough for two years – has 315 calories with good percentages of vitamins B-6 and C, potassium, iron, calcium, and magnesium.

FOOD & USES: These same leaves hung in a wardrobe can protect your clothes from moths. Bay enhances most dishes with a hearty flavor, like stews, beans, potatoes, or soups. For a different barbecue, try shish kabobs with bay leaves skewered between the fish, meat, and vegetable pieces. Add the bay leaves early in the cooking or marinating as it takes a while to give off its flavor, but always remove the bay leaves before eating.

RECIPES: CHICKEN STEW (NOT STEWED CHICKEN)

Ingredients: Half of a chicken skinned and chunked, 2 cups red beans cooked, 2 stalks celery, 1 carrot, 4 medium tomatoes, 1 medium onion, 2 ochro/okra, 1 sweet bell pepper - all chopped small, 2 TBS olive oil, 2 cups water, 2 TBS ketchup or tomato paste, 1 bay leaf, ½ TS fresh thyme, 1 TS fresh basil, ¼ hot pepper seeded (optional), salt to taste

Method: Boil chicken chunks in the 2 cups of water for five minutes. In a good-sized pot on medium heat sauté the celery, carrot, and onion pieces in the oil for about 5 minutes. Pour in chicken and broth. Add everything except the beans and simmer for half an hour with an occasional stir. Add beans and cook for 10 more minutes. The broth should now be thick. Remove the bay leaf and hot pepper piece. Serve hot. This is great for a rainy-day lunch or dinner.

HOT BAY BANANAS — This is easy and tastes great especially with fresh ingredients.

Ingredients: 8 ripe bananas (firm) sliced quartered longwise, 4 TBS butter (prefer unsalted),2 bay leaves, 1 cup fresh orange juice, 3 TBS fresh lemon juice, 1 cup brown sugar, 4 TBS brandy, cinnamon, and nutmeg to your taste, a pound package cake mix.

Method: Bake pound cake per package instructions. In a large frying pan melt butter until it bubbles and just begins to brown. Reduce heat and add bay leaves. Then mix in the orange juice, lemon juice, brown sugar, brandy, cinnamon, and nutmeg. Increase heat until everything begins to boil. Simmer stirring constantly until it has a thick, syrup consistency. Combine with banana quarters and stir gently to coat the pieces without breaking them. Once heated and coated, place bananas on sliced warmed pound cake. Cover with French vanilla ice cream. Enjoy.

BAY POTATOES

Ingredients: 6 large potatoes (or 1 per diner), 1/3 cup olive oil, 12 bay leaves, 1 TBS salt, 1 TBS powdered red (cayenne) pepper, herbs as sage, basil, oregano to your taste

Method: Slice the tops of each potato in a crisscross fashion without slicing through. Slide 2 bay leaves into each potato. Put potatoes in a baking dish or a bread pan. Mix oil with salt, pepper, and spices, and brush mixture onto potatoes. Cover dish with foil and bake at 350 for an hour. Uncover and put potatoes under broiler until they begin to brown. Serve hot.

BAY LAGER CABBAGE

Ingredients: 1 head cabbage cored and shredded, ½ lb. of ham, chicken, or beef chopped very fine (minced will work), 6 medium to large onions chopped fine, 3 bay leaves, ½ hot pepper seeded and minced (optional), 1 TS salt, ½ TS sugar, 1 bottle of lager beer

Method: In a large pot with a cover sauté the meat over medium heat. Then add bay leaves, onions, salt, pepper, and sugar. Cook stirring until onions lose color. Add cabbage and mix well. Cook until cabbage has wilted, usually less than 5 minutes. Add beer and lower the heat. Simmer covered for an hour. Remove bay leaves and serve warm.

EASY VEGETABLE CHILI delicious bean dish made as spicy as you can handle it.

Ingredients: 2 cups soya chunks, 5 cups red, black, or pink beans cooked (I prefer a combination of beans), 4 large onions chopped small, 6 medium to large very ripe tomatoes chopped, ¼ cup ketchup or tomato paste, 3 garlic cloves minced, 2 TBS ground cumin, 3 bay leaves, 1 hot pepper seeded whole or minced (optional), salt to your taste

Method: In a large pot with a cover mix all ingredients together. Add enough water so ingredients are covered. Bring to a boil and reduce heat. Simmer covered for at least an hour. Add more water if the sauce gets too thick. Remove bay leaves and pepper before serving hot.

DID YOU KNOW?

Bay leaves can also be crushed or ground before cooking. Crushed bay leaves impart more flavor than whole leaves, but are difficult to remove. It is best to use a muslin bag or tea infuser. In some cultures, the bay has a reputation as being a protective tree against lightning, witchcraft, and evil. A bay leaf tree is great for your yard as it is always green. Dry season, torrential rains, and hot winds do not affect it.

The Bay leaves are of as necessary use as any other in the garden or orchard, for they serve both for pleasure and profit, both for ornament and for use, both for honest civil uses and for physic, yea, both for the sick and the sound, both for the living and the dead; . . . so that from the cradle to the grave we still have use of it, we still have need of it.

~ Parkinson, 'Garden of Flowers' (1629)

BLACK PEPPER
Piper nigrum

With many local and exotic spices at hand, the most used is black pepper termed 'the king of spice.' Native to India it is now produced in the tropical East and West Indies, and throughout Asia. Black pepper, *Piper nigrum*, is a flowering vine whose 'heat' comes from the piperine, not capsicum found in hot peppers. This vine grows well on all the Hawaiian Islands and is commercially cultivated. As with almost all plants, try the social pages and nurseries on the NET for someone selling plants or ask at the garden supply shops.

There are a few different types of pepper vines. Their fruit are peppercorns. Black pepper is an unripe berry that has been dried for the sharpest and hottest taste. White pepper ripens more on the vine than the more common black variety and has a slightly less sharp taste. Green peppercorns are picked even earlier than the black or white for a fresher, less hot flavor.

Pepper has been an essence of East Indian cooking for more than four thousand years. Before petroleum, peppercorns were the original 'black gold' of trading. They were used as a form of money. Pepper was considered one of the five essential luxuries upon which foreign trade with the Roman Empire was based, the others being African ivory, Chinese silk, German amber, and Arabian incense. The term 'peppercorn rent' means small money. In the 14 and 1500s, a pound of pepper equaled a pound of gold, or up to three weeks' labor. It was the incredible value of pepper and a few other spices that led to the European efforts to find a sea route to India, and the eventual discovery of the Americas.

Upscale markets sell whole peppercorns, and grinders can be located in various kitchen stores. Peppercorns can be ground with a simple mortar and pestle. Fresh ground pepper has a superior taste to commercial powdered pepper. Whole peppercorns hold flavor almost indefinitely when sealed in an airtight container. Pepper is best when it's ground directly onto food. When cooking, add pepper at the end to preserve its aroma. White pepper is used in white sauces. Green pepper is best combined with garlic and other spices like cinnamon to make fresh sauces.

HOW TO GROW: The pepper vines are easy to grow in the tropics, but getting the seeds is the hard part. Local nurseries have pepper vines. With adequate water and some shade, pepper will grow almost anywhere on the islands. Plant the seeds ¼ inch deep rich potting soil 3 inches apart, Keep the soil moist. Peppercorn seeds can take thirty days to germinate at temperatures higher than 75 degrees F. Transplant seedlings at 4 inches to a partial shaded area.

It is a perennial climbing vine with aerial roots like an orchid. Plants can produce for thirty years. The vine can reach thirty feet long with wide glossy green leaves and will cohabit on citrus, cacao, and coffee trees. When it blooms each cluster will have about fifty small white blossoms. As the blossoms fade green peppercorn berries appear. They are green at first, maturing to a reddish. The pepper vine needs well-drained, humus-rich soil, and a hot wet tropical climate. Pepper is grown from cuttings or seeds cultured in a partially shaded area. Flooding or continually wet soil will kill pepper vines.

Plant in a raised grow box that can easily drain. Put large stones at the bottom, topped with smaller stones and sand. Then cover with compost material or very loose soil. The seeds or cuttings need to be watered lightly twice a day to keep the soil moist. Pepper needs poles or a trellis to support the vines. A long piece of the cut vine - two feet or more – is tied to a pole or trellis. Rough surfaces make these vines climb better. Plant two seedlings per post. Keep other vegetation cleared away, but keep enough trees to shade the pepper vines. Fertilize with

20-20-20 every third week. Pepper vines should bear within three years and bear heavily for a decade. Harvest all the peppercorns as soon as one ripens to red. You want the corns to be immature and that signifies that they have the sharpest taste. The corns can be sun dried. By the fourth year each pepper vine should bear a kilo of peppercorns. You can pickle or dry the green berries. The mature dark fruit should be boiled water for ten minutes immediately after picking. Then spread in the sun and dry to a dark brown or black. Try keeping ripe berries moist for a few days until the skin can be removed easily to make white pepper.

MEDICINAL: Black pepper is considered good for the digestive system and fights bacteria. Pepper soothes nausea and increases body temperature to fight fevers and chills. Its spicy hot flavor makes the nose and throat produce (water) a lubricating secretion and assists anyone who needs to cough up and clear their lungs. Pepper was also used as an ointment to re-lieve skin afflictions and hives. However, coarsely ground black pepper irritates the intestines.

NUTRITION: Black pepper, especially fresh ground, is almost a small vitamin pill. One teaspoon of black pepper contains 6 calories with manganese, copper, calcium, vitamins C and K, iron, phosphorus, potassium, and selenium.

FOOD & USES: For organic gardeners, one-half teaspoon freshly ground pepper to one quart of warm water sprayed on plants can be toxic to ants and moths. Sprinkle ground black pepper in doorways and windowsills to stop invading insects. Pepper is ranked the third most added ingredient to recipes, behind water and salt. Peppercorn is the most widely traded spice in the world, making almost a quarter all spices imported. This is by monetary value not weight because a greater weight of hot peppers is traded, but they have less value. Vietnam grows 30% of the world's peppercorns and Brazil grows 10%. Worldwide 350,000 tons are grown yearly.

RECIPES: OVEN-DRIED TOMATOES

Ingredients: 2 lbs. plum tomatoes cut in half and seeded, 4 TBS salt (coarse or sea preferred), fresh ground black peppercorns, 1 TBS dried marjoram, 2 TBS dried basil, 2 bay leaves, black peppercorns, 2 garlic cloves - sliced thin, 2 TBS olive oil

Method: Place the tomato halves, cut side up, on a baking sheet. Sprinkle with the salt, pepper, marjoram, and basil. Cover tray with foil. Bake at 300 degrees for an hour and remove foil and bake for another 45 minutes. Cool and pack into a sealing refrigerator container. Mix in olive oil, garlic, and peppercorns. This should make a quart. These are an excellent addition to any salad and will keep for several months refrigerated.

PEPPER STEAK

Ingredients: 1lb. thick strip steak, 2 TBS whole peppercorns, 1 TS allspice, salt, 1 TS cooking oil (canola preferred), 1 TBS butter or margarine

The sauce: 1 small onion chopped small, 2 TBS chive minced, 2 TBS brandy or red wine, ½ cup beef stock, 1 TBS soft butter or margarine, 2 TBS parsley chopped

Method: Trim the steak of any fat and slice in half and salt. Smash peppercorns using a rolling pin or mortar and pestle and roll the steaks on it until covered. In a large frying pan over high, heat the oil and the butter. When the skillet is very hot, place the peppered steaks and fry each side for about two minutes. (Longer if you detest rare or medium-rare meat as I do.) Making the pan sauce: Add onion and chives to the pan and fry for about a minute continually stirring. Add the brandy carefully as it should flame up and burn off the alcohol. Fry for a few minutes before adding the beef stock. Bring to a boil and stir in the soft butter. Pour sauce over the steaks and garnish with chopped parsley.

The disparity between a restaurant's price and food quality rises in direct proportion to the size of the pepper mill. ~ Bryan Miller

CARDAMOM
Elettaria cardamomum

My spice garden enlarged as my research progressed. I've added cardamom after reading about its qualities. Cardamom isn't difficult to locate as a spice, but locating quality seeds is another quest. Cardamom is also called Guinea grains, or grains of paradise. Guatemala is the largest commercial producer, followed by India and Sri Lanka.

Cardamom is often misspelled as *cardamon.* It is native to India and southeastern Asia. True cardamom has large leaves, white flowers with blue stripes and yellow edges, and grows 8 to 12 feet high. The fruit is a small pod with up to 20 dark brown seeds. It is the seeds that are used as a spice. Most used for cooking purposes is the true green cardamom, botanically *Elettaria cardamomum.*

Cardamom is one of the world's most ancient spices. It is native to southern India where it grows wild. Ancient Egyptians chewed cardamom seeds for clean teeth and fresh breath. Cleopatra was so enticed by the scent that her palace pleasantly smelled of cardamom smoke when Marc Anthony arrived. Ancient Greeks and Romans cooked with cardamom, and used it in medicines and perfumes. Ancient Indians used cardamom medicinally as a stimulant, to relieve indigestion, and as a cure for obesity, flatulence, and headaches. The Arabs attributed aphrodisiac qualities to it. It is a traditional flavoring in coffees and teas, essential in Arabic coffee. Freshly ground seeds are steeped with the coffee, or a few whole pods are put in the coffeepot. Only saffron and vanilla are more expensive spices.

HOW TO GROW: With correct attention, this spice will grow almost anywhere in Hawaii chain. This plant can enjoy some elevation and partial shade. Cardamom can be grown from seeds, but the most familiar fashion is like growing ginger, by spreading the roots. To try your luck, the easiest least expensive way is to locate some brown seeds (or root pieces), not green; or find the pods. The roots are separated into single pieces and planted. Seeds or root pieces should be sown an inch deep in rich soil prepared with well-rotted manure. Since car-damom requires plenty of rain, it is best planted in raised beds and kept moist. The cardamom plant likes a pH that ranges 5.0 to 6.5. Get your irrigation ready as cardamom requires 150 inches of rain yearly and an average temperature of 75 degrees. The cardamom root pieces, or seeds begin to germinate around 6 weeks. They are ready to transplant at a foot tall, after sprouting a couple of leaves. The cardamom plant grows best among trees in shady areas with just moderate sunshine.

Once the plants are started from either seeds or roots, they are best transplanted as they send up 6 to 8-foot leafy shoots. Mature plants send up flower spikes, which produce white or pale green flowers. Those produce green pod capsules, one to two inches long, containing the seeds. The seeds are in 3 double rows with about 6 seeds to the row. The larger cardamom plant is known as 'black' cardamom even though it is actually dark brown. The smaller cardamom is green. These seeds are small black and sticky. The best quality cardamom seeds are ripe, hard, and dark brown.

The premium green cardamom pods are picked by hand while still immature and sun-dried to preserve the bright green color. White cardamom has been bleached of color. It is often used in baking and some desserts that want to remain all white. Black cardamom is not a suitable substitute for the real thing. Cardamom is an attractive potted plant for the balcony or porch.

MEDICINAL: A curative, perhaps aphrodisiac, drink can be made by steeping seeds in hot water, or you can just suck a cardamom seed. This is good for the throat, respiratory tract, teeth, breath, and stomach. Green cardamom is used to treat tooth and gum infections (just like cloves), treat throat problems, lung congestion, and digestive disorders. Cardamom is one of the most effective remedies against halitosis/bad breath. Chewing the seeds eliminates bad odors. Cardamom contains an abundance of antioxidants, which protect the body against aging and stress, and fight common sicknesses.

NUTRITION: 100 grams of cardamom has 300 calories, 70 grams of carbohydrates, 10 grams of protein, and 7 grams of fat, with 12 grams of fiber. It is high in calcium, iron, magnesium, phosphorus, and potassium. It also has vitamin C, B, thiamin, niacin, and folate.

FOOD & USES: Cardamom is an ingredient in curry powders, and used to flavor sweets, liqueurs like Aquavit, and chewing gums. There are counterfeit cardamoms available today, which do not have all the culinary and medicinal properties. Java, Thailand, and Nepal all grow 'bastard' cardamom. Cardamom is highly aromatic and most commonly used in Asian, Indian, Arabic, and some Scandinavian dishes. Its complex flavor: slightly sweet, floral, spicy, slightly citrusy, and savory allows it to enhance seafood, sauces, meats, poultry, vegetables, and even desserts, pastries, and other baked goods. Its aftertaste is a warm, clean sensation similar to eucalyptus.

Cardamom seeds are available in some local markets, but it is better to buy the entire pods, as they will last longer, and storage is far easier. When buying cardamom make sure it looks fresh and smells with a strong aroma. To avoid spreading their aroma to other spices, cardamom should be stored separately in airtight containers in their pods until it is time to use. Put dry paper in the cardamom jar to absorb any moisture. If not, the cardamom will lose a lot of its aroma and will have less of an effect on the food. Pods should keep their aroma for a year. Keep in mind that when the spice is pre-ground its flavor and aroma are quickly lost. 10 pods are the equivalent of 1½ teaspoons of ground cardamom.

Generally, there are two ways people prepare cardamom. You can either use it as a whole pod, or remove the seeds and throw away the pod. Spices such as cardamoms can quickly over flavor the dish. Therefore, you need to be very careful when you add this spice; put in the exact amount in the right form as well. Cardamom is available in green and white pods with about 20 small black sticky seeds. You can either bruise them and toss them into your pot, or peel the skin off and use the seeds whole or ground. If you cannot find whole pods use the ground cardamom.

RECIPES: CARDAMON BLENDED SPICES

Ingredients: 2 TBS fresh ground cardamom, 1 TBS each: turmeric, chili powder, ground cinnamon

Method: Heat the mixture of all spices in a frying pan over medium heat stirring until the mix just begins to smoke. Allow to cool before storing in an airtight container. Use within 2 months and in the following recipe.

CARDAMOM CARROT SOUP

Ingredients: 1 TBS cardamom spice blend, 2 TBS oil, 2 large chicken thighs, 2 quarts of water, one onion chopped, 4 TBS fresh ginger peeled and chopped fine, 6 carrots peeled, 1 sweet bell pepper seeded, 1 potato peeled brown sugar, ½ cup fresh lime juice, ¼ cup parsley chopped, 4 TBS butter or margarine, salt, and pepper to your taste

Method: Simmer two chicken thighs in 2 quarts of water for 45 minutes until a nice broth is formed. In a large skillet heat the oil over medium and add onion, ginger, carrots, sweet pepper, and potato pieces. Stir in spice, blend and cook for twenty minutes. Put vegetables into chicken broth, add sugar. Boil then simmer for half an hour. Add lime juice, butter, parsley, salt and pepper. Cook for another 20 minutes.

Watching something grow is good for morale. It helps us believe in life.
—Myron Kaufmann

CAYENNE PEPPER – RED PEPPER
Capsicum frutescens

Islanders with 'hot mouths' love our spicy foods, and we owe it all to hot peppers. We know a variety of hot peppers by name, yet the world clumps them together under 'chilis,' which was the Aztec term. Cayenne peppers are not typical chili peppers and supposedly named after the Tupi Amerindian term for the Cayenne area of French Guiana. The botanical name is *Capsicum frutescens.* These peppers are another of the foods from South and Central America that have been grown for almost seven thousand years. According to some agronomists, hot peppers are a crossbreed of the potato and tomato families. Since the seeds stay fertile, peppers were easy to extend to the Hawaiian chain.

Columbus carried them to Europe as a replacement for black pepper. Magellan spread the hot pepper to Africa and Asia. Cayenne peppers are now grown on every continent except Antarctica. China, Turkey, Spain, and Mexico are the largest commercial cayenne pepper growers. Peppers add a unique flavor to cuisines with a certain amount of 'heat.' The heat is caused by capsaicin, thus the hotter the pepper the more capsaicin. The hottest are the scorpion, the habañero, Congo, and Scotch bonnet as well as cayenne peppers, and jalapenos. Cayenne pepper is a specific type of pepper, about five inches long, tapered, and slim, ranging from deep green to bright red when ripe. It is a larger variation of our difficult to grow 'bird' pepper. But cayenne pepper has become a term to mean any pepper ground into a fine dusty powder. The powder is red to red-brown, and some of the hotter versions include the seeds.

HOW TO GROW: Cayenne peppers are some of the easiest to grow and bear the longest. I start them from seeds in trays. Four plants are plenty for any household with many to give to friends. They should sprout in 10 days. Plant the seedlings about 2 feet apart and place a stick to later tie it to when it gets heavy with peppers. They will mature in about 70 days. Some trees bear for over 6 months with little attention except an occasional watering. Their only enemy is too much water. Pick when the peppers are about 5 inches long and turn red. The green ones are also very potent. To dry the peppers, be careful not to get them on your bare skin. The easiest way is to pull the whole plant out of the garden when it's filled with peppers and hang in a shaded breezy spot. To reduce to powder, carefully use a mortar and pistil, or a blender. I recommended long sleeves, full pants, rubber gloves, and a simple dust mask to prevent skin or nasal irritation.

MEDICINAL: *Back to Eden* written by Jethro Kloss concerns herbal remedies. Of all the herbs listed cayenne pepper seemed to cause the most spectacular cures. I grew it and offered some to our friend, a roadside vendor, to sell. He said islanders wanted only the usual long green or red 'chili' peppers, as that's what was usual. I offered to just give them away. Within days he wanted more, amazed that so many people had told him of the pepper's health benefits.

Cayenne is a food spice, but it also has been used as a miracle herb for the digestive and circulatory system. Peppers like cayenne have a reputation for causing stomach problems including ulcers. That is unfounded because hot peppers may help prevent problems by killing bacteria you may have eaten and stimulating the stomach to secrete a natural protective coating that prevents ulcers. Cayenne also helps the body secrete hydrochloric acid necessary for digestion. Once you have good digestion all the other organs of the body get the proper nutrients. Consumed as a tea or in a capsule, cayenne pepper lowers the effects of asthma, and clears congestion. Rubbed on the skin, mixed with a skin cream as a paste, will reduce the pain of arthritis or stop an itch. Studies have demonstrated if you consume a lot of 'pepper' your chances of having a heart attack or stroke are lowered. Cayenne reduces cholesterol. Gargling cayenne pepper tea helps relieve a sore throat.

Cayenne cleans the arteries, great for circulation, can rebuild blood cells, lowers cholesterol, and improves overall heart health. It rapidly equalizes blood pressure in your system, shrinks hemorrhoids, and heals the gall bladder. It has been reported a teaspoon of cayenne could bring a patient out of a heart attack. It is also reported to kill cancer cells in the prostate, lungs, and pancreas.

NUTRITION: Two teaspoons of cayenne pepper has about 10 calories. It is a good source of vitamins A and C, has the complete B complexes, and is rich in calcium, manganese, and potassium.

FOOD & USES: Cayenne pepper is typically used in cooking almost anything from seafood, to eggs, meat, and cheeses. It can be used in baking or barbecues. It makes stews, casseroles, and sauces tangy. It is an ingredient of English (Worcestershire) sauce. Cayenne will spice up any stir fry and where would our delicious curries be without pepper? For something entirely different add some to cocoa tea. Cayenne peppers combined with the juice of a lemon will enhance the flavor of simple cooked greens as spinach or bok choy.

RECIPES: CAYENNE OINTMENT - good for aches and sore muscles

Ingredients: 1 cayenne pepper chopped fine, ½ cup vegetable oil, 2 TBS natural beeswax grated. (Get this from a beekeeper or a hairdresser who waxes).

Method: Heat oil on medium heat in a small saucepan. Add minced pepper. Cook without boiling for 5 minutes. Remove from heat and pour through a strainer to remove pepper and seeds. Add beeswax and reheat until the wax melts. Pour into a suitable container or small jar and cool. Try on your next muscle ache.

JUMP UP JUICE - more spike than coffee and without caffeine

Ingredients: 2 TBS apple cider vinegar, 1/8 TS (a pinch) of baking soda, ¼ TS cayenne pepper powder, 1 glass of hot water

Method: First, turn your head away as you add the cayenne powder to the vinegar and water so you don't inhale it as the vinegar (acid) bubbles with the baking soda (alkaline). Sip slowly. It is a real energy boost, especially when I am hungry and tired in the afternoon.

MEXICANA KETTLE CORN

Ingredients: 6 cups popped corn, ¼ cup butter or margarine, 1/3 cup finely grated cheddar cheese, 1 TS cayenne red pepper powder or 2 dried peppers crushed, ½ TS ground cumin

Method: Melt butter in a saucepan. Add all ingredients to the butter and mix well before pouring over popped corn. Mix well, cover with foil and let sit for 5 minutes, and serve.

SPICED LENTILS

Ingredients: 2 cups brown lentils, 5 cups water, 1 TS turmeric, 2 cloves garlic minced, 2 TBS butter, 1 large onion chopped small, 1 small, sweet pepper preferably red color chopped small, 1 medium tomato chopped small, 1 TS ground cayenne pepper or 1 cayenne pepper seeded and minced, 1 TS coriander, 1 TS fresh ginger minced, 1 TS garam masala

Method: After washing lentils, combine them in a large pot with water, garlic, and turmeric. Simmer covered for half an hour. Remove cover, add remaining ingredients, and increase heat stirring constantly to reduce the liquid. Cook for 5 minutes. Serve warm with rice or pasta.

I actually start my day with a cup of warm lemon water with cayenne pepper. It jump-starts your detoxifying system in your body, jump-starts your liver, helps you eliminate the food that you ate the day before, and also just gets your body in an alkaline state ready to ward off disease.— Vani Hari

CHOCOLATE - KO KÃUA
Theobroma cacao

Few consumers of countless candy bars, understand what it takes to make that creamy sweet brown concoction called chocolate. The delicious sweet chocolate candies begin by removing the cocoa bean from the oblong cacao pod. A pod has from 30-50 seeds/beans.

Cocoa is known as hot chocolate or cocoa powder; the dry powder is made by grinding cocoa seeds and removing the cocoa butter from the dark, bitter cocoa solids. Cocoa is really the dried and fully fermented fatty seed of the cacao tree and not the cocaine plant .

In the 1830s, King David Kalakaua introduced cacao to the Hawaiian Islands. He treated friends, family, and dignitaries to the cocoa beans he grew. Hawaii, at 20 degrees north of the equator, is the coldest place in the world and the only place in the USA where is cacao commercially grown. Cacao/chocolate is a relatively new commercial crop on each of the main Hawaiian Islands, Maui, Kuai, Oahu, and Hawaii. Farmers harvest 13 tons of cacao pods annually, and work the beans to small, private, boutique chocolate factories.

The cacao tree is native to the Amazon in the lower Andes Mountains. The ancient Mayas may have brought it to Central America where it was cultivated by the Olmecs and the Toltecs. The Aztecs conquered both. Cocoa beans were used as currency before the Europeans arrived. The Aztecs were so powerful they received an annual tribute of 980 loads of cocoa beans; each load was exactly 8000 beans. As currency, 80-100 beans could buy a new robe.

Columbus encountered cocoa in 1502 when they captured a canoe off of what is now Honduras containing a quantity of mysterious-looking nuts. In 1519, Cortez witnessed Montezuma, the Aztec chief, drink enormous quantities of their variety of cocoa tea. Montezuma drank only chocolate from a gold goblet and ate with a gold spoon. It was flavored with vanilla and spices, and then whipped into a froth that dissolved in the mouth. The Spanish reported the Aztec chief enjoyed 50 servings of chocolate every day.

The Spanish carried chocolate to Europe and by the mid-1600s, it was a popular beverage. The Spanish also began cocoa plantations throughout the West Indies and the Philip-pines. The cacao plant was first given its botanical name by Swedish naturalist Carolus Linnaeus who called it *Theobroma,* 'the food of the gods' or 'cacao', or *Theobroma cacao.*

HOW TO GROW: Cocoa requires a humid tropical climate with regular rainfall and good soil. It grows best with some overhead shade. There are three main types of beans used to make cocoa and chocolate. The rarest and expensive is the Criollo, the cocoa bean used by the Maya. Only 10% of chocolate is made from Criollo, which tastes sweeter and smells better. The most popular bean is the Forastero, which is used to make 80% of chocolate. Forastero trees grow well in dryer African climates and produce cheaper beans. The Trinitario tree created in 1939 is a cross of the Criollo and Forastero, grown in Trinidad and Grenada, it is used in about 10% of chocolate. It grows best on sandy clay soil. Once planted cocoa trees require good drainage and regular weeding. They are topped to around twenty feet tall to make har-vesting easier. Local superstition says, 'Never prune a cocoa tree under a full moon or an evil spirit / ghost will live in that tree.' Cocoa trees bear at four or five years. A mature tree might have thousands of blossoms and only make about 20 pods. Ten pods, 300-600 beans, produce one kilogram of cocoa paste.

Here's where the labor problem arises. As the pods ripen, they are harvested with a curved knife on a long pole. The best pods are ready for harvest when green. Red, purple, or orange pods are considered of a lesser quality because their flavors and aromas are poorer, these are used for industrial chocolate. The seeds are extracted and carried to a place to ferment, known as 'sweating' for six or seven days, and then dried. This is important to increase the quality of the beans, which naturally have a strong, bitter taste. If the fermentation is too short the cocoa may be ruined.

Once fully fermented, the cocoa beans are sun-dried by spreading them out over a large surface and constantly raking them. Sun-drying permits no other tastes, such as smoke or oil, to taint the flavor. The beans are shuffled or 'danced' and clay mixed with water is sprinkled over the beans to obtain a finer color, polish, and protection against molds during shipment.

The beans are ground into a thick creamy paste, known as chocolate liquor, or cocoa paste. This 'liquor' is then further processed into chocolate by mixing in cocoa butter and sugar. Liquor makes unsweetened chocolate. The dried beans can also be separated into cocoa powder and cocoa butter using a hydraulic press, or the Broma process. This process produces around half cocoa butter and half cocoa powder. Standard cocoa powder has a fat content of 10–12%. Cocoa butter is used in chocolate bars, other confectionery, soaps, and cosmet-ics. Plain chocolate is made of cocoa powder, chocolate liquor, cocoa butter, and sugar. Milk chocolate has milk added. White chocolate is made of cocoa butter, milk, and sugar.

Dutch-processed cocoa powder is less acidic, darker, and smoother in flavor because they add an alkaline base. Another process that helps develop the flavor is roasting. Roasting can be done on the whole bean before shelling or on the nib after shelling. Low-temperature roasting produces a more acidic, aromatic flavor, while a high-temperature roasting produces a bitterer flavor. After the ingredients are mixed, the product is further refined to create chocolate suitable for solid bars and pieces. It's mixed, heated, and cooled very precisely in methods called 'conching' and 'tempering.' These processes take a week to make the finest chocolates.

MEDICINAL: Chocolate contains flavonoids, which act as antioxidants. Some protect from aging caused by free radicals, which can lead to heart disease. In moderation, it reduces blood pressure. Dark chocolate contains several antioxidants. Studies show consuming a small bar of dark chocolate every day can reduce blood pressure in individuals with hypertension. Dark chocolate has also been shown to reduce LDL cholesterol (the bad cholesterol) by up to 10 percent. Chocolate stimulates endorphin production, which gives a feeling of pleasure. It also contains caffeine and other substances which are stimulants. Dark chocolate has 65% cocoa content.

NUTRITION: Cocoa powder is actually good for you! At 12 calories a tablespoon, it has nearly twice the antioxidants of red wine, and three times the antioxidants in green tea. Cocoa contains magnesium, iron, chromium, vitamin C, zinc, and other minerals.

FOOD & USES: Cocoa beans, cocoa butter, and cocoa powder are traded on two world exchanges: London and New York. The London market is based on West African cocoa and New York on cocoa predominantly comes from South East Asia. Cocoa is the world's smallest soft commodity market. Cacao production has more than doubled in the last twenty years to 3.5 million tons, an increase due to the expansion of the plantations rather than increased productivity.

RECIPES: TOO EASY CHOCOLATE BAR PIE

Ingredients: 6 milk chocolate candy bars with almonds, 75 miniature marshmallows, ¼ cup milk, 1 pint whipping cream, 1 premade pie crust

Method: Melt candy bars and marshmallows with milk in saucepan over medium heat. A double boiler would be better. Cool for an hour before folding in half of the stiff whipped cream. Pour mixture into pie crust. Chill at least 6 hours. Cover top with remaining whipped cream.

CHOCOLATE CREAM PIE

Ingredients: ¼ cup cornstarch, 1 cup white sugar, ¼ TS salt, 2 cups scalded milk, 3 TBS butter,

½ TS vanilla, 1 ounce of unsweetened chocolate shaved thin, 3 egg yolks, 1 prebaked pie crust, spices like cinnamon and nutmeg, and nut pieces may be added.
Method: Combine cornstarch, sugar, and salt. Add milk and chocolate shavings. Cook over moderate heat, stirring constantly until mixture thickens and boils – about 2 minutes. Remove from heat. Whisk egg yolks into the hot mixture. Return to heat and cook for 1 minute, stirring constantly. Whisk in butter, vanilla. Let cool. Pour into baked pie crust. Bake at 350 for 20 minutes.

THE BEST COCOA TEA MIX-enough for a big family

Ingredients: 2 pounds instant chocolate drink, 1 lb. non-dairy creamer (flavored if possible as Hazelnut or French Vanilla), 1 cup sugar (powdered if possible because it dissolves better), 8 cups nonfat dry milk, 1 TS salt
Method: Combine ingredients well and seal in an airtight container. Refrigerate if possible. Add hot water to ¼ cup of this mixture for 1 cup of the best tasting cocoa tea.

CHOCOLATE LOVERS FUDGE

Ingredients: 3 cups semi-sweet chocolate chips, 1 can sweetened condensed milk, ½ TS salt, 2 TS vanilla extract, 1 cup chopped almonds or cashews (optional)
Method: In a medium pot melt the chocolate chips with sweetened condensed milk and salt in over low heat stirring constantly. Remove from heat; stir in nuts and vanilla. Get a square eight-inch pan and line with waxed paper and spread the mixture evenly into it. Refrigerate for at least 3 hours. Cut into squares and enjoy.

BROWNIES

Ingredients: 2 ounces unsweetened chocolate shaved thin, 6 TBS butter, 1 TS vanilla, 1 cup granulated white sugar, 2 large eggs, 1 cup all-purpose flour, ½ TS baking powder, ½ TS salt, ½ cup chopped almonds or cashews
Method: In a suitable pan combine chocolate and butter over low heat stirring constantly. Remove from heat and allow to cool. Whisk in vanilla and sugar. Add eggs one at a time whisking continually. In a bowl combine flour, baking powder and salt. Combine the flour mixture into the chocolate. Add nuts. Pour into a well-greased baking dish and bake at 350 for 25 minutes.

CHOCOLATE BISCUITS

Ingredients: 2 cups flour, ½ cup unsweetened baking cocoa, ¼ TS baking soda, ¼ cup brown sugar, 2 TS baking powder, ½ cup soft cooking margarine, 1 TS vanilla, ½ cup milk
Method: in a suitable bowl mix all dry ingredients, blend in margarine with a fork until flour mixture gets a crumb texture. Add vanilla and milk. The mixture should become a soft dough. Roll dough and cut into circles or squares. Bake at 450 for 15 minutes.

Cacao pods on tree may be yellow, orange, or maroon.

Strength is the capacity to break a chocolate bar into four pieces with your bare hands - and then eat just one of the pieces. ~ Judith Viorst

CORIANDER — CILANTRO
Coriandrum sativum

So, you've never heard of coriander? The bulk of most curry powder is ground coriander. To harvest coriander, you must plant cilantro. Cilantro is the leaves and stems of the coriander plant. When the plant flowers and turns to seed, the seeds are called coriander seeds. Cilantro is also the Spanish word for coriander. At maturity, cilantro seeds full and dried become coriander. Immature cilantro seeds do not have a pleasant odor, so the Greeks called the seeds bugs, 'koros' in Greek. Cilantro is part of the parsley family, often called Chinese parsley. Coriander's botanical name is *Coriandrum sativum*.

Cilantro's seeds are tiny, around an eighth of an inch, little balls slightly out of round. They stink if you work with them too soon. Wait till they are dried light brown. Coriander is available both whole and ground.

Coriander is a very old spice referred to in the Bible's Exodus. It is native to the Middle East, but has been grown in all parts of Asia for thousands of years. It grows wild in Egypt and in England where it was transplanted by the Romans.

HOW TO GROW: Cilantro will grow almost everywhere in Hawaii, and fresh is definitely best! I've tried this in my spice garden, and it grows well in slightly moist soil planted where it gets the morning light until about eleven or noon. Partial shade from a banana will also work. If cilantro gets a long, daily dose of tropical sun it will be bitter. In proper conditions, this plant grows like a weed. Make two plantings so you can have one for fresh cilantro leaves, the other will mature in about three months into coriander seed. You'll know when the seeds smell good they're ready to harvest. To grow plant the seeds about half an inch deep and about six inches apart and water regularly. This can also be grown indoors, but needs to get four to six hours of sunlight. Coriander seeds have a nice smell when they ripen, like fresh orange peel. East Indian, Mexican, and Middle Eastern cuisines depend on this herb. Its main produc-ers are North Africa, South America, and Russia.

MEDICINAL: Coriander seed, and especially the oil, is an appetite stimulator and will soothe a gaseous tummy, headaches, and arthritis. Cilantro can be used as a poultice against wound infections.

NUTRITION: 4 grams of coriander has 1 calorie with some vitamins A and C.

FOOD & USES: Whole coriander seed is easy to pound in a mortar. This will provide better flavor and aroma to your dishes. Whole seeds keep indefinitely, and light roasting will enhance their taste. Cilantro leaves can be chopped or minced before use and definitely lose taste if dried. This is one spice that has a wide variety of use. Coriander is an ingredient in ga-ram masala, pickling spice, and is used in cakes, breads, and other baked goods. Cilantro leaves should be used fresh in sauces, soups, and curries and sprinkled like parsley on cooked dishes. Even the cilantro root is used to spice meat and curries.

RECIPES: CORIANDER CARROT SOUP - good for a chilly, rainy day

Ingredients: 1 pound of carrots chopped small, 1 large onion chopped small, 1 bunch fresh cilantro leaves stripped from the stems and chopped - stems dumped, 1 TS ground coriander, 2 veggie stock bouillon cubes, 2 TBS oil, salt, and black pepper to taste, 1 quart of water, 1 cup cooked rice or small noodles

Method: In an appropriately sized pot heat the oil over medium and add the onion and carrots. Cook for 5 minutes until the carrots soften. Add ground coriander and cook for a few more minutes. Add salt and pepper and water. Reduce heat to low and stir in the veggie cubes. Simmer for a half an hour stirring occasionally. Add rice or noodles. Adjust salt and pepper to your taste. Stir in fresh cilantro leaves and serve with toasted bread.

CORIANDER BEEF OR PORK

Ingredients: 2 pounds beef steak or pork sliced into ½ wide strips, 1 cup water, ¼ cup cider vinegar, 2 TBS fresh lemon juice, ¼ cup canola oil, 2 TS ground coriander, ½ hot pepper seeded and minced (optional), 1 TS salt, 2 TBS cream (optional)

Method: Combine water, cider vinegar, half of the oil, pepper, salt, and coriander. Put the meat slices in a suitable dish, cover with mixture, and let sit overnight. Heat the rest of the oil in a large skillet with a cover and brown meat. Pour in the marinade and bring to a boil. Cover and simmer for about half an hour. If you choose, add the heavy cream to the juices remaining in the skillet and pour over meat before serving.

SIMPLE CURRY POWDER

Ingredients: 2 TBS ground coriander, 2 TBS ground ginger, 2 TBS ground cardamom, 4 TBS cayenne powder, 4 TBS ground turmeric.

Method: Mix well and store in a tightly sealed jar or refrigerate.

DID YOU KNOW?

Coriander is one of the oldest cultivated herbs. It is mentioned in the Bible and found in tombs of Egyptian pharaohs. It is also the stuff of dreams; *'while visions of sugarplums danced in their head.'* Sugarplums originally were sugar-coated coriander, a treat that initially tasted sweet with a spicy flavor aftertaste. That recipe later included small bits of fruit and became the sugarplum sweet we sing of today. Cilantro is men-tioned in the medical Papyrus of Thebes written in 1552 BCE. It was one of the plants of the Hanging Gardens of Babylon. Coriander is mentioned as an aphrodisiac in The Tales of the Arabian Nights.

CILANTRO GROWING

DRIED CORIANDER SEEDS

Coriander is good for diabetes. It can stimulate insulin secretion and lower blood sugar leverls. ~ Hartleys

CINNAMON - KINAMONA
Cinnamomum Zeylanicum Blume

Cinnamon is the inner bark of a tropical evergreen tree and one of the oldest spices. There are at least fifty different varieties and the best type is still grown in Sri Lanka (previously Ceylon) along the coast near Columbo. The Queen of Sheba gave cinnamon as a gift to King Solomon. The Chinese have used it as an herbal medicine for over four thousand years. In ancient Egypt, it was used as a drink, medicine, a preservative for embalming, and could be considered more valuable to the Pharaohs than gold. The Portuguese conquered Sri Lanka in 1536 to control the world's cinnamon supply. A century later the Dutch took control of Sri Lanka and started the planned cultivation of the spice that still exists. Instead of permitting the trees to just grow tall, the Dutch transformed the tree by continually cropping and topping it into almost a bush. The Dutch fought to control the valuable cinnamon monopoly for two hundred years until the English traveled cinnamon farming to South America and the Caribbean. Sri Lanka produces 90% of the world's genuine cinnamon *(Cinnamomum Zeylanicum Blume)*. That's where to get your tree.

The first recorded cinnamon tree planted in the Hawaiian Islands during the1870s on Oahu. The local name for cinnamon is 'kinamona.'

HOW TO GROW: In Hawaii, cinnamon trees grow best at elevations that receive more rain. You can make new cinnamon trees from stem cuttings. Take a cutting and strip off all but a few leaves. Plant the cutting in some well-draining soil from a nursery. Do not use soil from the garden because it may contain harmful bacteria and contaminate your tree and the soil's pH must be between 4.5 and 5.5. Cinnamon loves acidic soil, so this pH range is a must. Keep it in a warm, partially sunny window. Cuttings are slow growing and may not be ready to plant outdoors for several months. Decide if you want to plant the tree indoors or outdoors. Cinnamon trees will do well in either location if they receive full sun.

Other growing options are to purchase a cinnamon tree from a nursery or harvest fresh seeds yourself. If you choose to harvest the seeds from strong, healthy cinnamon trees with smooth, easy-peeling bark and a high oil content. Wait until the berries turn black first, then split them open. Set them out to dry in the shade for 2 to 3 days, then separate and rinse the seeds. Allow them to dry in the shade once more. You may be able to purchase fresh cinnamon seeds online, but you need to plant them as soon as possible.

A trimmed cinnamon tree will grow from ten to twenty-five feet with a thick, very scabby rough bark. They will grow tall but better to keep short. It grows better in sandy, well-drained soil. Water weekly or during severe droughts. The leaves are shiny and blooms small pale-yellow flowers that become dark purple berries. People we know who make cinnamon, skin fresh branch shoots of the tree, and the inner bark are left to dry and curl.

MEDICINAL: Cinnamon has many health benefits. It has shown promise in the treatment of diabetes, arthritis, high cholesterol, memory function, and even leukemia and lymphoma. It helps in removing blood impurities, is effective on external as well as internal infections, and is recommended for acne. It helps in destroying germs in the gall bladder and bacteria in staph infections. Cinnamon is considered a mild tranquilizer and relieves nausea, gas, and diarrhea. It is an antibiotic that fights some fungi and bacteria better than over the counter medication. One-half teaspoon of cinnamon each day may reduce blood sugar and cholesterol in Type II diabetes sufferers. Consuming cinnamon daily may increase insulin resistance, can help to control weight gain, and lower cholesterol, as well as to decrease the risk of heart disease. A tea of cinnamon and ginger is great for fighting a cold or flu and indigestion. Cinnamon is also anti-inflammatory, relieving pain and stiffness of muscle and joints and is recommended for arthritis. Cinnamon is a good brain tonic, boosts brain activity, helps remove nervous tension and memory loss.

NUTRITION: One tablespoon of cinnamon has 17 calories with only 1 calorie from fat and 5 from carbohydrates. It is a source of manganese, calcium, and iron.

FOOD & USES: Cinnamon should smell sweet when you buy it and will stay fresh in a sealed container. On cold rainy days make a tasty drink of warmed milk with cinnamon and honey or add it to coffee with cocoa powder. It pleasantly changes the 'usual taste' of bean and eggplant dishes. It is used in chewing gums because it is a good mouth freshener.

There is an ancient fable of the Cinnamon Bird that supposedly lived in Arabia and used cinnamon to build its nests. Greeks wrote that these birds flew to an unknown land and collected cinnamon and carried it to Arabia. The Arabians got the cinnamon by tempting the birds with pieces of raw meat. The birds carried the large pieces of meat back to their nests. The extra weight crashed the nests, and the people collected the cinnamon spice sticks.

RECIPES: EASY PAPAYA BREAD

Ingredients: 1 package yellow cake mix, 1 cup mashed papaya, 3 eggs beaten, 1/3 cup vegetable oil, ½ cup of sour cream, ¼ cup water, ½ cup applesauce, ½ TS ground cinnamon, ½ TS ground ginger, 1 cup raisins (preferably soaked in hot water), ¼ cup grated coconut (optional)

Method: Combine cake mix, eggs, oil, sour cream, applesauce, cinnamon, ginger, and water in a large bowl. Beat as smooth as possible before adding the papaya, raisins, and coconut. Pour into a greased, floured bread pan and bake for half an hour at 350 degrees. Insert a knife to see if it is baked through the center. If the knife doesn't stick and comes out clean, it is baked. If not continue baking for ten more minutes.

SWEET POTATO CASSEROLE

Ingredients: 3 sweet potatoes, peeled, boiled, and mashed, 2/3 cup all-purpose flour, 1 cup sugar, ½ cup butter, 1 TS ground cinnamon, 2 TBS orange juice, a pinch of grated ginger, ¼ TS ground nutmeg, ¼ cup milk, 1 TS vanilla extract, 2 eggs beaten.

Method: In a small mixing bowl combine half of the flour with the brown sugar. Blend in half of the butter until the mixture is flaky. In a second - larger bowl combine the remaining ingredients. Pour sweet potato mixture into a greased baking dish and sprinkle the top evenly with the flour/sugar/ butter mixture. Bake about 40 minutes at 350 degrees.

CINNAMON CHICKEN

Ingredients: 4 chicken breasts boned and chunked small, 1 sweet pepper cut into strips, 1 small pineapple chunked, ½ TS fresh ginger root minced, ½ TS hot pepper seeded and minced (optional), 2 leaves of culantro chopped fine, ½ TS ground cinnamon, 2TBS butter, ½ cup water

Method: In a frying pan cook chicken pieces in 1 tablespoon of the butter until browned. Remove from frying pan. In the same pan with the other tablespoon of butter quickly cook the sweet pepper. Add pineapple and water and remaining ingredients except culantro. Simmer for 15 minutes. Add chicken, heat, sprinkle with culantro, and serve with rice or pasta.

DID YOU KNOW? True cinnamon or Ceylon cinnamon is native to Sri Lanka, botanically *Cinnamomum verum* or *Cinnamomum Zeylanicum Blume.* Cassia is a related spice sometimes sold as cinnamon. It is not 'true cinnamon.' Most cinnamon powder sold in the United States and Europe is actually cassia. Cinnamon is mentioned in Chinese writings as far back as 2800 BC. Ancient Egypt used cinnamon in the embalming process. Cinnamon is one of the flavors used in 'cola' soft drinks.

Vegetables, herbs, and spices. If you can combine those ingredients, that would be the best dish you'd ever cook. ~ Rinrin Marinka

CLOVES
Syzygium aromaticum

Cloves are the flower buds of the clove tree, an evergreen medium sized tree with pale gray bark, and known botanically: *Syzygium aromaticum*. Cloves can be used whole or ground to a powder. This spice flavors beverages, liquors, culinary dishes, and pastries. They are called cloves because they're shaped like a nail and the Latin word for nail is clavus. Properly trimmed and maintained, a clove tree is a great addition to the home garden landscape. If the cloves buds aren't picked, they produce a beautiful scarlet red tubular flower with a great fragrance.

Cloves have always been an expensive spice from the ancient times. It is believed to be native to the Molucca Islands in Indonesian. From the 8th century until the Middle Ages, Arabs traded spices including cloves throughout Europe. At the close of the 1400s, the Portuguese gained control of the spice producing Maluku islands and brought large quantities of cloves to Europe. Then a kilo of cloves was valued at seven grams of gold. The Spanish and then the Dutch dominated the trade until the 1600s. The French introduced cloves to Mauritius in1770. Clove cultivation expanded to Guiana, Zanzibar, West Indies, and most of Brazil.

It's easy to grow a clove tree anywhere in the Hawaiian chain. All the islands already have trees and your effort will be to search them out through the NET social pages or agri-shops to get a sprout. Otherwise, buy seeds or trees online and follow the directions below. The rewards of beauty, fragrance, nutrition, and conversation topic are worth the efforts and expense.

HOW TO GROW: A clove tree is a long project. After planting, the clove tree will begin blossoming within 6-10 years and will reach full maturity (and produce the best harvest) at the age of 15 to 20. A fully grown clove tree will grow to 20 meters tall, but they are best when the center stem is cropped to keep them at 10 meters or less for easier picking. Cloves grow best in humid, tropical conditions to an elevation of 1,000m with a lot of rain; 70 inches a year is necessary. Cloves cannot endure a long dry spell. The secret to cloves is the alternating period of dry and wet weather increases the clove's oil content.

Clove is propagated through seeds, so find a tree and check for sprouts. Tree ripe fruits should be planted within 48 hours of collection. Seeds are obtained by removing the outer pulp by soaking them in water. Place clove seeds directly on the surface of the ground. The seeds do not need to be buried under the soil to put down roots. Since these trees require moist soil, you can improve soil conditions with rotted manure and compost and by covering the freshly planted seeds with a plastic sheet. However, growing a clove tree in a pot on the porch or balcony is possible with proper care in long cold or dry spells.

MEDICINAL: Cloves have many health benefits, including keeping blood sugar regulated and fighting bacteria. Clove oil is an important natural antibacterial drug. It's used in dentistry, pharmaceuticals, and aromatherapy. It's also used as an analgesic; clove oil is recommended for inhalation to treat sore throat, colds, coughs, and any breathing problems. One test-tube study found that clove extract helped stop the growth of tumors and promoted cell death in cancer cells. Clove oil boosts concentration. In addition, it revitalizes, energizes, and serves as an aphrodisiac. It's also a natural food preservative due to its antibacterial and antifungal effects. Studies show that the beneficial compounds in cloves could help promote liver health. The compound eugenol may be especially beneficial for the liver. Another study researched the effects of clove extract 'nigericin' and was found to increase the uptake of sugar from the blood into cells, increase the secretion of insulin, and improves insulin production both on human muscle cells and in mice with diabetes.

NUTRITION: 2 grams or 1 teaspoon of ground cloves contains 6 calories with 1 gram of fiber, lots of magnesium (55% of the daily value) and some vitamin K.

FOOD & USES: Cloves are largely used as dried whole buds. Ground clove powder is an ingredient in many curries. Clove oil is used to flavor food, in pickling, and the production of sauces and ketchup. This tiny spice is used in the pharmaceutical industry as a flavoring of mouthwashes, toothpastes, and disinfectants. Clove cigarettes, produced in Indonesia and distributed worldwide, are typically made up of approximately 60 to 80% tobacco and 20 to 40% ground clove buds and clove oil.

Clove oil is extracted by steam distillation from the tree's leaves, stem, and buds. Oil from the buds is clear while oil from the stems is a slightly yellowish liquid that: darkens with age. Clove leaf oil is a dark brown liquid obtained by the distillation of the dry leaves and used to produce eugenol.

RECIPES: MASALA CHAI TEA ~ a coffee alternative and great on chilly days

Ingredients: 4 whole cloves, 5 green cardamom pods, 2 or more thin slices of ginger, ½ cinnamon stick, 1-2 TBS loose black tea or 1-2 teabags, 1 cup of water, 1 cup milk, 2 TS sweetener, sugar, jaggery, honey

Method: In a mortar, lightly crush all the spices. Bruise them to bring out the flavor. Place in the pot of water on medium heat. Once it boils turn off and let steep for at least 10 minutes, the longer the stronger the flavor. Add milk, heat, and simmer to warm, add sweetener and strain. Enjoy.

MILK RICE

Ingredients: 1 cup long grain rice or any type, 2 cinnamon sticks, 3 whole cloves, 1 TBS lemon zest, 4 cups water, 1 egg, 3 cups milk, 1 can sweetened condensed milk, 1 TBS vanilla extract, ½ cup each: chopped raisins and almonds or grated coconut (optional)

Method: In a sizeable pot, soak the rice, spices, and lemon zest for at least an hour. Put on a high flame and bring to a boil, uncovered. Once it boils reduce heat and simmer for 15 minutes or until most of the water has evaporated. While rice is simmering, in a bowl beat the egg, add milk and combine. Then add this egg/milk mix with the vanilla and condensed milk to the rice and simmer, stirring, for half an hour until it thickens to the consistency you want. Remember that the rice will continue to thicken as it cools. The final cool sweet milk rice will be thinner than rice pudding.

SPICE ROASTED PUMPKIN

Ingredients: One 2-pound, unpeeled pumpkin or butternut squash—washed, halved, seeded and cut into 2-inch pieces, 8 garlic cloves, unpeeled, 1 ½ inch piece of fresh ginger, peeled and sliced thin, 1 hot red chili chopped, 1 TS cinnamon powder, ½ TS each ground cloves and nutmeg, ¼ cup light brown sugar, 1 ½ cups coconut milk, salt to taste

Method: In a suitable pan or wok, combine all ingredients and cook covered on medium about 45 minutes until squash/pumpkin is tender. Cook uncovered for 15 more minutes until half of liquid has evaporated.

DID YOU KNOW?

Cloves are dried flower buds picked before they open. The Island of Pemba off the coast of Tanzania produces the most cloves. The first reference to cloves in from China in the Han period about 100BCE and they are termed, chicken-tongue-spice. In Syria, archeologists found cloves in a ceramic vessel, with evidence that dates the find to within a few years of 1721 BCE.

CREPE GINGER
Cheilocostus speciosus

Crepe ginger is an edible and medicinal plant native to India and Sri Lanka with spectacular flower spikes. An erect plant can grow up to 3 meters tall, and it's related to ginger. Crepe ginger is mostly used as an ornamental throughout the Hawaiian Islands. It grows naturally on the edges of forests and in wetland areas and has been listed as an invasive species in Cuba. It's cultivated for its Ayurvedic-homeopathic value, but also is a beautiful ornamental for your home and great for your patio or balcony. The flower petals are quite sweet and nutritious, and this plant makes a great ground cover. IBotanically, it's *Cheilocostus speciosus.* It's known as crepe ginger because the pale pink flowers on top have a crinkled texture like crepe paper.

HOW TO GROW: Naturalized in Hawaii, crepe ginger can be found growing at the edge and inside shaded forests where the soil is moist at lower elevations. Crepe ginger grows similar to ginger or turmeric. It can be started from pieces of the root/rhizome, stem cuttings, or by seeds. The easiest way is to find one growing and dig a section of the root with at least two viable eye buds, usually seen at the tips or around the stem scars. Gently clean the root pieces; plant them separately 30cm (1 ft.) or more apart, to give it enough space to grow in a well-drained area. Crepe ginger thrives in fertile, moist soil with a pH of 5.7-7.5, in partial shade. This plant doesn't require special means of nurturing or harvesting. They have a high growth, and in some areas are considered weeds. Flowering begins after the rainy season. These have a tall center stems with one flower per plant that g rows f rom a very dense bright red center cone. The big flower resembles a white lily. The resulting fruit are shaped like flat teardrops and contains many black seeds.

MEDICINAL: Crepe ginger's most common therapeutic use for the prevention and treatment of diabetes and to maintain general good health. In contrast with most medicinal plants, crepe ginger tastes good and the medicinal benefits can be enjoyed raw, as a salad. Preparation is simple for medicinal purposes. All parts of the plants are used; leaves can stimulate appetite, treat skin issues, and as a bath for patients with a high fever. Juice pressed from leaves is used internally for eye and ear infections. The roots are edible and treat intestinal parasites, constipation, and have anti-oxidant properties. Juice from the root is a headache remedy. As a blood tonic, studies have shown that a diet with crepe ginger regularly reduces LDL-cholesterol and increases plasma insulin, tissue glycogen, HDL-cholesterol, and serum protein. Powdered crepe ginger leaves are available in some areas.

NUTRITION: This plant is a good source of nutrients and natural antioxidants. 100 grams of crepe ginger leaves has 19% protein and 12% fiber with plenty of vitamins C, E, and A. This plant contains lots of iron with calcium, magnesium, and potassium.

RECIPES: CREPE GINGER LEAVES SAMBAL - a South East Asian specialty

Ingredients: 20 crepe ginger leaves washed, 1 medium red onion and 2 hot peppers diced, ¼ cup fresh grated coconut, juice of two limes, ½ TS ground black pepper, salt to taste

Method: It is best to keep the crepe ginger leaves refrigerated until ready to serve. Then shred and in a proper bowl combine all the ingredients. Crepe ginger leaves tend to sour and gain a bad smell if cut and stored too long. Just toss a few times, don't stir much or it will be bitter.

DID YOU KNOW? Crepe ginger is mentioned in the *Kama Sutra* as an ingredient in a cosmetic to be used on the eyelashes to increase sexual attractiveness. It treats kidney problems and other urinary problems in Mizo Traditional Medicine. It was used as a traditional medicine by Malays when evil spirits have possessed a body.

NAMES: English-cane reed, wild ginger.

CULANTRO – SAWTOOTH CORRIANDER
Eryngium foetidum

One of my favorite seasonings is culantro and its botanical name is *Eryngium foetidum L., Apiaceae*. It is a biennial herb, which means it is a plant that lives only two years. It sprouts from seed the first season, but usually does not flower or fruit until the second year after which it dies. Culantro is native to the Central American tropics and the Caribbean Islands. Although widely used in cooking throughout the Caribbean, Central America, and the Far East, it's not well known in the United States and Europe. It is often misnamed for its close relative cilantro or coriander.

However you say or spell this great tasting leaf, it is one of the most used plants throughout the world. Culantro is part of the carrot family and is more commonly known in Hawaii as Mexican coriander, sawtooth coriander, or Chinese parsley. The entire plant, including the leaves, seeds, and roots are edible. Older plants usually have a stronger flavor. The smell supposedly will keep snakes away from your garden. This hardy plant will grow anywhere in the Hawaiian chain.

Fresh culantro will keep extremely well if wrapped in paper towels and then put in a plastic bag before storing in the fridge. Another method of keeping this herb fresh is to place it in a cup of water, uncovered in the refrigerator.

HOW TO GROW: This herb grows almost anywhere, especially in full sun with daily water. The seeds are tiny. To quicken the germination, wash the seeds in dish soap and only slightly dry before planting an inch deep and an inch apart. The plant should be watered and well-drained. Replanting every three weeks should keep your kitchen supplied with the fresh spice. A small four-foot by two-foot bed produces more than one kitchen can use. A well-cared for plant should have leaves of 8-10 inches. If you want to save seeds for future planting, wait until the leaves and flowers turn brown. Make rows 15 inches apart.

The easiest way to grow culantro is to cut the root from stalks bought from the market. Plant the roots in wet soil and keep damp. Trim the leaves with scissors and the plant will continue producing. Although culantro grows well in full sun most commercial plantings are partially shaded. This produces large plants with greener leaves, more marketable because of their better appearance and higher pungent aroma.

MEDICINAL: Culantro is used in traditional medicines for fevers and chills, vomiting, and diarrhea. Jamaicans use it to alleviate colds and convulsions in children. Bush medicine recommends the leaves and roots boiled and the water drunk for pneumonia, flu, diabetes, constipation, and malaria fever. The root can be eaten raw for scorpion stings. East Indians reportedly use this root to stop stomach pains. Research has demonstrated that leaves and stems from culantro help lower blood sugar levels in animals. To use culantro as a pain reliever; cut the leaves and boil them, apply the hot culantro water on your body parts, or drink it, or make a poultice with the warm leaves on the painful area.

NUTRITION: Other than taste, culantro has slight food value. A quarter cup of leaves has only 4 calories, with virtually no fat, fiber, cholesterol, or carbohydrates. The plant is reportedly rich in calcium, iron, carotene, and riboflavin. Leaves are a source of vitamins A, B 1, B 2, and C.

FOOD & USES: Culantro leaves are used in seasoning pickles, barbecue sauce, curries, and chutneys. In Central America, it's used in salsas and salads to burritos or meat dishes.

RECIPES: TRADITIONAL TRINIDAD GREEN SAUCE- can enrich most dishes
Ingredients: 12 culantro leaves, 1 head of fresh garlic, 1 stalk of full-grown celery, 2 leaves of Spanish thyme, ¼ cup vinegar
Method: Mince together with a blender and store in a bottle. Use to marinate.

PACIFIC SEAFOOD STEW – for extra-special occasions
Ingredients: ½ cup culantro, ¼ pound of shrimp/prawns, 2 pounds of fish, 1 cup white wine, ¼ pound butter, ½ cup chopped chives, 1 cup oyster juice (water poured off fresh oysters), 1 quart cream, 1 quart water, salt, and pepper to taste
Method: Boil wine in a large frying pan until only about 2 tablespoons remain. Add butter and lightly cook the green onions. Add the oyster water, water, and milk. Simmer as you blend in the culantro. Add salt and pepper levels to your taste. Add the oysters to the soup and cook for 5 minutes. Serve in bowls with rice. Serves 6. (Add a variety of garden vegetables to your taste)

SAWTOOTH CORRIANDER BARBECUE SAUCE
Ingredients: 3 cups culantro, 2 heads garlic, 1 cup parsley, ½ cup lemon juice, 2 TBS olive oil, salt, and pepper to taste
Method: Wrap garlic in foil and bake in a 400-degree oven for 45 minutes. Cool and squeeze the garlic into a food processor or blender. Add the remaining ingredients and blend. It is best to marinate the chicken, lamb, fish, or pork in this sauce overnight. While grilling, keep applying this sauce. You can also use this sauce over pasta, rice, or grilled vegetables.

DID YOU KNOW? Fresh culantro leaves give great flavor to mixed green salads, soups, and stews. Its distinct flavor enhances vegetable and meat dishes, chutneys, preserves and sauces. This tasty herb is an ideal substitute for cilantro. Unlike cilantro, culantro retains its good flavor when dried.
NAMES: Hindi: *bhand-hanya, broad dhanya, or coriander,* Thai: *pak chi farang (foreign coriander),* Malaysia: *ketumbar java,* German: *langer koriander,* Vietnam*: ngo gai,* Span-ish*: culantro, racao, recao,* English: *long leaf or spiny coriander*, Trinidad: *shado beni*, Dominica: *chadron beni,* Guyana: *fitweed,* Haiti: *coulante,* Puerto Rico: *recao*

Spice a dish with love and it pleases every palate.

CUMIN
Cuminum cyminum

Cumin is a seasoning readily available in the markets, yet still should be planted in your spice garden. It has sharp, almost overpowering, slightly bitter taste. Dishes cooked with it have a warm, spicy-sweet smell. East Indian and Mexican cuisines require cumin. To the unknowing, cumin's flavor is often confused for caraway. Some think there are dark seed and light seed cumin, but there's only light. The dark is from a totally different plant, the 'love in the mist flower.' Cumin is native to the areas surrounding the Mediterranean and Egypt.

HOW TO GROW: This is a spice that grows like a weed in the tropics. Cumin is a cousin of parsley and can grow a foot tall, and bend under the weight of the seeds. As the flower matures two seeds form in the head. The quarter inch or smaller seeds are shaped like a boat and slightly fuzzy. Cumin doesn't require much in your spice garden except for regular watering and full sunlight. In well-worked soil, plant the seeds about a half-inch deep and an inch apart. Don't permit the soil to dry out. The seeds usually mature in four months. Cumin plants will not all ripen at once, so wait until the first seeds are dry enough to crack when you pinch the pod between your fingers. Then cut the plant and hang over a clean cloth. After the pods thoroughly dry, put them in an old pillowcase. This will be used later to thresh them. Keep adding as they ripen and dry. Bang the case or bag against the floor to free the seeds. Sift to remove parts of the pods. Keep in a sealed container in a dark place or refrigerate until ready to grind.

MEDICINAL: Cumin could be called the 'breast spice' because it supposedly increases both lactation and size. Cumin is a stimulant and a great herb for digestive disorders as flatulence, indigestion, diarrhea, nausea, and morning sickness. A good recipe for one dose is one teaspoon of seeds boiled in one glass of water with a pinch of salt and a teaspoon of coriander or culantro leaf juice. The seeds are rich in iron and stimulate the secretion of enzymes from the pancreas which help absorb nutrients into the system. It boosts the liver's ability to detoxify the body.

NUTRITION: Six grams of cumin has about twenty calories of which half are from fat. It also has some iron and calcium.

FOOD & USES: This spice should be used minimally because its flavor can overpower other flavors in most dishes. A teaspoon is enough in a dish for four. Cumin is used to highly spice food. East Indian, Middle Eastern, Mexican, Portuguese, and Spanish cooks love it. Most curry powders and many savory spice mixtures have it as an ingredient. A pinch of cumin will invigorate plain rice, beans, and casseroles. It is a pickling ingredient for cabbage to make sauerkraut and is used in chutneys.

RECIPES: LATIN CHICKEN (OR TURKEY) SOUP

Ingredients: 1 TBS oil, 1 large onion chopped small, 2 stalks celery chopped small, 3 garlic cloves minced, 2 TS chili powder, 1 TS cumin for full flavor – ½ is adequate, 1 TS oregano, 1 quart water, 4 large tomatoes diced, 4 cups shredded cooked chicken or turkey, 2 TBS fresh parsley, 3 chicken bouillon cubes, 2 cups or one large can of black or red beans, 2 cups fresh corn cut from the cob or frozen corn, ½ cup sour cream optional, 2 TBS culantro chopped Method: In a large skillet heat oil, add onion and cook till soft and clear. Add garlic, chili powder, cumin, and oregano and stir for about a minute. Add water, diced tomatoes, celery, shredded chicken or turkey, parsley, and bouillon cubes. Bring to a boil, then simmer. Stir until the bouillon cubes dissolve. Add beans, corn, sour cream, and culantro. Simmer for half an hour

HANAUMA BAY PEPPER - SQUASH STEW

Ingredients: 1 cup lentils, 1 large onion chopped small, 4 stalks celery chopped small, 2 squash (Could be yellow, crookneck, zucchini) chopped, 3 large sweet pepper chopped small, 1 cup parsley chopped, 2 large tomatoes chopped, 1 can tomato paste or a ¼ cup of ketchup, 1 hot pepper seeded and minced (optional), 3 bay leaves, 4 garlic cloves minced, 3 TBS oil, 1 Spanish thyme leaf chopped, 1 TS each of ground cinnamon, ground cumin, and ground coriander, 1 can chickpeas, ½ cup plain yogurt (optional)

Method: Put water in a large 2-quart pot and bring to a boil. Add lentils, bay leaves, and half of the garlic. Simmer lentils about 10 minutes. Drain and place lentils in a bowl. Retain this water, but remove bay leaves. Toss with oil, thyme, and the remainder of the minced garlic. In a large skillet heat 1 tablespoon of oil and stir in spices. Add garlic, onion, celery, sweet peppers, and squash; simmer for 5 minutes. Add lentils, tomatoes, and paste, chickpeas, 2-4 cups of remaining lentil/ vegetable stock water. First add 2 cups and continue to add until you get the consistency of stew you desire from thick to runny. Bring stew to a boil; then reduce to simmer and cover for about half an hour. Stir occasionally. Add the chopped parsley. Serve in bowls topped with a spoon of plain yogurt, sour cream, or grated cheese.

PÃLOLO VALLEY BLACK BEAN BURGERS

Ingredients: 2 cups cooked black beans, 1 sweet pepper chopped very small, 1 small onion chopped very small, 3 garlic cloves minced, 2 stalks celery chopped very small, 1 egg, 1 TBS chili powder, 1 TBS cumin, 1TS hot pepper sauce, 1 cup breadcrumbs

Method: Mash black beans in a bowl with a fork or spoon. Add sweet pepper, onion, garlic, and celery. In a cup mix the egg with the spices and add to the bean paste. Add breadcrumbs until the bean mixture holds together. Make 4 patties. Put a tablespoon of oil in a skillet and fry the veggie burgers about 10 minutes on each side. If grilling, place patties on foil, and grill about 8 minutes on each side. If baking, place patties on a baking sheet, and bake about 10 minutes on each side.

DID YOU KNOW?

Cumin is the second most popular spice in the world. Black pepper is number one. A Middle Age superstition was that cumin kept yard fowl and lovers from wandering. It was also believed happiness would bless any bride and groom who carried cumin seeds at their wedding. Cumin's botanical name is *Cuminum cyminum.*

Unemployment is capitalism's way of getting you to plant a garden.

~Orson Scott Card

DILL
Anethum graveolens

Dill is not a traditional Hawaiian spice, but it is delicious. I keep four stalks growing and use it fresh when cooking fish fillets. Dill is another plant that is best used fresh as an herb - dill weed's wispy leaves, while dill seed is considered a spice. Yes, it is called a weed and it can grow that easily, perfect for a spice garden. Dill leaves have a crisp clean taste that is excellent with vegetable dishes, cucumber salads, and makes especially tasty potatoes.

Dill seeds have a much more potent flavor, like a blend of anise and celery. The seeds are eighth-inch ovals. Both seeds and leaves are used for pickling. Dill, botanically *Anethum graveolens*, originated around the fertile crescent in the Middle East. It was mentioned in Egyptian medical texts about five thousand years ago. Dill branches were found in Egyptian tombs. Roman soldiers burned dill seeds directly on to their wounds to promote healing. Dill is a favorite spice of northern Europe, particularly Germany and Russia.

HOW TO GROW: Dill is a perennial and self-seeding herb; one planting can last several years. Dill requires full sun, good drainage, and rich soil. The best pH is 5.8-6.5. Dill seed can be started indoors 4 to 6 weeks before transplanting to the garden, but seedlings form taproots that transplant poorly so dill is most easily started and grown in place. Place the seeds in shallow trenches ½ inch deep; thin successful seedlings from 8 to 12 inches apart. Dill is better grown in clumps, not rows. You'll only require 10 dill plants over the course of the season for cooking and culinary use.

Keep it weeded, occasionally water, and it will mature in about two months. Spread the seed over well-worked soil and cover with a half-inch of damp sand (Not beach sand). Sprouts emerge in about two weeks and should be thinned to six inches apart. Dill attracts beneficial insects that feed on aphids making it a good plant to protect your roses. Snip what you need with scissors and leave the remainder of the plant to keep growing.

Dill seed is harvested by snipping off the flat, yellow flower head as it ripens. Put the flowers in a brown paper bag and dry in the sun. Shake the bag a few times to separate the seeds. Store on a cool dark shelf or refrigerate. These seeds can be used whole or crushed in a mill or coffee grinder. These seed heads, combined with vinegar, garlic, sugar, salt, and pepper produce dilled pickles, or can be used in breads, stews, and rice dishes.

MEDICINAL: Dill weed contains carvone, which has a calming effect and aids digestion by relieving intestinal gas. Dill is believed to increase lactation in nursing mothers and is used in a weak tea for babies to ease colic, encourage sleep, and get rid of hiccups. Crushed dill seeds, mixed with water, are used to strengthen fingernails. Chewed dill seeds can cure bad breath. Dill tea is recommended to overcome insomnia. In foods, it will relieve that gassy feeling, stop hiccups, and ease digestion for children.

NUTRITION: 1 cup of fresh dill has only 9 calories and good source for vitamins A and C, manganese, folate, and iron, calcium, copper, magnesium, potassium, riboflavin, and zinc. Dill seeds are high in calcium, 1 tablespoon equals ¼ cup of milk. Dill is usually used by the tablespoon and the amount of nutrients you get from sprinkling it over your food will be considerably less.

FOOD & USES: Fresh or dried, dill's leaves and seeds are great additions to fish, lamb, potatoes, and pea dishes. Always add dill at the end of cooking, otherwise the heat will destroy most of its flavor. Use it sparingly or it will overwhelm other flavors.

RECIPES: PLENTY BEAN STEW

Ingredients: 1 large onion and the following chopped, 2 garlic cloves, 2 stalks of celery, 1 cup carrots, 2 cups potatoes, ½ cup of the following beans - pinto, kidney, black, and lentils, 1 bay leaf, 1 TS fresh dill leaves crushed, one TBS oil, 1 TS salt

Method: Sauté onion, garlic in oil. Combine ingredients except dill in a large slow cooker, crock pot, or large covered pot, bring to a boil and simmer for about 4 hours. Add dill before serving.

NA PALI DILLED FISH IN FOIL

Ingredients: 1-pound fresh fish fillets - salmon, grouper, or king steaks preferred, ¼ cup lemon juice, 1 TS fresh dill weed - leaves crushed, 1 medium onion chopped very small, 2 TBS butter or margarine, 2 TBS fresh chopped parsley, 1 TS salt

Method: Smear 1 TBS butter on 4 squares of aluminum foil, then put fish on each piece of foil. Melt the other TBS butter in a small sauté pan and add lemon juice, parsley, dill weed, and salt. Pour over fish. Top with onion. Fold foil so it doesn't leak and put the four pieces in another baking dish. Bake at 350 for twenty minutes or longer depending on the thickness of the fillets.

CHEESEY DILL BISCUIT BREAD

Ingredients: 2 cups bakers flour, 1/3 cup whole milk, ¾ cup cheddar cheese grated, 6 TBS cold butter or margarine chopped (must be cold), 2 TBS baking powder, ½ TS salt, 1 TS sugar, ½ TS baking soda, ¾ cup plain yogurt, 1½ TBS sliced fresh dill (or 1 TS dried.)

Method: In a suitable bowl whisk together all dry ingredients. Add cold butter pieces and continue to blend until the mixture is coarse. Add cheese and dill. Combine yogurt and milk into the flour-cheese mixture. On an ungreased cookie sheet, divide dough into quarter-cup mounds and keep about two inches apart. Bake at 400 for about fifteen minutes or until pale golden brown. Best to use the middle oven rack to keep bottoms from hardening and burning.

CABBAGE AND MUSHROOM PIE – absolutely delicious

Ingredients: Pie Crust: 3 cups flour, 1 TBS baking powder, a pinch of salt and pepper, 1/3 cup plus 1 TBS vegetable shortening, 6 large eggs beaten.

Filling: 3 cups cabbage shredded, 1 onion chopped, 2 TBS oil, ½ pound fresh mushrooms, 1 TS fresh basil chopped, 1 TS dried marjoram, ¼ TS dried tarragon, 4 ounces soft cream cheese, 4 hard-boiled eggs sliced, 1 TS fresh chopped dill, 1 egg - beaten, salt and pepper to taste

Method: Pie dough: Blend flour, baking powder, salt, and pepper. Whisk in shortening until mixture forms coarse crumbs. Beat in eggs until a soft, easy to work with dough forms. Cover and refrigerate while cooking the filling.

Filling: In a skillet heat oil over medium heat. Add cabbage, onion, and mushrooms. Cook cabbage till tender, about 20 minutes. Stir often. Add basil, marjoram, tarragon, along with salt and pepper.

On a lightly floured board, cut dough in half and roll out a top and bottom crust. Place bottom crust in a 10-inch-deep dish pie pan. Spread bottom with softened cream cheese and cover with a layer of the sliced hard-boiled eggs. Top with the cooled cabbage and mushroom filling. Sprinkle lightly with dill. Apply top crust, crimping and sealing edges. Cut four small slits in the center of the top crust in a decorative pattern to allow steam to escape. Brush lightly with the beaten egg. Bake at 350 for 30 minutes until crust is golden. Let sit 20 minutes before serving.

Store spices in a cool, dark place, not above your stove. Humidity, light, and heat will cause herbs and spices to lose their flavor.~ Rick Tramonto

FENNEL
Umbelliferae Foeniculum vulgare

In Hawaii, fennel bulbs and seeds are in the larger grocery stores on most islands, and it's being planted in home gardens around the state. I found my plant at a nursery online, and it is an attractive addition to my herb garden and spice shelf. Fennel is another plant considered an herb, a spice, and a vegetable. It has a thick, perennial root system and can grow to five feet tall. This herb is an erect cylinder of vivid green, smooth polished leaves. Every part of fennel is edible. The leaves can be used as a garnish. The swollen leaf base is eaten, and the seeds are used for flavoring.

There are two types of fennel that you may want to grow in your garden, depending on how you plan to use it. 'Florence Fennel' botanically is *Foeniculum vulgare var. azoricum.* This plant is an annual, grown for its thickened bulb-like structure just above the ground, up to 3 or 4 inches long, and oval. It has an aromatic and distinctive flavor and generally is used as a boiled vegetable. These fennel plants reach a height of 2 to 3 and look similar to celery.

Herb/condiment fennel is *Umbelliferae Foeniculum vulgare* and doesn't produce much of a bulb and is closely related to the vegetable Florence fennel. The plant is a perennial, which may be grown as an annual. However, it will yield more seed the following years. The leaves are arranged on either side of the stem with threadlike segments. Fennel blooms with a large, bright golden flower head. Fresh, young leaves and stems are minced and have an almost anise seed flavor. This herb is often used to season puddings, fish, and broths. The seed heads are cut and dried before threshing. The small, oblong seeds are the main spice value used in cuisine, confections, and liquors. A volatile oil pressed from the seeds is used in soaps.

Fennel originated around the Mediterranean and was well known to the ancient Greeks who used it as a digestive remedy. It is now grown worldwide where climate permits.

HOW TO GROW: Fennel prefers hot dry, sunny conditions, yet can adapt to partial shade. Since fennel grows so tall and truly can survive for years, plan your garden and place it where it will not shade or interfere with your other plants. A few plants about 6 inches apart are all that's necessary. Plant seeds an inch and a half deep in well-worked soil with and ideal pH of 5.5 to 6.8. Fennel doesn't require much attention or water, an inch of water every week, combined with rainfall and irrigation, is ideal. A warning, fennel will cross-pollinate with dill weed or coriander/cilantro; do not plant these herbs close together as both of their tastes will be dulled. After about 4 months collect the flowers before the seeds ripen. When dry, shake the seeds on to a white cloth. Pick fresh leaves as needed and harvest the bulb for salads. Fennel self-sows easily, so it's likely that if you plant it once, you'll see fennel popping up again.

MEDICINAL: Fennel is used as an eye wash and once believed to increase breast milk. Fennel is thought to curb eating and great for dieters. It will reduce gas and stomach cramps. It fights anemia, indigestion, constipation, colic, flatulence, diarrhea, respiratory disorders, menstrual disorders, and promotes eye care. With carrot juice, fennel is a good treatment for night blindness or to strengthen the optic nerve. Add beet juice to make a remedy for anemia resulting from menstruation. Fennel juice assists convalescence. The French use it for migraine and dizziness. Boiling fennel leaves and inhaling the steam can relieve asthma and bronchitis. Fennel is used after cancer radiation and chemotherapy treatments to help rebuild the digestive system. Ground fennel seed tea is believed to be good for snake bites, insect bites, or food poisoning. It increases the flow of urine.

NUTRITION: Three grams of fennel is about three calories, and this herb contains manganese, calcium, potassium, magnesium, phosphorus, and vitamin C.

FOOD & USES: Fennel is in mouth fresheners, toothpastes, desserts, and antacids. As an herb, fennel leaves are used by French and Italian cooks in fish sauces and to flavor mayonnaise. It is one of the best herbs for fish dishes. Fennel is an ingredient of Chinese Five Spices and of some curry powders.

RECIPES: BAKED FENNEL – POTATO

Ingredients: 4 TBS margarine or butter, 2 pounds Irish potatoes washed clean, 1 fennel bulb, a pinch of grated fresh nutmeg, salt to taste, ½ hot pepper seeded and minced (optional, 1 cup milk, ½ cup grated cheddar cheese.

Method: Slice the fennel bulb and potatoes very thin. Place half of the slices in a grease oven-proof dish and give a dash of nutmeg, salt and pepper if you choose. Cover with milk. Place more slices on top and cover with grated cheese. Cover dish with foil and bake at 350 for 45 minutes. Uncover and bake 15 minutes longer.

RICE AND FENNEL CAKE

Ingredients: 2 fennel bulbs peeled and chopped small, 2 cups milk, 1 cup cooked rice, 4 eggs, ½ cup brown sugar, 1 TBS butter or margarine, 2-3TBS bakers flour

Method: In a suitable pot boil the milk and stir in the rice and fennel pieces. Simmer for half an hour before stirring in the sugar. Mix in one egg at a time. Slowly stir in the flour. Spoon mixture into a greased cake pan and bake covered at 350 for 45 minutes. Uncover and bake 10 more minutes. Cool before serving.

ROAST GARLIC AND FENNEL

Ingredients, 3 heads garlic peeled, 2 fennel bulbs sliced, 1 bunch chives chopped, 1 TBS oil (prefer canola, ½ hot pepper seeded and minced (optional, a pinch of salt

Method: Place peeled garlic bulbs and sliced fennel on a piece of foil and add chives, pepper, salt, and oil. Wrap tightly and put in another oven-proof dish. Bake at 350 for half an hour. Serve as a spread for bread.

ROASTED FENNEL

Ingredients: 2 fennel bulbs (just use the base of the plant) sliced, 1-2 TBS oil (canola preferred), 1 TBS balsamic or another flavored vinegar

Method: In a bowl stir fennel slices with oil and vinegar until they are coated. Transfer to an oven dish and bake uncovered for 15 minutes at 400 degrees. Serve warm.

DID YOU KNOW? Ancients believed fennel seed improved eyesight, particularly curing nearsightedness. In medieval times this herb was hung over doors to ward off evil spirits. It is reputed to stimulate strength and courage and increase the eater's life span. Fennel seeds are used in India as an after-dinner breath freshener and to help indigestion. Flies are said to dislike fennel, and powdered fennel is used to chase flies away from dog kennels and horse stables.

Fennel is the spice for Wednesdays, the day of averages, of middle-aged people. Fennel smelling of changes to come.

~ Chitra Banerjee Divakaruni

FENUGREEK
Trigonella foenum-graecum

Fenugreek is an aromatic herb native to the Mediterranean region, southern Europe, and western Asia. It's best known as a popular seasoning plant. The dried or fresh leaves are used as an herb. The seeds are considered a spice and the fresh leaves and sprouts are vegetables. The seeds are used in cooking, to make medicine, or to hide the taste of other medicine. It has small round leaves and is one of the oldest cultivated medicinal plants. There's evidence the ancient Egyptians understood this herb's benefits because its seeds have been found in tombs, particularly King Tut's. Charred fenugreek seeds were recovered from the archaeological site at Tell Halaf, Iraq, and carbon-dated to 4000 BCE. In the first century A.D., the Romans flavored wine with fenugreek.

Mediterranean countries, Argentina, France, India, North Africa, and the United States grow fenugreek as a food, condiment, for medicines, and dye. India cultivates and consumes the majority, where it's also known as methi. This is the most fragrant member of the bean family, Fabaceae, and its botanical name is *Trigonella foenum-graecum*.

Fenugreek's strange, box-shaped seeds are yellow to orange. The chemical responsible for the distinctive 'maple syrup' smell is *Sotolon.* The seeds are used whole and powdered to prepare many Indian foods. Roasting reduces their bitterness and enhances their flavor.

HOW TO GROW: Fenugreek is an annual plant that reaches half a meter (19 inches) tall and has attractive three-pointed leaves. The blossoms are white and develop into long, slender, yellow-brown pods containing the orangish seeds. It's always a good idea to mix rotted manure or compost into the soil before sowing the seeds. This herb prefers a well-drained soil pH of 5.3 to 8. Fenugreek requires at least 4-5 hours of direct sun a day, and can tolerate afternoon shade. It can tolerate partial shade in warm climates, but in colder locations, it's best to grow it in a sunny spot. It takes fenugreek less than a week to sprout and about a month before it is ready to pick.

Once established, thin the seedlings to two inches apart. Water your crop regularly to keep it moist, particularly in dry weather. Do not over-water, as waterlogged soil will slow their growth. Pinch off the top third of mature stems periodically to encourage lush, branching growth. If you're not planning to collect the seeds, prune the top 6 inches of the mature plant to encourage more growth and prevent it from setting seed. As a member of the bean family, fenugreek needs little nitrogen fertilizer, and the plant will increase the nitrogen in the soil.

When sold as a vegetable, the young plants should be pulled with their roots attached and sold in small bundles. Fresh fenugreek leaves are an ingredient in some curries, like fenugreek-potato curry. Seed sprouts and fresh leaves are used in salads.

MEDICINAL: For thousands of years, fenugreek has been used in alternative Ayurvedic-homeopathic medicine to treat skin conditions and many other diseases. The plant treats bronchitis, fevers, sore throats, wounds, swollen glands, skin irritations, diabetes, ulcers, and some cancers, notably colon cancer. Fenugreek has been used to increase breast milk. One 14-day study in 77 new mothers found that consuming fenugreek herbal tea increased breast milk production, which helped the babies gain more weight. These seeds are also considered an aphrodisiac; men use fenugreek supplements to boost testosterone.

Fenugreek appears to slow the absorption of sugars in the stomach and stimulate insulin production. Both of these effects lower blood sugar in people with diabetes. The seeds have been used as an oral insulin substitute, and seed extracts have been reported to lower blood glucose levels in laboratory animals. In one study of Type 1 diabetics, researchers added 50 grams of fenugreek seed powder to the participants' lunches and dinners for 10 days. A 54%

improvement was found in 24-hour urinary blood sugar, and also reduced in the total and LDL cholesterol levels.

As a poultice, a paste of fenugreek seeds and-or leaves, are wrapped in cloth, warmed, and applied directly to the skin to treat local pain and swelling, aching muscles, gout, wounds, and eczema.

The downside of using fenugreek is that your perspiration and urine may smell like maple syrup. The same odor may happen with breast milk and/or a breastfed baby may begin to smell like maple syrup. If you are allergic to peanuts or chickpeas, fenugreek is in the same family and may cause an allergic reaction.

NUTRITION: The fenugreek plant is quite nutritious, high in proteins, ascorbic acid, niacin, and potassium. 100 grams (3.5 oz.) of fenugreek seed has 323 calories, 40g of protein, 58g of fiber 25 g with vitamins B2-riboflavin, B3-niacin, B-6, B6-folate (B9), and some vitamin C. It is packed with calcium, has loads of iron, magnesium, manganese, phosphorus, zinc, and some potassium. These strange seeds contain antioxidants, powerful phytonutrients, including choline.

FOOD & USES: For culinary purposes, fenugreek seeds and powder are also used in many Indian and Asian recipes for nutrition and their slightly sweet, nutty taste. In more recent years, it has become a common household spice worldwide, and is used as a thickening agent. Young seedlings and other portions of fresh plant material are eaten as vegetables again in some curries and salads. In foods, fenugreek is in some spice blends. It's used to flavor imitation maple syrups, foods, beverages, and tobacco. Fenugreek extracts are used in soaps and cosmetics.

Fenugreek tea: 1 teaspoon of whole fenugreek seeds in boiling water for 15 minutes. Sprouts are another way to eat fenugreek. Soak 1TBS of the seeds in water overnight. Drain the water the next day or drink it as fenugreek tea - and rinse the seeds with clear water. Place the seeds in a small, clear, shallow plastic container and place them on a windowsill with the cover partially open to get air. Rinse daily with fresh water. The seeds should sprout in around 5 days. They are great in salads and stir-fry.

RECIPES: CAULIFLOWER WITH FENUGREEK AND GINGER

Ingredients: 1 medium cauliflower washed and chopped, 1 cup canned or frozen peas, 1 medium sized red onion chopped, 2 tomatoes chopped, 1TBS fresh ginger grated, 1TS turmeric powder, 1TS chili powder, 1TBS fenugreek leaves or 1TS fenugreek powder, 1TS black mus-tard seeds, 2TBS yogurt, 2TBS oil, ½TS sugar, and salt to taste

Method: In a frying pan or wok, heat the oil on a medium flame for 1 minute. Add the mustard seeds, stir until they begin to crackle. Add the onion and the ginger and sauté for 4 minutes. Add the cauliflower, cover, and cook for 3 minutes. Stir in the peas, turmeric, chili powder, and salt. Reduce flame to low. Add tomatoes and cook covered for 5 minutes. Mix in the dried fenugreek, yogurt, and the sugar. Cover and simmer for another 5 minutes until the vegetables are soft.

FENUGREEK HAIR RESTORER

Ingredients: 2TBS fenugreek seeds and 1TBS coconut oil or olive oil

Method: Grind fenugreek seeds into a powder. Transfer to a bowl and add oil to it. Mix both ingredients and apply the paste on the roots of your hair. Let it dry for 10 minutes, and then rinse with water or a mild shampoo.

BE CAREFUL - DID YOU KNOW? Fenugreek seeds can cause diarrhea, flatulence, perspiration, and a maple-like smell to urine or breast milk. There is a risk of hypoglyce-mia particularly in people with diabetes. It may also interfere with the activity of anti-diabetic drugs. Because of the high content of coumarin-like compounds in fenugreek, it may interfere with the activity and dosing of anticoagulants and antiplatelet drugs. **NAMES:** Alholva, Bird's Foot, Bockshornklee, Bockshornsame, Chandrika, Egypt Fenugreek, Fenogreco, Fenugrec, Foen-ugraeci Semen. It is also spelled *foenugreek.*

HORSERADISH
A. rusticana

Living in the tropics we usually get our hot mouth through pepper sauces. At a friend's insistence, and perhaps humor, I tried horseradish in cocktail sauce. I loved its sharp bite and the different type of 'heat' that shoots up your nose and makes your eyes water. Horseradish is a root that can be bought fresh and easily cultivated in your herb garden. Get starter roots online.

Botanically horseradish is *A. rusticana.* Related to the mustard family, it is believed to be native to Russia or Hungary. The Egyptians seasoned food with it while the pyramids were being built. The 'horse' name may refer to the size of the root and its pungency. Horseradish was referred to as 'German mustard.'

Horseradish is a long, whitish-tan root and looks a lot like daikon, long-white radish. Thick roots are best. You can buy fresh, but it is regularly available grated, preserved in vinegar. Some are purple from added beetroot juice. Horseradish's 'heat' comes from isothiocyanate, a volatile compound when combined with air and saliva generates the sinus-clearing tang. Dried or powdered horseradish is more pungent than the vinegar preserved variety.

HOW TO GROW: I have three roots growing in my garden. Horseradish is a perennial, which means it will continue to grow. It can be planted anytime of the year in Hawaii. In fact, if you aren't careful, it will take over like a weed. When you see the large, long leaves you will know it is a cousin to mustard. The leaves can be cooked like spinach and have a nice, unique spicy taste. Before planting, I wrap the root pieces in wet cloth and place in my refrigerator for two days. It grows best in a place that gets only the morning sun. Sections of the roots are planted in soft well-worked soil about ten inches apart. Look for an 'eye' where a shoot is starting to form. As a beginner about 5 roots should be enough. After 4 months, carefully whisk away the soil from under the leaves to reveal the top of the root. If it is two inches wide of more, it's time to carefully dig it out. Save an 'eye' of the root to replant.

MEDICINAL: Horseradish dates back 3,000 years. It has been used as an aphrodisiac, a treatment for tuberculosis, a rub for low back pain, headaches, and a bitter condiment. Horse radish has been used to treat diabetes and circulatory problems. Consuming horseradish can relieve the symptoms of a sinus infection and treats water retention. Horseradish is a gastric stimulant that will help you digest rich foods. It is richer in vitamin C than an orange and works as an antiseptic. It has long been valued for its medicinal properties to help relieve respiratory congestion, and as a poultice to reduce aches from arthritis or rheumatism. Horseradish contains a chemical compound called 'sinigrin' that has been shown to help reduce inflammation.

NUTRITION: Horseradish has only 2 calories per teaspoon with vitamin C, calcium, magnesium, phosphorus, and potassium.

FOOD & USES: Before you grate fresh horseradish, the root should first be washed, trimmed, and peeled. Grate as you would cabbage or carrots. The outside layer has the most pungent taste. The whole root can be refrigerated for a few weeks. Grated horseradish may be kept in white vinegar or frozen in a sealed container.

This root's main use is in horseradish sauces, made most simply by mixing the grated root with sugar, spices, and vinegar. As a sauce, horseradish complements beef, chicken, seafood, and pork. Mixed with sour cream it is great on baked potatoes. Horseradish can be blended with butter for grilling. It is famous as a sharp 'seafood sauce' for shrimp cocktails. Served hot, horseradish loses its pungency and is quite mild.

RECIPES: HORSERADISH SAUCE

Ingredients: ¼ cup fresh grated horseradish, drained and squeezed dry (or ½ cup of the prepared with vinegar), ½ TS sugar, 2 TS Dijon mustard, 1 TS lemon juice, 1 cup heavy cream, ½ TS salt

Method: Combine horseradish, sugar, mustard, salt, and whisk until smooth. Gently fold in cream. Chill two hours

GERTI'S JAZZY HORSERADISH DIP

Ingredients: 1½ TBS fresh grated horseradish drained and dried, ½ cup plain yogurt, ½ cup sour cream, 1 bunch chives chopped, ½ cup cucumber chopped fine, salt, and pepper to taste

Method: Combine all ingredients in a small bowl. Chill dip overnight. Serve with raw vegetables or cooked shrimp.

HORSERADISH APPLESAUCE DIP

Ingredients: ½ cup applesauce, ¼ cup yogurt, 3 TBS fresh grated horseradish or prepared horseradish, 1 TS white vinegar, salt, and pepper to taste.

Method: Combine all ingredients in a bowl and chill.

WAILEA-MAKENA SPICY HERBAL PĀTÉ —appetizer, snack, or protein main dish.

Ingredients: 1 cup ground sunflower seeds, ½ cup cornmeal, ½ cup nutritional (brewers) yeast, 1 TBS parsley, 1 TBS basil, 1 TS thyme, ½ TS salt (sea salt preferred), ½ TS sage, 1 cup potato - finely grated, 1 and a 1/3 cup water, ¼ cup sunflower oil, 2 TBS soy, 1 TBS prepared horseradish

Method: In a bowl combine ground sunflower seeds, cornmeal, yeast, parsley, basil, thyme, salt, sage, and together. Grate potato and rinse, squeeze, and drain to remove excess starch. Add water, oil, soy, and horseradish; add potato last. Mix well. Grease an oven pan and spoon in the mixture. Bake at 350 for 45 minutes until brown. Let cool before serving with crackers or warm bread.

DID YOU KNOW?

In 1597 John Gerarde published a book of herbal medicinal plants and mentioned horseradish. Horseradish doesn't interest horses and it is not a member of the radish family. It is really a mustard. Germans called it 'meerrettich', or 'sea radish'. 'Meer' in German came out 'mare' in English. Perhaps 'mareradish' eventually became horseradish. The plant was known as in England as 'redcole' and as 'stingnose' in some parts of the U.S. Germans still brew horseradish schnapps and some also add it to their beer. Horseradish was rubbed on the forehead to relieve headaches. This root is still planted and harvested mostly by hand.

He said there were two kinds of bitterness: one that takes away the appetite and one that stimulates it. Pepper, he said, was of the first kind - it burns the tongue and nothing more. But horse-radish, though bitter, sharpens the hunger and makes a man impatient for the good things of the meal. So, he said, if a man becomes only bitter and downcast he goes no further. But a little bitterness, a little horseradish, may give one an appetite for perfection. ~ Guy Vanderhaeghe

LEMONGRASS – MAU ʻU LEMI

Cymbopogon flexuosus

Lemongrass is exactly what its name implies, a tall, thin grass with a purple stem that smells and tastes like lemons with a hint of ginger. Lemongrass will grow everywhere in Hawaii. Its health and aromatherapy benefits share this herb with the kitchen and the medicine cabinet. Lemongrass has a scent of flowers with mint. It is used as a culinary flavoring, an insect repellent, in perfumes, and most important, in natural medicines. This herb is a must-have for your garden.

There are over fifty different varieties of lemongrass. *Cymbopogon flexuosus*, or East Indian lemongrass, is also called Malabar or Cochin grass, saw grass, oily, or silky grass. It's native to Sri Lanka and Southeast Asia. The most recognized garden variety bears the scientific name *Cymbopogon citratus* or West Indian lemongrass, is a smaller variety than the one most often used for cooking and homeopathic medicine. For thousands of years, lemongrass has been utilized as a flavoring, fresh, dried, or powdered, and as a scented oil.

Lemongrass has been used continuously for over five millenniums as a food spice and a medicinal herb. Lemongrass probably came to the Hawaiian Islands with the European explorers. In the 1600s, the Spanish in the Philippines organized oil production. People of that time did not bathe as frequently as today and smelled bad. The upper class used scented oils, like citronella, as we use antiperspirants. The British imported lemongrass to Jamaica in 1799. The first recorded large-scale citronella oil production began in Florida in 1947. At the 1951 World's Fair in London, it was promoted to the public and in 1952 and at the Colombo Exhibition.

Today, India is the largest cultivator of lemongrass. It is ranked among the world's ten most profitable oil-bearing crops. China and most Asian countries produce lemongrass for export. *Cymbopogon nardus,* a different variety of lemongrass, produces citronella oil used to repel mosquitos. This grass will grow six feet tall and yields the antiseptics geraniol, and citronellol. It has a fresh lemony aroma is associated with cleanliness, and is a prime ingredient for soaps, household cleaners, disinfectants, and pesticides.

Asian cooking recipes often include lemongrass to flavor soups as Tom Yum Seafood broth and in various curries. Lemongrass is not only a cooking plant, but it's famous worldwide for its many medicinal benefits. This long grass is best known as an herbal tea to relieve fevers, nausea, anxiety, high blood pressure, bowel, and menstruation problems. It is an excellent natural insect repellent; the scent drives mosquitos away. Lemongrass can be used to repel mosquitos. Simply rinse, crush the grass blades in a mortar, and apply to your exposed skin.

HOW TO GROW: This grass is one of the many medicinal cooking herbs perfect for the home kitchen-spice garden. It is an attractive shrub that grows easily. Plant it along your porch and under your bedroom windows to keep mosquitos away. The easiest way to grow this herb is to buy complete fresh stalks at the open market. The stalks should be moist and firm, not dry or wilted. I chop off the blades to use for cooking and place the bulb-stalk upright in a glass filled with water. Change the water every day until roots sprout. Lemongrass is a nice, functional landscape plant.

Plant your lemongrass in a pot in full sun, water often. Use dry cow dung as a dry fertilizer mixed with water. Small new shoots will begin to grow off the side of existing stalks. In your garden, this grass will repel bad insects, even ants but will attract honeybees necessary for pollination. Use a knife or garden scissors to harvest a stalk of lemongrass.

MEDICINAL: Ever listen to an advert for modern medicine? The list of negative side effects is usually longer than the positive. The future of our planet is to 'go green.' Lemongrass – citronella oil - is being used in cleaning products, soaps, and shampoos. Using natural lemongrass presents little risk of reaction when compared to the chemicals used in commercial products. Lemongrass will safely moisturize the skin and treat damaged hair.

Before patent medicines, people of various cultures relied on lemongrass tea or oil. It was used for a variety of ailments, from respiratory issues to antifungal. As an antioxidant, lemongrass is essential in Ayurveda. It is commonly used to treat colds, congestion, nausea, indigestion irregular menstruation, diarrhea, and sore muscles and joints. India also used the oil to treat fungal infections of the skin. The essential oil of the plant is used in aromatherapy.

The Chinese use this herb in similar ways, while Cubans use the oil today to reduce blood pressure. Several studies indicate the lemongrass variety *Cymbopogon citratus* possesses anti-amoebic, antibacterial, antifungal, and anti-inflammatory properties.

Lemongrass tea is known as 'fever tea.' Chop fresh or dried leaves and use 1 teaspoon of lemongrass per cup of boiling water. Boil more leaves and steep overnight for a stronger, concentrated liquid infusion. Studies show drinking lemongrass tea infusions for 30 days can increase red blood cells/hemoglobin concentration.

Some people apply lemongrass as its essential oil directly on the skin at specific painful areas. Rub it onto the temples and behind the ears for headaches. Rub around the navel for stomach and abdominal pains. Anywhere a muscle aches apply as you would a menthol balm for a sprain. Smelling the essential oil of lemongrass is used as aromatherapy for relief from muscle pain, headaches, nausea, and anxiety.

Chewing the lemongrass blades can prevent mouth/oral infections by stopping the growth of bacteria cause cavities in the mouth and sore gums. Always try a small amount to see if you have a unique, adverse reaction. Don't use lemongrass if you're pregnant.

NUTRITION: This herb has a concentration of folate and is a source of synthetic vitamin A. Lemongrass oil contains a high percentage of citral, used in perfumes and ionone. Through distillation they are converted into synthetic vitamin A. Lemongrass is a good source of iron, zinc, and especially manganese. One cup has only 63 calories from 3 grams of fat and 2 grams of protein (if you swallow). Lemongrass is high in Omega 3 and 6 fatty acids.

FOOD & USES: Both the blades and bulbs of lemongrass have flavor, but the best taste is in the lower stalks. Usually, they are sold trimmed of the upper blades. The color should be pale green with a fat, juicy lower end. The upper leaves could be a bit dry, but don't purchase anything yellow or with brown bruises. It's best when it is chopped into very small pieces. Larger pieces should be removed from dishes before serving. Fresh may be stored in the fridge in a plastic bag for two weeks. It may freeze for up to six months. It's always best to use fresh.

Use lemongrass in meat, chicken, and seafood recipes. Fresh is preferred in dishes like stir fries as it gives a better combination of the flavors. If you plan to remove the stalks, make slices across the bulbs and gently pound the stalks before using. If the recipe requires the lemongrass to remain, slice as thin as possible or use a grater. Dried still provides a good, but duller flavor and best used in soups where it can rehydrate. Tom yum soup, shrimp and vegetable soup and Thai chicken noodle soup with lemongrass and coconut milk are great tasting favorites. Pow-dered lemongrass may be added at any point in the recipe. Fresh or dried lemongrass is fibrous, so thoroughly cook your recipe to soften it.

RECIPES: BASIC LEMONGRASS TEA: Slice stalks into 2inch pieces and steep in boiling water for 5-10 minutes, strain, and enjoy. Also, very thirst-quenching with ice on a hot afternoon. Using the tea to rinse your face will remove oil to help clear acne.

THE HULA GIRL'S LEMONGRASS - GINGER TEA: refreshing and relaxing. 6 cups wa-ter, 1/4 cup -1 stalk of lemongrass sliced to thin strips, 2 TBS fresh ginger peeled and chopped, 3 bags of tea of your choice. Ceylon black tea is good. In an open pot, bring all ingredients to a boil, let steep for five minutes. This may be served cool or warm.

SIMPLE LEMONGRASS FLAVORED RICE:

Ingredients: 2 cups rice, 2 big bulbs of lemongrass, 4 cups water, salt to your taste, 1 TBS-vegetable margarine if you choose. Wash rice and chop or grate lemongrass into small pieces, add all ingredients to your rice cooker, and set to cook. Enjoy both the taste and aroma.

Dr. CLAIRE'S TOM YUM SEAFOOD SOUP – A Thai Favorite

Ingredients: 4 cups of water, 12 medium to large shrimp or prawns peeled and deveined, or a pound of cleaned fish chopped into small pieces, 2-4 tomatoes depending on your taste, 2 green spring onions. 4 limes, 6 green chilies, 1/2 cup of oyster mushrooms (other types may be substituted), 1 large or 2 regular stalks of lemongrass, a piece of ginger as big as your thumb, 1 TBS fish sauce, 1 TBS coconut milk, 2 limes, 1 TS sugar optional

Method: Lemongrass, ginger, tomatoes, chilies, and mushrooms should be cleaned and chopped into small pieces. Bring a liter pot with water to a boil. Add ginger, lemongrass, and reduce heat and simmer for 3 minutes. Add shrimp and or fish and simmer another 3 minutes. Stir in chilies, tomatoes, and mushrooms and cook for 5 minutes. Remove from heat and stir in fish sauce, coconut milk, and squeeze the limes. If you add the lime juice while cooking, the broth will be bitter. Adjust to your taste, if too sour add the sugar. If not add more lime. To appeal to the eye, sprinkle the green pieces of the spring onion.

LEMONGRASS CARROT SOUP–for two

Ingredients: Everything should be chopped extremely small unless you have a blender. 1 stalk lemongrass peeled and chopped, ½ kilo carrots peeled and chopped, 1 medium red onion chopped, 2 cloves garlic chopped, a piece of ginger as big as your thumb peeled and grated, 3 cups coconut milk, 1 lime, 1 TBS oil for frying

Method: In a 2-liter pot add oil and over medium heat sauté onion, garlic, and lemongrass for 3 minutes or until you have a strong aroma, then add carrots and cook for 5 minutes. Add only 2TBS coconut milk. Let everything cook until carrots are tender. Pound with a pistil (use a blender if you have one) until smooth. Add remaining coconut milk, return to heat stirring until it bubbles. Remove from heat and squeeze in lime juice. For a garnish grate some lime zest.

DID YOU KNOW?

Lemongrass is used in India put between the pages of important manuscripts to preserve them. It is also called 'mosquito grass' because it is used as a repellent. Lemongrass was on a Qatar postage stamp. It may be burned as incense or as a religious offering to cleanse the spirit before a ritual. Burning may also protect a home against evil and bring good luck and love.

Great cooking is about being inspired by the simple things around you - fresh markets, various spices. It doesn't necessarily have to look fancy.

~ G Garvin

MACE

Myristica fragrans

According to my research, the Big Island, Hawaii, is the only place in the island chain that has nutmeg trees. The nutmeg tree is the only tree that grows two spices. Mace is similar to nutmeg with a slightly better aroma. Mace covers the nutmeg inside its shell. It is used to flavor baked goods, meat and fish dishes, sauces, and vegetables, and in preserving and pickling spice mixtures.

It's hard to find anything more purely red than fresh mace. When I carefully crack open a yellow -green nutmeg, see the red mace and get a faint whiff of nutmeg; it exhilarates me. Considering the universal popularity of these two spices historically, a lot of Europeans must have felt the same way. In England during the 1500s, one pound of mace was worth three sheep. High-quality mace retains an orangish-red color, but some types dry to light tan. Grenada is second to Indonesia in mace production. Mace is produced in a limited quantity on Hawaii, but if you are vacationing or on a Sunday drive around Hilo and along the Hamakua coast, search it out, along with fresh nutmeg. It's a chance of a lifetime to taste the freshest flavors of two incredible spices. You might get lucky and find them at some of farmers' markets on Oahu.

A whole dried mace is termed a blade. Preserved blades are preferred rather than ground mace, since fresh dried can be ground as needed. The flavor of an intact blade is much better than the powdered version. The flavor of mace is very delicate, so it should be carefully stored in a cool dry place and used quickly to maximize the flavor. Mace is a bright red lacy skin removed by hand after the nutmeg is harvested. Then it is left to dry flat in the sun. While it cures its intense aroma develops while its color fades. Nutmeg trees are native to Indonesia's Moluccan Islands. It is large tropical evergreen that can reach 60 feet. The trees are either male or female, and both are needed for pollination. Small, light yellow blossoms precede the pale-yellow fruit. As it ripens it splits to expel the seed.

HOW TO GROW: Nutmegs are grown from seeds, and after about six months they are ready to be transplanted. If you see trees during your island drives look for sprouted seeds. After five years the trees flower, and then can be sexed. The males are thinned to one male for every ten females. They bear after seven years, but reach full productivity at fifteen. Nutmeg trees continue to bear fruit for about fifty years. A single mature tree can produce 2,000 nut-megs per year. A pile of fruit large enough to make 100 pounds of nutmeg produces a single pound of mace. This naturally makes mace more valuable than nutmeg.

MEDICINAL: Mace and nutmeg are similar in culinary and medicinal properties. Both spices are efficient in treating digestive and stomach problems, relive intestinal gas and flatulence. It can reduce vomiting, nausea, and general stomach uneasiness.

NUTRITION: Mace features quite a different nutritional profile than nutmeg. It is less in calories, but has more concentrations of essential oils, vitamins A, B-2, and C, with caro-tenes, iron, copper, manganese, and calcium.

FOOD & USES: Dried mace pieces are not easy to crush. Ready-ground mace is easier to use, but the flavor and aroma will fade quicker. A trick is to dip the mace blade in a tiny bit of hot water. The blade and the liquid can be used in the recipe. One mace blade will season a dish for four. Mace should be added at the end of the cooking process, and the mace blade should be removed before serving. In baked goods and roasted meat recipes, mace is added at the beginning with the other ingredients.

Mace is used to flavor white sauces, lasagna, meat and vegetable stews, pastries, and some East Indian desserts. Add some to potatoes or sweet potatoes for something new. Hot chocolate drinks and tropical punches improve by adding a little mace. It is high in calcium, phosphorus, and magnesium. Five grams of mace has about twenty-five calories.

RECIPES: CHOCOLATE CHERRY PIECES

Ingredients: 1 cup seeded dried cherries minced, ½ cup butter - softened, ½ cup brown sugar, ¼ cup granulated sugar, 1 egg, 1 TS vanilla extract, 1 cup all-purpose flour, ¼ cup cocoa powder, ½ TS baking powder, ½ TS ground mace, one and ½ cups rolled oats, ½ cup chocolate chips, a pinch of salt

Method: In a large mixing bowl blend the butter and sugars until fluffy. Whip in the egg and va-nilla extract, add flour, cocoa, baking powder, mace, and salt. Whip until smooth before stirring in the oats, chocolate chips, and cherries. The final mixture will be very stiff. Drop tablespoons of dough onto greased or nonstick baking trays about an inch apart. Bake 10-12 minutes at 375 or until the tops appear dry, yet not browned. Remove and allow to cool.

HAMAKUA COCONUT MUFFINS

Ingredients: 1 cup grated coconut, 3 cups all-purpose flour, ¾ cup brown sugar, 1½ TS baking powder, 1 TS baking soda, ½ TS salt, ¼ TS ground mace, 1 egg, one cup milk, ¼ cup orange juice, 1/3 cup vegetable oil

Method: In a large bowl combine flour, sugar, grated coconut, baking powder, baking soda, salt, and mace. In a next small bowl beat the egg and combine with milk, vegetable oil, and orange juice. Slowly stir in flour mixture until dry ingredients are just moistened. Don't worry about the lumps. Spoon into greased muffin cups two-thirds full. Bake for 20 to 25 minutes at 425.

ALL DAY YAM BREAD – This takes a while and is worth the time and effort.

Ingredients: 1 cup cooked yam mashed, 2 packages dry yeast, 1½ cups very warm water, 6 cups unbleached white flour, 1 TBS salt, 1 TBS brown sugar, a pinch of ground allspice, ½ TS mace, 2 TBS soft butter, 1 egg for glaze. Sweet potatoes may be substituted.

Method: Dissolve yeast in the warm water. In a large bowl combine flour, salt, sugar, allspice, and mace. Stir in the yeast mixture. Add butter and mashed yam puree. The dough should now be moist and ready to knead for ten minutes by hand. Then put in a greased bowl, cover with a damp towel, and let rise in a warm place until doubled. Punch dough down and let rise again about 45 minutes. Punch down and shape into one large round loaf or divide. Let rise once more for another 45 minutes. Beat egg with a teaspoon of water and use as glaze for top of bread. Bake at 425 degrees for 45 minutes or until the loaf sounds hollow when tapped. .

DID YOU KNOW? One productive acre will yield 500 pounds of nutmeg, yet only 75 pounds of mace. This makes mace more valuable than nutmeg. Records show that in fourteenth century England one pound of mace was worth three sheep. Nutmeg has been cultivated for 1000 yrs. Nutmeg and mace are botanically *Myristica fragrans.*

Quotes are like cayenne pepper or some other strong spice: a little goes a long way, and too much is a disaster. ~ Ben Yagoda

MARJORAM
THE HERB OF HAPPINESS
Origanum majorana

Marjoram is oregano's sweet sister. Oregano is a Mediterranean spice with a zesty lemon-peppery flavor, while sweet marjoram is more delicate and fragrant. These two spices are almost interchangeable. They look almost identical except marjoram is usually a bright green and oregano a duller green. Both herbs are members of the mint family. Sweet marjoram is botanically known as *Origanum majorana*, while oregano is *Origanum vulgare*. Wild marjoram is better known as oregano, so think of marjoram as 'tame' oregano. Anyone who loves Italian food should grow marjoram. This herb will grow almost anywhere among the Hawaiian Islands.

The only way to discern the intricate differences between oregano and marjoram is to have a plant of each and take leaves; crush marjoram in your right hand while doing the same with oregano in your left. Oregano's aroma is a sharp pine smell of a commercial air freshener. Marjoram is more like a refined perfume with a clean, flavor. Sweet marjoram is a perennial, but to keep it from becoming woody, replant every year. Oregano is also a perennial that keeps growing with reasonable trimming.

Marjoram is native to North Africa and Western Asia, the Mediterranean area, and was known to the Greeks and Romans, who looked on it as a symbol of happiness. It was said that if marjoram grew on the grave of a dead person, he would enjoy eternal bliss. Egyptians used marjoram with other fragrant spices in their embalming process. To the ancient Greeks, marjoram was the herb of a happy marriage. Thought to be a favorite of the goddess of love, it was woven into wreaths that brides and grooms wore on their heads. Also, according to ancient folklore, sleeping with marjoram under your pillow may promote dreams of true love.

HOW TO GROW: Easy to grow from seed, marjoram prefers well-drained soil, full sun, and room to spread. This herb thrives in a rich, light loam soil with a pH range 4.9–8.7. The best is around 6.9. Cut the plant back if it becomes woody and it will regenerate. Its flowers run from pink to purple. It can be used fresh, or dried by spreading in a cool, well-ventilated place. This herb will grow to about a foot tall and needs to be spaced about eighteen inches apart from other plants in your herb garden. This plant's flavor usually peaks just before the flower buds form. To harvest, trim plant back by two-thirds its size.

MEDICINAL: Marjoram is considered the most fragrant essential oil among all herbs used in aromatherapy. It is also a warming and soothing massage oil for muscle aches. It fights asthma, headaches, and soothes digestion. Marjoram is used to loosen phlegm and is a decongestant used to fight bronchitis, and sinus headaches. It is useful as a tonic for the nervous system. Marjoram may be more calming than oregano, to soothe the nerves, reduce tension and stress. One component in marjoram is flavonoids, which relieve insomnia, tension headaches, and migraines. Oregano and marjoram have a high number of antioxidants, especially used fresh. A tea brewed from marjoram leaves may help with indigestion, headache, or stress. Externally dried leaves and flowers may be applied as poultices to reduce the pain of rheumatism.

NUTRITION: Marjoram is rich in magnesium and potassium. Marjoram is also a good source of folate and vitamins A and K. One teaspoon has 2 calories.

FOOD & USES: As with most herbs, fresh is best, but dried this herb holds its lovely fragrance and flavor much better than many other dried herbs. One tablespoon fresh equal one teaspoon dried. Marjoram is a natural with meat dishes, cooked or raw vegetables, fish, and chicken. It works well when it's added near the end of the cooking period. It's especially good along with other herbs in beef stew. Marjoram also is good on fresh tomato sandwiches, and it pairs well with eggs or cheese. A light sprinkling adds flavor to cream-based sauces or soups, especially potato soup. When growing this herb use the flowers to make an herbal vinegar. Marjoram is also used in body care products, including skin creams, lotion, body wash, and shaving gels.

RECIPES: MARJORAM PASTA

Ingredients: ¼ pound of your favorite type of pasta, 2 garlic cloves quartered, 3 cups fresh broccoli chopped small, 2 tomatoes chopped small, 4 ounces softened cream cheese, ½ TS salt, ½ TS black pepper, 1 TBS fresh marjoram leaves chopped small or 1 TS dried, 1 cup grated cheddar cheese,

Method: Bring a large pot of water to boil over medium heat. Add garlic and pasta. Just before the pasta is fully cooked add the broccoli. Remove from heat and drain keeping the pasta and broccoli in the pot. Add the cream cheese, salt, pepper, and marjoram. Stir to combine all the ingredients. Top with cheddar cheese and chopped fresh tomatoes.

EASY 'HOMEMADE' TOMATO SAUCE

Ingredients: 1-pound leanest minced beef (minced chicken or lamb could be substituted), 2 garlic cloves minced, 1 large onion chopped small, 2 eight-ounce cans tomato sauce, 1 TS dried oregano, ½ TS dried marjoram, 1 TS basil, 1 TS sugar, salt to taste

Method: In a large skillet on medium heat, brown and crumble the meat. As the meat begins to produce a liquid add the garlic and onion. When the meat is fully cooked and crumbled, drain off excess liquid. Add remaining ingredients. If you are using fresh herbs chop them just before adding. Return to heat. Bring to a boil and then simmer stirring for 10 minutes. Serve over pasta or rice.

LEMON MARJORAM ROAST CHICKEN

Ingredients: 1 whole chicken, 1 lemon halved, 1 TBS fresh marjoram, 1 TS black pepper, 1 TS salt.

Method: Wash chicken and rub inside thoroughly with salt and pepper. Gently squeeze the juice of the halves of the lemon into the chicken's cavity. Put the lemon halves inside and add the marjoram. Put the chicken in a baking or roasting dish and cover with foil. Cook at 350 degrees for an hour. Uncover and continue to cook for 15 more minutes.

TOMATO MARJORAM CHICKEN

Ingredients: 1 chicken cut into pieces, 3 large tomatoes (about four cups) chopped, ¼ cup oil (prefer olive or canola), 4 garlic cloves minced, ½ hot pepper seeded and minced, 2 TBS fresh marjoram chopped small, one TS salt

Method: Combine the tomatoes with the oil, garlic, minced pepper, salt, and 1 tablespoon marjoram in a large bowl. Place chicken in a baking dish. Pour tomato mixture over chicken. Bake chicken at 450 degrees for 40 minutes. Uncover and continue to bake chicken for 15 more minutes. Sprinkle with the remaining tablespoon of marjoram just before serving.

DID YOU KNOW? Oregano is often confused with marjoram, mainly because oregano translates to marjoram in Spanish. The tops and leaves of this herb are distilled to produce an essential oil. It has many components; one is camphor.

MINT
Mentha piperata

There's spearmint chewing gum and Peppermint Patties. There are many varieties of mint, each has a distinct taste and smell. Pennyroyal is probably the most commonly used in the kitchen. Spearmint is a sweet flavorthat seems to cool the mouth. Peppermint has a stronger mentholated taste, while pennyroyal has a strong almost medicine flavor. Mint grows almost anywhere with partial shade.I grow my mint in a window box by the door so if anyone, especially my dogs, brushes against, it produces the delightful smell. Mint is a great fragrance that symbolizes hospitality. Ancient Greeks and Romans rubbed tables with mint before their guests arrived. The Japanese learned to distill peppermint oil to produce menthol. Mint smell keeps mice and rats away and pennyroyal is an effective insect repellant for fleas and aphids.

Mint probably originated along the Mediterranean, but is now grown everywhere in the world. There are over a hundred different types of mint. Bowles has the best flavor for cooking or drinks. Spearmint plants have full flavor, apple mint has the best taste for fruit salads, while pennyroyal has the strongest flavor. Spearmint is an attractive perennial that can grow a meter tall with grayish-green leaves and clusters of small blue spiked flowers. Peppermint has purple spiked flowers on red-tinged leaves. Pennyroyal is the smaller mint that grows close to the ground with pink flowers.

HOW TO GROW: Mints thrive in cool and moist places, but grow almost anywhere. The best way is to divide growing plants, seeds take forever. Water and adjust your mint to enough sun, it should grow anywhere in Hawaii. Mint grows easily in either shade or sun with very little maintenance. It takes about ninety days for seedlings to mature. If not trimmed they will rapidly take over your garden. Just snip the necessary leaves when needed. Mint is another perfect plant to be grown indoors in pots. Don't overwater, or let mint dry out. The easiest method is to buy some at the market and soak the ends in a glass of water. Change water daily and within a week you'll see fine white roots. Carefully plant in a suitable pot is a partially shaded are for a month until the plants have stabilized. Then move to your herb garden. The best pH is 6-7.

MEDICINAL: Peppermint is one of the oldest and best-tasting home remedies for indigestion, headache, and rheumatism. After dinner mints are used to improve bad breath, ease gas, and aid digestion. Peppermint lessens the amount of time food spends in the stomach by stimulating the gastric lining to produce enzymes that aid digestion. It relaxes muscles lessen-ing stress, has antiviral and anti-bactericidal qualities, which clears congestion related to colds and allergies, and eases intestinal cramping.

NUTRITION: Two mint leaves have little calories with some calcium and potassium.

FOOD & USES: Mint combines well with many vegetables such as potatoes, tomatoes, or carrots. A few chopped leaves enhance salads and dressings. Peppermint is usually used in desserts like fruit salads, sorbets, and sherbets. Spearmint is used with grilled meats, stuffed vegetables, and rice. East Indians make fresh mint chutney. A mint julep is an American classic drink of bourbon whiskey and mint. The mojito cocktail uses rum or arrack with mint and lime.

RECIPES: MINT TOPPED EGGPLANT

Ingredients: 3 eggplants or about 2 pounds, 1 TBS sesame oil, 3 TBS fresh lime juice, 2 TBS water, 2 TBS sugar, 2 TS garlic minced, 1 hot pepper seeded and minced, 1 TBS fresh mint chopped

Method: First prick each eggplant all around with a fork or a knife so it won't burst as it roasts. Place the eggplants on a hot grill or rotate over a stove burner. Roast until its soft and blisters, about 3 minutes. Cool and peel under cool running water. Chop the eggplants into big pieces and place in a bowl. In another bowl combine the sesame oil, lime juice, water, sugar, garlic, and hot pepper. Pour over eggplant and sprinkle top with chopped mint and chives.

SUNSET BEACH MINT MANGO SALSA

Ingredients: 2 ripe mangos, peeled and chopped, 1 bunch chives chopped, ½ cup chopped mint, 2 TBS fresh lime juice, half hot chili pepper seeded and minced, salt to taste

Method: All ingredients should be chopped as small as possible. Combine all ingredients. Let stand at least an hour.

MINT JULIP COCKTAIL

Ingredients: 4 sprigs of mint, 1 TS powdered sugar, 2 TS water, 3 ounces Kentucky Bourbon Whiskey

Method: Combine everything in a shaker glass with a top. Fill with ice and shake. Add a sprig of mint to the glass. Enjoy.

FROZEN MINT DESSERT

Ingredients: 2 eight-ounce packages of cream cheese softened, ½ cup sugar, divided, one package lime flavor gelatin, 1 TBS lime zest, ¼ cup fresh lime juice, 1 eight-ounce tub whipped topping, 2 TBS fresh mint chopped fine, 1½ cups pretzels nely crushed, 6 TBS melted butter or margarine

Method: In a suitable bowl combine cream cheese, ¼ cup sugar until blended. Add dry gelatin mix, zest, and juice. Blend until creamy smooth and add whipped topping and mint. Pour into a 9-inch round pan lined with plastic wrap. Wrap must hang over pan. Sprinkle a mixture of pretzel crumbs, remaining sugar, and melted butter over the pudding mixture. Slightly stir it into the pudding. Cover dessert with ends of plastic wrap. Freeze until firm about four hours. Place frozen dessert upside down on a serving plate and remove plastic wrap and dish. Let soften for about 10 minutes before serving.

HANA HANDS FULL MOJITO

Ingredients: Juice of 1 lime, 1 TS granulated white sugar, small handful of mint leaves and 1 sprig still on the stalk for a garnish, 60ml white rum, and club soda

Method: Squeeze lime into a glass and remove seeds. Add sugar. With a long spoon or a pistil, muddle the mint leaves with the lime juice and sugar. I grind mine into the bottom of the glass to release as much of the mint essence as possible. Add rum while stirring then add ice and soda. Garnish, enjoy.

DID YOU KNOW?

Peppermint is a hybrid cross between water mint and spearmint. Spearmint results from a cross between apple mint and wild water mint. The common garden mint is spearmint, not peppermint as most people assume. In Mexico mint is known as yerba buena, the good herb. Botanically apple mint is *Mentha suaveolens*, spearmint *Mentha spicata,* and peppermint is *Mentha piperata.*

In Greek mythology, the goddess Persephone suspected her husband Hades, god of the underworld, of having an affair with a nymph named Minthe. In a jealous rage, she transformed the lovely nymph into a perennial herb coincidentally named mint.~ Wikipedia

MUSTARD
Brassicaceae

Mustard is a unique plant. Mustard is considered a fresh leafy herb and its seeds are a spice, and it makes some of the world's most used condiments. Of the many greens sold in the markets, mustards have the spiciest flavor of all the cooking leaves. They have a peppery flavor. I grow it because it is another excellent source of iron. Like its cousin spinach, it is extremely easy to grow. Mustard greens are high in vitamins A and C. Mustard will grow everywhere in Hawaii. Enjoy the mustard leaves and then it'll have yellow flowers followed by seeds.

Mustard is native to the Himalayan region of northern India more than five millenniums ago. Mustard is one of the oldest spices and one of the most widely used. The Ancient Greeks and the Chinese began using it thousands of years ago. Romans used it as a condiment and pickling spice. The usually yellow, sometimes brown, condiment was made by grinding mustard seeds into a paste and combining it with an unfermented wine called 'must.' Thus, it is named mustard. Mustard greens are used in Chinese, African, and Asian recipes.

HOW TO GROW: Mustard is a member of the *Brassica* family that includes brocco-li, cabbage, cauliflower, turnips, and radishes. The mustard family also includes plants grown for their leaves, like bok choy, as well as mustard greens. Three related species of mustard are grown for their seeds. Most mustard plants will reach a meter tall.

To grow plant seeds half an inch deep in well-worked soil with a pH of 5.5 to 6.8. Once they sprout, thin plants to six inches apart. With regular water and plenty of sun, mustard should grow nonstop. Cut the young and tender leaves and cook them first. As time progresses the leaf flavor will become stronger. Let the plants go to seed and cut off the seed pods before they burst. Put in a pillowcase and beat against the floor to separate the seeds, then make your mustard sauce.

MEDICINAL: Mustard was always important in medicine. Mustard oil can be so irritable that it can burn the skin. Diluted, it is used as a liniment for sore muscles. Powdered mustard is used in mustard plasters to fight bad coughs and colds. Various cultures have used mustard for snake bites, bruises, stiff neck, rheumatism, and respiratory troubles. It is delightful in bath water or as a foot bath to ease aches and pains. It is believed to rejuvenate hair growth when rubbed into the scalp. It is antibiotic and antifungal.

NUTRITION: A cup of cooked mustard greens has only 20 calories with vitamins K, A, E, and C, potassium, calcium, and iron.

FOOD & USES: Mustard oil is made from the brown seeds. It's used in India the same as ghee. Compared with other cooking oils, mustard is considered a low-fat variety, and its use may lower cholesterol. However culinary use of mustard oil is banned in some countries. It is also excellent as a massage oil.

Brassica alba also known as white mustard has bright yellow flowers that produce a round, light tan, very hard seed with a mild flavor. It is a good preservative and most commonly used in yellow mustard. In India, whole seeds are fried in ghee until the seed pops, producing a soft nut taste used to season some Indian dishes. White seeds are the base of yellow American-style mustard that has sugar, vinegar, and is colored by adding turmeric.

Brassica nigra or black mustard can grow to six feet tall and matures with a very fragile pod that produces a smaller round, dark seed, with a more pungent taste. Classic French Dijon mustard is made from these husked black seeds blended with wine, salt, and spices. *Bras-sica juncea* or brown mustard is the middle of the road seed with a sharper taste than the white, yet not as sharp as the black. Brown seeds are pounded and blended with other spices to create some curry powders. Mustard seeds are in most pickling spices. Mustard powder is used to flavor barbecue sauces, baked beans, many meat and fish dishes, deviled eggs, and beets. Mustard powder is made from white mustard seeds. It has no taste when dry, yet once combined with cool water (not warm) its sharp taste is created after about a ten-minute wait. A chemical reaction occurs, and warm or hot water can interfere. Once the strong taste has formed, other ingredients may be combined to further enhance the taste.

There are many commercial mustards with tastes from mild and sweet to sharp and strong. They can be smooth or coarse and flavored with a wide variety of herbs, spices, and liquids. Today our world consumes more than two hundred thousand tons of mustard yearly!

RECIPES: KIHEI COCONUT MUSTARD GREENS

Ingredients: 1 kilo of greens sliced, 1 onion cut into long strips, 2 garlic cloves chopped, 10 curry leaves, 3 green chilis chopped, ½ TS turmeric powder, 2 TBS grated coconut, 1 TBS ghee
Method: in a large, covered pot heat ghee and fry onion, curry leaves, and chilies until onions are soft. Stir in greens and turmeric on low heat and stir cook 10 minutes. Add coconut. Salt to taste.

SURFERS' MUSTARD AND PASTA

Ingredients: 1 bunch-two pounds, mustard greens sliced into 1 inch pieces, 1 pound pasta like elbow, springs or bow ties, 1 large onion chopped, 1 green sweet pepper seeded and chopped, 4 garlic cloves sliced thin, 1 thumb-size piece of ginger sliced, 4 TBS oil (canola preferred), 3 chilies chopped small, 2 TBS soy sauce, salt to taste
Method: Boil mustard green pieces in salted water for about 10 minutes to your desired consistency. Remove from heat, drain, and leave covered. Cook pasta, drain and keep covered. In a frying pan heat the oil and fry the onion, garlic, ginger, and pepper till soft about 2-3 minutes. Add hot pepper and soy sauce. Combine everything. Season to your taste. Serve warm.

STIRFRY BEEF or PORK & MUSTARD GREENS

Ingredients: 1 pound of boneless beef/pork sliced thin, 2 TBS sesame oil, ½ TS mustard seeds, 1 pound mustard greens, washed, stems removed sliced to inch wide pieces, 2 TBS chives finely chopped, 2 garlic cloves minced, 1 TBS fresh ginger peeled and sliced very thin, 1 cup beef broth, 2 TBS cornstarch or arrowroot, 2 TBS teriyaki sauce, 2 TS sugar, 3 cups cooked rice
Method: In a covered skillet heat 1 TBS oil and add mustard seeds. Cover and heat until seeds pop. Add sliced mustard greens and stir fry for 3-4 minutes. Remove from heat to a bowl. In the same skillet heat another tablespoon of oil. Add beef/pork, chives, ginger, and garlic. Stir fry until meat is well browned, about 3 minutes. Remove to bowl with greens. In the same skillet combine broth, cornstarch, teriyaki sauce, and sugar. Heat until bubbling and stir in rice. Lower heat and stir for 2 minutes. Combine all in a bowl. Enjoy.

MUSTARD - THE CONDIMENT

To make your own special tasting mustard all you need to do is either grind mustard seeds into a powder or mix prepared mustard powder with cold water. Wait about 10 minutes and you have simple mustard. You can add wine or hearty beer, spiced vinegars, or fruit juices for a truly unique flavor. Add spices to your taste. Salt, garlic, or onion powder, ginger, allspice, pepper, nutmeg, basil, and/or rosemary. Any herb-spice combo can make your mustard special.

DID YOU KNOW? Mustard was considered an aphrodisiac by the ancient Chinese. If a German bride wants to be the boss of the household, she sews mustard seeds into the hem of her wedding dress. Spreading mustard seeds around the exte-rior of the home will keep out evil spirits. Egyptian pharaohs stocked their tombs with mustard seeds to accompany them into the afterlife. Canada and Nepal grow more than half of the world's mustard.

NUTMEG
Myristica fragrans

If you have the room and a desire to gift future relatives and friends with a fantastic tree, be one of the few Hawaiians to grow a majestic nutmeg. Nutmeg makes a nice backyard tree that is easy to trim and shape. Nutmeg is an evergreen tree that originated in tropical Southeast Asia on Banda, the largest of Indonesia's Spice Islands. It is important to the world's spice cabinet for two spices derived from the seeds, nutmeg, and mace. Nutmeg was the spice that started wars and so important because of the medicinal properties of its seeds. It is an astringent and stimulant, as well as an aphrodisiac. At the height of its value in Europe, nutmeg was carried to demonstrate wealth. The fruit is light yellow resembling a small tennis ball. Nutmeg refers to the actual egg-shaped seed. Its reddish membrane covering is called mace. Nutmeg has a sweeter taste than mace. Mace has a more delicate flavor, like a mix of cinnamon with pepper. Indonesia and Grenada produce a combined 12,000 tons per year.

In the sixth century, Arab traders cameled nutmeg to Constantinople. In the 1300s, half a kilo of nutmeg was as valuable as a cow. The Dutch fought to control the nutmeg trade, including the extermination of the native people of Banda Island. By the mid-1700s the Dutch kept the price high by reducing the supply by burning supplies. The Dutch controlled Indonesia's Spice Islands until the 1940s. The British East India Company transported nutmeg trees to Singapore, India, Sri Lanka, throughout the West Indies, especially Grenada, where it is the nation's symbol. The trees on the Big Island probably came from Indonesia or Asia.

HOW TO GROW: Find a mature tree, search the surrounding area, and you should find some sprouted seeds. Carefully plant and coddle these shoots. Nurseries have trees available. The small trees do not take well to direct heat and sun. Plant with at least twenty feet of space on all sides from other trees, buildings, or fences. Make sure the soil drains. The optimum pH is 6-7. Give the sprouting nutmeg tree 12-24-12 starting-rooting fertilizer every 4 months. Start with broadcasting a ¼ cup around its roots when it is about 3 feet tall. As it grows increase to a cup. Be patient as it takes at least 7 years before the first spice harvest. When it starts to bear fertilize 3 times a year by broadcasting a cup of 12-12-17 fertilizer. Water during the dry season. As the fruit ripens, the outer covering splits showing the seed covered with red mace membranes. The nut must dry for at least 2 months.

MEDICINAL: Ground nutmeg is also smoked in India. Nutmeg oil is used in perfume, toothpaste, and cough medicines and in pharmaceutical drugs. The oil is used externally for rheumatism and gargled for bad breath and toothaches. Nutmeg oil is believed to have aphrodisiac qualities if topically applied to certain body areas. East Indians have long used nutmeg as a treatment for fever, asthma and heart disease. Arabs used it for digestive disorders and kidney diseases. Nutmeg oil stimulates the brain reducing mental exhaustion and stress. It improves the quality of your dreams, making them more intense and colorful. It is a good remedy for anxiety as well as depression. Nutmeg oil treats muscular and joint pain, and it is an excellent sedative. Nutmeg's relaxing aroma comforts the body and increases blood circulation.

NUTRITION: A grated nutmeg yields 2 to 3 teaspoons of ground nutmeg that has about 36 calories. This spice is high in riboflavin and thiamin with some magnesium, zinc, and copper.

FOOD & USES: Nutmeg makes a good combo with cinnamon, especially in hot beverages like cocoa, tea, or coffee. If possible, grate fresh nutmeg for your recipes and add it at the end of cooking since heat reduces the flavor. These spices should be kept in a sealed jar in a shaded corner or shelf of your kitchen. Mace is used in cheese, chicken, and fish entrees and especially chocolate or cherry desserts. Nutmeg works well with stewed fruit, custards, potato dishes, and is especially tasty with spinach. For all its great culinary and medicinal value, nutmeg has no notable vitamins or mineral content. Nutmeg and mace are botanically referred to as *Myristica fragrans.*

RECIPES: NUTMEG ICE CREAM

Ingredients: 2 cups milk, 2 cups heavy cream, 4 eggs, one cup sugar, 1 TBS fresh grated nut-meg, 1 TS vanilla extract, a pinch of salt

Method: Combine milk and cream in a suitable pot, bring to a boil, and remove from heat. Combine eggs, sugar, salt, vanilla, and nutmeg in a bowl. Add to the milk mixture and cook over medium heat with constant stirring. Put in an ice cream machine and crank. If a traditional ice cream machine is not available, put in a freezer until ice begins to form. Put in a blender and then refreeze.

SWEET SHEPHERD'S PIE

Ingredients: 2 pounds minced beef or chicken, 5 large, sweet potatoes peeled, boiled, and mashed, 1 medium onion chopped small, 2 garlic cloves minced, ½ hot pepper seeded and minced, ¼ cup all-purpose flour, 1 cup broth from the meat, 2 eggs beaten, ½ cup butter or mar-garine, 4 eggs hard-boiled, 1 bundle fresh spinach or bok choy chopped small, 1 TS rosemary, 1 TS salt, 1 TS nutmeg

Method: In a large skillet brown the meat and add enough water to produce a cup of broth. Re-move meat to a bowl. Ina small skillet butter, and sauté onions and garlic. Remove from heat and stir in flour slowly adding the broth. Bring to a boil, stirring constantly as it thickens add the beaten eggs. Pour this over the meat adding the spices and pepper. Press the mashed sweet potatoes into a greased baking dish forming a shell. Layer in the chopped spinach. Slice the hardboiled eggs and arrange on top of the greens. Cover with the meat mixture. Bake at 350 for 45 minutes.

LOW BUDGET CAPPUCINO MIX

Ingredients: ¾ cup instant coffee, 1 cup powdered cocoa tea mix, ½ cup sugar (less to your taste), 1 cup powdered non-dairy creamer (splurge and try hazelnut flavor), 1 TS ground cin-namon, ½ TS ground nutmeg

Method: Mix all ingredients and store in a sealed container. Refrigerate if possible. Use 2 ta-blespoons per cup of hot water.

PAUKAA BANANA-PUMPKIN SPICE BREAD

Ingredients: 4 cups bakers flour, 2 TS baking powder, 2 TS baking soda, 2 TS ground cin-namon, 2 TS ground nutmeg, 1 TS ground ginger, ½ TS salt, 1½ cup pumpkin cooked and mashed, 1½ cup very ripe bananas mashed, 1 cup plain yogurt, ½ cup cooking oil, 4 large eggs, 1 cup brown sugar, 1 cup grated coconut

Method: Combine flour, baking powder, baking soda, spices, and salt in a large bowl. In a second bowl mix pumpkin, bananas, yogurt, oil, eggs, and sugar until it is creamy. Slowly blend in flour mixture. Add half of the coconut. Pour into 2 greased bread pans and cover top with remaining coconut. Bake at 350 for 45minutes. Test center firmness with a toothpick.

If you've never experienced the joy of accomplishing more than you can imagine, plant a garden. ~ Robert Brault

PARSLEY
Petroselinum crispum

Parsley is the main ingredient of my homemade green seasoning. Parsley is a unique, bright green plant with a vibrant taste and fresh aroma. Even though parsley has a great taste, it may also be used as a medicinal herb. Originating along the Mediterranean Sea in Southern Spain, Italy, and Greece, parsley has been cultivated for over two millenniums. Originally parsley was grown for its medicinal properties to the extent the Greeks considered parsley to be a sacred gift from their gods. The queen of the Greek gods, Juno, grazed her horses in fields of parsley to keep them high-spirited. During the Middle Ages, the French popularized parsley as a kitchen herb. Parsley was introduced to England in the 1500s and brought to the Americas before 1800.

HOW TO GROW: Hawaii grows the two most common types of parsley, the curly fern leafed and the flat. We love the sharp smell and almost sweet taste of the flat-leafed variety. It is not difficult to grow from seedlings, yet very difficult to sprout directly in the garden from seeds. Work the dirt of a row until it is fine. Carefully plant the seeds every two inches. Carefully cover with about a half-inch of dirt. Water carefully so as not to wash away the seeds. Patience and constant watering are necessary as parsley seeds germinate so slowly it is said, 'Parsley goes to Hell and back again nine times before it comes up.' The easiest way is to look for parsley with the roots attached. Cut off the leaves and put the roots in a glass of water until they sprout fresh roots. Plant in pots and keep out of the hot sun until the plants are strong, about 4 inches tall.

MEDICINAL: Parsley is a natural diuretic and can be used by women to alleviate irregular menstrual cycles. Parsley eaten raw or as tea may ease the bloating that occurs during menstruation. Chewing parsley will help with bad breath from food odors such as garlic or onions. A tea made from parsley will stimulate as much as a cup of coffee, but watch out as parsley has been reported to be an aphrodisiac. A poultice of its leaves can be used for insect bites or stings. To make a parsley face tonic steep a quarter cup of fresh parsley in 2 cups of hot water for 2 hours. Then pass through a sieve. Drink before meals. Parsley's supply of vitamin A benefits skin tone and its vitamin E fights wrinkles. Parsley fortifies the immune system and benefits the liver, spleen, digestive and endocrine organs.

NUTRITION: Two tablespoons of parsley contain more than the recommended daily dose of Vitamin K. Parsley contains more vitamin C than lemon, orange, or any other fruit. It also has a good amount of potassium. Two tablespoons of parsley contain about three calories with antioxidants such as vitamin C, beta-carotene, and folic acid. To obtain the maximum food value from parsley it should be eaten raw. The heat from cooking greatly reduces the vitamin content and the taste.

FOOD & USES: If you are including parsley in cooked recipes use the flat leaf type and add it towards the end so it will retain most of its taste, color, and food value. In most upscale restaurants sprigs of parsley accompany the dinner entrée as a garnish. Don't push it aside, enjoy eating it, as it is not only tasty and cures bad breath, but can be very healthy for you. These antioxidants fight many diseases such as arthritis, asthma, diabetes, colon cancer, heart attacks, and strokes.

Parsley is fragile so wash it carefully in a colander or by swishing it in a bowl of water. Parsley makes an excellent addition to soups, salads, and baking.

RECIPES: TABOULAH – PARSLEY SALAD

Ingredients: ¾ cup wheat bulgur, oats, or cooked rice, 2 cups cold water, 2 cups chopped parsley, ¼ cup chopped mint leaves, ½ cup chopped onions, 2 medium ripe tomatoes chopped, ¼ cup olive oil, 2 TBS lemon juice, salt, and spice to taste.

Method: Soak oats or bulgur in a bowl with the cold water for a half-hour. Drain and mix grain with chopped onions. If substituting cooked rice just mix with chopped onions. Add parsley and mint. Mix lemon juice, oil, and spices then add to grain-parsley mixture. Add chopped tomatoes and chill for at least an hour before serving on a bed of lettuce. Serves six.

CAPTAIN DAN'S PARSLEY AND POTATO SOUP

Ingredients: 1 cup chopped potatoes, ½ cup chopped parsley, 1 bunch chopped chives, 1 medium onion chopped, 4 cups chicken stock or vegetable bullion, salt, and spice to taste

Method: Combine all ingredients except spices and parsley in a large pot and bring to a boil. Lower the cooking heat and simmer for an hour. Stir in salt and spices to taste. Just before serving add the chopped parsley.

WAIMANALO BEACH PARSLEY FISH CAKES

Ingredients: 2 cups mashed potatoes, ½ pound skinned salmon fillet, ½ pound skinned king mackerel fillet, 1-ounce melted butter, ¼ cup chopped parsley, ¼ cup chopped chives, 1 medium onion chopped, 1 egg beaten, ¼ cup crushed biscuits (crackers or breadcrumbs, 2 TBS oil for frying, salt, and spice to taste

Method: Mix mashed potatoes with butter and spices. Cover the bottom of a medium fry pan with a half-inch of water and poach the fish fillets until just tender and then flake the fish apart with a fork. In a suitable bowl mix the fish and potatoes together with the onion, parsley, and chives. First roll mixture into balls and then flatten with a spatula or spoon. Place beaten egg in one small bowl and biscuit/breadcrumbs in another. Dip fish/potato cakes First in the egg and then coat with the crumbs. Heat the canola oil in a skillet and fry the cakes until crisp. Drain on paper and serve with your favorite dressing.

CHINATOWN PARSLEY RICE

Ingredients: 2 cups boiling water, 2 cups rice (preferably brown rice), ¼ cup chopped parsley, 2 chives chopped, salt, and spice to taste

Method: Using a large pot, bring water to a boil, and then add rice and spices. Seal pot with foil, reduce heat, and simmer until rice is cooked. Stir in parsley and chives before serving.

PARSLEY COOKIES FOR THE PUPPIES—Does your pooch have bad breath???

Ingredients: 3 cups chopped parsley, ¼ cup chopped carrots, 2 TBS vegetable oil, 3 cups flour (prefer whole wheat), ¼ cup wheat bran, 3 TBS baking powder, ½ cup of water, more water may be necessary to get the correct consistency.

Method: Mix parsley, carrots, flour, bran, and baking powder with a half cup of water. Mix and knead. Add more water if the mixture feels too dry. Roll dough into half-inch thick logs. Cut the logs according to the size of your dog. A big dog should get four-inch pieces while small pompeks get only one-inch pieces. Bake pieces on a cookie sheet at 350 degrees for 30 minutes. This should make three-dozen dog biscuits. Store in a sealed container after they have cooled.

DID YOU KNOW?

Parsley has three varieties: curly, Italian flat-leaf, and Ham-burg. It can be bought fresh or dried. Curly is used as a garnish due to its appearance and bitter taste while Italian flat-leaf packs more flavor, thus adding more to a dish. Hamburg is not common, and the root is the part used rather than the leaves. Parsley's botanical name is *Petroselinum crispum*.

ROSEMARY

Rosmarinus officinalis

Rosemary and I are still getting acquainted, Rosemary, the herb, that is. It is one of the most attractive plants in my herb garden, but I'm still learning to enjoy its unique flavor. It has needle leaves like a pine tree. Its taste is like a piney and sweet-mint mix with a gingerish tang. Rosemary is native to the Mediterranean and is mostly used when roasting meats, fish, or poultry.

In ancient Greece, students wore rosemary head wreaths to improve their memory. During the Middle Ages people believed rosemary could ward off evil, so they slept with pieces of rosemary under their pillows. Rosemary symbolizes friendship and love and all brides carried it during wedding ceremonies. It was the main ingredient in various love potions.

HOW TO GROW: Rosemary is easy to grow in Hawaii. It is a perennial; it continues to grow year after year without replanting. Rosemary leaves look like miniature pine tree branches. It can grow to five feet or more, so make adequate room in full sun, and you'll have it around for a couple of decades. There are many different types of rosemary so be certain what type you are planting. It is slightly difficult to grow from seeds so cuttings from another mature plant is a suggestion. Seeds may take 2-3 months to germinate. Rosemary likes slightly alkaline, sandy soil with a pH of 6-7.5. Be careful not to let this herb dry out, or to overwater it. It adapts well to indoor container growing. Once or twice a year give it some fertilizer like special green, and it should be content to flavor many dishes for you. Seeds and plants are available at nurseries and online.

MEDICINAL: Rosemary stimulates the circulatory and nervous systems while it also calms indigestion. It will help fight severe headaches. Rosemary's essential oil will ease muscle strains and aches. Research has shown its ability to help prevent cancer and age-related skin damage, boost the functioning of the liver, and act as a mild diuretic to help reduce swelling. It is a usual ingredient in shampoos and hair conditioners that fight dandruff and (if you believe it) hair loss. It is thought that adding rosemary to your diet may reduce the chances of having a stroke or contracting Alzheimer's. Rosemary is a good herb to use as a mouth wash because it soothes sore throats and freshens breath. When put in a drawer or closet with clothes it will repel moths.

NUTRITION: A teaspoon of this herb has only 2 calories with slight amounts of vitamin K and folate with some calcium, potassium, and magnesium.

FOOD & USES: Rosemary's fresh, flavorful needles are chopped. A little rosemary goes a long way in most dishes. Try using it the next time you roast a piece of pork or a chicken. It improves the flavor of many vegetable dishes, especially tomatoes, squash, and spinach. It is a great addition to most marinades and light-flavored soups like potato or cauliflower. Rose-mary keeps extremely well when frozen in a sealed container.

RECIPES: BOUQUET GARNI – This is a restaurant chef's flavor trick. A classic herb combination secured in a cheesecloth pouch and used to flavor meat and vegetable dishes.
Ingredients: ¼ cup dried parsley, 2 TBS dried thyme, 2 dried bay leaves, 2 TBS dried rosemary
Method: Combine all ingredients. Store in a well-sealed container. Refrigerate. Cut a circle of cheesecloth about 6 inches in diameter. Fill with 2 tablespoons of herb combination. Tie tightly with string. Use in recommended dishes and remove intact after cooking.

LEMON ROSEMARY INFUSED OIL

Ingredients: 2 cups cooking oil (prefer either canola or olive), 1 sprig fresh rosemary, 1 bunch of fresh lemongrass, a fresh sprig of thyme, 1 garlic clove, 1 TS black peppercorns, 1 TS salt
Method: combine all ingredients. Put in a sealed bottle and refrigerate for 2 weeks before using.

Great on salads or fried vegetable dishes.

KOLOA ROSEMARY CHICKEN AND POTATOES

Ingredients: 1 chicken, 1 lemon quartered, 4 sprigs of fresh rosemary, 6 medium potatoes chunked, 3 TBS butter or margarine, 3 TBS oil (prefer canola), 2 TS dried rosemary, 1 TS dried thyme, salt, and pepper to taste, ¼ cup water

Method: Rub chicken with lemon inside and out. Leave ¼ of the lemon with 2 sprigs of rosemary in the cavity and the other sprigs under the chicken's loose skin wherever you can place it. Put the whole chicken in a suitable roasting pan and surround with potato chunks. Pour in water. Smear chicken and potatoes with butter, then cover with dried rosemary and thyme. Roast covered at 350 for an hour. Uncover and continue to roast for 15 more minutes.

KAHALA ROSEMARY STEAMED SHRIMP

Ingredients: 2 pounds fresh shrimp, shelled and deveined, 1 bottle of beer, 4 sprigs fresh rosemary, salt, and pepper to taste

Method: Place beer and rosemary in a steamer. Boil and steam shrimp for 5 minutes until pink. Add salt and pepper to your taste.

ROSEMARY CORNBREAD

Ingredients: 3 TBS fresh rosemary chopped fine, 1 medium onion chopped very small, 1 cup cornmeal, 1 cup bakers' flour, 2 TS baking powder, 1 TS baking soda, 1 TS salt, 1 TS cayenne pepper, 1 cup milk, ½ cup melted butter or margarine, 2 eggs, 1 TBS butter or margarine

Method: In a small skillet fry the onion in 1 tablespoon of butter until it becomes clear and soft. In a bowl combine all dry ingredients, rosemary, and cooked onions. In another bowl combine milk, melted butter, and eggs. Combine both wet and dry ingredients in a loaf pan. Bake at 400 for 45 minutes. Cool before cutting.

REFRESHING ROSEMARY LEMONADE

Ingredients: 3 sprigs of fresh rosemary, 2 cups fresh lemon juice, 2 TBS grated lemon skin, ½ cup sugar (more to your taste), 2 quarts of water, a pinch of salt

Method: Stir the sugar, one cup water, grated lemon, rosemary, and salt, in a pot and boil until sugar dissolves, about fifteen minutes. Cool the sugar mixture. Add it to the fresh lemon juice and the water. Serve chilled.

DID YOU KNOW?

Of the mint family, rosemary is not named after roses or a woman named Mary, but from the Latin *Rosmarinus*, which translates to 'dew of the sea.' Botanically rosemary is *Rosmarinus officinalis.* Chinese combine it with borax to treat baldness. Rub your pets with powdered rosemary leaves as a natural flea and tick repellent due to its antimicrobial properties. Rosemary hair rinse - boil 10 sprigs of fresh rosemary in 2 cups water for 30 minutes. Cool and store in a tightly sealed container. Refrigerate. Once a week combine one ½ cup rosemary water with a cup of warm water and work through your hair. You'll be amazed at the luster.

SAFFRON

Crocus sativus Linnaeus

Real saffron is the world's most expensive spice. What most people know as Indian saffron is really turmeric. Saffron comes from the dried stigmas of the saffron crocus flower, *Crocus sativus Linnaeus*. The stigmas are the female part of the flower. In a good year, each saffron crocus plant might produce several flowers. The only part of the saffron responsible for its therapeutic properties are the flower's three stigmas. With extreme care and planning, saffron will grow in the perfect micro climate on some of the Hawaiian Islands. But this would be a great project in a lanai pot. You won't get rich even if this is the best cash crop – a kilo of excellent saffron can bring $10,000 USD!

Saffron numbers: An acre of saffron flowers will yield only 10 pounds of spice. It takes 75,000 blossoms or 225,000 hand-picked stigmas to make 1 pound! There are 400-500 saffron stigmas to the gram. The stigmas are also called threads, strings, pieces, or strands. 1 gram equals 2 teaspoons whole, 1 teaspoon crumbled, or ½ teaspoon powdered. Don't buy ground saffron as it loses flavor quickly and is usually cut with turmeric.

According to a Greek myth, a handsome mortal named Krocos fell in love with a beautiful wood nymph named Smilax. But Krocus' affection was not returned by Smilax. Some versions tell that the Gods turned Krocus into a beautiful purple crocus flower, and other say Smilax did it to keep him close. Saffron is derived from the Arabic word 'zafaran', which means 'yellow.' Saffron was used to scent the baths and public halls of Imperial Rome. The Romans initially brought saffron to England, but it was lost to them during the Dark Ages. It is claimed today's English saffron comes from a 14th century pilgrim who smuggled one crocus bulb from the Holy Land.

Saffron's native to Western Asia, most likely Persia where Iran and Iraq are today. The crocus was cultivated in ancient Europe. The Mongols took saffron from Persia to India. In ancient time saffron was used medicinally, as well as for food and as a dye. It was cultivated in Spain around 700 AD. Spain now rivals Kashmir for the best saffron. Saffron is also cultivated in India, Turkey, China, and Iran. Saffron is prized for its taste and the way it colors food. In India, saffron orange is considered a beautiful color and is the official color of Buddhist robes.

Iran has over 90% of the world's saffron production, but the very best saffron is from Spain, and is termed 'coupe.' It is the least produced, thus, the most valuable. It is almost impossible to purchase. Mancha is the brand of the best that can be purchased. Its stigmas are deep red. Other Spanish types, Rio and Sierra, have stigmas that are lighter yellow.

HOW TO GROW: In warm climates, the saffron crocus flower grows to about a foot tall with long thin leaves. The blue-violet flowers contain the precious stigmas, delicate threads, each measuring about an inch. Superior saffron is a bright orange-red color. First locate the necessary bulbs from a supplier usually though the Internet. One bulb will cost about $5 USD; plan on buying at least 10. Prepare a spot in your garden protected from strong winds. Work the soil until it easily crumbles. Add sand and rotted compost. The optimum pH is 6-8. Plant the saffron bulbs pointy side up, 4 inches deep and 6 inches apart. Cover with cut veg-etation. They should sprout and can flower in 2-3 months. Commercially grown saffron is produced from corms or bulbs because the plants are sterile and don't produce seeds. The plant dies back once it has flowered. Each crocus bulb produces two to nine flowers per season, and each flower has three long red-orange stigma branches, attached together at the base. Saffron is perfect to grow in a porch or balcony pot.

Only the female stigmas of the flowers are used for the saffron spice. The stigmas are the three bright orange appendages attached to the center of the flower. You need to dry the stigmas in an airy spot away from direct sunlight. When the stigmas crumble easily when crushed, they are adequately dried for storage. Store in cool dark place in airtight containers.

MEDICINAL: Large dosages of saffron can be fatal. It is considered an excellent stomach tonic and helps digestion and increases appetite. It is a fact since antiquity, crocus was attributed to have aphrodisiac properties. Many writers along with Greek mythology sources associate crocus with fertility. People use saffron most commonly for depression, anxiety, Alzheimer disease, menstrual cramps, and premenstrual syndrome (PMS). Crocus in general is an excellent stimulant.

NUTRITION: One gram has 3 calories and is high in potassium, iron, magnesium, and calcium plus vitamins B-6 and C.

FOOD & USES: Saffron crocus becomes commercial saffron when cured properly. Each red stigma has complex chemicals that create the unique aroma, flavor, and yellow dye. *Crocus sativus Linnaeus* contains crocin, the source of its strong coloring property, and bitter-crocin, the cause of the distinctive aroma, taste, and essential oils. It should be kept in an airtight container in the fridge.

Due to its expense, unique taste, and strong dying properties, very little saffron is required for cooking. The trick is to mix the tiny bit of saffron evenly throughout the prepared dish. If too much is used it will overpower the dish and have a medicine taste. Saffron can be crushed to a fine powder in a mortar and pestle. The threads may be toasted in a heavy cast iron skillet over medium heat and then ground with a mortar and pestle before adding to the dish. With soups, or salad dressings crumble the threads and add directly to the dish.

The easiest way to get it thoroughly combined in your recipe is to steep saffron in hot water. Use only a pinch to a cup. Real saffron will expand in the hot water. A steeped cup should flavor a pound of rice. Saffron is used in Mediterranean and Asian dishes. Good yellow rice is made with saffron, not turmeric. It flavors and colors fish and seafood, especially the Spanish paella.

RECIPES: SAFFRON SCALLOPED POTATOES

Ingredients: ½ TS saffron threads, ½ cup whipping cream, 1 cup whole milk, 3 TBS butter, 1 tablespoon oil, 2 large onions sliced thin then chopped small, 2 garlic cloves minced, 6 large white potatoes sliced thin, 1 cup cheddar cheese grated

Method: In a suitable pot combine milk and cream and heat. Remove from heat and add saffron. Allow to steep for 20 minutes. In a frying pan with the butter and oil, sauté the onion and garlic until clear. In a baking dish, layer the potatoes, onion, and garlic. Cover with the milk - saffron mixture. Cover with foil and bake at 350 for an hour. Uncover and bake another 15 minutes.

INDIAN SAFFRON RICE

Ingredients: pinch of powdered saffron, 2 cups boiling water-divided, 2 TBS butter, 1 cup uncooked long-grain white rice - not rinsed, 1 teaspoon salt

Method: Steep the saffron in ½ cup boiling water. In a covered frying pan, melt the butter over medium heat. Stir in the rice and salt. Cook, stirring constantly, until the rice begins to absorb the butter and becomes opaque. Do not brown the rice. Add 1½ cups boiling water along with the saffron water. Cover tightly and simmer for twenty minutes until all the liquid is absorbed. Try not to peak or remove the lid while the rice is cooking.

DID YOU KNOW? All 7 of the world's most expensive spices can grow in the Hawaiian Islands: Saffron - kg: US$11,000, Vanilla - kg: US$440 Cardamom -kg: US$66, Clove - kg: US$22 Cinnamon- kg: US$14 Turmeric - kg: US$6-7 Black Pepper-kg $2.50

SAGE
Salvia officinalis

Poultry seasoning was almost impossible to locate during 2020, so I mixed my own from fresh sage and thyme. I use sage in soups, and especially with chicken. I have three plants in my garden, so I'm never without. Aromatic sage belongs to the mint family. Use it fresh for the best flavor.

Sage is native to countries surrounding the Mediterranean Sea and has been consumed in these regions for thousands of years. Botanically sage is known as *Salvia officinalis*. In medicinal lore, sage has one of the longest histories of use of any herb. The Greeks and Romans highly prized the healing properties of sage. The Romans treated it as sacred and created a special ceremony for gathering sage. Both cultures used it as a preservative for meat, a tradition that continued until the beginning of refrigeration.

Arab physicians in the 10th century believed that it extended life, while 14th-century Europeans used it to protect from witches. Sage was in much demand in China during the 1600s, because the Chinese yearned for sage to make tea. Dutch traders profited when the Chinese would trade three chests of their China tea for one chest of sage leaves.

HOW TO GROW: Throughout Hawaii sage is easy to grow in almost any well-drained soil or window pot in full sun. It can reach three feet tall, but trimming can keep it small. Two or three plants will keep you and friends supplied with this herb. Sage doesn't need much except sun and just enough water, don't overwater. It is a perennial, but I trim mine every year and sprout one or two of the cuttings to keep the flavor fresh. It can be grown from seed, but buy a plant when you can find it. Sage is great for your indoor spice garden and the best pH is 6-7. Never cut more than a quarter of a sage plant. Old leaves are strong flavored and are best for cooking; younger leaves have a lighter taste and are good with egg or cheese dishes. This is a good potted plant for your patio or porch.

MEDICINAL: Sage has the longest history of medicinal use of any herb. More than two millenniums ago, the Chinese were using sage tea as a mild tonic to invigorate the nervous system. The Romans named this herb *salvere*, meaning to save or cure. Sage stimulates the central nervous system, while reducing excess nervous energy. This mild tonic quiets the nerves and helps induce sleep. Sage will fight depression, mental exhaustion, trembling, and nervousness. Smoking sage helps relieve asthma, and sage tea will help fight a fever or a sore throat. Modern studies support its effects as an antibiotic, antifungal, and an astringent. Recent tests have confirmed sage is an outstanding memory enhancer. Sage has been studied to be effective in the management of mild to moderate Alzheimer's disease. This herb is effective as a food preservative because it inhibits the growth of bacteria.

NUTRITION: Sage is a good source of vitamins A and C with 6 calories in 2 grams with some calcium and iron.

FOOD & USES: When cooking, sage should be used in moderation as its flavor can overtake a dish. Sage combines well with rosemary, oregano, and thyme. Sage is a very powerful spice. It is sometimes combined with garlic and green pepper for frying meat. Because of its strong taste, a combination of sage with more delicate herbs does not make much sense.

When preparing any batter for frying for onion rings, chicken, or fish add 2 tablespoons of fresh sage for a unique and delicious taste. Sage goes with eggplant, pumpkin, tomato, and long bean.

Herbs are the friend of the physician and the pride of cooks. ~ Charlemagne

RECIPES: HICKAM FIELD FRESH HERB BISCUITS

Ingredients: 2 cups bakers' flour, ¼ cup mayonnaise, 1 TBS fresh sage minced, 1 TBS fresh thyme minced, 1 cup milk, ½ TS salt

Method: In a large bowl combine all ingredients. Use a large spoon, and carefully place the spoonfuls on a greased baking sheet. Bake fifteen minutes at 400 degrees.

SPRECKELSVILLE SQUASH SOUP

Ingredients: 4 pounds squash/pumpkin - peeled, seeded, and cubed, 1 medium onion chopped, 2 garlic cloves minced, ½ hot pepper seeded and minced, 1 TBS butter, 6 nice sage leaves minced or 2 TBS ground sage, 1 TBS brown sugar, 4 cups chicken broth, 1 TS salt

Method: In a large pot brown the onion and garlic in the butter. Add sugar, squash cubes, sage, salt and the pepper. Add the chicken broth, bring to a boil and simmer for 45 minutes.

SAGE PESTO - This is a great flavor to just blend with noodles, potato, or rice dishes. Ingredients: 2 cups sage leaves minced, 2 cups shelled cashews/almonds, ½ cup oil (prefer olive oil), 2 garlic cloves minced, 1 bunch chives chopped extremely small, ½ hot pepper seeded and minced (optional), ½ TS salt

Method: Place all ingredients in a blender and puree. Use a tablespoon with noodles or rice to your taste. Slowly add more to increase sharpness.

MOMMY'S EASY PUMPKIN SOUP

Ingredients: 1 small pumpkin peeled and cubed, 3 cups chicken stock, ¼ TS ground nutmeg, ½ TS ground sage, salt, and spice to taste

Method: In a suitable pot, add pumpkin to chicken broth and bring to a boil. Add remaining ingredients and simmer, stirring occasionally until pumpkin is tender and most water boils away. 4 tablespoons more water may be added to change consistency.

QUOTES OF SAGE'S MEDICINAL ABILITIES

Sage helps the nerves and by its powerful might palsy is cured and fever put to flight. ~ French saying.

DID YOU KNOW?

To hide gray hair put ½ cup of dried sage in 2 cups of water and sim-mer for half an hour. Cover and allow the sage water to steep overnight. Over a basin, pour the rinse over your hair at least 10 times catching and reusing the same sage water. On the last rins-ing, leave the sage water dry on the hair before rinsing with fresh water. Repeat every week until your hair is the desired shade. Rinse every month to retain your hair color. (Pioneer Thinking)

If one consults enough herbals... every sickness known to humanity will be listed as being cured by sage. ~ Varro Taylor Ph.D.

STAR ANISE
Illicium verum

Star anise grows easily in the warm climate of the Hawaiian Islands. This star shaped spice dates back 3000 years and is native to southern China and Southeast Asia. It's a fast grow-ing evergreen tree that can grow to 25 feet (6.6 m.) but is usually trimmed shorter to 10 ft. high and wide (3 m.) for easier picking. The reddish-brown fruit pods (not seeds) are an 8-pointed star that smells like licorice. It's used in Chinese cooking and in the five-spice powder mix. Historically in traditional Oriental medicine, star anise based teas and tonics are intended to improve digestion and protect against infection.

In 1578, star anise was introduced to Europe. Thomas Cavendish, a British sailor, car-ried it from Philippines. The spice's licorice flavor was used in puddings and fruit preserves. China produces 80% of world's star anise.

Star anise is often confused with aniseed, *Pimpinella anisum* - a relative of fennel. Both have a sweet, black licorice-like scent. Essential oils from both seeds contain the compound *anethole.* Herbal anise has been cultivated in Egypt for over 4,000 years and was brought to Europe for medicinal purposes. It is widely cultivated and used to flavor food, candy, and alcoholic drinks, especially around the Mediterranean. The main difference between anise and star anise is that anise seed is potent, with an almost spicy flavor, while star anise is subtly milder. Star anise fruit is used to make a highly fragrant oil that works well in cooking, making homemade pesticides, and body care products.

PLEASE TAKE NOTE: This article is about *Illicium verum*, and not other star anise plants that are toxic. Japanese star anise *(Illicium anisatum)* and swamp star anise *(Illicium parviflorum)* have the same common name but are highly toxic. The whole plant is toxic, es-pecially the fruit, which contains the poisonous anisatin. The 'good' non-poisonous star anise, *Illicium verum,* does have white or yellow blossoms. Japanese star anise *(Illicium anisatum)* is a similar tree that doesn't have the same pod shape as Chinese star anise. Swamp star anise *(Illi-cium parvi lorum)* has white flowers also and is more of a bush or a smaller tree. When buying a star anise tree be certain it is an *Illicium verum* and not a similar tree.

HOW TO GROW: Star anise is an evergreen tree that has large glossy green foliage, and white to yellow twisting blossoms about 1½ inches wide. Your tree should bloom in 5-7 years. The fruit pod has eight horns, each containing fruit. This tree won't grow where the temperatures drop below 15 F (-10 C). It will grow in almost any well-draining soil but the best type is rich and loamy with a pH of 6-7.5. It needs regular watering. In warmer climates, growing star anise in full shade is a possibility. Well-rotted manure is a suitable fertilizer. This is a fast growing tree that should be kept short. It has a spicy fragrance when it is trimmed and can be kept as a hedge. The star shaped fruit with 6 to 8 carpels (each contains a seed) is brown when ripe. Fruits are picked when still green and dried in the sun.

MEDICINAL: Ayurveda and traditional Chinese medicines use star anise to treat di-gestive ailments like stomach pain, infections, constipation, and indigestion. Star anise con-tains plenty of fiber - over 1 gram per tablespoon, which stimulates the digestive track and helps reduce constipation. Star anise seeds are used to produce sweet licorice scented oils that contain powerful ingredients: trans-anethole, limonene, caryophyllene, estragole, and linalo-ol. All possess antibacterial and anti-inflammatory properties. These compounds are used in aromatherapy and to combat respiratory infections. Star anise tea tonics stop the growth of dangerous fungi and bacteria, relieve swelling, muscle and joint pain, including rheumatism and arthritis. This tea also has a calming effect, relieving stress and anxiety. Star anise is a traditional remedy to normalize blood sugar levels, and improve blood circulation while lowering overall pressure. Star anise is also an effective sedative.

NUTRITION: 100 grams of star anise has 340 calories with 18 grams of protein and is packed with potassium, calcium, magnesium, and iron plus vitamins B-6 and C.

FOOD & USES: Star anise essential oils have a sweet scent and are used in aromatherapy, perfumes, toothpaste and cosmetics. This spice is also used in meat and poultry dishes as well as con-fections. It is one of the main ingredients in the traditional Chinese seasoning, five spice. The star-shaped fruit is still used today for mouthwashes, toothpaste, and body care products.

Insects do not like anethole, which is a major ingredient of star anise essential oil. Studies show anethole is 80% effective as a fumigant against German cockroaches. Such oils have also been used as all-natural pesticides protecting grains and other crops.

RECIPES: STAR ANISE MILK

Ingredients: 1 $^1/_2$ cups whole milk, 1 TS molasses, 2 whole star anise

Method: In a pot on low to medium heat warm the milk, molasses, and star anise stirring so it doesn't burn. Bring to a boil and serve hot.

CHAI SPICED TEA

Ingredients: 6 cardamom pods, 1/2 TS whole peppercorns, 5 cups water, 1/4 cup honey, 2 cin-namon sticks (3 inches), 8 whole cloves, 3 whole star anise, 1 TBS fresh ginger root minced, 5 black tea bags, 2 cups whole milk.1 TBS vanilla extract, ground nutmeg, optional

Method: In a mortar, pound/grind the cardamom pods and peppercorns into a paste. Add this mixture to the 5 cups of boiling water with honey, cloves, cinnamon, star anise and ginger. Simmer at least for 5 minutes, longer for a stronger flavor. Remove from heat and add tea bags, cover and let sit for at least 5 minutes. Strain and remove tea bags. Heat the milk in another larger pan and stir in the strained tea spice mixture. Add the vanilla. Top with ground nutmeg if desired.

CHINESE 5 SPICE POWDER

Ingredients: 6 whole star anise pods, 1½ TS whole cloves, 1 cinnamon stick (3 in.), 2 TBS fennel seeds, 1 TBS black peppercorns

Method: In a dry heated skillet lightly toast all spices for 2-3 minutes continually stirring so they do not burn. Place in a spice or coffee grinder or mortar and pestle and grind until smooth. Makes ¼ cup. Store in an airtight jar.

DID YOU KNOW? Star anise is 10 times sweeter than sugar because of the compound anethole. A large pharmaceutical company used to source star anise fruit for a popular flu medication and they consumed almost 90% of the world's crops. They have since switched to a bacterial source that contains the same compound. The seed pod's intensely sweet and warm licorice flavor is believed to have warming capabilities within the body. This is important in Traditional Chinese Medicine, or TCM, which uses warming and cooling ingredients to balance the qi or chi of the body, which is comprised of the cool yin and warm yang. These two must be in harmony for the body to function properly.

NAMES: Chinese anise, eight-horned anise, badiane

I grow plants for many reasons: to please my eye or to please my soul, to challenge the elements or to challenge my patience, for novelty or for nostalgia, but mostly for the joy in seeing them grow. ~ David Hobson

TARRAGON

Artemisia dracunculu

Tarragon is a small shrub, a perennial herb from the same family as sunflowers. It isn't seen much in the Hawaiian Islands, but garden shops have plants and the dried herb is available online. Two types are grown, the French variety with glossy sharp licorice smelling leaves and the much blander Russian type. Most dried tarragon comes from French tarragon. Only the leaves are edible. This herb is excellent with seafood, fruits, poultry, eggs, and most vegetables, as well as sauces - particularly béarnaise sauce. This herb can grow anywhere in Hawaii and great for a lanai or window. Mr. Google can help you find seeds or plants.

Native to remote areas of China and Russia, tarragon is believed to have been brought to Europe by the invading Mongols in the 13th century. Today, its primary producer is France. Tarragon is relatively new on the world's herbal scene. Unlike many herbs, it was not used by ancient peoples. Tarragon's name is from the French *esdragon*, which means 'little dragon' because of its winding root system. The tangled roots will strangle the plant if not often divid-ed. Botanically tarragon is *Artemisia dracunculu.*

HOW TO GROW: Tarragon is a delicate, green perennial herb with long, thin pointy leaves with an aromatic, distinct flavor with just a hint of anise. The French call French tarragon the 'King of Herbs'. When planting, be absolutely certain it is French tarragon, not inferior Russian tarragon, which is a completely different species. Like everything, plants can be bought from the Internet. French tarragon rarely, if ever makes seed, and so it must be grown from cuttings. This herb can grow to two feet across. To grow indoors it must have a good-sized, deep pot because its roots need plenty of room. It is best planted in sandy soil, pH 6.5-7.5 with the best possible drainage. Clay soils are bad for tarragon. Put gravel at the bottom of the pot before adding the soil to facilitate drainage. Keep this herb pruned so the plant is open to the breeze and try not to let it touch any other plant. Harvest tarragon when ready to cook with it Cut about a third of a branch, then chop the leaves fine, to fully release the full flavor.

MEDICINAL: French and the Russian tarragon are both well known to herbalists for their curative properties. Tarragon increases appetite, improves circulation of blood, and helps in the proper distribution of nutrients, oxygen, hormones, and enzymes throughout the body. It stimulates the brain, nervous, digestive, circulatory, and endocrine systems. This in turn stimulates the whole metabolic system and as a result, growth and immunity are stimu-lated.

NUTRITION: Tarragon is rich in iodine, minerals, and vitamins A and C, and was used to prevent scurvy. One teaspoon of ground tarragon has 14 calories with plenty of po-tassium.

FOOD & USES: Tarragon makes an excellent flavored vinegar, and an excellent herb butter, alone or in combination with other fine herbs. Tarragon vinegar is easy to make. Put fresh tarragon sprigs into a sterilized bottle of distilled white vinegar. Taste after a few days. Continue steeping until it suits your taste. Once the taste is strong enough remove the sprigs. Tarragon is also a good herb to use in infused oils. Heat greatly intensifies the flavor of tarragon, both fresh and dried. Care should be taken when using tarragon or it will overpower other flavors.

Tarragon is the main ingredient in Béarnaise Sauce and the French favorite herb mixture, 'fines herbes'. It is used with chicken, fish, and egg dishes. Tarragon flavors a popular carbonated soft drink in the eastern European countries of Armenia, Georgia, Russia, and Ukraine. The drink is made from sweetened tarragon concentrate and colored bright green. Commercially the fragrance is a component in soaps and cosmetics.

A half-ounce of fresh tarragon equals a third of a cup. One tablespoon of fresh tarragon equals one teaspoon of dried.

RECIPES: HERB OR SPICE INFUSED OIL

Ingredients: Favorite herbs and/or spices prefer sunflower, safflower, or extra-virgin olive oil
Method: Wash and dry your choice of herb branches and lightly bruise them to release flavor. Place them in a clean glass container that seals tightly. Cover with warm oil, and seal. Leave in a cool, dark place for 10 days or longer. If not strong enough for your taste add more herbs and remain sealed. If you do not strain the herbs out, the flavor will become stronger the longer it sits. If you infuse olive oil keep it refrigerated. Consider combinations with onion and garlic and a variety of herbs as rosemary, thyme, basil, tarragon, summer savory, oregano, culantro, marjoram, chives, dill, mint, parsley, and bay leaf. Use within 2 months and keep refrigerated.

TARRAGON SALAD DRESSING

Ingredients: 2 TBS fresh tarragon chopped fine or 1 TS dried, 1 TBS finely chopped parsley, 1 garlic clove minced, ½ cup sour cream, ½ cup mayonnaise, 1 TBS fresh lemon juice, salt to taste
Method: In a bowl combine tarragon, parsley, and garlic. Add sour cream, mayonnaise, lemon juice, and salt. Blend thoroughly.

FINE HERBS MIX

Ingredients: 1 TBS tarragon chopped, 1 TBS chives chopped, 1 TBS parsley chopped, 1 TBS chervil (French parsley) chopped
Method: Fresh Herbs: Combine tarragon, chervil, chives, and parsley. Add this mix at the end of the cooking process of any recipe to preserve their flavor.
Dried Herbs: Combine the herbs. Place in a glass jar and seal tightly. Refrigerate.

MY FAVORITE WAIKIKI HOTEL'S HERB MUSTARD SAUCE

Ingredients: ¼ cup unsweetened evaporated milk (must be chilled in the freezer for ten minutes), ¼ up tarragon vinegar, 2 TBS grainy Dijon mustard, ½ cup vegetable or olive oil, 2 TBS fresh tarragon chopped, 1 bunch chives chopped fine, salt to taste
Method: Use a blender, combine tarragon, vinegar, and Dijon mustard. Add chilled evaporated milk. Blender running, slowly pour in oil. Add chopped fresh tarragon, chives, and salt. Blend.

KONA EXQUISITE REFRIED POTATOES

Ingredients: 2 cups Irish potatoes cubed, oil for frying, 4 TBS butter or margarine, 1 TS tarragon leaves minced, 1 garlic clove minced, salt to taste
Method: Fry potatoes in oil until the color just begins to brown. Then melt butter in another skillet and add garlic and tarragon. Then refry the potato cubes until golden brown. Serve hot.

DID YOU KNOW? T a r r a g o n was considered 'the banishing herb'. Burn dried leaves while writing on paper what you want to banish from your life (bad habits or people). Then burn the paper with the remaining smoldering herb. It was also known to put guests at ease and make them feel welcome. It was carried in packet charms or sachets for love, peace, and good luck.

THYME
Thymus vulgaris

Fresh thyme is constantly used in my kitchen to flavor soups, stews, and casseroles. I use it with many meat and fish dishes. Thyme has a sharp smell that's spicy, but also very earthy. It has a warm flavor that remains after the meal and seems to freshen the breath with a faint clove aftertaste. It ranks as one of the fine herbs of French cuisine.

This is another herb first cultivated around the Mediterranean Sea. The Egyptians used it to embalm their mummies, and the Greeks used it smoldering to chase insects from their residences. In fact, that is where the name thyme originates. In Greek, 'thymon' translates 'to fumigate.' The Romans used thyme to make alcohol. The botanical name is *Thymus vulgaris*.

Although there are over 100 varieties of thyme, only common garden thyme and lemon thyme are used in cooking. The remainder are used as ornamental plants. One broad leaf variety is known locally as Spanish thyme. Garden thyme is a small perennial - once planted it's there as long as you care for it. Its appearance depends on soil type, care, and amount of sun, and water. Regular thyme is stiff and bushy with many thin, erect stalks covered with pairs of small, narrow dull green leaves. The blossoms are usually pale pink and loved by honey bees. Lemon thyme is smaller with bright green leaves that have a slight lemony taste.

HOW TO GROW: Thyme will grow almost anywhere with little effort as long as it is planted in well-drained soil with at least six hours of sun. It's usually planted from a cutting of a mature plant stuck in a pot of damp soil, but you can grow it from seeds. As long as you don't over water or allow it to dry out, thyme should catch in 2-3 weeks. Once it is about six inches tall it can be transplanted to your herb garden or to a larger pot. This is also a great plant for a box at your window. Thyme seems to be almost dry when picked.

I treat thyme like sage, oregano, and rosemary. I put it in a brown paper bag in indirect sunlight for a week. Then I force it through a strainer and store in a sealed bottle in the fridge. It is preferable to strip the leaves from the stems for your recipes when using either dry or fresh thyme because sometimes the stems can be woody.

MEDICINAL: In the past, nightmares were treated with thyme tea. Thyme oil was used during World War I to treat infection and to help relieve pain. Small amounts of this herb are sedative, but larger amounts are stimulant. Thyme is used against hookworm, roundworms, and threadworms. Thyme also warms and stimulates the lungs, expels mucus, and relieves congestion such as asthma. It also helps deter bacterial, fungal, and viral infections. Thyme has always been used as a poultice for wounds, insect bites, and stings. It is a good wash for sore eyes, and a hair rinse to fight dandruff.

NUTRITION: You'll only use a few teaspoons of thyme at a time. One gram of thyme has 1 calorie with higher than usual protein for a leafy herb; 100 grams of thyme has 6 grams of protein and high amounts of potassium, magnesium, and calcium. Thyme's an excellent source of vitamin C.

FOOD & USES: Thyme is traditional in bouquet garni combined with marjoram, parsley, and bay. Thyme works well with lamb, beef, poultry, fish, stews, soups, and bean and lentil casseroles. Use it with tomatoes, onions, cucumbers, carrots, eggplant, parsnips, leeks, mushrooms, asparagus, green beans, broccoli, sweet peppers, potatoes, spinach, corn, peas, cheese, eggs, and rice. Its flavor blends well with those of lemon, garlic, and basil. Store fresh thyme in a plastic bag in the vegetable crisper drawer of your refrigerator or stand sprigs in a glass of water on the refrigerator shelf. When cooking with thyme, be aware that one fresh sprig equals the flavoring power of one-half teaspoon of dried thyme. As with most leafy dried herbs, be sure to crush the leaves between your hands before adding them to your recipe. Dried thyme should be stored in a cool, dark place, in an airtight container for no more than six months.

RECIPES: RAINBOW BAY MARINA SPECIAL SHRIMP STEW

Ingredients: 2 lbs. (1 kg) fresh medium shrimp - peeled, cleaned, and boiled - save the water, 1 pound white potatoes - peeled and cubed, 1 lb. sweet potatoes - peeled and cubed, 3 cups of chopped onion, 1 cup of chopped chives, ¼ cup celery chopped, 4 medium ripe tomatoes chopped, four culantro leaves chopped, 6 garlic cloves minced, 1/2 cup vegetable oil, ½ cup flour, 2 TS salt, 1 hot pepper seeded and minced (optional), 1 TBS fresh or dried thyme, 3 bay leaves, 1 TS fresh lemon juice

Method: In a large pot heat the oil and whisk in a little flour at a time until all becomes a rich light brown color. Add chopped onion, chives, culantro, celery, and garlic. Keep stirring over low heat for 10 minutes. Add salt, pepper, bay leaves, and thyme. Mix well. Add the chopped tomatoes, 2 cups of the shrimp water, and lemon juice. Simmer for 1 hour with regular stirrings. Add the cubed potatoes and simmer for 10 minutes longer. Add the shrimp, cover, and simmer for half an hour. Serve warm with roti or fresh, crusty bread.

TROPICAL BIRD

Ingredients: 2 lbs. (1 kg) turkey chopped or chicken, 1 medium onion chopped small, ½ hot chili pepper seeded and minced (optional), juice from 1 fresh orange, 1 TBS corn starch, 2 TBS butter or margarine, 1 cup bread crumbs, 1 TS fresh or dried sage, 1 TS fresh or dried thyme crushed, salt to taste, 2 cups brown sugar

Method: Combine bread crumbs, sage, thyme, onion, salt, and pepper. Roll turkey or chicken pieces in crumbs. Place in a baking dish and cook in the oven at 400 degrees for 45 minutes. Com-bine orange juice and cornstarch and set aside. In a frying pan melt butter and stir in brown sugar until it melts. Stir in orange juice/cornstarch. Pour over baked poultry pieces. Serve with boiled banana.

FRENCH CREOLE SPICE MIX

Ingredients: 1/3 cup sweet paprika, 1/3 cup dried basil, 1/3 cup dried thyme, 3 TBS cayenne pepper, 2 TBS chili powder

Method: Combine paprika, basil, thyme, cayenne, and chili powder in a glass sealable container. Shake until well-combined. Best stored in the fridge.

GREEN BEANS AND FRESH HERBS

Ingredients: 2 lbs. (1 kg) fresh, canned or frozen green beans, 1 onion chopped small, 1 TS fresh parsley chopped, 1TS fresh thyme leaves chopped, 1TS lemon juice, ¼ cup butter, salt to taste

Method: Wash beans and chop into ½ inch pieces. Steam beans about 15 minutes and drain. In a skillet melt the butter and add the onion, lemon juice, parsley, thyme, and salt. Simmer for 5 minutes, stir in the green bean pieces. Serve warm.

DID YOU KNOW? The ancient Greeks used thyme in their baths and burnt it as incense in their temples, believing it was a source of courage. The Romans believed eating thyme either before or during a meal would protect you from poison. Emperors kept thyme close by. It was believed a bath in warm water with thyme could stop the effects of poison after it was inadvertently consumed.

TURMERIC – OLENA

Curcuma longa

Turmeric has the local Hawaiian name, olena. Turmeric, like ginger, is a root and not a flower. It will grow almost anywhere and continues annually. This root is another great addition to a home garden. The broad leaves make an excellent border, it's great for cooking and health.

Turmeric is also known as curcumin, and native to Southern Asia. It is a close relative to ginger. For more than 5,000 years, this root was used as a dye and a cooking spice in India. In medieval Europe, turmeric became known as 'Indian saffron' and used as an inexpensive substitute for real crocus saffron. Turmeric is one of the key ingredients in many curries, providing them both color and flavor. The root and rhizome (underground stem) of the turmeric plant are used medicinally. Turmeric's scientific name is *Curcuma longa.* This is one of the best things you can grow at home for taste, coloring, and health.

HOW TO GROW: I believe turmeric is such an important herb I have transplanted a row of turmeric along the fence in my backyard. It's easy to grow after you find some seed roots. Just plant it where it won't be too damp and leave it alone except for an occasional weeding. It's best to loosen the soil with a fork and work in lots of compost and rotted manure so the seed roots can expand. Turmeric thrives in well-drained sandy loam soils with a pH range of 4.5-7.5. Roots can be planted 6 inches apart. The bigger the seed root used, the greater the return you'll receive. Plant in full sun and water regularly. When the leaves fully wither, dig the roots, and replant some of the small 'knobs' attached. An attractive plant, ours grows about two feet tall, and turmeric has a large yellow and white flower spike surrounded by long leaves. The roots form bulbs and seed - rhizomes. Turmeric, like ginger, is perfect for porch or balcony pots.

Turmeric is famous for its roots' color, pale tan to yellow on the outside, but bright orange on the inside. It is used to color and flavor mustard, cheese, butter, pickles, relish, chut-neys, rice, and an important ingredient in curry powder. Turmeric's flavor resembles a combi-nation of ginger and pepper. Turmeric is a powerful coloring agent. Occasionally shredded and used fresh, turmeric is more often dried and powdered for use. The roots are boiled for hours, dried for days or weeks, and then ground into powder.

MEDICINAL: Turmeric keeps viruses from replicating, and kills staphylococcus and salmonella. It aids the digestive system. It's a blood purifier and regulates blood sugar. It corrects anemia and restores poor circulation. Turmeric removes oxidized cholesterol to prevent heart attacks. It keeps muscles and joints flexible and strong. Turmeric removes accumulation of cholesterol in the liver and promotes a healthy circulatory system. It is also a great stomach tonic for gas or indigestion. East Indian women with lovely, velvety skin often attribute it to consuming turmeric. It is a natural anti-inflammatory.

NUTRITION: Turmeric, per 100 grams, has 350 calories, with 8 grams of protein and 60 grams of carbs with 21 grams of fiber. This root is extremely rich in potassium and iron with magnesium and calcium. It has plenty of vitamins C and B-6.

FOOD & USES: Turmeric tea, pour 8 ounces of boiling water over a half-teaspoon grated or powdered turmeric, and let sit covered for 5 minutes, then strain. Drink two or three cups daily, as desired. Mix turmeric powder with cooking oil to make a thick paste. Put on the skin over wounds, bites, bruises, etc. Cover with a bandage and leave on for several hours. It washes off and the color disappears quickly. If you have a toothache, paint this on your face over the toothache.

RECIPES: EASY-EASY TURMERIC VEGGIE STEW – a surfers' delight

Ingredients: 1 squash seeded - peeled and chunked, 2 cups eggplant chunked with or with-out

skin, 1 cup lady fingers, 2 large tomatoes chopped small, 2 potatoes peeled and chopped small, 1/4 cup ketchup or tomato paste, 1 large onion chopped small, 1 carrot chopped small, 3 cloves garlic minced, 1 cup water, 1 whole hot pepper seeded (optional), 1 TS turmeric powder, 1/2 TS cumin powder, 1/2 TS cinnamon powder, salt and spices to taste
Method: In a large pot with a cover, combine all ingredients. Simmer for at least four hours. A slow cooker or crockpot is perfect for this dish. Serve with rice or pasta.

YELLOW VEGGIE CAKES

Ingredients: 1 carrot chopped small, 1/2 cup pigeon peas, 1 stalk celery chopped small, 1.5 cups potatoes - peeled, boiled, and mashed, 1/2 cup all-purpose flour, 1 hot pepper seeded and minced (optional), 2 leaves cilantro chopped small, 1 TS ginger root minced, 1 TS turmeric powder/, 1/2 TS salt, oil for frying
Method: Mix all ingredients together and roll into balls, about an inch to two inches in diameter. The bigger will take longer to fry. Drop into hot oil and fry till brown. dip in spicy sauce.

OLENA POMATOES

Ingredients: 4 potatoes peeled, chopped small, 1 stalk celery chopped, 1 medium onion chopped small, 2 medium ripe tomatoes chopped small, 2 cloves garlic minced, 1/4 cup oil, 1/2 hot pepper seeded and minced (optional), 1 TS turmeric powder, half TS cumin powder, 1/2 TS salt Method: In a frying pan with a cover, brown the onion and garlic in oil. Blend in all the spices, celery, and then add potato cubes. Cook for ten to fifteen minutes stirring intermittently. Add tomatoes and cook for another ten minutes.

KAANAPALI BEACH YELLOW SCRAMBLED EGGS AND POTATOES

Ingredients: 2 large potatoes either baked or boiled - peeled and cubed, 4 eggs, 1 small onion minced, 2 TBS evaporated milk, 1 hot pepper seeded and minced, 1 TS turmeric powder, 1/2 TS cumin powder, salt and other spices to your taste, oil, butter, or margarine for frying
Method: In a skillet heat 2 TBS butter or oil, lightly brown onion, and fry potato cubes. Add all spices Blend eggs and milk together. Add egg mixture and scramble. Serve with roti.

TOMATO CHICKEN

Ingredients: 1 whole chicken chunked (skinned if you like it so), 4 large tomatoes chopped small, 1 large onion chopped small, 3 cloves garlic minced, 1 hot pepper seeded and minced, 1/2 TS ginger root minced, 1 TBS oil, 1 TS turmeric powder, 2 TS cumin powder, 1/4 cup water, 2 bay leaves, 1/2 TS cinnamon powder, 1/2 TS nutmeg, salt, and other spices to your taste.
Method: Heat oil in a large frying pan with a cover and brown onion and garlic. Add spices including turmeric and cook for about two minutes. Add chicken chunks and cover with spice-onion mixture. Add tomato pieces and a quarter cup of water. Lower heat and simmer covered for up to two hours or until chicken is extremely tender. For thicker sauce uncover for last half hour.

DID YOU KNOW? As early as 3000 B.C. the turmeric plants were cultivated by a civilization in Punjab Pakistan. Kunkuma is a red powder made of turmeric and lime and worn by Hindus as the pottu, dot, at the point of the third eye on the forehead. 1280, Marco Polo mentioned turmeric in his travels in China: 'There is also a vegetable that has all the properties of true saffron, as well as the smell and the color, and yet it is not really saffron.' India is the world's largest producer and consumer of turmeric powder.

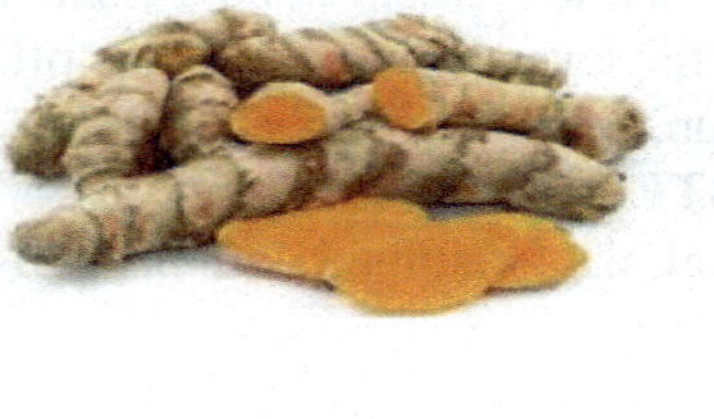

India has the lowest dementia rates in the world; they consume 25 to 50mg of turmeric every day which prevents inflammation of the brain.

Thetruthaboutcancer.com

VANILLA – WANILA
Vanilla planifolia

Vanilla is the only edible fruit of the orchid family, the largest family of flowering plants in the world. It's a tropical orchid, *Vanilla planifolia* (also known as fragrans) originally cultivated around the Vera Cruz area of Mexico. Another genus, the *Vanilla tahitensis*, cultivated in Tahiti, produces beans with stronger aroma, but less flavor. Like everything else these days, vanilla orchids are available anywhere from the Internet.

Hawaii is the only state where vanilla is commercially cultivated. Vanilla production is a perfect island venture and typical of the islands growing valuable crops in a small area. It also has potential for agro-tourism and is an excellent combination with macadamia nuts.

Only saffron and cardamom, are more expensive as spices than vanilla. Vanilla beans retail in some specialty shops for $2-3 US each. Like oil, the prices of valuable spices fluctuate. A good crop, coupled with decreased demand caused by the production of imitation vanilla, drops the market price. Some 1,800 tons were projected for natural vanilla for 2021. Around 65% comes from the northeastern region of Madagascar. Small-scale producers can't negotiate the price of their vanilla harvest. To produce just 1kg of vanilla beans, they must grow 600 orchid blossoms. Freshly plucked vanilla beans begin fermenting immediately and must be sold before they rot to middlemen for less than $50 US a kg. The real profit for natural vanilla at $875 US comes when beans are sold to curing facilities and exporters of vanilla around the world.

Vanilla is so valuable that in Madagascar vanilla theft was a major problem. Growers mark their beans with pin pricks before harvest for identification. Vanilla is expensive because it is the world's most labor-intensive agricultural crop. It takes 3 years after the vines are planted before the first flowers appear. The pods must be picked by hand 4-6 months after the fruit appears on the vines. The fruit is not permitted to fully-ripen, which would cause the beans to split, and reduce value. The green beans are then soaked in hot water, rolled in blankets to 'sweat' evaporating the water. Then they are stored in a ventilated room to slowly ferment and produce their unique aroma and flavor. They develop flavor and fragrance during this curing.

Then the beans are sorted for size and quality. Then they rest for a month or two to finish developing their full flavor and fragrance. By the time they are shipped around the world, their aroma is remarkable. The delicious flavor comes from the seed pod, or the 'bean' of the vanilla plant. The resulting dark brown vanilla bean is usually 7-9 inches long, weighs about 5 grams and yields about half a teaspoon of seeds.

Vanilla is native to Mexico (where it is still grown commercially), originally cultivated by the Totonaca tribe. They and vanilla were conquered by the Aztecs who in turn were conquered by the Spanish. Vanilla hit the big time when Cortez returned to Spain. It was valuable for both its fragrance and flavor. The term French vanilla is not a type of vanilla, but is often used to designate preparations that have a strong vanilla aroma and contain vanilla grains.

HOW TO GROW: Growing vanilla beans is time consuming and expensive. The first problem is to locate the proper orchid cuttings. Forget about trying from seeds. The flowers are quite large and attractive with white, green, greenish yellow or cream colors. Vanilla blossoms grow in bunches and open one by one, and only last about a day. Each flower opens in the morning and closes late in the afternoon on the same day, never to re-open. Only then, it's time to hand pollinate and to be scientific. Open the lip of the flower, take the pollen from the anther, and place it in the nectar, which is located in the stigma. (Look for a YouTube video before beginning.) Be careful because the white sap emitted by the plant can irritate your skin. After pollination, long green pods will begin growing. Vanilla flowers can only be naturally pollinated by a specific bee found only in Mexico. This rare bee graced Mexico with a 300-year monopoly on vanilla production. In 1841, a teenage French slave on Reunion Island discov-ered the plant could be hand pollinated with a thin piece of bamboo replicating the bees. This opened the global vanilla industry.

MEDICINAL: Vanilla is attributed with anticancer, anti-inflammatory, antibiotic, and high antioxidant properties. The vanilla aroma will soothe crying babies, calm adults, and treat insomnia and sleep-apnea. Vanilla as a substitute for sugar can reduce high blood glucose lev-els.

NUTRITION: A vanilla pod has 1 calorie with potassium, calcium, and magnesium.

FOOD & USES: Vanilla orchids are grown in tropic climates, primarily Mexico, Tahiti, Madagascar, Reunion, Mauritius, Comoro, Indonesia, Uganda, Sri Lanka, and Tongo, with most of the world's supply from Madagascar. Quality and aroma vary with location. The biggest buyer of vanilla is the United States. 97% of vanilla used is synthetic. The dairy industry uses a large percentage of the world's vanilla in ice creams, yogurt, and other flavored dairy products. Flavor chemists explain that artificial vanilla flavor is made from synthetic vanillin, which is usually synthesized from compounds found in clove oil, wood, and bark.

One-quarter teaspoon should be enough to flavor a recipe for 4 to 5 persons. Pure vanilla extract should have no sugar added and will last forever, aging like fine liquor. Vanilla extract is made by percolating alcohol and water through chopped, cured beans, somewhat like making coffee. Pure vanilla extract must contain 13.35 ounces of vanilla beans per gallon during extraction and 35% alcohol. Imitation vanilla is made from artificial flavorings, most of which come from wood byproducts and often contain chemicals. Twice as much imitation vanilla flavoring is required to match the strength of pure vanilla extract. The best rule for sub-stitutions is 1 teaspoon of vanilla extract equals 1 inch of a vanilla bean.

Choose nice fat vanilla beans with a thin skin for the most seeds. The pods should be dark brown, and just soft enough to wrap around your finger without breaking. To use the vanilla bean first split it lengthwise with a sharp knife. Scrape the seeds from the pod. A good way to store whole vanilla beans is to bury them in sugar. After two weeks the sugar will take the taste of the vanilla beans. This can flavor coffee and deserts. The beans can be removed and returned to the sugar jar. Keep refilling the sugar.

RECIPES: HOMEMADE VANILLA EXTRACT

Ingredients: 2 vanilla beans sliced open, 1 bottle vodka

Method: Place the beans in a 1 cup jar with a glass sealing top. Fill with vodka. Seal and put in a dark spot for 2 weeks. Replace what you use with more vodka until the beans are expired.

HAMAKUA COAST VANILLA PUDDING

Ingredients: 1/3 cup sugar, 3 TBS cornstarch, ¼ TS salt, 2½ cups milk, 1½ TS vanilla extract

Method: Combine sugar, cornstarch, and salt; then blend in milk. Cook over medium heat, stirring constantly, till mixture thickens. Cook 2 or 3 minutes more. Add vanilla. Pour into suitable bowls and chill until firm.

VANILLA RICE

Ingredients: 2 cups water, ½ TS salt, 1 TS sugar, 1/2 TS powdered cayenne pepper, 1 cup long-grain rice, ½ TS vanilla extract.

Method: Boil water in a suitable pan. Add the salt, sugar, cayenne pepper, and rice, and stir. Simmer until the rice is tender and the liquid has been absorbed. Remove from heat, sprinkle with the vanilla, and fluff the rice with a fork to separate the grains.

A garden is a delight to the eye and a solace for the soul. ~ Sadi

WATER HYSSOP – AEAE

Bacopa monnieri

Water hyssop, known as locally as 'aeae,' is an herb that likes moist areas and grows in muddy coastal areas. Its attractive flowers make it suitable for a potted house plant. Botanists believe this variety of *Bacopa* is native to eastern India, but now this plant is found naturally in wet marshlands throughout Hawaii, S.E. Asia, Central, and South America. It is a species of ground cover that grows into a mat that creeps across the ground, rooting as it advances from the nodes on its stem. Water hyssop can be found growing wild in most wet areas. Look for it growing in marshes, along brackish streams, almost anywhere that's moist. This is excellent and healthy ground cover. Seeds are available online.

The small white flowers can be kept in bloom almost constantly. Consuming water hyssop regularly is beneficial to the nervous system. Gotu kola (*Centella asiatica*), water hys-sop (*Bacopa monneiri*), and sweet flag (*Acorus calamus*), are considered memory-enhancing plants.

HOW TO GROW: Water hyssop is a short duration annual herb usually grown by stem cuttings. If you have seeds, it's best to plant about ½ inch deep in a nursery area or a big pot in the full sun. It is important to always keep the soil moist. They should germinate within 2 weeks. You can probably find plants near streams or rivers and in wetlands up to elevations of 1300 meters. Pull the roots and plant them or place 6-inch cuttings in a tall glass of water. Change the water daily and within 2 weeks you should see roots. Carefully transplant into a garden area in full or partial shade. The best pH is 5-.5. These plants need to be watered daily. Water hyssop can be grown totally in water, hydroponically, and can be used in an aquarium.

Water hyssop's white flowers grow 8 inches tall, spreads to 12 inches, and will look great as a border or from a hanging basket. The small oval seed pods are filled with tiny seeds.

MEDICINAL: *Bacopa monnieri* has been used in Ayurvedic/traditional medicine treatments for centuries. Water hyssop supposedly provides a cornucopia of treatments. But there are very few scientific studies to back up the Ayurvedic-homeopathic natural medici-nal claims. Taken regularly, this herb purifies the blood, strengthens the immune system, and may improve cognition and memory. It increases sex drive, cures impotence, is an antidepres-sant, lowers cholesterol, treats arthritis, insomnia, mental fatigue, inflammations, asthma, and rheumatism. A poultice of the boiled plant is placed on the chest to treat bronchitis and other coughs. Burns are treated with this plant's juice as a wash. A tea of water hyssop leaves (fresh or dry) is relaxing and relieves anxiety and stress. Take these supplements only after consulting a doctor or a health expert. **DON'T** use it while taking birth control pills or during estrogen replacement therapy.

NUTRITION: 100 grams of water hyssop leaves cooked has only 38 calories and contains 2g protein, 1g fiber with high quantities of vitamin C, calcium, phosphorus, and iron.

FOOD & USES: When water hyssop leaves are crushed, they give off a lemon scent. Leaves, flowers, and stems are eaten raw in salads, cooked, added to soups, or pickled.

RECIPES: WATER HYSSOP TAMARIND CHUTNEY

Ingredients: 100g water hyssop leaves washed and cleaned from the stems, 1TS coconut oil, ½TS curry powder, 2 dried red chilies chopped, 3/4 TS whole black peppercorns,1 cup split black lentils 2g tamarind paste (remove seeds), and salt to taste.

Method: Take a pan on a low flame and heat the oil. Roast the curry powder, red chilies, black lentils (urad dahl), peppercorns, and salt until it gives that great aroma. Mix in the tamarind and fry for another minute. Add the water hyssop leaves and stir-fry a minute, until it wilts. Put in a blender and grind until it's a stiff paste. Enjoy with biscuits, warm toast, naan, or roti.

THE HAWAIIAN HOME GARDEN GUIDE
FRUIT

What determines if something is classified as a fruit, vegetable, or nut? This question has been debated by botanists, chefs, and gardeners for centuries.

Fruits are the ripened ovaries containing the reproductive seeds of a plant. The plant produces flowers. The flowers are pollinated. Once pollinated, the flesh around the seeds often expands and becomes edible. Fruits are how flowering plants disseminate seeds. In cuisine, when discussing fruit as food, the term usually refers to just those plant fruits that are sweet and fleshy, like pomegranate, papaya, and the sugar apple. The 'ovary' definition also includes many common vegetables, as well as nuts and grains. They are the fruit carrying the seeds of their plant species. The enormous variety among plant fruit makes a precise definition impossible.

The differences between fruits and vegetables are confusing since many fruits are classified as vegetables by the public. A professional cook might determine a 'fruit' is sweet and a vegetable is not. Vegetables are the stalks, leaves, flowers, and roots of edible plants. Vegetables are drab in color - mostly green or brown; while fruits are bright vivid yellows, oranges, reds, and purples. A vegetable is a plant cultivated for the edible part. It can be a bulb, root, seeds, or the fruit itself. They are not sweet, and many have to be cooked to be edible. Fruits are not cooked in most cases, and almost all fruits can be consumed raw.

The fruit of a plant contains the seeds of the plant. Vegetables consist of virtually every part; seeds (peas and beans), stems (chives, celery), leaves (lettuce, cabbage, bok choy, and spinach), flowers (broccoli and cauliflower), and roots (carrots, malanga/ cassava, yautia/taro, and sweet potatoes). Nuts are tree seeds. Peanuts are a variety of beans. Peas and beans are vegetables, but peas in the pod are fruit. String beans, the green bean pod, and the beans inside, are fruit. Tomatoes, cucumbers, watermelon, and squash are fruit but are described elsewhere in this garden book. Some fruit, however, does not contain real seeds. Few modern varieties of bananas contain seeds. Strawberries have the seeds on the outside.

Finally, vegetables grow faster, while fruits are slower. A tree can take years to bear fruits. Vegetable plants grow quickly, and within months vegetables are ready to be harvested. The US Supreme Court ruled on fruits versus vegetables in 1893 with the Nix vs. Hedden case. Fruits and vegetables were taxed differently, and Nix who was a tomato importer wanted it to be a fruit, and Hedden, the tax collector, wanted it to be a vegetable. The Supreme Court ruled that because tomatoes were served with dinner and not after as a dessert. and that made tomatoes a veggie!

The most cultivated fruits worldwide are, in this order: bananas, apples, oranges, grapes, and mangos. Every fruit has many names. The botanical name is included to assist identification.

Knowledge is knowing a tomato is a fruit; wisdom is not putting it in a fruit salad. ~ Albert J. Schütz

PAGE:

Both photos courtesy 912crf.com

NOTES:

Gardening is a labor full of tranquility and satisfaction; natural and instructive, and as such contributes to the most serious contemplation, experience, health, and longevity.

~ John Evelyn, 1666

ACKEE

Blighia sapida

Ackee is a very strange fruit, almost a mistake of nature. It is delicious, and nutritious, yet can be poisonous if not picked when perfectly ripe. It is so plentiful in Jamaica it could be considered their national food dish, fried with salt fish. Through the NET, I've been lucky to locate a few trees on the Big Island west of Keaau and found the correctly ripened fruit. The ackee tree owner said he knew of a few on Oahu, Maui, and Kauai. *Blighia sapida* is the botanical name of the ackee.

Captain Bligh, of *'Mutiny on the Bounty,'* can be blamed for importing this tree from West Africa in the late 1700s. Ackee was a perfect food for the sugar plantations as it was cheap and nutritious. It is a distant relative of the lychee.

HOW TO GROW: Ackee grows throughout the Caribbean, and Central, and South America as an ornamental, but only Jamaica considers it an edible fruit. Ackee could make an excellent backyard tree to shade your hammock. It is a tropical evergreen, which will grow in most well-drained soils and loves plenty of sun. Leave plenty of space as it gets to 30 feet both high and wide. The pale green blossoms have a nice aroma. To me, the fruit resembles a cashew. As the ackee ripens it becomes red or orange. Mature fruit split open to reveal three black seeds in a creamy flesh. A good sized ackee is about a half pound.

Ackee may be propagated by seed; however, the seeds are short-lived and should be planted within a few days after extraction from the fruit. Seeds may take 2 to 3 months to germinate. Once the sprouted trees reach a meter, transplant to final location. It grows best in fertile well-drained light loams but tolerates slightly alkaline limestone sands, or deep clay soils. The best pH is 5.5-7.5. Seedling ackee begins to produce fruit in approximately 5 years.

MEDICINAL: How can a tree with poison fruit be Jamaica's national food? Picked before ripe, ackee contains an altaroid toxin that blocks the liver from releasing the natural supply of glucose to your body. We use glucose or blood sugar constantly for energy and maintaining body functions. Every few hours our body needs another burst of natural sugar to keep our blood sugar normal. Ackees must be properly picked and cooked. Ackee fruit or pods must fully ripen naturally and split open while on the tree. The ackee flesh must be cleaned, washed, and boiled. This water must be dumped and cannot be used again for cooking. The base membrane must be removed.

The illness resulting from eating bad, immature ackee is known as the 'vomiting sickness of Jamaica.' About two hours after eating unripe ackee fruit nausea begins followed by vomiting, dizziness, fever, convulsions, coma, and even death. This is caused by the lack of blood sugar or hypoglycemia, which can be corrected by an IV of glucose. Most cases of poisoning are young children of extremely poor families.

In traditional-homeopathic medicine, ackee's ripe arils are combined with sugar and cinnamon to treat fever and dysentery. The new leaves are crushed and applied to the forehead to ease a severe headache or migraine.

NUTRITION: Please don't be frightened away from this fruit. When properly prepared, the ackee is delicious, rich in vitamin A, zinc, iron, potassium, and calcium. A good-sized fruit usually has 150 calories. Ackee provides enough protein that it can be the center of a meal.

FOOD & USES: Even though this fruit can be life threatening, the ackee fruit is a major Jamaican export of more than a half a billion dollars a year! The US did not permit ackee imports until 2005 because so many people died from eating unripe ackee. Now Haiti is can-ning fully ripened fruits and exporting to the American market.

It can be consumed fresh, baked, boiled in milk, or in soup. They are delicious fried with onions, tomatoes, peppers, and saltfish. Ackee can be cooked with fish pork, or chicken. If you have your own tree, it'll be free.

When you buy fresh ackee, take the ackee out of the skin, remove the black seeds, and with a small knife remove the little pinkish - purple string membrane. This is the poisonous part of the ackee. Always drain the ackee after boiling. If you are timid, buy canned ackee. In Ghana, they use ackee blossoms for soap and perfume. In some African countries, the pods are crushed and thrown into the water where they have a narcotic effect on fish, making them rise to the surface where they can be easily caught.

RECIPES: ACKEE CHEESECAKE

Ingredients: The filling—a half dozen ripe ackee chopped small, a quart of French vanilla ice cream, 2 packages of cream cheese softened, 8 whole eggs

The shell: announce 8-ounce package of vanilla wafers or graham crackers smashed into small crumbs, ¼ lb. of soft butter

Method: Mix softened butter and biscuit crumbs. Press mixture evenly into a suitable pie pan. Combine sugar and cream cheese, add eggs and ice cream. Then mix in the chopped ackee. Pour this over the crushed crackers and bake at 350 for half an hour. Remove from oven, allow to cool before freezing. Serve direct from the freezer.

SIMPLE ACKEE AND SALTFISH

Ingredients: 10 ackees cleaned and boiled, 1 onion chopped, 1 sweet pepper chopped, 4 tomatoes diced, 2 stalks of celery chopped, 2 garlic cloves minced, 1 sprig of thyme, 1 hot pepper seeded-minced, 1 TS curry powder, 2 TBS cooking oil, 1 lb. of saltfish washed, boiled, and flaked apart

Method: In a large frying pan heat oil. Add curry, onion, thyme, and garlic stirring constantly. Then mix in the remaining vegetables. Add ackee last. Keep stirring and add saltfish. Cook covered for two minutes. Add dumplings, okra, or plantain pieces if desired.

ACKEE WITH OKRA

Ingredients: 12 ackee cleaned, 8 okra chopped (best if left to dry a few hours in the sun before preparation begins), 1 onion chopped, 2 medium tomatoes chopped, 1 bunch chives chopped, 1 sprig of thyme, 1 hot pepper seeded and minced, 1 garlic clove minced, 1 TS oregano, salt to taste, 4 TBS cooking oil.

Method: When cleaned put the ackee in a pot with water and salt and boil it for ten minutes until the ackee is almost soft. In another pot add the oil and sauté the onion, tomato, okra, garlic, chives, and hot pepper. Add about 2 tablespoons of water and the remaining ingredients. Cook until okra is tender. Add ackee to the vegetables and simmer stirring for 2 minutes.

ACKEE SOUP

Ingredients: 4 cups of cleaned boiled ackee, 2 cups chicken broth or vegetable stock, 2 large tomatoes cubed, 1 bunch chives chopped, 1 medium onion chopped, 1 clove garlic minced, ½ hot pepper seeded and minced (optional), salt to taste

Method: In a pot, combine ackees, chicken broth, and all ingredients. Simmer half an hour.

NAMES: The common name, ackee is derived from the West African Akan akye fufo. Also known as akee, or achee, and vegetable brains.

A society grows great when old men plant trees whose shade they know they shall never sit in.

This gardening saying is an old Greek proverb.

WI APPLE - AMBARELLA – GOLDEN APPLE
Spondias dulcis

The wi apple is an oblong, yellow-orange and tastes like a cross-breed of a mango and a pineapple. It is believed to have originated in New Guinea and native to Polynesia and Malaysia. The wi apple was first brought to the Western Hemisphere to Jamaica in 1782 by Captain Bligh.

The first tree in Hawaii was brought to Honolulu in the mid-1800s by the Hawaiian Kingdom Minister of Finance, John Walker. This plum-shaped fruit is popular in Asia and eaten at all stages of ripeness. Wi apple's main distinguishing feature is its spiny seed. These spines harden when the fruit matures, so the fully ripe fruit should be carefully sucked from the seed to avoid an unwanted pieced lip or tongue.

HOW TO GROW: Wi apple trees thrive throughout Hawaii in all types of well-drained soil. The trees can be grown from seeds, which take a month to sprout. After a few months, the sprouts can be planted in holes bedded with well-rotted manure. It is best to have the small trees at least partially shaded by mature bananas. They should be spaced at least fifteen feet apart and from fences or buildings. Wi apple trees can grow as much as six feet a year. The trees should be topped to keep to a reasonable height, otherwise, they will grow rapidly to forty feet or more, which makes the fruit difficult to harvest, and susceptible to damage from high winds. These trees should bear in 3-4 years. Dwarf types bear in one to two years at a height of less than six feet.

During the dry season, the leaves turn yellow and drop. Just as the rains begin, clusters form of small white blossoms of both sexes, which can self-pollinate. The fruit will appear in green clusters of ten or more ripening to a golden skin. Using a half cup of high nitrogen fertilizer mix once a month with regular watering, a mature tree should annually bear about two hundred pounds.

MEDICINAL: Tree sap can be used for medicinal poultices. A tea made from the bark and leaves is a supposed remedy for diarrhea and sore throats and coughs. The juice can be used to relieve diabetes, heart ailments, and urinary problems. Chopped fruits mixed with water are drank to treat coughs and fevers. The fibrous flesh helps clear the digestive tract and eliminate constipation. Boil the leaves and roots of wi apple and apply the liquid to the skin as a substitute for a moisturizer.

NUTRITION: This fruit has 160 calories for 100 grams because 10% is sugar, 85% is water, which is why it is great for juice. The fruit is a good source of vitamin C, potassium, carbohydrates, and fiber.

FOOD & USES: Wi apple leaves smell great, but are slightly sour and are used for flavoring, particularly curries. Indonesians make a dish with steamed leaves, salt fish, and rice. This tree's wood is adequate to make fishing floats or boats.

Wi apple has suffered by comparison with the taste and desirable appearance of the mango. However, if the wi apple is picked at the correct time, while still firm, it yields a delicious juice for cold beverages. As this fruit ripens the flesh changes from yellow to orange and becomes sweet with a sort of pineapple flavor. It can be frozen to make a tasty ice. Stewing the ripe flesh with a little water and sugar and then straining produces a rich apple-type sauce. By adding cinnamon or cloves, this sauce can be slowly cooked to a thick preserve similar to apple butter. Unripe fruits can be made into chutney or pickled. Unripe, green wi apple achcharu is made by peeling with a knife and sliced. Then add in a hot pepper, chili powder, 1 TS water, and salt. Combine and let sit for 1 hour. Wi apple is a good flavor ingredient for sauces and soups and can be used like papaya as a meat tenderizer.

RECIPES: WI APPLE CHUTNEY WITH RAISINS

Ingredients: 1 kilo wi apple fruit half-ripe, 1/2 kilo golden raisins, 2 garlic cloves minced, 2 TBS fresh ginger grated, 3 cups clear vinegar, 2 cups sugar, two TBS salt, 5 hot peppers cleaned of seeds, stems, and membranes (more can be added to taste), 1 TBS cinnamon, 5 whole cloves

Method: After thoroughly washing, slice fruit into sections and combine with all other ingredients in a stainless steel or cast-iron skillet. Aluminum will blacken the chutney. Bring to a boil and then simmer for an hour and a half until thick. Stir often to keep chutney from sticking to the skillet and burning. Fill jars, which have been sterilized by immersing in boiling water. It is best to use containers that do not have metal lids.

WI APPLE FRUIT SAUCE

Ingredients: 5 pounds fruit peeled with seeds removed, 1/2 cup water, 2 TBS cinnamon, 4 cloves, and other spices to taste

Method: Bring all ingredients to a boil and simmer for one hour with continuous stirring to mash up the fruit. Strain if the consistency is too thick. Continue cooking if the consistency is too thin. Serve hot or cold as a side for fish, beef, or chicken main dishes.

KAILUA WI APPLE CURRY

Ingredients: 1/2 kilo wi apple peeled and cut into small pieces, 3 garlic cloves, 1 medium onion, 2 green chilies, 1/4 turmeric powder, 2 TS chili powder, 1 TS curry powder, 1 TS mustard seeds, 2 TBS sugar, 2 cups coconut milk, 1 TS salt

Method: In a frying pan put wi apple, garlic, green chilies, onion, turmeric powder, chili powder, curry powder, and sugar. Add 1/2 cup coconut milk. Simmer for about 10 minutes, then add 1.5 cups of coconut milk. Cover and simmer stirring for 15 minutes. Add mustard seeds.

SOAKED WI APPLE

Ingredients: 12 full but not ripe wi apple. (The amount depends on the size of the container you are planning to fill.), 1/2 cup white vinegar, 1 TBS salt, 1/2 TS black pepper, 5 big cloves of garlic minced, 1/2 hot pepper to your taste

Method: Fill half of your container, jar, or bowl, with water. Add remaining ingredients. Peel and make cuts into the wi apple flesh. Put into container and let sit for at least two days.

WI APPLE KUCHELA - delicious and unique spicy - hot Indian relish

Ingredients: 12 green wi apples grated, 4 cloves of garlic minced, 1 hot pepper seeded and minced, 1 TBS salt, 1/4 cup garam masala, 4 TBS oil

Method: In a frying pan on low—heat oil, add garlic and pepper, wi apple, garam masala, and salt. Keep stirring until everything is evenly mixed. Remove from heat and put into sterilized bottles.

DID YOU KNOW?

Botanically this fruit is named *Spondias dulcis*. It has other callings such as the golden apple, pommecythere, Tahitian or Polynesian plum, Jew or Jamaican plum, mango jojo, or Tahitian quince. The name of the city of Bangkok, Thailand is derived from makok, the Thai name for wi apple. Sri Lanka is the world's biggest wi apple producer and exports 250 metric tons each year.

You are what you eat, so don't be fast, easy, fake, or cheap. ~ Além do Cabelo

AVOCADO - ‘ĀPALA

Persea americana

The avocado is not native to Hawaii, but the island chain boasts over 200 dfferent varieties. There are three main classffications. Immigrants brought and planted the seeds. Probably the Spanish horticulturist, Don Francisco Paula de Marin imported a few in the early 1800s. The fruit fall into three basic classiffications: Mexican with thin-skin, the Caribbean big and smooth-skinned with a smaller seed, and Guatemalan Haas with the hard, rough skin. Hawaii's initial cultivators were Japanese coffee farmers on the Big Island in the Kona area. Though avocados are plentiful, more than a million kilos, they are not an export crop, only enjoyed in the islands. This Hawaiian name is ‘āpala.

Avocado is a very nutritional tree fruit, an elder member of the laurel family that has been growing in Central and South America for an estimated ten thousand years. Spanish ex-plorers discovered the tree with long, egg-shaped leaves and greenish flowers without petals. Its name, avocado, is derived from the Spanish *aguacate*, but originally was known in the Mexican Aztec word, *ahuacatl*, which translates as ‘testicle tree’. Branch grafting has been the best method of cultivating this type of tree since the beginning of the twentieth century. Now avocados are grown throughout the world in temperate and tropical climates.

The fruit can be round or oval, with green, purple, or black skin. Avocados are usually 1-2 pounds, but can grow to double that. The fruits' flesh is yellowish green around a single large seed.

HOW TO GROW: Avocado trees are usually bought at nurseries. They grow to sixty feet with a spread of twenty feet, so they need ample space. A nice project for a young person or a newlywed couple is to sprout your own tree from seed. Select seed from a specific tree; a purple Pollock avocado is my favorite. Use the fruit, saving the seed. In a glass or cup of water, suspend the seed with toothpicks so only half of the seed is submersed. Watch as the seed sprouts roots.

Transplant the seed when the roots are about two inches long to a plant pot with soft soil and keep it moist. Increase pot size as the tree grows and plant in a well-worked two-by-two-foot hole when it is two years old or thirty inches tall. Avocado trees need well-drained soil and scheduled watering, especially in the dry months. Monthly broadcast a half cup of 12-24-12 for excellent root development. Once the tree blossoms broadcast a cup of 12-12-17-2 every other month. To keep the tree short – cut the top when it reaches about twenty feet, best done during the full moon. This will cause the tree to spread out and make picking this delicate fruit easier.

This fruit bruises easily, so it should not hit the ground. A pole picker or ‘kali’-a stiff wire hoop with a net - both knocks the fruit loose and catches it. Another method is for the picker to climb and throw or drop the fruits to a catcher with a large net usually made of a feed bag and two long straight branches, one for each hand.

MEDICINAL: Avocados are excellent for moisturizing and nourishing mature skin. This fruit helps prevent wrinkles when used as a mask. Avocados contain at least 14 minerals with vitamins A and E essential for skincare as we get age. Gently scrub your face before applying the avocado mask. Combine half a good-sized avocado with one egg yolk and a tablespoon of honey and blend. Apply the mixture immediately to your face, neck, shoulders, and hands. Relax in a cool, shaded place with the mask for half an hour. Rinse with warm water and apply a moisturizer.

NUTRITION: Avocados provide vitamin A, thiamine, and riboflavin. They contain the most protein of any fruit and more potassium than a banana. Although high in calories, eating avocado is good for the heart. They contain oleic acid, which helps lower cholesterol, potassium to lower blood pressure, and folate to lower the risk of heart attacks. One cup of avocado has 236 calories, but is also rich in vitamins K, B-6, C, and copper. Avocado contains a high percentage of lutein, an important nutrient for healthy eyes.

FOOD & USES: Avocados are used in salads, spread on sandwiches, and in soups. The common use is 'guacamole' where an avocado is blended with spices as a spread or a dip.

RECIPES: FIERY GUACAMOLE

Ingredients: 2 ripe avocados, ½ cup tofu, ½ onion minced, 1 garlic clove minced, 1TBS lemon juice, 1TS Worcestershire, 1 hot pepper seeded and minced, juice of 1 lemon or lime, salt

Method: Smooth tofu with a food processor or blender and then mash the avocados. Mix all ingredients. Place in covered bowl and the lemon juice should prevent darkening. Chill and serve.

CHILLED AVOCADO PRAWN SOUP

Ingredients: 2 ripe avocados, juice of 1 lemon and 1 lime, 1/2 bunch cilantro, 2 cups chicken broth, 1/2 TS Worcestershire, salt, and pepper to taste, ½ kg cleaned, shelled steamed prawns

Method: Peel seeded avocados and then mix with all ingredients except the prawns in a blender. Add prawns to the mix, chill the soup, and serve. Serves 4.

FETTUCCINE PASTA WITH WALNUTS AND AVOCADOS

Ingredients: 2 TBS oil (olive or canola preferred), ½ cup firm tomato diced, ¼ cup wine vinegar, 2 TBS chopped walnuts, cashews or almonds, ½ cup fresh or dried basil, 1 avocado diced, 2 TBS chives chopped, 1lb. dried fettuccine noodles, ¼ cup sweet pepper diced, ½ hot pepper seeded and minced (optional), and salt to taste.

Method: Cook the fettuccine pasta for 3 minutes and drain. In a large bowl combine all other ingredients with ½ the avocado. Combine with hot noodles and top with remaining diced avocado. Serve while the noodles are warm as a dinner entre' or chill for a salad.

HOLUALOA AVOCADO - BANANA BREAD

Ingredients: 1 ripe avocado-peeled and seeded, ½ cup rolled oats, 1 cup all-purpose flour, 1TS each baking powder and baking soda, ½ TS salt, 1TS cinnamon, ¼ cup oil-canola preferred, 1 cup brown sugar, 2 eggs, 2 bananas, ½ cup chopped almonds, cashews, ¼ cup sour cream

Method: In a large bowl mash the avocado. Add oil and brown sugar and mix –preferably with an electric mixer but a whisk will do – until creamy. Beat the eggs one at a time into the mixture. Add nuts and peeled bananas. Combine all dry ingredients in another bowl - oats, flour, baking powder, baking soda, salt, and cinnamon – and add to the avocado mixture. Stir in sour cream. Pour mixture into a greased nine-inch loaf pan and bake at 350 for one hour and fifteen minutes. Do the toothpick or knife test to see if the center is fully baked.

DID YOU KNOW? One tree can produce between 150 and 500 avocados per year. The average avocado contains 300 calories and 30 grams of healthy polyunsatu-rated and monounsaturated fat. Second, only to the olive, the avocado is comprised of about a quarter of oil. Mexico produces the most avocados, with California coming in second with 7,000 avocado groves. The Dominican Republic, Brazil, and Colombia also are top avocado producers. Once picked, an avocado takes at least a week to fully ripen. The refrigerator will slow ripening, but putting it in a paper bag with a ripe apple will speed up the process. Avocado's botanical name is *Persea americana.*

I love things that are indescribable, the taste of an avocado or the smell of a gardenia.~
Barbra Streisand

BAEL FRUIT - GOLDEN APPLE

Aegle marmelos L

This fruit is widely known as bael fruit. In Hawaii, it's being cultivated on Maui and Oahu on a limited scale. Botanically, it's *Aegle marmelos L.* It is often confused with the wood apple and in some places, they share that name, but the two fruits are different species. The wood apple is *Limonia acidissima.* Bael fruit is native to the northern part of India and spread to Southeast Asia. This is a nice tree for the yard. The fruit can be prepared may ways and parts of the tree are valuable for Ayurvedic and traditional medicines. Bael is believed to be a sacred plant and brings wellness, good luck, prosperity, and stability to your home. Bael seeds, like everything else are available through Amazon.

The bael tree is often found around Hindu temples. The leaves are offered to Lord Shiva. Its name in Sanskrit is bilva or sriphal or Shivadruma (the tree of Shiva) and bel, or bael in Hindi.

HOW TO GROW: This is a deciduous medium-sized tree and will grow to 40 ft. (13 m) tall with slender, drooping branches protected by ¾ inch (2cm) long thorns. Well-pruned, this can be a hedge; the thorny branches make an effective barrier. The bael fruit tree will grow where other fruit trees cannot. They'll grow in swampy, alkaline, or rocky terrain with a pH range from 5 to 8 and are very drought resistant. It will grow up to a 3600 ft. (1,200 m) altitude and can handle freezing to very hot temperatures. This is a hardy, well-adapted tree, but requires sun and will not tolerate shade or high winds. It prefers moist soil.

Trees may be grown from seeds and will bear fruit in six years. Grafted trees bear earlier. This fruit tree requires a pronounced dry season to give fruit. The full bearing potential will be in 15 years with an average crop of 200. The fruit ripens in the dry season when most leaves have dropped in anticipation of a bloom for the next crop. The tree will attract butterflies..

The bark is smooth, pale brown or gray. The grey-white wood is fine-grained, hard, but not durable; inclined to warp and crack during curing. It's used to make small objects such as tool and knife handles, pestles, and combs. This tree bleeds a clear sap from cuts. It leaks from nicked branches in long strands, slowly solidifying. It may look sweet, but will irritate your throat. The pale green or yellow blossoms are ½ inch (1.5cm), with a nice sweet aroma and usually appear with the onset of new leaves.

The round bael fruit 3-5 inches in diameter (8cm) and the skin or husk is thick and hard. It doesn't soften as this fruit ripens. It's smooth and green-gray until it's fully ripe when it turns yellow. Inside sections are filled with a nice smelling orange pulp. Each section has many seeds with each encased in transparent paste that becomes solid when dried. It takes about 11 months to mature on the tree and then 2-3 weeks after they are picked. It requires a hammer to open the shell. The flesh taste and has the consistency of sour orange marmalade.

MEDICINAL: The medicinal plants must be grown without chemical fertilizers or pesticides. The dried pulp of the bael fruit is astringent. It reduces irritation in the digestive tract and is an excellent remedy in cases of diarrhea. For hemorrhoid treatment, prepare a decoction of unripe fruit with fennel and ginger. As a sexual stimulant, prepare a decoction of only the unripe fruit. Ripe bael fruit is a laxative and relieves inflammation. It supports the healthy function of the stomach. Bael sherbet (fruits boiled in water) can improve your digestion. Don't indulge and eat too much; it can reduce your breathing rate, heart beats, and cause drowsiness.

Bael fruit has plenty of vitamin C to boost immunity and provide relief from the common cold. It's also effective in reducing high blood pressure and the risks of cardio-vascular diseases. The tree's sap, Feronia gum, reduces elevated blood glucose levels as an essential Ayurvedic treatment for diabetes. Research is needed to discover if prolong use could be a cure. Boil bael leaves in water and then wash your body to heal wounds and reduce body odors. However, eating leaves or drinking the tea may cause abortion and sterility in women.

The bark is used as a fish poison. The shell of the unripe fruit makes a yellow dye for calico and silk fabrics. The shell is rich in limonene and can be distilled to make a scented oil. An oil can be extracted from the leaves to make a useful herbal insecticide to fight the brown planthopper that harms rice crops. The gum is commonly used as a household glue and as a jeweler's adhesive. When the gum is combined with lime, it becomes a waterproof plaster. Artists use the gum as a protective coating on paintings. The fruit pulp has detergent action and used for washing clothes.

NUTRITION: 100 grams of bael fruit has only 140 calories with 60mg of vitamin C, 55mg of vitamin A, rich in potassium, and calcium. It provides niacin, riboflavin, and thiamine.

FOOD & USES: Bael fruit is eaten raw or made into marmalades, pickles, and drinks. The ripe pulp is a sweet-tasting and scented paste. The fruit may be cut in half, or the soft types are broken open, and the pulp with palm sugar is a breakfast treat. The blossoms can be made into a refreshing beverage. Its potent aroma will overtake your kitchen upon opening the shell and will predict the captivating taste. The scooped-out pulp from its fruits may be eaten with coconut milk and frozen into ice cream. It's also used in chutneys.

RECIPES: BAEL FRUIT RELISH

Ingredients: 2 ripe wood apples opened and the insides mashed, 1 TBS jaggery or brown sugar (more or less depending on the fruit), 1 TBS mustard oil, 4 green chilies (for spicier add 1 TBS red chili powder), 1 TBS coriander leaves, and salt to your taste.

Method: Break open the hard shell and scoop out the pasty insides with the seeds. Put in a ceramic bowl, don't use stainless steel. Combine all the ingredients. Cover with a lid and leave for 2 hours. Probably less because the aroma and anticipation will make you taste it! Enjoy.

BAEL CHUTNEY

Ingredients: pulp of 2 bael fruits, ¼ cup brown sugar, 1 cup of water, 1 TBS lemon juice, 1 TS Each of salt, roasted cumin seeds, black pepper powder, red chili powder, and fennel seeds

Method: Place pulp in a large bowl with 1 cup of water. Let sit for an hour and then work the pulp with your hands to loosen the seeds. Force through a sieve to strain the seeds. Heat the strained pulp in a frying pan, stirring for 10 minutes. Add sugar and all spices, and cook, stirring, for 10 minutes. Cool, stir in lemon juice. Test by dropping a spoonful on a plate. If it separates, it needs more time on the heat. Store in a sterilized glass container.

BAEL FRUIT TEA

Ingredients: 2 -4 slices of toasted bael fruit, 4 cups water, sugar to taste

Method: Slice bael fruit ¾ to an inch thick and heat in a frying pan or toast it in a toaster oven for only two minutes. Drop the toasted bael fruit slices into a pot. Add water and boil for at least 2 minutes. This method brings out the best flavor.

DID YOU KNOW? Evidence of the religious importance of the bael tree appears in **Shri Shuktam** of *Rig Veda* which reveres this plant as the residence of goddess Lakshmi, the deity of wealth and prosperity. In the traditional practice of the Hindu and Buddhist religions by people of the Newar culture of Nepal, the bael tree is part of a fertility ritual for girls known as the Bel Bibaaha. Girls are 'married' to the bael fruit; as long as the fruit is kept safe and never cracks, the girl can never become widowed, even if her human husband dies. This is a ritual that guarantees the high status of widows among the Newar.

BAKOTO PLUM
Flacourtia inermis

In Hawaii, the delicious bakoto plum is usually eaten out of hand, picked fresh. It's available in some village markets, but if you're a fruit lover, it's best to scout the trees and try to be there before the birds and bats get their share. This plum originated in the Philippine Islands and is well suited to Hawaii's climate and terrain. The red cherry-like fruits will make a nice tree for your yard. This tree is often confused with other plum trees because there are so many 'governor plums.' Its botanical name is *Flacourtia inermis*.

The Hawaiian Journal of History notes: In the 1930's plums were brought to Kauai and planted on lands that later became Kōke'e State Park. For years, plums have been declining due to weather, over harvesting, and damage to trees. This is a great reason to start a tree at home and continue this fruit in the island chain. Seeds and trees can be purchased online.

The bakoto plum is a beautiful, bushy tree that can grow to 10 meters. The trunk is seldom straight with a smooth light brown to gray bark. The leaves give this tree a holiday look and are very unique because they begin bright orange, to red, before maturing to dark green. The plums grow from yellow-green blossoms in bunches and are bright red, about an inch in diameter, and have a double layer of small grape-sized seeds in a star pattern. Bakoto are as sour and acidic as they are red, reminiscent of the Barbados/acerola cherry. As with many fruits, they're crunchy and more edible with some salt and chili pepper. The crunchy fruits are generally made into wine, jams, preserves, and syrups.

HOW TO GROW: Bakoto will grow in most soils even sandy coastal areas up to 1300 meters elevation. The best pH is 4.5 to 6. It is propagated through seeds and it's best to soak them pre-planting for two days in cold water. It likes a spot that gets a lot of sun and well-drained soil. For the best results, semi-annually fertilize with rotted animal manure as a mulch. Water heavily every two weeks. The tree does not produce fruits all year round. The wood is very hard and dense, usually orange or red. Besides being planted for its attractive foliage and fruits, the bakoto also attracts the Leopard and Rustic Butterflies. This is a very attractive tree and perfect to grow in a large pot on your patio/lanai.

MEDICINAL: The fruits have great nutritional value, but the bakoto is also known for ayurvedic -homeopathic treatments. The fruits treat jaundice and spleen problems. The juice and a leaf decoction are used to treat diarrhea, dysentery, and piles. The fruits, bark, leaves, and roots are used in concoctions for certain health conditions, especially for arthritis, some bacterial infections, nausea, sore throats, and to relieve chest congestion. The leaves are dried and powdered to relieve bronchitis and asthma. The bark and leaves are used to lessen the pain of toothaches and fight bleeding gums. A poultice of the roots will treat skin sores and boils. In Malaysia, a juice squeezed from the roots is used to fight herpes. In South India, they grind the seeds into a powder and combine with turmeric root and rub women after childbirth to lessen the pain.

NUTRITION: Bakoto fruit has all the Vitamins C, A, B, Thiamine, Riboflavin, and Niacin, which boost immunity. The fruit is rich in vitamins like ascorbic acid, carotene, and minerals like calcium, magnesium, and iron.

FOODS & USES: Fruits are consumed raw or used for making jelly, jam, chutney, pickles, preserves, pies, and confectionary. In Indonesia, cooked fruits are used in rujak, a dish with mixed fruit, chili sauce, and peanuts, and also in Asia, a mixed vegetable dish with chili flavor. Fruits are processed in the form of jam, syrup and sweets.

RECIPES: BAKOTO ACHAR RECIPE

Ingredients: 1kilo of bakoto not fully ripe, ½ cup chili powder, 6 red chilis chopped, 2TS mus-tard seeds, 1 TS roasted fenugreek powder, 6 cloves of garlic chopped, one handful of curry leaves, 2 TBS coconut vinegar, ½ cup salt, 1½ cup of water, ¾ cup ginger oil, a pinch of tur-meric powder

Method: Boil the bakoto in water until they become tender 5 minutes. If you boil longer the bakoto might spoil. The color will change. Strain but save the water. Let cool and then quarter each. Stir in the salt. Heat ginger oil in a pan or wok. Add mustard and stir until the pop. Sauté garlic, chilies, curry leaves, and spice powders. Combine with quartered, salted bakoto plums. Sauté for 2 minutes and then add the boiled water. When the achar/pickle mix becomes warm add the vinegar. Remove from heat and store in a sterilized jar. The longer it stores, the stronger the flavor.

BAKOTO EASY JAM

Ingredients: 1½ cups of bakoto, 2 cups brown sugar or jaggery, 1 TS cinnamon powder, ½ TS salt

Method: Quarter the fruit and remove the seeds. In a pot or a wok, add chopped fruit, and all ingredients on low heat, stirring constantly until fruit thickens and becomes darker. Remove from heat and cool. Store in a sterilized jar and refrigerate. Serve on coconut roti. **BAKOTO WINE** before you begin get the necessary jars and sterilize.

Ingredients: 1½ kilo plums, 1 kilo sugar, cinnamon, 10 cardamom, 1 inch of stick cinnamon, 6 cloves, juice of 4 oranges, 1½ TS vanilla essence, large wide-mouth glass jars, and liquor bottles.

Method: In a suitable pot, boil washed fruits for 20 minutes or until they are tender. Strain but save the water. Smash fruit with a spoon. Return smashed fruit to water and simmer. Stir in 1-kilo sugar. White or brown will alter the taste. I prefer brown. Remove from heat after sugar is fully dissolved. Let cool to lukewarm and stir in yeast. Pour into a wide-mouth glass bottle or ceramic. Seal with cheesecloth so it can breathe. Put somewhere it will not be disturbed for two weeks. After two weeks stir the mixture and let sit again for four weeks. With fresh cheesecloth strain the liquid twice. Go slow to remove as much particulate as possible. Funnel into liquor bottles and seal with wax the ones you aren't going to immediately drink. Enjoy.

DID YOU KNOW?

The genus Flacourtia was named after E. de Flacourt governor of Madagascar. The species *Inermis* means the tree has no thorns. He was named governor of Madagascar by the French East India Company in 1648. Flacourt restored order among the French soldiers, who'd mutinied. He wrote the *Histoire de la Grande Isle de Madagascar* in 1658. Unfortunately, he drowned on his voyage home in 1660. Flacourt was one of the few, if not the only, European to write about the elephant birds of Madagascar before they became extinct. That's why in English it is known as one of the many Governor Plums. In Indonesia, it's known as tome-tome. It's bakoto and lobi-lobi in Sinhala. Malaysians call it Rukam Masam. In India it's loika, lavaloikka, in Thailand: takhop-thai, and in the Caribbean, the Martinique Plum.

The hum of bess is the voice of the garden.~
Elizabeth Lawrence

BALATA - AUSUBO
Manikara bidentata

Some islanders know the tasty, sweet fruit balata with too little flesh, and too much seed. However, the ausubo tree is famous elsewhere for its wood. In its native Puerto Rico, balata (or ausubo, as it is known in PR) is that island's most valued tree for lumber known as bulletwood. From research, it seems for the true botanist, the tree is called the ausubo, and the fruit ausubo fruit. Ausubo's botanical name is Manilkara bidentata. Only the sap is known as balata.

Puerto Rican balata trees have lived for over four hundred years. That age is relatively incredible since strong storms, floods, and hurricanes often hit Puerto Rico. That is possible because the balata tree is a slow grower, very tolerant of shade, especially in the rain forest. This tree develops strong, deep roots that can withstand powerful winds. Since it grows so slowly, it is often consumed by grazing animals or wildfires. This tree is will grow anywhere in Hawaii if the terrain isn't excessively sandy.

HOW TO GROW: Balata trees are plentiful throughout the Caribbean chain, northern South America, Brazil, Peru, Florida, Hawaii, and Mexico. This tree is best grown from the seeds, and balata lovers know there are many seeds are in one fruiting season. Fresh seeds have a high germination rate. Be very patient as the seeds may take a year to sprout. It grows well in all types of soil, hillsides, floodplains, or along the sea. Balata only requires adequately drained soil. This tree doesn't do well in heavy sandy soil. Look for the biggest and juiciest balata varieties along the highlands and mountain ranges.

First and foremost, if you plan to grow a balata tree make certain you have enough space. These trees grow to more than 100 feet and about 4 feet in diameter. Once the tree is 3 years old give it starter fertilizer (12-24-12) mixed equally with calcium nitrate at 1 cup every 3 months. Balata grows slow, but needs water, so irrigate monthly in the dry season. Average height after a year is about 6 inches and after 5 years your tree should be almost 30 feet tall. If you want to top the tree it can only be cut when it is 3 years old. Wrap the trunk with aluminum foil so the whip string of a bushwhacker can't damage it.

MEDICINAL: Balata sap/gum is used in modern dentistry where it's ideal as a temporary filling for teeth and as a filling material inside tooth fillings. Drinking a cup of fresh milky sap will stop diarrhea. The leaves are used as both a tea and poultice to treat a paralysis of the arms or legs.

NUTRITION: A hundred grams of balatas, about a big handful, have sixty-two calories with two and a half grams of protein and ten grams of fat

FOOD & USES: The delicious fruits are most often eaten fresh. The main use of the tree is for commercial wood. The heartwood is light red when cut and turns to dark reddish brown with purplish shades when dry. Balata timber is used for railroad ties under the rails, heavy construction, furniture, and even pool cues. This wood can be bent in a steam box for boat frames and other curved pieces. Balata resembles mahogany and is resistant to the dry wood termites because the wood contains deadly toxins. If working with balata lumber, be sure to wear a dust mask. Mature trees are tapped for balata gum, which is similar to gutta-percha, a not very elastic rubber. Sap has been harvested from some trees for more than two decades. Originally, the latex was thickened by fire or dried in the sun, and souvenirs or novelties were fabricated. Now the latex is used for a non-elastic rubber needed for golf-ball skins, and ma-chine belts. It is waterproof and excellent for insulating underwater and underground electric cables and connections.

RECIPES: BALATA ICE: Work a quantity of balata fruits through a mesh strainer or a colander into a bowl to catch the juice. Put about six ice cubes in a blender with the juice, a literal pinch of cinnamon and nutmeg. Put the blender on chop until you gain the correct consistency for a snow cone. Be certain your blender is able to chop ice.

BALATA SHERBERT

Ingredients: juice from 5 pounds of balata fruits, ½ can sweetened condensed milk, 1package unflavored gelatin, 2 cups water.

Method: Juice the balata fruit by forcing it through a mesh strainer and collect the juice in a bowl. You need at least 2 cups of juice. In a pot on low, heat the water while stirring in the powdered gelatin. Stir in the milk. Once this is thoroughly mixed add the balata juice. Mix well. Allow to cool on the kitchen counter covered. Put in a covered container and freeze. After about 3 hours remove from freezer and blend to break up the ice crystals. Refreeze. This can also be made without milk as a tasty sorbet.

DID YOU KNOW?

Balata translates to 'sap' in Spanish. It is also called the 'cow tree' because of its latex sap. This may be folk lore. The sap from some of the species can be used to substitute cow's milk. The latex has the consistency and taste of cream, but overindulgence in it can result in severe constipation. Balata was often used in the production of high quality golf balls. Balata is the tree *Manikara bidentata* and it produces a natural non-elastic rubber obtained from the latex sap. Some trees have been tapped for sap for more than 25 years. It is often called 'bulletwood' since it is extremely hard. Balata wood is so dense it does not float in water. It must be pre drilled to drive nail.

The balata is a true rainforest tree. It is very tall, spreading when mature. Its long life is due to excellent root development and tolerance of shade. This enables balata trees to exist for three to four centuries. Inexperienced, and perhaps greedy foresters can kill this tree by slash-ing its bark to obtain sap. Some actually cut down the tree to obtain the latex.

Sweet Balata fruit

Ausubo tree being drained of sap.

Beneath these fruit-tree boughs that shed their snow-white blossoms on my head, With brightest sunshine round me spread, Of spring's unclouded weather, In this sequestered nook how sweet To sit upon my orchard-seat! And birds and flowers once more to greet, my last year's friends together. ~ William Wordsworth

BANANAS - MAI ‘A
Musa paradisiacal

Bananas aren’t trees, fruits, or vegetables, but a flower, part of the palm and lily family. Bananas grow the tallest of any flower on earth, some varieties to 40 feet, without a woody stem. Researchers believe bananas originated in Papua, New Guinea. India has cultivated the banana for at least 4,000 years. The earliest written reference is a Sanskrit text from around 600 BC. Alexander the Great discovered bananas in his conquest of India in 327 B.C. Nomadic Arabs cameled bananas to Palestine, Egypt, and Africa. The word ‘banana’ is derived from the Arabic word ‘banan’ meaning ‘finger.’ During the Crusades of the Middle Ages, both Moslems and Christians believed the banana was the forbidden fruit of paradise.

The banana is not native to Hawaii and its local name is ‘mai ‘a.’ Before 1300, Polynesian seafarers introduced the original *Iholenas*, *Maolis* and *Popoulus* with some *Fei* bananas to the Hawaiian Islands. Today the variety most grown bananas are Cavendish, including the ‘Hawaiian Apple and the Blue Java.’ The ‘apple’ banana has an apple smell, sand the Blue Java is also known and tastes like ice cream The Brazilian variety is another favorite. Hawaii grows more than 4 million pounds of bananas every year with more than half grown on both Hawaii and Oahu. Four corporations: Chiquita, Fyffes, Dole, and Del Monte control 80% of the world’s banana sales.

Although there are approximately 500 species of bananas, only 20 varieties are commercially cultivated. There are two main varieties of bananas, the sweet banana and the plantain. The fruit banana is eaten raw while the plantain is usually cooked. Plantains have lower water content, making them drier and starchier than fruit bananas.

HOW TO GROW: Bananas thrive in the hot tropics, with an average a humid 80 degrees Fahrenheit (27 degrees Celsius, and a minimum of 3 1/2 inches (75 mm of rainfall a month. The major exporters include Ecuador, Costa Rica, Colombia, Honduras, the Philippines, Panama, and Guatemala. Surprisingly, 80% of the bananas grown throughout the world are of the plantain or cooking variety. All varieties of bananas require rich soil with good drainage. Bananas do not grow simply from seed. Farmers start a crop by cutting growths (suckers, slips, pups, or ratoons from the underground stems of mature banana plants. These suckers are replanted and sprout three to four weeks later. In about nine months the plants mature to a height of about 15 to 30 feet. A 16 – 8 – 24 fertilizer mix is excellent for bananas and plantains. Bananas will grow almost anywhere in Hawaii, but need to be watered during the dry season.

When the banana bears fruit a large bud rises from the center of the bundle of leaves. The bud consists of small purple leaves called bracts. As the stem grows, the purple pulls back to reveal clusters of small flowers, which become tiny green bananas. Exactly how bananas ripen is a scientific mystery. Little bananas grow downward. Double rows develop vertically around the stem. As they ripens, they begin to turn upward against the force of gravity!

Bananas ripen three months after flowering. Harvested too early, instead of a sweet flavor, you get a floury pulp. Every bunch has many ‘hands’ or rows of bananas; while each bunch will yield about 200 ‘fingers’, or individual bananas. An average bunch of bananas can weigh between 80 and 125 pounds (35 to 50 kilograms. Bananas are not just green and yel-low; some bananas, are red. Banana’s botanical name is*Musa paradisiacal.*

MEDICINAL: The New England Journal of Medicine reported a banana a day can decrease the risk of death from strokes by as much as 40% in certain cases. Bananas have many medicinal uses. High in iron, bananas are great for treating anemia. The high potassium and low salt help reduce blood pressure. The high fiber can overcome constipation.

Bananas naturally contain tryptophan, which the body converts into serotonin – known to improve your mood and generally make you feel happier, fighting depression. Bananas neutralize burning stomach acid.

NUTRITION: Bananas not only taste good, but also are a healthy, quick energy food packed with vitamins, two grams of protein, and four grams of fiber. Bananas are rich in potassium, vitamins A, C, and the B complex. One banana is about 99.5% fat-free with usually 90 calories comprised of 75% water, 20% starch, and 1% sugar. Compared to apples, bananas have less water, fifty percent more food energy, four times the protein, half the fat, twice the carbohydrate, almost three times the phosphorus, nearly five times the vitamin A and iron, and at least twice the other vitamins and minerals as in an apple.

FOOD & USES: A banana's skin indicates the degree of ripeness. Green bananas can be used in soups and stews. When green bananas turn yellow, the starch becomes sugar. Partially ripe, yellow with green tips, may be broiled, baked, or fried. Ripe, all yellow, are eaten raw or in puddings, cakes, or pies. Full-ripe yellow with brown freckles again may be raw. Over ripe, all brown, are good if the flesh is firm. When they are the color you need, bananas can be refrigerated.

RECIPES: BANANA ICE CREAM — Makes a quart overnight.
Ingredients: 4 or more ripe bananas, 2 TBS coconut (either grated or powder, 1/4 cup pine-apple - chopped small, smashed, and drained, 1/4 cup milk, 1/4 cup of sweetened condensed milk.
Method: Mash bananas with a fork and mix everything in a small pot. Slowly bring to a boil while continuously stirring. Once cooled, pour into suitable freezing containers. Reusable ice cream containers are perfect. Add sliced bananas and nuts if desired. Freeze. Check after a few hours and whip again with a fork. This will remove most of the ice crystals. Freeze solid.

BANANA BREAD
Ingredients: 2 cups all-purpose flour, 2 TBS baking powder, 1/4 TS baking soda, 3/4 TS salt, 1/3 cup shortening, 2/3 cup sugar, 2 eggs, 1 cup mashed, very ripe bananas
Method: Blend flour with baking powder, soda, and salt. Cream shortening and sugar until light and. Add eggs one at a time and beat well after each addition. Add bananas and mix. It will be easier to add the flour in four portions and beat until smooth after adding each portion. Turn into a well-greased pan (8x4x2-1/2 inches) and bake in a moderate oven (350 degrees F) for 50 minutes, or until bread tests done in the center with a knife or toothpick. Cool on a rack before slicing. Yield: 1 loaf

BANANA FRITTERS
Ingredients: 1.5 cups flour, 1 TS baking powder, 1/4 TS baking soda, 1/4 TS salt, 3/4 cup water, four firm bananas, 5 cups vegetable oil (for frying
Method: In a large bowl combine one cup of flour, baking powder, soda and salt. Gradually blend in the water and beat with a whisk until smooth. Cut each banana crosswise into 3 pieces. You should have 12 pieces altogether. Coat bananas with the remaining half cup of flour. Heat oil in a large skillet over high heat. Dip banana pieces in the flour-water mixture, coating well. Cook 4-6 pieces at a time until golden brown, about 3-5 minutes. Drain on paper towels.

DID YOU KNOW?

Bananas float in water because they are less dense in comparison. Bananas grow on plants that are officially considered an herb and they are classified as a berry. The bananas we eat today, the Cavendish, are different from pre-1960s' bananas, the Gros Michael, as those have been wiped out by Panama disease.

You cannot taste me, until you undress me. Sincerely, banana. - Anonymous

BILIMBI

Averrhoa bilimbi L.

The bilimbi is a great, unique, medicinal tree, perfect for your Hawiian backyard. The bilimbi tree is a native of Malaysia and the Indonesian Moluccas. It's common in the Philippines and India in the home garden. The green, round bilimbi resembles a cucumber. Botanically *Averrhoa bilimbi,* L., it's related to the golden, sharp-angled carambola.

The fruit was taken from Timor to Jamaica in 1793, supposedly on Captain William Bligh's second breadfruit voyage, and was distributed widely in the New World. It's believed that the first wave of Filipino immigrants in 1906 brought this tree to the Hawaiian Islands.

HOW TO GROW: The attractive, tree is long-lived, reaches 16 to 33 ft. (5-10 m in height; has a short trunk that branches to about 10 meters if not topped. The trees prefer full sun and a pH of 5.5 - 6.5 in a well-drained soil up to 1,000 meters elevation. The small yellow-green or purple blossoms have a sweet fragrance. The unique fruits, lime green, five-sided cylinders, are directly on the trees' main branches in clusters. Unripe fruit is crisp, glossy bright green to yellowish green. The soft flesh is green, jelly-like, juicy, and acidic with a half dozen flat seeds. They easily bruise because of the thin skin and cannot be kept more than a few days. Seedling trees will fruit in 4–6 years. Healthy, well-pruned 10-year-old trees can produce 100 pounds per season.

MEDICINAL: In the Philippines, the leaves are applied as a paste or as a poultice on insect bites, and to reduce the swelling and pain of rheumatism. An infusion made from the leaves or blossoms treats coughs and is taken as a tonic after childbirth. A decoction made from the leaves treats hemorrhoids. Bilimbi juice treats hypertension and regulates blood sugar. CONSUMED IN EXCESS, bilimbi's oxalate acid can lead to a risk of kidney stones and even kidney failure.

NUTRITION: 100 grams of bilimbi has 30 calories with vitamins A and C, calcium, phosphorus, and potassium.

FOOD & USES: Bilimbi's are not a popular commercial crop, as the shelf life is less and difficult to transport long distances and considered too acidic to eat raw. In Costa Rica, a relish is made from the fruits to accompany beans and rice. The bilimbi is used to add a sour taste to sambal curries, and soups and to make chutney. Because of its oxalic acid content, bilimbi juice bleaches stains from the hands and from white cloth, and brightens brass.

RECIPES: EASY BILIMBI CHUTNEY

Ingredients: 6 bilimbi, 1 small onion chopped, 4 green chilies chopped, 1 TBS coconut oil, salt, and spices to taste

Method: Grind all ingredients together except coconut oil, into a coarse paste. Transfer to a bowl and pour coconut oil over it

BILIMBI JUICE

Ingredients: 6 bilimbi, 1TS chopped ginger, ½ TS salt, ¼ cup jaggery or sugar, 3 cups water

Method: Cut fruit into rounds. Put into a blender with all dry ingredients. Blend into a paste and strain. Rinse blender, add water to paste, and blend again. Best served chilled.

EASY BILIMBI JAM

Boil 6 or more fruits in 1 cup water with the juice of 1 lime and ½ cup sugar until all the water is gone. Use as marmalade.

NAMES: English - cucumber tree or tree sorrel, Malaysia- billing-billing, Indonesia - *balimbing, blimbing,* Jamaica - *bimbling plum,* Cuba - *grosella china,* Philippines – *kamias*

BRAZIL CHERRY - GRUMICHAMA

Eugenia brasiliensis Lam.

The Brazilian cherry /grumichama tree is native to southern coastal Brazil, can be pruned to about 6 ft tall making picking easier. This elegant ornamental tree is perfect home garden landscaping. In Hawaii, it produces fruit to 2000 feet of elevation. According to Don Francisco de Paula Marin's journal, the first Brazilian cherry tree was planted in Hawaii circa the late 1700s. This tree has been grown in Jamaica since 1880. Although the fruit is small and susceptible to the fruit fly, its great taste resembling the North American sweet cherry, makes it perfect for a yard.

HOW TO GROW: Seeds should be sown as fresh as possible because it loses its viability in about 6 weeks. Plant in individual containers, several seeds to a pot, and move away from the hot afternoon sun to a light shade. Germination rates are usually low, with the seed sprouting within 1-2 months. Seedlings are slow growers. Soil should be kept moist with a pH of 5.5 - 6.5. This tree will dry out of not regularly watered. The Brazil cherry grows well in full or partial sun, but needs protection from high winds. It matures slowly from seed and the trees bear in about 4 years. In the tropics blossoms and harvest last several months.

In Hawaii, the trees bloom and fruit from July to December, with the main crop in the fall. There are two distinct fruits yellow or red-skinned. The red turns from bright red to nearly purple/black is sweeter with a thin skin, soft pulp with a mild taste. The skin is thin, firm, and the red or white pulp is juicy.

MEDICINAL: In Brazil, an infusion of 10 g of leaves or bark in 1½ cups of water treats rheumatism.

NUTRITION: 100 grams has 60 calories with vitamins A and C, with calcium, iron, and phosphorus.

FOOD & USES: These cherries are usually eaten fresh, but can also be used for juices, preserves, and pies. The bark and leaves contain an essential oil with a pleasant aroma.

RECIPES: BRAZILIAN CHERRY POPSICLES

Ingredients: 6 cups cherries, seeded and halved, 1 cup sugar, ½ cup water

Method: Blend the cherries into a pulp. Boil the water and sugar into a syrup. Put the cherry pulp in molds. Fill mounds with sugar syrup and freeze.

BRAZIL CHERRY PRESERVES

Ingredients: 6-8 cups seeded cherries, 2 cups sugar or less according to your taste, zest strips of 1 lemon, juice of 2 lemons, 12 cloves.

Method: In a large pot or wok, stir all ingredients on low heat. Simmer 15 minutes until fruit is soft. Bring to a boil. Strain the cloves and lemon zest/peel. Bottle in sterilized jars.

Change is a continuous process. You cannot assess it with the static yardstick of a limited time frame. When a seed is sown into the ground, you cannot immediately see the plant. You have to be patient. With time, it grows into a large tree. And then the flowers bloom, and only then can the fruits be plucked.

~Mamata Banerjee

BREADFRUIT - 'ULU

Artocarpus altilis

The Hawaiian name for breadfruit is 'ulu.' Breadfruit originated in Micronesia in the Pacific Ocean. It is the fruit of a beautiful tree that grows to a hundred feet. The tree acquired its name because explorers found the fruit could be eaten before it is ripe, tasting and feeling like fresh bread. Native Hawaiians believe the god Ku changed into an 'ulu/breadfruit tree and spread the fruit trees on every island to bring food to his people. Breadfruit is one of the Polynesians' 'canoe plants that arrived before 1000 AD. European explorers first encountered breadfruit about 1600 and quickly realized its nutritional value to cost-effectively feed their enslaved labor. Breadfruit's botanical name is *Artocarpus altilis* of the mulberry family.

Breadfruit is not an important commercial crop anywhere because it doesn't ship well, but it is important forfilling bellies where it grows. It will grow anywhere up to 900 meters in elevation if there is enough rainfall. Developed countries weren't interested in breadfruit until 'gluten-free' became important. Properly packaged, dried breadfruit will keep for 2 years.

The real cause of the famous Mutiny on the HMS Bounty wasn't just the beautiful Tahitian women, but breadfruit! The sailors didn't enjoy the thought of living thirsty while the breadfruit trees got watered. After all the problems, when the trees finally reached the Caribbean Islands, the slaves didn't like the taste and refused to consume it! It took decades until their descendants learned delicious recipes, and now appreciate the breadfruit.

HOW TO GROW: Breadfruit needs space to grow and deep fertile well-drained soil. However, breadfruit has adapted to various climate and soil conditions throughout the world. Some Paci ic island varieties grow along rivers while others thrive on sandy coral soils. One variety is tolerant of the salty seawater environment. Transplant a sucker of the seedless bread-fruit variety. It should be partially shaded and watered daily. These sucker trees should bear fruit in about ive years. Mealybugs and ants are their enemies. Breadfruit is fragile and easily bruised. The fruit must be used within a week of picking. One means of preserving is to keep excess fruit underwater until they are desired. All parts of the breadfruit tree including the fruit are rich in milky latex. Jamaicans partially roast their breadfruit to congeal or thicken the latex for export markets.

MEDICINAL: Breadfruit roots, leaves, and its latex sap are used in natural-homeopathic medicine. Breadfruit root and leaves are chewed for arthritis, asthma, back pain, diabe-tes, fever, gout, high blood pressure, liver disease, and toothaches. The latex sap is eaten for diarrhea and stomach pain. It's reported that research has shown breadfruit leaf extract inhibits the enzyme that assists the development of prostate cancer. These leaves also kill cancer-causing cells.

NUTRITION: 200 grams of breadfruit has 200 calories; 2 grams are protein with less than 1 gram of fat. It has 25 grams of carbs and 11 grams of fiber. Breadfruit is a good source of vitamin B and with more vitamin C in riper fruits. It contains calcium, phosphorus, and iron.

FOOD & USES: Breadfruit may be consumed before it ripens. Generally, unripe fruits are green, turning yellowish-green as it ripens. When it is fully ripe the fruit is a yellow-brown. Unripe breadfruit can be chunked and boiled with seasonings and other vegetables as a type of chowder. Breadfruit can be steamed, boiled, roasted, or fried. Ripe fruits may be quartered and steamed with seasonings or it may be rolled in lour and fried. The pulp from ripe breadfruits can be mixed with coconut milk, salt, and sugar to create a pudding. In Barbados, breadfruit has been dried and made into lour as a 'gluten free' substitute. Soft overripe fruit is best for frying as chips.

Gardening is full of mistakes, almost all of them pleasant
and some of them actually instructive. ~ Henry Mitchell

RECIPES: SIMPLE BREADFRUIT DESSERT

Ingredients: 1 very ripe breadfruit mashed, 2 TBS butter, 2 eggs beaten, 1/4 cup brown sugar, and spices as cinnamon, nutmeg, and cloves to your taste, two TBS brandy (optional)
Method: Boil all ingredients and blend well. Serve warm or cold.

TWO-DAY BREADFRUIT CHIPS

Ingredients: 1 green breadfruit, ice cubes, vegetable oil for frying, salt, and spices to taste
Method: Scrub, peel and core the fruit. Put pieces in ice water in the fridge overnight. Slice breadfruit pieces as thin as possible and replace in ice water. Heat the oil and fry until golden brown. Drain and spice as you like it.

LAUPAHOEHOE CURRIED BREADFRUIT

Ingredients: 1 large breadfruit peeled and cut in pieces, 2 cups water, onion sliced, 2 cloves of garlic sliced, 2 green chilies chopped, 10 curry leaves, 1TS of the following: turmeric, un-roasted curry powder, chili powder, fresh ground black pepper, mustard seeds, cumin seeds, cinnamon powder, 3 dried red chilies crushed, milk from one coconut, oil, salt to taste
Method Boil breadfruit pieces in water with turmeric for half an hour. Drain off water. In a small frying pan heat oil add mustard seeds, cumin seeds until they pop. Then add dried chili, garlic, onion, curry leaves, and cinnamon. Stir fry till onions are brown. Heat the breadfruit in the saucepan. Add coconut milk, curry powder, chili powder, pepper and green chili. Add the oil mix and salt to taste, stir well. Cover with a lid and simmer for 20 minutes or till gravy thickens.

KAUAI OLD-TIME TOURIST BREADFRUIT PANCAKES

Ingredients: ½ of a soft breadfruit, cooked and mashed, 1-pound fresh tuna or 2 cans, 2 beat-en eggs, ½ cup crushed pineapple drained, 1onion chopped, 2 cloves of garlic minced, 1 hot pepper minced and seeded (optional), salt and spice to taste, one cup bread crumbs, and oil for frying.
Method: Mix all ingredients (except the oil and breadcrumbs) into ten, four-inch cakes. Dip in breadcrumbs and fry until light brown on both sides.

BAKED STUFFED BREADFRUIT

Ingredients: 1 breadfruit, 1-pound minced beef, chicken, or fish, 1 onion chopped, 1 tomato chopped, 2 garlic cloves minced, 1 bunch chives, 1TBS butter or margarine, salt, and spices to taste.

Method: Boil the breadfruit for 10 minutes in salted water. It will not be fully cooked. Fry onions, garlic, chives, and minced meat, then add tomato and spices. Peel and core the bread-fruit. Stuff the fruit with the minced meat mixture. Brush the outside of the fruit with butter or canola oil. Bake in a 350-degree oven for 40 minutes. Every 10 minutes brush with butter or oil. Serve hot.

CHEESY BREADFRUIT

Ingredients: 1 breadfruit, peeled, cored, and chunked, ¼ cup butter or margarine, 2 onions chopped small, 1 clove garlic minced, 2 cups whole milk, 2 TBS flour, 1 cup cheddar cheese grated or sliced very thin, ½ a hot pepper seeded and minced (optional), salt to taste
Method: Cook breadfruit chunks in salted water until tender. In a skillet heat the butter and sauté the onions and then add the minced garlic. Add flour and milk stirring constantly. Re-move from heat and place breadfruit chunks in an oven-proof dish or pan. Add milk/onion/flour mixture. Stir in the grated cheese and minced hot pepper. Bake for half an hour at 350. Serve hot or cold.

DID YOU KNOW? The male breadfruit flower is used to repel mosquitoes. Breadfruit tree bark can be harvested for strips of fiber without killing the tree and used to make mosquito nets, and paper. The sap can be used as chewing gum.

BREADNUT - CHATIAGNE - THE GREEN CHESTNUT
Artocarpus camansi

Breadnut was introduced to the Hawaiian Islands within the last century probably Filipino immigrants. It's a food that really fools you by its appearance, closely resembling its cousin, the always sort of bland breadfruit. Breadnut looks as if a breadfruit had received a serious electric shock. The seeds and husk make breadnut unique and flavorful. In Trinidad and Guyana, it's an East Indian wedding dinner delicacy. This is a dish you seldom find in any restaurant and must be prepared differently depending on this fruit's ripeness.

The breadnut tree, breadfruit's relative, is botanically *Artocarpus camansi*. Breadnut trees originated in New Guinea, or possibly on the Molucca Islands in Indonesia. They are believed to have been brought to the Caribbean, Jamaica, and St. Vincent, with breadfruit trees by Captain Bligh in 1793. There is another tree also named the breadnut that exists in Central America and bears one-inch fruits. *Brosimum alicastrum* is commonly known as the breadnut or ramon and was cultivated by the pre-Columbian Mayan Amerindians.

HOW TO GROW: Breadnut trees are beautiful and eventually grow tall and full. This tree can grow 5 feet every year for 10-12 years and the diameter of canopy can be half the height. The only place it is presently cultivated on a large scale is in the Philippines. Mostly it is a backyard tree. A breadnut tree makes a great addition to any garden or shade for a hammock area. Breadnut succeeds up to 1500 meters, in well-drained soil with a pH of 6.1-7.4, and likes the full sun. The easiest was to get a tree is to locate someone that has a healthy tree and check around the base of the tree for shoots, or try your luck with seeds. This tree cannot be grown from root cuttings. Leave about twenty-foot radius clear and plant this tree away from septic tanks as the roots will travel.

Fresh seeds must be used and will germinate in moist potting soil within two weeks. Sprouts should get shade for the late afternoons. Seedlings grow rapidly and can usually be transplanted within 6-8 months when they are 3 ft. tall. Plant in a hole worked about a foot and a half deep. Water weekly during the dry months, and every two months broadcast some 12-24-12 fertilizer. The breadnut shoot can also first be potted and used both indoors and out-doors, treated like a large shrub. This tree is very adaptable and should begin to bear fruit in 8-10 years.

The breadnut fruit is a bright green pod usually 5 to 8 inches in diameter. A mature tree can produce 600 fruits every season. Each fruit can have as many as a dozen to 150 seeds, each seed weighing from 7-10 grams The spiny fruit has very little pulp and is grown for the seeds. Immature fruits are sliced thin and fried. The seeds are high in protein and low in fat. Unless picked the seeds -nuts will continue to grow until the fruit bursts open for the birds to enjoy. The fruit is very perishable with a shelf life of not more than two days and refrigeration doesn't help.

MEDICINAL: A tea can be made of yellowing leaves to lower blood pressure, treat respiratory problems, and fight diabetes. A decoction of the bark is used to heal wounds and treat dysentery. Crushed leaves are used for thrush and juice from stems of leaves for ear infections.

NUTRITION: 100 grams of breadnuts has 360 calories, of which 13g protein, 6g fat, 76g carbs, and 14g fiber, with vitamins C, B6, calcium, phosphorus, potassium, iron, thiamin, folate, and niacin. In 100 grams, there are good quantities of four valuable amino acids: 3g of methionine, 2.6g of leucine, 2.4g of isoleucine, and 3.1g of serine.

FOOD & USES: When roasted they are similar to chestnuts in texture and flavor. They can be canned in brine, or processed into nut butter or nut paste, flour, or oil. Dried male flowers can be burned to repel mosquitoes and other flying insects. The wood is light in weight, flexible, and easy to work and carve into statues, bowls, and fishing floats.

The easiest way to enjoy breadnuts is to buy the seeds already cleaned in the market or from a roadside vendor. Then boil in salted water for about forty minutes. Peel and eat. For something different peel the boiled seeds, chop, and sauté with butter, garlic, and culantro. Eat over spring or bow-tie pasta.

RECIPES: RED BREADNUT

Ingredients: 1 cup breadnut seeds - boiled, peeled, and chopped, 1 cup onion chopped small, 2 garlic cloves minced, 3 cups tomatoes chopped, 2 cups cooked chickpeas, 1 hot pepper seeded and chopped small, 1 bunch culantro chopped small, 1 TS lemon juice, 1 TBS oil, peanut preferred, 1 TBS garam masala, ¼ TS more or less of each of the following - depending on your taste - cumin, mustard seeds, turmeric, and salt

Method: In a large frying pan heat the oil, with the cumin and mustard seeds until they start to pop open. Add the onion and fry till brown. Add the chopped tomatoes and simmer until it be-comes a thin paste. Add culantro, pepper, breadnuts, and chickpeas with the garam masala, the other spices, and 2 cups of water. The amount of water depends on your desired consistency, less for thick, more for thin. Simmer for a half an hour. Serve with rice or roti. For an Italian flavor to this dish use olive oil and substitute oregano, basil, and thyme for the herbs, and add more garlic. Eat over pasta.

AUNTY'S BREADNUT

Ingredients: 1 good size breadnut, 1 cup coconut milk, 1/ 2medium onion chopped, 4 garlic cloves minced, 2 TBS oil, 1TBS curry powder, ½ TS achar masala, ½ TS roasted cumin, ½ TS salt

Method: Cut breadnut in half, remove core, cut into one-inch strips, and peel. Remove seeds and peel them. Strip the surrounding husk into thin pieces. Wash and set aside. In a good sized pot heat the oil, add curry powder and achar masala. As the powder starts to sizzle add ½ cup of water and stir until it thickens. Add garlic and onions, and simmer until it becomes a thin paste. Add breadnut husk, seeds, and salt. Lower heat and cover. Cook until mixture just begins to stick. Add coconut milk with 1 cup of water and increase heat until it boils. Reduce heat and cover. Simmer for half an hour, until the seeds are soft. Cook off all extra liquid. Add cumin and cook five more minutes. Cool and serve.

DID YOU KNOW? Breadnut is often considered to be a form of seeded breadfruit. However, it's a separate species and the ancestor of seeded and seedless breadfruit (*A. altilis.*)

These trees are primarily grown for seeds; which are a good source of protein and low in fat compared to other nuts such as almonds, Brazil nuts, and macadamia nut The fat extracted from the seed is a light yellow, viscous liquid at room temperature with a ristic odorsimilar to peanuts. It has physical properties similar to olive oil. Its seeds are a good source of minerals and contain more niacin than most other nuts. The seeds comprise half the weight of the fruit.

The only way to keep your health is to eat what you don't want, drink what you don't like, and do what you'd rather not. ~ Mark Twain

CANISTEL - EGGFRUIT

Pouteria campechiana

The canistel is a rare, delectable tropical fruit often called the eggfruit, or yellow sapote. This evergreen tree is usually short, seldom higher than eight meters, with fragrant blossoms. I've only enjoyed canistel a few times when I was shopping in the farmers' markets in the Hilo area. Scientists believe that the canistel is native to Mexico's Yucatan area. Trees are available at nurseries on most of the Hawaiian Islands.

HOW TO GROW: Because of its rarity and small size this tree is another good one for the backyard. It'll grow in most soils and usually will do well where other citrus trees won't. Canistel can be planted from seeds, but may take 4-5 months to germinate. Cuttings take longer to root. Grafted trees will produce fruit quicker. Combine mulching with a half cup of 12-12-24 fertilizer every other month for the first year to promote root growth. After 2 years and the tree begins to blossom, use a cup of a bearing fertilizer like 12-12-17-2 every four months. Water regularly, especially in the dry season. Canistel has few pests, but a regular spray of an insecticide as Fastac mixed with a foliar spray at three-month intervals produces better fruit.

This fruit is usually almost round, but can be slightly oval. As it ripens its smooth, glossy skin turns yellowish. The yellow flesh is firm, but the center is softer and pasty. It has been often likened in texture to the yolk of a hard-boiled egg. The flavor is sweet like a baked sweet potato. The sweet musky flavor and the unique texture of canistel flesh have turned a few away. It is an individual taste.

MEDICINAL: Almost all parts of canistel trees are used for medicinal properties. The oil from the seeds is used as a hair dressing, thought to prevent hair loss. The seed kernel oil is also used to treat indigestion, ulcers, toothache, eye, and ear diseases, and as a skin ointment. The residue remaining after oil extraction is applied to treat painful skin afflictions. A decoction of the bark can also help to heal skin disease and taken to reduce fever. The leaves are reported to be anti-inflammatory. Fungus infections, as athlete's foot, can be treated by applying the latex sap.

NUTRITION: Canistel is a very rich fruit. 100 grams has 130 calories, 2g protein with 13g fat. It has a good amount of vitamin C, thiamine, and niacin with calcium, iron, and phosphorus.

FOOD & USES: Canistels are usually enjoyed raw and plain. Some locals eat them with salt and pepper with a sprinkling of lime juice. I have a friend who slightly cooks them smeared with mayonnaise on a barbecue grill.

RECIPES: CANISTEL SMOOTHIE

In a blender mix the skinned, seeded pulp of 6 canistels with 2 cups of whole milk, 2 TBS brown sugar, 1 TS vanilla extract, ½ TS cinnamon, and ½ S nutmeg. Fill blender with ice cubes and blend.

CANISTEL PIE

Ingredients: 2 cups mashed canistel pulp, ½ cup sugar or honey, ½ TS salt, ¼ TS nutmeg, 1 TS vanilla, 1 TS lime juice, 2 beaten eggs, 2 cups evaporated milk, and 1 ready-made pie crust. Method: Blend ingredients and pour into pie crust and bake for one hour at 300° F. Let cool for at least an hour before serving.

CANISTEL SPREAD

Ingredients: 2 cups of canistel skinned seedless pulp, 1 TBS brown sugar, 1 TS lemon zest

(grated lemon peel, and 1 TS lemon juice.
Method: Mix pulp and sugar in an electric blender. Then cook in a covered pot over medium heat stirring constantly. Add lemon juice and zest. Remove from heat and refrigerate in suitable container. Use for topping on toast, ice cream, or pancakes.

BAKED CANISTEL CUSTARD

Ingredients: 2 cups mashed canistel pulp, 1 cup sugar, ¼ TS salt, 3 eggs, 1 TS vanilla, 2 cups whole milk
Method: Combine dry ingredients in a small bowl. In a bigger bowl whisk the eggs and add dry mix. Whisk in canistel pulp, milk, and vanilla. Pour mixture into a suitable oven baking dish and bake at 350 degrees for 45 minutes or until firm. Test with a toothpick. Cool and serve.

CANISTEL-COCONUT BREAD

Ingredients: 2 cups ripe canistels skinned and mashed - seeds removed, 2 cups flour, ½ TS baking soda, ½ cup soft butter or margarine, 1 cup sugar, 1 cup whole milk, ½ TS vanilla extract, 2 eggs, 1 cup grated fresh coconut, a pinch of the following - salt, ground cloves and cinnamon
Method: Combine in a suitable bowl flour, salt, baking soda, and spices. In another bowl blend butter and sugar, slowly stirring in the eggs. Whisk in milk, vanilla, and canistel before combining with the flour mixture. Finally add the grated coconut. Pour into a greased bread pan and bake at 350 for 45 minutes. Check with a toothpick before removing from the oven. This is best if baked on the center oven rack.

CANISTEL CHICKEN SOUP

Ingredients: 1½ cups canistel seeded, 1 package chicken soup mix, 1 onion chopped small, 1 garlic clove minced, 1 TBS butter or margarine, 2 cups condensed milk, 2 cups steamed chicken pieces, 2 cups of the water the chicken was steamed in, ½ hot pepper seeded and minced, 1 bundle chives minced, salt to taste
Method: In a frying pan, melt the butter before adding onion and garlic. Fry until onion is tender. Add canistel, milk, chicken, chicken stock, soup mix, chives, pepper, and salt. Cook for 15 minutes, stirring constantly.

BAKED CANISTEL— easy and different

Cut 4 fruit in half and scoop out the flesh removing the seeds. In a bowl combine flesh with 2 TBS onion chopped small. Add ½ a hot pepper seeded and minced (to taste, 1 TBS mayonnaise, and 1 TBS lemon juice. Return mixture to fruit shells and bake in suitable oven ware at 300 for fifteen minutes. Cool and enjoy.

DID YOU KNOW?

Canistel's botanical name is *Pouteria campechiana.* This delicious fruit has many names including egg-fruit, yellow sapote, siguapa, zapotillo custiczapotl fruta de huevo, zapote amarillo, kaniste, limoncillo, mamee ciruela, zapotillo de Montana, huevo vegeta, mammee sapota, and many more.

There are blessings in being close to the soil, in raising your own food even if it is only a garden in your yard and a fruit tree or two. Those families will be fortunate who, in the last days, have an adequate supply of food because of their foresight and ability to produce their own. ~ Ezra Taft Benson

CEYLON OLIVE
Elaeocarpus serratus

The Ceylon olive is a true wonder of nature and will grow and produce in most areas of Hawaii. One tree is at the Wahiawa Botanical Gardens in Honolulu. These trees produce fruit similar in appearance to a Middle East / European olive. The Ceylon olive's botanical name is *Elaeocarpus serratus* and the European olive is *Olea europaea.* This tree is supposedly native to Sri Lanka and India. Growing this tree would be a worthwhile family project for both land-scaping beauty and healthful fruit. The ovoid, green fruits have both nutritive and medicinal value used to treat various diseases and taste good. Shampoo and conditioners are made from Ceylon olives.

This evergreen tree is grown in home gardens across the country and there are some commercial orchards, but not enough. The tree usually measures 12 meters tall and 8 meters wide. The plant will grow anywhere in Hawaii. Lacy, fragrant white blossoms, like snowflakes, precede the locally famous Ceylon olive. The fruit is green to blue, oval-shaped when ripe with a big brown seed. The flesh is sour, but pasty like an avocado.

HOW TO GROW: First, with the Ceylon olive, you need patience. It can take up to 2 years for the seed to germinate. Wow! That's because this seed has an extra hard, woody shell. To accelerate germination, it's recommended to make cracks vertically through the seed coat using a seed cutter followed by soaking in distilled water for 24 hours at 27 °C. But don't file away too much and damage the inner pulp. What the inner core needs is moisture to sprout. So be patient. Plant several seeds in pots and keep them in partial shade. You don't want the potting soil to dry out or become too moist. When they sprout and are about 6 inches, separate to individual pots. You can try cuttings, but it may be best to look under olive trees for sprouts. The seeds have a small window of fertility and must be planted quickly. The final transplant should be in well-drained, loamy soil with a pH of 5.5 to 7.5, in full sun. It grows fairly rapid-ly, becoming a sturdy, hardwood tree, with branches capable of safely supporting adult olive pickers. Trees start bearing fruit in three to four years, up to 50 kilos of fruit per season. The wood is pale yellow. The blue occurs in some olives from anthocyanin, an unstable pigment affected by the soil pH.

People are the biggest threat to these olive trees. The tree reproduces by seed and the big, 2 gram brown seeds are used as prayer beads in Hinduism, especially in Shaivism. They are sometimes called blueberry beads. The seeds are associated with the Hindu Lord Shiva, called his tears, and worn for protection while chanting mantras such as *Om Namah Shiva-ya*. The seeds are primarily used in India, Indonesia, and Nepal as beads for organic jewelry and malas (bead necklaces; they are valued similarly to semi-precious stones. Non-Hindus wear 'rudraksha' seed bracelets because there is a belief cosmic energy can be absorbed. Amazon sells them. Increased seed collection has shrunk the natural seed bank. This tree is in the 'threatened' category.

MEDICINAL: Ceylon olive seeds also produce oil. 100% pure oil extracted from seeds is a dietary supplement. Two drops of oil once a day are good for internal healing. It's also used as hair oil, will end dandruff and condition hair, and reduce acne and pimples. It also can help heal skin and stop itching in conditions like eczema and ringworm. It's also used as a massage oil and can treat burns and scars.

Ayurveda homeopathic medicine refers to the Ceylon olive seeds with its bark and leaves as a cure for various ailments: headache, fever, skin diseases, as a blood purifier, for coughs and breathing problems, treat high blood pressure, and heart diseases. It is used in cosmetics to make the skin glow and remove facial wrinkles. For improving memory power:

The paste of the leaves has been known to be a cure for ulcers. Ground leaves are also applied to the scalp before bathing to prevent dandruff, banish lice, and moisten and condition the scalp to promote healthy, silky hair. Local supermarkets sell veralu shampoos and conditioners.

The Ceylon olives possess anti-inflammatory, antibiotic, anti-anxiety, analgesic, antidepressant, and antihypertensive properties and are also used for rheumatism and an antidote for poison. For centuries, Ayurvedic medicine has valued the Ceylon olive. It is time for more clinical trials to support its therapeutic use. Herbal personal care/appearance products are in demand by the global markets. The Ceylon olive could be a perfect home garden business for anyone with land, a love of nature and traditional medicine, and extreme patience.

NUTRITION: These fruits are high in starch and sugar and have low amounts of protein and iron. Per 100 grams of edible flesh: 28 calories, ½ g fat, 1 gram of fiber, and ½ gram of carbs. They have 5mg of calcium, 3mg phosphorus, with 8mg vitamin C. Even the dry leaves have a vitamin C content of 257mg per 100 gm.

FOOD: Raw or cooked can be used in curries and also pickled like olives. There is an ice cream flavor made from Ceylon olives. Another combo is the olives stuffed with dates.

For a nice healthy snack, pick out the softer olives that are ripe. Peel o the esh, add jaggery-palm sugar or regular brown sugar and combine. This is a sweet and sour taste. If you have a sour mouth, try nil veralu, especially one referred to as young nil veralu, soak or use fresh, chop the olives into pieces, and add pepper and salt. Do not use hot water because that will harden the skin and the flesh will become tough. The water should be lukewarm and soak the olives for only 10 minutes.

RECIPES: PICKLED CEYLON OLIVES –ACHCHARU

Ingredients: 500g ripe olives, 2 TBS coconut oil, 2 cloves garlic chopped, 1TBS chili flakes, 10 fresh curry leaves, 1 TBS coconut vinegar, 2TBS mustard paste or sauce, TBS brown sugar, ½ TS ground black pepper, 1 TS salt.

Method: Boil olives in salted water, drain, and keep aside. Heat oil, add garlic, and fry to a golden brown. Add other ingredients stirring for 2 minutes. Add boiled olives and stir well. Transfer to a large sterilized glass jar. Seal and store in a cool, dark place for 2 weeks before eating. Once opened, the olives will keep in the fridge for 3 weeks.

CEYLON OLIVE CHUTNEY- JALPAI CHUTNEY

Ingredients: 150g Ceylon olives, 10g raisins soaked in water, 1 clove garlic minced, 1 dry red chili, ½ TS chili powder, 2TBS mustard oil, ½ TS of the following, mustard, fennel, and cumin seeds, 2 TBS brown sugar.

Method: Boil the Ceylon olives in a pressure cooker for 3 minutes, remove but save the water separately. Remove seeds and smash the flesh. Heat oil in a wok and add smashed olives, stir in the spice seeds, garlic, salt, sugar, and raisins. Sauté for 5 minutes. Add boiled water and stir until it reached your desired consistency.

DID YOU KNOW? The Ceylon olive has many names: Indian olive or wild olives in English. Also, rudraksh, mas veralu eralu, pol veralu, thiththa veralu & gal veralu, and jalpai in Bengali.

Nutrition is so important. It can't be stressed enough.
~ Dwayne Johnson AKA The Rock

CHENETS - GENIPS – MAMONCILLO – SPANISH LIME
Melicoccus bijugatus

The Spanish lime trees are a great backyard trees throughout the island chain. These trees grow on every Hawaiian island but are seldom grown in orchards. The Spanish lime/chenet/genip/mamoncillo – whatever you call them – the season is always eagerly awaited. After washing the fruit, cracking the crisp skin with a bite and manipulating to suck the flesh from the skin and then spit out the big seed sounds complicated, but worth every drop of the sweet juice.

These sweet, unique green fruits have the Latin botanical name *Melicoccus bijugatus* are similar in taste and related to lychee and several other edible tropical fruits including the longan and pulasan but native to the Western Hemisphere. Chenets grow everywhere in the Caribbean and every place has a different name, so I'll stick with chenet. I wonder if anyone has a chenet plantation?

They look like a small key lime, yet grow in bunches like cherries or plums. Inside the fruit's crisp leathery skin is a succulent and tasty pulp with a flavor of lemonade spiked with grapefruit juice. The pulp ranges in color from pinkish with an orange tint to yellow. The fruit is almost all seed, and these seeds can be roasted and eaten like nuts. This fruit is related to the lychee and classed as a drupe. A chenet fruit has a tight and thin, yet rigid layer of skin that experienced eaters crack with their teeth. Inside the skin is a sweet to tart jelly like pulp. It is sucked by putting the whole fruit inside the mouth. The fruits seed is big so the flesh layer is only about a quarter inch thick. Be careful to lean forward when you are indulging in as the juice makes brown stains.

WARNING! Be careful eating chenet fruit because the large seed is slippery and can accidently lodge in the throat and can cause choking, especially in young children.

HOW TO GROW: Great for the home garden, this tree provides shade and after a chenet tree is a meter tall, it is almost maintenance free. The downside of chenets: to get fruit it's necessary to grow both male and female trees. A chenet tree's usually started from cuttings and fresh seeds take a month to germinate. Try the easy way: find the one tree that produces the biggest fruit. Some chenets are as small as three-quarters of an inch while others are an inch and a half wide. Snip about three or four young branch tips and dip in a rooting compound. This will 'almost' guarantee your tree will bear fruit like its parent. Plant in the soil at a place where it can spread to a 30 ft. diameter. The trees require well-drained soil with a pH of 5.5-8. These trees adapt to their surroundings and after 3 years can withstand a 3 month drought. This tree wants sun, not shade. With the best conditions, tropical humidity and rainfall inter mixed with a dry season, sandy loam soil, regular fertilizer and weeding, this tree's girth can grow about a half inch wider every year. The largest tree recorded in Puerto Rico is more than 1 meter (39 in. in diameter.

Keep the weeds pulled from around this young tree and water during dry spells. The tree grows slow and it will be 5-10 years before it bears fruit. Grafted trees bear earlier than seedlings. Chenet trees grow to seventy feet tall and forty feet wide, if not topped and pruned. Use a half cup of starter 12-24-12 fertilizer every other month until the tree is about three years old. Then apply a cup of bearing fertilizer 12-12-17-2 every other month. Once it starts to flower increase the fertilizer to every month.

MEDICINAL: In Venezuela, roasted chenet seeds are smashed, combined with honey and eaten or brewed into a tea to curb diarrhea. An astringent decoction of chenet leaves is taken as an enema for intestinal complaints. In Nicaragua, seed extracts treat parasites. Adding chenets to your diet will help prevent diabetes, cardiovascular disease and gastrointestinal disorders. Look out, eating a lot of chenets will demonstrate its laxative properties.

Chenets contain phenolics, flavonoids, and sugars that have medicinal and healthy diet properties.

NUTRITION: 100 grams of chenets has 60 calories with some calcium, phosphorus, iron and niacin. The fruit is a source of fiber, vitamin A, vitamin B, vitamin C, calcium, iron and the amino acids tryptophan and lysine.

FOODS & USES: The #1 use is eating chenets out-of-hand. The size of the fruit limited their commercial success in international markets. The removal of the pulp from the seeds can be challenging and low yielding. If you are lucky to find a tree with extra-large chenets, you can scrape the pulp from the seeds to make marmalade or jelly, but this is a lot of scraping. Refrigerated, chenets can keep for 2 weeks. Peeled fruits can be boiled and juice is a fantastic cold drink and especially for daiquiris. Colombia and Venezuela sell canned chenet juice.

The peel of the fruit is also a promising source of bioactive components (extra-nutritional ingredients in small quantities in foods including lycopene, resveratrol, and tannins. A dye is made from the juice. The blossoms are rich in nectar that makes a dark, great-flavored honey. The heartwood is beautiful, fine-grained, and yellow with dark lines. It is a heavy hardwood, but will rot outdoors. Chenet wood is used for rafters, indoor framing, and cabinetwork and to make furniture. Amerindians in the deep Orinoco grasslands eat cooked chenet seeds as a substitute for cassava.

RECIPES: KICKING SPANISH LIME SHERBERT

Ingredients: 2 cups chenet juice (probably will take about 3 gallons of chenets!, 4 TBS ginger water –see below, 1 package unflavored gelatin, 2 TBS rum (optional, 1 cup ginger beer, 1 egg white

Method: Wash ginger root and slice thin. Put in a small pot and add just enough water to cover pieces and boil for 5 minutes. In a bowl add 1 TBS chenet juice to unflavored gelatin before adding the hot ginger water. Combine and whisk all ingredients except egg white. Put into a suitable container and put in freezer until it begins to freeze. Remove from freezer, blend and add egg white. Fully freeze in a sealed container.

CHENET WATER - Perfect for a hot afternoon. Peel a pound of chenets, avoid staining any clothes with the juice. Put in a large pitcher and let sit for an hour in the shade. Pour into ice filled glasses. Add sugar as desired. This raw unsweetened juice will curb your thirst and cool you off.

DID YOU KNOW?

Chenets are not a citrus fruit. According to Caribbean folk tales, girls learn the art of kissing by eating the sweet flesh of this fruit. More folklore has it that if a woman finds two seeds in her chenet, she'll have twins. Depending on size, there can be 28-50 chenet fruits in 1 pound. This tree's leaves when spread on a floor of a stable or fowl pen are reputed to attract and then kill flies.

Growing food was the first activity that gave us enough prosperity to stay in one place, form complex social groups, tell our stories, and build our cities.

~ Barbara Kingsolver

COCO PLUM – FAT PORK
Chrysobalanus icaco

Why does every tropical fruit that is considered a 'plum' have too many names? Another specialty fruit is the coco plum that's grown throughout the Hawaiians Islands. Its scientific name is *Chrysobalanus icaco*. This tree is also known elsewhere as the icacos plum, and bears what we call 'coco plum fruit or apple.' The coco plum is native to Mexico, Central America and South America, to Ecuador, northern Brazil, and throughout the Caribbean chain. This type of 'plum' is usually found wild, and seldom cultivated. I only know of one coco plum tree. This almost unknown, pinkish fruit is about the size of a big plum.

My research indicates that it's known as the coco-plum, the cotton plum, icacos plum, icaque ponne, pork-fat-apple, or zicate. Coco plums are usually a flushed pink color, but they can be white or purple. The flesh is never crisp, but spongy, whitish, and slightly sweet, but usually tasteless when ripe. This tree's availability to water determines the thickness of the fruits' flesh.

References to the coco plum usually refer to it as a small shrub growing to 10 feet. My research describes it as a slow-growing, small evergreen tree with a twisted trunk that can grow prostrate along the ground that's usually found where the soil is moist or flooded. The coco plum tree I know of is at least 30 feet and usually bears fruit twice a year.

Why is it called fat pork? I believe the name 'fat pork' comes from a successful pig farmer who was lucky to have a couple of these trees close to his pens. When this fruit dropped to the ground, his pigs got fat. Another name, 'cotton plum' is derived from someone referring to the consistency and taste of the fruit as 'sweetened cotton.'

HOW TO GROW: This is a shrub, 'rarely a tree' that has very striking bright green leaves in contrast to a reasonably attractive - reasonably tasty pink to purple fruit. It can be grown in gardens, as a hedge, and even in big porch pots. Fresh seeds will germinate usually within a month and can be planted at their permanent location at about 6 months. It can be cultivated from seeds and is a relatively slow growing tree. Thus, its area must be cleaned of weeds at least monthly. Once this tree is 6 feet tall it can withstand damp soil or very dry conditions. It does best in soil with a pH range of 6.6 to 8.4 with regular watering/irrigation.

This is a good tree for a river bank, or to stop soil erosion close to a beach, even in sand dunes, because the coco plum loves the hot, humid tropical lowlands. Coco plum can survive wind storms, salt spray, and floods. Shade is its biggest enemy. Coco plum shrubs can grow as small, individual, multi-stemmed trees, or spaced 3-4 feet apart, they can be pruned to form a hedge. They should bear fruit in 3-4 years.

MEDICINAL: A tea made from coco plum leaves is reported to help control type II diabetes. A combination of fruits, bark and leaves boiled in water will produce a tea that will fight severe diarrhea. A tea made only from the bark may help kidney ailments. The teas can be used externally as washes to treat skin problems.

NUTRITION: These fruits are so rich that 100 grams has 50 calories with some vitamin C. calcium, iron, and phosphorus. The chemical features of the coco plum include flavonoids, terpenoids, steroids, and tannins. Scientific medical interest in flavonoids has increased because many of them exhibit anti-inflammatory, antiviral, antibacterial, as well as anticancer abilities.

FOOD & USES: Coco plums are a rich fruit with a consistency close to that of an ackee. They are edible raw, stewed in sugar, dried like prunes, or made into jams and jellies. The purple or reddish fruits are reported to have a better flavor than white skinned coco plums. To get a better taste of these bland fruits, cover in sugar water over night before using them in any recipe.

To dry and preserve the fruits, pierce the plums through the center and the seed. This method permits the plum's juice to seep into the seed. After separation from the dried shell/ skin, the dried fruit kernel surrounding the nut is sucked and the dried flesh is eaten.

The seeds are reported to have a good, nutty flavor, eaten either raw or cooked. An oil suitable for human consumption can be extracted by pressing the seeds.

RECIPES: COCO PLUM SHERBERT

Ingredients: 30 ripe coco plums halved with seeds removed, ½ cup sugar, ¼ quarter cup water, 1 cup fresh orange juice, ½ cup corn syrup, 2 TBS grated orange peel, 1 TS fresh lemon juice, 1 TS vanilla extract, 1 TS salt, 1 cup whipping cream.

Method: Combine coco plum, sugar, and ¼ cup of water in a large skillet. Bring to a boil and then reduce to a simmer always stirring until coco plum starts to fall apart, usually about 10 minutes. Set aside until cool. Pour into a blender and puree until smooth. Add remaining ingredients and blend. Freeze for 3 hours and then blend again to break down ice crystals.

COCO PLUM STUFFING – for a unique chicken dinner

Ingredients: 1 whole roasting chicken – 4-5 pounds, 12 coco plums - seeded and peeled, 1 garlic clove minced, 1 small onion chopped very small, 1 TS salt, 2 TBS brown sugar, 2 bay leaves, a pinch of cinnamon, and or nutmeg optional.

Method: Rub the cavity of the chicken with salt. Combine the remaining ingredients and fill the chicken. For a more nutty taste use the cinnamon and nutmeg. Cover with foil and bake for an hour at 300 F. Remove foil and continue to bake for 30 minutes more or until skin has started to brown. Cool slightly before serving.

COCO PLUM POTATOES

Ingredients: 2 dozen coco plums seeded and peeled, 3 large potatoes boiled and chunked, 1 bunch chives - chopped small, 2 culantro leaves minced, 1 TS salt, 2 stalks celery chopped small, ½ a small onion chopped small, and 1 cup grated cheddar cheese.

Method: In a suitable greased baking dish or bread pan, combine all the ingredients except cheese. Cover with foil. Bake at 250F for 30 minutes. Uncover and spread cheese on top. Continue to bake for 15 more minutes. Allow to cool slightly before serving.

COCO PLUM AND RAISIN COMPOTE

Ingredients: 2 pounds of coco plums (about 60 peeled and seeded, ½ cup raisins, ½ cup brown sugar, 1 TBS fresh lemon juice, ½ TS salt, 1 cup water, 1 TS cinnamon, and a pinch of nutmeg

Method: Boil the water in a medium sized pot. Add sugar and stir until a syrup begins to form. Mix in the coco plum pieces, raisins, and salt. Reduce heat and simmer for about 10 minutes. Add lemon juice, cinnamon and nutmeg. Stir over heat for 2 minutes. Remove from the heat and add cover. Cool to room temperature. Eat over ice cream or cake.

DID YOU KNOW?

Icacos point on the southwestern tip of the island nation of Trinidad, in the southern Caribbean is named after the coco plum aka icacos plum. A British surveyor in 1797 named the area 'Marsh of Icaque.' The coco plum shrub covered the land. The name originated from Haiti where the Taino Amerindians called the fruit icaco. It was considered a subsistence food, only if you are starving.

Growing your own food may be one of the most powerful steps you can take for the health of yourself, your family, and your planet.

~ Lindsay Oberst

CUSTARD APPLE - CHERIMOYA

Annona cherimola

I beg a friend for a few of the delicious cherimoya/custard apples that grow near his new house. This year he gave me a newly sprouted tree. To me, spooning into a chilled custard apple is a true natural delight. It is a heart-shaped fruit, usually yellow-tan with edges of pinkish brown. The creamy insides are as close to real custard as Mother Nature can make. A sweet custard apple feels like it has hard sugar inside because the creamy flesh crunches as you eat it. Grow your own.

Scientists believe the custard apple originated in the Caribbean before Spanish explorers carried the seeds throughout the tropical world. Custard apple's botanical name is *Annona cherimola*. The *Annona* family includes soursop and the sugar apple.

HOW TO GROW: A custard apple tree grows from seeds or grafts. These trees grow best in rich soil, especially along rivers between 800 and 5000 meters of elevation. It matures quickly if it is mulched, fertilized, and watered regularly. The first two years use a starter-rooting fertilizer mix such as 12-24-12, a half-cup bi-monthly. After blossoming begins, use the same amount, but switch to a bearing fertilizer mix as 12-12-17-2. Prune to shape the tree to your yard. Hand pollinating can increase the yield. The chalcid fly is the custard apple's insect enemy. Spray monthly with an insecticide as Fastac or Pestac combined with a soluble foliar fertilizer. Bats also will damage these fruits. The cherimoya/custard apple will grow anywhere throughout the Hawaiian chain.

The trees are wide with large leaves, usually about twenty feet tall. The leaves hang over the delicate fruit to protect it from burning in the tropical sun. The pale-yellowish, slightly fragrant blossoms hang in clusters, but never seem to fully open. This fruit is usually about four inches in diameter and irregularly shaped like a heart, almost round or oval with a depression at the base. The skin is thin, but tough. It can be yellowish-brown with a brownish-red tint when ripe. The custard apples usually have more than fifty seeds, but some varieties have a single seed or as many as seventy. It has a thick, white creamy, somewhat grainy flesh. The flavor is sweet and agreeable though without the sugary sweetness of the cherimoya, sugar apple.

MEDICINAL: Crushed leaves or a paste of the fruit's flesh may be applied to boils, abscesses, and ulcers. The bark is very astringent and when boiled in water, it is drunk for a tonic and a remedy for diarrhea and dysentery. Pieces of the skin of the tree's roots will relieve a toothache when put against the bad tooth. Central Americans roast and smash the seeds, then combine with water or raw milk to induce vomiting, or purge waste from the body. This is used for poisonings. This same roasted seed powder is mixed with cooking grease or oil to kill skin parasites as lice. A tea made from custard apple skin is used to fight pneumonia.

NUTRITION: One hundred grams of custard apple has about a hundred calories with good amounts of calcium, phosphorus, and vitamin C. Custard apples are a well-balanced food having protein, fiber, minerals, vitamins, energy, and little fat. They are a good source of dietary fiber, vitamin B6, magnesium, potassium, with some B2 and complex carbohydrates.

FOOD & USES: Check the ripeness of a custard apple just as an avocado, when you gently squeeze, it gives under your fingertips. Custard apples can be purchased ready to eat, or hard to the touch— to ripen in a few days. They should never be black or pulpy. Fully ripe it is soft to the touch; and the stem and attached core can be pulled out. Custard apples are only eaten when soft, and only the flesh is eaten. To eat, simply cut in half and scoop out the white flesh. The custard apple should be moist with a pleasantly sweet aroma. They are absolutely the best when chilled.

RECIPES: CREAMED CUSTARD APPLE

Ingredients: the pulp of two seeded, pureed custard apples, one TS lemon juice, two cups cream or one cup whole milk, three TBS clear gelatin, one package softened cream cheese, one-half cup powdered sugar, one-quarter cup boiling water

Method: Dissolve gelatin into boiling water. In a bowl whisk the soft cream cheese gradually adding the gelatin mix, lemon juice, powdered sugar, and cream. Add custard apple and whisk as smooth as possible. Chill and serve with cake or pastry.

CUSTARD APPLE SORBET

Ingredients: pulp of 6 custard apples - peeled and seeded, ½ cup powdered sugar, 1 cup boiling water

Method: Dissolve powdered sugar in boiling water stirring until a syrup forms. Let cool before blending in the custard apple pulp. You can combine both in a blender or food processor. Freeze until it is a stiff slush. Blend again and refreeze.

CUSTARD APPLE SAUCE - different for the barbecue grill especially for fish steaks and fillets.

Ingredients: 1 custard apple - skinned, seeded, and pureed, 1 bunch chives chopped tiny, 2 gar-lic cloves minced, 1 TBS butter or margarine, ½ cup fish or chicken stock, ½ cup white wine, ½ hot pepper seeded and minced (optional, salt, and spices to your taste

Method: Cook chives and garlic in butter. Add wine and simmer until it thickens before adding the stock. Simmer and add custard apple. Add spices and salt. Apply to fish steaks while they are either baking or grilling.

CUSTARD APPLE CAKE

Ingredients: 4 ripe custard apples peeled, seeded, and minced, ¼ pound butter, ¾ cup powdered sugar, 3 eggs, 1 TS vanilla extract, 2 cups self-rising flour, 2 TS cinnamon, ½ TS nutmeg Method: With a whisk or electric mixer beat a half cup powdered sugar and vanilla into the butter in a medium bowl. Add eggs one at a time. Beat until the mixture is pale and creamy. Add custard apple puree to the mixture. Mix till combined. Divide flour in half and carefully fold into the mixture so not to get any lumps. When flour is fully added, and the mixture is smooth pour into a greased oven pan. Sprinkle remaining sugar, cinnamon, and nutmeg over the mixture. Bake for one hour at 350 degrees. Let cool and serve.

DID YOU KNOW? The atemoya is a delicious crossbreeding of a custard and a sugar apple. Custard apples are also called cherimoyas, bullock's heart, or bull's heart, all due to its shape and a reddish tint. The skin color is reflected in the Bolivian name, chirimoya roia. This fruit is juicy with creamy white flesh and large, black seeds, and tastes like a combination of pineapple, mango, papaya, and vanilla.

In our society growing food ourselves has become the most radical of acts. It is truly the only effective protest, one that can —and will — overturn the corporate powers that be. By the process of directly working in harmony with nature, we do the one thing most essential to change the world — we change ourselves. ~ Jules Dervaes

DRAGON FRUIT – PITAYA

Selenicereus undatus

The dragon fruit is one of the most beautiful and widespread members of the cactus family. You can find plants at local nurseries and fruit in the farmers' markets throughout the Hawaiian chain. It can be a beautiful garden landscape ornamental with fragrant night-blooming flower that bears delicious expensive fruit. But this is a rapidly growing, vining cactus, usually with sharp spines and will climb over everything, trees, walls, and rocks, if it is not kept under control. In many countries, dragon fruit is considered an invasive species. There are 19 edible species of cactus that produce pitayas; dragon fruit are the biggest and most delicious. Pitaya usually refers to fruit of the genus *Stenocereus.*

The dragon fruit looks like it came from outer space with bright red, purple, or yellow-skinned varieties. The fruit is oval or pear-shaped up to 5 in. /12cm long by 3in. /9cm wide and can weigh more than a kilo. The flesh is either white or red, dotted with edible black seeds. The taste is a combination of watermelon, berries, and kiwi. The cactus is also very different as it vines and climbs over almost any terrain. At your home, it would be best to keep it under control by running it up a strong wide trellis or fence. It would be nice around a swing seat because the flowers bloom huge, 9 inches wide, and have a nice aroma, but only for one night and then they fall off. They're called moonflower or queen of the night.

The two main varieties of dragon fruit are probably native to the highlands of Central and South America. Sweet dragon fruit come from the most cultivated genus *Hylocereus*, of the *Cactaceae* family, while sour dragon fruit are from the *Stenocerus* genus. Pre-Columbian Amerindians carried the seeds to tropical areas of the Americas and the Caribbean. According to legends in the 1200s, Aztec emperors ate fresh pitaya carried 500 km from Western Mexico by runners. Spanish explorers brought the fruit to the Philippines in the 1500s. European explorers gave it to the Chinese in Taiwan in 1645 and it spread throughout the country. French missionaries, while they were also in Mexico, were the first to export the fruit from Central America to South East Asia around 1860. Dragon fruit was only known as pitaya before it was renamed by Asia marketers. This name is derived from a legend that the fruit was the last breath exhaled by a dragon defeated in battle. This fruit has become an important fruit crop throughout South-East Asia and is now cultivated widely in the tropics and subtropics. Under prime cultivation, this fruit can produce 4 times a year.

HOW TO GROW: Dragon fruit is a fast growing, spreading triangular shaped cactus with frequent sharp black spines and doesn't tolerate extreme temperatures or full sunshine. Yellow dragon fruit are the sweetest, but red are also tantalizing. Fresh seeds germinate easily, usually within a few days. Plant 1/2" deep in moist, sterile soil. It's very important to keep soil temperature warm, 70-85F. The best pH is 6-7 and in most soils use high amounts of organic material. These vining cactus can be grown in big 15-gallons pots with a strong T inserted at planting for the vine to climb. I had one growing, didn't pay attention, and it crawled up the side of my house. Repainting didn't remove its trail. Many places consider it an invasive species.

Dragon fruit are self-sterile and need to be pollinated by nocturnal moths or bats. The large dragon fruit flowers require pollination during the night as they generally whither in the day and only last up to 24 hours. In 1-2 months, depending upon climate and elevation, the fruit develops and is ready for picking. All varieties become more intense in color as they ripen, but nature gave these fruits a timer. When ready to pick, the spines or needles, drop off around the head of the fruit. It's cultivated commercially in Israel, Thailand, Philippines, Okinawa, Japan, Taiwan, Sri Lanka, southern China, Malaysia, Vietnam, Indonesia, and northern Austra-lia. Vietnam, where it is called *thanh long*, is the world's leading exporter with revenues from dragon fruit making up 55% of the country's fruit export revenue.

MEDICINAL: Dragon fruit stems and flowers are used to treat diabetes, as a diuretic, and promote healing wounds. The fruit lowers cholesterol, boots immunity, and helps digestion. In Taiwan, it's used by diabetics as a substitute for rice to add fiber to their diet.

NUTRITION: Dragon fruit are rich in vitamins C, E, B1, B2, B3, phosphorus, fiber, magnesium, iron, niacin, potassium, and calcium. There are only 60 calories per 100 grams. Like everything, consume dragon fruit in moderation because of fructose levels. The tiny seeds are a good source of omega-6 and omega-3 fatty acids. The fruits' phosphorus promotes anti-aging properties and lycopene fights cancer. Eating it helps to promote the growth of two types of healthy bacteria in your stomach: lactic acid bacteria and bifidobacterial.

FOODS & USES: The peel can be used to produce betacyanin and coloring pigments, and a gum in the food and the cosmetics industries. The red or purple pulp can be blended as a drink, used for sherbets and salads, to make syrups, and juice or wine. Frozen pulp can flavor ice cream, yogurt, jams and jellies, candy, and cakes. The edible unopened flowers are steamed or cooked as a vegetable. The flowers are added to soups or made into a tea. Same as with most fruit, chilling improves the taste. Then cut it in half and spoon out the flesh or mix the juicy flesh with sugar or milk into a smoothie. Dragon fruit goes well with milder salad greens and spinach.

RECIPES: PINK DRAGON SYRUP SODA

Ingredients: 1 dragon fruit cut in half, the juice of 1 lime, sugar to your taste, club soda.

Method: Scoop the flesh and blender until smooth. Strain and discard the solids. Add lime juice and sugar to your taste. This produces a syrup you can refrigerate. In a tall glass with ice add 3 ounces and then gently stir in club soda and liquor (vodka or gin if you choose.

DRAGONFRUIT TROPICAL PUDDING

Ingredients: ½ cup water, one packet unflavored gelatin, 1 dragon fruit peeled and cubed, 5 rambutans-peeled, seeded, and diced, 4 TBS brown sugar, 1 cup coconut milk

Method: Put the water into a bowl and sprinkle the gelatin on top and let it sit and absorb for 5 minutes. In a saucepan, combine the sugar and coconut milk and simmer on low. Smash dragon fruit with a fork. Then add it and the rambutan to the sweetened coconut milk. Stir in the gelatin for 1 minute. It's best to pour into the serving bowls and then refrigerate for at least 4 hours.

DRAGON FRUIT SALSA

Ingredients: 1 cup 2 dragon fruits chopped into small cubes, 2 green onions chopped small, 5 stalks of cilantro, juice of one lemon or lime, 2 hot pepper seeded and minced (optional)

Method: Combine and let sit so the flavors meld. Enjoy with crackers, biscuits, roti or chips.

DID YOU KNOW? Botanically dragon fruit has two names *Selenicereus undatus* and Hylocereus *undatus*. It's the strawberry cactus pear in Latin America, *pithaya, pitajaya,* or *pitahaya,* S.E. Asian refer to it as the dragon pearl fruit. In Mandarin, it's long guo, in Vietnamese: thanh long. The flowers are called the belle or queen of the night, the Cinderella plant. In the Spanish speaking Caribbean Islands, it is *Flor De Caliz*, and in Thailand it's *Geow Mangon.* The pithaya roja variety of dragon fruit has the brightest red flesh, almost magenta. Eating too much red dragon fruit can give rise to a harm-less condition called pseudo hematuria, which turns your urine reddish. The rare tyalgum purple dragon fruit is considered superior quality. Its creamy white flowers, as large as an automobile headlamp, produce fruits that resemble cricket balls with brownish skin.

DURIAN – THE KING OF FRUITS

Durio zibethinus

The durian fruit is either absolutely loved or absolutely hated, but it is the delicious dessert fruit of Southeast Asia. The creamy flesh is delicious, but the aroma can be noxious. Durian looks like a breadfruit that suffered an intense electrical shock. Usually, 9 inches in diameter, big ones are double that with green, spiky skin. If you love the sweet, unique taste of the durian's custard flesh, you condone the nauseating smell that has been compared to sewage combined with rotting garlic, and honey. This delight grows sparsely on all the Hawaiian Islands in private gardens. The big Island and Oahu have commercial gardens.

King David Kalākaua is credited with bringing durian to Hawaii in 1881. It was a gift from the king of Thailand when he stopped in Bangkok during his world circumnavigation. The Thai King respectfully supplied the Hawaiian travelers with mangosteens, coconuts, and durians. The king saved seeds and they were later planted at The Grove Homestead, on Kauai.

The durian has a thick thorny-spiked gray to dark green skin. As this fruit ripens, it splits open into a half dozen sections. The milky-white flesh has a thick consistency that may taste slightly like strawberries, or almonds with caramel. Each section may have ten to fifteen seeds that may be roasted similar to chestnuts. The sickening smell may suggest rotting meat. Eating a durian is similar to enjoying Limburger cheese; hold your nose!

I recommend wearing gloves or wrapping the durian in a dish towel before attempting to open the hard, spiny rind. Chop it open and pry the shell apart and remove the flesh. It's usually white or pale-yellow, but some varieties have red or green flesh. Usually, one fruit provides two cups of pulp. Durian is eaten raw, or used for juice, candy, and most often to flavor ice cream.

Scientists believe the durian originated in Malaysia. In the early 1400s, a Chinese explorer Admiral Zheng discovered Malaysian natives eating a 'foul-smelling fruit named tuer-wu.' During the Portuguese explorations of SE Asia, they found the durians of Malacca to be the Orient's best fruit. They created durian plantations in India, Sri Lanka, Zanzibar, and Africa.

HOW TO GROW: This is an easy tree to grow and suitable for most backyards. A tropical climate below 3,000 meters is the habitat of the durian and the temperature is never below 50 degrees with the usual sticky 80% humidity and with rain around 100 inches annually. It grows easily from seeds, but the seeds cannot be dried for transport and must remain moist for planting. Within two weeks the seeds will germinate. The sprouts can be ten inches tall before they drop the original seed-shell. It does well in soil with a pH between 5.5 and 7.5. The trees blossom twice a year, and the fruit matures in 4-5 months. Usually the fruit load is so heavy and the root system so small, it is best to prop a durian tree to keep it from falling over. Also, it is best to tie the fruits to the branches because they tend to drop before they are ripe.

MEDICINAL: The leaves, roots and fruits are used in homeopathic medicine to treat fever, jaundice, and skin eruptions. Durian quickly produces energy. Durian builds muscle tissue and improves blood pressure. It may slow or stop the spread of cancer cells, lower cholesterol, and has antibacterial properties. The downside of durian is to never combine it with alcohol. Research found durian prevents alcohol from dissipating and increases alcohol levels in the blood causing nausea and vomiting. Durian may cause you to feel heated.

NUTRITION: Durian is nutritious, but fattening at 150 calories per 100 grams. Diabetics beware; durian has a high sugar content. Eat small portions: per quarter pound, it has 150 calories, with 3 grams of protein and 35 grams of carbohydrates, and 6 grams of fat. It is high in vitamin C, vitamin B-6, thiamine, potassium, and manganese.

FOOD & USES: Durian is used in desserts and side dishes. It is also sold frozen, which makes the flesh fall apart and more fibrous. Durian fruit is considered in Malaysia to be an aphrodisiac. There is a saying, when the durians come down, the sarongs come off!

RECIPES: DURIAN SHAKE

Ingredients: 1 cup durian flesh seeded, 2 bananas, ½ cup milk, ¾ cup water. ½ TS cinnamon, ½ TS nutmeg, (optional 1ounce rum

Method: Blend until smooth. Sprinkle nutmeg and cinnamon on top.

THAI COUNTRY PIE

Filling Ingredients: 1 lb. minced chicken, 10 fresh bird chilies or 3 regular long green chilies, 6 garlic cloves chopped, 1 TBS coriander root chopped (coriander seeds may be substituted, 2 TBS cooking oil, 2TBS oyster sauce, 1 TS sugar, a ½ pound of holy basil (tulsi leaves.

Method: In a pistil, pound chilies, garlic, and coriander to a smooth paste. Coat wok with cooking oil and fry paste until a good aroma. Add chicken and fry until done. Then season with oyster and fish sauces, and sugar. Stir in holy basil leaves and remove from the heat. Boil durian and garlic in water for about 5 minutes. Peel garlic and mash it with the durian. Stir in egg yolks, salt, and pepper. In a separate saucepan, fry a handful of holy basil leaves until crispy.

Durian topping ingredients: 300g unripe, seeded durian, 6 cloves of garlic – leave the skins on, 2 egg yolks, ½ TS salt, ½ TS pepper, garnish with fried crispy holy basil leaves

Finish: Spoon chicken filling into a sizable ovenproof dish and top with mashed durian. Bake at 180C/350F for 20 minutes. Remove from oven and garnish with crispy holy basil leaves.

FANCY DURIAN BRUNCH SANDWICH

Ingredients Durian filling: 1lb. durian, 2 TBS granulated sugar, 1/3 cup condensed milk, ½ water

Ingredients for dipped toast: 4 slices whole-grain bread, 2 eggs, ½ cup milk, 2 TBS butter or margarine for frying, cinnamon, and nutmeg powders

Method: In a pot on medium heat, stir durian with sugar, milk, and water until smooth – 10 minutes set aside. In a bowl, combine egg, spices, and milk. Dip bread slices and fry them in a skillet, the equivalent of French toast. Remove from skillet to plates and spread with creamy durian mixture. Sprinkle with cinnamon and nutmeg.

DURIAN LOLLIES or ICE CREAM

Ingredients: 1 pound ripe durian, 1 TBS confectioner's (powdered) sugar, 1 cup whole milk

Method: Blend ingredients until smooth. Freeze with sticks to make popsicles. If you don't have any sticks, enjoy in a bowl.

DID YOU KNOW? There is an international agriculture-tourism industry offering Southeast Asians and Chinese travel packages to enjoy various durian producing areas. Some famous durian areas are Penang Island in Malaysia and Chanthaburi in Thailand. A study on the aromatic compounds in durian found 44 active compounds, including some that contribute to scents of skunk, caramel, rotten egg, fruit, and soup seasoning. Durian has been forbidden to be transported by many airlines and some subway train systems because of its aroma. Thailand is the largest producer with 650,000 metric tons. There are more than 500 varieties of the durian, and the most common is *Durio zibethinus*. The name 'durian' comes from the Malay word for thorn. In Thailand, it is *turien,* and in Vietnamese, *sau rieng*. The Dutch named it '*stinkyrucht*,' which translates to stinky fruit.

ELEPHANT APPLE - CHALTA

Dillenia indica

Elephant apple or elephant apple, has the botanical name *Dillenia indica*, and is one of Hawaii's strangest fruits. It is both unique and unattractive, yet tasty. Plants are available at nurseries. Elephant apple may be called chalta, the wood-apple, monkey fruit, or curd fruit. It is common to India and throughout Southeast Asia, and a favorite fruit of elephants as well as the Hindu God Lord Vinayaka. I learned to pick ripe elephant apple by climbing the tree and dropping the tough-skinned fruit. Ripe elephant apples don't bounce.

The English name, elephant apple, came from the fact that in its native range of northern India, this fruit is desired by the local wild elephants. The elephant apple is usually four to five inches in diameter, with a hard, woody, light grey skin, a quarter-inch thick. The inner brown pulp has a weird grainy texture, with almost a sour smell and plenty of small, whitish seeds. Depending on the ripeness, elephant apple tastes run from sweet to acidic if overripe. There seem to be two types of elephant apples. One has large, sweeter fruits; the other has smaller more acidic fruits.

HOW TO GROW: The elephant apple tree prefers full sun or very light shade. It has dark green, toothed, leathery leaves and grows up to 25 feet. The elephant apple tree has a standing core reaching up with long drooping branches returning to the earth. The bark is scaly. The flowers of the tree are white to pale red in color and have a distinctive very pleasant odor.

Elephant apple trees need plenty of rain, but also require a distinct dry season. It grows best in light soils with a pH of 5.5-7. Trees grown from seeds usually don't bear fruit for ten years or more. Grafted dwarf elephant apple trees bear within five years. Hearty trees, elephant apple needs only a half cup of starter fertilizer (12-24-12) every other month until the tree bears. Then, twice a year A cup of bearing fertilizer (12-12-17-2). When picking elephant apple first check its outer husk. Shake it gently to see whether the fruit has become dislodged, which will indicate it is ripe. When you open the fruit, the pulp should be a rich brown. The skin or husk can be cracked with a hammer. To discern the ripeness of an elephant apple sniff for a sweet smell. If they are not ripe enough set them in the hot sun for a day or two days till they ripen. If these fruits are not ripe the pulp will not come out of the shell when scooped, and it will taste slightly bitter.

MEDICINAL: A tea made from elephant apple leaves helps avoid repeated colds and related respiratory conditions, cures a sore throat, and treats chronic coughs. Fifty milligrams of elephant apple juice mixed with warm water and sugar will assist blood purification. Regular consumption of elephant apple is recommended for people with kidney problems. Scientists believe an extract of elephant apple fruits may fight types of human leukemia. Elephant apple is a high energy food as 100 grams produces a hundred and forty calories, and benefits diges-tion. All parts of the elephant apple tree are used to heal snake bites. Tea made from the flowers is used to cleanse eye infections. In India, elephant apple is consumed as a tonic for the liver and heart. Consumed unripe, it will cure diarrhea and hiccough, sore throat, and gum diseases. A poultice of the pulp is used on insect bites. Dried leaves are used as a sandpaper substitute to polish ivory.

NUTRITION: Elephant apple pulp is a third carbohydrate with a small bit of protein. One fruit has about fifty calories with beta-carotene, thiamine, and riboflavin.

FOODS & USES: Elephant apple fruit pulp is also used for glue; mixed with lime and plaster it is used as a sealant; and added to watercolor paints. In the cosmetic industry, limonene is extracted from the rind as a rich oil used to scent hair products. The husk makes a yellow dye. The seed pulp is also used as a household glue.

RECIPES:

ELEPHANT APPLE - DAHL 1– Dahl (daal is dried lentils, peas, and beans that don't need pre-soaking.

Ingredients: ½ pound of pigeon-peas, 6 elephant apple slices, 3 green (not ripe hot peppers seeded and minced, 1 whole red congo pepper or 2 red chilies, 1 cup water, 1 TBS mustard powder, 3 TS mustard oil, 1 TS ghee or butter, 1 TS salt, 1 TS sugar (more to taste, 1 TS cumin, 2 bay leaves.

Method: Slice elephant apple and mash. Boil pigeon peas over medium heat. After a few minutes, add salt, green hot peppers, and sugar - according to taste. After boiling remove the surface foam. In a large skillet fry red hot pepper, cumin, mustard, mashed elephant apple slices, and bay leaves in the ghee/butter. Add water and boil. Pour boiled peas into the mixture. Cook 2-3 minutes. Remove from heat.

ELEPHANT APPLE - DAHL 2

Ingredients: 1 elephant apple, 3 green unripe hot peppers, 1 TS cumin seeds, 1 TBS salt, a pinch of powdered turmeric, 1 TS oil for seasoning, 1 TS each of black lentils (urad dhal, sugar, mustard seeds, cumin seeds, ground chickpea powder, fenugreek seeds, and dried red peppers.

Method: Remove elephant apple pulp. Combine by grinding together pulp with cumin seeds, turmeric, salt, and green peppers to a paste. In a skillet heat the oil and add seasoning ingredients. Add the paste to the seasoning and simmer for two minutes. Serve with rice or roti.

ELEPHANT APPLE CHUTNEY

Ingredients: 1 elephant apple, ½ cup brown sugar, 3 TBS achar masala, 1 entire head of garlic minced or grated, 1 TS salt, 2 TBS oil, 1 hot pepper (optional, ¼ cup water

Method: Break the shell of the elephant apple and empty the flesh content into a bowl. Chop elephant apple into one-inch chunks and boil till tender. Heat oil in a frying pan and add garlic, achar masala, sugar, pepper, and salt and pound. When everything is cooked a few minutes add water and bring to a boil. Add elephant apple pieces and mix thoroughly. Cover and simmer for 5 minutes. Serve warm or cold.

ELEPHANT APPLE SWEET

Remove the flesh of 4 elephant apples to a bowl and combine with 1 cup coconut, ½ cup grated coconut, a pinch of nutmeg, and cinnamon. Add sugar to taste. Blend and drink on ice or freeze to make a sorbet.

DID YOU KNOW? The fruit pulp is used as a hair rinse. The leaf juice is applied to the scalp to treat dandruff and to prevent baldness. The wood ash is added to clay bricks to increase fire resistance. It can also be used as a large container plant and will attract several bird and bees. Prefers sunny position, a well-drained slightly acidic soil rich in humus. The tree can be easily grown by seeds or semi-ripe cuttings.

NAMES: Bengalese: elephant apple, outenga in Assamese, wood-apple, chalita, elephant apple, monkey fruit, curd fruit, bael fruit, matoom, Bengal quince, golden apple, holy fruit, stone apple, velakkaya, chulta, hondapara tree, ma-tad, Indian simpoh, and kath bel.

My green thumb came only as a result of the mistakes I made while learning to see things from the plant's point of view. ~ Fred Dale

GIANT GRANADILLA – BARBADINE
Passiflora quadrangularis

Giant granadilla is the sweet, good tasting fruit of a climbing vine perfect for growing on a fence and also known as barbadine. The flesh of the giant granadilla can be cooked as a vegetable, a dessert, or strained for the juice. This fruit is native to tropical Central and South America. Trinidad and Barbados have grown giant granadilla since the mid 1700s. Trinidad sent the first giant granadilla seeds to the United States in 1909. It survived only in southern Florida's warm climate. This tasty vine fruit was reported in Hawaii as early as 1888. It quickly naturalized.

HOW TO GROW: Giant granadilla's blossoms are incredibly beautiful, reddish-green on the outside with white, pink, and purple inside. The fruit sprouts from the blossoms like a weird shaped, pale green barbell. It will grow to a light yellowish-green, foot-long fruit weighing up to a kilo. It is easily grown from seeds, which sprout in two weeks. This fruit grows between 700 and 1,500 ft. of elevation and thrives best in soil with a pH of 5.5-7.5. Seeds from one mature fruit should produce at least ten plants. First fork a hole a foot square and deep to produce fine dirt. If available, mix in some well-rotted chicken manure into the hole to keep the soil porous. Plant four seeds about an inch from the surface. A distance of ten feet is necessary between vines. Keep the plant watered daily. Drought is one of giant granadilla's biggest enemies. As it matures weave the vine into a fence or a trellis. Use a tablespoon of a fertilizer mixture high in nitrogen at the roots once a month. Every two weeks spray the vine with a mild insecticide as Malathion. When the blossoms open, spray the entire vine with a soluble 20-20-20 fertilizer every other week. A giant granadilla vine will climb trees to 50 feet if permitted. That is out of easy picking range. On the island of Java one vine was reported to have grown to one hundred and fifty feet.

Giant granadilla's enemies, birds and boring insects, will appear with the fruit. It is wise to protect the hanging fruit by putting them (while on the vine into plastic or brown paper bags until they mature. Do this carefully and don't pull or damage the immature fruit on the vine. One vine properly watered and fertilized can produce 2-3 fruits every month. It takes ap-proximately 2 months for the fruits to ripen after the appearance of blossoms. Vine growth and fruit production, quality, and size benefits by yearly pruning. This produces young branches, with more flowers. Giant granadilla should not be permitted to grow on any tree since it will compete aggressively for sunlight, and it can damage or kill the support tree. The fruit is melon-like, has a delicate skin with a thick layer of white flesh, which tastes similar to a pear. It is ready to harvest when the skin becomes translucent and glossy, and turns slightly yellow. Giant granadilla requires careful handling to prevent bruising.

MEDICINAL: The fruit is valued in the tropics an appetite enhancer. In Brazil, the flesh is used as a tranquilizer to relieve nervous headaches, asthma, diarrhea, dysentery, neuras-thenia, and insomnia. The seeds are reported to have a narcotic effect in large doses. The leaves boiled in water are used for bathing skin eruptions. Poultices made from the leaves are applied to liver ailments.

NUTRITION: Giant granadilla is a good source of vitamins B-6 and C, with magne-sium, calcium, and potassium. 100 grams has 60 calories with 2g of protein and 10g of fiber.

FOOD & USES: Immature fruit can be boiled, breaded, then fried as a vegetable, or cubed and stir-fried. Its juice is great chilled or as a flavoring for shaved ice. Boiling the un-peeled flesh and the pulp separately is the first step in making giant granadilla jelly. The juice is strained from both and combined with sugar and lemon juice. Boil the combined juices again until it jells. Australians make giant granadilla wine by crushing the entire fruits with sugar and warm water and permitting it to ferment for three weeks. The Aussies fortify the mix with a quart of brandy and hide it in a dark place for a year. Jamaicans bake the roots of old vines as a survival food, a substitute for yam.

RECIPES: GIANT GRANADILLA COLADA

Ingredients: 1 giant granadilla, 1 can condensed sweetened milk

Method: Peel the giant granadilla and cut out the pulp. Force seeds and pulp through a strainer. Add the condensed milk and blend. If it's too thick add water or regular milk. Add sugar, cinnamon, and nutmeg to your taste. This may also be made with chilled water and ice. Use 2 cups of water and 2 trays of ice. Add sugar and other juices as orange or passion fruit to your taste. Put everything in a blender and push 'chop' mode. Hit the button about 4 times until the icy mix is grainy.

GIANT GRANADILLA EASY CAKE

Ingredients: ½ giant granadilla fruit, 1 yellow cake mix, 3 eggs, 1/3 cup cooking oil or margarine, cinnamon or nutmeg to taste

Method: Peel giant granadilla and put in a blender to get juice. Strain to remove pulp and blend all ingredients. Pour into a greased cake pan. Bake for 10 minutes at 350F degrees. Then lower temperature to 300F for half an hour. Check the center with a toothpick. Allow the cake to cool before slicing.

GIANT GRANADILLA ICE CREAM

Ingredients: ½ giant granadilla fruit, 1 tin evaporated milk, 1 tin sweetened condensed milk, cinnamon and or nutmeg to taste

Method: Crush or blend giant granadilla to get the juice. Strain to remove pulp. Blend all ingredients and put in a suitable container for freezing. The refrigerator cube tray may be used to make delicious ice blocks.

GIANT GRANADILLA TART

Ingredients: 1 giant granadilla skinned and seeded, about 2 dozen sweet biscuits or cookies of your choice, 4TBS orange marmalade, 2 TBS sugar, 1 cup full whipping cream (this must be chilled at least overnight, 1 cup hot water, 1 cup cold water

Method: Combine 2 TBS marmalade first with hot water and then cold water, chill in fridge. Whisk cream until thick adding giant granadilla pulp. Add sugar and marmalade water. Crush cookies/ biscuits with a rolling pin and cover the bottom of a cake or pie pan. Add a layer of giant granadilla pulp mixture and then another layer of crushed biscuits. Refrigerate until thor-oughly set.

Names: barbadine in Trinidad, grenadine in Haiti, *tumbo* or *tambo* in Peru and Ecuador, *kasa ora* and square-stemmed passion flower in the Philippines, *markeesa,* or *manesa* in Indonesia; *timun belanda, marquesa* or *men-timun* in Malaysia, it's *sukhontha-rot* in Thailand; *dua gan tay, or giant granadilla* in Vietnam.

No occupation is so delightful to me as the culture of the earth, and no culture comparable to that of the garden. - Thomas Jefferson

GOVERNOR PLUM – CERISE PLUM

Flacourtia indica

The governor plum is a small berry-like fruit resembling an English plum, but is often called the cerise plum. This small plum tree is on every Hawaiian Island and considered invasive on Oahu. That's good because it can become invasive due to birds eating the fruit. Farmers have used it to mark boundary lines, pen cattle, and protect gardens. It's a sour sweet nutritious fruit consisting of a yellow or white flesh and a tough central core containing 6-10 seeds. Different from other fruit, it's at its sweetest after a good massage. They are hard and need to be rolled between the palms of your hands, or thumb and forefinger and then squeeze the pulp-jelly into your mouth. After the tenderizing rub, you'll taste the sweet-tart flavor.

Because there are so many fruits referred as plums or cherries, and many governor plums, it's best to know and refer to the botanical name, *Flacourtia indica.* The Martinique or bakoto fruit is *Flacourtia inermis* and also known as another of the many governor's plums. But these are two slightly different fruits. English colonists in India named it, the *'indica'* tree, ramontchi, governor, or acutes plum. It has sharp long spikes on the trunk of the tree and branches and makes a barrier hedge for cattle, unwanted visitors, or bad neighbors.

Governor plum is a small berry type of fruit up to an inch across. At first, its skin is tough and green. As it begins to mature, it becomes dark red. The flesh is a pale yellow and the taste is both sweet and sour, with an acidic tang.

HOW TO GROW: The governor plum is an easy evergreen tree to grow, so easy it may become an invasive species. Once it sprouts from fresh seeds, it doesn't require much care. Seedlings like semi-shade. You can transplant sprouts to large patio pots in 3 months. This is a slow-growing tree with drooping thorny branches that will only reach 15 meters. It is best to keep it trimmed. It is drought resistant, and loves humid tropical areas. They will grow up to 2400 meters in elevation with the best pH 5-6.5. It likes limestone, clay, or sandy soil, but cannot tolerate shade. The tree has rough, pale, powdery, gray to brown bark and scalloped leaves. Once the small white to pale yellow blossoms appear, it will be six months until the fruits are ripe.

The tree can be planted as a windbreak, living fence to hold cattle in, or keep out strangers. Great for around a garden plot. Although slow growing, it responds to trimming. The branches that are trimmed are good for charcoal or firewood. The hardwood is heavy and used for rough lumber, and good for small wooden tools such as plow handles.

MEDICINAL: The decoction from the bark is used to relieve arthritis pains. Most parts of the plant are used for cough, pneumonia, and bacterial throat infection. It has also been used for diarrhea. In 2010, *The American Eurasian Journal of Scientific Research* published a study that demonstrated governor plum leaves contain potent antioxidants, which may slow down the signs of aging and reduce stress. A juice pounded from the leaves treats asth-ma and bronchitis. Governor plum fruit stimulates the appetite, fights jaundice and enlarged spleens. The tiny seeds are ground into a paste and used to ease rheumatic pain. A decoction of the root in combination with the leaf sap treats malaria and relieves body aches. The bark is used for rheumatic pain and as a gargle for hoarseness. In 2011, *The International Journal of Drug Development and Research* published a study demonstrating compounds in the leaves contain significant antimicrobial and antibacterial qualities useful to treat infectious diseases and inflammation. As a diabetic treatment, boil governor plums with water and strain.

NOTE: Eating governor plum 72 hours before a carcinoid tumor exam may result in a false positive. This is because of the fruit's high level of serotonin. Doctors check for cancerous tumors by measuring serotonin levels in urine. If higher than normal, doctors give the prognosis of cancer.

NUTRITION: 100 grams of governor plum fruit has only 94 calories and contains many minerals. It is an excellent source for iron, potassium, calcium, niacin, and vitamin C.

FOOD: Fruits are perishable and best to eat when picked or within two days. Enjoy a handful of governor plum like a large grape but spit out the seeds and the skin or first slice it in half and remove them. Governor plum can also be stewed as a desert as a topping for pastry or ice cream or a flavoring for Italian ice or sherbet. The fruits are used to make jellies and jams. Ripe fruits can be dried and stored as a high-potency natural snack.

RECIPES: PICKLED GOVERNOR PLUM

Ingredients: 500 grams of governor plum, a piece of ginger as big as your thumb, 3 garlic cloves chopped, 1 medium onion chopped, 1 green chili (optional) 3 TBS chili oil, ½ TS cinnamon powder, ½ TS fenugreek

Method: Sauté the whole governor plum with ginger, garlic, onions, and chili oil. Once onion and garlic become transparent remove from heat. Cool and store in the fridge. This is an excellent and unique appetizer.

GOVERNOR PLUM COMPOTE

Ingredients: 500 grams governor plum seeded, (Try to keep them as whole as possible.), ½ cup sugar, 2 sticks cinnamon, 6 cloves, ½ pint boiling water

Method: In a suitable saucepan, heat the water, and when boiling carefully stir in the sugar until it forms a syrup. Add the governor plum, cinnamon sticks, and cloves. Simmer stirring for 10 minutes. Spoon out the plums to individual serving bowls (Best on top of a slice of cake.), and continue to simmer the sauce until very thick, but doesn't stick to the pan. Remove from heat and pour over plums. At the very end you can stir in an ounce of brandy (optional).

SPICY GOVERNOR PLUM SAUCE – excellent for chicken or pork

Ingredients: 2 kilos-4 pounds of governor plum, seeded and quartered, 1 medium onion chopped, 1 garlic clove peeled, 2 cups sugar, 2 hot chili peppers whole, 2 cups vinegar

Method: Blend governor plum, garlic, onion, and chili peppers until smooth. Transfer to a suitable pot add sugar and vinegar. Bring to a boil and then reduce heat and simmer until the liquid reduces by half. Stir frequently. Remove from heat and carefully transfer to sterilized jars.

DID YOU KNOW?

These trees were used to build the Great Governor plum Fence of India. Governor plum (*Flacourtia indica*) can be cropped and maintained as a bushy shrub or rise to a tree with thick spikes on its trunk and branches. It can grow up to 25 feet as a shrub or rise to 50 feet as a tree. The British Raj in India resourcefully chose it as the main shrub to lay a 1,500-mile long living thorny fence across India, to prevent the natives from engaging in the lucrative salt trade without paying a tax. It was officially called the Indian Inland Customs Line and was in existence from 1840 to 1879 and was a huge success in raising revenue for the British Empire

The primary reason diseases tend to run in families may be that similar diets tend to run in families.

~ Michael Greger

GRAPEFRUIT

Citrus paradise

The grapefruit tree is perfect for any yard, orchard, or garden in Hawaii because they are small and well-adapted to the soil and climate. Any citrus makes a nice home garden tree. The grapefruit is a large citrus fruit related to the orange, lemon, and pummelo. Grapefruits are categorized as white, pink, or ruby - the color of their flesh. Historically, it is a very young fruit. The grapefruit is a new citrus and has been around only for 300 years. In Barbados, it was accidentally crossed between a fruit called the pummelo and the orange. The Trinidadian name, 'shaddock', comes from the seafarer Captain Shaddock who carried the pummelo seeds from Indonesia to the West Indies in 1793. The grapefruit has many names including the 'forbidden fruit' in Barbados and the 'chadique' in Haiti. The Dutch call it 'pampelmoose,' which translates to pumpkin-sized citrus. Grapefruit's botanical name is *Citrus paradise.*

Because of the isolated location, Hawaiian citrus has been initiated by seeds. The first grapefruit seeds arrived before 1850. Once the trees were stable, propagation was increased by grafting. Today, the islands' commercial grapefruit orchards are small and sell locally.

At first, the grapefruit was too bitter to be enjoyed, but cross-pollination (again in Jamaica) between the grapefruit and the tangerine, created the 'ugli fruit.' Correctly named, the ugli has wrinkled skin and a flat bottom. Most people didn't care for the thick-skinned, sour fruits. The name grapefruit comes from how the fruits grow in clusters like grapes. The grapefruit was first commercially cultivated in the US in the 1880s. Presently the United States grows over 40% of the world's grapefruit.

HOW TO GROW: Grapefruit love Hawaii's sun and will grow up to a 2000 ft. elevation. This fruit likes soil with a pH of 6 to 7. To grow a tree, first, find either a tree vendor or a fruit you like and sprout the seeds. Seeds are available from Amazon. Trees should grow to twenty feet and bear fruit in five years. Grapefruit will develop in most soils, but require regular watering. A little high nitrogen fertilizer mix should be applied monthly. A single tree can bear over a thousand pounds of fruit yearly. Ripe fruits may be stored for 2-3 months on the tree permitting extra growth.

MEDICINAL: Grapefruit lowers some cholesterol, helps digestion, and reduces gas. It is also a treatment for water retention, urinary, liver, kidney, and gall bladder problems. Grapefruit can be rubbed directly on the skin to alleviate pimples and greasiness. The leaves have antibiotic properties. The grapefruit may cure many of the world's ills. It's claimed the fruit's enzymes will burn body fat. Grapefruits' essential oil has a stimulating trait. A diet including half a grapefruit, or a large glass of grapefruit juice should melt away ten pounds in twelve days. (Wouldn't that be great!) Check with a qualified doctor or pharmacist before drinking grapefruit juice if you're taking pharmaceutical medicines. Certain medicines when combined with grapefruit juice become more potent. Compounds in grapefruit juice slow the normal detoxification and metabolism processes in the intestines and liver, which hinders the body's ability to break down and eliminate these drugs.

NUTRITION: One grapefruit has 80 calories with 1 gram of protein, 18 grams of carbohydrates, and 3 grams of fiber. It is an excellent source of vitamins A and C, most B vitamins with folic acid, pectin, calcium, potassium, and magnesium.

FOODS & USES: The grapefruit industry created special 'grapefruit knives and spoons' to easier eat the juicy fruit. Most of the commercial grapefruit production is aimed at canning the juice, yet very little of the fruit is wasted. Cooking oil is pressed from the seeds.

Farmers revitalize their soils by using the seed hulls. The pulp is dried and fed to cattle. The seed extract is used as a remedy for foot fungus, and a concoction prepared from the blossoms is used as a blood tonic and as a cure for sleeplessness. Grapefruit is more than a fruit or a juice.

It can be used in a multitude of recipes. Grapefruit sections chopped with cilantro, sweet onions, sweet peppers, and half a hot pepper makes an excellent salsa. To make a great salad, mix grapefruit sections with cooked shrimp and chopped avocado placed on a bed of lettuce. Mix the juice with club soda and perhaps your favorite liquor. Grapefruit ice can be easily prepared by mixing grapefruit juice with some pulp and freezing. After two hours machine blend it or stir well and return to the freezer.

RECIPES: HIDDEN REEF'S GRILLED CITRUS FISH

Ingredients: 2 pounds king mackerel or any fish, 1 lemon, 1 orange, 1 grapefruit (or sour orange), 1/2 cup olive oil, 1 TS thyme, 2 cloves minced garlic, 1 onion minced, salt, and spice to taste

Method: Season the fish. Light your grill. Section the fruit and place in a small skillet with olive oil, thyme, salt and spices, onion, and garlic. Bring to a boil and simmer for about three minutes. Grill the fish until well done. Cover with the citrus sauce and let stand for 5 minutes before serving.

CITRUS SHRIMP STEW

Ingredients: 2 pounds medium or large shrimp peeled and deveined, 2 cups each of orange and grapefruit juice, 1/2 cup chopped onion, 1 orange and 1 grapefruit peeled and cut into sections without the membrane, 1 hot pepper, seeded and minced, 3 bunches of cilantro chopped, 2 TBS brown sugar, 2 medium tomatoes chopped, 3 potatoes boiled and cubed small, 1 nice taro/eddo sliced, 1 small giant taro/taro boiled and chopped, 2 green cooking bananas chopped, 1 bunch celery chopped, salt, and spices to taste

Method: In a small skillet combine juices, onion, 2 bunches chopped cilantro sugar, ½ minced hot pepper, and bring to a boil, then simmer for 10 minutes. Use ½ of this citrus mixture to marinate the shrimp for 2 hours in the fridge. (The longer you marinate the shrimp the better - up to 2 days.) In a large pot bring remaining citrus marinade to a boil before adding potatoes, tomato, celery, taro/eddo, giant taro/taro, and banana, 1 bunch chopped cilantro and the remaining ½ minced hot pepper. Cook for half an hour. Add shrimp and cook for 15 more minutes. Serve with rice.

CUBAN BLACK BEAN GRAPEFRUIT ENSALADA

Ingredients: 2 grapefruits-peeled, seeded, and sliced thin, ½ cup grapefruit juice, 1 bunch cilantro chopped, 1 pound boiled or canned black beans, 1 nice cucumber sliced, 1 cup papaya seeded, peeled, and cubed, 125 grams cheddar cheese grated, 1 head of lettuce, ½ TS ground cumin, 2 TBS honey, salt, and spice to taste

Method: Cover four salad plates neatly with lettuce and border each with grapefruit slices. Spoon each plate with equal amounts of the beans, cucumber slices, and papaya cubes. Cover with grated cheese. Mix the grapefruit juice with cilantro, cumin, honey, salt, and spices. Pour mix over salads.

DID YOU KNOW?

Most of the world's grapefruit are either grown in the USA in Florida or Texas. A grapefruit is 75% juice, and 1 fruit will give 2/3 cup of juice.

A grapefruit is only a lemon that saw an opportunity and took advantage of it ~ Oscar Wilde

GUAVA – KUAWA

Psidium guajava

Guava is another fruit native to the Western Hemisphere that has over a hundred species. Guavas grow in many forms and colors; pear-shaped, round, or oval; with yellow to green skins, and creamy or grainy yellow, pink or red flesh. The botanical name *Psidium guajava* refers to the most common type, the apple guava, which we call the Chinese guava. The smaller, sweet-smelling lemon guava and strawberry (*Psiddium cattleianum*) are more common among the world's islands. The Spanish explorers probably planted it on Hawaii during the 1500s. The strawberry guava was brought to Hawaii in 1825 on the HMS Blonde from England. Guava can become an invading pest as birds randomly distribute the seeds.

All guavas have rows of small hard seeds with a strong aroma and taste. Guava is used green or ripe in punches, syrups, jams, chutneys, ice creams, and a paste called 'guava cheese.' Scientists believe the guava was first cultivated in the mountains of Peru thousands of years ago, but man and birds have spread the seeds through all the tropics. The European voyagers carried the guava from the West Indies to the East Indies, Asia, Africa, and Egypt. India now invests over 100,000 acres to produce over 25,000 tons of guava annually.

The Hawaiians say 'kuawa' for the white guava. The lemon guava, kuawa-lemi, has sour pink pulp, kuawa-momona has bigger seeds and thicker skin, kuawa-ke'oke'o is like the lemon guava but has whitish pulp. The Beaumont and Ka Hua Kula varieties are grown commercially.

HOW TO GROW: The guava is usually a small tree growing to thirty feet, but new grafted types seldom reach fifteen feet. It is a type of evergreen with smooth brown bark. These trees can be grown from seeds, but better results are delivered from the grafting-budding process. Guavas prefer full sun and can grow in almost any soil type. They flourish in well-drained soil with a pH of 6. Mature guava trees need a half-pound of nitrogen-rich urea a year, but should also be fertilized monthly with a quarter cup of the mix 10–4–10 plus 5% magnesium. Pruning will increase blossoms and larger fruit. Red alga is a parasitic problem. Spraying with a copper-based algaecide at the first appearance should control this problem. Mealybugs and fruit flies can also be problems. Where fruit flies are a problem, the immature fruit is cov-ered with paper bags for protection to assure prime quality produce for the markets.

MEDICINAL: Guava is one of the best sources of dietary fiber. Its seeds are excellent laxatives. Guava fights diabetes, protects the prostrate, and reduces the risk of cancer. Guavas are rich in vitamins, proteins, and minerals, but with no cholesterol, and easily satisfy an appetite. The guava tree leaves are also a natural astringent that are used to stop diarrhea. They can be pounded into a poultice for wounds, boils, and aches. Guava leaves can be chewed to relieve a toothache. Amazon Indians use a tea of the leaves as a remedy for sore throats, nausea, and to regulate menstrual periods. Tender leaves are chewed for bleeding gums and bad breath. If chewed before drinking alcohol, it is said to prevent hangovers. A poultice of guava blossoms will relieve sun strain, conjunctivitis, or eye injuries.

NUTRITION: Guavas are high in vitamins A and C, phosphorus, and niacin. Some types of guavas have four times the vitamin C of an orange. A quarter pound of guavas is 60 calories.

FOODS & USES: Fine-grained guava wood is valued in India for carvings. It is also good for charcoal. Guava bark and leaves are almost 25% tannin, which is necessary to process animal hides. Asians use the leaves as a dye for cotton garments. The common way of preparing guavas is to remove the center pulp and stew them in the shells. Cooking will usually reduce the strong odor associated with guavas. Guavas are used in jams, jellies, and paste, canned, or frozen.

Straining the liquid after boiling seeded guavas makes guava juice, one of the main ingredients of Hawaiian Punch.

RECIPES: GUAVA SAUCE – a great addition for fish, pork, duck or chicken

Ingredients: ¼ pound of guava, 2 cups orange juice, sugar, and spices to taste

Method: Place guavas in a large pot, cover with orange juice, and simmer until cooked. Strain and add sugar and or spices.

PRINCEVILLE GUAVA BREAD PUDDING

Ingredients: 12 small guavas, or 3 large guavas – boiled and strained or a ½ package guava paste, 4 cups scalded milk, 2 cups bread cubes, 4 beaten eggs, ½ cup sugar, one TS vanilla extract, half TS nutmeg, half TS cinnamon, salt to taste

Method: Soak bread cubes in scalded milk for 5 minutes. Mix in sugar, salt, vanilla, and eggs. Pour into a baking dish. Cut paste into half-inch cubes and spread out evenly through the dish. Sprinkle top with cinnamon and nutmeg. Bake at 350 for one hour.

CHINATOWN GUAVA CAKE

Ingredients: 12 small or 3 large guavas boiled and strained, or ½ pound guava paste sliced a ¼ inch thick, ¾ cup butter, 1 cup sugar, 2 eggs, 2 cups bakers flour, 1 TS baking powder, 1 TS vanilla extract, ¼ TS salt

Method: In a medium skillet melt butter and slowly mix in sugar. Add eggs individually and vanilla. Separately combine flour, baking powder, and salt. Then combine the flour and melted butter/egg mix. Pour half the mix into an eight-inch baking pan. Cover with guava slices, cover with remaining batter. Bake at 350 for an hour.

POACHED GUAVA

Ingredients: 8 very ripe guavas—peeled, seeded, and halved (save seeds and pulp), 1½ cup water, 4 cups sugar, 3 TBS lemon juice

Method: Slice the guavas into ¼ inch strips. Place seeds and pulp in a skillet with the water and boil for five minutes. Use a wire mesh strainer to strain the liquid to another saucepan. Add sugar, guava strips, and lemon juice. Boil for 3 minutes or until fruit strips are soft. Serve as a topping for cakes or ice cream.

A garden returns 50x the investment you put in to it.
Not just food, but joy, peace and a real connection with creation.
A spiritual retreat from a noisy world and hurried people. Grow a garden

Everything that slows us down and forces patience,
everything that sets us back into the slow circles of nature, is a help.
Gardening is an instrument of grace.. ~ May Sarton

HOG PLUM — YELLOW MOMBIN

Spondias mombin L.

The yellow mombin is great shade tree for home garden landscape. This tree is native to Brazil and central South America to Peru and has been naturalized in Hawaii. Portuguese explorers spread this yellow plum tree to parts of Africa, India, Nepal, Bangladesh, Indonesia, and the Caribbean. Net research showed the red mombin as *Spondias purpurea,* the purple mombin as *Spondias purpurea L*, while the hog plum is *Spondias lutea* or *S.mombin. Spondias mombin L.* is a genus of fruit trees that comprises 18 species native to tropical America and Asia, and Madagascar. I've also seen it referred to as *Spondees monsoon*. The yellowish hog plum is related to the mango, cashew, and ambarella. (Beware, some call the ambarella hog plum because of the prickly surface of the seed/pit.) In South East Asia there is another hog plum that bears a resemblance to the Western hemisphere's tropical version, but botanically it's *Choerospondias axillaris.* Yellow mombin is also botanically Spondias mombin lutea. When I typed Spondias mombin L. into the search box for *Useful Tropical Plants,* it returned 128 pages with 100 species listed on each page!

Yellow mombin, hog plum, is the fruit of a fairly large tree that can grow to about 20 meters tall and almost a meter thick. This is twice as large as the red mombin tree. It has pale yellow blossoms that produce oval, bright yellow fruit, which are the size of a small plum. The hog plum has an acidic-sweet tangy mushy pulp around a large, fibrous pit/seed much like a small arambella. The fruit hang along the branches in numerous clusters of a dozen or more. Because the taste is just so-so, and the seed so big, very few people cultivate these trees. Most of the trees I know of are wild usually near a river. When they bear the surrounding ground is littered with yellow fruit. It is one of the few fruit trees that islanders don't rush to. The reason this is named the hog plum is that it is excellent feed for pigs. This fruit is also excellent to cook with pork.

HOW TO GROW: Hog plum trees it can be found at elevations up to 1,000 meters. They require full sun; prefers a medium to heavy, well-drained, fertile soil with a pH in the range 5.5 - 6.5, but can tolerate 4.3 – 8. A tree can be started from seed. Fresh seeds germinate well, usually within 2 months. First, find a wild hog plum tree and search for some sprouted seeds. If you are not successful, then enjoy some of the juicy plums and plant the seeds. Or you can take some cuttings from young branches, coat them with a rooting compound, and stick them in potting soil. These will root quickly in loose soil kept moderately moist.

Once the cutting has rooted, probably in 6 months, transplant it to an area where it can spread as it matures. Be tender with the seedling as hog plum is known for a long tap root. A hog plum tree needs about 6 meters (18ft.) on all sides. The tree is fast-growing and needs plenty of water during the dry season. Seedlings will bear in about 5 years, and grafted trees will bear sooner. This can be a great shade tree.

MEDICINAL: Another reason to have this tree in your home garden is the hog plum tree is very therapeutic in bush medicine. Tea can be made with the blossoms and leaves to relieve indigestion and sore throats. The juice from crushed leaves combined with the powder of dried leaves are used as poultices on wounds and inflammations. A decoction of the leaves and young stems is used as an eyewash. Boiling the bark in water is drunk to cure ailments of reproductive system, a remedy for diarrhea, and hemorrhoids. A decoction of the blossoms treats laryngitis. The plant leaves reportedly contain antiviral and antibacterial qualities, and a tea made with the leaves is used to rid the body of parasitic worms. The leaves are fed to pregnant domesticated animals to hasten littering or to expel the placenta after successful littering. **WARNING!** Eating many hog plums can cause diarrhea.

NUTRITION: 100 grams of hog plums has 52 calories and 1g of protein with some vitamin A and lots of vitamin C. They contain some iron, thiamin, niacin, calcium, manganese, potassium, and phosphorus.

FOOD & USES: The yellow mombin is less desirable than the purple mombin, and usually eaten out-of-hand, or stewed adding sugar. The juice can be mixed with other citrus to make a tasty punch, or used to flavor ice creams and sorbets. Immature green hog plums can be pickled with culantro and hot pepper as chow, or eaten with salt and pepper sauce, same as the chili plum. The tree is occasionally planted to shade coffee plants. Ashes from the roots are used in making soap.

The heartwood is cream to pale yellow in color. The wood is lightweight, soft; making it easily attacked by insects. The trunks are occasionally used for dugout canoes; branches are cut for posts, boxes, matches, and tool handles. Its hardness, density and light color make the tree useful for wood pulp to make paper.

RECIPES: STEWED HOG PLUMS

Ingredients: 2 pounds fresh ripe hog plums (about 50), ½ cup sugar, 1 cup water, ¼ TS salt, 1 TBS fresh lemon juice, 1 cinnamon stick

Method: Wash and cut plums in half and remove seeds. In a medium pot, combine sugar, water, salt, lemon juice, and add a cinnamon stick. Bring to a boil stirring constantly. Place plum pieces in the boiling syrup. Continue boiling for 5 minutes. Remove from heat and refrigerate.

HOG PLUM SORBET

Ingredients: 3 cups of washed sliced plums, pits removed, ¼ cup plus 2 TBS sugar, 1 TS fresh lemon juice, ¼ TS salt, 2 TBS orange juice

Method: Blend the sliced plums, sugar, lemon juice, and salt at purée speed until it becomes very smooth. Then force this puree through a fine mesh sieve to remove the hog plum skin. Add the orange juice and blend well again. Freeze in a suitable container. After 3 hours in the freezer, take out and blend again to break down the ice crystals. Freeze and serve.

HOG PLUM MARMALADE - Hog plum trees bear an abundance of fruit. Ingredients: 4 pounds of hog plums, washed with the seeds, 1 pound golden raisins, 2 cups water, 4 cups brown sugar, juice of 2 fresh lemons, and the grated peel of 1 lemon

Method: Put the hog plums, water, sugar, raisins, and salt in a large heavy skillet, cast iron preferred. Bring to a boil, stirring for 20 minutes or until the seeds float. Remove from heat and spoon out the seeds. Grate the peel of one of the lemons and add with the lemon juice. Simmer uncovered for about an hour stirring frequently so it doesn't stick to the bottom of the pan and burn. Remove the marmalade from heat and put 1 tablespoon of marmalade on a chilled plate. If the plum marmalade makes a crinkly track, the marmalade has set. When cooled pour into warm, sterilized jars and seal the jars with wax. Store in a cool dark place.

NAMES: June plums, yellow mombin, or monkey mombin

Change the food in the schools and we can influence how children think. Change the curriculum and teach them how to garden and how to cook and we can show that growing food and cooking and eating together give lasting richness, meaning, and beauty to our lives. ~ Alice Waters

INDIAN JUJUBE - DUNKS

Zizyphus mauritiana

I never thought an apple could be grown in the tropics until my neighbor, Rambo, showed me his jujubes. The green fruit was about the size, shape, and color of a small green, sour crabapple, but Indian jujube's taste is very tart and bitter sharp. As I talked to elders, and researched, I found a grafted jujube tree is rare and is believed to have originated in China.

The Indian jujube tree is cultivated on every Hawaiian Island. Oahu considers it invasive. This makes a fantastic tree for your home garden and is the fruit is great for your health. The jujube tree can grow to about forty feet. It will adapt to almost any soil or climate condition and is very drought resistant. These should be grown in full sun. It grows quick because its tap root descends to locate water and nutrients. The yellow, five-petal blossoms are self-pollinating and produce varied sizes of fruit. Most fruit are small, about an inch in diameter.

A well-cared-for grafted tree, as my neighbor's, produces three-inch diameter fruit. Jujube fruit can be round or oblong with thin skin. Just like an apple it ripens from a yellow-green to a full red. The flesh is white and crisp and even smells almost like an apple before it fully ripens. It is slightly acidic and perfect for making spicy chow-chow. Ripe flesh is less firm and pulpy. Overripe fruits have wrinkled skin and are soft to the touch.

HOW TO GROW: Indian jujubes are easily grown from seeds of fruits that are fully ripened on the tree. Each fruit has a center pit with two seeds. A trick for sprouting the seeds is to put them into water. Discard those that float. Of the rest carefully try to delicately split the shell. If you are careful and successful, the seeds should germinate within two weeks. However, grafting produces better fruit. With warm temperatures and direct sun, jujube will thrive. Regular watering and fertilizing can not only increase the size of the fruit, but triple the harvest. Fruit flies are this tree's main pests that can be controlled by a regular spray of insecticides mixed with a foliar fertilizing mixture containing zinc and boron.

MEDICINAL: Jujubes have been used by Chinese and Indians for many millenniums, for their many medicinal uses. This fruit has more vitamins A and C than apples. In fact, Indian jujubes have twice the vitamin C than citrus and are used as a tea to cure sore throats. These fruits are used to calm digestive and intestinal ailments. A tea of the bark will help fight diarrhea. It is believed that a diet of jujubes will cure baldness. Beware and don't eat too many raw fruits as they have a laxative quality. This fruit improves stamina and strength, stimulates the immune system, helps liver functions, is sedating, and serves as an all-purpose tonic. Raw jujubes are wrapped or poulticed over cuts and ulcers. They are used against blood circulation ailments and fevers. Combined with salt and hot peppers they ease indigestion and gas. Dried ripe fruit is a mild laxative. Jujube seeds are sedative and are taken, sometimes with whole milk, to halt nausea, vomiting, and abdominal pains, especially during pregnancy. Pulverized roasted seeds combined with cooking oil are rubbed to relieve rheumatic areas. Leaves are applied as poultices to assist liver ailments, asthma, and fevers. A decoction of the bitter, astringent bark will fight diarrhea and dysentery, and relieve gingivitis. A paste made from the boiled bark is applied on sores. Jujube root decoction will make a good purge. The powdered root is dusted on wounds. Juice of the root bark is said to alleviate gout and rheumatism. BEWARE - Strong doses of the bark or root may be toxic. An infusion of the flowers serves as an eye lotion.

NUTRITION: 100 grams, about 3 Indian jujubes, has only 80 calories with 1 gram of protein and 10 grams of fiber. Dried fruits have more calories than fresh ones. They are high in vitamin C and potassium. The fruit is often eaten to reduce anxiety. Research studies indicate the jujube may improve memory and help protect brain cells from damage by nerve-destroying compounds. Studies with mice even suggest that extracts of jujube seed may treat dementia caused by Alzheimer's. The seeds themselves are not usually consumed.

FOODS & USES: Indian jujubes are eaten fresh or dried and used for pickling like chow-chow or achar. Jujube vinegar, juice, marmalades, and honey are commonly sold in Asia.

RECIPES: JUJUBE WATER

In a clean gallon bucket mash a dozen jujubes. Fill with water and cover for 4 hours. Strain and chill. This is very refreshing over ice.

JUJUBE BUTTER

Wash, seed, and quarter about 30 ripe jujubes. Put in a large pot and cover with water. Add 3-6 cups of sugar depending on your sweet tooth, a TBS each of cinnamon and nutmeg, and 1 TBS cornstarch. Simmer uncovered for 2 hours. Let stand overnight. Eat on bread,or biscuits.

JUJUBE BREAD Ingredients: 2 cups minced fully ripened jujubes, 1 cup brown sugar, ½ cup butter or margarine, 1 cup water, 2 cups whole wheat flour, 1 TBS baking soda, and 1/2 TS salt, and 1 TS cinnamon.

Method: In a large frying pan combine the jujubes, water, sugar, and butter and bring to a boil. Once thoroughly blended remove from heat and allow to cool. Then add the dry ingredi-ents till it is a stiff mixture. Spoon into greased ovenware and bake for one hour at 350 degrees. Check with a toothpick before removing from the oven.

SPICED JUJUBE PIECES

Ingredients: 3 pounds of jujubes - quartered seeds removed, 4 cups brown sugar, 5 cups of water, 1 TBS cornstarch, 1 TS nutmeg, 1 TBS cinnamon, 1 TBS salt

Method: Wash jujubes; drain and prick each several times with a fork. In a kettle, bring to boil **water, sugar, spices, and corn starch.** Add pieces and simmer, uncovered, stirring occasionally, for half an hour. Remove from heat, cover, and chill overnight. The following day, bring syrup and jujubes to a boil and simmer, un-covered, for half an hour. With a slotted spoon, lift jujubes from syrup and place slightly apart on cookie sheets or trays suitable for the oven. Dry in oven at 250 for two hours. Check fruit pieces frequently and turn fruit occasionally. Turn oven off, but do not remove pieces. Remove from oven the next morning and store in the refrigerator.

DID YOU KNOW? There are two types of jujube trees farmed for their fruit, the Indian and the Chinese jujube. There are 700 varieties of the Chinese jujube, and 90 varieties of the Indian. The Chinese jujube has been cultivated for 4,000 years. In Jamaica, jujubes are called coolie plums or crabapples. Jujube's botanical name is *Zizyphus mauritiana*

Courtesy Wikipedia

NAMES: *English:* jujube, or Chinese date (which leads to confusion with the hardier species), Indian plum, Indi-an cherry and Malay jujube; *Jamaica*: coolie plum or crabap-ple; *Barbados:* dunk or mangustine; *Trinidad and Tropical Af-rica:* dunks; *Queensland:* Chinee apple; Venezuela: ponsigne or yuyubo; *Puerto Rico:* aprin or yuyubi; *Dominican Republic:* perita haitiana; *French-speaking West Indies*: pomme malcadi, pomme surette, petit pomme, liane croc chien, gingeolier or din-doulier; *Philippines:* manzana or manzanita ("apple" or "little apple"); *Malaya:* bedara; *Indonesia and Surinam:* widara; *Thai-land:* phutsa or ma-tan; *Cambodia:* putrea; in *Vietnam:* tao or tao nhuc. In *India* it is most commonly known as ber, or bor.

You are also far less likely to waste food when you have nurtured it from a seed into a plant.
~ Darina Allen

JACKFRUIT – NĪNEKA
Artocarpus heterophyllus

Jackfruit in Hawaii is 'nīneka.' This fruit is two kilos or larger, with odd, bright green skin that's tough with rough points all over it. The oblong fruit has seedy flesh with a sort of a sweet custard taste. It has a pleasant aroma. The white jackfruit pulp tastes strange, yet the pale-yellow pulp around the seeds tastes good, like a pineapple smoothie. The seeds also have tasty insides. Every part is sticky. Jackfruit is an acquired taste. For good nutrition, every home should have a jackfruit tree.

Jackfruit has East Indian origins back at least three millenniums. Supposedly the way the fruit got known in English as 'jack' is from the Portuguese 'jaca,' from the Malay 'chakka,' from the Indian 'Katah.' Jackfruit arrived in the Hawaiian Islands before 1890. Jackfruit season is August through December.

HOW TO GROW: Raising your own jackfruit tree is not difficult, but will take pa-tience and maybe years. Get seeds that are fresh because they lose fertility within a month after harvesting the fruit. Soak the seeds in water overnight and then plant in the soil. In 3-8 weeks, the seeds will germinate. It is necessary to transplant a jackfruit seedling when it has 4 leaves because after that the taproot of the seedling will be difficult to transplant. That root is long, delicate, and easily broken. The jackfruit tree grows in elevations to 5000 ft. and can adapt to various soil types, but it must drain well. The optimum pH is 5-7.5. Plant it in full sun. Wet soil will kill the tree. This tree matures quickly and should bear fruit in 3-4 yeas. Grafted trees bear in 3 years.

I went to see our friend's tree and found the big fruits growing directly from the trunk of the small fifteen-foot tree. Their tree was at least forty years old. They had other younger trees grown from this tree's seeds, and all took about three years to bear fruit. The biggest fruit they'd seen was about thirty pounds! The jackfruit bears the world's largest fruit. The heaviest jackfruit according to the Guinness Book is 42.73 kg and 57.15 cm long set in Pune, India in 2016. When ripe it should thump hollow, as a melon. Hawaii's jackfruit are usually 35-40 pounds.

MEDICINAL: An extract of roots treats skin diseases, asthma, and diarrhea. An extract from leaves and the latex sap cures asthma, prevents ringworm infestation and heals the cracking of feet. An extract from the leaves is given to diabetics as a control measure. Heated leaves are used in poultices to cure wounds, abscesses, earaches, and as a pain reliever. An infusion of mature leaves and bark treats gallstones. Ripe fruits are laxative.

NUTRITION: A one-cup serving of raw jackfruit has about 160 calories, with 4 grams of fat and 2 grams of protein. It provides 10% of the daily requirement of vitamin A and 20% of vitamin C with folate, niacin, potassium, magnesium, calcium, iron, zinc, and phospho-rus. It's one of the few fruits that's high in B vitamins. The benefits of the antioxidant seeds fight cancer, hypertension, aging, and ulcers. A powder ground from the seeds relieves indigestion.

FOODS & USES: When working with jackfruit recipe wipe your knife and cutting board with cooking oil because the fruit has a white, very sticky, milky latex goo. Always remove the skin. Fresh ripe jackfruit has brown spots on the green skin. Saw the fruit across the width. Core and toss the inedible center. The cut open jackfruit has large seeds encased in tasty flesh surrounded by white 'stuff.' Chop into sections and remove the seeds. Eat the flesh. There may be a hundred or more seeds. The nut-sized seeds are wrapped in a tasty skin. Inside they are crisp like a nut and can be fried, roasted, or boiled for five minutes, then roasted. It is high in pectin and has white latex goo used in India and Southeast Asia as glue.

Some believe jackfruit goo is the secret base for the flavor of Juicy Fruit chewing gum.

Jackfruit can be eaten green or ripe; raw or prepared. It is a staple food boiled or curried. In many countries, it is considered a 'poor man's food.' The younger the jackfruit, the milder the flavor. Ripe jackfruit is used mostly for desserts.

RECIPES: JACKFRUIT DELIGHT-sweeten to taste

Ingredients: 1 jackfruit peeled and cut into small pieces, 2 cups grated coconut, 1 to 2 cups powdered sugar, ¼ cup small pieces of coconut, 1 cup milk, 2 to 4 TBS brown sugar, 1 TS cinnamon, 3 TBS butter, 10 chopped cashews

Method: Fry jackfruit pieces in two TBS of butter for 6 minutes. Boil grated coconut and powdered sugar in 2 cups of water for 10 minutes before adding the fried jackfruit. Continue boiling for two minutes. Fry cashews and coconut pieces in 1 TBS butter. As they brown add sugar and cinnamon. Mix with jackfruit and add milk. Simmer till everything is warm. This can also be served chilled.

HAWI JACKFRUIT CHICKEN

Ingredients: 1 chicken cut up, 2 pounds jackfruit, 5 large tomatoes chopped, 1 medium onion chopped, 1 TS minced garlic, 1 TS minced ginger, 1 TBS minced thyme leaves, 2 TBS culan-tro, five bay leaves, 7 TBS curry powder, 3 TBS cooking oil, 2 cups water

Method: Cut jackfruit into one-inch cubes. Keep in ice water to reduce browning. Heat oil in a large frying pan. Add chicken and brown. Add onions, garlic, ginger, thyme, and tomatoes. Add water if too thick and starts to burn. Add jackfruit chunks. Stir before adding the curry powder, bay leaves, culantro, and one cup of water. Simmer covered for half an hour. Remove bay leaves before serving.

WAIKIKI BARBECUED JACKFRUIT

Ingredients: 1 medium onion, 1 sweet pepper, 1 small hot pepper seeded, 4 cups jackfruit chunks, 1 TBS teriyaki, 1 cup BBQ sauce, salt, and spices to taste.

Method: Sauté onion and peppers in a sizable skillet. Add sauces and simmer for 20 minutes. Add water if necessary. You want this thick. Serve on toasted French bread.

JACKFRUIT ICE Ingredients: 1 jackfruit peeled, trimmed and seeded, 2 cups sweetened condensed milk, 2 TBS lemon juice, sugar to taste

Method: Blend ingredients and freeze until it starts becoming icy. Blend again and refreeze. Grated coconut and fried jackfruit seeds may be added with cinnamon and nutmeg for a variation.

DID YOU KNOW? Jackfruit is the largest of all tree grown fruits. Similar to the pineapple, jackfruit is made up of fused individual fruitlets. Archeological discoveries in India reveal jackfruit was cultivated in India 3000 years ago. Jackfruit is a cousin to the breadfruit in the mulberry family, *Moraceae*. Its botanical name is *Artocarpus heterophyllus.* Jackfruit matures eight months after flowering and contains as many as five hundred one-inch seeds that can be boiled or roasted and used for seasoning. The seeds are valued more than the pulp. The flavor of this fruit's flesh tastes like melon, mango, papaya, and banana com-bined. This fruit is only viable for a few days.

BE LIKE A JACKFRUIT, HARD ON THE OUTSDE
BUT SWEET TO THE CORE.

JAMAICAN PLUM – JOCOTE - PURPLE MOMBIN
Spondias purpurea L

Islanders love strange things like the Jamaican plum. We eat it before they ripen with salt and some pepper. I love plum season so much, I know where every tree is close to my house. They are also called Governor or the Jamaican / purple / June plum. I eat them green, but Jamaican plums turn a gorgeous deep red or burgundy color, with a beautifully contrasting yellow juiciness inside and are very juicy when ripe. Few make it to be fully ripe between the birds and kids sucking the plum pit. With care and regular maintenance, these trees will grow anywhere in Hawaii.

Jamaican plums originally grew from southern Mexico to northern Peru and parts of north-coastal Brazil. As with so many fruits and vegetables, the Spanish and Portuguese planted this fruit throughout the tropics worldwide. This fruit tree has been naturalized where the explorers had colonized: the Caribbean, Africa, India, Hawaii, SE Asia, and Indonesia.

HOW TO GROW: The Jamaican plum is an attractive tree that can reach a height of 25 meters (82ft., but is more likely to be in the range of 7 - 10 meters (32ft. tall. Grafted varieties are usually without thorns, self-pollinating, and easy to grow. Like every fruit tree, this requires rich, well-drained soil with a pH of 6-7 and will grow at elevations up to 2,000 meters. The Jamaican plum is tolerant of a wide range of soil conditions, and also will tolerate some light salt spray. This is a good tree to plant in areas to halt soil erosion. For a good fruit set, it requires growing conditions with a marked dry season of up to 6 months.

Plants grow rapidly, often 5 feet a year, and produce large quantities of red, or purplish-red, and sometimes yellow cylindrical fruits about 1inch in diameter. The fruits are 25-30mm (1½in. long. Trees prefer full sun for best growth. Plants generally need pruning at least once or twice a year to keep from becoming too overgrown. Seedling plants can start producing fruit when about 4-5 years old, but grafted trees bear in 2-3 years. Every other month for the first two years sprinkle a cup starter fertilizer 12-24-12 around the tree. Water if there is a long drought. A well-maintained plum tree can bear for15 years.

MEDICINAL: This plum is used as to increase urination and to relieve stomach spasms. A decoction made from the plums is used to bathe wounds and heal mouth sores. Syrup prepared from these plums is taken to stop chronic diarrhea. A bark decoction is a remedy for, ulcers, dysentery, for bloating caused by intestinal gas in infants, and animal mange. The gum-resin of the tree is blended with pineapple or soursop juice for treating jaundice.

WARNING! The Jamaican plum is a member of the sumac family and if you contact with the sap it may cause a skin rash.

NUTRITION: Jamaican plums are a good source of vitamins A and C and potassium, they have very little protein and a trace of fat, but contain more antioxidants than any other fruit.

FOOD & USES: This variety of plum can be repeatedly pruned as a living hedge. It is also used for making jams, and ice cream. Unripe fruits are pickled or made into a tart green sauce. The young shoots and leaves often are conspicuously colored with red and purple. They have a rather agreeable acid flavor, and often are eaten raw in salads or cooked. The leaves contain 5.5% protein.

RECIPES: SPICY PLUM JAM

Ingredients: 5 pounds ripe plums, ½ cup water, ½ cup fresh lemon juice, 1TS ground cloves, 1 TS ground cinnamon, 5 cups of sugar, a half bottle Certo

Method: Remove seeds from plums. In a small pot combine water, cinnamon, cloves, and lemon juice. Cover and simmer on low for 5 minutes. Add sugar (more for sweeter) and bring to a boil for 1 minute stirring constantly. Remove from heat and add Certo, stirring for 5 minutes. Spoon into sterilized - boiled - jars to ¼ inch from the top and seal with melted paraffin. Should make about 8 pints.

JAMAICAN PLUM WINE

Ingredients: 5 pounds ripe plums, 4pounds of sugar, 4 liters of boiled water (boiling should remove all impurities), 1 package dry yeast.

Method: First, wash all plums and then mash the pulp in a spotless 5-gallon bucket with a top. Pour in water, mix in sugar, and stir until dissolved. In a cup of warm water, dissolve the yeast, and pour into the plum mixture. Cover and let sit in a cool place for a week, stirring once each day. Using organdy or cheesecloth, strain and place in another spotless bucket with a cover. Check daily to see if fermentation bubbling has finished. Strain again with cheesecloth and funnel into sterilized (boiled) preferably brown bottles. Seal caps with melted wax-paraffin. Store in a dark place for at least 6 months. Makes 5 liters.

PLUM CRISP

Ingredients: 1 cup very ripe plum pulp - skinned, peeled and seeded, 1 cup flour, 1 cup brown sugar, ½ cup butter or margarine, 1 TS ground cinnamon, ½ TS ground cloves, 1 TS ground nutmeg, 1 TBS Cornstarch, and ½ TS salt

Method: Combine the flour with half of the sugar and the spices. Blend in the butter until the flour mixture is flaky. Mix the plum pulp with the remaining sugar and cloves and put in a greased baking dish or bread pan. Cover with flour mixture. Bake for 1 hour at 350F.

PLUM DUMPLINGS: A unique dessert or appetizer!

Ingredients: 2 cups ripe plums peeled and seeded, 2 cups flour, 3 large potatoes peeled boiled and mashed, 2 TBS butter or margarine, and 1TS salt

Method: Mix ingredients together, except plums, into a dough about a ½-inch thick. Slice dough into 3-inch squares. Place a tablespoonful of plums in the center of each square. Pull up the corners to cover the plums and pinch shut. Carefully drop in boiling water and cook for 15 minutes. Adding sugar or slight pepper sauce to the plums will change the flavor and intention of these dumplings.

DID YOU KNOW? The name 'plum' is derived from the old English word 'plume,' which comes from Latin. The fruit is grown on every continent except Antarctica. China grows the most plums. Plums can grow in many colors such as purple, reddish purple, yellow, red, green, or white. In Chinese culture, plums symbolize good fortune. It is unknown how many species of plum exist. Some say there are 19 species while others say there are 40 species.

NAMES: As with most fruits and vegetables, they are known by local names. Someone gives a fruit a moniker and it's known by that name to that family in that village. In Mexico and Central America, it's known as jocote, derived from the Aztec/ *Nahuatl* word *xocotl*, meaning any kind of sour or acidic fruit. It's known as ciruela, ciruela, or ciruela which is Spanish for plum, traquead in *Panama*, ciriguela, cirigüela, cirguela, cirguelo in *Ecuador,* and siniguelas in the *Philippines.* Huesito in *Colombia*, Other names are: red mombin, plum, purple mombin, hog plum.

When the old plum tree blooms, the entire world blooms.~ Dogen

JAVA PLUM–INDIAN BLACKBERRY– BLACK PLUM

Syzygium cumini

The Java plum is a plum-like fruit botanically known as *Syzygium cumini*, and sacred to both Buddhists and Hindus. It is native to India and Java plum trees will grow almost anywhere and tolerate most conditions except extreme salt air. Botanists believe, more than three millenniums ago, the valuable, tasty oblong, purple fruit and wood was so desired, traders intentionally carried the seeds to be cultivated in Bhutan, Nepal, Queensland, South Wales, Zanzibar, and the southeast coast of Africa. After arriving before 1900, it's been naturalized throughout the Hawaiian Islands. If not maintained, it's considered an invasive species. Search out the delicious java plum fruit during June through August.

This is a perfect home garden tree. The java plum tree was so desired because it had many uses: the delicious fruits, sturdy wood for building, and firewood. The seeds are good for Ayurvedic-homeopathic medicine and were a valuable commodity traded internationally until 1800. It is used to shade coffee trees and as a nest tree for the tasar silk worm. Its nectar makes excellent honey.

Today, this fruit tree grows in almost every tropical and sub-tropical country world-wide. It is in various countries of SE Asia, the Caribbean, and Eastern Africa such as Nepal, Pakistan, India, Surinam, Trinidad, Bangladesh, Tobago, Puerto Rico, Indonesia, Florida, and California.

HOW TO GROW: The java plum tree can grow to 25 meters (82 ft.) but can also be trimmed into a suitable fruit-bearing hedge. Since the trees tolerate strong gusts, a row of these trees makes a good windbreak. If you know of a java plum tree, look underneath as they self sow. In some countries, this makes the java plum undesirable because it can take over an area and kill other vegetation by blocking sunlight with its dense canopy.

It is best to plant fresh seeds an inch deep in fresh loamy soil either during the rainy season or keep the pots moist and they should germinate in 2 - 4 weeks. The sprouts will die if the soil dries out. In nine months, the sprouts can be moved from the nursery beds to a location with full or partial sun. The sprouts require frequent watering. Java plum trees prefer pH from 5.5-7, but will tolerate more alkaline conditions. These trees will grow in areas up to a 1400-meter elevation. Seedlings may reach a height of 3 meters in only 2 years and full height in 40 years. This tree, like most, prefers moist soil and regular rains but can withstand a period of drought. Unless for a hedge, trees should be spaced 40 feet apart. These trees can survive for more than a century.

The java plum tree has white to pale pink, ½ inch blossoms that usually bloom in May. Fruits transform as they mature from green to white, to deep purplish, to black as they ripen. The white to pale lavender flesh is acidic with one large green seed. A 5-year-old tree can easily produce 700 fruits. A large tree can litter an area with fruits that have a sour odor as they rot, attracting insects. Unless topped and trimmed, the java plum is a large evergreen tree that provides dense foliage and shade, great for a hammock or swing. The bark is light gray and flaky; heart-wood is red to gray with a straight grain. It is termite resistant and durable in water. In India, java plum wood is used to construct posts and beams for homes, masts and oars for boats, water troughs for animals, and handles for farm implements.

MEDICINAL: The Java plum is respected for its medicinal contributions. The fruit juice can be applied to the skin as an astringent. It will stimulate the appetite, relieve gas, prevent scurvy, and cause the body to excrete water. It is also a treatment for diarrhea, indigestion, asthma, sore throat, and skin diseases as ringworm. Extracts from the seeds are used to lower blood pressure. The fruit is loaded with vitamin C and iron, increases hemoglobin. It is astringent to fight acne. It has 55 mg of potassium per 100 g keeps your heart healthy. Seeds, bark, and leaves are used to treat dysentery and reduce glucose in the blood and excess sugar in the urine of diabetic patients.

NOTE: Because it lowers blood sugar, one should not consume java plum two weeks before and after surgery. It's not recommended to eat this fruit on an empty stomach or with dairy products. And like every fruit, eat too much and you'll get sick with fever and body aches.

NUTRITION: Per 100 grams, java plum has 250 calories. These fruits are extremely high in potassium, and has plenty of calcium, magnesium, phosphorus, with vitamins C and A. Fruits have some of the highest levels of natural folic acid and recommended for pregnant women. They also are an excellent source of vitamins B3 and B6. They have no cholesterol, but some sodium.

FOOD & USES: Java plum is usually eaten raw, fresh from the tree. Extremely acidic fruits can be soaked in salted water or eaten with salt and chili powder. They make a suitable juice close in taste and appearance to grape juice. These fruits are good for jams. The whiter the flesh indicates a higher concentration of natural pectin, and the jam will be very hard unless the prep on the stovetop is kept to a minimum. Purple flesh fruits need additional pectin or combined with guavas to make jams and jellies. The juice can be made into a frozen sherbet or syrup for ices. Java plum makes decent wine and excellent flavored vinegar. An oil distilled from the leaves is used to scent soap and is combined with other scents to make perfumes. The bark is high in tannin and used for tanning leather and to create a brown dye.

RECIPES: JAVA PLUM AND RICE SALAD

Ingredients: 100g of java plum seeded and chopped, 2 cups cooked brown rice, 1 cucumber, 2 tomatoes chopped, 2 spring onions chopped, the juice from 2 lemons, 1 TS ground black pepper, 1 TS brown sugar, 2 TBS coconut or olive oil, salt to taste.

Method: Combine all chopped vegetables with the cooled, cooked rice. Combine the lemon juice, oil, sugar, salt, and pepper, and add to the rice and vegetables mix.

JAVA PLUM JAM

Ingredients: 1 cup java plum pulp, ½ cup sugar, 1 TBS liquid glucose, ¼ TS cinnamon powder, ½ TS clove powder

Method: Slowly cook sugar with ½ cup of water until it forms a syrup. Stir in glucose. Add java plum pulp and combine with cinnamon and clove powder. Cook until it thickens. Cool and transfer to sterilized jars.

JAVA PLUM CHUTNEY

Ingredients: 1 kilo of java plum seeded and halved, 1large onion purple preferred - chopped, 1 cup raisins, ½ cup white sugar, 1 TS crushed chili flakes, ½ TS dry ginger powder, 1 TS salt, 2TBS coconut or cider vinegar, ¼ TS cinnamon powder, 1 TS clove powder.

Method: In a suitable pot or wok, combine java plum with sugar and cook until it liquefies. Stir in all ingredients except vinegar. Stir often to keep from sticking and burning. When it thickens, spoon a drop into a bowl of cold water. If it thickens, it's finished. If not, continue to cook and test again. Once it sets, add vinegar and cook for 5 minutes. Place in sterilized jars.

DID YOU KNOW? This tree is planted near Hindu temple because it's sacred to the Hindu gods Krishna and Ganesha. According to some sources, the java plum tree, also known as jambu, was the tree that kept constant shade over Siddhartha during his first mediation. This fruit has many names: black plum, Indian blackberry, ma-dan, jaman, jambu, jambul, jambool, Java plum, Malabar plum, duhat, and the Portuguese plum.

Maybe a person's time would be as well spent raising food as raising money to buy food. ~ Frank A Clark

KUMQUAT
Citrus japonica or Fortunella japonica sp. Swingle

A kumquat isn't much bigger than a grape, but packed with sweet-tart flavor. The edible peel is sweet, and the juicy flesh is tart. This is another unique, easy to manage tree for your home. The kumquat came to Hawaii in the early 1800s brought by Chinese or Japanese immigrants.

Kumquats are believed native to China. They were described in Chinese literature in 1178 A.D. They have been grown in Europe and North America since the mid-19th Century, mainly as ornamental dooryard trees and as potted specimens in patios and greenhouses. Kum-quats will grow on most of the Hawaiian Islands, but not commercially. China is the largest producer of kumquats, with more than 18,000 tons yearly.

HOW TO GROW: The kumquat will to 5000 ft. in elevation. They require full sun, regular watering, and prefer a pH 6-7 in well-drained soil. They are also perfect for patio or lanai pots. The kumquat tree is slow-growing, shrubby, compact, 15 ft. tall and 6 ft. wide. They can be grown as close as 3 feet apart for a unique, decorative hedge. The branches occasionally having a few thorns, glossy leaves with clusters of dainty white blossoms. The oblong or round fruit is gold to reddish-orange, 1½ in. wide. Kumquats bear late in the year. The Hawaiian kumquat varieties are Meiwa (*Fortunella crassifolia*), and Nagami (*Fortunella margarita*).

Grafted trees can bear at 3 years and those from seed can take much longer. It's best to start kumquat seeds in a pot of moist soil. Place the seeds on top and cover with about 1/4 inch of coarse sand. Place the pot in a clear plastic bag, and close it. The seeds need a warm place to germinate, like a window. Germination should be within a month. Transplant to larger pots when the seedlings are 6 inches tall. Gradually harden the trees by moving to locations of more sun. Don't damage your tree picking the fruit, use scissors or snips. Kumquat fruits don't store well. They have thin, delicate peels. Eat fresh or keep in the fridge.

MEDICINAL: Kumquats are high in fiber and aid digestion. The skin contains several useful oils that fight cancer. The juice reduces the chances of kidney stones. Because they con-tain pectin, vitamin C and oils, they are antibacterial and anti-inflammatory. Eating kumquats is good for treating respiratory ailments.

NUTRITION: 100 grams have 3.8 grams of protein, a great source of vitamin A with calcium and potassium

FOODS & USES: Fresh kumquats can be eaten raw. The fruits are easily preserved whole in sugar syrup. Taiwan exports kumquats canned. Kumquat marmalade is excellent

. RECIPES: CANDIED KUMQUAT

Ingredients: 4 cups of chopped seeded kumquats - 1½ lbs., 1 cup water, 2 cups sugar

Method: Heat the water and sugar until it boils. Reduce heat and simmer for 4 minutes. Add the kumquats and simmer for 10 minutes. Drain the kumquats through a colander on top of a bowl. Return the drained syrup to the pan and simmer for 5 minutes to reduce the syrup to about ¼ cup. Combine the kumquats and the syrup. Serve or jar and refrigerate.

DID YOU KNOW? Kumquats were botanically known as *Citrus japonica* until 1915 when Dr. Walter Swingle gave them their own family, genus *Fortunella,* which embraces six Asiatic species. Kumquat, cumquat, or comquot, is Chinese for gold orange. The Japanese name for the round is *kin kan* or *kin kit*, and the oval type *too kin kan*. In Southeast Asia, the round is *kin, kin kuit,* or *kuit xu,* and the oval, *chu tsu* or *chantu*

LANGSAT - DUKU

Lansium domesticum

Throughout Hawaii, this fruit keeps its English name, the langsat, but more often it is known by duku. This fruit is native to western Malaysia and the botanical name is *Lansium domesticum.* That's important because there are three common varieties of this fruit. They are the same, but different – if that makes sense. This fruit appeared in Hawaii around 1930. Trees are available at nurseries and seeds can be purchased online.

The differences in the three varieties of langsat: gaduguda/duku is the largest, round, and pale yellow with a thick, rough skin with soft fuzz and a strong aroma of grapefruit and a big seed. Longkong/dongkong is round with rough, thick, pale yellow skin, no sticky latex, and few seeds. Langsat has a thin skin with some fuzz and some sticky latex, oval-egg shaped only 1.5 inches long, not as sweet, and a small seed. Yes, it's confusing.

Langsat's skin is yellow to brown and often spotted, with a smooth and waxy texture like a young potato. The skin is not stuck to the fruit as an orange, making peeling easy. The skin does have a sticky white latex when peeled so have a cloth ready to wipe your hands. To reduce the skin's gumminess, dip the fruit briefly in boiling water. Once the peel is removed there are usually five segments that can be pried apart and consumed like orange slices. Most segments have large, bitter seeds, some varieties have small seeds, and a select few types are seedless. The smaller, darker langsat are sweeter than the larger fruit. The flesh has a texture similar to grapes with the sweet-sour flavor and an aroma similar to grapefruit. Don't eat the seeds as they contain a small amount of toxins.

Lansium Pubescens, the typical wild langsat, which still grows wild on Sumatra, is a thin tree producing round, thick-skinned fruits loaded with sticky latex. *Lansium domesticum (*more confusion because it is also listed as *Lansium parasiticum),* langsat/duku is a broader tree with oblong, not-so-sticky fruit. It is a tropical evergreen that will only grow in areas of high rainfall, and below 750 meters of altitude. It will not survive in dry, arid environments. It requires humid conditions.

HOW TO GROW: If you're lucky to know where a langsat tree is located, look for seedlings beneath the tree. Otherwise, langsats are grown from seeds that are only fertile for 1 or 2 days after removal from the fruit. It prefers well-drained, but moist soil in a pH range of 6 - 6.5. They do best in semi-shade during the early years. And there will be many years because this tree bears fruit in 12-15 years. However, when the tree reaches the age of 20, the yield jumps to 100 kg of fruit per year. In orchards, the trees should be spaced 10 meters apart.

The leaves look like frizzy hair and the small blossoms will be pale yellow to white. The fruits grow in bunches of 20 and sprout directly from the bark of the tree. As they ripen, they develop a brown crust on its thin skin to a pale-yellow hue with a pleasant aroma. The fruit will be the sweetest when golden brown. Overripe fruits will have brown, pliable skin with black flecks. The flesh of a langsat/duku should be white and semi-translucent. If it turns purple or brown, it's overripe. The wood is light-brown, medium-hard, fine-grained, tough, elastic, and durable. It's used for rafters, tool handles, and small utensils.

The main growers are Indonesia, Thailand, the Philippines, and Malaysia. In the west-ern hemisphere, there are successful commercial orchards in Surinam. In the Philippines, a productive tree averages 1,000 fruits per year, where it is grown in half shade interplanted with coconut.

Throughout the Hawaiian Islands, langsat/duku is grown in home gardens with a few small orchards on every island. When in season, fruit is commonly found in farmers' markets.

MEDICINAL: Langsat/duku bark can be boiled and drank for diarrhea and malaria. A paste made from the seeds is used to reduce fevers. A poultice of the bark is used to fight the poison of a scorpion sting. Dried peels are burned to repel mosquitos and smells good enough to double as incense. Juice pressed from the leaves is used as eye drops for inflammation. Langsat should not be consumed by diabetics. The sweet juicy flesh contains sucrose, fructose, and glucose.

NUTRITION: 100 grams of langsat/duku has only 60 calories with good amounts of calcium, phosphorus, vitamin A, thiamine, and riboflavin. The high content of riboflavin helps reduce the pain of migraines. White langsat skin brightening body scrub is formulated from the extract of langsat fruit.

FOOD: This fruit is usually eaten out-of-hand or served as a dessert, and may be cooked in various ways. The taste of the fruit blends very well with chicken and pork. When cooking with it, should be placed last in the dish because it will dissolve in long periods at high temperatures. It is mixed with pineapple, mango, and rambutan into fruit salads. It works well with green salads, particularly those with cucumbers, carrots, mandarin, tofu, sesame seeds and a soy-based dressing. Great for juice or blended into drinks, or cooked into a thick syrup and added to ice cream and pastries. Freeze the juice to make popsicles and ices, add orange juice for a good complement. Peeled, seeded fruits are canned in syrup or sometimes candied. The fruits will keep for 3-4 days at room temperature and up to one week when stored in the refrigerator.

RECIPES:BON BON COCKTAIL

Ingredients: 60ml/2oz rum or vodka (optional), 4 peeled and seeded langsat/duku, 1 TS of fresh lemon/lime juice, ice

Method: Combine all the ingredients in a blender and process until smooth.

DID YOU KNOW?

An arrow poison is made from the fruit peel and tree bark. Both possess a toxic property, lansium acid; when injected it stops the heartbeat in frogs. Langsat still grows wild in the forests of Sumatra. It's a symbol of one of the Thai province - Narativat. Langsat blossom is the official flower of South Sumatra. Malaysia is the largest producer of langsats, followed by Thailand, the Philippines, and Indonesia.

It has many **NAMES**: the langsat, lansa, langseh, langsep, lanzon, lanzone, lansone,duku or kokosan. domesticum, called the duku, doekoe, or dookoo, Bon Bon, duku, doekoe, or dookoo, longdong, gaduguda, dongkong.

Fruit is nature's candy. — Unknown

Healthy eating is a way of life, so it's important to establish routines that are simple, realistically, and ultimately livable. ~ Horace

LEMON - LIMONA

Citrus limonum

In Hawaiian, lemon is known as 'limona.' Everyone enjoys the taste and the health value of the versatile lemon. Hawaii grows less than 50 tons and must import 2,000 tons. A lemon tree makes a great addition to any home landscape. Lemon's botanical name is *Citrus limonum.*

The lemon is one of the world's oldest fruits and was probably first grown on the lower slopes of the Himalayan Mountains near Assam in northeastern India. A DNA study of the lemon indicates it is a hybrid descendant of the sour orange and citron. Citrus fruits have been cultivated in southern China and Southeast Asia for approximately 4000 years. By 200 A.D. the lemon had reached southern Italy and spread to Iraq, Egypt, and China during the next five centuries. Today, the biggest lemon producers are India, Argentina, Iran, and Brazil. Almost 14 million tons of lemons are grown worldwide.

HOW TO GROW: Lemon trees are easy to grow from seed and will do well in most soils. Fill a pot with moist potting medium. Cut open your lemon and remove a seed. Scrape the seed clean of any pulp. Do not delay to plant. The seed must still be moist. Plant the seed about half an inch deep in the middle of the pot. Cover the pot with clear plastic wrap, seal the edges with a good rubber band, and poke small holes in the top with a pencil. Place the pot in a warm, sunny location. Keep the soil moist but not wet. Take the plastic covering off when the shoot emerges after about two weeks. Put it where it will get full sun. Plant where water will not be a problem during the rainy season as lemon trees are subject to root rot. Spray every tree monthly with an insecticide to controls mites mixed with a soluble nutrient and throw a cup of 12-12-17-2 (blue) fertilizer. This and regular watering in the dry season produces large juicy fruit.

The very thorny tree can grow large and should be pruned after about two years to keep it less than fifteen feet tall for easy picking. The rough-skinned variety is named 'Jambiri' and is used for a rootstock to graft most citrus such as sweet orange and grapefruit.

MEDICINAL: Lemons, rich in vitamin C, strengthen the body's immune system as an antioxidant, protecting cells from damage. For a cough, sore throat, or cold; make a syrup by heating one tablespoon lemon juice with two tablespoons honey. Try lemon juice mixed with salt and ginger as a tonic for a cold. The aroma of lemon oil has been tested to reduce stress in aromatherapy. Lemon juice mixed with water will aid digestion and pat an insect bite with raw lemon to stop the itch and reduce swelling. Lemons help cleanse the body through perspiration and as a natural diuretic. Lemon juice is believed to cleanse the liver of toxins. It was the juice of the lemon originally picked in the Mediterranean countries, not limes, first used aboard Brit-ish sailing ships to prevent scurvy.

NUTRITION: A lemon has only 25 calories with 1 gram of protein, 8 grams of carbs with calcium, iron, potassium, and phosphorous. 100 ml of lemon juice has 50 ml of vitamin C (ascorbic acid).

FOOD & USES: In Mexico and India, fine-grained lemon tree wood is used for carving toys and other small articles. A sprinkling of lemon juice on sliced apples, bananas, and avocados acts as a short-term preservative and keeps them from turning brown. The sour juice has been used for bleaching freckles, lightening hair, and is used in some facial cleansing creams. Use a slice of lemon dipped in salt to clean the copper bottoms of scorched cooking pots and pans. It will remove the dried soap and stains in the shower and washbasin. Cut a lemon in half and dip in baking soda for scrubbing dishes and stained surfaces. Oil from the lemon peel is used in furniture polish or make your own by mixing a cup of olive oil with a cup and a half of lemon juice. Its 'clean' scent is widely used in soaps, shampoos, and perfumes. Mixed with water the juice will help potted plants (except chrysanthemums) to hold their flowers longer.

A lemon contains 3 tablespoons of juice. Although not anti-septic, lemon juice will clean most meats and fish of anything on their surface. If left in the tart juice, beef or lamb gets tender. Lemons, with or without oranges, are used to make marmalade. Lemon zest is the grated rind used to flavor many dishes. Try something new and stuff lemon slices inside a whole chicken and bake for a unique flavor. Lemon juice squeezed on vegetables while steaming keeps colors bright. Rice cooked with lemon juice will be fluffier.

RECIPES: LEMONADE:

Lemonade may have originated in medieval Egypt, but a popular thirst quencher everywhere. Mix the juice of 1 lemon with 3 cups pure water per every 2 glasses of beverage desired. Add sugar to taste. Serve over ice.

LEMON LADY FINGER SOUP

Ingredients: 1 chicken cut up, juice of 2 lemons, 2 cups sliced okra/lady fingers, 6 cups water, 1 large onion chopped, 4 medium tomatoes chopped, 1/4 cup tomato paste (or ketchup, 1/3 cup uncooked rice (prefer brown, 2 TS salt, 1 TS turmeric, pepper, and other seasonings to taste

Method: Rub chicken parts with lemon. Put chicken and lemon juice in a large pot and add water. Bring to a boil and then simmer covered for fifteen minutes. Add all other ingredients and cook for half an hour or until chicken is tender and rice is ready.

MEDITERRANEAN POTATOES

Ingredients: 7 medium potatoes peeled and quartered, 1/4cup olive or canola oil, 2 TBS ketch-up or tomato paste, the juice of 2 lemons, 1.5 cups chicken broth, 1 TBS oregano

Method: Bake, microwave, or boil potato pieces until tender. Mix oil, tomato paste and all other ingredients in a saucepan and simmer for five minutes stirring so it doesn't burn or stick. Pour mixture over potatoes and let stand for at least a quarter of an hour before placing uncovered in an oven at 350 for twenty minutes.

LEMON RICE

Ingredients: 2 cups cooked rice, 1 TBS lemon juice, 1 TS lemon zest (grated lemon skin, 1 TS each of the following: turmeric powder, mustard seeds, fennel seeds, 3 dried chilies chopped, salt to your taste, 1 TBS oil

Method: Heat oil in a frying pan. Add mustard and fennel seeds until the pop, all dried chili pieces. Stir in turmeric powder and salt. Add rice to pan. Stir in lemon juice and lemon zest.

LEMON SORBET DESSERT

Ingredients: 2 cups water, 1/2cup sugar, 2 cups lemon juice, zest of 6 lemons

Method: Heat water enough to completely dissolve the sugar and pour into a bowl over the lemon zest. This will enhance the lemony flavor. If you have an ice cream maker, use it or just put it into a freezer until it just begins to freeze. Then put into a blender to smooth and reduce the ice crystals. Return and fully freeze. Adjust the amount of sugar used according to your family's sweet tooth and add any type of other citrus available.

DID YOU KNOW? India produces approximately 3 million tons of lemons per year. Lemon trees can produce up to 600lbs of lemons every year. Lemon trees produce fruit all year round. Lemon tree leaves can be used to make tea. Lemon zest, grated rinds, is often used in baking. There are 2 types of lemons – Acidic and sweet. Acidic lemons are those that are grown commercially Sweet lemons are those that are homegrown by gardeners.

We are living in a world today where lemonade is made from artificial flavors and furniture polish is made from real lemons. ~ Alfred Newman

LIME - LEMI

Citrus aurantifolia

Every Hawaiian Island grows juicy limes. Like most of the other citrus fruits, the lime comes from Southern Asia. Spanish explorers and the Portuguese sailed the lime from India to the Western Hemisphere. Perhaps early missionaries brought seeds to the islands.

There are three basic types of limes with various names. The Key, or Mexican lime has many seeds and is smaller than the big seedless Tahiti variety. The Tahitian may be a genetic hybrid between the true lime and citron that arrived in California around 1850 with fruit imported from Tahiti. There is also a Southeastern Asia variety called the Kaffir lime.

The lime was medicine for ancient peoples. During the Middle Ages fragrant limes were used to ward moths from hanging clothes just as today's mothballs. Sailors loved the lime since it prevented the weakening disease of scurvy. Picking limes ashore was considered the best duty of the British sailors, who became known as 'limeys.' The term 'liming' or hanging out is derived from sailors' lime picking, which always included resting under the lime tree.

HOW TO GROW: The rough-skinned Kaffir or the Tahitian Lime are the best choices to grow for your home use. These trees need good, well-drained soil, and planted in full sun. When planting, carefully place the tree and water after refilling half the soil in the hole. Pile the second half of the soil higher to prevent puddles after rains that could cause root rot. Mound the soil about two feet around the tree's trunk to create a three-inch high dam. This will hold water for the fine outer roots during the dry season. Water every few days for the first month, then weekly for the next four months if it is the dry season. Pull or gently hoe all weeds and lawn grass from inside the water ring so the young tree doesn't have to compete for water. Do not use herbicide.

A half-cup of urea sprinkled is the recommended fertilizer every three months. After the tree starts to blossom, alternate between urea and 12 –12 17 –2 every two months. Limes are sturdy trees that have few natural enemies, so little chemical spraying is necessary. Fruit that ripens to yellow on the tree will soon turn brown at one end. Brazil and Mexico lead the world's lime production with a combined 1.2 million metric tons.

You can also grow a dwarf lime variety in a container. Dwarf Key Lime is preferred for a container and should only grow to six feet and bear fruit within three years.

MEDICINAL: Limes also contain flavonoids, with antioxidant and cancer-fighting properties. It helps prevent diabetes, constipation, high blood pressure, fever, indigestion, and improves the skin, hair, and teeth. Lime oil smells sweet and blends well with citronella, lavender, or rosemary. It also increases blood circulation, relieves arthritis, reduces high blood pressure, and fights colds and flu. It can be rubbed on areas to reduce acne and relieves insect bites.

NUTRITION: One lime has only 20 calories with absolutely no fat, sugar, or cholesterol. Each fruit is packed with vitamins C, B6, A, E, folate, niacin thiamin, riboflavin, pantothenic acid, copper, calcium, iron, magnesium, potassium, zinc, phosphorus, and protein. One lime contains a third of the daily requirement of vitamin C as ascorbic acid.

FOOD & USES: Citrate of lime and citric acid are also derived from this fruit. Lime is also used in pepper sauce and chutneys. Lime juice is used in marinades, salad vinaigrettes, fish dishes, cocktails, and ceviche (pronounced sa-vi'-chee or sa-vich'). In ceviche style cooking, lime juice mixed with chopped hot and sweet peppers, tomatoes, culantro, and onions not only flavors fish or seafood; its acid actually cooks the flesh firm and opaque.

If it came from a plant, eat it; if it was made in a plant, don't. ~ Michael Pollan

RECIPES: NAWILIWILI CEVICHE SHRIMP

Ingredients: 1-pound large shrimp, 6 limes (the juice from 4 and 2 sliced thin), 1 large ripe tomato chopped, 1 medium sweet onion, 1/2 hot pepper minced, 1 cup fresh orange juice, 1 bunch cilantro chopped fine, salt and spice to taste

Method: Bring a two-quart pot of water to a boil. Turn off and then add shrimp for only one and a half minutes, then remove from water, drain, and place in a glass bowl. Add all chopped vegetables, the minced hot pepper, lime and orange juices, and lime slices to the shrimp. Let sit covered in the fridge for at least four hours. Serve on biscuits, roti, or warm garlic bread.

GRILLED LIME CHICKEN

Ingredients: 3 limes, 4 large chicken breasts, 3 TBS canola oil, 1 onion minced, 6 garlic cloves minced, 1/2 hot pepper minced (optional), 1 TBS cilantro chopped fine, salt and spices to taste

Method: In a bowl mix oil, lime juice, culantro, and hot pepper. Add chicken. Let marinate in the fridge for 4 hours. Grill over hot coals for 20 minutes basting with the marinade.

PICKLED LIMES

Ingredients: 15 ripe limes, 2 TS roasted chili powder, 1TS hot red pepper smashed, 4 cloves, 1/2 TS cardamom powder, 1 TS ground black pepper, 1 TS sugar, clear vinegar, water, lots of salt

Method: Boil water. Put 10 limes in boiling water for no more than 1 minute. Dry and make a deep 'X' cut on each end of each fruit. Push as much salt into the cuts as possible. Place on a foil-lined bowl. In a separate bowl squeeze out the juice from the remaining 5 limes and mix with salt. Place both bowls in a sunny spot to dry. The juice/salt will dry quicker and then store it in a cool place. Dry the salted limes till the skins are tan. Place the dried limes in a bottle. Add all remaining ingredients plus the dried lime juice/salt. Fill the bottle with vinegar, cap, and shake. Seal and place in a cool dark place to age, the longer the better.

WAINIHA LIME BALLS

Ingredients: juice of 3 limes, grated peel of 1 lime, 2 cups grated green papayas, 1/2 cup water, 1 cup brown sugar

Method: Boil grated green papaya for 1 minute, cool, and strain. Press papaya to remove all excess liquid. Place sugar and 1/2 of water in a saucepan and bring to a boil. Stir until thick then add grated papaya, lime juice, and grated lime peel. Boil for two minutes stirring constantly. Remove from heat and slightly cool before rolling spoonfuls into balls. Dry on wax paper.

THE EASIEST LIME PIE

Ingredients: 1/2 cup lime juice, 5 egg yolks beaten, 1 can sweetened condensed milk, 1 pre-made pie shell.

Method: Bake pie shell in a 175c/350f-degree oven for ten minutes. Blend egg yolks and milk in a bowl. Mix in lemon juice. Boil mixture in a double boiler stirring constantly until it thickens, or it will stick and burn. Pour into pie crust and put in the fridge for at least an hour before serving.

DID YOU KNOW?

Limes are believed to be native to Indonesia or Southeast Asia. The English word 'lime' is derived from the Arabic word 'līma.' The fruit was unknown to Europe until the Crusades. By the mid-1200s, it was cultivated in Italy and France. The Spaniards carried lime seeds to the Caribbean islands and Mexico and were commonly being grown in Haiti in 1520. Portuguese traders brought the Persian lime.

Eat healthy day and night, to keep your future looking bright. ~ Unknown.

LONGAN – DRAGON'S EYE

Dimocarpus longan

Longan is a fruit highly valued in Hawaii and SE Asia, especially in China, both as a tasty food and for Ayurvedic-homeopathic medicines. This is a perfect tall tree for your landscape or potted for a lanai or balcony. The longan has been growing throughout the island chain since the early 1900s. Hawaiians developed the 'Kohala' longan that has a sweet butterscotch flavor, sweeter and larger with smaller seeds than a typical Asian variety.

This tree provides shade, and the small light-yellow blossoms have a sweet fragrance. Trees can be purchased from nurseries. Longan can be grown on patios and balconies in pots. Its botanical name is *Dimocarpus longan*. Longan originated in southern China, between 150-450m in elevations (500 and 1,500 ft.). The earliest record is from 200 BC during the Han Dynasty. The emperor proclaimed lychee and longan trees be planted in his palace gardens. Longan was introduced into India by the British in 1798. China is the biggest producer with 2,000 tons annually.

The fresh, delicious longan fruit is small, but bigger than a berry, and grows in bunches with a hard shell encasing a translucent sweet white flesh. A fully ripened, fresh-picked longan fruit is round, about 2 cm (1 in.), and has a thin, firm, yellow-brown to light reddish-brown rough shell. It's easy to peel by squeezing the pulp out as if one were splitting a sunflower seed. If the shell is soft, the fruit was usually picked too soon. Longan fruit must be hand-picked and remain fresher when attached to the branch. The flesh is a pale translucent white with a musky, drier sweetness similar to dates. Longan's don't have the tropical, grape-like sweet-sour of a lychee, and have less aroma. The seed is round, jet-black, shining, with a circular white spot at the base, giving it the aspect of an eye.

HOW TO GROW: Longan is a good-looking, fast-growing, evergreen tree with a many-branched crown, usually growing to 15 meters. Longan is a picky tree suited for higher elevations. A definite change of seasons is necessary for blossoms and fruiting to occur. High-volume blooms only happen after a cool 2-3 month winter season with temperatures of 10 degrees C (50f). After that cool season, hot temps are good. It prefers some shade from the hot sun. This tree also thrives in sandy soil with a 5.5 - 6 pH. Rich humus soils will produce branches rather than blossoms.

Usually, longan trees are started from seed. After drying in the shade for 4 days, the seeds should be planted only 2 cm deep, otherwise, they may send up more than one sprout. Within 2 weeks you should see growth. In six months, transplant the seedlings to a shaded area. Longan saplings are big enough to set in place at 2-3 years in holes with compost and manure. Longan bears fruit irregularly; one good year is followed by 1 or 2 poor years. To combat this, the trees should be fertilized with high nitrogen after every harvest and with a smaller dose when blossoming. It's best to add fresh, rich soil around the tree's base every year. Longan requires regular water and can withstand a short flood, but not a long dry spell. Irrigation is necessary for dry periods. These trees should be topped as they can grow to over 30m (100 FT.). Longan can bear fruit in drooping clusters within 5 years. In a good year, a longan tree can produce 200 kilos (440 lbs.) of fruit.

MEDICINAL: The longan has been used medicinally for relaxation, and to produce 'internal heat,' which is thought to improve sex and cure skin problems. The flesh of this fruit and seeds are used medicinally. The Chinese also use the leaves and the flowers as herbs. The fruit flesh enhances appetites, reduce fevers, expel intestinal parasite worms, and an antidote for poison. A decoction of dried fruit is a treatment for insomnia and general exhaustion. In North and South Vietnam, the 'eye' of the seed is pressed against a snakebite in the belief that it will absorb the venom. Pulverized seeds in tea counteract heavy sweating and the powder, which contains saponin, tannin, and fat, will stop a wound from bleeding or help heal a burn quicker. The leaves contain quercetin which has an antioxidant and anti-inflammatory effect that reduces inflammation, kills cancer cells, controls blood sugar, and helps prevent heart disease. Preliminary research on the extracts of dried longan seeds has shown treatment possibilities using their en-zymes to treat colon cancer.

NUTRITION: 100 g (3.5 oz.) of longan fruit has only 60 calories, 15g of carbohy-drates, 1g protein, with vitamins B1, B2, B3, loads of vitamin C, manganese, phosphorus, and potassium.

FOOD & USES: Longan fruit is delicious eaten raw and can also be dried, preserved in syrup, cooked in sweet and sour dishes, or in dessert soups. At room temperature, longan fruit remains in good condition for several days. Because of the firmer rind, the fruit is less perishable than the lychee. The fruit's flavor may be improved by cooking. In China, most of this fruit is canned in syrup or dried. Longan fruit can be preserved by drying with or without the shell. The resulting dried leathery, black flesh is used to flavor drinks and make beverages, including liquor.

To dry the fruits, they are first heated to shrink the flesh and facilitate peeling off the shell. The seeds are then removed, and the flesh dried over a slow fire, giving them a sweet smoky flavor. Longan fruit are canned in their own juice with little or no sugar to preserve the fruits. Canned longan fruit retain their flavor better than canned lychees.

The reddish-brown heartwood is strong, very hard, difficult to split, highly durable, but slow to dry. It's used for posts, agricultural implements, furniture, construction, textile looms, and rifle butts.

RECIPES: DRIED LONGAN FRUITS

Dried longan fruit is naturally sweet and chewy. If your tree grows a lot of fruit, drying can make fruit candy for all year round. BUT you must have an oven. First, peel the shell with a quick twist. Seed the fruits by slipping a finger alongside the seed and the longan's flesh should split in half easily. Place the peeled and pitted fruits on a baking tray or cookie sheet. Slide the sticky fruit into a low-temperature oven and let it bake-dry for hours. As long as the tempera-ture is kept low, the fruit should not dry out. Check the fruit every hour. When the fruit is dry to the touch and chewy on the inside, it is ready. If you wait too long to remove the longan fruit it will become hard. Hard is okay, but chewy is better.

LONGAN – COCONUT ICE CREAM

Ingredients: 500g longan fruit peeled and seeded, ½ cup shaved coconut, 1½ cups whipping cream, 2 cups sweetened condensed milk, 1 TBS vanilla extract, and a blender and hand mixer. Method: First, place a mixing bowl and the beaters in your freezer for 15 minutes. Place the longan fruit in a blender and pulse. Blend so that chunks remain. Pour the whipping cream in the chilled bowl and beat at medium speed for 2 minutes. Add the condensed milk, the blended but still chunky longan fruit along with its juices and the vanilla extract. Beat this combination for 5 minutes. Freeze overnight.

DID YOU KNOW? Longan's Asian name may be derived from the Mandarin 'long yan,' (龙眼), or the Cantonese *lùhng-ngáahn* 龍眼, literally meaning 'dragon eye' because the peeled fruit resembles an eyeball as the black seed shows through the translucent flesh like a pupil/iris or a dragon eye. They are also called lungan, dragon's eye, or cat's eye. Longan is of the soap-berry family wth the litchi and the rambutan.

LOQUAT – JAPANESE PLUM – PIPA
Eriobotrya japonica

A relative of the rose, the loquat, grows small, round fruits in clusters that taste like a blend of peaches, citrus, and mango. Loquats are classified as subtropical fruit. It's a large ever-green shrub or a short tree grown commercially for its orange fruit and leaves for tea. Botanists credit Chinese immigrants bringing the loquat to Hawaii.

The loquat is native to southeastern China. For more than a millennium, it's been cultivated in Japan. Loquat has become naturalized in India. Japan is the leading grower with 17,000 tons annually. Brazil has 150,000 loquat trees. Over 800 varieties of loquat exist in Asia.

HOW TO GROW: Loquat fruit, growing in clusters, are oval, rounded, or pear-shaped, 1–2 in. long, with a smooth yellow or orange skin. The juicy pulp can be white, yellow, or orange and tangy sweet. Loquat does better in the cooler hill regions. Loquats are normally bushy, rather dense trees, and will grow to 30 ft., but usually topped at 10-15ft. Trimmed shorter, it's a shrub and perfect for a porch or balcony pot. In Taiwan, because of the hazard of strong typhoons, the loquat is grown as a dwarf, 3 ft. high and wide.

Growing from seed presents some problems, as you don't know exactly how the fruit will turn out. Seeds come from grafted trees and you may get the rootstock's product. It may not produce fruit. Either way, it'll make a nice ornamental plant. Purchasing a small tree from a nursery is better for growing fruit. Water the seeds lightly once a day in their pots until sprouts appear. After that, water the pods when they're dry. At 6 inches, transfer the sprouts to larger single pots. Loquats do well in most soils, but prefer slightly acid soil with a pH 5.5-6.5. Grafted trees fruit in 3 years, seedlings in 8-10 years. The 1-inch white blossoms have a nice aroma.

Wait for the fruit to ripen on the tree. These fruits don't ripen well off the tree. Look for the proper color. Loquats are not easy to pull off the tree, even when ripe. Clip the stalk of the fruit with shears. Loquats are unusual among fruit trees in that the flowers appear in the autumn or early winter, and the fruits are ripe at any time from early spring to early summer. Mature loquat trees may yield more than 100 lbs. per year depending upon tree size and care. For the best fruit, water regularly and four times a year apply ½ pound of 6-6-6 fertilizer.

Loquats need to have the seeds removed to eat. The big seed is most of the fruit and slightly poisonous in large quantities.

MEDICINAL: Loquats have been used in traditional medicine for thousands of years to treat diabetes, arthritis, rheumatism, and nausea. Powdered loquat leaves treat diarrhea, depression, and even help with a hangover. The leaves are used in Japan to make biwa cha, a beverage that treat skin conditions, bronchitis, and other respiratory illnesses. In 914, the first Chinese medical textbook was translated to Japanese telling how to use loquat to obtain clear lungs.

NUTRITION: 100 grams of loquat have 47 calories, 1½ g protein with vitamin A, calcium, magnesium, phosphorus, and potassium.

FOOD & USES: Fruits are eaten fresh, and delicious simply stewed with a little sugar added. Taiwan exports loquats canned in syrup. Loquats are used commonly as a natural sweetener for many different types of food and are used to make marmalade and jelly in various locales. Loquat wood is used to make rulers and other writing instruments.

DID YOU KNOW? For centuries only the Chinese royalty could eat the fruit, as it was thought that loquat fruit falling into the rivers gave the koi, or carp, the strength and desire to swim against current and up waterfalls and be turned into mythical dragons. Japan got the trees from China as early as 700 AD.

LYCHEE

Litchi chinensis

Like so many fruits, the lychee tree is great for the home yard and garden. Trees are found in Hawaii, but they are not as popular as its hairy cousin, the rambutan. A lychee tree can grow to 30 meters, but you want to keep it shorter. It's very classy with a wide top of dense green leaves and the green, yellow or white blossoms have a pleasant aroma. Keeping it pruned short will make it easier to pick the delicious fruit. The outside of the fruit is pink-red, roughly textured and the shell/casing is inedible. The lychee's sweet, almost clear or white flesh with one seed is eaten raw or in many desserts. It's added to ice creams, cakes, and beverages, because of its succulent taste and pleasant fragrance.

This is a unique fruit grown everywhere in the tropics and subtropics, it carries only one name, lychee, litchi, or lichi. This is an evergreen tree and a member of the soapberry family. Its botanical name is *Litchi chinensis* because it originated in China. Wild trees still grow in parts of southern China and on Hainan Island. The fruit was used as a delicacy in the Chinese Imperial Court and in such demand that a special fast horse - pony express-type courier service brought fresh lychees from the far southern Guangdong Province. Records in China refer to lychee as far back as 2000 BCE, but the cultivation of lychee began in1059. The lychee was described and introduced to the West in 1656 by Michal Boym, a Polish Jesuit missionary.

Lychee cultivation spread through S.E. Asia and Indonesia. In the 1600s, it was carried to Burma, and a century later to India. It arrived in the Caribbean in 1775 and was being greenhouse-grown in England and France during the early 1800s. Lychee was imported to Hawaii in 1873 from China and became known as 'The Afong Tree,' in Oahu. It reached Florida in 1883. Annually, China produces at least 200,000 tons of lychees, followed by Taiwan with 130,000 tons.

HOW TO GROW: Save your seeds and grow a tree. Seeds have very short fertility and should be planted as soon as taken from the fruit. Sow the seed a half-inch deep in good-sized pots. The seeds germinate without pre-treatment when sown fresh. The best places to grow lychees is where it gets 16c/60f or lower. The soil should be kept moist with a pH range of 5-5.5. The bark is grey-black, the branches are a brownish-red. Trees will bear in three to five years. Lychee is a sub-tropical fruit and can be grown at low elevations up to 600 meters. In the hot months water weekly. In cooler temperature, water once a month. The hundreds of small blossoms grow in clusters of ten or more, a foot or longer.

Once the blossoms appear, it usually takes 90 days for fruits to ripen. This differs with varieties. The maturity of the fruit is judged by the color that changes from green to pink when mature. For extended freshness, it is best to pick the fruits in bunches along with a portion of the branch and a few leaves. A well-cultivated tree can yield 50 kilos of fruits a year.

Lychee fruit varies. They can be round, ovoid, or heart-shaped, up to 2 in, by 1.5 in., weighing approximately ¾ oz. The skin is thin, tough, and green when imma-ture, and ripen to red or pink-red. The skin can be smooth or rough. The husk is in-edible, but easily removed to expose the pearly white flesh that has a floral scent with a sweet flavor. The husk turns brown and dry when left out after harvesting.

MEDICINAL: These fruits are anti-hyperglycemic and can lower blood sugar levels in diabetics. Consult with your doctor before eating many to avoid side effects. These fruits are loaded with vitamin C, which builds the immune system, protects against colds, infections, and fights inflammation. Tea is sometimes made from the lychee husks/skins to treat smallpox and diarrhea. The seeds are ground in India to treat stomach ailments. Sore throats are treated with a decoction of bark, root, and blossoms.

NOTE: In 2019, dozens of children in India, Bangladesh, and Vietnam died of an unknown illness, hypoglycemia, or low blood sugar. Because the illness occurred in a cluster with-in a lychee-producing area, scientists are considering it was caused by a toxin found in the ly-chee; higher in unripe fruit, which the children could have consumed. Other investigators think that the pesticides used to grow the lychees might also be to blame. ***There's no way to prevent low blood sugar when consuming lychees, doctors suggest not to only eat lychees at a sitting.***

NUTRITION: This fruit is 70% water and a great source of hydration for the skin and has a powerful compound 'oligonol' to protect the skin from UV light and free radicals that can damage the skin cells. Lychee is perfect for dieters, 100 sweet grams has only 66 calories, and because it has only a tiny amount of saturated fats and no cholesterol. It is rich in vitamins C and A with many B vitamins, potassium, thiamin, niacin, folate, and copper.

FOOD & USES: Lychees are commonly sold fresh in Asian markets. Smooth-skinned fruits grown on lowland, the 'water lychee' is preferred to the rath-er highland grown prickly fruits of the 'mountain lychee' The red husk turns dark brown when the fruit is refrigerated, but doesn't change the taste. It's also sold canned. The fruit can be dried in the husk, and the flesh will shrink and darken.

Fresh lychees have a sweet taste like a combination of strawberries, watermelon, and grapes, and can be kept in the refrigerator for up to 2 weeks. They can also be pickled, canned in syrup, used in jams, sauces, ice creams, and fruit salads. Dried lychees, called lychee or litchi nuts, are also popular, are similar to raisins used as a snack or chopped into fruit or green salads. Chinese use dried lychee to sweeten their tea. The tough outer skin is peeled like a grape to reveal a pearly white interior pulp. If the fruit is very ripe, you can tear off the end of the skin and then push the fruit out. If not, use a serrated knife to cut lengthwise through the skin and around the seed. Then just peel away the skin and the inner membrane to reveal the fruit.

RECIPES: Make a lychee smoothie by blending yogurt, honey, chopped ly-chees, fresh lime, powdered cardamom, and ice cubes, until smooth and frothy.

LYCHEE JUICE

Ingredients: 4 ripe lychees, 2 TBS jaggery or brown sug-ar, 1 TBS fresh orange juice, 2 cups water, a pinch of salt

Method: Clean lychee, seed, and chop. Machine blend lychees and water, then strain through a cheesecloth or sieve. To the resulting clear juice add the remain-ing ingredients and blend again. Refrigerate for at least an hour. Serve chilled.

LYCHEE AND OATMEAL FACIAL TREATMENT - brightens and rejuvenates your skin

Ingredients: 2 lychees peeled, and seeded, ½ cup basic oatmeal, ½ cup water

Method: Blend ingredients into a paste. Apply to your face and let sit for 10 minutes, wash off.

LYCHEE SORBET

Ingredients: 1 pound of fresh peeled, seeded, and chopped lychees, ½ cup white sugar, 1.5 cups water, juice of a half of a lime,

Method: Bring water to a boil stirring until sugar dissolves. Allow to cool, pour into a blender, add chopped lychees, and lime juice. Blend until an even consistency. Pass through a sieve. Freeze until solid, break apart and blend again for five minutes. Freeze overnight and serve

Love is a fruit in season at all times.

MAMEY SAPOTE

Pouteria sapota

The mamey sapote or mamey Colorado is a great tree for your yard. Native to Mexico and Central America, the mamey sapote is widely cultivated for its large, edible fruits. It's attractive, stately, has many culinary and medical uses, but it is also unique, kind of rare. This evergreen has bright green leaves that are two feet long and the fruit resemble a cross between a big mango and a coconut with rough, light brown shin. The creamy flesh is deep orange with the texture of a baked sweet potato. The taste is compared to apricots. Some have fruit all year.

HOW TO GROW: It can succeed in the lowland moist tropics up to 1400 meters. It's large, not clustered with branches giving it good wind resistance. Grafted trees from a nursery are best. If you want to try planting seeds, nick or crack them first for quicker germination. Plant a tree 30 feet from anything, as the mamey, if not topped, can grow to 75 feet. It requires a rich, deep, well-drained soil, including sandy and clay, with a pH of 5– 7 and full sun. Requires a very soil. Seeds which have a hairline crack in the seed coat appear to germinate more Grafted trees bear in 3-5 years while seedling trees take at least 7 years or longer. After the small pinkish-white blossoms appear, it takes from 13 - 24 months for fruit to reach maturity. Kumu Aina Farms on the Big Island has a hundred plus mamey trees.

This tree is unique as it may have flowers, immature, and mature fruit at the same time. The sapote can bear 200 to 500 fruit per year. The fruit can weigh up to 3 kg and usually con-tains only a single, large seed, but can have up to four. A ripe mamey should give when pressed like an overripe avocado. The ripe fruit are reddish, soft, and sweet and eaten raw or processed into a variety of beverages or desserts.

MEDICINAL: The oil pressed from the seeds is used as a skin ointment and as a hair dressing believed to stop hair loss. The oil is also reputed to be diuretic and is used as a sedative in eye and ear ailments. The pulverized seed shell is mixed with wine as a reputed remedy for coronary ailments, kidney stones, and rheumatism. The seed residue, after oil extraction, is used for poultices on skin problems like boils. Drinking tea made from the bark and leaves treats arteriosclerosis and hypertension.

NOTE: The raw seed has stupefying properties, and this may be due to its HCN (*Hydrogen Cyanide*) content. The milky sap of the tree is highly irritant to the eyes and caustic on the skin. The leaves are reportedly poisonous.

NUTRITION: 100 grams has 120 calories with 1.5g of protein. The fruit is an excellent source of vitamin B6 and vitamin C, and is a good source of riboflavin, niacin, vitamin E, manganese, potassium, and dietary fiber.

FOOD & USES: The fruit is eaten raw or blended into milkshakes, smoothies, ice cream, and fruit bars. It can make marmalade and jelly. It's best to eat the fruit as soon as it's ripe, but you can also freeze it. Ground seeds are mixed with cornmeal made into a confection. They can be brewed with sugar and cinnamon to make a nutritious beverage called pozo. Pressed, the seeds yield a white, semi-solid, vaseline-like oil which is edible when freshly extracted and refined.

RECIPES: MAMEY ICE CREAM

Ingredients: 3 cups mamey diced, ¼ cup evaporated milk, 1 cup sweetened condensed milk, 1 TBS vanilla extract, pinch nutmeg and cinnamon, ¼ cup whipping cream. Method: Beat the whipping cream until it doubles. Blend the mamey with the evaporated and condensed milks and spices. Add the whipped cream. Freeze for at least 8 hours.

MANDARIN CITRUS

Citrus reticulate

The mandarin orange tree is a slightly smaller tree than the regular orange tree and perfect for the home garden. The mandarin is native to China and Southeast Asia. Chinese and Arab traders were the first to spread this fruit tree throughout Asia, India, and the Middle East. It's the largest and most diverse group of tasty citrus. I included mandarins, in addition to regular oranges, because mandarins are different with loose, easier to peel, skin, and separate into segments.

It may have been the Dutch, Portuguese, or Spanish explorers who brought the first mandarins from Asia to Hawaii. During the 18th and 19th centuries, the British brought many species and varieties of trees, especially citrus from Asia to the Caribbean to start plantations to feed England and Europe.

Two Chinese varieties of mandarin oranges were growing in Italy by the mid-1800s. Today, world commercial production of mandarins is more than 12,500,000 tons, or 13% of all citrus production. Spain and Japan each grow 16% of the world's production.

HOW TO GROW: Mandarin orange trees are the most adaptable citrus, and are more cold and drought tolerant than the sweet orange. If there is a cold blast, the fruits must be picked because they are tender and will be damaged by the cold. The climate can affect the fruit's taste and juiciness. Although a wide range of soils will produce good fruit, the ideal fertile soil must be well-drained with a pH of 6-6.5. Protect mandarin trees from high winds. These oranges can be grown from seed. Keep a few seeds from a large, juicy mandarin. Plant it in moist soil and keep in a sunny warm place. It should germinate within a month.

It can take 3-5 years for grafted mandarin orange trees to bear fruit but will bear for 20 to 30 year. I have a mandarin tree, but it's grafted. With regular water and ¼ cup of 12-12-17-2 fertilizer bi-monthly; the tree bears so heavily I must prop the branches or they'll snap.

MEDICINAL: For hundreds of years, Chinese traditional medicine has used mature mandarin orange peel to improve digestion, relieve intestinal gas and bloating, and resolve phlegm. Immature mandarin orange peel treats liver and stomach problems.

NUTRITION: Per 100grams, mandarin oranges have 47 calories with vitamin C, calcium, magnesium, potassium, phosphorus, and iron.

FOOD & USES: Mandarin oranges are best eaten out-of-hand. The sections are delicious in fruit salads, puddings, or on bakery goods and pastries. Small oranges are sectioned and canned in syrup. The essential oil flavor from the peel is used commercially in hard candy, gelatins, ice cream, chewing gum, and carbonated beverages. At room temperature, mandarins will last about 1 week. Refrigerated in a bag they should last 2 weeks to 1 month.

RECIPES: EASIEST AND TASTIEST MANDARIN PUDDING

Ingredients: 2 mandarins peeled, sectioned, and seeded, 2 juiced, 1 package vanilla pudding mix

Method: Follow pudding directions, but use mandarin juice as liquid. Add sections, chill, serve.

DID YOU KNOW? A ripe mandarin is firm to slightly soft, heavy for its size, and pebbly-skinned. The name may relate to the yellow color of some robes worn by mandarin nobility. During Chinese New Year, mandarins are considered traditional symbols of good luck because.when spoken in Mandarin, it sounds similar to the word "wealth". The orange hue of the fruit is also said to symbolise 'gold', which makes it a very auspicious fruit

MANGO – MANGA

Mangifera indica

Hawaii has incredible views, friendly people, and the food tastes fantastic. But nothing tastes better than the first mango of the season. Local mangoes are ripe from May until October. June is the best month. Due to the climate and elevations, some mangos are always available. Try fresh, in chutney, or pickled. After pineapple, mango is Hawaii's most cultivated fruit.

About four thousand years ago the delicious mango originated in East India in the foothills of the Himalayan Mountains. Mangos were one of the first fruits humans farmed. Asian kings had vast orchards of trees. An Emperor of Delhi, Akbar (1556 to 1606, had a hundred thousand hectares of mango trees. The word 'mango' is derived from the Tamil East Indian dialect word, man-kay, which Portuguese explorers changed to 'manga.' India is still the largest producer of mangos with very little exported. More mangos are eaten throughout the world than any other fruit. There are over 20 million metric tons of mangos grown yearly. Mango's botanical reference is *Mangifera indica*. Honolulu is the major market for mangos in Hawaii. In 2019, Honolulu sold 60,000 kg, 75% of which came from Oahu. The other 25% were from Hawaii, Kauai, and Maui.

Captain Meek, originally from Marblehead, Mass., is given the usual credit for first mango in 1824 to Hawaii from Manila. Three small mango trees were brought to Honolulu from Valparaiso, aboard the HMS Blonde in 1825. From Mexico, some mangos were imported by Don Marin around 1800. Today, the island chain has over 500 mango varieties. Arguably, the most popular are the Haden, Rapoza, and Pirie.

HOW TO GROW: First find a very ripe mango of the variety you enjoy the best and carefully cut the fruit from the seed. Do not damage the seed. The thinner area on the seed is called the eye. In a pot with drain holes filled with moist soil, plant the seed, eye up, and cover with ½ in/1.27 cm of soil. The seed should sprout within a month. Mangos don't require much but water when the soil is dry. Transplant outdoors when the young tree has several leaves.

Unless you choose a dwarf, grafted mango, allow at least a circumference of 50 ft. for the tree. It should bear about the third year. Plant in an area that drains well to avoid root rot. Keep adequately watered. In the dry season use the hose water for at least an hour a month for excellent fruit. Tend with a cup of high nitrogen fertilizer mix monthly and watch for pests. It takes about 4 to 6 years for a tree to bear fruit. Their flowers are pollinated by insects, but less than 1% of the flowers will mature. It takes about 4 months for the fruit to mature and each one is harvested by hand once a year. Your tree should give you a lifetime of shade and delicious fruit. Trees can still bear fruit after 30 years. I just sprouted two mangos accidentally in my compost pile. It's easy.

MEDICINAL: It is believed mangoes strengthen and invigorate nerve tissues in muscles, heart, and brain. Mangoes have enzymes such as magneferin, katechol, oxidase, and lactase, which helps clean bowels. Mangoes are an antidote for all toxic effects inside the body. Mangoes also contain pectin, a soluble dietary fiber, which has been shown to lower blood cholesterol levels. Chinese medicine uses mangoes to treat anemia, bleeding gums, constipation, cough, fever, nausea, seasickness, and to help with weak digestion.

NUTRITION: Mango is a true 'comfort food' because, like papaya, mango contains a stomach-soothing enzyme. One large mango has about a hundred calories with no cholesterol and half of the necessary daily fiber. Mangos not only make you feel good, provides a quarter of the recommended daily allowance for vitamin C, two-thirds of vitamin A, some vitamin E and K, fiber, phosphorus, and magnesium. Mangoes are particularly rich in potassium, which can help reduce high blood pressure.

FOOD & USES: Mango wood is hardwood, and it takes 15 to 20 years to mature to use for furniture. The natural color is golden brown. In the northeastern Thai region of Isaan, krok (mortars) made of mango wood are particularly prized as they are considered inherently aromatic and of the perfect hardness with which to pound the region's famous som tum, or grated salads. Do not burn mango wood or leaves as it produces toxic fumes that can cause serious irritation to the eyes and lungs.

RECIPES: KANEOHE MANGO CHUTNEY - great with everything Ingredients: 3 nice mangos, washed, peeled, seeded, chopped, 3 cloves of garlic minced, 3 thin slices of fresh ginger, 2 TS mustard seeds, 2 cloves, and 2 cardamom seeds ground, 1 TS hot red pepper flakes or 1/2 habanero minced with seeds, 1/3 cup vinegar, 1/4 cup sugar, salt to taste
Method: Combine mango pieces and sugar. Let stand in a bowl for 3 hours. In a mortar, crush garlic, ginger, cloves, cardamom, pepper, and mustard seeds. Mix together with vinegar and the remaining sugar. Simmer for 15 minutes. Add mango-sugar pieces and cook for another 15 minutes. Adjust sugar and pepper to your taste.

MANGO CRISP

Ingredients: 5 cups ripe mangos peeled and sliced, 1/2cup brown sugar, 1/2 cup flour, 1/2 cup soft butter/margarine, 1 TBS lemon juice, 1/4 TS nutmeg, 1.5 TS cinnamon
Method: Mix mangos, nutmeg, one TS cinnamon, and lemon juice and place in a greased ten-by-ten baking dish. Blend flour, brown sugar, a half TS cinnamon with butter until mixture is crumbly. Drop the flour mix over the mango. Place in a 175c/350f-degree oven for 35 minutes.

MAUI WOWWEE MANGO SALSA

Ingredients: 1 cup diced mangos, 1/2 cup diced red sweet peppers, 1/2 cup red Spanish onion, 1 small hot pepper minced, 1 TBS fresh lime juice.
Method: Combine all ingredients and chill. Use as an accompanying sauce or marinade with meat, chicken, or fish dishes. Eat with roti, or chips.

PICKLED MANGO

Wash, peel, and slice 4 half-ripe mangoes into thin strips. Place in a bowl with a pinch of salt, one hot pepper - seeded and minced, 1 TS of the following: black pepper, chili powder, sugar, vinegar, 12 cilantro leaves minced. Combine, cover, and let sit for an hour. Enjoy.

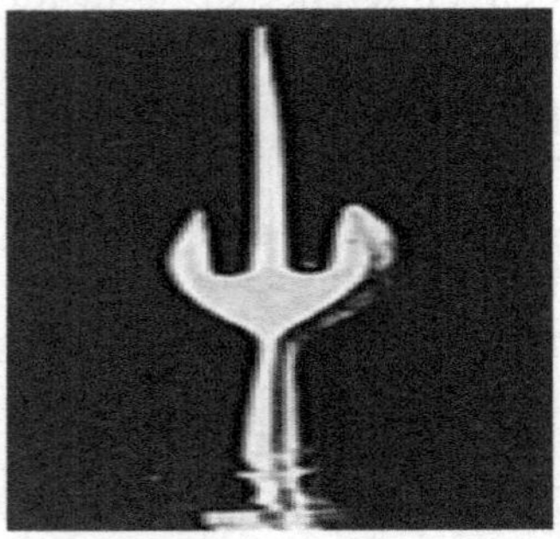

DID YOU KNOW? This luscious fruit is mango in English and Spanish, and only slightly different in French (*mangot, mangue, manguier*), Portuguese (*manga, mangueira*), and Dutch (*manja*). In some parts of Africa it is called *mangou, or mangoro*. India produces more than half of the world's mango crop nearly a billion tons on two and a half million acres. There are more than 500 different varieties of mango in India where it is the national fruit. Mango is from the same family as poison ivy, and the cashew, pistachio, and Jamaican plum tree.

The long mango fork is designed to pierce a mango through the seed. With the peel removed, the most flavorful flesh around the seed can be enjoyed like a lollipop.

A king asked a sage to explain the Truth. In response, the sage asked the king how he would convey the taste of a mango to someone who had never eaten anything sweet. No matter how hard the king tried, he could not adequately describe the flavor of the fruit, and, in frustration, he demanded of the sage, "Tell me then, how would you describe it?" The sage picked up a mango and handed it to the king saying "This is very sweet. Try eating it!" Hindu Teaching.

MANGOSTEEN

Garcinia mangostan

Hawaii is in the early stages of mangosteen cultivation, but this tasty fruit can be found on every island. The Big Island has the most and the oldest trees. Look to the Asian farmers' markets during the summer for this unique fruit. The mangosteen is no relation to the mango. The mangosteen is a tropical evergreen tree with an aril fruit, native to Indonesia. The tree is almost twice the size of the usual orange. The ripe fruit is deeper, more reddish-purple than the canistel. The segmented white flesh has a combination of flavors. Mangosteen's botanical name is Garcinia mangostana.

HOW TO GROW: Search for them with the grafted tree vendors. You might get lucky as they are few. It can be grown from seed, but do multiple plantings as the success rate is low. Keep the seed moist until it is planted. In damp potting soil, seeds should sprout within a month. This tree has a long, fragile taproot and that makes transplanting difficult. It is advised to transplant this tree before it gets more than 18 in. /47 cm tall. This will keep the taproot at a manageable length. It may take two years to grow to this size. Plant a mangosteen when you have a baby. It may bear fruit when your child can drive! Young trees need to grow in the partial shade, while mature trees require plenty of sun for the fruit to properly develop. It can produce fruit twice a year. Depending on the age of tree, mangosteen can produce from 200 to 3,000 fruit per season. Older trees produce more fruit. Mangosteen is a perennial plant that can survive more than a century.

Be certain to dig an exceptionally deep hole, thirty inches, and have someone suspend the mangosteen seedling while crumbled earth is gently put around the long root. The mango-steen tree is a very slow growing. It does well in deep rich, sandy soil. Keep the surrounding area weeded. It needs good drainage, but can handle and may thrive in damp soils. It does not do well in an area that has sea blast. The mangosteen needs about twenty feet on all sides. Every other month sprinkle a cup bearing fertilizer 12 –12-17- 2. The tree may bear fruit within five to seven years, or longer. Keep this tree watered! Mangosteen trees, with the roots almost con-stantly wet, as around a pond or along a stream, bear the most fruit. To combat the dry season, use a mulch of coconut husks or palm fronds around the base of the tree to help hold moisture. Spray with an insecticide combined with a foliar fertilizer once a month. Keep ants away from this tree as they will decimate the tender leaves. Keep watch for the mangosteen caterpillar. Neem extract has worked well controlling these pests in Thailand where it is the national fruit.

The exquisite taste of mangosteen's juicy, snow-white, soft flesh is well worth the ef-fort and wait. The fruit may be seedless or have multiple seeds that cling to the flesh. The flesh is slightly acidic, but an exquisitely delicious mix of peach, strawberry, and pineapple flavors. Young mangosteen are pale light green. Ripe mangosteen are bright purple and just slightly soft. These fruits must be harvested by hand and not be allowed to damage with a fall. One taste and you know why they call mangosteen, 'The Queen's Fruit.' Explorers to the Far East re-turned with tantalizing stories of the mangosteen. The return journey by sail took too long and the queen never got a taste. There is a legend Queen Victoria offered a sizeable reward, even a knighthood, to anyone who would bring her a mangosteen. In reality, trees were introduced from India in 1833 and were grown in heated glasshouses in England, and fruits were harvested in 1855. The queen did get a taste.

To open mangosteen use a knife to score the peel. Grab the fruit with both hands. Take care not to get any of the juice on your clothes as it will stain. Twist gently along the score with the thumbs until the rind cracks. Pull the halves apart along the crack and remove the fruit.

MEDICINAL: Mangosteen are dried to use medicinally throughout Asia. The powder is made into an ointment for skin disorders. A piece of this fruits peel put into boiled water and soaked overnight is a bush remedy for chronic diarrhea. Be aware, the juice will stain clothing. Mangosteen's peel contains the highest level of nutrients, while the actual fruit pulp is one of the world's best tasting fruits. Mangosteen has compounds that have antioxidant, anti-bacterial, anti-fungal, and anti-tumor activity. It also has antihistamine and anti-inflammatory properties.

NUTRITION: Mangosteen have 60 calories per 100 grams of flesh w ith calcium, phosphorus, folate, manganese, and iron.

FOOD & USES: Mangosteens are non-climacteric ripeners, meaning that after picking, they do not ripen further. Mangosteen should be picked only when ripe and eaten soon after. Mangosteen is also used for the preparation of ice creams, sorbets, yogurts, smoothies, cocktails, and salad dressings. In Indonesia, a natural batik dye is extracted from mangosteen peels for shades brown, dark brown, purple, or red hues. Mangosteen twigs are used as chew sticks in Ghana. The wood is used to make spears and furniture in Thailand. Mangosteen skin is used to tan leather in China.

RECIPES: MANGOSTEEN SORBET

Ingredients: 1 cup mangosteen chopped as small as possible, 1 cup sparkling grape juice, 1 egg white, one half cup white sugar, six slices of peeled firm limes, six slices of a firm peeled orange
Method: Put mangosteen pieces in a bowl and add the grape juice. Stir the effervescent fruit juice into the puree. Whip the egg white until it thickens adding the sugar. Combine this with the mangosteen and grape juice. Gently stir in the lime and orange pieces. Freeze.

WAIKIKI SUNSET MANGOSTEEN COCKTAIL

Ingredients: 2 ounces mangosteen juice (squeeze one fruit), a splash of fresh lime, orange, or lemon juice or a combination, 3 ounces of either vodka, arrack, or light rum, 1 twenty-ounce soda. Method: Combine ingredients and pour into two glasses with ice. Top off with soda.

MANGOSTEEN PUDDING

Ingredients: 2 cups mangosteen segments chopped small, 2 cups milk, 2 cups whipping cream, ¼ cup white sugar, ¼ cup hot water, 2 tablespoons gelatin
Method: In a coffee cup dissolve gelatin in hot water. In a 2-quart pot combine the sugar and mangosteen. Constantly stirring, bring mixture to a boil over low heat. Stir in milk, whipping cream. Bring this mixture to a boil again stirring constantly over low heat. Add gelatin and leave it until it boils one more time. Remove from heat and pour into a suitable container.

DID YOU KNOW? Mangosteen is perhaps the finest flavored of all of the fruit in the world, hence the nickname 'Queen of Fruits'. The King? Another beloved Southeast Asian fruit called durian, also known as the 'King of Fruits.' According to Chinese philosophy, the primary forces that influence the body's health are elements of cold and hot, or yin and yang. Following this concept of dualism, Mangosteens supply the cool element to offset the heat of durian, so this royal pair is often eaten together in Asia. Look in the positioning of fruit in tropical Asian markets. Durian (tropical fruit covered with thorny husk) and mangosteen are usually placed next to each other in the markets, because durian produces a strong heating effect which mangosteen easily neutralizes. Many compare the taste to a combination of strawberries and oranges. Mangosteen skin is dried and used to prepare an excellent, healthy herbal tea.

MOUNTAIN APPLE - 'Ô HI'A 'AI

Syzygium malaccense

Nothing is quite like the taste of a chilled mountain apple / Malay rose apple on a hot day. The Hawaiians call these delicious fruit 'ō hi'a 'ai. Our tree gets overloaded with fruits twice a year. This fruit has a different name in every country. Its botanical name is *Syzygium malaccense,* and it is part of the Myrtle family. It's called the Malay rose apple because most botanists believe this fruit originated in the lowlands of the Malaysian Peninsula.

The mountain apple is one species cultivated since prehistoric times by the Austronesian peoples. This is one of the ancient canoe fruits. It spread by Polynesian seafarers to far islands and countries where it's been naturalized. Portuguese explorers carried it from Malacca to Goa in India; then they introduced it into East Africa and Brazil. Captain Bligh, on the British ship, Providence, brought 3 varieties from the islands of Timor and Tahiti to Jamaica in 1793.

HOW TO GROW: Mountain Apple trees grow fast to 50 feet. It is best to top them at twenty-five feet to make the picking easier. The trunk usually is straight for 15 feet and then branches out. The bark is rough, light brown, and flaky. It is a perfect backyard tree with colorful flowers and shiny, waxy fruits, although there is some cleanup when the blossoms drop. The fruits are up to 4 inches long and 3 inches wide and can be brilliant red, sometimes white, or what we call the 'Chinese,' white with red or pink streaks. The juicy white flesh is crispy and mildly sweet. Usually, mountain apples have one seed, but some types have two.

The fruit matures two months after the flowers open. They are apt to fall and get damaged when they reach full ripeness. They should be handpicked. The recommended fertilizer is a time-released 8-3-9 spread in a circle below the branch ends before watering. When watering, soak the soil, but do not water again until the top two inches under the tree have dried. A good tree, fertilized with high nitrogen and given regular water, can produce more than a hundred pounds of delicious fruit.

Fresh seeds germinate and may sprout from rotted fruit under the tree. Do not dry these seeds; instead, plant in potting soil 1.5 inches deep as soon as you remove them from the fruit. You should see sprouts within a month, and transplant at eight months. Cuttings can be transplanted six weeks after roots have formed. Mountain apple trees adapt to most well-drained soils except if highly alkaline or salty. These trees do well with a pH of 4.5 to 7.5 where they receive full to partial sun and start flowering and fruiting within six to eight years. Cuttings have been rooted in sand on some of the Hawaiian Islands.

MEDICINAL: The mountain apple tree is valuable as a natural medicine source. Cambodians use a decoction of the leaves and seeds to break a fever. In Borneo, the twigs and bark are boiled and drank to fight diarrhea. Hawaiians squeeze juice from the bark and drink it for sore throats. Bush doctors in Brazil use various parts of this tree as treatments for headaches, coughs, and diabetes. The roots are known to be a diuretic and an effective preparation for combating skin irritations and edema or swelling. The leaves can fight acne because they have anti-inflammatory, antibacterial, and antioxidant properties.

NUTRITION: Mountain apple has slight protein and fats but is a good source of calcium, phosphorus, iron, Vitamins A and C, riboflavin, thiamine, and niacin.

FOOD& USES: Wood from the mountain apple is hard and heavy, reddish to light brown, but tends to warp, making it difficult to work. The wood is used for beams for the thatch-roofed sheds, rural houses, serving bowls, and cutting boards.

The ripe fruits are eaten fresh, out of hand, in fruit salad, fruit cocktail, stewed with spices as a dessert or eaten dipped in sauces. In Indonesia, the flowers are eaten in salads or are preserved in syrup and so are baby leaves and shoots before turning green, are cooked and eat-en as greens with rice. A tasty dessert is peeled, and sliced mountain apples stewed with cloves, cinnamon, and nutmeg served warm as a topping over ice cream. This fruit can also be sliced and candied. Half ripe fruits are used for pickles, jelly, and preserves. The red blossoms, fresh off the tree, can be eaten in salads.

RECIPES: MOUNTAIN APPLE WINE

In Puerto Rico, both red and white table wines are made from fresh mountain apples. The fruits are picked from the tree, not fallen/damaged fruit, as soon as they are fully ripe. Immediately dip in boiling water for one minute to remove any bacteria. Then remove the seeds. Before beginning, to ensure all this work will produce good tasty wine, every utensil, pot, container, and bottle must be sterilized. If not, bacteria will grow.

Ingredients: 5 kilos of mountain apples scrubbed, seeded, and sliced, 9 cups sugar, 2 TBS yeast, 5 liters water, 1 cup raisins (dark raisins for red wine, golden for white wine), 10 cloves, and 2 cinnamon sticks or 2 TS ground cinnamon.

Method: Bring water to a boil and carefully transfer to a sterilized bucket or buckets. Leave space for the mountain apple pieces, add yeast, cloves, cinnamon, and 4 cups of sugar. Cover with cheesecloth and a lid. Leave in a cool dark place for three weeks. Then strain through cheesecloth and blend with 5 cups of sugar. Transfer to sterilized bucket and distribute the rai-sins evenly. Let sit in the same cool dark place for two more weeks. Then transfer to sterilized liquor bottles and seal with wax.

HAKALAU SPICED MOUNTAIN APPLE

Ingredients: 15 mountain apples diced and seeded, 12 culantro or cilantro leaves minced, 5 cloves garlic minced, 1 hot pepper minced (more to taste) 1.5 TS salt or more to taste, juice of two large limes or 2 TS white vinegar.

Method: Combine all ingredients in a suitable container. Chill in the fridge for at least 1 hour. For more spice, add more pepper and let sit longer.

MOUNTAIN APPLE JAM

Ingredients: 5 kilos 12 pounds of mountain apples, washed, seeded, and chopped into small pieces, 4 cups sugar- white or brown (more to your taste), 2 TBS ground cinnamon and 6 cloves, 1 cup dark rum - optional

Method: Put mountain apple pieces in a large pot on medium heat. Mountain apple is more than half water, so it will slowly reduce in its own juice. After the chunks have reduced to about half add sugar, cinnamon, and cloves. Boil on medium heat for 1 hour stirring frequently otherwise, it will stick and burn. Reduce heat and simmer for another hour until most of the liquid disappears. Stir in rum and bottle in sterilized jars. If you have heat seal caps great, but you can seal with melted wax.

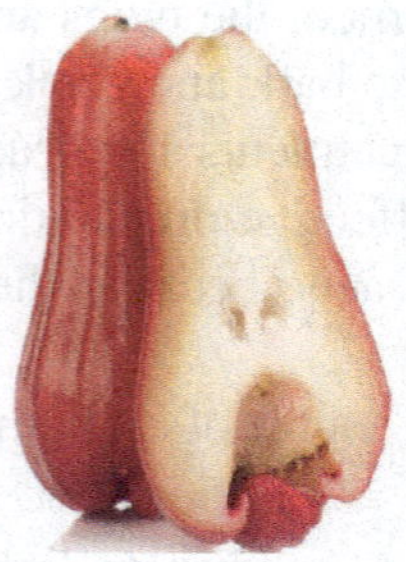

DID YOU KNOW? The Mountain apple is also known on some islands as the rose apple. There is another quite similar fruit, also known as the rose apple. Mountain apple is *Syzugium malaccense* and is a medium-sized tree with a straight trunk with very pale brown bark, pink blossoms, usually bears bright red or white fruit. The rose apple, *Syzugium jambos,* is a small tree with a twisted truck with reddish-brown bark, white blossoms, and bears yellowish-green fruit. Alternate names for the mountain apple are loku jambu, wax apple, Malay rose-apple. In Thailand it is *chom-phu-daeng,* in Tahiti: *ahia,* Vietnam: *man hurong tau,* Puerta Rico: *Malaya,* Costa Rica*: manz*

Healthy cooking is the best preventive medicine. ~ *Unknown*

NAMNAM

Cynometra-cauliflora

Namnam is this strange fruit's name, and it comes from an unusual tree that grows sparingly on every Hawaiian Island. This ugly fruit grows on the lower part of the gnarled and twisted trunk. The tree is attractive tree because it's good for conversation, and produces fruits that promote good health. Namnam is perfect for the home garden or on the patio in a large pot. Botanists believe the tree originated on the eastern Malaysian Peninsula, or on Sulawesi and the Moluccas. The tree has a long history of cultivation in India, Sri Lanka, and Indonesia for food, and as an ornamental with many medicinal uses, but no commercial production. It is believed to have been brought to Hawaii in the early 1800s. Namnam can also be grown on a smaller scale, as a bonsai variety.

This is a shrub or small tree with a twisted trunk. It will grow to 8 meters with dense yellow-white foliage on its many branches. The botanical name is *Cynometra-cauliflora.* Namnam is a wrinkly fruit that almost looks like fungus shaped like a kidney pasted to the tree trunk. The taste has a refreshingly sharp, sour apple flavor. The fruit and leaves have many Ayurvedic - homeopathic uses. It can be referred to throughout Hawaii as namnam, nam-nam, or naminam. In Indonesia: name, niam niam, namu namu, Malaysia: salah nama, and in Thailand: hima or nang-ai.

HOW TO GROW: Namnam grows best from a fresh seed soaked in cool water for two days. The seeds are flat, kidney-shaped and will take up to 3 months to sprout. In the first year, it may sprout to a foot, send branches out with close leaves, and then it will be dormant. This period is the namnam's dwarf version and will acquire many hard knots on the stem-trunk. These knots will eventually have beautiful pinkish-white blossoms in the sixth year. The small, 4-inch fruits mature in 2 months. There are sweet and sour types of this tree. The unattractive namnam is a fleshy fruit with a rough, brownish-green skin hanging directly from the trunk. Each has only one seed and will yellow as it ripens. The more sunlight, the sweeter, and bigger the fruits. Namnam grows well in the tropical wetlands in full sun or partial shade up to 400 meters and prefers a pH of 5-6. These trees do not produce a lot of fruit which keeps them more of a curiosity. The namnam is very fragile and will perish after only 2 days and that prohibits commercial cultivation. Search this fruit out and keep salt handy to enhance its sweet-sour (masam-manis) taste.

MEDICINAL: This is another one of those obscure fruits that boasts a long list of traditional medicinal uses. This fruit is not readily available; you either own a tree, or know someone who owns a tree. The fruits are reported to have useful medicinal properties and are used in various folk medicinal preparations. In some places in India, namnam fruit are sold in the market only for medicinal purposes. It's used to treat loss of appetite. The seeds yield an oil used in India to treat skin diseases. In Indonesia, namnam is used to treat diarrhea.

Little research has been done to discover the true medicinal value of this fruit. Namnam fruit contains vitamin C, vitamin A, and antioxidants to increase natural immunity. The leaves can be prepared as a tea, which also increases natural immunity, helps the flow of urine, and treats kidney stones. Though studies have thus far been inconclusive, namnam has great potential in fighting leukemia.

This seasonal fruit has a short shelf-life and producing a tonic-wine can preserve health-boosting properties and reduce post-harvest losses that treat diabetes and some cancers.

NUTRITION: Per 100 grams: vitamin A 150–500 I.U., vitamin C 25 mg, and extremely high in flavonoids.

FOOD & USES: The young fruit may be eaten raw or cooked with a pleasant, sharp flavor. Sri Lankans use young namnam for sambal. The sour-acid content decreases as the fruit matures. It's best to pick the fruit when the skin turns yellowish-brown. Namnam is used in fruit salads, pickled, and can be cooked with sugar into compotes. Indonesians cook namnam to flavor curries and slice the ripe fruit for a salad with peanut sauce.

RECIPES: SPICY NAMNAM FRUIT SALAD

Ingredients: 1 namnam peeled and seeded cut into chunks, ½ pineapple chopped,1 curry mango peeled and cubed, 1 cucumber chopped, ¼ cup peanuts roasted and chopped (almonds or cashews may be substituted), 1 TBS toasted sesame seeds DRESSING: 1 TBS each of the following; soy sauce, coconut vinegar, brown sugar.

Method: All fruit peeled and chopped into the same size pieces. To make the dressing, combine all ingredients in a saucepan and simmer until the sugar dissolves. While warm, pour over fruit pieces and serve.

NAMNAM FRITTERS:

Ingredients: 2 not yet ripe namnam peeled and seeded, 1 carrot peeled, 1 apple, 1 clove of garlic, 1 hot chili minced (optional), 2 eggs, ½ cup self-rising flour, 1 TS salt, 1 TS ground black pepper, 2 TBS oil for frying

Method: Grate the namnams, apple, and carrot into a bowl. Add chili pepper, eggs, flour, salt, and black pepper. Stir the mixture until stiff. Add more flour if necessary. Heat the frying pan with oil and drop in tablespoons of the fruit-flour mix. Fry on one side and flatten with a spatula. Then carefully flip. Fry until brown. Serve with yogurt or chili sauce as a dip.

NAMNAM CHUTNEY:

Ingredients: 2 cups namnam peeled, seeded, and chopped, ½ cup onion chopped small, 2 cloves of garlic minced, 3 green chilis chopped, ½ cup dates seeded and chopped, ½ cup brown sugar, ¼ cup coconut vinegar, ¼ cup water, 1 TBS coconut oil, salt to taste

Method: In a suitable frying pan or wok heat oil and fry onion and garlic until transparent. Add all ingredients and bring to a boil while stirring. Reduce heat and simmer for half an hour or until the mixture thickens. To test, put a spoonful on a plate and divide it with a spoon. If no liquid runs, it is thick enough. If not, return and simmer longer. Store in sterilized jars and refrigerating is recommended.

NAMNAM SAMBAL – spicy exotic Indian condiment

Ingredients: 1 namnam peeled, seeded, and grated, 1 cup grated fresh coconut, 1-2 hot chilies (to your taste) minced, 1 medium onion chopped, 1 clove garlic minced, 1 lime for juice, ½ TS sugar, salt to your taste

Method: Combine all ingredients, except lime juice in a bowl. Put suitable portions into a mortar and crush until it is all smashed. (You could use a food processor, but the flavor is different). When you are satisfied with consistency spoon into a bowl and add the lime juice. Adjust the chilies and salt to your taste.

We're only truly secure when we can look out our kitchen window and see our food growing and our friends working nearby.

~ Bill Mollison

OLIVE
Olea europaea

The olive is native to Asia Minor and spread from Iran, Syria, and Palestine through the Mediterranean basin more than 6,000 years ago. Humans grew olives before they developed an alphabet. In the early 1800s, President Thomas Jefferson, imported French olive trees and planted in coastal Georgia and South Carolina. Olive trees are relatively new to Hawaii. The beautiful foliage varies from grayish green to silver. In Kona, white blossoms appear in March or April.

Plant an olive tree as an ornamental and your descendants will enjoy it. These trees live for hundreds of years! An olive tree can live an average of 300 to 600 years. On Crete, the Olive Tree of Vouves is 3,000-plus years old and still produces fruit. The Al Badawi olive tree, located in Bethlehem, in the Palestinian West Bank, is somewhere between 4,000 and 5,000 years old. Over 800 million olive trees are planted worldwide. The olive branch is a symbol of peace.

Olive trees are hardy and thrive in poor conditions. Spain produces half of the global olive oil supply; other major producers are Italy, Tunisia, Greece, and Turkey. Per capita consumption is highest in Greece, followed by Italy and Spain. Olive oil production has tripled in the last 60 years, reaching 3,400,000 tons. Olives are pressed for oil or preserved in brine.

HOW TO GROW: Olive trees are very hardy. They can grow in the desert and are drought resistant, but too much water is a killer. Buy two different types of olive trees and plant close together to benefit from cross-pollination. The tree requires chilling for good fruiting, but is injured when temperatures drop below 20 degrees. Plant your olive trees in a sunny spot with well-drained sandy, unimproved soil. It's best to buy trees from a nursery. They aren't expensive. Two types that do well on the Big Island are Spanish varieties, Arbequina and Arbosana.

If you want to try to sprout olive pits, it's not easy. You need raw olives. Scrape off all the flesh and soak overnight. Clean the seeds with scouring pads. With a pair of pliers, nick the pointed end but don't crack the hull. Soak again for a day. Plant the pits in a mix of half sand half potting soil 1 inch deep. Put olive pots in a shaded location on a germination mat set at 60F for 30 days. Keep each pot moist. Don't over water. Increase the germination mat to 70F. Within 60 days you should see sprouts. Reduce the mat's temperature by 15 degrees F each week until it matches the outside temperature. Keep them in a lightly shaded area during the hot summer months and then transplant them in mid-autumn when the weather is again cool and moist. (From: *Gardening Know How.*) They should bear in 5 years.

For the olive to successfully fruit it requires 200 to 300 hours below 47 degrees each year for flower bud development. Olive fruits begin green and will generally become a blackish-purple when fully ripe. Some varieties remain green or turn a copper color. Shape, size, and flavors vary. Olives are usually too bitter for eating right off the tree, but are better preserved and eaten later. Others are better for oil

MEDICINAL: Olive tree leaves are used against high blood pressure, gout, arteriosclerosis, and rheumatism. Eating the fruit or oil reduces blood sugar, cholesterol, and uric acid. It also treats inflammation, diarrhea, respiratory and urinary tract infections, stomach and intestinal diseases, asthma, and hemorrhoids. They are a laxative, mouth cleanser, and dilates blood vessels. Olive oil is very good for your hair and it can be used as a hair mask conditioner.

NUTRITION: 100g of olives has 112 calories with 1g protein, 11g fat, and iron.

FOOD & USES: The olive is pickled in brine or squeezed for oil. It takes 11 pounds of olives to yield 1 quart of olive oil. U.S. olive oil consumption has increased to 80 million gallons. Over 800 million olive trees are planted worldwide. The olive branch is a symbol of peace.

ORANGE – 'ALANI
Citrus sinensis

I know, I write this many times, but I love chilled, fresh fruit. On a hot day, very few fruits taste better than a peeled, fresh orange. 'Alani is the local Hawaiian name for orange. The orange was first cultivated in the area between Southern China and northeastern India thousands of years ago. This fruit slowly migrated to the Middle East about 800 BCE and made the trip to Europe seven centuries later. In 1792, the seeds of the Hawaiian or Kona orange were carried to Hawaii from Cape Town, South Africa by Captain George Vancouver, the famous British naval explorer. Before efficient steamship transport, outside of the tropics, oranges were considered a 'holiday' treat because of their cost. The great oranges Hawaii grows are consumed locally. Living in Hawaii? Grow your own sweet oranges.

Brazil leads a list of the world's orange growers including the US, Mexico, Spain, and Israel. Yearly, Hawaii grows more than 150 tons of oranges. Botanically sweet orange is *Citrus sinensis.* Perfect, name-brand, brilliant orange fruits are injected with a dye – citrus red #2 – into the skins (not the pulp). Also, the oranges grown in the temperate climate of Florida tend to become more orange than their tropical cousins. Partially green oranges or even those with brown rusty spots may be as delicious. Oranges with smooth textured skins, firm, and heavy are the better choice. Small, heavy oranges with thinner skins are the best for juice.

Oranges have both sweet and bitter varieties. Most of the oranges grown are round like king, osbeck, or dansi. The Mediterranean naval orange actually has a second fruit set in the blossom end. Washington Navel Oranges are the most widely grown in Hawaii. Other orange varieties include the Cara cara, Fisher and Lane navel, Moro blood, and Valencia. More varieties of the orange are sweet 'Parson Brown', seedless 'Hamlin.' 'Marrs,' and the pineapple orange. The thin-peel, seedless Jaffa was developed in Palestine, while the very desirable, large seedless Valencia came from Spain, and the smaller mandarin came from China. Other members of the orange family are the everhard, kumquat, calamondin, and some tangelos. The tangerine is a type of mandarin first grown in Tangiers Morocco in 1841. Bitter oranges make marmalade and liquors as Triple Sec, Grand Marnier, and Cointreau.

HOW TO GROW: To grow oranges first select the variety you love. You can try planting seeds, but this will usually have disappointing and time-wasting results. This is because the bitter orange is often used as rootstock for grafting. It is better to buy a tree created by budding. All types prefer well-drained loamy soil with a pH 6 to 8. Continually damp or dense clay soil will result in poor growth, low production, and a short life for the tree. Oranges need full sunlight for growth and production. They should be planted at least ten feet from buildings, driveways, walkways, and fences. Keep trees twenty feet apart. When planting orange trees do not leave any indentation where water can settle and cause root rot. The soil around the tree should be higher with an outside ring of soil constructed to make efficient watering. Inside the ring should be barren of grass. The orange tree should be watered daily when planted, then weekly. A good way to protect the new orange tree is to wrap some aluminum foil around the young trunk. The foil protect sagainst herbicide spray and bushwhacker damage.

Fertilize infant orange trees with a cup of ammonium sulfate (urea 21 – 0 – 0) every three months during the first year. Increase to two cups the second year and three cups the third. Once the tree begins to blossom add 12 – 12 –17 –2 at the rate of two cups every three months per tree plus the ammonium sulfate. Always keep the grass away from the tree. Leafminers are a problem, but indiscriminate spraying of pesticides will destroy the natural predators. If you feel spraying chemicals is the answer, first identify the problem by speaking with someone at an agricultural center, and then select the appropriate chemical. Apply it properly, at the appropriate time to control the pest minimizing the amount of chemical used.

MEDICINAL: Remember how our grannies had dried orange peels hanging for use in flavoring tea? Seems they knew more than we do today. Orange peels contain compounds known as 'polymethoxylates (PMF)', which has been tested to lower cholesterol significantly – as much as some prescription drugs! Just dry the peel and add to tea or boiling water. Chinese use dried mandarin peel to regulate their chi or flow of energy.

NUTRITION: Each orange has about 60 calories packed with vitamins C, and B-1, fiber, and folate. One orange contains about 50mg. vitamin C; or two-thirds of our daily need.

FOOD & USES: Cooking with oranges is easy. Sauté onions and ginger slices in butter and then pour in orange juice. This is a great sauce for fish or chicken. Mix peeled and seeded orange pieces with beet root slices for a tasty salad. Sweet potatoes or squash can be simmered in orange juice for a wonderful taste.

RECIPES: SUN SOUP

Ingredients: 1 cup orange juice, 1 medium onion chopped, 1 cup 'soaked' red lentils, 1 cup pumpkin-boiled and mashed, juice of 1/2 a lemon, 1 sweet red pepper chopped, 1/2 cup sliced mushrooms, 1.5 cups vegetable broth, 1 TBS butter or margarine, 1 TBS grated ginger, 2 TBS sugar, 1/2 TS cumin, 1/2 TS coriander, 1 TS cinnamon, 1/2 TS salt

Method: In a soup pot brown the onions in the butter adding the sweet pepper and mushrooms. Slowly add vegetable broth with lentils with spices. Stir in pumpkin, and then add lemon juice. Bring to a boil, add orange juice, and simmer until lentils are tender.

SWEET AND SOUR CABBAGE

Ingredients: 1 cup orange juice, 4 cups shredded cabbage, 1 cup shredded carrots, 1 sweet pepper chopped small, 1 TBS cornstarch, 1 TBS vinegar, 1 TBS brown sugar, salt, and spices to taste Method: Mix brown sugar, vinegar, cornstarch, and a pinch of salt. In a large pot heat orange juice until it begins to boil before adding cabbage, carrots, and pimentos. Mix in sugar cornstarch combination. Cook for fifteen minutes stirring to keep from sticking.

ORANGE SWEET POTATOES

Ingredients: 1/2 cup orange juice, 1 pound sweet potatoes-peeled and boiled, 1 TBS brown sugar, 1 TBS butter or margarine, 1/2 TS cinnamon, 1/2 TS nutmeg, salt to taste.

Method: Combine orange juice, sugar, and spices. Pour over sweet potatoes and boil uncovered for five minutes. You may have to add a few tablespoons of water if the mix gets too low. Stir occasionally.

KONA ORANGE CAKE

Ingredients: 1 1/2 cups sifted self-rising flour, 1/2 butter, softened, 1/4 cup milk, 1/4 cup orange juice, 2 eggs, 3/4 cups powdered sugar, 1/4 cup orange zest

Method: Combine all ingredients and beat thoroughly for about five minutes. Pour cake batter into a greased cake pan or muffin tray. Bake in the oven at 180C / 350 F for 40 minutes. Enjoy.

DID YOU KNOW? Oranges originated around 4000 BCE in Southeast Asia, from which they spread to India. Oranges are unknown in the wild. They are a hybrid of the pomelo/jabong fruit, or 'Chinese grapefruit' (which is pale green or yellow), and the tangerine. There are typically ten segments inside an orange. There are over 600 varieties of oranges. Spain has over 35 million orange trees. With a high resistance to disease, more oranges are killed by lightning than by plant diseases. The color orange is named after the orange fruit. Before orange made its way from China to Europe, yellow-red was called simply yellow-red, or just red.

PAPAYA – PAPAIA
Carica papaya

It is always papaya season in Hawaii. A papaya a day should keep the doctor away. Many people swear to good health from consuming this fruit, usually with the seeds. Papaya is native to the tropics of the Americas. Columbus called papaya 'the fruit of the angels.' Circa 1550, Spaniards carried seeds from Mexico to the Philippines and the papaya traveled from there to Malacca, Sri Lanka, India, and probably Hawaii. There is a debate if papayas were introduced before or after Captain Cook arrived. The pear-shaped, orange papayas can grow to twenty pounds with soft, sweet flesh and great for health. The seeds can be eaten, but they have a peppery taste.

Basically, there are two types of papayas, Hawaiian and Mexican, and they have many sub-vareties. Garritt Wilder, a botanist, introduced the very popular Solo papaya to Hawaii from Barbados in 1910. The Hawaiian Solo type are pear-shaped fruit generally weigh about 3 pounds and have yellow skin when ripe. The flesh is bright orange or pinkish and are easy to harvest because the plants seldom grow taller than 8 feet. The Maradol/large Mexican papayas originated in Cuba after a long breeding process. Mexican papayas can weigh up to 10 pounds and be more than 15 inches long. The flesh may be yellow, orange, or pink. Genetic modifications saved the Hawaiian papaya industry from ringspot virus (PRV). The most common type is Red Lady which produces a succulent fruit in 7-8 months. Papaya's botanical name is *Carica papaya.*

HOW TO GROW: Papaya trees like sunshine and reflected heat, so the hottest place, against the house where nothing else will grow, is a perfect location. Papayas dislike wind, but can withstand some stiff breezes. Papayas can be grown from dried seeds. Papayas do not handle transplanting well. For the best results start them in large containers, such as a gallon paint pail, so they only must be transplanted once. Do not damage the root ball and tap root when planting in the soil. Papaya trees are either male or female. The female has flower blossoms close to the trunk and males have blossoms on branches. Males are not necessary for pollination and they do not bear fruit. Plant at least six plants to ensure of having females. Like bananas, papaya is a good all-year crop for a small farmer. Plant a papaya tree every 6 months and that will keep you supplied.

In a well-drained area, fork a two-foot square hole a foot deep. I add a shovel full of rotten manure and then add a layer of dirt building a mound before placing in the new tree. It is necessary to water often during the early weeks. Almost all transplants enter a 'shock period' as they become accustomed to their environment. Too much rain is the biggest enemy of the papaya. In the rainy season with hard winds, these trees may need to be propped or staked with ropes to keep the roots from pulling out of the soil. If you are planning to grow papaya commercially, plow the land so it drains well. Use a 12 – 24 –12 fertilizer for the first two to four months switching to 12 –12 – 17 –2. Papaya trees can bear fruit two years or longer. Papayas do not need to be pruned, but some growers pinch the seedlings or cut back established plants to encourage multiple trunks.

Bunching is a disease where the top leaves do not spread, and black spot rot is another disease that occurs on the fruit. Papayas are ready to pick when the skin is yellow-green. Bring inside to protect against bird damage and store at room temperature for two or three days until they are almost fully yellow and slightly soft to the touch. If you want to speed this process, place them in a paper bag with a banana or ripe apple. Ripe papayas should be stored in the refrigerator and consumed within one or two days, so you can enjoy their maximum flavor.

MEDICINAL: The most common use of papaya is to aid digestion. Papayas are the only natural source of papain, an effective natural digestive aid, which breaks down protein and

Like the papaya, let your roots grow deep and your fruits grow sweet.

cleanses the digestive tract. Less food settles into the metabolism and becomes fat, making papaya's natural digestive properties an advantage to people trying to lose weight. Eating papayas helps to prevent diabetic heart disease. Nutrients in papayas prevent cholesterol from clogging arteries reducing the possibility of strokes. Papaya may lower cholesterol, prevent colon cancer, and reduce inflammation caused by rheumatoid arthritis. Researchers believe papaya may be great for your sense of sight. The high vitamin A (300 percent of the daily need) content of papaya will help smokers prevent emphysema or lung inflammation from second-hand smoke.

NUTRITION: A cup of papaya has 120 calories with good doses of vitamins A, E, and C, folic acid, potassium, and fiber.

FOOD & USES: Dark green fruit will not ripen properly off the tree, even though it may turn yellow on the outside. Papayas are often sliced and eaten alone. They can also be cooked to make chutney or various desserts. Green papayas should not be eaten raw because of the latex they contain, although they are frequently boiled and eaten as a vegetable. In the West Indies, young leaves are cooked and eaten like spinach. In India, seeds are sometimes used as fake black pepper.

RECIPES: SIDE STREET PAPAYA CURRY

Ingredients: 1 half-ripe papaya peeled and chunked small, 1 onion sliced, 3 green chilies chopped, 1 TBS curry powder, 1 TS each of: fenugreek seeds, turmeric powder, chili powder, ground black pepper, 3 cloves garlic, 1/2 TS ground mustard seeds, 1.5 cups thick coconut milk, salt to taste

Method: Add all ingredients except coconut milk into a suitable frying pan and cook until papaya softens in about 10 minutes Add coconut and cook over medium heat stirring constantly so it doesn't stick for 10 more minutes. Serve with rice.

PAPAYA STUFFED WITH SHRIMP - a great lunch

Ingredients: 2 ripe papayas cut in half lengthwise and seeded, 1/2 lb. cooked shrimp cleaned and cut into one-inch pieces, 1 lime, 1 small onion minced, 1 stalk celery minced, 1/2 cup may-onnaise, pinch curry powder, salt, and pepper to taste

Method: Combine mayonnaise with juice of the lime, curry powder, salt, and pepper. Add shrimp, onion, and celery. Mix and chill for one hour. Spoon shrimp salad into the papaya halves.

THE EASIEST THAI GREEN PAPAYA SALAD - shred the papaya like matchsticks

Ingredients: 2 TBS roasted peanuts, 1-2 hot peppers to your taste, tiny bird peppers recommended, 3 garlic cloves, 1 1/2 tablespoons sugar - palm sugar jaggery recommended 2 yard-long beans green beans/green beans chopped to ½ inch, 1 tomato chopped small, 2 TBS fresh lime juice, 2TBS Thai fish sauce or 1 TBS salt, half a medium green papaya, seeded, peeled, and shredded, 1 carrot, shredded

Method: With a mortar and pestle, grind the peppers, and garlic into a coarse paste. Add sugar, beans, and tomato. Lightly pound and add lime juice. Add to shredded papaya and carrot. Stir in fish sauce or salt. Sprinkle with roasted peanuts when ready to serve. Boiled shrimp maybe added.

DID YOU KNOW? Papayas have 33% more vitamin C and 50% more potassium than oranges with fewer calories, 13 times more vitamin C and more than twice the potassium than apples, and four times more vitamin E than both apples and oranges. Tea made from papaya leaves is consumed in some countries as protection against malaria. Papaya leaves are steamed and eaten in parts of Asia. The bark of the papaya tree is often used to make rope. Modern confirmed that unripe papaya is a natural contraceptive but large quantities can also induce a miscacriage.

PASSION FRUIT – LILIKO'I
Passiflora edulis

Liliko'i is the Hawaiian name for passion fruit. The tasty and fragrant passion fruit vine is native to southern South America. The early Spanish missionaries in Brazil frst witnessed the beauty of the vines' scented blossoms for the duration of Lent and Easter and named the fruit after the 'Passion of Jesus.' Passion fruit belongs to the family *Passifloraceae* which contains 12 different classes with more than 500 species. Annually, Hawaii produces only 13 tons.

Passion fruit can be purple, red, but Hawaii's most common variety is yellow. To start your crop, buy fruit from at least two different sources. The purple-fruited species is supposedly self-fertile and the yellow fruited species despite claims to the contrary is self-sterile and requires another for pollination. The two types of passion fruit have clearly differing exterior appearances. The bright yellow variety, also known as the 'golden passion fruit,' can grow up to the size of a big orange and has a smooth, glossy skin. The dark purple (mauve) passion fruit is usually smaller than a lemon, but the purple passion fruit also has a higher flesh proportion, a richer taste and aroma, and is less acidic than the yellow.

HOW TO GROW: Plant the seeds while they are fresh. Remember, two different fruits are necessary for future pollination. The vine is fast growing and perfect to provide privacy on the fence, especially in direct sun. Soil pH should be from 6.5 to 7.5. If the soil is too acid, lime must be applied. And good drainage is essential. Beautiful passion fruit blossoms can appear after the first year. The fruit ripens slowly taking almost three months. The vines will grow best in sandy soil, yet can adapt to almost any soil that drains well. The roots do not pene-trate the soil very deep, so they should be molded with some rotted manure. The most important part of successfully growing passion fruit is keeping it regularly watered. With constant water and monthly fertilizing with 10-5-20, the vines can almost constantly bear fruit. Pruning is nec-essary to keep the vines from getting out of control. This plant is a vine with great clinging and climbing capabilities. It binds to almost any support it can find. It can grow twenty feet a year.

Passion fruit usually only lives from 5-7 years. The main pest is nematodes. Passion fruit is easily picked as when it ripens it drops. Ten pounds of fruit per vine is a good harvest. Each fruit makes a small amount of very pungent, concentrated juice. The flavor of the passion fruit is sweet-tart, guava-like, and musky. The fruit should stay good in the fridge for two weeks while you gather more to make the tasty juice.

MEDICINAL: In Madeira, the juice of passion fruits is used to stimulate appetites and as treatment for gastric cancer. In Puerto Rico, where the fruit is known as 'parcha,' it is widely believed to lower blood pressure. Fresh passion fruit is high in beta carotene, potassium, vitamin C, dietary fiber, and polyphenols. They are antioxidant and anti-inflammatory effects plant compounds that may reduce your risk of chronic inflammation and heart disease. Passion fruit is usually safe to eat and good for you, but some people are allergic to it. This is more likely if you're allergic to latex. Passion fruit seeds are packed with piceatannol, another polyphenol that may improve insulin sensitivity in obese men and potentially reduce the risk of type 2 diabetes.

NUTRITION: A cup of juice has 100 calories with one gram of protein and 11 grams of carbohydrates. It is a good source of vitamins A and C with potassium and magnesium.

FOOD & USES: Look for passion fruit that has thick skin and feels heavy for its size. Wrinkled skin means it's drying out. It's always a good idea to wash it well, because cutting into it, the knife can spread bacteria from the peel to the flesh.

RECIPES: PASSIONATE ICE

Ingredients: 2 cups milk, ½ cup strained passion fruit juice, ½ cup sugar, ¼ cup fresh lime juice. Sugar content can be adjusted to taste.

Method: In a small skillet on low heat, stir the sugar into the milk until it completely dissolves. Permit to cool before adding the juices. Pour into a suitable container and freeze for 3 hours. Remove from freezer and blend until smooth. Return to container and freeze. This blending removes the ice crystals. Rum, arrack, or vodka may be added before blending and refreezing.

PASSION AND CASHEW SALAD DRESSING

Ingredients: ¼ cup passion fruit juice, 1/3 cup olive oil, 2 TBS clear vinegar, ¼ TS minced ginger, ¼ TS mustard, ½ TS pepper sauce, ¼ cup chopped cashews (or peanuts), salt, and spices to taste

Method: In a container that will seal, mix passion juice, pepper sauce, cashews, ginger, spices, and oil. Shake vigorously and use over a salad of greens or watercress.

PASSIONATE FISH FROM THE CHARCOAL BARBECUE

Ingredients: 1 cup passion fruit juice, ½ cup soy sauce, 2 TBS sugar, 1 medium 1-2 kg fish cut in 4 pieces - king or tuna, 1 cup finely sliced onions, ½ cup chopped chives, ¼ cup sliced celery, ½ cup sliced carrots, 2 TBS minced garlic, ½ TS pepper sauce salt, and spice to taste, 8 banana leaves

Method: Light your barbecue grill or coal pot. In a small skillet on medium heat mix the soy sauce with sugar stirring for three minutes. Lay out a banana leaf and place the fish on it. Coat the fish with the soy sauce and cover with onions, carrots, ginger, and garlic. Using only half a cup, sprinkle each piece of fish with passion fruit juice. Completely wrap the fish with the banana leaf and then wrap it in a second leaf to seal in the flavor. If the grill is very hot, give each side 10 minutes. Uncover and sprinkle the remaining passion fruit juice on the cooked fish. Serve on a bed of rice.

KAUNAKAKAI'S VALENTINE'S DAY LOVERS' CHEESECAKE

Ingredients: 1 cup crushed passion fruit peeled with seeds removed, 1 pie shell, 1½ lb. cream cheese, 1 cup sugar, 2 TBS corn starch, 1 TS real vanilla essence, 3 eggs, ¾ cup sour cream,

Method: Blend cream cheese with the sugar until it is velvety. Thoroughly blend in cornstarch and eggs. Mix in the sour cream and vanilla before adding the crushed passion fruit. Pour mixture into the pie shell and bake at 300 for 70 minutes. Turn off the oven but leave the cheese-cake in the oven, with the door slightly ajar, for 2 hours or until cooled completely to prevent the cake from cracking. Cool before refrigerating. This must be well chilled, best overnight, to set. Cover so it does not acquire tastes or smells from the fridge.

ROMANCE CAKE

Ingredients: 1 package cake yellow or white follow directions, but add 1 TS vanilla extract, 3 passion fruit seeded, skinned, and crushed.

Method: Combine all ingredients and bake per box directions. Test with a toothpick or knife to be certain it is thoroughly baked.

DID YOU KNOW? The passion fruit has had a religious association as reflected by the name 'passion' given to it by Catholic missionaries who thought certain parts of the fruit bore some religious connections. These missionaries used the fruit to illustrate the crucifixion to the Amerindians. Botanically it is *Passiflora edulis.* **NAMES:** *Spanish*: granadilla (granadilla means little pomegranate because of the numerous seeds), parcha, parchita, parchita maracuyá, *Portuguese*: maracuja peroba, *French:* grenadille, or couzou; *Sri Lanka:* vael dodang, and *Jamaica:* mountain sweet cup. The purple form may be called purple, mauve, red, or black granadilla. The yellow fruit is widely known as yellow passion fruit. The seeds are edible, and you can eat the orange pulp straight from the shell.

PEACH PALM - PEJIBAYE

Bactris gasipaes

The peach palm/pejibaye has been planted on all the Hawaiian Islands, but few people have tried the tasty fruits. This palm was introduced to the Hawaiian chain circa 1940. The seeds came from Brazil and Peru with a high rate of success. There are more than 200 names for the species of palm scientists label as *Bactris gasipaes*. The fruit is commonly known as the pewa. In Brazil the fruit is 'pununha,' Costa Rica it is 'pejibaye,' Peru it's 'pijuayo,' 'cachipay' in Colombia, and 'macana' in Venezuela. This palm can be utilized in four ways: cooked fruit and palm hearts for people, animal feed, flour for bread, and vegetable oil. Do not confuse with the cabbage or palmetto palms that also produce palm hearts.

HOW TO GROW: The palm can grow to about fifty feet, but it's usually shorter as most pewa can be picked without a ladder. Peach palm originated in the Western Hemisphere, probably native to the Amazon rainforest, and is presently grown from Nicaragua and Honduras to northern Bolivia. This type of palm thrives at low altitudes, especially at sea level. You'll see beautiful palms with bunches of red, two-inch fruits hanging beneath the palm leaves. Once you find a tree, carefully free a shoot from the parent palm. This must be done with a broad sharp chisel as the parent's roots can be tough. The only concern is to plant in soil that doesn't hold water. However, it will do fine in the rich soil along drains or rivers if the roots don't constantly stay wet. The ideal pH is 5-6.5. Dampness is one of the peach palms' main enemies. Loosen the soil about a foot in diameter and a foot deep, but only cover the roots of the shoot. Weeds are another enemy, so clear the area around the palm three or four times a year. I recommend wrapping aluminum foil around the trunk, so a bushwhacker doesn't damage the trunk. Peach palm trees make good shade cover for coffee, bananas, and cocoa.

Although this palm is not so bushy, it needs to be about ten feet from other trees and walls. It will take about five years before developing shoots at the base of the fronds, which form a clump of pewa/fruit. A peach palm tree can live sixty or more years. Pewa is a fragile fruit that lasts only four or five days, so only pick what you can use. Various colors of pewa indicate different varieties of the peach palm. Many species of birds feed on this palm, especially parrots and parakeets. Once the fruit is ripening, I put pieces of aluminum foil or pie plates on the palm leaves to chase the birds. A great cluster has a hundred or more pewa and will weigh up to ten kilos. A peach palm makes a great tree for the home, the fruit are a very colorful red.

MEDICINAL: Pewa is recommended as a remedy for headache and a bellyache. The oil from the seeds is used as a rub to ease rheumatic pains.

NUTRITION: The pewa contains high amounts of carotene, iron, calcium, phosphorus, and are especially high in vitamins A and C. A 100 grams of pewa have 150 calories.

FOOD & USES: Pewa is probably the most nutritionally balanced of tropical fruits with twice the protein content of the banana and produces more carbohydrate per acre than maize. An edible oil is obtained from the seed called oil of macanilla. Pewa cannot be eaten raw, it must be boiled with salt and spices, and eaten only after the skin is peeled off. We boil pewa in salted water, often with salt beef or fish for seasoning. The fruit is peeled the seed removed and eaten plain or with a dip of mayonnaise and pepper sauce. The pewa flesh can be deep-fried or roasted. The fruits are also ground into flour for baking bread, cakes, or roti. The leaves provide a green dye for coloring fabrics. Agri scientists are breeding spineless trees that bear seedless fruit with a higher carotene content. The trunk is valuable timber.

The peach palm can be harvested for heart of palm. This is the tender, growing tip or of the palm, the core of the plant. This has good commercial farming potential as the first harvest can be from 18 to 24 months after planting. In Brazil, heart of palm is a big agri-business. There is a growing demand for the heart of palm internationally for gourmet salads and dishes.

Don't toss the seeds, as they are rich in protein and fiber. They make good food for ducks and swine. Like other palms, the seed is rich in saturated fatty acids, and could be used to manufacture cosmetics and soap.

The following recipes might seem like a lot of work for us pewa lovers who just want to peel and suck. Try them, they're delicious.

RECIPES: CREAMED PEWA

Ingredients: four pounds pewa shelled and seeded, two medium onions chopped small, three TBS cornmeal, two-thirds cup milk, one TBS butter, half TS fresh grated nutmeg, salt, and other spices to your taste

Method: In a skillet fry the onion in the butter and add the pewa flesh. Increase the heat and sauté quickly. Add milk and thicken with cornmeal and season to your taste. Serve on rice with grated nutmeg.

FRIED PEWA NUMBER ONE

Ingredients: 3 pounds of pewa cleaned and seeded, ¼ cup all-purpose flour, 2 TS baking powder, ¼ cup milk, 1 egg separated, oil for frying, half TS salt, pepper, and spice to taste

Method: In a bowl combine flour, salt, and baking powder. Mix egg yolk with milk and grad-ually add this mixture to the dry ingredients. When mixture is stiff blend in egg white and add pewa flesh. Carefully drop spoonfuls of this mixture into hot oil. Fry on both sides and eat while hot.

FRIED SPICY HEARTS OF PALM

Ingredients: heart of palm sliced into 1/2-inch rounds, spices, and salt to your taste, 1 TBS oil

Method: Coat both sides of palm rounds with spices. Fry in oil 2 minutes until light brown. Enjoy.

DID YOU KNOW?

Fruit should be boiled within 2 to 4 days of picking. They are eaten boiled in salty water, often with salt pork for seasoning. The fruit is then peeled, and the seed remove. Flavor varies with the carotenoid content and may be bland or have a strong nutty taste. Boiled pewa is eaten plain or with a dip of mayonnaise or cheese, or deep-fried, or roasted. It may be ground into meal, mixed with egg and milk, and fried as tortillas. Raw fruits may be kept for several weeks in a cool, dry place and cooked fruits may be held in the refrigerator for 5 or 6 days.

To forget how to dig the earth and to tend the soil is to forget ourselves.

~ Mahatma Gandhi

PINEAPPLE – PAINA

Ananas comosus

Pineapple is one of God's best tasting, and strangest creations. Pineapple is Hawaii's largest fruit crop. They are easy to grow. Don't waste your tops! We saved our cut off tops and planted them above the front yard drain. After a year we had sweet pineapples with little effort. Although it looks like a cactus, the pineapple, botanically *Ananas comosus,* is the fruit of hundreds of individual flowers that cluster on the barb of the plant. When mature, all the pineapple's fleshy tissues swell with juice. The pineapple grows on a 'bromeliad' with stiff leathery leaves around a center spike, which we call the core. Pineapples are usually 4 - 9 pounds, but the Giant Kew can be more than 20 pounds. In 1922, Dole bought the Hawaiian Island of Lanai. That pineapple plantation grew 75 % of the world supply. The local Hawaiian name is paina.

The pineapple originated in South America where Indians named the fruit 'anana' meaning excellent fruit. These Amerindians planted pineapples with their sharp picker leaves surrounding their villages to keep out intruders. In 1493, Columbus discovered the Carib Indians growing pineapples on the island of Guadeloupe. His sailors created the present name because the exterior appeared as a pinecone, yet the core tasted as an apple. By the mid-1500s, pineapples were being grown in the West Indies for export to Britain. Two centuries later, pineapples were being grown throughout the Caribbean chain. French King Louis 14th loved the sweet fruits' taste so much he forgot his manners and cut his mouth trying to bite an unpeeled pineapple. He treasured the fruit so much glass 'greenhouses' were created to grow them. Pineapples became a status symbol as a party decoration and as a dessert.

No one is certain who brought the pineapple to Hawaii. In 1820, missionaries found them already growing. A Spanish shipwreck in the early 1500s may have accidentally deposited pineapples. Don Francisco de Marin may have brought them from Mexico in the late 1700s. The founder of Hawaii's pineapple industry was Captain John Kidwell, who tested a variety of pineapple and finally selected the Smooth Cayenne in the 1880s. It took another century and the efficiency of steamship transport to make commercial production feasible. Pineapples were made available to the world when Dole began canning them in 1903. Today, Thailand exports the most canned pineapples. Costa Rica is the world's largest producer of pineapples, 3.42 million-metric tons of 28 million tons grown worldwide! Yearly, Hawaii makes over $3 billion from pineapples.

HOW TO GROW: Pineapples rarely have seeds, but are grown from cuttings, which are either the tops or the suckers that appear close to the base of the fruit. The soil must be worked a foot wide and deep. Some rotted manure can be placed deep in the hole. Plant two feet apart, water regularly, and fertilize monthly with a high nitrogen mix. Harvest when they become a bright gold and can be twisted from the plant. Cut and leave on part of the stem so they keep better.

MEDICINAL: Pineapple contains micro-nutrients believed to prevent cancer and breaks up blood clots. Pineapple juice kills intestinal worms, relieves intestinal disorders, and soothes the bile. The juice also stimulates the kidneys and aids in removing toxic elements in the body. Pineapple contains a mixture of enzymes called bromelain that helps reduce swelling by arthritis, gout, sore throat, and acute sinusitis. This also helps accelerate the healing of wounds due to injury or surgery. For the medicinal benefits eat pineapple between meals. Eaten with meals the enzymes are used digesting food.

NUTRITION: Two slices of a regular-sized pineapple should be about 100 grams and has 60 calories with no fat or cholesterol. Pineapple is a good source of vitamin C.

FOOD & USES: Pineapple tastes good and is good for you. Pineapple juice is chemically close to stomach juices. Consumed moderately pineapple aids digestion. It has plenty of fiber and helps the body relieve fluids, especially mucus from nasal passages. Never consume an unripe pineapple, as it can be poisonous causing throat irritations and diarrhea. The juice is an excellent cooking marinade tenderizing the meats while adding a tropical flavor. When you buy a ripe pineapple should be firm, smell sweet with fresh green leaves. Reduce the acid content, let the pineapple to sit for three days before using. To increase its sweetness, salt, and let sit before eating.

RECIPES: SCALLOPED PINEAPPLE

Ingredients: 1 medium pineapple peeled and cut into half-inch chunks, 1 cup butter or marga-rine, 2 cups sugar, 4 beaten eggs, ¼ cup milk, 1 TS vanilla essence, 4 cups white bread cut into cubes (Sugar may be reduced to your taste.)

Method: In a frying pan over medium heat; blend butter, sugar, and eggs into a cream then add remaining ingredients. Pour into a greased baking dish and bake at 175c/350f-degrees for an hour.

WAHIAWA PINEAPPLE PIE

Ingredients: 1 pineapple peeled and crushed, 1 can sweetened condensed milk, ½ cup fresh lemon juice, 1 prepared pie crust, 1 container (½ pound) non-dairy whipped topping

Method: Mix sweetened condensed milk with lemon juice. Add pineapple and whipped top-ping. Pour into pie crust. Chill at least 4 hours before serving.

MAMA'S PINEAPPLE RICE

Ingredients: 1 fresh pineapple, 2 TBS canola oil, 2 cups cooked rice, 1 cup medium shrimp peeled and cooked, 1 chicken breast cooked and cubed, 2 bunches chives chopped, ¼ cup carrots chopped, ¼ cup peas (Canned peas and carrots are fine).

Method: Cut pineapple in half and clean out the shell. Place shell under broiler or on a grill until thoroughly heated. Chop a half cup of the pineapple flesh as small as possible. In a large skillet heat oil and stir in rice cooking for 2 minutes. Stir in remaining ingredients including the chopped pineapple. Spoon into heated pineapple shells.

SPICEY PINE FRITTERS

Ingredients: 1 fresh peeled pineapple chunked small as possible, 1 hot pepper seeded and minced, 2 bunches chives minced, 1 medium onion chopped as small as possible, 2 garlic cloves minced, ½ TS turmeric, 1 ½ cup flour, ½ cup milk, 2 beaten eggs, ½ cup canola oil for frying, salt, and spices to taste

Method: Mix flour, milk, eggs, salt, and spices until smooth. Refrigerate covered for 4 hours. Mix with pineapple, pepper, chives, onion, and garlic, and blend into batter. Drop batter from a large spoon into a skillet with hot oil and fry about 3 minutes a side. Remove fritters and drain.

DID YOU KNOW? Next to bananas, pineapple is the second most popular tropical fruit. Pineapples originated in a place between today's Paraguay and Brazil. More than 200 flowers, that have different colors ranging from purple to red and even lavender, are produced by a single pineapple plant. These flowers have their individual fruits which look like scales. These individual fruits fuse around this central core that individual berries together to form the fruit. Canned pineapple wasn't financially feasible until Henry Ginaca, an engineer, invented a machine in 1911 that could remove the outer shell, inner core and both ends of 100 pineapples in less than a minute. The world's largest pineapple ever recorded was grown by Christine McCollum from Bakewell Australia in 2011. It measured 32cm long had a 66cm circumference and weighed 28kg.

PLANTAINS - LAUKAHI

Musa x paradisiaca/ P. hawaiensis

Hawaiian Islanders love bananas, and they are the state's most widely consumed fruit. Plantains are a starchier variety. Hawaiian plantains are called 'laukahi' and shorter and fatter than traditional plantains. The Island of Hawaii has almost 800 acres devoted to banana cultivation.

Different from usual plantains, they are sweet and easier to peel. These starchy bananas are found in every market and delicious cooked in various ways.

Plantains are in the genus *Musa*, and likely native to Malaysia or India and what is now the Indonesian islands. The present Hawaiian plantain may have evolved from Polynesian canoe plants or evolved from those brought by European explorers. Alexander the Great's army carried plantains to southern Europe during his world conquest.

HOW TO GROW: Hawaii's unique variety of plantain is botanically *P. hawaiensis* and easy to grow. I know of three types of plantain, but there are probably others, or different names. The smallest, most curved plantain is the French. Horn plantain is bigger, but the 'giant' are the biggest. To grow, first, you must locate some healthy suckers. You do not need anything except a small grapefruit-sized ball termed 'the eye.' You do not transplant the big plants. The bigger the banana tree you attempt to plant will have less success. The big ones will die and then sprout. In plantains and bananas, small suckers adapt best. Carefully chop off enough of the 'eye' that will grow without hurting the parent plant. This can be done best with a sharp tool; a shovel will work. It is recommended to soak the shoots for an hour in a pesticide solution like Malathion to kill any worms. Dig a hole about two-foot deep and break up the compacted soil as loose as possible. Farmers recommend putting a piece of pitch in the hole before placing the eye. Make sure that the eye has a bit exposed to the air. Water, use 12-24-12 fertilizer at the base, and watch it grow. This will be a big tree, so give it adequate space. Bananas cannot crossbreed.

Plantains mature in a year or less. They should be spaced about six to ten feet apart; air must circulate. Plantains can grow to twenty feet tall are excellent garden shade and attractive landscaping. They take little work except for trimming, watering during the dry spells, and bi-monthly broadcast of a half cup of high potassium fertilizer around the base. Once the plant sends out the flag that announces it is ready to bear, feed it with a half cup of high-potassium fertilizer. Water for bigger fruit. If the bunch gets too heavy and the tree starts to tilt, carefully prop, or tie it for support. Once the bunch begins to form it may take two months to develop. Keep the birds away from the ripening plantains. Pick before they ripen and store. Plantains are perennials, always making more shoots. Bananas must be constantly trimmed, or within a few years, they will crowd themselves and produce only small bunches.

Presently Southeast Asia produces 35 million tons annually, India produces 11 million, Africa 7.7 million tons, Central America 8 million, and South America 16 million tons. They are a staple crop in much of the tropics.

MEDICINAL: Ash plantains contain high volumes of dietary fiber that helps reduce bowel-related problems such as constipation. The tender leaves, smeared with coconut oil, are used as a cool and soothing bandage burns, blisters, and ulcers.

NUTRITION: Plantains are belly fillers, not for the dieters, as they are mostly carbo-hydrates - approximately 40 grams per half banana with 180 calories. Plantains have twenty times the vitamin A, about three times the vitamin C, double the magnesium, and twice the potassium.

FOOD & USES: The plantain is a banana that's eaten cooked rather than raw and are drier with lower water content, making them starchier than fruit bananas. Plantains can be cooked at varying ripeness. Green plantain is starchy like a potato and can be boiled or fried.

RECIPES: BOILED GREEN PLANTAINS – a side dish.

Ingredients: 3 green plantains, ½ cup diced onion, 1/3 cup olive oil, 1 TBS vinegar, 1TBS salt, ½ cup grated cheese, white or cheddar.

Method: Peel the plantains. Beware green plantains have a slimy goop when they are peeled. A trick is to let dishwashing detergent dry on your hands. After peeling the plantains, the slime should easily rinse from your hands. Rinse any soap from the plantains before cutting them into one-inch pieces. Boil for ten minutes in salted water or until tender. Sauté the onion in oil and vinegar. Mash the plantain pieces with a potato masher/ricer or a sturdy fork and fold in the sautéed onions. Cover with grated cheese. To make this into a casserole; add pieces of cooked carrots, green beans, beef, or chicken, and bake for 20 minutes at 150c/300f.

FRIED GREEN PLANTAINS – Tostones – an appetizer

Ingredients: 4 green planltains peeled and sliced into one-inch rounds, 4 cloves of garlic smashed, 1TBS salt, 1 quart water, oil for frying.

Traditional Method: smash garlic with salt to a paste and mix in water. Soak plantain pieces in garlic water for an hour. Drain and fry pieces in vegetable oil until golden brown. Be careful of the oil splatter. Flatten fried plantain pieces by pressing them with a large spoon on wax paper and return to hot oil for 2 minutes. Great served warm with hot sauce and cold drinks.

PUNA PLANTAIN CURRY

Ingredients: 4 green plantains peeled and sliced into rounds, 1 medium onion chopped small, 1 green chili chopped, 1 TS of each of the following: turmeric powder, curry powder, mustard seed, cinnamon powder, 1 cup thick coconut milk, 1 cup water

Method: After slicing the plantains immerse in a bowl of salted water with turmeric powder to halt the browning. Put all ingredients except coconut milk in a frying pan and add water. Cook 20-30 minutes stirring until plantains are soft. Add coconut milk and cook 10 minutes.

PLANTAIN FRY SAMBAL

Ingredients: 3 plantains, 6 roasted dry red chilis, 1 medium onion chopped small, 2 green chilies chopped, 1TS mustard seeds, 1 TS turmeric powder, 1/2 TS sugar, 2 TBS vinegar, salt to taste, 4 TBS thick coconut milk, cooking oil

Method: Blend dry chilies, mustard seeds, sugar, salt, and vinegar into a paste. In a deep fry pan heat oil and brown plantains pieces. In a bowl, combine all ingredients - onion and chili pieces, spice paste, and coconut milk. Add salt and sugar to your taste. Enjoy.

RAW PLANTAIN WITH GRATED COCONUT

Ingredients: 4-6 plantains peeled and cubed, 1 small onion chopped small, 2 green chilies chopped, 4 dry red chilies, ½ medium sized onion sliced, 1/2 TS each of the following: mustard seeds, fenugreek seeds, curry powder, red chili powder, turmeric powder, cinnamon, salt (adjust spices to taste), ½ cup grated coconut fresh is best, 2 TBS oil for frying

Method: Blend fresh coconut with green chili, red chili, curry powder, and turmeric powder into a paste. Keep aside. In a frying pan, heat oil, add mustard seeds, and fry until they pop, then add onion and dried red chili. Cook a few minutes until onion's tender. Drain ash plan-tain pieces and pat dry. Add plantain pieces and cook for 3 minutes. Add fresh coconut paste and stir in fenugreek seeds, and cinnamon. Cover and cook stirring for 5 minutes. Add water if necessary.

PLUMS

Spondias etc.

I added this page about plums because they're an easy-to-grow tree. All tropical plum trees can be pruned, shaped, and kept to a reasonable height. Some can be worked into a living and fruiting hedge. The different varieties, with reasonable maintenance will bear an abundance of healthy fruit for more than a decade. The plum trees in this book will adapt to soil and climates as long as it's tropical. Traditional / Ayurvedic / homeopathic medicines use parts of the plums mentioned for various treatments. Plums may have been one of the first fruits cultivated by humans. With olives, grapes and figs, archaeologists have discovered plum remains in Neolithic age sites.

NAMES CAN BE CONFUSING. Plums are one of my many favorite fruits. Researching the different types of governor, Jamaican, and chili (Chile), has driven me up the proverbial wall. A friend at the agricultural ministry says there are three distinct types of plum, the Japanese, the west Asian, and the European. . Other friends provided red plums they said were governor. I still haven't found the chilies! I decided, if they haven't ripened and green, they must be called chili. I'm certain a reader can differentiate. There is also the hog plum, but that is definitely different.

One reference confirmed my theory. From the horticultural website of Purdue University: *One of the most popular small fruits of the American tropics, the purple mombin, Spondias Purpurea L., has acquired many other colloquial names: in English, red mombin, Spanish plum, hog plum, scarlet plum; purple plum in the Virgin Islands; Jamaica plum in Trinidad; Chile plum in Barbados; wild plum in Costa Rica and Panama; red plum.*

There are three main types of plums, the Japanese, the west Asian, and the European. There are thousands of varieties of plums and all are family to the rose. Purple mombins / Jamaican plums are native to Central and South America. Purple mombin plums have large seeds compared to its thin flesh. The flesh is sour, but sweetens if you add salt. I believe we eat them green just to beat the birds to the tree.

It's now believed plums originated in China. The first written reference is 470 BC. The Assyrians first cultivated wild plums over 2,000 years ago The Crusaders are credited with bringing plums to Europe in 1369. In ancient Roman times, 300 varieties of European plums were mentioned. Japanese plums originated in China rather than Japan about 3-4 centuries ago. They are sometimes referred to as the Japanese apricot.

Internet research showed the red mombin as *Spondias purpurea*, the purple mombin as *Spondias purpurea L*, while the hog plum is *Spondias lutea or S.mombin*. These plums have been botanically researched with published scientific documents since 1753. Of the18 species in *Spondias*, ten are native to the New World, distributed from Mexico to southern Brazil, one is native to Madagascar, and seven are native to Asia and the South Pacific, from Malaysia through tropical China, Sri Lanka, Indochina, Thailand, India, Myanmar / Burma, Solomon Islands east to Polynesia. *Spondias* dulcis is cultivated in tropical America and the Caribbean. European explorers introduced *S. mombin* and *S. purpurea* to Tropical West Africa, Asia, and the Caribbean.

The reader may wonder, as the writer did, how to they tell one species from another… scientifically. To provide a glimpse as to how all the varieties/species of plums are classified, I borrowed from *The Flora of the Lesser Antilles*, Bornstein (1989) who researched *S. mombin, S. purpurea,* and *S. lutea.* This was the criteria.

1) Wood anatomy: gray

2) Outer bark: densely to broadly fissured, sometimes thick, usually rough, often with raised lenticels, rarely (some *S. mombin* and *S. purpurea*) with large, corky, tooth-like projections.
3) Inner bark: usually broadly marked with lines (white and rose, red, orange, or brown)
4) Sap or resin: viscous and usually clear or less often cloudy.
5) Leaf architecture / shape: Leaves alternate or opposite, aggregated toward branch tips, sometimes deciduous.
6) Breeding systems: blossoms are usually hermaphroditic.
7) Blossoms: petals yellowish-white; stamens, flower, and fruit anatomy.
8) Pollen: round or elongated, and size.
9) Hybridization: implicating *S. mombin* **as o**ne of the putative parents
10) Seedlings – with or without a tap root

After all those critical, great, and minute differences, various species remain referred to with local names as this: The common name for the ramontchi fruit in India is governor's plum. The ramontchi is native to tropical Africa, Madagascar, India, parts of Malaya and Southeast Asia, and much of Malaysia including the Philippines. It has been planted in Florida, Puerto Rico, Trinidad, Guatemala, Honduras and Venezuela. The fruit is round, half to one inch thick, smooth, glossy, dark red-purple, with light-brown, acid to sweet, astringent, slightly bitter, flesh, with six to ten small, flat seeds. Ramontchi is not what the Caribbean Islanders call governor plums.

Names don't matter as long as the reader realizes that the plum tree provides good fruit, when regularly included in a diet, it will keep you healthier and also treat physical ailments. All human cells protect themselves against free radical damage by antioxidant compounds such as ascorbic acid / vitamin C, and tocopherol, and glutathione. During our lives these protective mechanisms are sometimes disrupted and wear down. Since limited pharmaceutical drugs are available, the treatment of degenerative diseases has always been assisted by natural, nontoxic, and affordable fruits and vegetables. Plums are a great source of inexpensive preventive medicine.

Spondias is a genus of flowering plants belonging to the cashew family (Anacardiaceae). This genus comprises 18 species distributed across the world's tropical regions. Plums have abundance of bioactive compounds such as phenolic ac-ids, anthocyanins, carotenoids, flavonoids, organic citric acid, fiber, tannins, aromatic substances, enzymes, and minerals as potassium, phosphorus, calcium and magnesium, with vitamins A, B, C & K.

A variety of bioactive phytochemical constituents were isolated from different plants belonging to the genus *Spondias*. Diverse pharmacological activities were reported for the genus *Spondias* including slowing cell damage, antioxidant, ulcer protection, liver cleansing, anti-inflammatory, anti-arthritic, and anti-dementia effects. Eating plums helps to prevent heart disease, lung and oral cancer, lowers blood sugar and blood pressure, Alzheimer's disease, muscular degeneration, improves memory capacity, boost bone health, and regulates the functioning of the digestive system. These attributes indicate the plums potential to treat various degenerative diseases. From: *Ge-nus Spondias: A Phytochemical and Pharmacological Review by the Faculty of Phar-macy,* Ain Shams University, Cairo, 11566, Egypt.

Remember, all fruit is better for you when you grow it yourself and know that little or no chemicals have been used.

The author relied on information from:
A revision of Spondias L. (Anacardiaceae) in the Neotropics
by John D. Mitchell, Douglas

POHA - CAPE GOOSEBERRY

Physalis micrantha

Poha is a unique fruit that's native to the South American countries of Chile, Ecuador, and Peru. It is also known as the wild tomato and grows as a weed in Hawaii's tropical wet low elevation areas. It may have been brought to the islands on purpose or accidentally by European explorers and colonists before 1825. Because poha is easily cultivated, this plant is perfect for the home garden grown for its tasty, unique fruit, medicinal purposes, and as an ornamental. Its botanical names are *P peruviana* and *Physalis micrantha.* It is also considered invasive.

Known as the Cape gooseberry, poha resembles a Chinese lantern with a small, round golden fruit wrapped in a papery case. It is another fruit that tastes both sweet and sour somewhat like a mixture of a pineapple, a mango, and a tomato.

HOW TO GROW: This is an attractive small tree that usually reaches only a meter but can grow taller with good care. Poha will grow almost anywhere in any conditions from seeds or sprouted cuttings from sea level to 3000 meters in almost any soil type with a pH range from 4.5 to 8. Poha grows better with only a little daily water and planted in sandy gravel soil with partial shade and protected from high winds. Plant 6 seeds to each hole about 6mm (½in.) deep and 55cm (2ft.) apart. Thin and keep the best plants. With reasonable care, regular water, and a little fertilizer, a plant can have an annual yield of 200 or more fruits. This shrub should blossom within 3 months after planting and fruit in 6. The tree will produce a good crop for 2 years. The marble-sized, 2cm (¾in.) fruit turn golden as the papery husk dries.

MEDICINAL: In Ayurvedic-homeopathic traditional medicine, poha treats ailments from malaria to toothaches and rheumatism. It is considered a diuretic and relaxant. Infusions treat indigestion, fever, and jaundice. Poultices of steamed leaves treat inflammations and sore muscles.

NUTRITION: 140 grams of poha has 74 calories with 2g of protein, and has high quantities of vitamins A, B1, B3, and C with good amounts of iron, phosphorus, and calcium.

FOOD & USES: Poha can be eaten raw, cooked, or dried. The papery husks aren't edible, but keep the berries in their husks until ready to use. In the husks, poha can garnish a dish or as a beautiful dessert dipped in melted chocolate. This fruit is delicious as a snack, a flavoring for puddings, and in pies, sauces, jams, jellies, *ice cream,* and chutneys.

RECIPES: POHA/GOOSEBERRY PRESERVE:

Ingredients: 3 cups poha husk removed and halved, 3/4 cup natural orange juice, ½ water, 2/3 cup sugar, 1TS vanilla extract, 1TBL butter (optional)

Method: In a proper pot, combine orange juice, water, and poha halves and bring to a boil. Reduce heat and simmer until the fruit is soft. More water/juice may be added. Stir in the sugar and vanilla until it becomes a thick jam consistency. Remove from heat, stir in butter. Enjoy.

POHA SMOOTHIE

Ingredients: 18 poha, husked, washed, and quartered, 2 bananas peeled, 1 cup milk (real, or soy), 1 bunch spinach washed, 2 sprigs mint, and ice

Method: Wash all ingredients. Put them into a blender. Chop and blend until smooth.

NOTE: The unripe poha fruit has been reported to be poisonous to some people.

DID YOU KNOW? During the 1800s, poha was commercially grown near the Cape of Good Hope in South Africa. That is how it became known as the cape gooseberry. **NAMES:** Aztec berry, Barbados gooseberry, bladderberry, cape gooseberry, Chinese lantern, goldenberry, Peruvian ground berry, poha berry, strawberry tomato, winter cherry

POMEGRANATE

Punica granatum

The pomegranate is another passionate fruit. Like the sugar apple, you must passionately want to enjoy its unique flavor, to deal with the seeds. The round, usually reddish-maroon three-inch fruit has a distinctive royal crown at the blossom end. The pomegranate is technically a big, tough-skinned berry with many seeds in juicy, transparent, jellied membrane compartments. This tree is perfect for your home, both for fruit and medicinal uses. If you do not like seeds, this fruit is not for you. A juicer or food processor can be used to extract the juice from the pulp. Strain the juice to remove any seed sediment. Seeds are more than half of the weight of this fruit. Spanish traders or missionaries probably brought seeds to the Hawaiian Islands.

The pomegranate has every reason to wear a crown. It is an ancient fruit native to an area in the Himalayas north of India, but has been cultivated throughout the entire Mediterranean region since ancient times. Greek mythology explained the four seasons with pomegranates. Demeter, who was the goddess of the harvest, had a daughter Persephone who was kidnapped by Hades, the lord of the underworld. Demeter refused to permit anything to grow on Earth until Persephone was returned. Zeus, the lord over all the lesser gods, ordered her to be reunited with her mother, but she had eaten four pomegranate seeds. The rule was if you ate anything while in the underworld you were forced to spend eternity there. Each seed equaled one barren month on Earth, or winter.

The Moors brought the pomegranate to Spain where it is the national emblem. The Spanish city of Granada is named for it. Kandahar in Afghanistan is famous for pomegranates.

HOW TO GROW: The unique pomegranate is a perfect backyard tree. It grows to between fifteen and thirty feet. The pomegranate will grow in any well-drained soil, even rocky. For a small tree, it has almost as many branches as seeds. Some trees at France's Versailles Gardens have lived for two hundred years. Every three months feed young, not yet bearing trees, a half cup of 12-24-12 starter fertilizer around the roots and water during the dry season. Once it bears change to a cup 12-12-17-2 or another bearing salt twice a year. Pomegranate trees are on every Caribbean island. Seeds are available online.

Pomegranates will develop suckers at the roots. These can be used to plant, or trees can be raised from seeds. The tree can be evergreen or deciduous. Pomegranate seeds sprout easily, but better trees are developed from cuttings. The pomegranate may begin to bear in a year after planting, but two to three years is more common. The fruits ripen six months after blossoms appear. Too much sun exposure will dull the usual reddish skin to a burnt brown and toughen the skin. The pomegranate is equal to the apple in having a long storage life of more than six months if refrigerated. This fruit improves with time as it gains juice and flavor. How to eat a pomegranate without a mess: cut out the crown blossom end and remove some of the white membranes while trying not to break the red pulp around the seeds. The entire seed is consumed raw, though the juice is the tasty part. With a sharp knife cut slits in the fruit's skin making quarters. Break the pomegranate apart on the slits and bend back the skin to remove the seeds. It is also possible to freeze the whole fruit, making the red arils easy to separate from the white pulp membranes.

MEDICINAL: Pomegranate juice has also been shown to lower blood pressure, inhibit viral infections, and may destroy dental plaque. Pomegranates can thin the blood, increase blood flow to the heart, reduce blood pressure, reduce plaque in the arteries, and reduce bad cholesterol while increasing good cholesterol. A seed decoction has been used to treat syphilis. The juice treats jaundice and diarrhea. Dried, powdered blossoms treat bronchitis.

NUTRITION: A hundred 100 grams of pomegranate has 70 calories with plenty of potassium, vitamins C and B_5. Pomegranates are considered one of the best fruits for fighting illnesses due to its high levels of vitamins, minerals, and antioxidants.

FOOD & USES: Pomegranate adds a distinctive flavor to sorbets, icings, salad dressings, soups, and puddings. The juice provides a fresh, unique flavor to marinade fish, chicken, pork, and beef. Grenadine is a thick red syrup made from pomegranates and often used in cocktails such as the Tequila Sunrise. An ordinary home kitchen orange-juice squeezer can extract the juice. In northern India, pomegranate seed sacs are dried in the sun for two weeks and sold as a spice. Avoid using aluminum pots or carbon steel knives with pomegranates as they can turn the juice bitter.

Pomegranate juice has been used as a natural dye in many countries and will stain fingers and clothes. Ink can be made by steeping the leaves in vinegar. The Japanese make an insecticide from the bark. The pale-yellow wood is hard and excellent for walking canes.

RECIPES: POMEGRANATE CHUTNEY - Seeds in the following recipes include the juicy membrane.

Ingredients: 1 cup pomegranate seeds, ½ cup red currant jelly, 1/3 cup chives including tops chopped fine, 1 TBS fresh ginger minced, 1 TBS hot pepper seeded and minced, 1 TS ground coriander, 1 TBS fresh lemon juice, salt to taste

Method: Chill the seeds overnight in an airtight container. Soften currant jelly in the sun or in a sauté pan on high for about a minute. Add chives, pomegranate seeds, ginger, hot pepper, coriander, and lemon juice. Add salt to taste. Let stand 15 minutes before serving.

POMEGRANATE VINEGAR

Ingredients: 1 cup fresh pomegranate seeds, 2 cups white vinegar

Method: Place pomegranate seeds in a clean wide-mouthed bottle with a lid. Rough the seeds up with a spoon then cover with vinegar and seal tightly. Place jar in a window in full sunlight and let steep for two weeks. Use through a strainer or strain vinegar through a cloth.

POMEGRANATE ROAST CHICKEN

Ingredients: ¼ cup cooking oil, 2 cloves garlic minced, 1 medium chicken-quartered, 1 pomegranate-halved, juice of a lemon, ½ TS cinnamon, 1 TBS brown sugar, salt to taste

Method: Mix oil and garlic and coat chicken pieces. Put pieces in a baking dish with any garlic oil that remains. Bake at 190c/375f-degrees for thirty minutes. Then combine pomegranate lemon, cinnamon, sugar, and salt and baste the pieces. Return to the oven for another twenty minutes until the skin is browned. Serve with the juices.

POMEGRANATE CAKE

Ingredients: seeds from a large pomegranate, ¾ cup sugar, 6 TBS margarine, 2 large eggs, 1 large egg white, ¾ cup milk, 1 TS lemon zest, 2 TS vanilla extract, 1 TS baking soda, 3 cups bakers flour, ½ TS salt

Method: Blend sugar and butter until creamy – ten minutes of whisking. Add eggs and egg whites one at a time and continue to blend. Combine milk, lemon zest, vanilla, and baking soda. Combine flour and salt, with the milk-butter mixture. Add pomegranate seeds. Spoon into a greased bread loaf pan and bake for an hour at 350 degrees. Cool before slicing.

DID YOU KNOW? Botanically pomegranate is *Punica granatum,* derived from punicus, meaning from Carthage where this fruit was discovered. Pomegranate is called granada in Spanish and grenade in French. Buddhists believe the pomegranate to be a bless-ed fruit, and some scholars believe the pomegranate, not the apple, was the forbidden fruit in the Garden of Eden. A pomegranate tree may live for more than a century. Once the pomegranate fruit is picked it stops ripening, but will become more flavorful after being in storage.

POMELO – JABONG

Citrus maxima

The pomelo is the largest citrus fruit and the ancestor of the grapefruit. Unfortunately, most of a pomelo fruit is a spongy rind with white inedible pulp. It took a lot of evolution to produce the relatively thin-skinned grapefruit. The pomelo is a delicious fruit with a reddish core with succulent tangy-sweet juice. These fruits can reach two kilos in weight and are almost the size of a volleyball. The largest pomelo on record weighed twenty-two pounds!

This fruit was probably brought to Hawaii Spanish or Portuguese seafarers. The USDA began experimenting with pomelo varieties from Thailand and the Philippines circa 1900. Like many tropical fruits this one has numerous names pumelo, pummelo, Chinese grapefruit, Lusho Fruit, jabong, pompelmous, and shaddock. The pomelo is native to southeastern Asia and probably originated in Malaysia, Indonesia, New Guinea, or Tahiti. The original trees were grown from seeds, but today they are grafted. The pomelo is not specifically grown for export but a small farm crop at this moment. The fruit's flesh may be white or red.

HOW TO GROW: You can try to start a tree from seeds, but I recommend finding a garden shop that has grafted trees. A tree sprouted from grafted seeds may produce the root-stock rather than the graft. Plant it in an area that gets full sun and is well-drained. These trees will not survive any standing water, or seriously damp soil. Give it a cup of 12 -24-12 starter fertilizer every three months for the first two years. Once you see blossoms give it a cup of 12-12-17 -2 bearing fertilizer every six months. In the dry season give this tree plenty of water, five gallons, every week. This will produce succulent fruit. Twice a year spray the tree with an insecticide like Fastac mixed with a foliar mix with trace minerals. Watch your tree for damage to the new leaves from ants. A grafted pomelo is a nice backyard tree. It doesn't have a lot of small branches, so it doesn't make deep shade. After five years our grafted tree is still only six feet tall and seems to be always bearing. It began to bear huge fruits after the second year. These trees are unique sights with a bunch of big fruits hanging on a small tree.

MEDICINAL: This fruit fights atherosclerosis and helps regulate blood pressure. Pomelo is great for dieting because it quenches your appetite and has only thirty-five calories per hundred grams. Limonoids found in pomelo are being studied for their cancer protective properties. Eating a pomelo is invigorating and increases stamina and lifts your spirit.

NUTRITION: 1 cup of pomelo has only 70 calories and is loaded with vitamins, rich in vitamins C and B complex, and beta-carotene. It is a great source of folic acid which is especially good during pregnancy. It contains a lot of heart-friendly potassium.

FOODS & USES: The pomelo is eaten by peeling it, skinning the segments, and enjoying the juicy pulp. Because the fruits are so large, usually about ten inches in diameter and the flesh center much smaller – about four inches across – it is difficult to determine when the fruit is ripe. If the fruit stays too long on the tree the pulp dries out. The fruit rarely changes from its green color as it matures, but some do become slightly yellow.

Pieces of the pulp can be segmented for salads and desserts, or just eaten raw. The juice is delicious. The pomelo tastes like a sweet, mild grapefruit with little of the bitterness common among grapefruit. The peel can be candied. Be careful if you must peel a lot of Pomelos because there are chemicals in the peel that can irritate your skin. The Chinese believe that a bath with water from boiling pomelo husks will cleanse a person and repel evil.

RECIPES: POMELO CHICKEN SALAD

Ingredients: 1 pomelo peeled, seeded, and broken into pieces, 1 cooked chicken breast shredded, 1 cup cooked medium shrimp cleaned and deveined, 1 TS chopped hot red pepper seeded and minced, 1 TS sugar, juice from one lime, 1 TS chopped fresh culantro, small head red leaf lettuce. Garnish with ¼ cup roasted peanuts, and julienned fresh red sweet pepper.

Method: Add shredded chicken and shrimp with the pomelo. In another bowl combine the chopped red chili, sugar, lime juice and culantro. Mix with the pomelo, chicken, and shrimp. Spoon this over the lettuce. Top with peanuts and garnish with sweet red pepper slivers.

POMELO ICE

Ingredients: 1 pomelo peeled, an ice cube tray, and fresh pure water.

Method: Peel and core the fruits over a bowl and squeeze to catch the juice. Combine with water and pour into the ice cube tray. Freeze overnight in the freezer.

SUNSET BAY POMELO FISH

Ingredients: 1 cup of pomelo juice, 1 pomelo peeled and sliced a half-inch thick, 2 pounds of fish steaks or fillets like king, tuna, or salmon, 1 small onion chopped small, 1 clove garlic minced, 1 TBS culantro

Method: Combine juice, onion, and garlic and culantro. Marinate fish for at least an hour, but 4 hours will be better. Save marinate. Line a baking dish with the pomelo pieces. Lay the fish on the Pomelo and pour the marinade over the fish. Cover tightly with foil. Bake at 350 degrees for thirty minutes. Uncover and bake another five minutes.

CHICKEN POMELO

Ingredients: 1 whole chicken, 2 pomelos—one quartered and one juiced, ½ hot pepper seeded and minced (optional), salt, and seasoning to taste.

Method: Season inside of chicken cavity with salt and hot pepper. Place pomelo quarters halves inside chicken. Place chicken in a pan. Pour half of the pomelo juice into cavity and the remainder on the chicken. Cover with foil and bake at 400 degrees for forty-five minutes. Remove foil and continue to bake for 10-15 minutes till brown.

FRIED POMELO

Ingredients: 2 or more pomelos seeded and halved, 2 TBS butter or margarine, 2 TBS brown sugar, cinnamon, and nutmeg to taste

Method: Heat a large frying pan and melt butter over medium heat. Put pomelos face down and cover. Fry for four minutes. Remove and sprinkle with sugar, cinnamon, and nutmeg.

PENNILESS SURFERS' BROILED POMELO CRISP

Ingredients: 1 pomelo seeded and halved, ¼ cup old-fashioned oats, 1 TBS brown sugar, ½ TS ground cinnamon, a pinch of salt, 2 TBS butter

Method: In a bowl combine oats, sugar, and cinnamon. Put pomelo halves on a tray and cover with oat mixture. Put under broiler for eight to ten minutes. Serve warm.

DID YOU KNOW?

Pomelo is the largest citrus fruit identified as *Citrus maxima Merr., (C. grandis Osbeck; C. decumana L.).* The common name is derived from the Dutch pompelmoes, which is rendered pompelmus or pampelmus in German, pamplemousse in French. Alternate names include shaddock, limau abong, limau jambua, Bali lemon, and pummelo.

Growing your own fruit at home is a wonderful treat and a huge money saver. Homegrown food of any kind tastes delicious and being able to go outside and pick fresh fruit from a tree is something everyone should experience.
~ Julio Belson

RAMBUTAN

Nephelium lappaceum

The rambutan is one of the world's strangest looking fruits. This golf ball-sized, pinkish-red fruit looks as though it's a children's cartoon character who received an eletric shock. Rambutan's name comes from the Malay word *'rambut'* for 'hair,' because of the fruit's surface. This is a delicious fruit once you understand how to get at the sweet part inside the hard, fuzzy shell. Its unmistakable appearance is often compared to a sea urchin.

To open a rambutan, find a groove between the soft, fuzzy spikes with your thumbs. Then push downwards and then outwards, prying it apart to get at the lychee-like fruit. They are usually eaten out of hand. Suck it whole in your mouth, but spit out the seed. The flavor's similar to the lychee, longan, and grape. Its translucent white flesh has a sweet creamy taste. Consuming raw seeds may have narcotic and analgesic effects, causing sleepiness, coma, and even death.

Rambutan is believed to have originated on Borneo. Arab traders who dominated the seas from the 1200s to 1400s spread the fruit to India, Asia, and Africa. European explorers carried trees to the Caribbean, Central, and South America. In 1906, rambutan was successfully grown in Puerto Rico and introduced to the Philippines in 1912 from Indonesia. This tropical fruit is of the same family as the lychee and there are over 200 varieties of rambutans with varying colors, flavors, and appearances.

Today, Thailand produces the most rambutan, with 588,000 tons - 55% of the world's supply. Indonesia grows 320,000 tons - 30%, and Malaysia 126,300 tons 12%. These three countries grow 97 percent of the world's supply. It is grown commercially within 12–15° of the equator and at elevations up to 500m. Hawaiians love rambutans; most fruit comes from small home gardens. Commercial orchards began circa 1970 on several islands. Rambutan was one of the three top tropical fruits throughout the chain in the late 1990s. The season is from February thru March. It's related to several other tropical fruits as lychee, longan, pulasan, and mamoncillo.

HOW TO GROW: The rambutan is an evergreen tree and loaded with bright red fruit it makes a great addition to the home landscape. It usually grows to 20 meters and spreads, so it needs to be spaced at least 7 meters from any buildings or other trees. This fruit does best in deep soil, clay loam or sandy loam rich in organic matter with a pH of 5.5 to 6.5. It thrives on hilly terrain with good drainage. Rambutan seeds should be planted within two days after removing from the fruit. They should be planted with the at side down so the sprout grows straight and builds a strong root system.

Rambutan trees have sexes. Male trees are rare and they won't produce. There are fe-males, and more often hermaphrodites, which have a majority of femalef lowers with a small per-centage of male owers. 500 greenish-yellow flowers occur in each hermaphrodite branch cluster. In an orchard, separate the trees at least 10m in every direction. The wood is reddish-white, or brownish, suitable for construction but must be carefully dried or it will warp and split.

The fruit is a round, 6cm/2.5 inches in diameter, with a single seed and hang in a loose cluster of 10–20. The thick skin is reddish, sometimes orange or yellow, covered with soft spines They ripen only on the tree, Rambutan can bear twice a year. Interestingly, the Rambu-tans remain fresh longer when harvested with the branch attached.

MEDICINAL: Rambutan is nutritious yet low in calories and may aid your digestion, boost energy and the immune system, ght anemia, and aid weight loss. All of the rambutan tree is used, from the root to the skin of the fruit, the seed, and leaves to relieve such ailments as dysentery, diabetes, and fever. The leaves can be made into a poultice and placed on the temples to ease a headache. A decoction of the astringent bark is a remedy for throat problems, while root decoction will reduce fever. A study in Thailand discovered many flavonoids in rambutan, which act as anti-inflammatories with anti-cancer and antioxidant properties. The light brown seeds are high in certain fats and oils, valuable to industry, and used in cooking, and the manufacturing of soap.

NUTRITION: Rambutans are a good source of copper to help maintain bones and nerves. It provides vitamin C, an antioxidant that strengthens the immune system and reduces inflammation. Rambutan fruit contains many nutrients, but only in small amounts, except for manganese with 16% of the RDA. Per 100 g consumed, with 75 calories you get 10mg of calcium, 12 of phosphorus, and 30 mg of vitamin C.

FOOD & USES: Rambutans are most often eaten fresh by tearing the shell open. It does not adhere to the flesh. They're stewed as dessert and canned in syrup. The seed yields a 40% solid, white fat or tallow resembling cacao butter. When heated, it becomes a yellow oil with a sweet aroma.

RECIPES: RAMBUTAN CURRY

Ingredients: 1 peeled and seeded rambutans, 1TS turmeric powder, 4 cloves of garlic chopped, 3 green chilies chopped, 1 stalk lemongrass, smashed, (only the inner core is used, sliced thin), one onion chopped small, 10 curry leaves, ½ cup pineapple cubes, 1 TBS fish sauce, 2½ cups coconut milk, 2 cups chicken broth, 500g chicken breast deboned and sliced thin, 1 TBS brown sugar 2TBS coconut oil

Method: In a mortar, grind the garlic, turmeric, lemongrass, curry leaves, chilies, into a paste. In a big pot or wok, heat on medium flame 1TBS oil and add pineapple cubes. Stir until brown, about 5 minutes. Carefully spoon out cubes to another bowl, then add the other TBS of oil and the onion. Cook until brown, then stir in the paste and stir for 2 minutes. Add the coconut milk and chicken broth. Bring to a boil, constantly stirring, and then reduce heat to a simmer. Add chicken and cook for 10 minutes. Add the already cooked pineapple and the rambutan and cook for 2 minutes. Remove from heat and stir in the fish sauce. Serve with rice.

RAMBUTAN SMOOTHIE

Ingredients: 3 rambutans peeled and seeded, 2 cups fresh grated coconut, 1 banana

Method: Put all ingredients in a blender and process until a smooth liquid.

RAMBUTAN MARTINI

Ingredients: 8 fresh rambutans peeled and seeded, 2TS lime juice, 2TBS white sugar, ¼ cup water, ½ TS vanilla extract, 1 cup ice, 4-6 ounces of vodka

Method: In a saucepan on medium, make simple syrup - heat the sugar and water until the sugar is dissolved. Pour into a small bowl to cool. Place the rambutan in a cocktail shaker and press with a big spoon to get the juice out. Add the vodka, lime juice, vanilla, ice, and cooled simple syrup. Cover and shake for a minute. Strain into martini glasses and garnish with pieces of the rambutan.

TO MAKE A RAMBUTAN DAIQUIRI use the same ingredients but blend and don't strain. Use tall glasses.

DID YOU KNOW?

The botanical name for rambutan is *Nephelium lappaceum*. In Viet Nam, the rambutan is termed 'chom chom,' which means 'messy hair.

A rumbutan has an agressive shell hiding a vuknerable delicious soul. ~
Anonymous

RED BANANAS

Musa acuminate

The dwarf red, or Cuban red banana is reported to be grown everywhere throughout the Hawaiian chain. They are a heartier tree than a regular yellow Cavendish. They make a great fried banana for breakfast or as a dessert. This is a variety with reddish-purple skin. They are smaller and plumper. The ripe banana flesh is creamy yellowish and even pink when fully ripe, but they are certainly sweet. They taste the sweetest when fully ripe. The redder a fruit, usually means it contains more carotene. Perhaps the red banana is nutritionally better than the yellow varieties.

Red bananas are known as Jamaican bananas, red Spanish, red Cuban, Colorado, Macaboo, or Klue Nak. In India where it may have originated, red bananas are 'Lal Kela.' Since Alexander the Great found a liking for red bananas in 327 BCE they are grown the world over. They are a strain of the Cavendish banana. The botanical name is *Musa acuminate.*

The red banana plant is large and is highly resistant to disease. The entire plant is elegant for a productive landscape theme as it is almost all maroon. As the bunch evolves it goes from deep maroon to a bright orangish-red when fully ripened. As the fingers are small the bunch may be large and the tree will probably need to be propped in the latter stages of growth. These bananas have a thick peel, but not as thick as a plantain. The flesh is firm and has a nice aroma and flavor quite different from other bananas. This tree is uniquely red maroon with nice stature and is a great addition to a back or front yard landscape.

HOW TO GROW: First, find someone who has red bananas and get some suckers. Soak in a recommended mixture of a mild insecticide such as Malathion and water for an hour before planting to rid it of any worms. Dig a deep hole and refill as your crumble and soften the earth. Plant the sucker so a bit is exposed. This will be a bigger than usual tree and needs a space of eight to ten feet from any neighbor or building. Water regularly, and every other month hit it with a quarter cup of some high potassium fertilizer. When it shoots up the signal leaf indicating it is about to bear, keep it moist and broadcast a cup of high potassium salt around the base. This tree should bear in sixteen to eighteen months, but it may take two years to make a bunch. This strain of banana appears to be more resistant to the diseases prevalent today.

I've had red bananas since they were recommended by a friend who had eaten them in Mexico. Once they start bearing you should have a bunch every three months from the same stool. The bunch bears about six nice hands, the fruits are thick, about six inches long. They are beautiful to look at and delightful to eat, especially fried for breakfast. There is a white variety, but that seems to be silly. Why would you want a white version of a red banana?

MEDICINAL: Bananas are said to contain everything a human body needs including all eight amino acids, which our bodies can't self-produce. Bananas are a good source of fiber and potassium. The fruit is a mild laxative, good for cardiovascular health, protects against strokes and ulcers. It reduces water retention and is preferred for anemic patients because it's rich in iron. Red bananas have more vitamin C than the usual yellow types. The redder the fruit the more nutritious. Eat at least one banana a day. It comes in a truly germ-proof package because its thick peel is an excellent protection against bacteria and other contamination. The flower of this plant is used to treat ulcers, dysentery, and bronchitis and cooked flowers are good for diabetics.

NUTRITION: Every red banana has about 115 calories with 400 mg potassium and 15% of your daily requirements for vitamins C and B6 and one gram of protein. Red bananas are excellent as an energy snack. High in potassium, they can help reduce cramping after exercise or stress. Vitamin B-6 in red bananas helps break down proteins and form red blood cells.

FOOD & USES: Red bananas are eaten in the same way as yellow bananas, by peeling the fruit before eating. They are most frequently eaten whole raw or chopped and added to desserts or fruit salads. They can also be baked, fried, or toasted. Red bananas are one of the varieties commonly used for store-bought dried bananas. Ideally, red bananas should be eaten when they are soft, not mushy. They will ripen in a few days at room temperature, and refrigeration is generally not advised because it can make them extremely mushy. A red banana will also emit ethylene gas that quickly ripens other fruits.

RECIPES: RED BANANAS WITH CARDAMOM

Ingredients: 6 red bananas peeled and cut long ways, ¼ cup butter, ½ cup brown sugar, 2 TBS fresh lime juice, ½ TS fresh cardamom seeds ground, a quart vanilla ice cream

Method: Melt butter in a sauté pan over medium heat. Add brown sugar and stir until dissolved. Add bananas, lime juice, and ground cardamom. Cover and simmer for five minutes, stirring occasionally, until bananas are tender. Spoon bananas with sauce over ice cream.

FRIED RED BANANAS WITH FRUIT SALSA

Ingredients: 6 red bananas-peeled, sliced long ways, 2 TBS butter for frying, 2 TBS brown sugar, 2 tangerines or tangelos - peeled, 1 grapefruit or pummelo/jabong, and 1 mangoes- all fruit peeled, seeded, and chopped as small as possible, 2 TBS honey, 1 TS vanilla extract, 2 TBS grated fresh coconut

Method: In a sauté pan brown the red banana slices in the butter over medium heat. After turning the slices twice sprinkle the brown sugar in the pan and stir until the slices are covered. Remove from heat. In a bowl combine all fruit pieces, vanilla, and honey. Let stand in the fridge for an hour or until chilled. Put banana slices in bowls and cover with fruit salsa. Sprinkle each bowl with fresh coconut and serve.

RED BANANA FACIAL MASK

One way to prepare an easy and effective face mask is to mix mashed red bananas with a half cup of steel cut oats, 1 TBS each of lemon juice and honey. Mix into a paste. Apply it on your face and let it dry and then wash it off.

DID YOU KNOW?

Red bananas are often used to make dried banana chips. Dried bananas contain have five times more calories than fresh. Bananas are fermented to make beer in east Africa. India grows the most bananas. Wild banana varieties carry much bigger seeds that would make it difficult to eat, but necessary for the plant's reproduction. The definition of fruit is 'a mature ovary containing seeds.' In fact, a banana is more precisely referred to as a berry! The modern banana is cultivated to be seedless. The yellow bananas that we eat (*musa*) do have seeds, but they're so small they're not functional, meaning they are not useful for the plant's reproduction. According to banana giant Dole, Americans eat thirty-three pounds of bananas annually.

Life is full of banana skins. – Proverb

SAPODILLA – CHICO

Manilkara zapota

The sapodilla or it's AKA 'chico of chikoo' is round, about three inches in diameter, with a flat base and a thin brown leathery skin. Chico is short for chicozapote. Some Hawaiians call it the 'chewing gum' tree. It resembles a small potato. It is about 2-4 inches in diameter. The flesh varies from yellow to reddish-brown, and usually has a grainy texture. Sapodilla tastes sweet, like a pear dipped in crunchy cinnamon-brown sugar. Fruits usually have from 3 to 12 hard, black, shiny flattened seeds about 3/4-inch-long at the center of the fruit. Ripe sapodillas have a great aroma. This fruit is believed to have originated in Southeast Mexico. Sweet-toothed Amerindian tribes as the Mayans and Aztecs, and the European explorers spread sapodillas throughout the tropical Americas to Hawaii, southern Florida, and the West Indies. The Spanish, perhaps Don Marin, probably brought the sapodilla to Hawaii.

HOW TO GROW: Sapodillas are usually grown from seeds and take at least five years to bear fruit. It is wise to use the seeds from the largest, sweetest sapodillas you are lucky to locate. A grafted tree will bear fruit in about three years. This tree will grow easily in most well-drained soils with a pH of 6-8 and a place that gets full sun. They can be grown in a big pot on your porch or balcony of you keep it trimmed. Every month fertilize with an 8-4-8 mix for the best results. Once the tree blossoms it takes about five months for the fruit to ripen. Sapodillas are difficult to decide when they are ripe enough to harvest. When the fruit is brown and pulls easily from the stem without leaking any of the latex, it is fully mature yet should be kept at room temperature for few days to soften. Wash off the sandy scruff before setting the fruit aside to ripen. It should be eaten when they just start to get soft, before it gets mushy. It is an ideal dessert fruit as the skin (not to be consumed) serves as a 'shell.' Do not accidentally swallow a seed, because the protruding hook can cause you to choke erasing all the enjoyment from this delicious fruit. Ripe, firm sapodillas can be stored for several days in your refrigera-tor. Frozen fruits can be kept perfectly for a month.

MEDICINAL: Young sapodilla fruits can be boiled and eaten as a bush remedy for diarrhea. A tea made from the old, yellowed leaves is a reputed remedy for coughs and colds. Eating crushed seeds are claimed to help heal the bladder and expel kidney stones. The liquid essence of the crushed seeds is used throughout the Yucatan as a tranquilizer. The seeds can be crushed into a paste to soothe insect bites. Chicle, the tree sap, is used in Central America as a primitive dental filling.

NUTRITION: 100 grams of raw sapodilla has about 83 calories with calcium, potassium, phosphorus, vitamins A and C, and folate.

FOOD & USES: The sapodilla produces strong and long-lasting timber. This type of wood is so strong and durable that the timbers used as beams in ancient Mayan temples have been found intact in the ruins. It has also been used for railway crossties, floor planks, carts, and handles. The wood's reddish core makes archers' bows, furniture, railings, and intricate cabinets. Due to the latex content, sapodilla sawdust irritates the nostrils.

The sap of the sapodilla tree is called chicle. Containing 15% rubber, chicle is harmless and tasteless. It was dried and chewed by the Mayans as a primitive chewing gum to ward off hunger. The Mexican General Santa Ana introduced chewing chicle to America in the mid-1800s beginning the chewing gum industry. The demand for sap became so great that it was 'tapped' from sapodilla trees throughout Central America delivering over seven tons a year.

Most often sapodillas are eaten raw, but a sauce can be prepared from peeled seeded fruit forced through a strainer, and then mixed with orange juice and heavy cream. The fruits' flesh can be mixed with egg and cream to make a delicious custard. Crushed boiled fruits can be strained to create a sweet syrup. Mashed sapodillas can be added to a batter to make dessert fritters. If you add sugar when cooking sapodillas, the flesh will turn red.

RECIPES: SIMPLE SAPODILLA PIE

Ingredients: 2 cups sapodillas - peeled, seeded, chunked, ½ cup golden raisins, ½ cup lemon or lime juice, ½ cup brown sugar, 1 pie crust

Method: Fill pie crust with sapodilla pieces, then top with raisins. Pour the lime juice over the pie to prevent the sapodilla from becoming chewy. Sprinkle sugar over the pie and cover with top crust with plenty of fork holes to release the steam. Bake at 350 degrees for 45 minutes.

EASY SAPODILLA RICE

Ingredients: 2 cups cooked rice, 2 sapodillas - peeled, seeded, and chunked, 1 TBS lemon zest, 1 TBS ginger peeled and minced

Method: Mix everything into the warm cooked rice. Let sit for fifteen minutes covered and serve.

SAPODILLA CUSTARD

Ingredients: 1½ cup sapodilla - peeled, seeded, and mashed, 1½ cup milk, 4 eggs whisked slightly, 1 ripe banana sliced, 4 TBS brown sugar, slight salt

Method: Bring milk to a boil in a medium pot and turn off before adding all ingredients. Pour into a well-greased (with butter) baking dish. Bake at 350 degrees for half an hour. Top with fresh banana slices before serving.

SAPODILLA SOUFFLE`

Ingredients: 1 cup sapodilla - peeled, seeded, and mashed, ½ cup heavy cream, ¼ cup milk, 1 TS cinnamon powder, 1 TS lemon juice, 2 TBS brown sugar, 1 TS salt, 4 whipped egg whites, ¼ cup melted butter

Method: In a skillet combine sapodilla, milk, and cream. Simmer for 10 minutes constantly stirring. Add lemon juice, sugar, cinnamon, and salt. Remove from heat and cool for 2 hours. Then when it has reached room temperature, fold in the egg whites. Carefully fill small ovenproof bowls and bake at 300 degrees for half an hour. Brush tops with melted butter. Serve without delay.

BRIAN'S BAR SAPODILLA COLADA

Ingredients: ½ cup sapodilla peeled and seeded, ½ cup milk, 1 TBS honey or brown sugar, 2 TBS brandy or dark rum (optional), 4 cups ice

Method: Put everything in a blender or processor until smooth. Sit back, relax, and enjoy.

DID YOU KNOW? Sapodilla is a distant relative of the canisteland starfruit. Many believe the flavor bears a strik-ing resemblance to caramel. Others think it tastes like a combination of cinnamon, apple, and pear. An excellent backyard tree, the sapodilla usually bears fruit twice a year, but may have blossoms year-round. There are several varieties of sapodillas including the brown sugar with grainy flesh, prolif-ic with pinkish flesh, the early ripening Tikal, and the large Russel, which can grow to five inches in diameter.

NAMES: The sapodilla's botanical name is *Manilkara zapota*, but is known as Chico in *Mexico* and the *Philippines*, Chi-kuu in *India*, and Chicozapote in *Venezuela.*

To plant a garden is to believe in tomorrow. ~ Audrey Hepburn

SOURSOP

Annona Muricata

Occasionally I'll see a soursop in the market, but it is not always available. It is another unglamorous, yet very tasty fruit that originated in the Western Hemisphere. It is native to northern South America. Spanish explorers carried the soursop across the world. It is 'guanábana' in Spanish-speaking countries. Soursop comes from the Dutch and means 'sour sack.' Spanish explorers or settlers probably introduced this fruit to Hawaii. This tree is not yet planted commercially in Hawaii, and the demand far exceeds the supply.

Soursop has a weird irregular shape with a greenish skin covered with short stubs that look like pickers. Its skin almost makes you afraid to touch it until you taste the delicious flesh. A ripe soursop feels soft to the touch. The foot-long prickly green fruit can weigh up to five pounds. The thick, inedible skin hides a white pulp that is a bit fibrous, grainy, with an exceptional taste, like a combination of pineapple and strawberries, or coconut and banana. Soursops may have a few seeds, or over a hundred. Soursop is usually juiced rather than eaten directly. Eating it raw is a bit difficult because of the many large seeds, and the sections of soft pulp are held by fibers.

HOW TO GROW: The soursop tree is perfect for a backyard garden. There are many classifications of these fruit from sweet to acidic taste, small to big, round, oblong, and angular shapes, juicy to dry. So, choose your seeds from a tasty variety. The soursop is usually grown from seeds. They should be sown in containers and kept moist and shaded. Germination takes from 15 to 30 days. Soursop will grow almost anywhere in the tropics. It grows best in rich, deep, well-drained soil with a pH of 5-6.5. Most are bushy evergreens with low branches and only mature to twenty feet tall.

Mealybugs, moths, and wasps are the main pests and can be prevented with a regular spraying of an insecticide like Fastac or Pestac. Use a cup of starter fertilizer like 12-24-12 every other month until it blossoms then switch to 15-15-20. Water regularly, especially in dry spells. Don't expect a big crop as most soursop trees only produce less than two dozen fruits.

Soursops should be picked when firm, just slightly soft, and starting to yellow. If permit-ted to tree ripen either the bats or birds will get more than you, or they can fall and smash on the ground. A bruised soursop will blacken like a banana and should be refrigerated. Soursop with all its valuable nutrients and Ayurvedic-health benefits is not grown commercially and remains an underutilized fruit. Have your own tree!

MEDICINAL: Enjoying a soursop is not only about excellent tastes and aromas, but consuming this fruit can better your mood. Soursop juice will fight fevers, and supposedly increases mother's milk after childbirth. Crushed seeds can be pounded, and the result used as a body wash to against ticks and lice. The leaves are considered to be a sedative, helping to reduce hypoglycemia and hypertension.

The root bark is used as a supposed antidote for poisoning. The juice of the fruit can be taken orally as a remedy for liver ailments. A decoction of the young shoots or leaves is regarded as a remedy for gall bladder trouble, as well as coughs, catarrh, diarrhea, dysentery, fever and indigestion. To speed the healing of wounds, the flesh of the soursop is applied as a poultice unchanged for 3 days. The seeds have emetic properties and treat vomiting.

NUTRITION: 100 grams of soursop has 60 calories with good amounts of calcium, phosphorus, and amino acids. The fruit contains significant amounts of vitamins C, B1, and B2.

FOOD & USES: Soursop is usually pressed through a colander or strainer to extract the juice from the pulp. The juice can be blended with milk or water. Do not consume the seeds.

The seeds contain 45% of a yellow oil, poisonous, which severely irritates the eyes. The wood is pale, aromatic, soft, light in weight and not durable. The bark, as well as seeds and roots, are used as fish poison.

RECIPES: BASIC SOURSOP JUICE

Remove the seeds from a soursop, strain the pulp, and blend the juice with sweetened condensed milk. Chill and enjoy.

KEOKEA SOURSOP FREEZE

Take the soursop juice combined with sweetened evaporated milk and pour into a suitable container or ice cube tray. Stir a few times while it is freezing to break up the ice crystals.

SOURSOP JUICE EXTRORDINAIRE

Ingredients: 2 cups soursop pulp and juice, 2 TBS lime juice, 1 TS vanilla extract, 1 large can of sweetened condensed milk (or 2 small)

Method: Combine soursop with the other ingredients. It is best to use a blender. Pour into a suitable container and freeze till slushy and blend again. Refreeze.

5-STAR SOURSOP CHEESECAKE

Ingredients: 2 cups vanilla wafers crumbled, 4 TBS butter melted, 1 eight-ounce package of cream cheese soft, 1 small can of sweetened condensed milk, ¼ cup fresh lemon juice, 1½ cup soursop pulp blended or whipped, 3 TS plain gelatin dissolved in ¼ cup hot water, 1 TBS fresh mint chopped

Method: In a pie pan combine melted butter with crumbled vanilla wafers. In a suitable bowl combine the remaining ingredients whipping until smooth. Pour into crumb lined pie pan and chill for 4 hours before serving. Sprinkle with mint and enjoy.

DID YOU KNOW? The soursop, also known as the prickly custard apple, was one of the first fruit trees carried from the Americas to the tropical Far East. It is popular from southeastern China to Australia, throughout lowland Africa, and Malaysia. The tastes of soursops are divided into three classifications, sweet, slightly acidic, and acidic. Then they are classed by shapes as round, heart-shaped, oblong, and angular. Finally, they are classed by flesh texture from soft and juicy, to firm and comparatively dry.

HEALTH NOTE: Soursop, botanically *Annona Muricata*, gained attention in the 1970s as a natural cancer cell killer. In the few laboratory studies that have been performed, extracts from soursop can kill some types of liver and breast cancer cells usually resistant to chemotherapy drugs. One study conducted by the Catholic University of South Korea found that soursop was 10,000 times more effective at killing colon cancer cells than chemotherapy. Its anti-tumor effect holds the most interest. But there haven't been any large-scale studies in humans, so there is no conclusive proof it can work as a cancer treatment. However, some tests did discover that in some people, soursop can cause nerve damage similar to Parkinson's Disease. Soursop also fights bacteria and fungus infections, is effective against internal parasites, and lowers high blood pressure. The flesh and juice fight depression, stress, and nervous disorders. It is reported crushed soursop leaves mixed with water and the juice of two limes will sober a drunk when rubbed on his head. A tea made from this tree's leaves will have a calming effect. A poultice of mashed leaves will fight skin problems as eczema and soothe rheumatism. Chewed leaves combined with saliva put on bad cuts will prevent scarring. A compress soaked with a decoction of leaves will reduce inflammation and swollen feet. This same decoction will kill head lice and bedbugs

Life without love is like a tree without blossoms or fruit. ~ Khalil Gibran

STAR APPLE - CAIMATE

Chrysophyllum caimito

Throughout the state of Hawaii, the star apple, also known as caimate/caimito, is another unique fruit grown on sparsely on every island. Trees are available through nurseries online. The green or purple fruit has a soft, sweet, milky pulp with flat seeds. It is called star apple because when cut, the seed groups radiate from the core like a many-pointed star. The skin shouldn't be eaten; enjoy the pulp only after cutting this fruit either in half or quartered. Don't let any of the harsh tasting sap of the skin contact the milky flesh.

Spanish explorers in Peru first recorded star apple in the mid-1500s. It is common on most of the Caribbean Islands, Polynesia, Asia, and throughout the world's tropical countries. Botanists believe the star apple originated in the West Indies. Trinidad, Puerto Rico, and Cuba have the most star apple trees.

HOW TO GROW: It is a nice tree for the garden as its leaves are two-toned with a shiny blue-green upper side and a coppery underside. It can be topped and trimmed to stay at about twenty feet tall. Star apples will grow in almost any soil that is well-drained. This fruit prefers a pH of 5.5-6. These trees are easy to start from seeds or cuttings from mature trees, but they may take five years to bear. Grafted varieties have been known to bear fruit the first year after being planted. Young trees need regular watering and monthly need a half cup of starter 12–24–12 fertilizer. After the trees blossom, use a foliar spray combined with an insecticide-miticide once a month, and feed them a cup of bearing fertilizer 12-12-17-2. Leave star apples on the tree until fully ripe. It is best to pick star apples leaving the stem in the fruit, so they keep fresh longer.

MEDICINAL: In addition to having tasty fruits, the star apple has varied uses in bush medicine. A regimen of drinking tea made from the golden leaves steeped in boiling water is believed to cure diabetes, rheumatoid arthritis, and prevents cancer. A tea made from the tree's bark is used to alleviate coughs. It is also considered a total physical tonic. All bush remedies must be properly prepared and used in the proper dosage only after consulting a doctor.

NUTRITION: Per 100 grams, Star apples have 62 calories, more than 2 grams of protein, with vitamins C, B1, B2, and B3. This fruit is a good source of calcium, iron, and phosphorus.

FOOD & USES: Star apple is seldom used as lumber although it is an attractive and durable wood. The bark is rich in tannin used to tan leather and is thought to fight cancer. A drink made from the boiled bark is used as a stimulant tonic. Cooked star apple is used to reduce a fever. The leaves are grated and applied to cuts to reduce infection. The leaves may be boiled and the resulting liquid drank to combat hypoglycemia.

Star apples are usually eaten fresh, best when chilled, and may also be used as an ingredient of ice cream and sherbet. This fruit mixes extremely well with coconut. A hundred grams of star apple has sixty calories with calcium and phosphorous. In Jamaica, the flesh of this fruit is mixed with sour orange or lemon juice and the mixture is called 'matrimony.' The flesh is also combined with regular orange juice, a pinch of sugar, nutmeg, and eaten as a dessert called 'strawberries and cream.' However, too much star apple can cause constipation.

RECIPES: HALAWA VALLEY FROZEN FRUIT MIX

Ingredients: the seeded pulp of 6 star apples, 1 cup chopped pineapple, 2 mangos skinned seeded, and chopped, 1/2 cup shaved coconut, 1/2 cup currants or raisins, 1/2 cup shaved almonds, 2 cups coconut water.

Method: Mix, and freeze. After 4 hours stir or blend again to remove ice crystals. Freeze..

STICKY RICE PUDDING

Ingredients: seeded pulp of 6 star apples, a pinch of cinnamon and nutmeg, 1 cup milk, 2 cups cooked rice

Method: Mix all ingredients in a pot. Bring to a boil carefully stirring to avoid sticking and burning. Cool and refrigerate.

POIPU BEACH STAR APPLE SHAKES

Ingredients: seeded pulp of 2 star apples, 1 TBS brown sugar, 1/2 TS vanilla, ice

Method: Combine ingredients in a blender. The amount of ice and the duration of blending will determine the thickness of the shake.

STAR APPLE CHICKEN

Ingredients: 6 nice star apples the riper the better, 1 chicken—cut up and steamed -fully cooked, 4 cups cooked rice, 2 TBS cooking oil, 1/2 cup sweet pepper chopped small, 2 bunches of chives chopped small, 2 cloves of garlic minced, 1/2 cup condensed milk, salt and spices to taste

Method: Heat in a large frying pan. Simmer sweet pepper, chives, garlic, and star apples until tender, maybe ten minutes. Add condensed milk and seasonings. Continue to simmer and stir so it doesn't stick for another ten minutes. Put chicken and rice in a suitable oven-proof dish and cover with caimate –milk mixture. Bake at 350 for half an hour.

DID YOU KNOW?

The star apple is also known as cainito, caimito, caimate, golden leaf tree, abiaba, pomme du lait, estrella, milk fruit, and milky apple. Star apple's botanical name is *chrysophyllum cainito*. The skin contains an unpleasant sticky latex. Do not bite directly into one because the white sappy latex will make your lips very sticky.

Odd as I am sure it will appear to some, I can think of no better form of personal involvement in the cure of the environment than that of gardening. A person who is growing a garden, if he is growing it organically, is improving a piece of the world. He is producing something to eat, which makes him somewhat independent of the grocery business, but he is also enlarging, for himself, the meaning of food and the pleasure of eating. ~ Wendell Berry

STAR FRUIT - CARAMBOLA
~ FIVE FINGERS ~ *Averrhoa carambola*

I love juicy star fruit. It has many different regional names. The star fruit is believed to have originated in Sri Lanka (Ceylon and was cultivated for centuries before Spanish explorers brought trees to the Caribbean and the Americas. The islands have more than thirty commercial orchards growing this fruit. Star fruit season in Hawaii is from September to April,

The star fruit title comes from the five-pointed star that appears when you slice it across the width. Some areas also call it five fingers because it also looks like a hand, with the fingers extended. *Averrhoa carambola* is the botanical name. The Portuguese word *carambola*, first known use 1598, was taken from Marathi karambal derived from Sanskrit karmaphala. Its first name is from a famous 12th century Arabian physician, Ibn Rushd, who was called Averroes.

There are two varieties of starfruit, sour and sweet. The fruits with narrow spaced fingers or ribs are less sweet than the fruit with thick fingers. This fruit starts green, and yellows as it ripens, though it can be eaten in both stages.

The star fruit is all about its shape. They can grow to about a foot long always with five triangular sectioned ribs. Slices cut in cross-section become star-shaped. It has a thin skin from pale to dark yellow with a waxy smooth texture. The flesh is a paler yellow to almost clear. A great star fruit is crisp and very juicy, without any fiber. It has a sharp acidic aroma from the oxalic acid. The taste ranges from lip-clenching tart to almost sweet. Even the sweet variety seldom has more than 4% sugar. Usually, a star fruit has ten to twelve flat, half-inch long brown seeds. The color of star fruit slightly intensifies as they ripen. This is a good container tree for the balcony or patio.

HOW TO GROW: To grow a tree from seeds is difficult since the seeds become infertile in a few days after removing them from the fruit; so, have your potting soil ready. Nice fat seeds should sprout in a week to ten days. Grafted trees can be purchased at garden shops. Dwarf trees are available for your patio or balcony. A star fruit tree is perfect for a garden as they seldom grow more than 15-20 feet tall. Keep them at least twenty feet apart and away from walls. Young trees should be sheltered from a full wind. Only light pruning is ever required. Make certain the area you plant is drained, as water is this tree's biggest enemy. The best soil pH is 4.5-7. However, star fruit must be watered regularly during the dry season to produce a juicy crop. Run a hose for half an hour every week during a dry spell. Spray with a pesticide-miticide and foliar fertilizer once a month. Sprinkle about a half-cup of 12-12-17-2 every month around the base. Star fruit should flower a few times a year, with the best crop during the heated dry season.

MEDICINAL: In bush medicine, the star fruit tree is a virtual pharmacy. In India, ripe star fruit is used to halt hemorrhages and to relieve a bleeding hemorrhoid. Dried fruit or the juice may be taken to fight a fever. A heated or cooked version of this fruit will fight intestinal gas, diarrhea, and is believed to relieve an alcohol hangover. A salve made of star fruit will aid eye afflictions. Star fruit is used by Brazilians to fight kidney and bladder problems. It is also used to treat eczema. A decoction made from the fruit will overcome severe nausea and vomiting. Star fruit tree leaves can be plastered on the temples to soothe headaches. A poultice of crushed leaves will eradicate ringworm. This tree's roots, combined with sugar are believed to be an antidote for poison. Hydrocyanic acid has been detected in the leaves, stems, and roots. Powdered seeds serve as a sedative in cases of asthma and colic.

NOTE: Individuals with kidney problems shouldn't eat star fruit because of oxalic acid. Juice made from star fruit is very dangerous because this acid is concentrated.

Star fruit intoxication can cause hiccups, vomiting, numbness, insomnia, weakness, confusion, and muscle convulsions. Symptoms 30 minutes to 14 hours after eating.

NUTRITION: One nice star fruit has about twenty calories and is high in carbohydrates, calcium, phosphorus, vitamins A, B, and C, plus amino acids.

Star fruit is a good source of dietary fiber and can help to reduce blood pressure and cholesterol. Ripe star fruit are great eaten just picked and washed, and of course the green or slightly ripe make excellent chow. They can be stewed with cloves and raisins. Star fruit cooked or raw are a great accompaniment for seafood. Star fruit juice will clean brass and silver. It will also remove rust stains from white clothes.

RECIPES: CHINATOWN STAR FRUIT CHOW

Ingredients: 4-6 nice sized starfruits sliced about ¼ thick seeds removed, 2 culantro leaves minced, 1-2 cloves of garlic minced, 1 hot pepper seeded and minced, and a pinch of salt

Method: Combine all ingredients in a bowl. Let sit for at least 2 hours before serving. Best chilled.

STAR FRUIT BREAD

Ingredients: 4 medium starfruits washed and minced, ½ cup sugar, ¾ cup milk, 1 egg, 1 cup hole wheat flour, 1 cup white baker's flour, ¾ TS salt, 1 TS baking powder, 1 TS baking soda, 1 TS powder ginger, ½ TS cinnamon, ¼ cup currants or raisins (Increase the sugar to your taste.)

Method: Combine the fruit, sugar, egg, and milk in a large bowl. In another bowl mix the remaining ingredients and then slowly add to the fruit combination and mix until all the flour is moist. Don't over blend. Pour into greased bread pans and bake at 350 for forty-five minutes.

FRIENDLY BEACH STAR FRUIT CHICKEN

Ingredients: 1 chicken chunked, 4 star fruit cut to ¼ star slices, 1 large onion sliced thin, ¼ cup olive oil, 2 TBS honey, ¼ cup fresh lime juice, 2 TBS lime zest (grated lime peel), one bunch culantro / cilantro chopped fine, 1 TBS fresh ginger minced, ½ cup raw almonds (cashews or peanuts may be substituted), ½ hot pepper seeded minced, salt, and spices to your taste

Method: Combine chicken, honey, lime juice and zest, onions, ginger, and pepper in a large bowl preferably with a tight cover. Refrigerate for a day or two, stirring occasionally. Put chicken mixture in a baking dish and cover with the nuts. Cover with foil and bake at 375 for about half an hour. Uncover and bake for another 20 minutes. Add culantro / cilantro before serving.

STAR FRUIT SAUCE – great on fish or chicken and on cake, or ice cream.

Ingredients: 4 ripe starfruits, 2 cups either orange, passion fruit, or guava juice, 2 cups sugar, 2 TBS fresh ginger grated, 2 bay leaves best boiled in a cheesecloth bag, 1 stick of cinnamon or 1 TBS powdered, ½ TS nutmeg, a ½ cup of apple, or cranberry sauce or a ¼ cup of orange marmalade may be added for a taste twist, 1 TS cornstarch to thicken.

Method: Remove seeds before dicing 3 starfruits. In a 2-quart pot combine the sugar, juice, and corn starch, bring to a boil stirring constantly. Reduce heat and simmer for 5 minutes with occasional stirring. Add diced starfruit, other fruit sauces if desired, bay leaves, cinnamon, and nutmeg. Simmer for 10 minutes and cool before serving.

Eating well is a form of self-respect.~ Anonymous

STRAWBERRY – QUEEN OF BERRIES

Fragaria × ananassa

The strawberry is a widely grown hybrid cultivated worldwide and even in Hawaii. People love this fruit because of its great red color, unique smell, and juicy, sweet taste. Roman literature mentions strawberries for medicinal purposes. The French began cultivating this sweet red fruit in the 14th century. The modern strawberry is a crossbreed of a North and a South American variety created in France in 1750. The legend behind the name strawberry is because English children picked fresh berries and tied them onto stiff grass straws. They sold them as 'straws of berries.' 10 million tons of strawberries are produced every year; China grows 40%. Strawberries are not classified as berries. Blueberries and raspberries have seeds inside while strawberries have their seeds outside. The botanical name is *Fragaria × ananassa.* Strawberries belong to the family of rose, along with apples and plums.

HOW TO GROW: Temperature and sun have a lot to do with growing strawberries in the tropics, but it can be a fun and delicious hobby plant for your garden or in porch pots. Choose varieties that have been adapted for warm climates. Containers are the best way to initially test your garden skill with the fragile strawberry. Big pots, 18 inches deep and wide, will allow the roots to stay cool and you'll be able to adjust the plants' position for optimum sun and temperature. Strawberries grow best in well-draining soil with a pH ranging from 5.5 to 6.5, in partial sun. In hot tropical climates, strawberries require regular watering and are planted in a soil/compost mixture that will retain water. Covering pots or planting areas with white polyethylene helps stop evaporation and the white color reflects the sun's heat. Black plastic will absorb heat.

Keep the soil mix moist. A trick is to soak the plants and then reduce watering for 2 days then soak again. If the plants' leaves turn pale green or yellow, they are getting too much H2O. Use rotted manure in your soil mix but monthly hit the plants with a just a pinch of high potassium fertilizer. Beware of snails which also love this ripe red fruit. When you are successful, the ripe berries must be picked daily or every second day with the green caps attached and half an inch of stem. Strawberries need to remain on the plant to the fully ripen. They do not ripen after being picked. Don't wash the berries before you're ready to eat them.

There are several commercial strawberry farms on the Big Island and Maui. Hawaiians have been growing strawberries vertically since the 1920s. It takes adapted strains, especially created soil mix, and shields from heavy sun and rains to produce good berries on a large scale and be cost effective. Knowing that, support the local strawberry farmers by buying and enjoying their harvest. Try to grow your own sweet red berries. The recommended warm weather varieties are: Pocahontas, Tioga, Sequoia, Headliner, and Tioga. *This is from a UH at Manoa bulletin.*

MEDICINAL: The entire strawberry plant was used to treat depressive illnesses. The leaves and roots were brewed into a tea. Strawberries help to regulate and stabilize blood sugar. In 2011, doctors discovered eating 35 strawberries a day reduced the complications of diabetes, such as kidney disease and neuropathy. Strawberries are an excellent source of folic acid, an essential for pregnant women to protect against severe birth defects. Ancient Romans used strawberries to treat fainting spells, fevers, throat infections, gout, and bad breath.

NUTRITION: Per 100grams of strawberries, there are only 33 calories. They are high in vitamin C with trace amounts vitamins E, Band 1, B2, B3, B5, B6, B9 (folic acid), and K. The mineral content is high in manganese and potassium, with trace amounts of calcium, iron, magnesium, phosphorus, and potassium.

FOOD & USES: The same as every fruit and vegetable, fresh strawberries, that you've grown or been recently harvested provide the best nutrition. They should be bright red, feel firm, have a light pleasant, sweet aroma, and be wearing the green caps. Small strawberries are sweeter, with a more intense flavor. Eat them as soon as possible, within two days or they will lose flavor and nutrition. The natural red color, anthocyanin, is sensitive to heat and will brown your berries if not kept at room temperature or refrigerated. The strawberries should not be washed until you are ready to eat them, and it's best to keep them in a covered container in the fridge.

In addition to being consumed fresh, strawberries can be frozen, made into preserves and jellies, as well as dried and used in a multitude of prepared foods such as cereal and energy bars. Strawberries and strawberry flavorings are a popular addition to dairy products as yogurts and ice creams.

RECIPES: EASY STRAWBERRY DUMPLINGS

Ingredients: 4 cups of strawberries sliced ¼-in-thick, 1cup all-purpose flour, ½ cup sugar, 2TS baking powder, 1/4 TS salt, 2TBS unsalted butter - cold and cut into chunks, ¾ cup whole milk

Method: In a big frying pan or wok with a good sealing lid/cover on low heat, combine the strawberries and sugar stirring until it becomes syrupy. Remove from heat and let sit for 15 minutes. In a bowl combine the flour, baking powder, and salt. With your fingers knead the cold butter into the flour blend until it is a coarse mixture. You do not want the butter to be smooth, but in small pieces. Stir the milk into the flour-butter mixture to make a thick batter. Bring the strawberries again to a boil over medium heat. Slowly stir and add the flour batter to the boiling berries. Stir once or twice and then simmer on low and tightly cover the frying pan or wok. Let cook for 15 minutes. The dumplings should expand but appear dry on top. Let dumplings sit for 5 minutes before serving warm with ice cream.

SIMPLE STRAWBERRY JAM:

Ingredients: 1kg (2.2lbs) of cleaned strawberries, ½ cup sugar, 2TS fresh lemon juice

Method: Use an electric blender and chop the strawberries until they become a coarse mixture. Pour into a large skillet or wok; stir in sugar and lemon juice. Cook over medium heat, stirring constantly, until the thick mixture bubbles in 10 minutes. Transfer the strawberry jam to steril-ized jars and let cool. Note: this can be thinned and used as strawberry sauce.

STRAWBERRY- AVACADO SALAD

Ingredients: Dressing: 2TBS granulated white sugar, 2TBS olive or coconut oil, 1TBS honey, 1TBS vinegar, 1 TBS lemon juice.

Salad: 2 cups green salad leaves, 10 strawberries chopped, 1 avocado, seed removed, peeled, and sliced, ½ cup cashews or almonds chopped

Method: Dressing: combine all ingredients in a bowl and stir. Salad: Add strawberries and av-ocado slices to the salad greens in a bowl. Pour dressing and cover with chopped nuts.

DID YOU KNOW?Botanically, strawberries aren't berries**.** They are derived from a single flower with more than one ovary, making them an aggregate fruit, yet bananas are considered a berry, with one, giant seed. There are more than 600 varieties of strawberries that differ in flavor, size, and texture. The strawberry was a symbol for Venus, the Goddess of Love because it's often heart-shaped and has a rich, red color. In the 13th century France, strawberries were believed to be aphrodisiacs and were served in soups to newlyweds.

Remember when your body is hungry, it wants nutrients not calories.

~ Nourishing Children Org.

SUGAR APPLE – CHIRIMOYA
Annona squamosa

Everywhere throughout the islands, adults and children love sugar apples also known as chirimoyas. Sugar apples are what I call a 'busy fruit.' They keep you occupied sorting and spitting out the seeds. This is another crazy fruit that is so sweet, but the seeds are poisonous. **DO NOT SWALLOW THE SEEDS.** I love this fruit, the creamier the better. You never forget the location of a good sugar apple tree. This popular fruit originated in Central or South America and is now grown in the tropics around the world. Sugar apple seeds keep extremely well, up to four years.

The Spanish Conquistadors were able to carry viable seeds and transplant sugar apples to the Philippines, and it quickly spread throughout Asia. The Portuguese supposedly brought the sugar apple to India in the late 1500s. Sugar apples are widely cultivated in southern China, Queensland-Australia, Polynesia, Hawaii, tropical Africa, Egypt, and even the lowlands of Palestine. India probably grows the most sugar apples. It is also one of the most important fruits of the interior of Brazil. Cuba developed a seedless variety in the 1940s.

HOW TO GROW: Sugar apple trees are nice for the small backyard garden as they seldom get taller than 6 meters (20 ft.) and are extremely easy to grow from seeds. The seeds germinate better after being dried for at least a week after removal from the fruit than when perfectly fresh. They should sprout within 2 months. When the sprouts have 3 leaves, it's time to plant in well-drained soil with a pH of 6. Pick a breezy spot that gets 8 hours of sun. The main enemy is too much water and associated fungus. It should be planted at least 10 feet away from other trees or walls.

It usually take 4 years to bear, but grafted trees should bear sooner. The blossoms are very fragrant. For spectacular fruit, spray monthly with a pesticide that also works for mites, and mixed with a foliar booster. Young trees to two years old should get a half-cup of 12-24-12. Older trees should get a half-cup of 12-12-17-2. When the blossoms appear, water the tree every other week. Fruits ripen in about 4 months. A luscious, perfectly ripe sugar apple is worth the effort. Chilled is better.

MEDICINAL: A tea made of sugar apple leaves is considered an excellent tonic, especially for colds and diarrhea. A bath in the leaves relieves severe arthritis and rheumatism.

BEWARE: Sugar apple seeds are poisonous. Powdered seeds are used as fish poison and as insecticide in India. Powdered seeds made into a paste will kill head lice, but must be kept away from the eyes because it is highly irritant and can cause blindness.

NUTRITION: 100 grams of sugar apple has 95 calories with 2g of protein, but less than a half gram of fat. They contain plenty of vitamins C and B-6 with good amounts of potassium, iron, and magnesium.

FOOD & USES: A big sugar apple is a cone about 4 inches long with a thick, knobby, segmented grayish green skin. Once you pull it apart the creamy white flesh smells as sweet as it tastes, like a prepared sherbet. The flesh can be pressed through a sieve to eliminate the seeds, and then added to ice cream, or blended with milk to make a tasty 'colada' drink. Ripe sugar apples can be frozen and eaten like ice cream. They are best served chilled, cut in half, or quartered; and eaten with a spoon.

RECIPES: SUGAR APPLE CHICKEN

Ingredients: 1 cup seeded and pureed sugar apple, 1 cup plain yogurt, 1 chicken quartered or chunked, 3 medium onions (prefer white) sliced thin, 3 garlic cloves minced, ½ cup of chickenbroth, oil for frying

Method: combine puree, yogurt and let sit. Brown the chicken pieces in a large frying pan with 2 tablespoons oil and remove. In the same oil sauté the onions and garlic, then add the chicken, the broth, and cover. Cook at medium heat until the chicken is tender. Drain the liquid from the pan with ½ cup remaining. Lower heat and stir in the sugar apple puree and yogurt. Stir briskly for 3 minutes and remove from heat. Serve warm.

SUGAR APPLE CUSTARD

Ingredients: 2 sugar apples seeded and pureed, 1 pound of softened cream cheese, 2 TBS sugar, ¾ cup cream, 1/3 cup boiling water, 3 TS gelatin (plain preferred, but flavored will slightly alter the taste)

Method: Dissolve the gelatin in 1/3 cup boiling water. Whip the cream cheese adding the gel-atin and water, sugar and cream. Add the sugar apple puree and beat until smooth. Refrigerate for 4 hours before serving.

SUGAR APPLE CAKE

Ingredients: 2 ripe sugar apples seeded and pureed, ½ cup sugar, 1 TS vanilla, 2 eggs, ¼ cup of soft butter or margarine, 1½ cups of bakers flour, 2 TS baking powder, 2 eggs, ½ TS ground cinnamon, ½ TS nutmeg

Method: Whip the butter, sugar, and vanilla until creamy. Add eggs one at a time to mixture, then sugar apple puree, and spices. Continue to whip while adding the flour. Pour into a greased bread pan and bake at 350 for an hour. Check consistency with a toothpick. Cool and serve.

DECADENT SUGAR APPLE PIE

Ingredients: 2 cups sugar apples seeded and pureed, 1 pack of cookies of your choice smashed to crumbs, ½ cup strawberry jam or chocolate syrup, 1 large banana sliced thin, ½ pint whipping cream, 1TS lemon or lime juice

Method: Layer the bottom of a pie dish, or medium sized bowl with the cookie crumbs. This is kind of sticky so you might want to keep your fingers moist. Pour into this crust just enough of the sugar apple puree and then carefully stir in swirls of the jam or syrup. Mix banana slices with the lemon /lime juice and top the pie with the slices. Repeat this until the pan is filled. Refrigerate for at least an hour. Top with whipped cream when serving.

DID YOU KNOW? This fruit's scientific name is *Annona squamosa*, but it is called chirimoya throughout South America. The name originates from the Andes mountain Indian language word 'chirimuya', which means "cold seeds', because the seeds will germinate and the plant grows at higher al-titudes. It is named pomme cannelle in Guadeloupe, French Guiana, and French West Africa. It is rinon in Venezuela, and sweetsop in Jamaica and the Bahamas. It is known as sitaphal, custard apple, or scaly custard apple in India.

Maybe we should stop asking why real food is so expensive and start asking why processed food is so cheap.

~ Anonymous

SURINAM CHERRY
Eugenia uniflora L.

The Surinam cherry is native to the northeastern coast and central areas of South America. This tree is adaptable to most soils and elevations. Spanish and Portuguese explorers spread this tasty fruit around the tropics. Trimmed, this is a great ornamental plant for your home garden. Birds and other wildlife love the fruit and the seeds easily germinate, so it's considered an invasive species forming dense thickets stifling native plants. It's often used as a living hedge/fence.

It usually produces two crops and may blossom and fruit all year in tropical climates. The ½ to 2 inch wide, very juicy, thin-skinned cherry turns bright red or almost black when ripe. When partly ripe they are juicy and tart, but sweet when fully ripe.

HOW TO GROW: The Surinam cherry is considered a shrub or small evergreen tree up to 20 feet. They are usually started from fresh seeds. When the seedlings are about a foot tall, transplant to an area of full sun. They take 2-6 years to fruit. Grafted trees usually take 2 years. Like most trees, this cherry grows best in moist, well-draining soil with a pH 5.6-7.5. This tree is adaptable to soil types. It can withstand some drought and is somewhat salt-tolerant. Abundant water increases fruit size and sweetness. Because this is a slow grower, don't prune until 6-7 years old, then shape to improve harvesting or to grow as a screen or hedge. A well-maintained tree can produce 6 pounds of fruit.

MEDICINAL: A leaf infusion treats fevers and increases appetite. A leaf decoction is a cold remedy. The leaves contain citronella and are used in South American homes spread over the floors to repel flies.

NUTRITION: 100 grams of Surinam cherries has 50 calories with vitamins A and C, calcium, and phosphorus.

FOOD & USES: If seeded and sprinkled with sugar before placing in the refrigerator, they will become mild and sweet served with cake or ice cream. They are an excellent addition to fruit cups, salads. and custards. They are made into jam, jelly, relish, or pickles. Brazilians ferment the juice into vinegar or wine. Surinam cherries last one day at room temperature and up to one week refrigerated.

RECIPES: SURINAM CHERRY SAMBAL – hot relish

Ingredients: 2 cups ripe Surinam cherries seeded, 1 TS sugar, ¼ cup coconut oil, 3 shallots diced, 1 TBS lemon or lime zest, 6 hot chilies, seeds removed, 2 garlic cloves chopped, 1 TBS fish sauce.

Method: Sprinkle sugar over the cherries and place them in the fridge overnight. In a wok or frying pan, (non-stick better), heat 2 TBS of the oil on low heat. Add shallots, zest, and chilies. Sauté, stirring for 5 minutes. Add the garlic and cook for another minute. Transfer the sautéed vegetables with the cherries to a mortar and pestle or a food processor. Form a thick paste. Heat the remaining oil in the wok/skillet over medium. Sauté the paste while stirring until it turns darker red, about 5 minutes. Add the fish sauce and keep stirring for another few minutes. Transfer to a jar and use as a condiment.

NAMES: Brazil cherry, Cayenne cherry, pitanga, and Florida cherry. The Brazil cherry mentioned earlier in this book, *Eugenia brasiliensis Lam.,* ripens to a deep purple with 2 seeds. It's a relative of the Surinam cherry.

TAMARIND

Tamarindus indica

Tamarind trees are planted throughout Hawaii and used in many types of tasty curries and chutneys. This tree is great for the home as it gives plenty of shade but also grows wild. The tree has long been naturalized in the Pacific Islands. It's reported that the first tamarind trees were brought to Hawaii in 1797 by Don Francisco de Paula Marin and planted in Pauoa Valley.

This fruit is strange. Once you taste tamarind, your lips pucker. You love the taste, yet can't decide why, or even describe the lavor. The six-inch pods can have as much as twelve seeds in a sticky brown paste. The shells are brittle and break easily when the pods are fully ripe and can remain ripe on the tree for months.

The tamarind is the only spice to originate in Africa, although India also tries to claim the delicious fruit harvesting 300,000 tons every year. Ancient Egyptians and Greeks enjoyed tamarind. Its botanical name is *Tamarindus indica*. 'Tamarind' translates as 'Indian date'.

HOW TO GROW: Plant a tamarind tree for your children's children. Tamarind is a slow-growing tree and can mature to the grandeur of seventy-foot height and thirty-foot width; so, give it a lot of space. It's a hearty tree that can adapt to most conditions, and it can survive for more than 300 years. Soak tamarind seeds overnight in warm water to speed up germination. Sow seeds a half-inch deep. Germination occurs within 2 weeks. Tamarind trees from seeds may take 7 years to bear fruit. It's recommended to buy a plant from nursery or grow from grafted cutting for quicker results. Once the tree is almost a meter tall, dig a hole twice the size of the root ball of the plant. Firm the soil and water the plant thoroughly.

It grows in pH level around 4.5 – 9, but prefers well-drained, deep loamy soil. Tamarind can also withstand some salt spray. Its red or yellow flowers that bloom in spring are not very showy. To get the best results in both the number of pods and the amount of pulp, fertilize young trees to three years old with a cup of 6-6-3, three times a year. Bearing trees should get 8-3-9, at a cup for every five years of age at the same intervals. East Indians shake the branches to make ripe fruits fall. Pickers should not use long sticks to knock the fruits because this could damage both the fruits and tree.

MEDICINAL: A hundred tons of tamarind are imported yearly by US drug companies. Tamarind preparations are known as coolants for fevers, and as laxatives. Tamarind leaves and lowers, dried or boiled, as poultices will relieve swollen joints, sprains, and boils. A thick tea made of the fruit or leaves is used in cases of gingivitis, asthma, and eye inflammations. Fresh pulp can be applied directly on inflammations and used as a rinse for a sore throat. Pets infested with leas or ticks can be washed and then rinsed with strong tamarind water. Let it dry on them as a repellent. A tea from the tree's bark makes an excellent tonic.

NUTRITION: Tamarind raw pulp has 280 calories per cup with a little protein and fiber Tamarind is high in calcium, phosphorus, and potassium.

FOOD & USES: Hard tamarind heartwood makes the best hoe handles, also mortars, and pestles. Green pods can be used as a seasoning, boiled with rice. The tamarind pulp is used in a variety of sauces including Worcestershire (English Sauce). Tamarind balls are probably the most common way of eating the pulp. To separate the pulp from the seeds, work the goo on a colander and keep adding powdered sugar. Shape the strained goo into balls and roll in sugar. It's messy to make, yet delicious to savor. Tamarind water is another easy treat. Put as many shelled fruits as you choose (more will make a stronger lavor) in a bottle of water overnight. Cloves, ginger, and even a hot pepper may be added to enhance the lavor. Add sugar to your taste. In Thailand, they grind the dry tamarind seeds for a co ee substitute. These seeds have a property that makes things gel better than pectin and are used as a stabilizer for ice creams.

Tamarind-ade is a popular tropical drink and is now bottled carbonated in Guatemala, Mexico, and Puerto Rico. The easiest method of preparing this drink is to first shell the fruits and put six in a bottle of water. Let stand overnight before adding a tablespoonful of sugar. Shake vigorously. For a stronger drink cover about thirty tamarinds with hot sugar syrup. Add spices of your choice such as cloves, cinnamon, allspice, ginger, pepper, and-or lime slices. Let stand in a dark place for several days before straining. Dilute with water and chill to serve.

RECIPES: TAMARIND PASTE

Shell 1 pound or more of tamarind and remove stems, fibers, etc. Put into a pot and cover with boiling water to soften the pulp. Push softened mixture thru a sieve. Place in a pot with one and a half cups of apple cider vinegar, a half cup of brown sugar, plus spices you may choose. Boil until a thick paste. Push thru a sieve again. Cool before freezing.

BEACH LOVERS TAMARIND BLACK BEANS

Ingredients: 2 cups soaked black beans, ¼ cup tamarind paste, 1/3 cup oil, 2 onions chopped, 1 tomato-chopped small, 2 TBS grated ginger, 2 TBS minced garlic, 1 hot pepper-seeded and minced (optional), 1 TS roasted ground cumin, ½ TSP turmeric, 2 TS roasted cumin seeds, 1 TS garam masala, 2 bundles culantro/cilantro chopped, salt to taste

Method: Soak tamarind paste in 2 cups of hot water for an hour. When it has cooled work the paste between your hands to produce a strong juice. In a large skillet heat oil and sauté the onions until they brown. Add garlic and ginger and cook for five minutes. Stir constantly or it will stick. Add the hot pepper, cumin, and turmeric; cook for half a minute and remove from the heat. Stir in the tomato, the beans, and 2 cups of water, Cook covered until beans are tender. Add water as necessary to provide a good sauce. Stir in the tamarind paste and simmer for 10 minutes. Add the roasted cumin seeds, garam masala, and culantro/cilantro. Cook for five minutes and remove from heat; let sit for ten minutes before serving.

MOMMY'S TAMARIND CHUTNEY

Ingredients: ¼ cup tamarind paste, ¾ cup water, 1/3 cup seedless raisins, 1 TBS sugar, 1 TS roasted ground cumin seeds, 1 TS lemon juice, 1 hot pepper seeded and minced, salt to your taste

Method: Bring water to a boil before adding tamarind paste and raisins. Remove from heat and soak for an hour. Work the tamarind paste to a creamy juice while adding the spices, pepper, and lemon juice. Blend it very smooth. Cover and let stand at room temperature overnight.

DID YOU KNOW?There is a superstition it is harmful to sleep or to tie a horse under a tamarind tree. Some African tribes regard this tree as sacred. Burmese believe the tamarind tree is the house of the rain god. Hindus marry a tamarind tree to a mango tree before enjoying the mango fruit. In central Africa, corn soaked with tamarind bark is fed to domestic fowl; if they stray or are stolen, they will always return home. Malaysians combine tamarind and coconut milk and feed it to infants at birth to hope-fully make them wise. Tamarind pulp has more sugar and fruit acid per volume than any other fruit. Marco Polo, the explorer, claimed the pirates of the Malabar Straits forced the people they captured to swallow a mixture of saltwater and tamarind. That would cause their victims to vomit and reveal if they'd swallowed any gems, or pearls..

TANGELO
Citrus X tangelo J.

Okay, so you thought you knew of every type of citrus grown in the Hawaiian Islands, but you never heard of a tangelo? If that's the case, then you have missed one of the best tasting fruits. I discovered tangelos by accident one day at a roadside market. I frequent one that usu-ally has a variety of fruit like canistels and starfruit. The vendor offered me a taste of a unique fruit, which I thought was a miniature grapefruit, until I tasted it. The tangelo is very sweet, but without the usual acidic bite of typical citrus and has a slight spicy aftertaste.

Tangelos may be accidental hybrids of a variety of mandarin oranges and the grapefruit or pummelo. The first reported deliberate hybrids were crossed by Dr. Walter T. Swingle at Eustis, Florida, in 1897, and Dr. Herbert J. Webber at Riverside, California, in 1898. The fruit these botanists created are so unique that a separate botanical classification was created, *Citrus X tangelo* J.

Or tangelos were created by Mother Nature over the centuries of citrus coexistence. The most common citrus fruits are lemons, limes, grapefruits, oranges, and tangerines. Most of these have varieties that have adapted over time to a variety of climates, altitudes, and regional elements. The tangelo is a cross of two plants, the tangerine, more commonly known as the mandarin orange, and the grapefruit, a relative of the pummelo. Some botanists are convinced tangelos are be originated in Southeast Asia some 3,000 years ago. A tangelo is the size of an orange, yet oblong instead of perfectly round. Some tangelos have a neck on them much like the mandarin orange. A ripe tangelo is very juicy and tastes like a sweeter, orangish pomelo/jabong. The peel is fairly loose, and easily removed like a tangerine. Even though this hybrid already existed naturally, scientists didn't quit manipulating the tangelo until they did nature one better.

The two main types of commercial tangelos are the Minneola tangelo, which was cre-ated in 1931, and the Orlando tangelo, created in 1911. Each of these 'breeds' is the hybrid of one specific type of tangerine and one specific type of grapefruit. Minneola tangelos are a Duncan grapefruit crossed with a danci tangerine.

HOW TO GROW: The tangelo is a perennial evergreen fruit tree that can also be enjoyed as an ornamental plant when grown as a dwarf tree or bonsai in lanai or balcony pots in tropical and subtropical areas. Whether planted permanently in soil or in pots, tangelos want sun.

Personally, I don't recommend attempting to grow any hybrids from seeds because the end results are variable. But if you want to experiment and save some money, get seeds from several mature tangelos and soak in water for at least 8 hours. Then place in moist potting soil. It's recommended to keep the seeds in a warm sunny area and they should sprout within 2-3 weeks. They should bear in 3-5 years.

My tree is a grafted variety purchased from a local garden shop. The tangelo needs plenty of sun and well-drained soil with a preferred pH of 6-6.5. When it is 5-6 ft. tall, every second month, sprinkle a cup of bearing fertilizer, 12-12-17-2, around the base. In the dry season it needs a good drenching every other week. On the full moon, I check every grafted tree and trim unwanted branches from the original root stock. They are easy to tell because those branches should be the only ones that have thorns and they usually grow straight up. My tangelo tree began to bear fruit in its second year. You can expect 50-70 lbs. of fruit from a well-maintained tree.

MEDICINAL: The tincture of the peel is used as a pleasant bitter tonic to treat coughs and choking.

NUTRITION: A tangelo weighing about 100grams has 100 calories with 1g of protein, plenty of vitamin C, folate, and some potassium.

FOODS & USES: The tangelo is excellent for eating fresh or adding to fruit or vegetable salads. Its segments will liven up coleslaw or tuna salad. It's excellent as a dessert fruit.

RECIPES: TANGELO SORBET

Ingredients: 4 cups fresh tangelo juice from about 10 fruits, 1 cup sugar, 1 cup water

Method: In a sauce pan combine water and sugar, bring to a boil and then simmer. Add grated rind (zest) of 1 tangelo. Simmer for 10 minutes stirring frequently. Add 4 cups of the strained tangelo juice. Remove from heat to a bowl. Cool before putting in the freezer. After 2 hours in the freezer, remove and put into blender, or vigorously beat with a slotted spoon. This is to break up the ice crystals. Refreeze. For very smooth sorbet wait another 2 hours and blend again before freezing solid.

TANGELO RICE SPECIAL

Ingredients: 1 cup rice - prefer brown whole grain, 3 cups vegetable stock, ½ cup fresh squeezed tangelo juice, ¼ TS salt, one-half cup shelled pigeon peas, 1 TS grated tangelo rind (zest), ¼ tangelo peeled and chopped small, 1 TS ground coriander, 2 TS chopped raw almonds, peanuts, or cashews.

Method: In a suitable pot combine rice, salt, vegetable stock, and tangelo juice. Boil, cover and simmer, occasionally stirring, for half an hour. Add peas, tangelo rind, and coriander. Stir thoroughly and cover again, let simmer another 20 minutes until the rice is cooked to a nice texture. Remove from heat and let stand, covered, for 5 minutes. Uncover, stir in chopped nuts, and serve immediately.

TANGELO APPLE SALAD

Ingredients: 4 tangelos – 2 peeled and sectioned and 2 juiced, 6 apples peeled, cored and sliced, 1 cup yogurt plain or orange lavored

Method: In a bowl combine all ingredients. Chill and serve.

TANGELO CAKE

Ingredients: ¼ cup butter or margarine, ¾ cup baker's lour, ½ cup milk, zest and juice from 1 tangelo, ½ cup powdered sugar, 2 eggs, 2 TBS brown sugar

Method: Combine butter, milk, lour, and sugar. Beat with a mixer or whisk for ive minutes. Pour batter into a cake pan and bake at 350F for 40 minutes. In a sauce pan on medium heat combine brown sugar, tangelo zest, and juice until a thin syrup forms. Remove from heat and pour over cake.

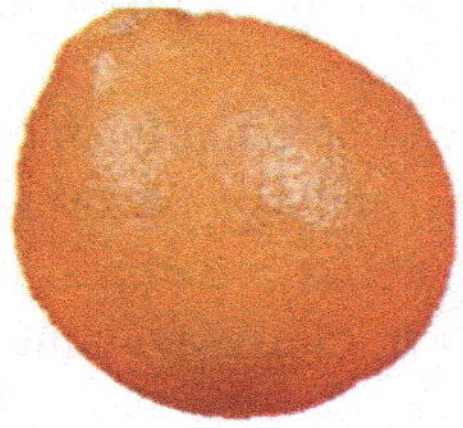

DID YOU KNOW? Natural occurring tangelos probably are caused by insect cross pollination between the Mandarin orange and the pomelo/ jabong, the ancestor of the grapefruit. They are so unlike other citrus fruits that they have been set aside in a class by themselves designated Citrus X tangelo J. Tangerines are deliberate or accidental hybrids of the mandarin *(Citrus reticulata)* and the sweet orange. The Jamaican 'Ugli' fruit is believed to be a chance hybrid between a mandarin orange and grapefruit. UGLI® is the registered trade mark under which Cabel Hall Citrus Ltd. markets its brand of tangelos from Jamaica.

Nutrition isn't just about eating, it's about learning to live.

~ Patricia Compton

TANGERINE
Citrus reticulata

Tangerine is the name of a variety of citrus fruit and categorizes some varieties of the mandarin orange family. This is a perfect tree for any Hawaiian porch or backyard. All varieties of tangerines are related to varieties of the Chinese mandarin orange, but not all mandarins are tangerines. In general, 'tangerine' refers to mandarins with a reddish-orange skin. Because of the fruits' similarities, tangerines are sold labeled mandarin oranges, and vice versa.

Mandarins originated in SE Asia and tangerines developed in North Africa. This ex-plains their names: tangerines because they developed near Tangier. Oranges are the result of cross-pollination between pomelos/jabongs and mandarins. Tangerines are the result of cross-pollination between different varieties of mandarin oranges.

Often called baby oranges, tangerines are smaller, pocket-sized, somewhat flattened and generally oblong. Ripe oranges are firm and heavy when ripe; ripe tangerines are softer to the touch and are reddish-orange. Tangerines have a very thin, loose skin, making it easy to peel. The tangerine's flavor tends to be just a bit sweeter than the orange's flavor is, though both fruits are sweet. Tangerines have a stronger citrus flavor than an orange and the aftertaste will be shorter with a tangerine. Oranges are believed to have originated in northern India, southern China, or SE Asia. Around the mid-1400s, Italian traders brought oranges to the Mediterranean area and it was cultivated for medicinal purposes. The Spanish explorers brought the orange to the western hemisphere.

HOW TO GROW: From seeds, taste a few tangerines and keep the seeds from those you like. Scrub the seeds clean and dry them before planting 2-3 together in moist potting soil. Once seedlings have a pair of true leaves, transplant them to large pots, about 4-6 inches in diameter. Tangerines can either be grown directly into the soil or in pots or containers but want sun and warm temperatures above 70F. Keep watered, but not wet. They should bear in 3-5 years. Grafted trees bear in 2-3 years. During the first 2 years, give a soda cap full of 12-24-12 fertilizer bimonthly. Once blossoms appear change to a bimonthly cap full of 12-12-17-2 fertilizer.

MEDICINAL: Tangerine peel/skin is chewed to treat asthma, indigestion, and clogged arteries. They possibly help prevent cancer, especially lung, colon, and rectal cancer. The es-sence of tangerine peel helps control the side effects of chemotherapy.

NUTRITION: One tangerine, approximately 90 grams has 50 calories with 1g pro-tein, loads of vitamin C and potassium with traces of vitamin B-6 and magnesium.

FOOD & USES: Tangerines have sections that easily separate and can be eaten fresh or canned. They are used in fruit salads, flavorings for baked and pudding deserts, and as a flavoring for meat and fish marinades. The skin can be peeled and grated for zest used as a garnish. If cooking with tangerines, use low heat and a short time, so the flavor is not lost from over cooking. Shelf life is 2-3 days or a week in the fridge.

RECIPES: TANGERNE BUTTER: Grate a tangerine peel and 2 garlic cloves minced and mix it into a soft butter for seafood and vegetables.

DID YOU KNOW? Tangerines are named after Tangiers in Morocco. They were shipped to Europe in 1848 (perhaps they were mandarin oranges). The name tangerine is applied to the port of Tangiers the same as Florentine applies to cuisine from Florence, Italy.

TROPICAL APPLES

Malus pumila – Malus domestica

There's nothing quite like the tasty crunch when you bite into a juicy apple, and it's ten times better when it's home-grown. Apples, like peaches and other stone fruit require a 'chill' period, a certain number of hours of cool temperatures produce fruit. In Hawaii, they grow best at elevations above 2,000 feet. There has been limited success at lower elevations throughout the island chain. Even having a few apple trees that don't bear yearly, is still worth working for. Multiple trees are necessary for cross-pollination.

The accepted botanical name is *Malus pumila*, but is also *Malus domestica*, *Malus sylvestris, Malus communis*, and *Pyrus malus (ITIS website)*. Apples are members of the Rose family, with many other great tasting fruits: pears, plums, peaches, cherries, strawberries, and raspberries. DNA analysis indicates that apples originated in the mountains of Kazakhstan in southern Russia and western China. Since apples need cross-pollination, the countless varieties have inherited different forms of particular genes from the different parents. It's difficult, but not impossible to grow an apple from seeds.

HOW TO GROW: 'Dorsett Golden' and Anna" varieties require a low number of chill hours and have been successful in Hawaii, mostly on the Big Island. It is not always location, location, location, but also perseverance and maintenance. When buying a sapling, always check the required chill hours and USDA growing zones that match your location. To be sure of low chill trees and not wasting time, effort, and money, buy the recommended saplings from a nursery. Trees can grow to 40 feet in height, but better to prune to keep under 20 feet. It takes more than 3 months from blossoms to fruit.

Even though some varieties are listed as self-pollinating, get a second tree. Cross-pollination is essential for apples, pears, most sweet cherries, and most Japanese plums producing more fruit on regular cycles. Pollen is primarily transferred by honeybees when temperatures are above 65°F. Cool weather, rain, or winds can keep bees in their hives. Pesticides should not be used during blossoming. Keep your trees close together; distance can cause poor pollination.

Dwarf apple trees have many advantages over standard apple trees. You can grow them in small spaces, in large pots, easily prune, spray and harvest. After planting, dwarf apple trees produce fruit much earlier, usually after three years, than standard apple trees, which may take a decade.

An apple tree can be grown from seeds. It takes time and patience, but what's worthwhile that doesn't? Remove many seeds from apples; clean and dry them. The more you begin with, the better the chances a few of them will sprout. Because of genetics, apple seeds need a cold period snap into regeneration. Get a plastic container with a lid, like for margarine. Fill a ½ inch with moist sand. Place the apple seeds and cover with another ½ inch of moist sand. Punch several holes in the lid, seal and put in the fridge for at least 50 days. When the time is up, remove the seeds from the fridge, plant ½ inch deep in quality potting soil. Plant several in a pot and carefully drip water so the seeds aren't disturbed. Within a month sprouts should appear. Remember, be patient because apple trees grown from seed won't produce fruit for about 6 to 10 years.

Choose a location in full sun with good air circulation so leaves dry quickly after a rainfall or watering to avoid leaf and fungus diseases. The soil must drain well, and the preferred pH is 6.0 to 6.5 but apples can adapt to 5.5 to 7.0. Clay and water retaining soils will cause root rot. Dig a hole twice the diameter of the sapling's root ball and 2 feet deep. Loosen the soil on the bottom and sides. Unwrap and spread the apple roots, so they're not twisted. Other apple trees that bloom at the same time must be planted nearby, preferably within 100 meters or closer. One application of a high potassium at the start of the season, regular watering, look for pests, don't prune until they bear fruit, thin the crop a month after blossoming, is all it takes to have nice apples.

MEDICINAL: An apple a day may keep the doctor away and so will a tablespoon of good raw apple cider vinegar. Apple cider vinegar with its sour taste and strong smell is mostly apple juice with yeast added. That ferments the juice into alcohol and then into acetic acid. Raw apple cider vinegar (not plain cider vinegar is known to be a good source of acetic and ascorbic acid (vitamin C, mineral salts, amino acids, and other key components of good nutrition, but it is also a well-loved folk remedy thought to ease digestion, fight obesity and diabetes, wash toxins from the body, kill lice, and reverse aging. Add 1 to 2 tablespoons to water or tea. One researcher found vinegar improves blood sugar and insulin levels in people with type 2 diabetes. Apples also contain some chemicals that seem to be anti-bacterial, anti-inflammatory, and kill cancer cells. Apple peel contains ursolic acid assumed to build muscle and increase metabolism.

NUTRITION: 100 grams of raw apple has 52 calories with significant levels of vitamin C and potassium.

FOOD & USES: Apples can be eaten cooked or raw, or enjoyed as juice, pie, cider, apple wine, butter, and apple sauce. Pruned limbs and twigs can make chips used in hot or cold smoking of meats and cheeses. Apples are used in cosmetic lotions and shampoos. Apple wood is good for furniture.

RECIPES: APPLE CRISP FROM THE FRYING PAN

Ingredients: APPLES: 2 cups apples - peeled, cored, and chopped, 1 TS ground cinnamon, 1 TS brown sugar, 2 TBS butter. TOPPING: 1 TS unsalted butter, 1 TS brown sugar, ¼ cup rolled oats, ¼ cup nuts, almonds, walnuts, peanuts

Method: For the apples: Melt the butter in a frying pan, add apples, cinnamon, and brown sugar. Stir until well combined. Cover and cook on medium, stirring until apples soften – about 5-7 minutes. For the topping: In a saucepan, melt the butter. On low heat, add sugar, oats, and nuts. Brown for a few minutes until crunchy. Watch closely, it can stick and burn. Serve apples in a bowl and sprinkle the topping on. Top with ice cream or milk.

CINNAMON APPLES

Ingredients: 4 apples, cored, peeled, and chopped, 1 TS lemon juice, 2 TBS brown sugar, 1 TS ground cinnamon, 2 TBS butter, 2 TBS water, ½ TS vanilla extract, pinch of salt

Method: Combine apple chunks, with lemon juice, sugar, and cinnamon. In a frying pan or wok on medium flame, heat the butter. Add apples and water, stir until apples begin to soften. Stir in vanilla extract and salt. Serve warm.

DID YOU KNOW? There are more than 7,000 varieties of apples worldwide. The apple tree originated in an area between the Caspian and Black seas. The number 5 is a magical number for apples. The word "Apples' has 5 letters, apple blossoms usually form in clusters of 5, and blossoms have 5 petals. Worldwide production of apples in 2018 was 86 million tons, China grew 41 million tons. The largest apple weighed three pounds. The US grows more than 20 million pounds of apples.

And when you crush an apple with your teeth, say to it in your heart: Your seeds shall live in my body, and the buds of your tomorrow shall blossom in my heart, and your fragrance shall be my breath, and together we shall rejoice through all the seasons. ~ Khalil Gibran

TROPICAL APRICOT – CEYLON GOOSEBERRY
Dovyalis hebecarpa

The tropical apricot is a rare and very healthful fruit available to grow anywhere in Hawaii. Seeds were brought from Florida to Hawaii in 1920. It was introduced to Cuba and Puerto Rico in 1930 and is now known as Puerto Rican cranberries. It is a small tree or shrub that will grow to 20 feet (6m This fruit is usually used in drinks, as a flavoring, or jellies, and jams. The acidic taste is not for everyone. This fruit usually isn't eaten out of hand because the juicy flesh is acidic-sweet and the ripe purple skin has small, fine hairs and extremely bitter. The taste is compared to an apricot, or a cranberry. Every fruit contains roughly 6 or more small, seeds. The tropical apricot is medium-sized, about an inch (2.5 cm, orange when unripe, and purple when mature. The fruit is rich in vitamins and minerals and high in antioxidants.

HOW TO GROW: The tropical apricot is a nice tree for the home to make tasty tonic juices. The short tree is very attractive and manageable, with wide sweeping branches, and does well from sea-level to 2,600 ft. (800 m in either dry or moist climates, although proper fruit development requires regular watering. It prefers a pH of 6.5-7.5 and grows best in a fertile, humus-rich, well-drained soil. Trees bear within 5 years. In some areas, these trees bear twice a year. A 15-foot tall shrub can produce over 100 pounds of fruit per year.

It is best grown from seeds in individual pots. Seedlings are often found under mature trees. The trunk and lower branches have long, sharp 1½(4cm inch spikes. The spikes make this a protective-productive hedge. The blossoms are greenish-yellow. This tree will bear many, many small juicy fruits that resemble grape tomatoes.

MEDICINAL: The tropical apricot is packed with antioxidants (fights cancer and heart disease, polyphenols, (help digestion and the metabolism, fights diabetes, and anthocyanins (fight viruses, cancers, and inflammation In a 2007 study published in the *Journal of Agricultural and Food Chemistry,* tropical apricots have 10 anthocyanins and 26 carotenoids. In anthocyanin-rich juice, like gooseberry, scientists found that the compounds reduced the risk of heart attacks. The fruits' skin must be utilized because all anthocyanins had higher concentrations in the skin than in the pulp.

NUTRITION: 100 grams of tropical apricot provide: 1.2 mg of iron, 26 mg of phosphorus, 13 mg of calcium, 98 mg of vitamin C, 0.4 mg of vitamin B2, 0.3 mg of vitamin B3, 0.02 mg of Vitamin B1. A cup has 63 calories.

FOOD & USES: The tropical apricot offers a unique, sour taste. Some cut the fruit in half and scoop out the flesh. Some remove tropical apricots skin before consuming because of its irritating, small hairs. Removing the fruit's hairy seeds isn't necessary, as they're soft and have a negligible taste. Use them quick, tropical apricots will quickly spoil even in the fridge. For syrups and jam recipes, blend and strain the pulp to achieve the smoothest consistency. Every pound of fruit will require approximately 1/4 cup of sugar and 3/4 cup of water.

RECIPES: EASY TROPICAL APRICOT SPREAD

Ingredients: 1 kilo (2.2 lbs.) washed fruit, 1 cup white sugar, and 1 cup water

Method: In a large pot on medium heat combine tropical apricots, sugar, and 1 cup water. Boil until the fruits are soft and then carefully whisk to crush as many as possible. Lower heat and stir until liquid thickens as it reduces by half. The fruits should shed their skins. Let cool and then blend and strain to remove the seeds and skin. Let cool and then chill overnight. This is a nutritious spread for bread or toast, biscuits, or an ice cream topping.

TROPICAL PEACHES

Prunus persica

Peaches growing in the tropics is now becoming a reality. It takes a reasonable amount of work and the right conditions, but what doesn't? The peach is native to northwest China and then spread westward through Asia to the Mediterranean countries and later to other parts of Europe. The peach first appeared In India circa 1700 BCE. The Spanish explorers took the peach to the New World, and as early as 1600 the fruit was found in Mexico.

The botanical reference *Persica* refers to Persia (Iran having orchards. Trees were transplanted to Europe. Peaches belong to the same family that includes the cherry, apricot, almond, and plum, in the rose family. Deciduous fruit trees such as apples, pears, peaches, and plums, originating from the temperate zones, have been grown in the subtropics and tropics for several centuries. The peach is a stone fruit. Stone fruit are fruits that all share in having a single large seed, or pit in the middle. These fruits usually have thin skins, tree ripen only and must be picked on the perfect ripe day. Peach blossoms have a sweet light aroma.

There are hundreds of varieties of peaches, but they are all divided into two categories, freestone, where the flesh easily separates from the pit, and clingstones where it doesn't.

HOW TO GROW: You can plant a peach tree anywhere and it will usually survive, but they need specific conditions to produce fruit. Peaches have a chilling requirement. Chilling is the number of hours between 45 and 32 F. Standard peaches have chilling requirements of 600 to 1,000 hours and we know that doesn't happen in the subtropics or tropics.

Having trees produce peaches in the tropics successfully depends on choosing a variety of tree that has a low chill factor, elevation of where you planted, and total rainfall. If your location gets chilly at night and not overly wet, and that continues for several years, you should be able to grow peaches and other 'stone' pitted fruit.

In the tropics, on an island with different elevations, you are looking for what's termed a 'micro-climate.' We've all traveled around the islands and stayed where it's been cold and damp on an average night. Hawaiian researchers have found varieties that produced fruit on the island of Molokai with no chill period, tropic sweet and tropic beauty. Midpride and Bonita varieties need as little as 100 chill hours.

The Hawaiian Islands are blessed with many different climates in the central highlands and can grow a variety of international crops. Low elevations produce the best coconuts, mangoes, and papaya. The best elevations in Hawaii are 600m to 1200m (2000 – 4000 ft.). There are varieties of peaches that produce at 200m with no chill hours. Rain is another factor. Hard rains will knock off young flower buds and increase diseases. Less than 80 inches of rainfall per year should be ok, with less rain being better. Stone fruits like peaches thrive on the Big Island in areas like Volcano and Waimea on Hawaii Island or Makawao on Maui. These have all the right elements.

Peach trees are available online. Peach trees require full sun with sandy loam well-drained soil with a pH of 6.5. Compost should be worked into the soil 5 feet in diameter before planting. Fertilize once a year, at the end of fruiting, with one pound of 10-10-10 fertilizer per 100 square feet or by topdressing with 2 to 3 inches of compost. Peaches require regular water and regular pruning while it is dormant. The open-center system of pruning is recommended for peaches. For larger, sweeter fruit remove excess fruit about four weeks after flowering so that fruit is spaced about eight inches apart on the branches. Thin the fruit so they are not touching. This will allow them to grow larger and increase their flavor. A well-maintained tree bears 35 kg or more yearly.

If you want to try to grow a peach tree from seed; carefully crack open a peach pit and remove the kernel. Place the kernel into a plastic bag. And put that in your fridge. Hopefully it will germinate within 2-3 months. Then carefully transfer the sprout to a pot.

MEDICINAL: Chinese traditional medicine uses the peach to aid digestion, promote circulation and reduce fatigue. A tea of peach leaves treat fever, headache, skin problems, ulcers, and malaria. A tea made from the bark and dried leaves treats coughs. Root bark is used for dropsy and jaundice.

NUTRITION: 100 grams of a juicy peach has only 40 calories with considerable amounts of vitamins A and C and calcium, phosphorus, and potassium.

FOOD & USES: A juicy peach is best eaten raw, cooked, dried, or frozen for later in pies, jam, or ice cream. The leaves will produce a dye and remove strong odors from containers as garlic or cloves. Gum from the stems can be used as an adhesive.

RECIPES: PEACH SALSA

Ingredients: 2 cups of peeled diced peaches, 1 large red onion chopped small, 2 or more green chilies, 1/3 cup cilantro, juice of 2 limes, salt to taste

Method: Combine all ingredients in a bowl and refrigerate for an hour before serving with chips.

PEACH CRISP

Ingredients: 4 cups sliced fresh peaches, ½ cup all-purpose flour, 1 cup rolled oats, 2 TBS brown sugar, 2 TBS butter, ½ TS ground cinnamon, 1 TBS lemon juice

Method: In a skillet or wok, melt 1 TBS butter. Add peaches, sugar, and lemon juice. Bring to a boil and then reduce heat and simmer uncovered for 5 minutes. Stir in flour and oats. Let cool and refrigerate an hour before serving with yogurt or ice cream.

SPICY HOT PEACH JAM

Ingredients: 1½ kg peaches, peeled and chopped, 1½ cup sugar, 2 green chilies, 5 strips of lemon peel, ¼ cup fresh lemon juice, 1 TS salt

Method: Mix peaches and sugar and let rest for half an hour. In a skillet or wok, combine all ingredients and simmer for half an hour. When cool, remove the lemon peel. With a potato masher or fork, mash the fruit. Fill sterilized jars. Good in the refrigerator for 3 months.

GRILLED PEACHES

Ingredients: 4 ripe peaches cut in half with the pit removed, 4 TBS butter or margarine, 2 TBS honey or sugar, 1 TS vanilla essence, ½ TS cinnamon

Method: In a saucepan, combine everything except the peaches and cook until brown – 4-5 minutes. Grill peach halves face down for 5 minutes until they have good char marks. Flip and grill for 3 minutes. Immediately drizzle with the browned butter. Serve with ice cream or cake

DID YOU KNOW? Peach blossoms are highly prized in Chinese culture. The ancient Chinese believed the peach to possess more vitality than any other tree because their blossoms appear before leaves sprout. When early rulers of China visited their territories, they were preceded by sorcerers armed with peach rods to protect them from spectral evils. Reported to be the world's best tasting peach is the Water honey variety from China. It has green skin and white flesh. The largest peach on record weighed 25.6 ounces (716 g) from Michigan, USA. The nectarine is a peach with a smooth skin.

An apple is an excellent thing -- until you have tried a peach. ~ George du Maurier

WAX APPLE – WAX JAMBU – ROSE APPLE

Syzygium samarangense

Hawaiian fruit lovers know this fruit as the wax apple. The wax apple is not an apple, but because it is usually a red fruit and crunchy when bitten, the English applied the name apple. Because it is juicy and the skin shines, it's the wax apple. This fruit is a thirst quencher. The botanical name is *Syzygium samarangense* and that's important because there are many types of similar 'apples,' rose, water, mountain, wax, water, all with names specific to a locale on the Indian subcontinent. The thin skin is a glossy green or red, two to three inches long. They are usually seedless, but different varieties may have 1 to 6 seeds. An Indonesian variety has white skin. The flesh is white or pink, and usually faintly sweet. There are so many types, some varieties can be tart, crisp or soft, juicy, or dry. It's considered a good thirst quencher.

The wax apple is native to the area that includes the Greater Sunda Islands, Malay Peninsula, the Andaman, and Nicobar Islands. Prehistoric migrations carried this to a wider area of S.E. Asia and India. Now the wax apple is cultivated worldwide throughout the tropics. Because of their similarity in appearance, it is often confused with Syzygium aqueum, another wax apple, although the latter is more commonly cultivated.

HOW TO GROW: The wax apple is an evergreen tree best grown in lowlands where there is a separate and definite rainy and dry season, but will tolerate 1000-meter elevations either in full or partial sun. The tree grows best in well-drained, but moist soil with a pH of 5.6 to 7.5. It will quickly reach 15 meters. The leafage is very thick and dark green. Its blossoms vary from white to pale yellow to pink. The wax apple is excellent for a windbreak. It has a coarse dark brown bark, and the wood is very hard.

This tree is easy to start from fresh seeds. Find a fruit type you like and plant the seeds. Pink fruits have more juice and flavor. Green fruits are crisper, but juicy and sometimes cooked. For planting purposes, do not permit the seeds to dry. The young sprouts need shade and should bear fruit in 5 years. Usually, these trees bear twice a year. A cultivated and fertilized wax apple can bear 700 fruits. Wax apple fruit are bell-shaped and range in color from white, pale green, deep green, or red/crimson, lavender to deep purple, to black. As it ripens, the fruit expands outwards, keeping a slightly concave inner curve at the base, or 'the bell.' The skin of healthy wax apples shines.

Despite its English name, a ripe wax apple only resembles an apple on the outside in color. It does not taste like an apple, and it has neither the fragrance nor the density of an apple. The liquid-to-flesh ratio of the wax apple compares to watermelon, but with softer flesh. The seed is in an edible, but flavorless cotton-candy-like mesh. The color of the juice depends on the type of fruit, transparent to purplish.

MEDICINAL: Medicinally the astringent blossoms are boiled and used to treat fevers and diarrhea. A decoction of the astringent bark is a local application on thrush. It boosts good HDL cholesterol, increases metabolism, and prevents constipation. Because it is mostly water with some potassium, it helps prevent muscle cramps. The leaves of wax apple can be used as a tea as a possible supplement for type II diabetes patients. This fruit is being studied for numerous pharmacological anti-diabetic and antioxidant properties. Experiments with various parts of the tree are being conducted for anti-inflammation, wound healing, antibacterial, and anticancer properties.

NUTRITION: Wax apples are good for most pregnancies with vitamin A and iron. According to early writings, a wax apple salad is a ceremonial dish for new mothers. These bland fruits help keep expectant mothers hydrated during morning sickness. The leaves have an analgesic property that can help relieve post-birth pain. One hundred grams or one fruit has 25 calories, with some thiamine, riboflavin, niacin, vitamins A and C, slight calcium, magnesium, phosphorus, iron, and no cholesterol. This fruit is 90% water.

FOOD & USES: Generally, the paler or darker the color, the sweeter the fruit. In S.E. Asia, the black fruit are 'Black Pearls' or 'Black Diamonds,' while the very pale greenish-white ones are called 'Pearls.' This fruit is often served uncut, but cored, removing the inner mesh around the seed and keeps the unique bell-shape. The fruit is used in salads, also sautéed. It is mainly eaten as a fruit, but also used to make achar pickles. The wax apple is mainly eaten by children, to relieve thirst. In Indonesia, the fruits are sold in markets in piles or skewered on slender bamboo sticks. Superior types are sometimes served sliced in salads.

RECIPES: WAX APPLE JUICE

Ingredients: 15 wax apples cored and seeded, 2 TBS sugar, lemon juice to taste, water to suit you

Method: Blend all ingredients and puree till smooth. Strain and chill. Enjoy.

WAX APPLE JAM

Ingredients: 1 pound of wax apples, 1liter water, 2 cups white sugar, ½ TS cardamom powder, 1 TBS lemon juice or more to your taste, 2 cloves.

Method: Scrub the fruit and slice it into small pieces. Put in a suitable pot and cover it with water. Boil for 5-10 minutes or until the fruit slices are soft. Drain and mash with a spoon and blend in the sugar. Cook over low heat, stirring, and mashing. Add the cardamom, cloves, and lemon juice. This will be the flavor, so adjust to your taste. Keep stirring. To test if the consistency is good, drop a spoonful on a plate. The water should not separate. Color could be added if you choose.

WAX APPLE ACHAR PICKLE

Ingredients: 2 cups fruit cut in small pieces, 1 TBS ginger-garlic paste, 1 green chili pepper or one hot pepper depending on your taste, 3 TS red chili powder, 1 TS mustard seeds, a pinch of each turmeric and fenugreek powder, sesame oil, and salt to your taste, ½ cup vinegar

Method: Heat the mustard seeds in the sesame oil until they pop. Stir in ginger-garlic paste, green chili, and sauté, stirring in the wax apple pieces. Add turmeric and fenugreek powders. While stirring, add more oil if needed. Add salt and remove from heat. Let cool and mix in the vinegar. Store in a tight glass container. Best to refrigerate.

Every pleasure I've had, I owe in some way to the sweetness of fruits. ~ Marjory Sheba

DID YOU KNOW? This fruit as many names, wax apple in the Caribbean, water jambu, wax apple, wax jambu, Semarang rose-apple. In the Philippines, it had the ancient name, dambo, but now is known as macopa. In Southern Indian Tamil, it's jambakka or chambakka. In Hindi, it's 'pani seb,' jambu, or 'panneer naval.'

Every time you eat a fruit for the first time that year you need to make a wish.

— André Aciman

WEST INDIAN CHERRY – ACEROLA
Malpighia emarginata

The most popular tropical cherry is the acerola cherry, also known as the Barbados or West Indian cherry. Acerola is the one tree that should be growing in your Hawaiian home garden. It is a relatively small fruit tree and may be trimmed to a convenient sized shrub. This fruit is nature's vitamin pill.

The acerola has recently become famous since it has the highest vitamin C content of all fruits, and is now cultivated for medicinal purposes. A hundred grams of cherry flesh usually has about 2,000 mg of Vitamin C. It's amazing that immature green cherries actually have twice the vitamin C than the fully grown red. Botanists believe this fruit tree originated in Mexico's Yucatan Peninsula. Acerola cherries can be found in all of Central America, throughout the Caribbean, parts of South America, India, Madagascar, and Southeast Asia.

HOW TO GROW: The acerola cherry is grown from seed, cuttings, or grafts. It can adapt to most environments found in tropical regions and is able to withstand a severe drought. Its shallow root system can be uprooted by strong winds. For the tree to produce abundant fruit, regular watering is necessary with adequate drainage. This tree can be grown in sizable pots where it will get sun. The pots should have good drain holes and the potting soil kept slightly alkaline. The soil is best with a pH of 7-7.5.

The acerola is an attractive tree with dark green leaves and pink or white blossoms. It will grow to about 15 feet and can be shaped by pruning. Keep the tree to a height that can be easily picked, or the birds and bats will get most of the cherries. It is self-pollinating and loves the sun. Acerola fruit are brilliant red and juicy, sour with a slightly sweet aftertaste. The cherries mature from blossoms in about a month. Fertilize immature trees every 3 months with 12-24-12 and water regularly. Once blossoms appear, use 12-12-17-2 and increase water until the fruits appear. For a better harvest, lime the soil once a year. Nematodes are a problem for immature trees. Spray the trees monthly with an insecticide such as Fastac combined with a foliar mix containing boron. This is an excellent backyard tree because with proper fertilizing and watering, an acerola cherry tree can bear in its third year. At 5-6 years, it will fruit 3 times a year, and can bear fruit for 20 years.

MEDICINAL: Acerola cherries with such a high vitamin C content are beneficial for patients with liver ailments, diarrhea, and dysentery as well as those with coughs or colds. The juice may be gargled to relieve a sore throat. These cherries can also reduce joint inflammation from arthritis and prevent infections. A steady daily dose of acerola has shown reductions in the blood glucose levels of mice, which may prove to be a useful anti-diabetic treatment.

NUTRITION: One acerola cherry is a natural vitamin pill; 100 grams has 32 calories. In 1945 the West Indian cherry was found to be extremely high in ascorbic acid. One tiny acerola cherry has 81 milligrams of vitamin C, 25 percent more than the recommended daily allowance, more vitamin C than an orange. These little round bright red fruits also have vitamins B1, B2, B3, and B5 with high amounts of vitamin A with calcium, potassium, and magnesium. Acerolas also contain antioxidants as flavonoids, anthocyanins, and polyphenols.

FOOD & USES: Because the nutrients acerola contains, it's used in cosmetics as an anti-aging skin treatment. The cherry wood is surprisingly hard and heavy, and won't ignite even when treated with flammable fluid unless perfectly dried. Eat these cherries fresh from your tree. They bruise easily. Once the cherries are picked they have a shelf life of only 3-5 days and lose their vitamin content. You can also preserve fresh acerola cherries by freezing them as soon as you pick them. Even the juice will spoil easily, unless preservatives are added.

They are sour, the very high vitamin C content makes them acidic. It's great for flavoring ice cream or topping puddings or cakes. It flavors drinks, cocktails, and other desserts. Add another favorite juice such as orange, pineapple, or a banana. Add ice and blend until smooth. Pitted cherry pulp can be frozen with additional sweeteners and blended into a sorbet. Milk may be added to either make a shake or to refreeze into ice cream.

RECIPES: CHERRY JUICE: Strain seeded cherries until you have 2 cups of juice. Mix with an equal amount of water. It is slightly acidic so add sugar to taste. Pour over ice and enjoy.

CHERRY CONCENTRATE: Strain seeded cherries until you have 2 cups. Put into an uncovered pot over medium heat. Stir constantly as half of the liquid evaporates. Remove from heat and cool before pouring into containers for freezing. Keep as a cold remedy or as flavorings for drinks, cakes, and frozen desserts.

TROPICAL FROZEN BLEND: Strain seeded cherries until you have a cup of juice. Mix in a blender with a cup of orange juice, a quarter cup of pineapple, and two ripe bananas. Add ice to the blender's capacity. Cover and blend until smooth. Guava and apple are also good combinations with acerola cherries.

NAHIKO CHERRY SORBET: Get 2 cups of seeded cherry pulp. Add 2 tablespoons of sugar, 1 teaspoon of lime juice, and blend with a tray of ice cubes. Freeze until slushy, blend again and refreeze. A cup of milk may be added, but the amount of ice should be reduced by half.

EXQUISITE CHERRY GINGERBREAD

Ingredients: 1 cup cherry pulp, 1 cup milk, 3 TBS butter, 3 TS active dry yeast, ¼ cup firmly packed brown sugar, ¼ cup fresh lime juice, ¼ cup fresh cherry juice, 1 TS salt, 4 cups baker's flour - divided, 1 cup toasted coconut, 1 TS grated lemon peel, 1 TBS minced fresh ginger

Method: Combine milk and butter in a small pan and heat just until butter melts. Remove from heat, pour into a large bowl, and stir in yeast. Let sit for 5 minutes before adding brown sugar, lemon and cherry juices, salt, and 2 cups of the flour. Combine thoroughly, electric mixer preferred. Add remaining 2 cups of flour and mix until the dough doesn't stick to the bowl. Add tablespoons of water if too dry. Blend in cherry pulp, a ¼ cup of coconut, grated lemon peel, and ginger. Knead dough until it is elastic and let sit in a warm place until it doubles in size - usually about an hour. Punch it down and place in a greased baking pan and let it sit for another hour. Rub grated coconut into the top. Bake at 350 for an hour or until a toothpick or knife pulls out without sticking.

DID YOU KNOW?

Acerola cherries, by weight, have 20 times the vitamin C of an orange. During WWII, acerola seedling trees were given to houses because the vitamin C would keep people healthy and help the war effort.

NAMES: *Malpighia emarginata* is generally accepted as the correct botanical name for the acerola. They are also known as *M. Punicifolia L.* and *Malpighia glabra L.* Barbados cherry, which is also called the West Indian cherry, sweet, Puerto Rican, Jamaican, the native cherry, garden cherry, or French cherry. In Spanish, it is named acerola, cereza, or cereza colorada.

Plant the trees just for beauty. If flowers bloom or fruits ripen, enjoy it as a gift and appreciate nature as a universal giver. ~ Debasish Mridha

THE HAWAIIAN HOME GARDEN GUIDE
ROOTS – CORMS – RHIZOMES

Edible roots, taro. and others, are a crucial and cultural part of the original and present day Hawaiian diet. Easy to grow in a perfect climate, roots are abundant on every island. Most were carried in canoes by the early Polynesian seafaring explorers. Taro, because the roots stored well, was and remains one of the most important foods not only in the Pacific region, but for the entire world as a subsistence crop. Taro and kalo, the most common names of the roots, includes hundreds of varieties. Some roots grow and multiply in wetlands and swamps with an ease and intensity that makes them 'invasive.' Other roots require well-drained soil.

Roots aren't only important for the underground food, but also for the plant leaves. Kalo may be defined as the entire, incredibly valuable plant that has both the taro root and the 'luau leaf.' Steamed, or stewed in soups, this nutritious leaf is delicious.

The root is the part of the plant usually below the soil's surface. Roots, which are edible for humans are 'enlarged' roots that store food for the plant, and from which the above ground plant grows. The major functions of roots are to provide stability and anchor the plant's body in the soil, absorb and store water and nutrients to keep the plant alive. In some cases, they are a means of asexual reproduction.

Beets, carrots, and radishes are examples of true root vegetables. They are the taproot of the plant, which is formed as the very first root from the seed. However, 'roots' also include corms, rhizomes, and tubers. Tubers are underground stems. They grow wide instead of long. Irish and sweet potatoes, cassava/manioc, tannia/malanga, and yams are tubers. Corms are the underground swollen base of the solid tissue of the stems. A corm grows vertically and is a mass of stored water and nutrients to feed the plant. Roots are attached to the bottom of the corms. They do not have separate fleshy scales as in a bulb, although there is usually a dry papery covering, skins that are really modified leaves. Eddoes and sweet corn root are corms.

Rhizomes are knobby underground stems that usually grow horizontally relished for their pungent and flavorful flesh. A rhizome is similar to a corm, but new rhizomes are not produced annually. Instead, the older parts die off and the tips of the rhizome grow longer. The plants grow rhizomes, and rhizomes grow roots plus nourish the above-ground parts of the plant. A rhizome may be propagated by division when the resting buds grow and produce leaves for a new plant. Ginger and turmeric are rhizomes. Confused: Peanuts are 'technically root vegetables,' because they are tubers that grow off a rhizome underground, but are treated as a nut.

Apologies if the roots have unfamiliar or incorrect local names.

PAGE:

Courtesy R. Douglas

Everything that pushes up out of the earth I love. Everything under the earth, root vegetables, I love to cook. — Alain Ducasse

AIR POTATO – HOI

Dioscorea bulbifera

The Ancient Polynesians introduced the air potato to the Hawaii. It is one of the 'canoe plants' named Hoi or Pi'oi. There is some confusion if it is poisonous or edible. Be certain which type of air potato you are growing. In the early 1800s, it was naturalized on Barbados and within a century, the air potato was growing in gardens on almost every tropical island. Few people know of it, but this vegetable has been cultivated for millenniums. Air potato is native to Asia, and seafaring explorers carried it to all parts of the globe. Asia, Africa, the Pacific Islands, Northern Australia, Florida, and in the West Indies, all grow the air potato.

This starch is a perennial, climbing aerial yam with smooth, glossy leaves. A properly maintained air potato can spread to 10 meters (39 ft.) over the ground and wrap around trees or supports. This plant is often cultivated in tropical areas, mainly for its edible aerial bulbs and for its roots. Before cooking, air potato is as hard as a rock with rough, thick tan to brown skin or it can be light grey and smooth. The orange/brown flesh is firm, starchy, and slimy when peeled. Air potatoes have a mild, earthy, and sometimes bitter taste. The fruits range in size from 2cm (¾ in.) to 15cm (6in.) **TAKE NOTE**: **Edible varieties of *Dioscorea* have opposite leaves, and poisonous varieties have alternate leaves.** The fruit of this species contain toxic substances, including the altaroid dioscorine that can be destroyed by thorough cooking. Some varieties are low or free of toxins. They wrap around anything counterclockwise while yams wrap clockwise.

HOW TO GROW: In Hawaii, the air potato does best in moist low attitude gardens. Fruits will sprout dormant buds. Each section with a bud can be chunked and planted. The cut fruit should remain in the sun for a day to seal the wounded surface, to promote wound healing and reduce the possibility of rotting. Usually, air potato plants take two seasons to produce a full-size fruit. Air potato prefers a well-drained, sandy loam with a pH in the range 6-6.7, and full sun. Fruits are produced within 6 months from planting, though some forms can produce a crop in as little as 3 months. Big air potatoes weigh 2kg (4½ lbs.), but they average about half a kilo (1lb.). Roots are usually around 0.5kg, though they can be up to 1.5kg. The air potato can grow quickly, 8 inches a day. This is listed as an invasive plant because each fruit can fall and sprout another vine.

MEDICINAL: The air potato variety with white spots on its large, round leaves and also on the fruits are not eaten but used in traditional medicine. The juice of the air potato is used externally to sanitize wounds, usually as a poultice, to treat cuts, sores, boils, and inflammations. Juice squeezed from the roots expels parasites and when combined with palm oil, it's rubbed to soothe rheumatism. Fruits are used to treat boils and fevers. The leaves are steamed and applied for conjunctivitis/pinkeye. Air potato contains the steroid diosgenin, which produces synthetic commercial steroidal hormones for birth control pills and steroids used to boost men's testosterone and used in bodybuilding.

NUTRITION: Air potato has 100 calories per 100 grams. It's 7.5% protein, and very rich in potassium, niacin, vitamin C, magnesium, and calcium, but high in sodium.

FOOD & USES: The fruits must be cooked thoroughly, usually boiled first, to destroy the toxic altaroids. Raw, they are bitter and slimy. Once boiled, the skin should rub off in your hands. Air potato has an agreeable taste; locals boil it for 20 minutes in the skin, clean it, and eat it plain. Boiled air potato can be prepared like a yam, fried, baked, or roasted. They can be added to soups and curries. In a cool, dry, dark place, air potatoes store for a couple of weeks.

RECIPES: MASHED AIR POTATO CURRY

Ingredients: 8 air potato boiled, skinned, and crumbled, 1 medium onion chopped, 4 chilies chopped, 1TBS each: turmeric powder, yellow lentil, black gram, black mustard seeds, cumin seeds, plus 4 curry leaves, 2TBS oil, and salt to taste

Method: In a suitable pan or wok heat the oil and add, stirring, the lentils, gram, mustard, and cumin seed. Then the onion and chilies. Then stir in the turmeric powder and salt. Add crumbled/mashed air potato. Stir and serve.

NAMES: air potato, air yam, aerial yam, bitter yam, cheeky yam, potato yam, parsnip yam, Varahi Kand in Hindi, up-yam in Nigerian Pidgin English

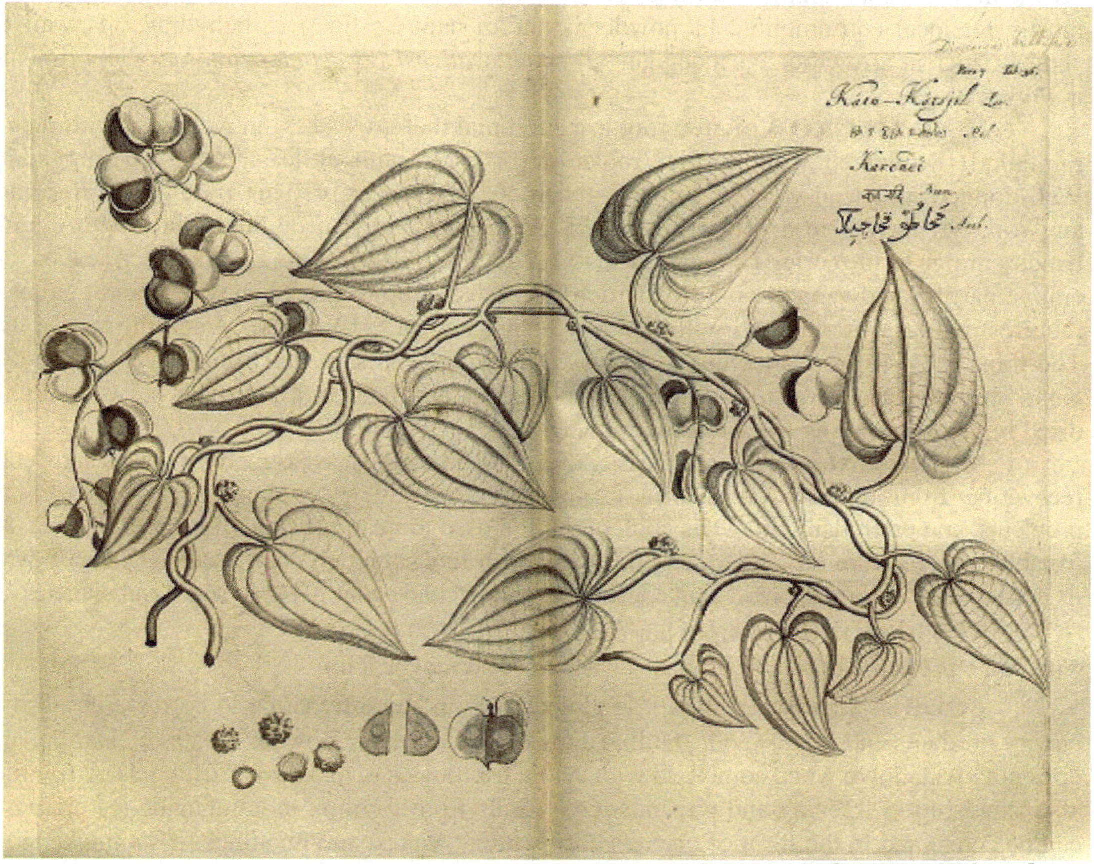

Courtesy Wikipedia

Your brain is your garden, what you sow is what you will reap. Seed it with hope,positive energy, good manners, and creativity. Irrigate it with faith, hard work, and good company. ~ Zeyad Massry

Sow a thought, and you reap an act; Sow an act, and you reap a habit; Sow a habit and you reap a character; Sow a character, and you reap a destiny.
~ Charles Reade

The roots of all goodness lie in the soil of appreciation for goodness. ~ Dalai Lama

ARROWROOT
Maranta arundinacea

West Indian arrowroot is a perennial that grows tall, 1½ meters shigh, and produces long, fleshy roots that taste like corn when boiled. In the 1600s, Spanish, Portuguese, and Dutch explorers spread arrowroot from the Western Hemisphere. It's native to northwestern South America and one of the earliest plants domesticated with evidence of cultivation dating to 8200 BCE. Some archaeologists believe that arrowroot was used not as food initially, but as a poultice for wounds. The Caribbean island of St. Vincent produces 95% of the world's supply. It is grown in S.E. Asia, China, Australia, Hawaii, Puerto Rico, and Florida in the United States, mostly for local consumption. Its powder is one of nature's finest carbohydrates. Hawaii has another, ancient arrowroot, a canoe plant, *Tacca hawaiiensis,* or *leontopetaloides* that originated in Polynesia.

HOW TO GROW: Arrowroot is a perennial that grows best in wet areas with regular rainfall. It is planted with smaller rootstock or roots with at least two sprouting nodes. Well-drained and well-worked soil lets the roots spread easily. This root grows in almost any soil, but thrives in a pH of 5.5 to 6.5. Arrowroots are narrow and white with circular bands similar to turmeric. The plant grows 1-2 feet tall, and has long, dark green leaves, and enjoys partial shady conditions. Arrowroot is a perfect crop to grow beneath coconut palms. It requires regular watering and will be ready for harvest in 8-12 months when the leaves wilt. The roots grow from 20cm (8in) to 45cm (18in) long. The skin must be peeled for use as it has a stronger taste. Wash several times before beating them in a mortar. The wet starch is dried in the sun, leaving the arrowroot powder.

MEDICINAL: Arrowroot is easily digested and can be used as food for babies and people recovering from illnesses or medical treatments, especially heart ailments. It helps uri-nary problems, and eaten daily, it lowers cholesterol. It's used to treat intestinal disorders, diarrhea, and irritable bowel syndrome. The root's fiber helps push foods through your system efficiently while its nutrients are absorbed, preventing constipation, and helping control blood sugar and diabetes.

NUTRITION: Arrowroot has 65 calories for 100 grams with some protein, potassium, iron, vitamin B-6, and magnesium.

FOOD & USES: Arrowroot is gluten-free, and an alternative to corn starch used in bakery products and infant foods. Unlike cornstarch, arrowroot powder creates a clear gel and does not break down when combined with acidic fruit juices to make clear fruit gels. Arrowroot also stands up to freezing and prevents ice crystals from forming in homemade ice cream. It can be consumed in the form of biscuits, pasta, bread, cakes, and puddings. It's a thickener in Asian sweet and sour sauce and makes a great binder in meatloaf and veggie burger mixtures. Arrowroot is an herbal replacement for talcum powder and skin care products that often contain carcinogens. It's a perfect alternative for baby powder. It's also used in hair dye because of its thickening ability.

RECIPES: ARROWROOT MILK PORRIDGE

Ingredients: 2 TS arrowroot powder, 2 cups milk, 1 cup boiling water,
1TBS brown sugar, a pinch of salt, ½ TS Vanilla extract, ¼ TS of nutmeg and ½ TS cinnamon
Method: In a pot, boil the water and cinnamon. Add the salt. Reduce heat to a simmer. Mix the arrowroot powder with just enough cold water to make a paste. Combine the sugar and milk. Constantly stirring, add the arrowroot paste to the pot of boiling water. Then add the sugar and milk to the pot. Simmer for 10 minutes and add vanilla and grated nutmeg. Serve hot.

NAMES: obedience plant, India: desi arrowroot; Spanish: arraruz, Chinese: zhu yu, Puerto Rico: amaranta, Cuba: raíz americana; sagú, Thailand: sakhu, Vietnam: củdong

BEETROOT

Beta vulgaris

Beetroot as a side dish never excited me when I was a kid. My grandmother served curried beets and roti. I loved beets when I had them baked with spicy mustard. Eating beetroots dates to prehistoric times. Ancient civilizations along the shores of the Mediterranean Sea first cultivated beets for medicine. Later, as the cultivated root grew plumper and more succulent, they were desired for their pleasant taste and texture. Beetroot can be a nutritious crop for the Hawaiian home garden. Protect against high summer temperatures with frequent watering.

Both the root and leaves are eaten, and dried beet powder is used as a coloring agent for many foods. Some frozen pizzas are colored by using beet powder added to the tomato sauce. The most common garden beet is a deep ruby red, but yellow, white, and even candy-striped (red and white are farmed. Beets to botanists are *Beta vulgaris*.

HOW TO GROW: The most grown varieties in Hawaii are Detroit Dark Red and Baby Ball. Both mature within 3 months. To plant beets, you must work the soil with a fork until it is loose to about ten inches deep. This is necessary for the beetroot to form to a good size. Plant seeds about a half-inch deep and one inch apart. Keep a foot between rows. It may take 10-14 days for the beets to sprout. If the seeds are planted too deep, or the soil gets hard after a heavy rain, the beets will sprout irregularly. Placing a board or a piece of plastic as a cover over the row of beet seeds may help keep the moisture in and protect from damage from hard rains. Remove any cover as soon as sprouts appear. The sprouts should be thinned to about two inches apart.

Clean the beet rows of weeds often but do it by hand. This is because using a hoe may damage the beets' shallow roots. Beets will not develop if they are overcrowded or are forced to compete with weeds. About a month after the beets sprout, carefully mold, pulling soil up to the leaves. Every month give the beets a light sprinkle with a 12 – 24 – 12 fertilizer mix. Keep the soil moist for the best results.

In about 2 months the beets should be about 2 inches in diameter and harvesting can begin. With proper thinning, light fertilizer, and water, local beets can grow to 3 inches in diameter. Larger beets tend to taste woody. Planting a row of beets every month will keep you supplied with this healthy food.

MEDICINAL: Red beet juice is high in iron, and it regenerates and reactivates the red blood cells, supplies fresh oxygen to the body. It is useful in treating anemia. Beet juice is a good treatment for jaundice, hepatitis, and nausea. Add a teaspoonful of lime juice to increase its medicinal value. Fresh beet juice mixed with a tablespoonful of honey taken every morning before breakfast will help heal a gastric ulcer. Beet juice combined with carrot and cucumber juices, is one of the finest cleansing agents for the kidneys and gall bladder. Beet juice acts as a natural tonic for dry, tired skin. Blend one beetroot with a half cup of cabbage; set some juice aside. Add one tablespoon of mayonnaise or olive oil with the beets and cabbage that remain after juicing and apply to your face. Mix the beet juice with water and freeze to ice cubes. Use one cube to cleanse and tone your skin every morning and evening.

NUTRITION: Beet juice is considered to one of the best vegetable juices. It is a rich source of natural sugar. It contains sodium, potassium, phosphorus, calcium, sulfur, chlorine, iodine, iron, copper, vitamins A, B1, B2, C, and P. This juice is rich in easily digestible carbohydrates, but the calorie content is low. Beets are rich in folic acid, which fights heart disease and anemia. Beets are high in fiber. Fiber is good for the intestines and helps regulate blood sugar and cholesterol. One cup of fresh, cooked beets has only seventy-five calories of which one and a half grams are fiber, with the same amount of protein, and eight grams of carbohydrates.

FOOD & USES: When preparing, always wash beets, and leave on the skin to contain the nutrients. Be careful because beet juice will stain anything it contacts. Beets can be microwaved, steamed, boiled, pickled, roasted, or eaten raw. Beets are delicious roasted in a hot oven. They are done when easily pricked with a fork. Raw beets also can be grated, diced, or sliced. Beets pickled in salt-vinegar brine are a tasty addition to salads.

RECIPES: BORSCHT - BEET SOUP

Ingredients: 5 nice-sized beets, 1 TBS vegetable oil, ½ carrot chopped, 2 garlic cloves, 1 small head of cabbage, 8 cups chicken broth, 2 large potatoes, 1 large onion, salt and pepper to taste

Method: Steam beets until skins pull off. Chop beets and potatoes into small cubes. Chop garlic, and onion, then sauté in oil. Add beets and carrots and stir constantly. In a large pot heat the chicken broth, add the potatoes and cabbage and cook for 10 minutes. Add the beet, garlic, onion mixture. Simmer for 20 minutes. Add salt and pepper to taste. Lemon juice may also be added if you like tart soups. Serves 6.

SPICY SHREDDED BEETS

Ingredients: 3 nice beets washed, peeled, and grated, ½ cup grated onion, 3TBS margarine, ½ cup water, 2 TBS lemon juice, ¼ TS each of cinnamon and nutmeg, salt and pepper to taste

Method: In a covered pan cook all ingredients over low heat for half an hour. Beets should be tender. Do not stir the mixture often. Serves four.

ROASTED BEETS

Ingredients: 5 nice-sized beets, 4TBS olive oil, 1 TBS Dijon Mustard, juice from 1 orange, and 1TBS vinegar, 1 garlic clove

Method: Place cleaned beets in a large bowl and mix with one tablespoon olive oil, salt, garlic, and pepper. Place in ovenware and into a 400-degree oven for 45 minutes. While beets are roasting, in a bowl mix mustard, orange juice, 3 tablespoons olive oil and vinegar. Remove beets and cool. Rub off the beet skins and slice or cube. Mix with dressing. Serves six.

CURRIED BEETS

Ingredients: 6 nice beets - cooked and chopped small, 1 large potato chopped small, ½ cup pigeon peas, ¼ cup of chopped onion, 3TBS butter, ½ TS cumin seeds, 1 sprig curry leaf, 2 green chilies chopped, ½ TS garam masala, salt, and spices to taste

Method: In a suitable frying pan heat butter and fry cumin seeds, bay leaf, chili, onion, pepper, and garam masala for 1 minute. Add potato, peas, and beets. Cook, stirring, for 2 minutes. Add salt, spices, and curry leaves with a little water. Cook gently until the potato is tender. Serve over rice or with roti.

DID YOU KNOW?

The Romans considered beet juice an aphrodisiac and grew beets throughout their empire. Australians put pickled beets on hamburgers. Homemade wine can be made from beets. Beet greens are even better for you than beetroots.

Breathe Properly, Stay Curious, and Always Eat Your Beets!
— Tom Robbins, *Jitterbug Perfume*

CALAMUS ROOT - SWEET FLAG

Acorus calamus

Calamus root is a perennial plant with stemless leaves on reeds that grows in moist, wet soil, scarcely on every island. It makes a clump of sword-shaped leaves pushing up to 150cm (5ft. tall. The roots double every year, so it seems to creep in all directions from where it was originally planted. It is often grown as a home garden ornamental because the stems, or reeds, are attractive and fragrant. It also has pale yellow flowers. A marshy area near a stream or pond is ideal. It's also known in English as sweet flag. The botanical name is *Acorus calamus.*

This is a water-marsh-wetlands plant that has a variety of uses as food and traditional medicines. Calamus root is probably native to India and traveled east and west along the Silk Road. The ancient Egyptians and Chinese considered it an aphrodisiac. The Romans and Greeks used the center reeds for weaving baskets. In the Middle East, warriors carried the root to treat battle wounds. A century ago, Dutch children chewed calamus root roots as natural chewing gum and enjoyed it candy-coated with caramelized sugar.

HOW TO GROW: First, find some calamus root roots/rhizome; look online. This is a water plant that thrives in continuously wet, moist soil or the edge of a pond. Planting is a trick. To begin, fill a small plastic pot that has drain holes at the bottom with rich compost. Put that container inside another bigger, watertight pot that has 3 cm (1in.) of water. That will permit the smaller pot to stay moist. Push the calamus root seed root into the surface of the potting soil, but let the surface of the seed root lay on top, open to the air and sun. Do not bury the seed. Keep the soil very moist and they should sprout within 2 weeks. When the sprouts grow to 12cm (4in.) transplant to separate pots and water them often or plant them in shallow water at the edge of a pond. Calamus root wants full or partial sun and mucky soil. It will grow in moist regular soil, but it won't get tall.

In 2 years, calamus root will multiply and you can divide the roots to plant other areas. In 3 years, there should be enough extra firm, large roots to harvest. The skin will become brown as the root matures. Harvesting is a dirty job because the roots and reeds are tangled in the moist, muddy soil about 30cm (1ft.) down. Once stripped of the reed leaves, the roots must be washed. Calamus root should not be peeled because in the outer section near the surface is where the cells of the aromatic volatile oils are located. Store in a cool, dry area.

MEDICINAL: The calamus root is used in traditional Ayurvedic-homeopathic medicine to revitalize the brain, improve the memory, and nervous system. This root simultaneously boosts energy, and you feel focused and alert. It's also a remedy for digestive disorders like gas, nausea, and loss of appetite. Chewing the root will kill the pain of a toothache and is supposed to break the addiction to tobacco. For two millenniums, calamus root has been used as an aph-rodisiac throughout the Middle East, India, China, and S.E. Asia.

NOTE: Chewing large amounts of the leaves and/or roots, and especially as a distilled essential oil, can cause mild hallucinations.

NUTRITION: Calamus root is a vegetable multivitamin and mineral. 100 grams has only 120 calories, but 15g are protein, 6g of fiber, a very high content of vitamin E with vitamins C and A. This root has very high amounts of potassium, sodium, magnesium, and calcium, with lesser amounts of phosphorus, copper, and zinc.

FOOD & USES: Peeled and washed, and after removing the bitter outer layer, it can be eaten raw with a taste similar to cinnamon and nutmeg. Its tasty roasted dried and powdered. Calamus root has a spicy flavor and can be a substitute for ginger and used in tea. This root is sautéed with sugar into a sweetmeat. The young stems are peeled, and the insides are eaten raw.

Young leaf buds are also added to salads. The leaves can be used fresh with milk to flavor custards and puddings the same way as a vanilla bean or cinnamon quill. The leaves are used in basket making or woven into mats or used to thatch a roof. All parts of this plant can be dried and used to repel insects or to scent linen cupboards. Growing this plant is said to repel mosquitoes. It has a fragrance reminiscent of patchouli oil. Rich in starch, the root contains about 1% of an essential oil that is used as a food flavoring.

RECIPES: CALAMUS ROOT CANDY – SORE THROAT LOZENGE

Ingredients: 500g calamus root peeled and chopped into ½-inch pieces, 4 cups of sugar, plus ½ cup sugar or powdered sugar (icing sugar or confectioners' sugar). 4 cups water

Method: In a pot, boil the calamus root pieces for 10 minutes. Drain, rinse the roots, and boil again. You can save 1 of these waters to make calamus root syrup. Boil and rinse two more times, 4 times total. Then bring only 4 cups of water to a boil and slowly stir in the sugar until it dissolves. Hold back 20 pieces of the root. Add the calamus root pieces and simmer for 10 minutes. Remove from the heat and again drain and save the sugar syrup. Roll the root pieces in the extra ½ cup of sugar. Lay the sugared calamus root pieces on waxed paper to dry for a full day. Store in an airtight container. The sugared pieces are bitter and spicy hot until they sit for a few days.

CALAMUS ROOT SYRUP: continued from above

Ingredients: Boil the 20 remaining pieces of calamus root with the extra water and sugar syrup, stirring until it is the thickness you desire to treat coughs and sore throats.

HAIR REJUVENATING OIL

Ingredients: 100ml each sesame, coconut, and castor oil, 5 calamus roots, 1TBS fenugreek seeds

Method: Slightly pound the calamus root. Don't peel or smash the roots, just rough it up so that the skin is broken, and the inner flesh is exposed.

Method: In a saucepan, bring the oils to a boil. Then add the pounded calamus root pieces and fenugreek seeds. Reduce heat and simmer for a few minutes. Pour into a sterilized jar. Use as needed. To prevent hair loss, rub it into your scalp at night twice a week.

DID YOU KNOW? Benedictine and Chartreuse liqueurs were flavored with this root, as well as many liqueurs, beers, tonics, and gin as late as the 1960s. It was an ingredient in the original recipe for Dr. Pepper soft drink. The famous Stockton herbal bitters include calamus and gentian roots. It has been used by many cultures since biblical times. In Sanskrit, it was kalamas, the Roman used calamus and the Greek used kalamos to name the reed of the plant. In the 17th century, calamus root had so many uses everywhere in the world that it was over-harvested and nearly extinct. Since ancient times, *Acorus calamus* has been used as a hallucinogen. It may have been one of the ingredients of the Holy Oil that God commanded Moses to make (Exodus 30). Calamus oil is one of the ingredients in an ancient recipe for anointing oil found in the Bible. In medieval churches and palaces, the reeds were placed on the floor because of its sweet, spicy fragrance for freshening the air. This plant was once used in toothpowders. It has been used magically for luck, healing, money, and protection. It is said that if you place the root in the corners of the kitchen, you will never be hungry. Calamus earned its nickname 'Singer's Root' for its ability to numb the vocal cords of tired singers so that they could continue. If your voice is strained or raspy, or you get laryngitis, try gently chewing just 1 cm (½in. of this root and leave it between your cheek and gum throughout the day. **NAMES:** calamus root, sweet flag, rat root, sweet sedge, flag root, sweet calomel, sweet myrtle, sweet cane, sweet rush, beewort, muskrat root, pine root, myrtle grass, sweet cane, sweet grass, sweet rush, and wild iris

CARROT – KĀLOTI

Daucus carota

We take the delicious, crunchy carrot for granted. Carrots can grow all year throughout the Hawaiian Islands, but those grown during the cooler months seem to taste better. There never seems to be any shortage even though most are imported. My favorite is to pickle firm carrot pieces with hot chili peppers. The carrot, botanically *Daucus carota*, is one of the world's most popular vegetables grown for the thickened orange root.

Carrots originated some 5000 years ago in the hills of Punjab and Kashmir in India and slowly spread into the Mediterranean area. The first carrots were white, purple, red, yellow, green, or black - not orange, and its roots were thin. Ancient Egyptians highly regarded the me-dicinal properties of carrots. Drawings containing information about carrots used in medicinal treatments and even carrot seeds have been found in pharaoh tombs.

HOW TO GROW: Always reasonable in the market, carrots are easy to grow locally. Although carrots will endure some heat, they grow best planted in a shaded, cool area with sandy soil with a pH of 6-7. The best growing temperatures are 15° to 20°C. Romance is a good carrot seed, but it's best to try a few different varieties of seeds. The soil should be prepared to a depth of ten inches with clumps smashed completely so the carrots can fully develop. Two or three seeds should be planted every inch, at a half-inch deep, with rows 12 to 18 inches apart. Thin the seedlings when they are one inch tall to one seedling per 1 to 2 inches. It is essential to keep weeds under control for the first few weeks. Carefully pull loose dirt up and mold the rows when the carrot roots begin to enlarge, usually forty days after planting. Fertilize with a higher phosphorus mix when the plants are about a month old. Carrots can be harvested when the root tops are at least ¾ inch in diameter, about 70 to 90 days after planting. Dig to remove the roots without damage. They may be harvested over a 3 to 4-week period. Big, overgrown carrots are less tasty. Stored properly carrots keep 4 to 6 months.

MEDICINAL: Carrots are rich in beta-carotene, which the human body converts to vitamin A. Beta-carotene fights some forms of cancer, especially lung cancer. Eating carrots may also protect against stroke and heart disease. Carrot juice strengthens the eyes and is a great treatment when rubbed on dry skin. Carrot juice drank daily may cure colic, colitis, and peptic ulcers. Raw carrots are good for fertility. Eaten after a meal, a raw carrot destroys germs in the mouth and prevents bleeding of the gums and tooth decay. Carrots speed up digestion, and when eaten regularly prevent gastric ulcers and other digestive disorders. Carrot soup is a natural remedy for diarrhea. Carrots fight all body parasites including intestinal worms.

NUTRITION: A half-cup serving of cooked carrots contains four times the recommended daily intake of vitamin A, with only 35 calories, 1 gram of protein, 8 grams of carbohydrates, with 2 grams of fiber. Carrot is rich in alkaline elements that cleanse and rejuvenate the blood. Carrots also supply vitamin C, calcium, phosphorus, sodium, sulfur, and chlorine with traces of iodine. The mineral contents in carrots lie very close to the skin and should not be peeled or scraped off.

FOOD & USES: Always wash carrots first. Raw carrots are naturally sweet, yet slightly cooked carrots may be sweeter. Carrots lose very little vitamins and minerals during cooking. Carrots are a versatile veggie and can be shredded, chopped, juiced, or cooked whole. They are delicious roasted, boiled, steamed, stir-fried, grilled, and they team up beautifully with almost any vegetable companion. Carrots boost the nutritional value of soups, stews, and salads.

RECIPES: FRESH HERBS WITH BRAISED CARROTS – a tasty side dish Ingredients: 500g (1 lb.) carrots chopped 3 inches long, 1cup canned or fresh beef broth, 1TS

honey, 1TBS butter or margarine, 2 TBS fresh parsley, chopped (or 1TBS dried)
Method: In a medium saucepan, bring beef broth to a boil, add carrots, honey, butter, and parsley. Cover and simmer for 10 minutes. Serves 4.

MARINATED CARROTS — This is a great healthy salad or snack.
Ingredients: 500g (1lb.) carrots, 2TBS lime juice, ½ TS brown mustard, ¼ cup olive or virgin coconut oil, 2 spring onions chopped, 1 TBS chopped parsley, 1 clove garlic crushed, and ¼ TS salt
Method: Scrub and peel carrots. Cut into ½ inch-wide strips. Boil carrots for 3 minutes or until tender, but still crispy. Drain. In a small bowl, stir together the lemon juice and mustard. Using a fork mix in olive oil a little at a time. Add chives, parsley, and garlic. Pour over warm carrots. Season to your taste with salt and pepper. Refrigerate 6 to 8 hours or overnight. Makes 3 cups.

HEALTHY CARROT DESERT
Ingredients: 3cups carrots grated, ½ cup wheat germ, 1cup dried fruit - raisins, or dates (seeded, 1 cup chopped nuts or grated coconut, ½ cup honey, 1 TS cinnamon, ½ TS nutmeg
Method: Combine all ingredients. Press into a bowl and chill. Turn and tap out onto a plate. Serve with plain or vanilla yogurt.

CARROT-PINEAPPLE CAKE
Ingredients: 2 cups shredded carrots, 3 cups flour, 1 small pineapple chopped and mashed saving the juice, (You can use an 8 oz. can of pineapple drained. 1½ cups brown sugar, 3 large eggs beaten, 1½cups canola or virgin coconut oil, 1cup chopped cashews (or grated coconut, 2 TS cinnamon, 3TS baking powder, 1TS baking soda, 1TS vanilla extract, 1TS salt
Method: Mix flour, sugar, baking soda, baking powder, cinnamon, and salt in a large bowl. In a separate bowl, add beaten eggs and stir in oil and vanilla extract. Add this and the crushed pineapple, shredded carrots, and nuts to the flour and stir until the mixture becomes moist. . This will make a large cake or can be halved. Pour into greased and floured pans. Bake 50 minutes to an hour at 350 degrees. Use the toothpick method to tell when done in the center. Cool completely before serving.

BAKED CARROTS
Ingredients: 2cups carrots sliced thin, 3TBS butter, ½ TS nutmeg, 1TBS brown sugar, ¼ cup water, salt, and spices to taste
Method: In suitable ovenware combine nutmeg, sugar, salt, and your spices with water. Add carrots and dot top with butter. Bake at 350 - 20 minutes. 10 minutes before serving

DID YOU KNOW?Carrots are a distant cousin of cilantro. Ancient Greeks believed eating carrots made both men and women more amorous and recommended women eat carrot seeds to prevent pregnancy. Dutch carrot growers cross-bred pale-yellow carrots with red carrots and created today's orange carrot to honor of the House of Orange, the Dutch Royal Family. China is the world's top carrot grower, producing 35% of the world's carrots. Carrots are about 87% water. The longest carrot ever recorded was almost 17 feet long and the largest carrot ever recorded weighed 18 pounds. Unlike most other vegetables, carrots are more nutritious when eaten cooked than eaten raw. Carrot greens are high in vitamin K, which is lacking in the carrot root and its skin contains 10% of all nutrients found in carrots.

It's a sad donkey that runs out of carrots.
~ Anthony T. Hincks

CASSAVA – MANIOKA

Manihot esculenta

Cassava must be one of God's greatest gifts to this earth. It takes little effort to plant; this root grows easily and tastes so great. Cassava is a perennial bush with an edible root. The root is a tasty food and a great source of carbohydrates. Scientists rate cassava after rice and wheat in the importance of feeding the world. The root is the main source of nutrition for about 500 million people. In Hawaiian, it is manioka or manioc. This root grow's easily throughout the Hawaiian Islands but not heavily planted. Cassava's botanical name is *Manihot esculenta.*

HOW TO GROW: Cassava is an important food source because it is easily grown in almost any type of soil with a pH that may vary from 5 to 9. Cassava can withstand almost any conditions from infertile soil to droughts. Only 20 inches of annual rain is necessary for the crop to mature. Cassava bush grows 6-8 feet tall with large green reddish veined leaves. The stalk has bumpy nodes from which new plants are grown. Cassava plants may be bitter or sweet. Consult the local agriculture station and try to get some MX or butter-stick variety planting sticks. The roots grow in clusters and can be up to five inches thick and over eighteen inches long, covered with a thin brown bark. Since the bark may contain toxins, it must be removed before cooking.

Some varieties of cassava are poisonous without proper preparation. The poison is linamarin, which is chemically similar to sugar. When eaten raw, the human digestive system converts it to cyanide poison. It only takes two cassava roots to contain a fatal dose of poison. This cyanide is a natural protection for the roots as it repels most bugs and borers. Primitive tribal cultures grind the roots into a paste, which releases many poisons. The paste is placed in a wicker tube (called a tipitipi), stretched on a frame. The juice containing the poison drains out. The paste is made into loaves and sun-dried for better storage. The poison is used to kill river fish..

To plant cassava, first you need to locate a suitable variety already growing, preferably MX. Cut the stem of the growing plants into one-foot-long pieces. Dig a two-foot circle about a foot deep with a fork, and then hoe the clumps until the soil is smooth. Insert the stems, root side down, about four inches. This is the most important aspect of growing cassava. The thicker end of the stem must be planted as it produces the root. Cassava is a long crop, taking eight to ten months. Some new varieties mature in less than six months. Watering every other day is necessary at the beginning, but can be reduced to once a week after the new leaves have sprout-ed. Water heavily twice a month during the dry season. Cassava's main enemy is too much water, which will rot the roots. Use a high-potassium fertilizer monthly. The only problem with cassava is it spoils relatively quickly. Stored properly in a cool dark place, the roots should last three weeks.

Uncovering the cassava roots takes time and patience, or the roots will be damaged. To harvest, carefully clear the soil from around the tree to determine where the roots run. Carefully fork and gently pry until the tree is free enough to shake. Continue working the roots loose. If a root is pierced with a fork or broken it must be cooked, otherwise it will spoil and rot. Digging cassava is a lot like unwrapping a gift; you never know what you are going to get!

MEDICINAL: Cassava leaves and roots have been a folk remedy for tumors and can-cers, which may be due to the B17 content, also known as laetrile. Vitamin B17 is also found in some seeds like apricots, peaches, and apples. B17 stimulates hemoglobin red blood cell count. The bitter cassava variety is used to treat diarrhea and malaria. The leaves are used to treat hypertension, headache, and pain. Cubans commonly use cassava to treat irritable bowel syndrome, the paste eaten in excess during treatment.

NUTRITION: Cassava is an important food source because it is high in starch, which produces energy. A half-pound of cassava has 215 calories, of which a quarter is carbohydrates, with one gram of fat, and three grams of protein. It also has vitamin C.

FOOD & USES: Cassava can be processed into lour, pastes, and granules, or fermented. Some African countries boil cassava leaves as a vegetable. In Central and South America cassava is boiled, fried, or made into cassava lour bread. Cassava is a great addition to soups and stews.

RECIPES: CASSAVA CAKE

Ingredients: 2 cassava roots and 1 dried coconut–both grated, 1 cup milk, 1 egg, 1 cup sweetened condensed milk.

Method: Mix cassava, coconut, and milk and place in a greased baking dish. Bake in a 350-degree oven for half an hour. Remove. Beat the egg into the condensed milk. Spread this mixture over the cassava cake and replace it in the oven for twenty minutes. Let cool before slicing.

CASSAVA PONE

Ingredients: 4 pounds cassava, 2 pounds pumpkin, 2 dry coconut, 2 small cans of evaporated milk, 1 TBS vanilla, 1¼ pound margarine or butter, cinnamon, and nutmeg to taste.

Method: Grate cassava, pumpkin, and coconut meat and mix well with other ingredients. Pour into a well-greased baking dish and bake at 350 degrees for 1 hour.

ASTOUNDING CASSAVA

Ingredients: 2 pounds boiled, mashed cassava, 1 egg, ¼ cup lour, 1-pound minced beef, chicken, or fish, 1 bunch chives, 1 bunch celery, salt, and spices to taste, oil for frying

Method: Spice the minced meat to your taste and fry till brown. Mix the egg, some salt, into the mashed cassava with the lour. Flatten this mixture and divide it into two parts. On one part, place the minced meat, celery, and chives. Cover with the other part. Cut into squares and pinch the ends. Fry until golden brown. This could also be baked in a casserole dish at 350 for an hour.

CASSAVA BANANA CAKE

Ingredients: 4 pounds fresh cassava-peeled and grated, 2 TBS sugar, 16 ounces of coconut cream (canned Coco Loco), 1 ripe banana mashed,

Method: Place sugar and coconut cream in a large bowl and mix until all sugar is dissolved. Stir in mashed bananas. Blend the grated cassava into the mixture and mix until fairly smooth. Pour the mixture into the greased cake pan and bake at 250F for 30-40 minutes until golden brown.

CASSAVA PUDDING

Ingredients: 2 pounds fresh cassava - peeled and grated, 3 ounces shredded coconut, 1TBS freshly grated ginger root, ½ pound of sugar, 4 cloves lightly crushed

Method: In a large bowl, combine all ingredients and blend well. Divide the mixture in four and place each portion in the center of four pieces of foil or traditionally on banana leaves. Wrap into parcels and steam for 40 minutes. Serve hot.

My friend CB harvesting my cassava roots

DID YOU KNOW? It is reputed Columbus was introduced to cassava when he first visited the Island of Hispaniola. The native Arawak Indians fed the explorer a type of fried cake prepared from the root. Portuguese and Spanish sailors brought cassava to Europe, Africa, and Asia – mainly as a food source for slaves. In 2000, worldwide cassava production was more than 175 million tons, with more than 50% grown in Africa. Cassava is also used for animal feed, laundry starch, and has become a source for ethanol fuel production. Thailand and Indonesia are making biodegradable grocery bags from cassava.

COCOYAM - MALANGA – YAUTIA - TANNIA
Xanthosoma sagittifolium

Cocoyam is the local Hawaiian name for a tasty root with a large spearpoint-shaped leaf. Cocoyam as with taro, names other roots depending where you are standing and who you are talking with. The local names for Hawaiian roots would be an excellent master's thesis. This delicious root is also known as tannia, yautia, malanga, elephant ears, Mickey Mouse ears, or cocoyam tanier. Botanically, it's *Xanthosoma sagittifolium.* I have to choose one name and that will be cocoyam. This root originated in tropical South America. English and Spanish explorers found it cultivated throughout the Caribbean chain of islands. They carried it to Asia and Africa, where it became a major food source. Some kind person brought it to Hawaii. Many consider cocoyam's flavor and texture superior to kalo, cassava, potatoes, and yams. This root is a local favorite root and the grown on every island in hot uplands and swamps and heavily cultivated in Kana, Hawaii. Another slightly different variety is termed bun-long. This is a beautiful big, green leaved plant for landscaping.

HOW TO GROW: First, you must find some cocoyam root seeds to plant. Look in the markets and you might be lucky to find one sprouting a little pink head. The Hawaiian Islands have a few types of cocoyams, just as there are several types of manioc/cassava. The one cocoyam that has a faint red line along the leaf stem is tasty. This root grows best in open sun, but needs consistent water. Don't bother planting if your soil is all clay. This root grows best in moist, loose soil with a pH of 5.5-6.5.

To plant cocoyam, first work the ground with a fork so that it's soft and free of hard, dirt clumps. Blend in some well-rotted chicken/cow manure, a quarter-cup of crushed lime-stone, and form mounds. If you are using a head, split it, and put it in the mound, split side up, and cover with about 5 inches of loose soil. If you are planting a shoot attached to the seed, slide it in so the attached root is facing up, and cover with about 5 inches of loose soil. If you plant shallower, it will produce many small side shoots rather than one big root. Cocoyam's big ele-phant ear leaves don't like neighbors, so space your plant about a meter apart. After two weeks use 12-24-12 fertilizer at about an eighth cup per plant and again every two months when you remold. Keep the weeds down until the leaves develop. Fungus and bacteria are cocoyam's main enemies; spray monthly with Ban Rot or Rizolex. This is a long crop and can take from 8-12 months. In the sixth month fertilize with a 13-13-21 mixture. Keeping these roots well-drained during the rainy season is the best solution. During the severe part of the dry season, water your cocoyams heavily, once a week. Harvest carefully with a garden fork. It's wise to wash the roots clean of dirt and rinse with a mild disinfectant if you wish to store them for months. Save the small seeds, cormels, to replant. Some areas consider this an invasive plant, but it is an invasive with tasty roots, but requires cultivation.

A seasoned farmer recommended planting cocoyams in the ascending moon and to get long fat roots, but if he plants in the descending/waxing moon, he gets long thin roots. Cocoyam is in demand because of its taste, and unlike kalo/taro/eddo, it always will boil/soft-en. West Africa, Cuba, Puerto Rico, and the Dominican Republic are the biggest producers of cocoyam.

MEDICINAL: This root is used in Brazilian traditional medicine to prevent osteoporosis. Eating delicious cocoyam regularly will reduce hypertension and rheumatism. Used as a poultice it will make boils form a head. The juice is used for skin care. Research is being done if this root inhibits the proliferation of leukemia cells.

NUTRITION: 100 grams of cocoyam has 105 calories with 2 grams of protein. It contains high amounts of potassium, phosphorus, calcium, and magnesium with trace amounts of copper, zinc, manganese, and selenium. It has amounts of vitamins A, C, B9 (folates), and beta carotene.

FOOD & USES: Malanga makes a great landscaping plant. It is tropical-exotic and will grow to a more than a meter tall with full sun and plenty of water. The best thing is it can be easily groomed so there are no weedy areas, just lush tall malanga that look like they belong in Jurassic Park. Plan your backyard landscaping, so your cocoyams won't shade a large area.

RECIPES: COCOYAM FRITTERS

Ingredients: 500g (1lb.) cocoyam - peeled and grated, 2 stalks of celery chopped, 1 medium onion minced, 4TBS flour, 1TS cayenne pepper, 1TS salt, 1TS lime juice, 1 egg, 2TBS milk, spices to your taste.

Method: Combine everything in a bowl. Beat the egg in well. Carefully drop spoonfuls of this cocoyam mixture into hot oil until both sides are brown. Drain and serve hot.

WAIKOLOA COCOYAM PIE

Ingredients: 500g (1lb.) cocoyam - peeled, boiled, and mashed, 1pound cooked minced meat seasoned to your taste, 1 medium onion chopped, 2 garlic cloves minced, 1 medium sweet pepper chopped small, ½ hot pepper seeded and well minced, ½ cup cheddar cheese grated, ¼ cup milk, 2 TBS flour, 1 TBS butter, salt, and seasonings to your taste

Method: Mash the cocoyam with milk, butter, your seasonings, and flour into a pie shell shape. Then fill with the minced meat (lamb, beef, or chicken) combined with onion, garlic, and pepper. Bake at 350 for a half-hour and then cover the top with grated cheese. Bake for another half-hour. Serve hot.

KAPAPALA COCOYAM SOUP

Ingredients: 500g (1lb.) of cocoyam- peeled and chunked, 2 cups vegetable stock, 1 medium onion chopped fine, 2 bunches chives chopped, 2 garlic cloves minced, 2 TBS butter or margarine, ½ cup milk, salt, and spices to your taste

Method: Lightly brown cocoyam pieces with the onion and garlic in a large pot with the butter. Add stock and spices, and boil for 5 minutes. Simmer for 20 minutes. Then blend in the milk and chives and serve.

COCOYAM BEDUMA – delicious, exotic Sri Lankan stir-fry

Ingredients: 250g (½ lb.) Cocoyam– peeled and sliced thin, 1 onion chopped, 1TBS chili flakes, 1TS turmeric powder, 10 curry leaves, piece of pandan/rampe, salt to taste, oil for frying

Method: Put cocoyam slices in a bowl and coat with a mixture of all the ingredients. In a suitable pot or wok, fry the slices until they float golden brown.

DID YOU KNOW? 400 million people depend on cocoyam for food and The *Spanish Caribbean* calls it yautía or malanga. In *Thailand* it is kradat-dam, *Vietnam* it's khoai sáp or hoàng thu, *Papua New Guinea* it's kong kong taro. In *India* it's palhembu or seemavhembu. In *Hawaii* it's widely known as Chinese taro, Mickey Mouse Ears. There are two types of cocoyam: white and yellow. West Africa is now the major cocoyam producer. According to the National Library of Medicine, cocoyam is one of the world's six most roots.

One who plants a garden, plants happiness. ~ Unknown

COLOCASIA – WILD TARO – EDDO

Colocasia antiquorum

Most islanders enjoy eating different roots and all have many nicknames. This particular root can be called colocasia, little, hairy, or wild taro, or eddo. Most Hawaiians call it great food. In the Hawaiian-English dictionary (*https://ulukau.org/*), there's no translation for this specific root, as Mr. Google wants to rename it *Colocasia esculenta,* but this is a very different, smaller root. There were 300 varieties of what's termed kalo/taro. Almost 90 still exist of the dry and wet land kalos.

I know this tasty root n the Caribbean as the eddo, but for Hawaii, its name is colocasia. This will grow almost anywhere throughout the tropics. Once planted, the colocasia can be harvested every six months. It's a starch with a hairy outer coating like a cross between an old red onion and a hairy coconut, a good substitute for potatoes. The starch molecules in colocasia are among the smallest in the plant world and easy to digest.

Colocasia is a very old food, with a history of cultivation longer than that of wheat. First grown in Southeast Asia, this root was first recorded by the Chinese about the time of Christ. It's mentioned on one of the bamboo slips found in the two-thousand-year-old tomb of the Lady Tai. Colocasia were grown around the Mediterranean long before the potato. Who brought this specific root to Hawaii; was it the Asian, or the European explorers and settlers?

HOW TO GROW: Growing colocasia is very easy in moist soil and best with a pH of 5.5 to 7. First, locate some mall roots to plant. For a nice sized row, you'll need a gallon bucket of starter colocasia. Fork the row about 12 in. deep and the same wide. Pull dirt up so the row is about 8 in. high. Plant the starter colocasia so the green stem points upward. Space the roots about 6-9 inches apart. Keep colocasia watered and they should sprout new stems in 2 weeks. For a small vegetable plot, or even a flower garden, colocasia make a nice border, but initially will take daily watering. A soaker hose is a useful tool to get the colocasia growing. Colocasia grow up, not down as most roots, so dirt must be carefully pulled around the protruding roots. This molding will cause the colocasia to start more of the small clusters. Once a month fertilize with diamonium sulfate and phosphorus. Colocasia must be harvested in the dry season when their leaves yellow, wilt, and disappear. This is usually a 5–6-month cycle depending on the occurrence of rain. Use a fork and carefully pry the soil from the clusters. Then wash and store in the sun to dry. Once dry, remove to a cool dry place and colocasia will last for a few months. In the United States, colocasia are considered an invasive plant. The only problem with planting colocasia is getting rid of them. Left alone, colocasia will naturally multiply, especially if the area is often wet.

Clusters of smaller brown hairy roots surround the central 'head' root. It is the smaller roots, to harvest. The flesh is usually white; but this root can also be yellow, pink, or orange, and can weigh up to five pounds. The taste is like an Irish potato, but with a pleasant, slightly nutty flavor. Raw colocasia should never be eaten, all varieties contain calcium oxalate crystals. These crystals can cause considerable discomfort, but disappear during cooking. Even peeling can irritate the skin.

MEDICINAL: Since ancient times, the colocasia root has treated various ailments as asthma, arthritis, diarrhea, and skin problems. The corm juice is a laxative, and pain reliever. The leaves of the plant are anti-diabetic, anti-inflammatory, will expel intestinal parasites/worms. A decoction of the leaves is drunk to promote menstruation. The leaf juice treats conjunctivitis. In New Guinea, the leaves are heated over a fire and wrapped in a poultice to draw boils.

NUTRITION: 100 grams of colocasia root has 112 calories, 2 grams of protein with no cholesterol. The roots contain vitamins A, C, B-6, and E, folate, niacin, potassium, and calcium. The leaves are rich in vitamin C, phosphorous, calcium, iron, riboflavin, thiamine, and niacin.

FOOD & USES: Colocasia have a creamy, nut-like flavor. They can be fried, baked, roasted, boiled, or steamed. Colocasia absorbs large quantities of liquid while cooking adding bulk and flavor. Casseroles, soups, and stews benefit from these roots. Select tubers that are firm, hairy, with no wrinkling. Store roots for up to one week in a cool and dry location. Don't let them dry out.

RECIPES: COLOCASIA ROOT CURRY

Ingredients: 500g (1 lb.) peeled and chopped colocasia, 3 garlic cloves chopped, 1 red onion chopped, 3 green chilies chopped, 10 curry leaves, 1TS curry powder, 1TS turmeric powder, ½ TS mustard seeds, 1TS chili powder, 1TS or 1 stick of cinnamon, ¼ cup grated coconut, 1/2 cup of coconut milk, and salt to taste

Method: In a suitable pot or wok, mix all spices, salt, and onion with the colocasia pieces, add enough water to cover, bring to a boil and then reduce heat and simmer. In a mortar, grind the garlic, chilies, and coconut into a paste and stir into the simmering colocasia. Add coconut milk, increase the heat until it boils. Reduce heat and simmer for 5 minutes. Enjoy.

CREAM OF COLOCASIA ROOT SOUP

Ingredients: 2lbs. colocasia - peeled and diced, ¼ cup celery chopped, 4 cups chicken broth, 2 garlic cloves minced, 1 medium onion chopped, 2 TBS butter or margarine, 1 bunch culantro chopped, salt, and spices to taste.

Method: In a large pot on low heat, melt butter, stir in onion, garlic, and celery; then add colocasia. Cover for fifteen minutes. Add chicken broth and boil until the colocasia are soft. Add culantro. This may be served it 'as is' or pureed in a blender.

COLOCASIA ROOT SHOESTRINGS

Ingredients: 2lbs. colocasia, salt, and spice to taste, oil for deep frying (3-4 cups)

Method: Peel and slice colocasia into thin strips (1/8 inch thick), Place strips in ice water for half an hour then towel dry. Drop into heated oil and fry until golden brown. Turn carefully with a wire skimmer. Drain on newspaper or paper towels. Sprinkle with salt and seasonings.

COLOCASIA CAKES

Ingredients: 1-pound colocasia grated, 2 culantro leaves chopped fine, 1 medium onion chopped, 1 sweet pepper chopped, 3 TBS butter or margarine, 2 TBS flour, salt, and seasoning to taste.

Method: Cook onion, pepper, in butter until browning. Blend grated colocasia, flour, spices, and culantro. Drop large spoonfuls into the hot skillet on the onions and peppers. Fry until golden brown. Cakes should be 3 inches in diameter and less than 1 inch thick. Flip with a spatula.

DID YOU KNOW? Cultivated for 6,000 years, colocasia has traveled all around the world. The botanical name for colocasia is *Colocasia antiquorum*. Researchers believe it originated in the Bay of Bengal region. Traders are credited with taking colocasia root to Japan, China, and the Mediterranean over 2,500 years ago. Historians say it reached the South Pacific around the time of Christ. It was many centuries later when New Zealand and Hawaii received the colocasia. This root is a staple in Asia, the Pacific Islands, Africa, the Caribbean, and parts of South America. Soviet astronauts took dehydrated colocasia to the space station!

Colocasia antiquoram also known as eddoes and *colocasia esculenta* aka dasheen, look alike but *esculenta* is bigger above and below ground.

Gardens are an autobiography, and still a friend you can visit anytime.~ Unknown

FINGER ROOT – CHINESE GINGER

Boesenbergia rotunda

In English, this Asian root is called finger root, because the roots grow in appendages similar to fingers from a central orb or palm. The mature root looks like a multi-fingered hand. Finger root is grown for its edible roots used for nutrition and as a spice. For centuries it has been for traditional-homeopathic medical treatments. It is perfect for the home garden as a functional ornamental hedge plant because its white or pale pink flowers are fragrant. The roots have a light brown skin over pale-yellow flesh that has a pleasant aroma.

Finger root is considered a distant relative of ginger and galangal. It's believed to be native to the tropical rain forest areas from southern China to Malaysia. Finger root's taste is ginger's sweet/warm bite combined with the mild bitter heat of black pepper. Its flavor is milder, and earthier than galangal. It's grown on a limited scale on every Hawaiian Islands; try to locate some.

HOW TO GROW: It's a perennial root, termed a rhizome, usually grown as an annual. Once planted, its roots will continue to creep and multiply, but the above-ground green plant wilts at the end of every growing season. Finger root requires well-drained organic rich soil with a pH of 6-7 and grows best with morning sun and afternoon shade and regular water. To plant, find some finger roots and wrap in damp paper in your fridge for a few days. Carefully separate the roots, plant 2 inches deep one foot apart. Pineapples look great planted between the finger roots.

The green tops will grow to 2-3ft. tall with long dark green leaves. The white or pale pink flowers appear one at a time. Well-maintained roots can grow to a foot long in soft rich soil and usually mature within 3-4 months. This root grows well in humid, hot areas up to 1,200m elevation. These roots also grow well in porch/balcony pots.

MEDICINAL: Tonics based on finger roots are used in post-partum teas. It is eaten raw, boiled, or steamed to treat stomach ulcers, indigestion, nausea, and to relieve gas pains. It also treats coughs, and mouth ulcers. Eating finger root fights the bacteria, H.pylori, associated with peptic ulcers and colon cancer. Externally, it is used to treat ringworm. Finger root is believed to be an aphrodisiac. Research is ongoing that finger root's anti-bacterial properties fight E. coli.

NUTRITION: 100grams of finger root has 80 calories with almost 2g of protein. It contains potassium, magnesium with vitamins B-6 and C.

FOOD & USES: Finger root is an herbal spice with a unique flavor and pleasant smell used in place of ginger or galangal in fish curries and coconut recipes. It contains essential oils that are the base for various fragrances. It can be dried, pickled, or frozen and made into a powder. It's an ingredient in some Thai-style curry powders. These tasty roots are principal ingredients in panang curry, drunken noodles, and many soups.

RECIPES: FINGERROOT AND LEMONGRASS TEA

Ingredients: 4 -6 finger roots scrubbed, sliced thin, 4 lemongrass bulbs peeled and sliced thin, 4 cups water, honey, or sugar to taste.

Method: Put lemongrass and finger root pieces in a pot with the water and boil for 5 minutes. Reduce heat and simmer for 15 minutes. Stir in sweetener if desired. Enjoy warm or chilled.

NAMES: *English:* Finger root or Chinese-keys, galingale, tumicuni, Chinese ginger, *Thai:* krachai, *Indonesian:* temu kunci

GALANGAL
Alpinia galanga

Galangal root is not well known in the Hawaiian Islands, but you can find it on every island. This is another Asian root. It's native to Southwest China and throughout Southern Asia. It's another rhizome/root related to ginger and turmeric, with a creeping stem that sends out shooters to create new plants. For many centuries, all three have been used in Ayurvedic and traditional Chinese medicine.

This root is also known as Thai ginger or Siamese ginger. There are a few different varieties of galangal. Greater galangal has orange-brown skin with white/pale yellow flesh. Lesser galangal has orange flesh and sharp aroma.

HOW TO GROW: Galangal is an attractive plant that will give your home garden a tropical flair. This makes a good border or hedge. It can grow to 6 feet in height with small light green flowers with a red-veined tip. The flowers produce red berries. Growing it is easy and it's a low-maintenance plant. Pick a location with full sun in well-draining soil, but keep the soil moist. It's recommended after you locate your galangal seed roots, wrap them in a damp paper towel and put in your fridge's veggie drawer for a few days. This encourages the growth of 'eyes' that will sprout into other roots. Cover it with several inches of soil. Space the roots a foot apart. As the eyes sprout, it will become a bushy plant. It takes about a 10-12 months to reach maturity. The roots are usually pale yellow to white. Some varieties are blue-tinged, giv-ing it another name of blue ginger. When dried they are light brown. You don't want to overly shade nearby garden plants, but you may use the shade to your advantage. Once established it is relatively easy to move.

MEDICINAL: It can add flavor, antioxidants, and anti-inflammatory compounds to your dishes and provide many health benefits. These include protecting you from infections and potentially from certain forms of cancer. Test-tube studies suggest that the active compound in galangal root, known as galangin, may kill cancer cells or prevent them from spreading Emerging evidence suggests that galangal root may boost male fertility.

NUTRITION: 100 grams of galangal contains 45 calories and is also a source of sodium, iron, and vitamins A and C.

FOOD & USES: Similar to ginger and turmeric, galangal can be eaten fresh or cooked and is a popular addition to many Chinese, Indonesian, Malaysian, and Thai dishes. Ginger offers a fresh, sweet-yet-spicy taste, while galangal's flavor has a sharp citrusy, almost piney flavor. Galangal is not typically regarded as synonymous with ginger in traditional Asian dishes. Galangal's usually available in Asian markets as whole fresh root, or in dried and sliced, or powdered form. It can't be grated like ginger, so it must be sliced.

RECIPES: FERMENTED CARROTS WITH GALANGAL AND LIME

Ingredients: 1lb. of carrots, peeled and sliced 1/4-inch thick, 1 TBS galangal peeled and thinly sliced, 1 TBS grated lime zest (absolutely no white pith), 2 TBS salt, 2 cups water

Method: Pack the carrots, galangal, and lime zest in a 1-quart jar leaving 1 inch of space at the top and set aside. Dissolve the pickling salt in 2 cups of hot water. When cool, cover the carrots in the jar. Seal with a top and store in a cool dark place for 5 days. Remove the cheesecloth, cap the jar with the regular lid, and place in the fridge. Use it within a week.

DID YOU KNOW: During the Middle Ages, the spicy and pungent root known as galangal was a hot commodity. It was widely traded from its native Indonesia and China throughout Europe. It's known as 'The Spice of Life,' Laos Root, Kah,or Galanga.Today, Russians use it to flavor liquor.

GINGER – AWAPUHI

Zingiber officinale

Ginger was probably brought to Hawaii by Chinese immigrants who worked on the sugar plantations. This root will grow in every part of the Hawaiian chain. Commercially, Hawaii annually grows more than 2,500 tons with most of the original farms on the Big Island. Ginger has a true tropical appearance, great for landscaping, kitchen use, and better health. The strange looking, brown, knobby ginger root is both a unique, sharp-tasting flavoring and medicinal herb. Ginger originated in Asia and belongs to the same spice family as cardamom and turmeric. The Greeks and Romans used the root for medicine, yet its use almost disappeared. The Italian adventurer Marco Polo rediscovered ginger in far eastern Asia and India where it has been growing for 5000 years. India, Nigeria, and China are the world's biggest producers of ginger root.

The name 'ginger' means horned root in Sanskrit. Botanically it is *Zingiber officinale*. Ginger is used medicinally because it stimulates and strengthens the stomach, breaks colds, and coughs, diarrhea, rheumatism, and especially for nausea. The root can be boiled and pounded into a paste applied to the forehead to ease headaches or made into a poultice to soothe arthritis. As a flavoring, ginger is used for beverages as ginger beer and for gingerbread. Although Queen Elizabeth I is reputed to have invented gingerbread for Christmas; the ancient Greeks created it to aid digestion. The Spanish brought ginger to the Caribbean and by 1550 they were exporting it. Jamaicans and early American settlers made beer from it

HOW TO GROW: First, at the market, purchase some fat ginger roots with many buds or fingers. Wrap these roots in damp paper towels or wet newspaper and place them in your fridge's vegetable drawer or cool shade, until sprouts appear. Fork a row about ten inches deep and wide. Create a raised bed that will drain, otherwise the ginger root will rot. The best pH is 5.5-6.5 and at altitudes less than 1,500 meters. Carefully separate the root 'eyes' or buds. Plant the ginger root pieces eight inches apart and two inches deep with the buds or fingers turned upward. Water regularly and fertilize with 12-24-12. It takes approximately nine months for ginger to mature. Ginger is ready to harvest after the green tops begin to wilt. Break up the clumps of roots and wash extremely well, and completely dry in the sun before storage.

MEDICINAL: For thousands of years, Chinese medicine has used ginger to help cure and prevent several health problems. Ginger is available fresh, dried, preserved by pickling, tinctures from distillation, or candied. Dried ginger can be ground into a powder. Ginger tea is a safe remedy for motion sickness and the nausea of morning sickness during pregnancy. It is also very helpful in reducing nausea from chemotherapy. Ginger slows the feedback from the stomach to the brain to prevent the feeling of nausea. Ginger may inhibit blood clotting and reduce heart attacks. Ginger tea may also relieve the pain of your arthritis and cure a common cold. To prepare the tea, just slice some thin slivers off the root and boil. Let stand for ten minutes before drinking. Scientists suggest that only a gram a day of ginger should be ingested. Ginger is an antiemetic, anti-nausea, anti-clotting agent, anti-spasmodic, anti-fungal, anti-inflammatory, antiseptic, antibacterial, antiviral, analgesic, circulatory stimulant, carminative, expectorant, increases blood flow, promotes sweating, and relaxes peripheral blood vessels. Ginger is good for your health and has been said by some to be a plant directly from the Garden of Eden. Always check with your physician before using any herbal remedies.

NUTRITION: Nutritionally ginger doesn't have much. 1 TBS of ginger has 5 calories with only trace amounts of vitamins B-3, B-6, and C, iron, potassium, magnesium, phosphorus, zinc, folate, riboflavin, and niacin. ***Ginger is like a warm hug in a spice.***

FOOD &USES: To candy ginger, thinly slice peeled ginger root and place in a pot withjust enough water to cover. Boil for half an hour until tender. Drain and place ginger slices in a saucepan with an equal amount of brown sugar and three tablespoons of water. Cook over low heat, stirring continuously until all liquid is gone. Sprinkle ginger pieces with more brown sugar.

RECIPES: GINGER RICE

Ingredients: 1 thumb-size piece of peeled ginger root, 2 cups chicken broth, 2TBS soy sauce, 2TBS olive (or any) oil, 1 ½ cups rice, 2 garlic cloves minced, salt, and spice to taste.

Method: On a cutting board, thinly slice peeled ginger and smash slices as best as possible in a spoon. Scrape all pieces and juice into a medium size saucepan with the broth and soy sauce. In a skillet, heat the oil and add rice and garlic. Stir until rice browns in about 5 minutes. Remove from heat and dump the rice into a covered casserole dish adding ginger broth. Place in a 400degree oven until all liquid is dissolved in about 25 minutes.

EASY GINGER ALE

Ingredients: 2 cups peeled ginger chopped as small as possible, 4 cups water, 3 strips of fresh lemon peel 4 inches long, ½ cup sugar, 3 liters of club soda

Method: In a large pot mix water, ginger, and lemon. Boil uncovered for 10 minutes. Add sugar and boil again for 15 more minutes. Pour the mixture through a fine wire strainer. (Note: This ginger may be reused to flavor baking or ice cream.) Cover and cool the liquid for at least 2 hours. Add 1 cup of the ginger syrup to a liter of club soda or moderate to your specific taste.

SPICEY GINGER FISH

Ingredients: 2 pounds of filleted kingfish, tuna, salmon, or dolphin, ½ of a hot pepper - seeded and minced, 1 ripe pineapple chopped, 1 TBS fresh ginger-chopped small, ¼ cup chicken broth, ½ cup diced sweet pepper, 2 TBS cilantro, salt, and spice to taste

Method: Season the fish fillets with salt, hot pepper, and any other spices you like; then sauté for 3 minutes on each side. In another skillet on high heat, cook pineapple slices, sweet pepper, and ginger until brown. Add cilantro and the chicken broth, then scrape the pan clean of pine-apple and pepper pieces. Place the fish on individual plates and cover with the ginger-pineapple sauce.

GINGER PUMPKIN CAKES

Ingredients: ½ cup brown sugar, 2 cups all-purpose flour, 1TBS baking powder, 1TS cinnamon, ½ TS salt, 5 TBS butter or margarine, 1 egg, ½ cup boiled and mashed pumpkin, 4 TBS minced ginger, ¼ cup sour cream

Method: Mix all ingredients (except 1TBS of butter) until a soft dough forms. Then roll out on a floured surface. Knead at least ten times. Using a rolling pin, make the dough into a 10 by 5-inch rectangle. Cut dough into 5 sections 2-inch long, then cut these diagonally. Place these 10 triangles on an ungreased baking sheet. Brush triangles with melted remaining tablespoon of butter. If desired sprinkle with reserve sugar. Bake 15 minutes at 400 degrees or until golden brown. Cool for 10 minutes.

DID YOU KNOW? Traditional Chinese and Ayurvedic medicine have used ginger for over 3,000 years as a 'carrier' herb, one that enables other herbs to work better in the body. Ginger was one of the first spices exported from Asia.

NAMES: ***In China:*** jing, ***India:*** aale, adu,aduwa, alha, suhnthi ***Thailand:*** khing, khing-daeng, ***Philippines:*** basing, luya, ***Vietnam:*** cay gung, ***German:*** ingwer, **French:** ginembre, Spanish: gengibre, ***Netherlands:*** djahe, gember, ***Italy:*** zenzero

Every time I feel like something is missing from a dish, I think, 'Oh, I know, I'll add a pinch of dry ginger.' If it's not salt and it's not vinegar, it's probably missing dry ginger. ~ Alex Guarnaschelli

GUINEA ARROWROOT – TOPPEE TAMBU.
Calathea allouia

Guinea arrowroot or toppee tambu is another easy to grow, delicious root for your home garden. Guinea arrowroot has many names, topiambour, or sweet corn root. It is probably best known as Guinea arrowroot, which is not the thickening agent - arrowroot from St. Vincent.

HOW TO GROW: Guinea arrowroot is very easy-to-grow anywhere in the Hawaiians and makes a nice landscaping border for gardens, another edible hedge. The broad green leaves can reach five feet tall and a few will bear white flowers. The almost round root can be up to two inches in diameter and resemble 'new,' first harvested small potatoes, or water chestnuts. Guinea arrowroot is seldom affected by any insects or worms, and only needs watering in a drought. These roots like moist soil and humid air. To locate some seed roots (rhizomes) find someone selling them along a roadside. They either have some for planting or will direct you to where they bought the Guinea arrowroot. To prepare for planting find a well-drained area, fork and mix in well-rotted chicken manure. Harvests will be less in clay soil than sandy. The preferred pH is 6-7. Plant the 'seeds' about a foot apart. This plant will even grow in shade, which makes it perfect for interspersing between fruit trees, cassava, or plantain. It is a long crop, taking about nine months to harvest.

This root isn't cultivated on a large scale in most countries. However, in the few plantations where it's grown, no pest attacks or diseases have caused significant damage during the last 15 years. With the impact of fungus and pests, and the escalating costs of preventative garden treatments, Guinea arrowroot may be a crop that can make a nice profit.

MEDICINAL: It is a good food for a non-irritating recuperation diet, and for infants as a replacement of breast milk. It can be eaten in the form of jelly seasoned with sugar, lemon juice, or fruit. The leaves are also used in a broth as a diuretic and to treat cystitis.

NUTRITION: Guinea arrowroot is more than a starchy, belly filler. Per 100 grams these roots have 60 calories, 9 grams of carbs, a half gram of protein, with calcium, phosphorus, and some vitamin C.

FOOD & USES: Guinea arrowroot roots are usually boiled for 15 to 20 minutes and can then be fried, grilled, or mashed; and especially are great for stir-fry. Its flavor is similar to that of cooked young corn. Guinea arrowroot remains crisp long after cooking. Tender new leaves can be boiled in soups and stews. Flour made from the dried roots contains 15% starch and 6.6% protein. Guinea arrowroot may have a future as a starch as the world searches for new food sources.

Guinea arrowroot roots are unique because after they are cooked their texture remains crisp. It is best to cook the Guinea arrowroot before using them in any recipes. As well as being eaten on its own, Guinea arrowroot can used in salads, soups, and fish dishes.

RECIPES: GUINEA ARROWROOT STEW

Ingredients: 2-4 lbs. (1-2kg) fresh Guinea arrowroot washed, chop into spoon-size pieces, 1 sweet green pepper seeded and sliced, 3 tomatoes quartered, 1 large onion sliced thin, 2 garlic cloves minced, 2 TBS oil for frying, a pound of stew beef, chicken, or pork (if not, keep it vegetarian), a pound of cooked red beans, 3 cups water, 2 leaves culantro minced, salt and spices to taste

Method: In a deep skillet or big pot fry the onion in the oil till it becomes clear, add meat (if using), cook till no red remains inside. Add beans and remaining vegetables then water. Cover and simmer for two hours. Add culantro and cook for 10 more minutes

Deep in their roots, all plants keep the light. ~ Theodore Roethke

GUINEA ARROWROOT FRENCH STYLE—CREAMED

Ingredients: 2 lbs. (1 kg) Guinea arrowroot boiled and skin removed, then sliced ¼ inch thick. 1 onion sliced thin, 1 garlic clove minced, 2 TBS oil for frying, 1 cup evaporated milk, 3 leaves culantro or ¼ cup parsley, salt and spices to taste

Method: Sauté onion and garlic in the oil in a frying pan. Add Guinea arrowroot and brown a bit before adding condensed milk. Simmer and stir frequently so the milk doesn't burn. Add culantro or parsley at end of cooking, about 15 minutes. Great side dish with chicken or fish.

SCALLOPED GUINEA ARROWROOT

Ingredients: 2 lbs. (1kg) Guinea arrowroot, boiled and skinned, 1 onion sliced thin, 1 garlic clove minced, 1 cup evaporated milk, 1 cup cheese grated, ½ TBS nutmeg, salt, and spices to taste.

Method: Combine all ingredients in an oven ware pan. Bake at 350 for forty minutes.

TOPPEE SHISH KABOB

Ingredients: 6 metal skewers, 2 lbs. (1kg) Guinea arrowroot boiled and peeled, 2 medium onions quartered, 2 sweet peppers quartered, 3 medium tomatoes quartered, 2 ripe plantains skinned and cut into 2-inch pieces, curry powder or cumin, salt, and spices to taste

Method: Put all ingredients on to metal skewers, roll or dust with either curry or cumin. Cook over a coal pot or gas grill until all veggies just start to brown. Serve with pasta or rice.

DID YOU KNOW?

Guinea arrowroot is also called sweet corn root. During my research, I became confused with the arrowroot thickener grown mostly in St. Vincent, and the Guinea arrowroot. Guinea arrowroot botanically is *Calathea allouia,* while the thickener arrowroot is *Maranta arundinacea.* The *English* say either Guinea arrowroot, or sweet corn root. *Caribbean islanders* call it topee tampo, topi-tamboo, or topinambour. In *Spanish* it is dale dale, especially in *Peru.* It is agua bendita, cocurito in *Venezuela.* Lerenes in *Puerto Rico.* Brazilians say it in *Portuguese* as ariá, or láirem. *French* say touple nambours particularly in *Santa Lucia,* but in the French islands it is alléluia. Also known as curcua d'Amérique/turmeric of the Americas.

As the garden grows, so grows the gardener. ~ Unknown

HAUSA POTATO – CHINESE POTATO

Plectranthus rotundifolius

The hausa potato, like so many plants and vegetables, has many names including the local potato, country potato, potato mint, or the black potato. The egg-shaped roots look like small, black Irish potatoes and are sweeter than other local yams. Instead of developing one big root like a yam, the hausa vine develops many round roots bigger than a chicken egg, but smaller than a tennis ball. Botanists believe the hausa root originated in central or east Africa, but many millenniums ago it spread into South-East Asia, India, and the western hemisphere. This root can be found in almost every tropical country around the world. Today, the hausa potato is cultivated in the Hawaiian Islands for local consumption. It's a perennial, edible plant scientifically known as *Plectranthus rotundifolius, Solenostemon rotundifolius, and*/or *Coleus rotundifolius*. This is one of the few roots crops that is a member of the mint family (*Lamiaceae*). Yes, the hausa's related to peppermint!

HOW TO GROW: The hausa potato is another very attractive edible plant for your home garden. This root doesn't have a vine, but is a 60cm (2ft.) tall plant, with mint-shaped leaves that have a nice minty smell and small purple blossoms. In the islands, the best method to grow a hausa is to locate some existing roots to use for seed. The seed roots should be cut into 2-inch pieces and let dry for two days so they don't rot once planted. Or you can root cuttings about 20cm (8in.) long. The soil should be prepared by digging a hole the size of a gallon bucket or larger and fortified with rotten manure and/or compost. Loose, rich soil where the hausa pieces are planted equals a greater harvest. Plants should be 50cm (18in.) apart. It prefers full or partial sun with a pH of 6.5 - 7.5. The stem area must be weeded, so hausa gets plenty of sun and must be regularly watered for the best results. These roots mature in 6 months and you'll see the leaves begin to dry. Carefully dig and separate the roots. They are usually 4cm (2in.) long in clusters of 5. If the roots stay in the soil after the tops whither, the roots will begin to rot. It's best to store the extra roots in a box of dry sand or a basket filled with straw in a breezy area. Harvested, the hausa potato loses flavor after two months, so keep replanting in fortified soil. The white, starchy, slightly aromatic tubers become dark with age.

MEDICINAL: Hausa potato leaves are used in Ayurvedic-homeopathic medicine to treat dysentery, and bloody urine. The juice from the leaves is used for disorders. The roots treat hemorrhoids, nausea, throat infections, and are used to kill parasites. A paste of the root treats skin problems as burns, wounds, sores, and insect bites.

NUTRITION: Per 100g, hausa potatoes have 40 calories, 1g protein, 1g fat, high in calcium and iron with lesser amounts of potassium, phosphorus, manganese, and magnesium.

FOOD & USES: The hausa is a staple food. The leaves can be cooked as a green vegetable. This root is best used as a substitute for potatoes in recipes and are usually cooked in a curry. They can be boiled, roasted, baked or fried. The hausa root is 16% starch.

RECIPES: CHINESE POTATO CURRY

Ingredients: 250g of hausas peeled and diced, 4 garlic cloves chopped, 1 large onion chopped, 3 green chilies chopped, 1 sprig of curry leaves, 1 small stick of cinnamon, ½ TS each of fenugreek seeds and ground mustard seeds, 1 ½ TBS curry powder, 1 TS turmeric powder, 1 cup coconut milk, 1 cup yogurt, salt to taste, 1 TBS oil

Method: Put all ingredients in a suitable pot or wok and cover with water. Bring to a boil and simmer covered stirring occasionally until hausa are tender. Add the coconut milk and again boil. Stir in curd or yogurt adjust salt and simmer for 5 minutes. Enjoy with rice or roti.

NAMES: Chinese or hausa potato, local potato, country potato, black potato, Sudan or Zulu potato

JERUSALEM ARTICHOKES - SUNCHOKES

Helianthus tuberosus

Jerusalem artichokes are native to North America and were grown by Amerindians when the Puritans landed in the 1500s. Botanically, they are *Helianthus tuberosus.* The French explorer, Samuel de Champlain, considered these roots' taste was similar to an artichoke and sent the first of these roots to France in 1612. Today, China is the main world producer followed by Mexico and Spain. The Jerusalem artichoke is a member of the sunflower family, and its flower is very similar. Why it's named Jerusalem is a question. Italian immigrants to America named the root 'girasole,' the Italian word for sunflower. It's possible that this word was corrupted over decades by Americans to Jerusalem. These grow easily throughout the Hawaiian Islands, but because of the number of seeds and the seeds' ability stay viable for more than 5 years, this plant is considered invasive. The flowering head resembles a sunflower, so be careful with the seeds. Perhaps after harvest, sprout the seeds and eat them in a salad.

HOW TO GROW: This is another exotic easy-to-grow plant for the home garden or in large pots filled with good potting compost. Jerusalem artichokes thrive in sun or shade and even grow in poor soil. To produce better, the soil should first be tilled, adding compost. Plant small roots, as you would ginger or turmeric, in well-prepared soil, 10cm (4in) deep and 30cm (1ft) apart. Once they sprout and grow well, these roots can be left alone and only watered if a severe drought. Weeding isn't necessary, because the plants produce a dense cover of foliage. When the stems reach 30cm (12in) high, the bases should be molded to provide stability against rain and winds. When the foliage starts to turn yellow in autumn, trim the stems to 7.5cm (3in). Harvest carefully, finding the tubers with a small garden fork. Some can remain in the ground for the next season, but the plants need to be dug up and replanted in fertile soil.

Jerusalem artichokes have fibrous roots as long as 50 inches. Stems grow as tall as 4 meters (12 ft.). Stems will become woody over time. The edible roots resemble ginger, knobby clusters from 7.5 to 10 cm long, 3–5 cm thick.

MEDICINAL: The Jerusalem artichoke is used in traditional medicine to treat diabetes, heart disease, and rheumatism. The presence of inulin stimulates the growth of bifido bacterium to fight harmful bacteria in the intestines reducing concentrations in the intestines of possible cancer-causing enzymes, prevent constipation, and boost the immune system. Potassium-rich Jerusalem artichokes improve bone health and reduce the risk of osteoporosis.

NUTRITION: 150 grams of raw Jerusalem artichoke has 110 calories, 3g of protein, and is an extreme source of potassium, phosphorus, iron, magnesium, and important trace minerals of zinc and copper plus vitamins A, b1, B2, and B3. This root is an excellent source of vitamin B9, folate, and beta-carotene.

FOODS & USES: The Jerusalem artichoke can be eaten raw, sliced in a salad or boiled, roasted, braised, sautéed, or stir-fried, as a potato substitute with the same consistency and texture, but much more nutritious. These roots have a delicate sweet nutty flavor similar to an artichoke.

RECIPE: SPICED JERUSALEM ARTICHOKE

Ingredients: 500g (1lb.) Jerusalem artichokes, 1 garlic clove minced, 1 medium onion chopped small, 2 green chilies chopped, lemon, 2 TBS oil, 2 TS cumin seeds, 1 TS garam masala, salt to taste, and a bunch of fresh coriander chopped.

Method: Peel the artichokes and chop them into cubes. In a wok or frying pan, heat the oil and add cumin seeds. Add garlic and onion. Once the seeds crackle, add the chilies and the chopped artichokes. Stir until the cubes are coated with the spices. Add salt. Cook on low heat for 5 minutes, stirring occasionally. Cook longer for a softer texture. Remove from the heat and squeeze a lemon over the dish. Sprinkle with the coriander.

DID YOU KNOW? Jerusalem artichoke produces many miniature sunflowers at the end of the branches. They are used as a source of fructose and in the production of ethanol as a biofuel.

NAMES: German turnip, groundnut, cane truffle, American potato, or topinambour

Jerusalem artichokes have a great affinity with nuts.
I love them with chopped walnuts or almonds, lemon juice,
garlic, herbs and plenty of olive oil.
~ Yotam Ottolenghi

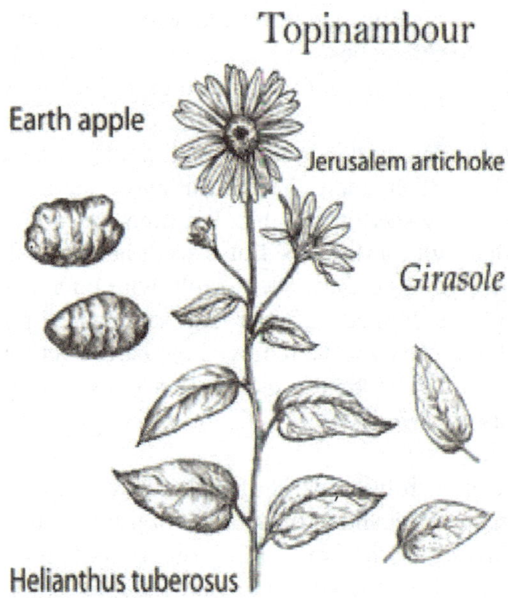

Courtesy of Wikipedia

We can make a commitment to promote vegetables and fruits and wholegrains on every part of every menu. We can make portion sizes smaller and emphasize quality over quantity. And we can help create a culture - imagine this - where our kids ask for healthy options instead of resisting them.
~ Michelle Obama

LOTUS ROOT - LOTUS YAM

Nelumbo nucifera

Lotus root originated in India and has been eaten to aid good health for millenniums. The plant is long revered for its beautiful flowers, but is mainly grown for the edible stems. Lotus root is a vegetable that roots underwater and will grow more than a meter long and resembles a long, thin squash. For centuries, this flower has been used to portray purity, innocence, and godlike beauty. The seeds stay viable for millenniums. Lotus roots found in ancient Chinese excavations have been carbon-dated to the fifth century AD and have been germinated! The lotus was brought to China and Egypt thousands of years ago, and other parts of Asia, Japan, and northern Australia. The botanical name is *Nelumbo nucifera.* The water lily is *Nymphea lotus* transplanted to Egypt and Europe, a different plant. Lotus grows in the Hawaiians. I'll bet you didn't know it was food.

HOW TO GROW: The home garden should be a beautiful place to relax and enjoy nature's beauty and bounty. Growing lotus is perfect for that. Asian countries such as China, India, and the Philippines are the main cultivators of lotus roots. This can be eaten stir-fried, steamed, or grilled, and it's always available. It's an aquatic perennial plant about 2-inches (5cm wide, white or brown spotted, and will grow to 20 inches (50cm or longer. The beautiful pink or white lotus flower remains on the surface, but is rooted to the muddy bottom of shallow ponds, marshes, lagoons, and flooded fields.

If you want a gorgeous relaxing area for your home garden, build a small pond, 2 meters in diameter. As long as it's filtered, you can have koi, goldfish, and two beautiful lotus plants. To grow lotus, look for someone who has plants. Growing by seed is possible, but time-consuming. Even if a seed has all the ideal conditions, it can lay dormant for decades. Lotus seeds are about the size of a small olive and are very hard. It's recommended to hold the seed firmly with pliers, but don't crack it, and rub the concave end (the other end is pointed against rough sandpaper or a concrete block. Sand it until a white oval appears. That means you've removed the hard layer, but the hard work is just beginning. To sprout the seed, it must be kept in clean water (no chlorine, changed every other day. The temperature cannot vary from 70 degrees F for several days, maybe weeks, before the seeds swell and sprout. It's best to buy roots.

Plant the seedling in a large container of water with two inches of a clay mix or aquatic soil dredged from a river or lake. The water must remain warm and lit by sunlight or grow bulbs. It may take weeks for the seedling to sprout small leaves, which signify it's growing roots and a stem. Give it time, it will get bigger. That's why you should also have fish in your small lotus pond to watch. The rule is the longer the stem, the thinner it is. Stems can be harvested in 4-6 months depending on the climate. In warmer areas, lotus grows quicker. At har-vest, the water is drained and the roots and stems are carefully removed from the bottom silt, but in a small, home pond you can feel along the bottom and yank the roots. The roots grow in segments from 2 to as many as 6. Segments are what you buy at the market. Commercial yields can be 4000 lbs. per hectare. Removing the flowers makes a bigger crop.

Lotus stems stay crisp when cooked. There are many internal circular passages in each segment. Where the stems are broken, you'll find a sticky secretion. Starch from the lotus stems is a thickening agent. Immature leaves are consumed as greens and mature seeds, and the flower's core is eaten and reported by the Chinese to have medicinal qualities. The lotus root industry is expanding; lotus root chips, tea, lotus root soup, and lotus root powder are available. Lotus root should never be consumed raw as it tends to contain parasites that may be detrimental to your health. Peeled, sliced, and boiled, it resembles a spoked wheel with a flavor like coconut.

MEDICINAL: Lotus root increases the blood pressure, maintains proper enzyme activity, can uplift your mood, and possibly prevents some cancers. The combination of copper and iron serves to improve blood circulation in the body and increase energy. The potassium in lotus root also serves to help regulate blood pressure. It relaxes the blood vessels and improves blood flow. A large amount of dietary fiber helps to improve the digestive process and reduce constipation. It's also a low-calorie food that helps you to maintain and lose weight while still getting some protein and nutrients in your diet. Vitamin B contains pyridoxine, which helps to improve your mood and relax your mind. It even helps to reduce headaches and serves to lower stress. It's another reason the lotus flower is associated with peace and tranquility. Pyridoxine also helps to lessen the risk of a heart attack and improves heart health overall. Ayurvedic-homeopathic medicine uses lotus root, the flowers, stamens, rootstock, seeds, and leaves to treat bleeding piles, snake bites, diarrhea, coughs, and skin diseases. Powdered lotus root is sold as an herbal medicine.

NUTRITION: 100 grams of lotus root has74 calories, 0-fat, 40 mg sodium, 556 mg potassium with 17.4g carbohydrates, 4.8g dietary fiber, 0-sugars, and 2.5g protein. This vegetable is very nutritious and contains large concentrations of vitamins and minerals including Vitamin B, Vitamin C, iron, potassium, copper, thiamin, and zinc, among others.

FOOD & USES: Fresh lotus root is a long, ugly, brownish-gray, vegetable, with a woody texture, easily prepared by slicing it into pieces of the desired shape and size. The unique shape of this root with several holes in the middle, like a spoked wheel, adds visual beauty to any dish. Keep lotus roots in the refrigerator until you're ready to use them. Before cutting into lotus roots, wash your hands, and rinse the outer skin under running water. Remove the outer skin and chop it into cubes or thin slices. Pieces can be added to soups, stir-fries, or many other dishes. It can be deep-fried, boiled, braised, pickled, or used as a garnish and flavoring for salads, snacks, and rice. Lotus leaves can be used for wraps. Select fat, firm roots without cracks or bruises. The outer skin of lotus root tastes bitter, so peel it.

RECIPES: LOTUS ROOT CURRY

Ingredients: ½ lb. (225 g) lotus root washed and sliced into small pieces, 2 whole tomatoes, 2 green chilies, 1 medium onion chopped, 10 curry leaves, ½ TS each of the following spices: fenugreek seeds, turmeric powder, chili powder, salt, 1TS roasted chili powder, 1 cup (250ml) thick coconut milk, 3 TBS oil or ghee

Method: In a frying pan or wok, with the oil, fry all the spices, salt, and curry leaves until there's a good aroma. Add the lotus root pieces, sliced tomatoes, chilies, curry leaves, and onion. Fry a few minutes until onion is browned. Add the curry powder and stir for 1 minute.

LOTUS ROOT SPICY STIR-FRY

Ingredients: 1 ½ cup (350g) peeled and crosswise sliced lotus root, 4 cloves of garlic sliced, 4 green chilies sliced, 1 medium onion chopped, 2 TBS soy sauce, 1 TS brown sugar, salt to taste, and 2 TBS oil.

Method: In a small saucepan, combine soy sauce and brown sugar. Remove from heat. Drop lotus root slices into a pot of boiling water for 2 minutes to blanche. Rinse under cold water, drain, and pat dry. Heat the oil in a frying pan or wok, fry lotus root slices until brown. Add garlic, chilies, and onion. Then coat with soy-sugar sauce. Enjoy as an appetizer or side dish.

DID YOU KNOW? Lotus is the national flower of India and Vietnam. 1,300-year-old lotus seeds recovered from a dry lakebed in northeastern China were successfully germinated. **NAMES:** Botanical names include *Nelumbo nucifera* or *Nelum-bium nelumbo.* lotus root, East Indian lotus, Egyptian lotus; lian, sacred lotus, bean of India.

PARSNIPS
Pastinaca sativa

Parsnips are the royalty of roots and offer a great fresh taste. They resemble the size and shape of carrots, but the skin and flesh of parsnips are white to cream-colored. Parsnips are the result of continuous cultivation. And to grow this root in the tropics, it may take several tries. Because it's kind of rare in the Hawaiian Islands and a very healthy root, it can be a good crop for the home garden. Used as a staple food like carrots, parsnips are also a natural sweetener for breads, cakes, and jams. Parsnips are native to Europe and Asia and were used by the ancient Greeks and Romans for medicinal and food purposes. Early colonists spread this white root to every temperate part of the world. Plant in September and harvest in March.

If you haven't tried parsnips, they are a tapering white root, with a green leafstalk stem 1-2 ft. tall. When fully ripe, at the ends of the stems are yellow flowers. The green tops are toxic, and handling requires gloves because they can cause allergic reactions. Parsnips are biennial, but are grown as an annual. They are cultivated mainly in temperate regions and in the cooler parts of the tropics, including East and southern Africa.

HOW TO GROW: First, you must have fresh seeds. That is of great importance. In the tropics, parsnips should only be cultivated above a 900 m altitude, but patient, innovative people have grown them at sea level. It can grow in full or partial sun. Plant in a cool location since rising temperatures can cause the quality and quantity of the root to decline. They grow well in sandy loose fertile soils that can hold water with a pH of 6 or above. This is necessary to grow long, straight roots. Stony soils are not suitable for parsnips. Good parsnip seeds will germinate within a month. Once you have the soil ready, make a nice, straight groove. It's best to plant the seeds directly, ½ in. deep, 2 inches apart. They may need to be thinned to 4 inches. Harvest within 4-6 months. Since this is a long crop, you could interplant with beans, peppers, or salad greens.

MEDICINAL: Parsnips were used in herbal medicine and considered an aphrodisiac. Parsnip tea is an excellent diuretic and helps to cleanse by stimulating the production of urine. Parsnips are rich in potassium to keep the heart healthy. Eating parsnip also helps to make your bones stronger, lowers the chances of developing diabetes, helps to reduce cholesterol, and prevents the onset of depression.

NUTRITION: Per 100 grams, parsnips have 75 calories, 36mg of calcium, 29mg of magnesium, 375mg or potassium, 17mg of vitamin C, and 22.5mg of vitamin K.

FOOD & USES: Parsnips can be eaten raw, but cooking brings out their flavor. They can be baked, roasted, boiled, fried, or pureed. They give a rich flavor to soups and stews.

RECIPES: CURRIED MASHED PARSNIPS

Ingredients: 2lbs. parsnips chopped into 1-inch rounds, 2 TBS butter or margarine, 1 TBS curry powder, 1 TBS honey or brown sugar, 1½ cups milk, salt to taste.

Method: Melt the butter/margarine in a saucepan, add parsnips, and cook for 5 minutes. Add the curry powder and honey/sugar; cook for 2 minutes. Add the milk, bring to a boil, cover, and simmer for 15 minutes or until the parsnips are ten-der. Mash with a potato masher and season to taste.

NAMES: Botanically: *Pastinaca sativa,* chirivía, grand chervis, panais, parsnip root, pastenade, pastinacae herba, past-inacae racine-blanche.

There is a southern proverb - fine words butter no parsnips. ~ Walter Scott

'UALA KE 'OKE 'O - POTATO

Solanum tuberosum

The potato is a root that grows from a green vegetable plant from 30cm tc a meter tall (1ft. to 3ft. that produces a brown-skinned root that is the world's largest fooc crop. English-speaking people say 'white potato' or 'Irish potato' to separate it from the sweet potato. In the 1500s, the Spanish found the potato at an elevation of 3000m in the Bolivian Andes and brought it to Europe. From there, potato became another of the world's starchy belly-filling food staples, the same as rice or cassava. Some type of 'white' potato is now grown in all temperate countries and in tropical highlands. The potato, with an annual production of 368 million tons, is the fourth most important food crop in the world after corn, wheat, and rice. Its botanical name is *Solanum tuberosum.*

The first recorded successful potato cultivation in Hawaii was in the mid-1800s. Then Hawaii exported large quantities of potatoes grown on Maui to the continental US. The best growing areas were the Makawao and Kula regions on Maui, and the Waimea and Waiki districts of the Big Island. Varieties to grow today, include Spunta, Gold Rush, and Red Pontiac. Grow them for yourself and friends. Commercially, in the tropics, the potato cannot compete with the other native root crops, but work at growing many types.

HOW TO GROW: Climate and elevation are important for potato cultivation, but with the proper planning, you can grow a good supply almost anywhere. The best tropical elevation is 2000m (6500ft. Regular watering is the most important factor. That problem at home is solved with a hose. Potatoes need about 70cm (28in. of rain during their growing season. It must be even and not a torrential deluge. In the tropical lowlands, potatoes should have partial shade, getting sun in the cool mornings, then a good soak of water, and shade in the heated afternoons. Pick a proper spot to plant and then work the soil so it's loose and fortified with rotted manure and/or compost. The best pH is 5.5-6. To get a good crop, check your plants regularly that the water hasn't eroded soil from the top of the roots. **NOTE: If that happens as a result of exposure to the sun, any potato with a green discoloration, contains solanine, and has been known to cause fatal poisonings.** Not enough water or sun, or too much water and sun will ruin your crop.

It's best to fertilize every two weeks with NPK, with high 'N' nitrogen-120, 'P' phosphorus-80, and 'K' potassium-80. Every garden shop should have that mixture. It is equally important to revitalize the soil before and after planting with rotted manure.

Find potatoes with eyes, yes, eyes, buds, little bumps that grow on potatoes. Put some in your vegetable drawer wrapped in slightly dampened newspaper and check regularly until you see the eyes grow. You need at least one eye on each chunk of potato you plant. Make mounds or ridges and plant the chunks about 10c (4in. deep and 30cm (1ft. apart. Weed control is very important and mulching, covering the area but not the green potato tops with cut grass-straw, rice straw or husks, palm fronds will help prevent unwanted evaporation or erosion. Mulching immediately after planting will help the shoots to develop quicker. Potatoes and tall crops grow well together. Corn planted early followed by potatoes will provide needed shade and double the crops from the same area. The green tops are usually 1 ft. tall with white to blue 1in. blossoms.

Potatoes should be harvested 2 or 3 weeks after the foliage has withered naturally, usually 4-5 months after planting. It's wise to check after 100 days by carefully digging one root.

MEDICINAL: Potatoes are reported to be a general remedy, a laxative, a sedative, increase urine flow, and increase breast milk. It's a traditional remedy, externally applied boiled or raw, for burns, corns, and warts. Eating potatoes may worsen arthritis.

Potato sprouts/eyes are considered toxic.

NUTRITION: Per 100 grams, potatoes contain 80 calories, 2g protein, and good amounts of calcium, phosphorus, iron, potassium, and magnesium, with some vitamins A, B-1, B-2, and C.

FOOD & USES: Potatoes are eaten: boiled, baked, fried, stewed; a versatile food you can prepare it any way you can imagine. Potato flour is the oldest commercially processed potato product and can be used for baking. Ripe potato juice is excellent for cleaning cottons, silks, and woolens. Surplus potatoes are used for animal feed and making alcohol. Boiled with weak sulphuric acid, potato starch transforms into glucose, and then is fermented into alcohol, to yield 'British Brandy.' The Netherlands, Denmark, and the USA use large quantities of potatoes to make a large-grained starch used by the food, paper, and textile industries, and manufacturing adhesives.

RECIPES: POTATO LEEK SOUP

Ingredients: 1k (2lb.) of potatoes cubed, 3 garlic cloves minced,3 large leeks, chopped, 2 green chilies(optional), 10g (¼cup) fresh chives chopped, 1.5L (6cups) vegetable broth, 480ml (2cups) water, 2TBS butter, salt, and pepper to taste

Method: In a large pot on medium heat, melt the butter. Add the chopped leeks and stir until coated with butter. Cover the pot and lower heat, cook for around 10 minutes until the leeks are just browning and soft. Add garlic, potatoes, salt, and pepper. Cook for 1 minute, then add vegetable broth, water. Bring to a boil. Lower heat, cover again and simmer for 15 minutes, until potatoes are tender and easily speared by a fork. Stir in chives.

POTATO PANCAKES – LATKAS

Ingredients: 4 large potatoes peeled and grated, 1 garlic clove minced, 1 large onion grated, 1 green chili chopped small(optional), 1 egg beaten, 2 TBS all-purpose flour, salt, and pepper to taste, 2 TBS oil for frying

Method: Combine potatoes, garlic, chili, and onion into a large bowl. With a cloth, press and drain off/absorb any excess moisture. Mix in egg, salt, and black pepper. Add enough flour to thicken the mixture, 2 to 4 TBS. Heat oil in the bottom of a heavy frying pan or wok over high heat. Drop large spoonfuls of the potato mixture. Press with a spatula to flatten to make 1/2-inch-thick pancakes. Fry, turning once, until golden brown. Transfer to paper towel covered plates to drain the excess oil. Enjoy for breakfast, lunch, or dinner.

POTATO CURRY

Ingredients: 500g (1lb.) potatoes peeled and cubed, 4 garlic cloves minced, 1 red bell pepper chopped, 1 large onion chopped, 2 green chilies, 1TBS red chili powder, 1 TBS curry powder, 1 TS turmeric powder, 1TS cumin powder, salt to taste, 2 cups water, 2 TBS oil

Method: In a frying pan or wok on medium heat, add the cooking oil, garlic, onions, and chilies. Cook for 30 seconds and then add the potato pieces. Cook for 2 minutes. Stir in the spices. Cook, stirring for 1 minute, and then add the water. The potatoes should be almost completely submerged in water. Simmer on low heat for 20 minutes or until the water has evaporated and the curry has thickened. Add sweet red pepper and stir occasionally. The potatoes should be cooked until soft, if not add one more of cup water and continue to simmer. When the curry is the thickness you want and the potatoes soft, add the coriander and cook 1 more minute. Serve with roti or rice.

DID YOU KNOW? There may be 20 distinct species of cultivated potatoes. There are yellow and even blue potatoes. European farmers planted potatoes instead of grains because no army would camp long enough to dig up all the potatoes because you are too vulnerable to counterattack.

NAMES: English potato, Irish potato, white potato, alu or aloo-*India*, jaga-imo - *Japan*, kartoffel-*German,* papas-*Central America*, patatas-*Spanish*, pomme de terre - *French*

RADISH
Raphanus sativus

Agricultural literature says not to plant radishes in hot weather. Radish seeds are easier to find than the round globe root vegetable in the tropic markets. I enjoy the spicy crunch and have learned to grow them in partial shade away from most, if not all, of the afternoon heat. Daily watering is imperative They grow better at higher elevations above 300 meters. The early Scarlet Globe radish grows throughout the islands and is one of the quickest to mature.

Radishes are root vegetables that belong to the mustard family. The name radish is derived from the Latin word radix for root. There is some confusion about where the radish originated. Southeast Asia is the only region where truly wild radishes have been discovered. India, Central China, and Central Asia appear to have been secondary locations where multiple shapes and colors developed. It was cultivated in China and SE Asia 3000 years ago, but there is evidence the first radish was discovered growing in Egypt in 2780 BCE and the radish was eaten by the slaves building the pyramids. Greeks and Romans grew radishes in the first century AD. In another chapter, the author has included the long white radish, usually known as the Japanese daikon. The 'salad' radish of numerous varieties, varying sizes, flavors, colors, and lengths deserves mention.

The radish was one of the first vegetables cultivated by Europeans in the Western Hemisphere. The Spanish were growing them in Mexico in the early 1500s and in Haiti by 1565.

Radishes with their mild to hot peppery flavor and crunchy texture are still popular today. There are four basic types according to the seasons when they are grown. They grow in a variety of shapes, lengths, colors, and sizes, such as red, pink, white, gray-black, or yellow radishes, with round or elongated roots that can grow longer than a parsnip. They grow quicker in cooler climates, and can be eaten raw, pickled, boiled, or fried. The leaves can be eaten fresh as salad or stir fried. Seeds can be used as spice. These roots are useful in traditional medicine, and as a bio-fuel. The world produces about 7 million tons of radishes. China, Japan, and South Korea, are the main growers.

HOW TO GROW: I always allow at least 6 weeks for radishes. Sow radish seeds in rows one foot apart in loosened, light, sandy loams, with a soil pH 6.5 to 7. I recommend the bed get full sun until noon and then full or partial shade until 4pm. If you are a radish lover as I am, make your own shade. First, soak the bed and add water-retaining coconut husk pieces if available. Seeds should germinate in 3-5 days. Cover them 1/2 to 1-inch deep. The depth the seeds are planted will affect the size of the root. A half inch deep is recommended for small radishes to an inch and a half deep for large radishes. Until you see sprouts, water them gently so the seeds don't wash away. A week after the radish sprouts appear, thin 2-3 inches apart. Seedlings too close will jam each other and not get plump. Small salad radishes will mature nicely with 2-inch spacing, but 4-inch spacing is necessary between daikons and other big radishes. Keep the bed damp. Once successful, plant the same as salad greens, a row every two weeks. The soil must never dry out. Radishes can be useful as companion plants for many other crops because their pungent odor deters insect pests.

After harvesting, radishes can be stored for 2-3 days at room temperature, and about 2 months refrigerated.

MEDICINAL: The radish can be used to relieve stomach aches, ease digestion, as a diuretic, and to help regulate blood pressure. It contains a substance known as sulforaphane, which has the potential to prevent cancer development. Radishes also contain coenzyme Q10, an antioxidant that helps block the formation of diabetes. Radishes in your diet will help flush out toxins from the liver and kidneys and help to heal any damage. They are also used to treat bronchitis, fever, colds, coughs, and for high cholesterol. In Britain, the radish is used to treat kidney stones, intestinal parasites, and skin problems. Radish seeds are used as a carminative to relieve gas, as a digestant, and expectorant. The Spanish black radish is highly regarded for health and natural medicine.

NUTRITION: Radish roots and leaves have antioxidant properties. When compared with roots, the leaves possess higher levels of proteins, calcium, and ascorbic acid. Per 100 grams, radishes have 16 calories with reasonable amounts of vitamins B-6, C, and folate, plus minerals as potassium, calcium, magnesium, copper, and manganese. Raw radish is 95% water.

FOOD & USES: The root of the radish is the part most commonly eaten, but the entire plant is edible. The green tops can be used as a leaf vegetable used in potato soup or as a sautéed side dish. The seeds can be used as a spice or sprouted and eaten similar to mung beans. Roots can be pickled, boiled, and fried. The seeds can be pressed to extract radish seed oil. Wild radish seeds contain up to 48% oil, but are not suitable for human consumption. The oil is a potential source of biofuel.

RECIPES: SPICED RADISHES

Ingredients: 500g -1 lb. radishes, leaves removed, halved/quartered 100g unsalted butter, 2 green or red chili peppers chopped thin or 1 TBS chili flakes, and the juice from ½ a lemon

Method: Heat the butter in a saucepan on low until it foams and becomes golden brown. Stir in the radish slices with the chili pieces or flakes and coat in the butter and peppers, about 10 minutes. Increase the heat to high for 2 minutes. Add the lemon juice, let it sizzle. Season to taste and serve.

SAUTÉED RADISH GREENS

Ingredients: 4 cups radish greens (from 1 bunch of radishes), 4 garlic cloves minced, 2 TBS olive oil, 1-2 hot chili peppers chopped small to taste, and a pinch of salt

Method: Remove the stiff stems where the greens start to grow from the roots. Soak the radish leaves in a large bowl of cold water for a few minutes after washing the leaves. Remove the leaves and lay flat to dry for about 15 minutes. Heat the oil in a frying pan or wok over medium heat. Add the minced garlic and cook for 1-2 minutes until it browns and become fragrant. Add the chopped chili pepper and cook about 1 minute more. Do not overcook or burn the pepper. Add the radish greens and constantly stir until they are covered in oil and completely wilted, about 2-3 minutes. Enjoy with crushed hard-boiled eggs.

DID YOU KNOW? Round radish varieties are usually 1 inch wide. Cylinder types are 7 inches long, while carrot-like varieties grow to the size of 24 inches. Giant radishes were described by a German botanist in 1544 weighing 100 pounds. The longest radish was grown by Joe Atherton of the UK, measuring 6.703 m (21 ft. 11.89 in) long. According to Official Guinness Records, the world's heaviest radish was grown by Manabu Oono of Japan and weighed 31.1 kg (68 lb. 9 oz.). Most radishes have a crisp texture and a sharp, peppery flavor, caused by glucosinolates and the enzyme myrosinase, which combine when chewed to form allyl isothiocyanates, also present in mustard, horseradish, and wasabi. Oilseed radish is a variety cultivated because its oil is used as biofuel.

Photo courtesy 912crf.com

RUTABAGA

Brassica napobassica

In the late Middle Ages, rutabagas in nature are believed to have developed from turnips crossing with wild cabbage in Russia or Scandinavia. The root resembles a turnip except the flesh is yellowish. The leaves are edible only when the roots are very young, bluish green and smooth like cabbage. Rutabaga is derived from Swedish 'rotabagge,' meaning thick root and ram's foot.

Rutabagas aren't a common vegetable in the markets, so you should grow some. They're delicately sweet, reminiscent of both cabbage and turnip flavors. Young roots have a pleasant taste.

HOW TO GROW: Growing them isn't much different from turnips, but rutabagas take at least a month longer to mature. In Hawaii's tropical climate, it's best to grow them when the temperatures are cooler. You can plant them in a place that gets shade from the hot afternoon sun. If you work the soil bed with mulch it should keep the soil cooler. Growing rutabagas will be a labor of your love for that specific root. Grow a short row to test the site. A recommended rutabaga variety for warm climates is 'American Purple Top' a 3-month wonder. Plant the seeds directly into well-worked, loose soil with a pH of 6-7 that will drain when the rains come. After they sprout, thin to 6 inches between roots. Cook the sprouts you remove. Constantly weed and work the soil loose between the remaining plants to 3 inches deep. This will aerate the soil and make the roots expand. Roots of both turnip and rutabaga are milder flavored if the soil is kept moist; they become more pungent under drier conditions. Harvest rutabagas when they are 3-5 inches wide.

MEDICINAL: Rutabagas have folic acid and antioxidants to fight inflammation, prevent premature aging, and reduce cancer risks. They are a natural diuretic and treats chronic constipation. A cold remedy is to peel and blend a rutabaga into a paste and mix with honey. Use a teaspoon 3 times a day. The oil from the seeds can be mixed with camphor as a remedy for arthritis and rheumatism.

NUTRITION: Per 100 grams of rutabaga, there are only 38 calories, with lots of vitamin C, with potassium, calcium, and magnesium.

FOOD & USES: Rutabagas can be grated raw, in salads, steamed, boiled, baked, mashed, creamed, glazed, or fried, and are tasty in soups and stews. They have great flavor, easy to prepare, and are very nutritious. The color deepens during cooking. Add a tablespoon or two of sugar to the boiling water to subdue any bitterness. Rutabaga skin is edible, with a strong flavor. If purchased at a market, it's probably covered in wax, so it's best to remove the skin before cooking the vegetable.

RECIPES: BOILED RUTABAGA

Ingredients: a 2 pounds of rutabaga peeled and diced into ¾ in. cubes, 1 bay leaf, spice and salt to taste, 2 TBS olive oil, and fresh ground black pepper.

Method: Bring the water to a boil. Reduce heat, cover, and cook for about 20 to 25 minutes until rutabaga is fork-tender. Carefully drain the water. Remove the bay leaf. Place in a bowl and add olive oil and spices.

MASHED RUTABAGA- same as above. As you mash blend in butter or sour cream.

NAMES: swede, Swedish turnip, yellow turnip, neep, or snagger.

UALA - SWEET POTATO

Ipomoea batatas

Hawaiians know the sweet potato as 'uala pukui,' a staple root vegetable that requires a long growing season to produce mature roots. Sweet potatoes are often confused with yams; however, yams are quite different. These roots are grown on every island, but almost 75% are grown on Oahu. The average annual sweet potato harvest is under 400 tons. The local demand is so great that Hawaii must import 40% more. One of the best varieties to grow is Waimanalo Red. In the Hawaiian language dictionary, (https://ulukau.org/), I found hundreds of different terms relating to sweet potatoes.

The sweet potato, botanically *Ipomoea batatas,* is native to Central America, where it was cultivated 5,000 years ago. Yams originated in Africa and are not very sweet and grow as large as 100 pounds. Before Europeans landed in the Western Hemisphere, the sweet potato was already well-traveled. This root had already passed through South America and Mexico, then it was carried by boat to far away Pacific islands and farther on to New Zealand. In many countries of the Pacific, sweet potato is a prime food source, especially if the rice crop fails.

Columbus brought back to Spain many new foods he discovered during his first voyage in 1492. Sweet potatoes were among Columbus' treasures. The Portuguese seafarers carried them to Africa, Asia, India, and Sri Lanka. When it arrived in Hawaii is a mystery, because it was already being cultivated by the Hawaiians in 1778, when Captain Cook arrived.

HOW TO GROW: The sweet potato is a perennial vine that bears an edible root, usually with purple or brown skin. The flesh can be a variety of colors, from white to bright orange. This tuber loves the sun and rain and is best cultivated in well-drained sandy loam soil with a pH of 5.7-6.7. Banking, pulling the soil into ridges, is the favored cultivation method. Fresh cuttings, called slips, from existing sweet potato vines are pushed about 10cm (4in.), into the loose banked ridge and 30cm (1ft.) apart. Weeding is important in the first month, so the sweet potato vine doesn't have to compete for moisture and sun. The row should be molded at least 8 inches high. Add another inch of cover soil when the slip begins to grow. Keep 3 feet between rows, because healthy plants will vine and spread. The rows must be kept wet in the development stage. Carefully pull out any weeds. Once the vines are mature water sparingly, perhaps once a week. Sweet potatoes prefer hot dry weather but need occasional watering. If too arid, the bushes will die or produce small roots. A heavy rain, or over-watering will cause the roots not to form properly. To harvest dig them carefully with a fork and dry in a shady, cool spot for a week. After curing, sweet potato roots can be stored for 3 months.

MEDICINAL: Sweet potato has an aphrodisiac property that helps to correct male sexual dysfunction and increase testosterone. Eating these roots helps to lose weight by reducing cravings. To stop aging wrinkles, make a paste of ½ cup of boiled sweet potatoes with 1 TBS of honey and apply to your face every evening for 30 minutes. Then rinse with only water, no soap. Sweet potato promotes hair growth when rubbed onto the scalp. Combine ½ cup of mashed sweet potato with 1TS of each: olive oil, coconut oil, and honey. Rub onto a wet scalp, let sit for ½ hour, and then remove with mild baby shampoo.

NUTRITION: 200 grams of cooked sweet potatoes has 180 calories with 4g of protein and 6g of fiber. The sweet potato is very nutritious and one of the best sources of vitamin A. It also has good amounts of vitamins C and B6, potassium, manganese, niacin, and copper. The orange and purple varieties are richer in antioxidants Sweet potatoes may be shredded raw and added to salads or to top soups. Sweet potatoes can also be juiced.

FOODS & USES: When buying, select firm roots without bruises. DO NOT RE-

FRIGERATE; it will reduce the good taste. Store in a dry area at 13C (55F). Sweet potatoes are delicious: baked, grilled, fried, or boiled. If boiled, drain immediately. Sweet potatoes can be fried like regular potatoes to produce chips. Young leaves and shoots are sometimes eaten as greens. It's also a good source of starch, glucose, sugar syrup, and industrial alcohol.

RECIPES: SLICED BAKED SWEET POTATO

Slice peeled sweet potatoes ¼ inch thick. Place on a piece of foil or baking sheet and brush with vegetable oil. Bake at 400 degrees for half an hour.

FRIED SWEET POTATO CAKES

Ingredients: 3 sweet potatoes, 2 eggs, ½ cup flour, 2 TBS cooking oil, salt, and pepper to taste

Method: Peel and grate the raw sweet potatoes. Mix in eggs and flour. Season to taste. Form into cakes about 1-inch thick. Heat oil in a frying pan or wok and place cakes. Cover and fry till cooked through and the cake breaks easily. Uncover and brown. Serves four.

SWEET POTATO SOUP

Ingredients: 4 sweet potatoes peeled and chopped into 1-inch pieces, 1 large onion chopped, 1½ cups coconut milk, 3 cups vegetable stock, ½ TS turmeric, 1TS ginger, salt, and seasoning to taste.

Method: In a large pot put the vegetable stock, sweet potatoes, and onion. Boil until the potato is soft. Put pieces and liquid into a food processor with turmeric, ginger, and coconut milk. Blend, puree, or force through a sieve. Return mixture to stockpot and heat. Add salt and seasonings.

SPICY SWEET POTATOES – makes a great side dish or a unique appetizer.

Ingredients: 3 large sweet potatoes, 2TBS butter or margarine, ¼ cup chopped or sliced almonds (Other nuts - even coconut - can be substituted), ½ cup of flour, a pinch (to taste) each of salt, pepper, cloves, and cinnamon, 2 cups vegetable oil for frying

Method: Peel and mash boiled sweet potatoes, then add nuts and spices. Blend until sweet potatoes can be rolled into small balls. Roll the balls in flour. Deep fry golden brown and serve hot.

SWEET POTATO CASSEROLE

Ingredients: 4 medium sweet potatoes – boiled and mashed, 4 ripe bananas – mashed, ½ cup chopped nuts, 6 TBS butter or margarine, ½ cup brown sugar (or less to taste), ½ cup coconut, 1 cup crushed corn flakes

Method: In an appropriate pot or bowl combine mashed sweet potatoes with nuts, 4 TBS butter, a ¼ cup of the sugar, and the coconut. Pour half of this mixture into an oven casserole dish. Spread mashed bananas and cover with the remaining sweet potato mixture. Cover top with crushed corn flakes remaining butter and sugar. Bake at 350 degrees for 15 minutes. Uncover and bake for another 10 till top is browned and crunchy.

DID YOU KNOW? Sweet potatoes were a staple of the Maori in New Zealand before the arrival of Europeans. George Washington Carver found 300 uses for peanuts. From the sweet potato, Carver developed 73 dyes, 17 wood fillers, 14 candies, 5 library pastes, 5 breakfast foods, 4 starches, 4 flours, and 3 types of molasses. Carver's research also demonstrated the value of soil regeneration by planting sweet potatoes as a rotation crop. It ranks as the world's 7th most important food crop—after wheat, rice, maize, potato, barley, and cassava. China grows 90% of the 13 million tons that are produced globally per year. Hawaiians make sweet potato beer. **NAMES:** In New Zealand sweet potatoes are also known as kumara, batatas or boniatos in South America, and in India it is called sakharkanda.

Perhaps no crop has larger possibilities for quick increase of food production for both men and animals than the sweet potato. ~ David Houston

TARO – KALO – LEHUA MAOLI

Colocasia esculenta

Among the Hawaiian Islands, many roots answer to the name 'taro / kalo.' Root names are an island/district/village -even family determination. Kalo is one of the original canoe plants carried over a millennium ago, and is traditionally considered one of their most important crops equal in prominence to the sweet potato. *Colocasia esculenta* is a fast-growing tropical plant cultivated for its roots. Another name is 'lehua maoli' which loosely translates to the common/usual taro root that fed the million Hawaiians who existed before being 'discovered' by the Europeans. Taro and the broad heart-shaped leaves are called taro bush. In the Caribbean, it's dasheen bush. Taro leaves can easily be confused with the *Colocasia antiquorum* root also named colocasia or eddo. The taro leaf stem attaches to the center of the leaf and does not touch the 'heart notch.' These leaves are not very tasty until cooked and drained to remove the oxalic acid. These big roots are the main ingredient of the Hawaiian dish 'poi,' made from steaming or boiling taro before mashing it into a paste. Information was gathered from *Roots* by Diane Morgan and *Kalo* by Kawika B. Winter.

Botanical-anthropological evidence suggests that taro originated probably in India, Malaysia, or New Guinea. It's still growing wild in S.E. Asia. As an easy to grow, delicious food starch, this root spread throughout China, Japan, and the Pacific Islands. This root also traveled west to the Middle East, Arabia, and the Mediterranean region. Two millenniums ago, 100 B.C.E., it was growing in China, Egypt, and the east coast of Africa. Slave ships carried it to the Caribbean.

HOW TO GROW: Kalo/taro/dasheen roots comprise one or more large central heads, sometimes called a 'mommy,' which may grow to eight pounds. Around the taro head is a cluster of smaller corms, usually 110g (4oz.) in size. The growing season for taro roots is about seven months. These big green leaves that can grow to 2 meters (7ft.) make a good tropical landscaping effect and contribute to tasty dinners. *The author is more familiar with this root's Caribbean name.*

Taro roots left in the ground usually remain in good condition, and they will sprout again. First, fork the soil a foot deep and wide. Add either a phosphate fertilizer or rotted manure to the bottom of the forked ditch. Taro roots or suckers (small roots) are planted whole, three inches deep, two feet apart in rows spaced four feet apart. Taro requires moist soil. Along an irrigation or drainage ditch is perfect. Every 2 months use a phosphate-rich fertilizer mix, as 12-24-12.

More sun produces bigger corms, and more shade produces better tasting leaves. So, it's best to get the morning sun and afternoon shade. Taro does not compete well against weeds before its big leaf canopy is formed. The best method of controlling weeds is to pull them by hand. This should not be too difficult as the soil should be moist. Water is the key ingredient to making kalo/taro produce. Using insecticides can control most insect pests, such as leafhopper and aphids. Since these roots are near to the surface, beware that watering may expose them, and ants may attack the roots. Malathion, a mild pesticide, should stop them. Taro is ready to harvest when the leaves turn yellowish, and the corms protrude. This may be 6-8 months de-pending on location, soil fertility, and wetness.

MEDICINAL: The leaf juice of kalo/taro is considered a coagulant, stimulant, and increases blood circulation. It's a useful treatment for internal hemorrhages, earaches and dis-charges, and swollen lymph nodes. The root's juice is a pain killer and a laxative. Eating taro stimulates the appetite and helps cure constant thirst.

NUTRITION: Kalo/taro is a belly-filling starch with 250 calories to a cup. It has low fat, no cholesterol, one gram of fiber, but 60 grams of carbohydrates with healthy Palmitic, oleic and linoleic acids and adequate levels of the other essential amino acids. Taro is low in protein, high in potassium, but contains moderate quantities of calcium, phosphorus, and magnesium.

FOOD & USES: Kalo/taro can be boiled, steamed, and baked, or fried to make chips. Islanders grow these starchy roots to use like potatoes, but sweeter, to thicken soups and stews. The leaves are cooked as greens similar to spinach. The stock or leaf stem can be peeled and boiled to a taste like asparagus. Also, blanched young shoots obtained by growing in heavy shade (usually under bananas) supply a tender vegetable having a flavor like mushrooms.

RECIPES: TARO GNOCCHI – (gnocchi is the Italian version of potato dumplings)
Ingredients: 2 pounds taro peeled and chunked, 1 medium onion chopped very fine, 2TBS olive oil, 2TBS fresh basil herb, 2 garlic cloves minced, 1 bunch cilantro chopped small, 1 cup dry white wine, 2 medium ripe tomatoes chopped small, ½ cup prepared spaghetti sauce, 2 large eggs beaten, 1 cup bakers flour, ½ TS nutmeg, 1 cup grated Parmesan or cheddar cheese.
Method: Boil taro pieces in salted water until tender, approximately 30 minutes. To make the sauce, use a large skillet, heat the oil adding basil, cilantro, garlic, and onions. Cook for about three minutes. Next, lower the heat before adding the white wine, tomatoes, and marinara sauce. Simmer for one minute. Then season with salt and pepper to your taste. To make the gnocchi, drain the taro before passing them through a ricer or sieve into a large bowl. Add eggs and nutmeg. Season to taste Slowly add flour evenly, stirring constantly to prevent lumps. Bring a pot of slightly salted water to boil. Carefully spoon the taro mixture into as neat balls before dropping into the boiling water. Cook these taro gnocchi balls until they float for 5 minutes. Drain and add to the skillet with the sauce. Serve warm with grated Parmesan or cheddar cheese.

KALO CHIPS
Ingredients: 1 pound raw kalo/taro peeled and sliced 1/8-inch-thick, oil for frying, salt, and spices.
Method: Place taro slices in a large strainer or colander and blanch them with boiling water. Drain and dry. Sprinkle with spices and cook in hot oil until golden brown. Drain and eat warm.

KALO SALAD
Ingredients: 3 pounds kalo/taro peeled, boiled, and diced, 1 cup mayonnaise, 3TBS vegetable oil, 3bunches of parsley, 2 bunches of spring onions chopped, 1large onion chopped, 2 garlic cloves minced, 3 TBS white vinegar, 12 green or black olives sliced thin.
Method: Combine onions, garlic, parsley, chives, and all seasonings with the oil and vinegar. Add taro cubes, olive slices, and mayonnaise. Cool and serve.

KALO PUFFS
Ingredients: 2 pounds peeled, cooked, and mashed taro, ¼ cup milk, 2TBS butter or margarine, 1 egg beaten, 4TBS breadcrumbs, ¼ cup all-purpose flour, 1 cup frying oil, salt, and spice
Method: Combine mashed taro, butter, milk, and the egg into a soft mixture. On a smooth flat surface mix breadcrumbs and flour. With your hands, form balls of the taro mixture and roll in the flour and breadcrumbs. Fry in hot oil until golden brown. Serve hot.

DID YOU KNOW? Kalo/taro/lehau maoli is also known as callaloo, elephant ears, tropic potato, dasheen, cocoyam, or blue food. Botanically it is *Colocasia esculenta*.

The author apologizes for this root's confusion of names.

TURNIP
Brassica rapa

This is another hard-to-find root in Hawaii. The turnip is thought to have originated in middle and eastern Asia and is grown throughout the temperate zone. Turnips are one of the oldest known crops; people have eaten them for 4,000 years. European explorers and the British brought this root to all their colonies. The tops and the roots of turnips are equally tasty. The common turnip is white skinned except for the purple, red, or greenish above the ground sunburned top.

Turnips are a quick-growing crop. The green tops are fuzzy with succulent stems simi-lar to mustard greens but not as curly. The roots are rounded, white or white with a purple top. The inside the flesh is smooth, crisp, and white. Sometimes they are confused with rutabagas; they are similar, but two different vegetables. China grows 20 million tons yearly.

HOW TO GROW: Turnips grow best at 27C/80 F. Some turnip varieties as Purple Top White Globe, Golden Ball, De Croissy, and Marteau can be grown in tropical areas. Tur-nips are easy to grow. They require 8 hours of sunlight, regular watering, and rich, well-drained soil tilled a foot deep and wide with some compost mixed in to keep it loose. Make a narrow furrow. Sow fresh seeds ½ in. deep and 2 in. apart. 18 inches between each row. Seeds should germinate in a week. Once seeds have sprouted and reached about 3 inches tall, you can thin them out to where there are 3 inches between the turnips in each row. Don't discard those seedlings — they're great in a salad or on a sandwich as micro-greens. If you only grow turnips for the green tops, you can plant them closer together. Harvest in 6-9 weeks, when they're 3 inches in diameter or less — any larger and they become pungent, pithy, and stringy. A good turnip can weigh a kilo.

MEDICINAL: Turnips have been used to treat rheumatoid arthritis, edemas, headaches, jaundice, hepatitis, and sore throats. Turnip is rich in glycosylates and isothiocya-nates that have anti-tumor properties.

NUTRITION: Turnips greens are very nutritious; 100 grams of boiled leaves have only 20 calories, 1g protein, loaded with vitamin K, and vitamins A, C, folate, and lutein. The root per 100 grams has 28 calories with vitamins C and B6, folic acid, calcium, and potassium.

FOODS & USES: Use turnips like potatoes, baked or boiled in stews, soups, stir-fries, or lightly steamed with some butter, salt, or lemon juice for flavor. Enjoy shredded turnip in-stead of cabbage in your next coleslaw salad. Use turnip greens instead of spinach. They're delicious sautéed or steamed as a side dish with garlic, onion, and lemon, or in soups, stews, and pastas.

RECIPES: ROASTED TURNIPS

Ingredients, 1½ pounds turnips peeled and chopped small, 1 TBS olive oil, 1 TBS butter or margarine, salt, pepper, and spices to taste.

Method: In a wok or frying pan on high, stir-fry turnip pieces with olive oil, salt, and pepper until tender. Remove from heat and toss with butter and spices.

DID YOU KNOW? Irish legend says that the first jack-o-lantern was carved from a turnip. It wasn't until the tradition was brought to America that pumpkins replaced turnips. The heaviest turnip weighed 17.7 kg (39 lb. 3 oz.) and was grown in Alaska in 2004.

WHITE RADISH – DAIKON

R. Sativus var. L

Hawaiians prepare the white radish, daikon, in many ways from kim chee to soups. This spicy radish is easy to grow, very tasty, nutritional in both the root and leaves. It is an Asian favorite and can be grown all year throughout the Hawaiian Islands.

Thousands of years ago radishes were first cultivated in China 5000 years ago. Along the Mediterranean, the Greeks grew a primitive type of radish, but the Egyptians farmed radishes to a great extent for the Pharaoh's tables. The radish did not reach Europe or Britain until the mid-1500s. Radish gets its name from the Latin 'radix' meaning root. Daikon means 'great root' in Japanese. In a cool shaded area, daikon will grow rapidly. A mature white radish can weigh 5 or 6 pounds. There are several varieties of radishes. Some are red, white, or black, thin, and long, while others are short and round. All radish top greens are edible.

HOW TO GROW: Daikon is spindle-shaped and can be grown wherever there is sun and moist, fertile soil, even on the smallest backyard garden if there's enough water. A good long variety to try is Minowase or the Japan Ball variety that's white and round. Create a raised bed (4 in.) because radishes need loose soil that drains well to develop the root. Fork the soil a foot deep and remove all stones from the radish bed. Spread a cup of 10-20-10 fertilizer mix for every ten feet of row or add a five-gallon pail of well-rotted chicken manure. Spread 2 tablespoons of Sevin Powder throughout each row. The more the soil is worked, the easier the radishes will grow. The following day wet the bed thoroughly and carefully plant radish seeds ½ inch deep and 1 inch apart. Within a week you should see radish sprouts. Water regularly, but do not soak the bed. Some radish types are ready to harvest within a month. To ensure a continuous supply, make successive plantings of short rows every 2 weeks. Radishes may be planted in spaces between vegetables like peppers or tomatoes.

MEDICINAL: Radishes are rich in antioxidants and potassium that help lower high blood pressure and reduces the risk fofheart disease. The radish is also a good source of natural nitrates that improve blood flow and helps regulate blood sugar level.

NUTRITION: All radishes are very flavorful and low in calories with a good balance of fiber, potassium, and folate. 100 grams of white radish has only 18 calories because they are 94% water. They are a good source of vitamins A and C with calcium and phosphorus. A ½ cup of white radish daily will help prevent kidney or gall stones.

FOOD & USES: Red radishes are usually eaten raw and are in demand by the hotel industry. White radish can be pickled and processed in various ways. Always wash radishes. There is no need to peel or remove the skin from red radishes except at the top and root end. Radishes may be sliced, diced, shredded, or served whole. However, white daikon radishes should be peeled because the skin is bitter. Those smaller than six inches can be eaten raw, but the longer ones are excellent when added to stir-fry or cut up and simmered in stews and soups. Daikon aids in the digestion of fatty fried foods.

RECIPES: WHITE RADISH CURRY

Ingredients: 250g white radish peeled and chopped into thin sticks, 1 garlic clove minced, 1 medium onion sliced thin, 2 dried red chilies chopped, 10 curry leaves, 1/2TS each of turmeric, cinnamon, and mustard powder, 1TS curry powder, 2 TBS oil, 1 cup coconut milk, 1 TS salt

Method: In a frying pan or wok, heat the oil and fry the garlic, onion, chili, and stir for 2 minutes. Stir in all the spices and then add the white radish sticks. Stir-fry for 5 minutes. Stir in the coconut milk. Simmer stirring. Serve with rice or noodles.

PASTA WITH RADISH

Ingredients: 4 white radishes with green tops, 2 TBS olive oil, 1 onion chopped small, 1 pound pasta cooked (shells, springs, or elbows) - keep a quarter of the cooking water, ¼ cup grated parmesan cheese or cheddar, salt, and pepper to taste

Method: Wash radishes and greens several times before separating tops. Slice radishes very thin. In a large frying pan, heat the oil. Fry onions for 2 minutes until soft. Add radish slices and greens. Cover and cook for 5 minutes until the greens have wilted. Season and cool before adding drained pasta. Add the remaining water from cooking the pasta, stirring while adding grated cheese.

HOT AND SOUR SPINACH RADISH SOUP

Ingredients: 5 cups chicken broth, ¼ cup rice wine vinegar, 2TBS sugar, 2 green chilies minced, 1 thumb-sized piece of ginger peeled and minced, peeled shrimp, 1 ½ cup sliced radish, 1 ½ cup shredded spinach leaves, ½ cup chopped chives.

Method: In a large skillet, bring chicken broth to a boil. Add rice vinegar, sugar, hot pepper, and ginger. Then add shrimp and cook a few minutes until they become pink. Mix in radish, spinach, and chives. Cover and let sit for 5 minutes before serving.

CHINESE RADISH SALAD

Ingredients: one large white radish – sliced into large match sticks, 2 TBS rice or clear vinegar, 1 TBS soy sauce, 1 TS sugar, 2 TS sesame oil, 2 cloves garlic minced.

Method: Mix all ingredients in a large bowl.

DAIKON SALSA – great on grilled fish, pork or beef. Use it on scrambled eggs or as a chip dip. Ingredients: 2 white radishes, 2 medium ripe tomatoes, 1 medium onion, 1 bunch of cilantro, 1 garlic clove minced, juice from 1 lemon, 2 green chilies minced, salt, and spices to taste. Method: Chop the ingredients fine, mix and place in the fridge for an hour to let the flavors mix.

HEALTH NOTE

Radishes and their green tops are an excellent source of vitamin C. The leaves contain six times the vitamin C content of their root and are also a good source of calcium. Red radishes provide the trace mineral molybdenum and are a good source of potassium. Daikon provides a very good source of potassium and copper. Historically radishes were used as a medicinal food for liver disorders. They contain a variety of sulfur-based chemicals to increase the flow of bile, which helps to maintain a healthy gallbladder and liver, and improve digestion.

DID YOU KNOW?

Radish's botanical name is *raphanus*. There are two types; the spring usually red *(R.sativus)* commonly in salads, and the Asian or winter white daikon radish *(R. sativus variety longipinnatus)*.

What is life if no laughter and love, caring and compassion, fresh bread and crisp radishes? ~ James Kavanaugh

WHITE TURMERIC – ZEDOARY
Curcuma zedoaria

Throughout the Hawaiian Islands white turmeric is a rare, but available root. White turmeric is a perennial herb, also known as zedoary, and indigenous to Bangladesh, Sri Lanka, and India. Zedoary was an ancient food plant that spread during prehistoric times to the Pacific Islands and Madagascar, 3-5,000 years ago. This root's foliage produces a beautiful yellow and purple flower, excellent as a hedge or in rock gardens. Its use as a spice and in Ayurveda/traditional medicine is becoming better known. The root is white inside and smells like a mango. It tastes like ginger, except with a bitter aftertaste. White turmeric is cultivated in India, Sri Lanka, and China for its starch-rich tubers, zedoary root.

HOW TO GROW: White turmeric will grow almost anywhere with proper attention. Zedoary likes wet lowland forests, close to drains, streams, and rivers. First, find some root pieces to plant. Roots are available online. I put a few in a damp cloth inside my fridge for a few days until I see some sprouts. Zedoary grows best in well-drained, sandy soil, rich with compost with a preferred pH 6.0 - 6.5. Soak the soil prior to planting. Put the roots about 1 inch deep with the sprout or pointed end up and plant about 8 inches apart. Plant in full or partial sun, water regularly and every other month lightly fertilize with 10-10-10 or organic manure.

Once it flowers, the roots will be ready to harvest in 8 months. Zedoary makes an attractive fragrant, yellow/purple flower about a meter tall. It should bloom at the start of the rainy season as the leaves develop. The roots are ready for harvest when the plant wilts.

MEDICINAL: Various parts of the white turmeric plant are used in Ayurveda for the treatment of different ailments. Traditional medicine claims it treats cancer, ulcers, diar-rhea, dysentery, and even toothaches. Preparation methods begin with careful washing with lots of water to remove most of the exterior protein and water-soluble nutrients. The rinsing is supposed to remove a poison that is yet to be identified. Zedoary is used for stomach pain, loss of appetite, indigestion, as a mosquito repellent, and many other conditions, but there is no good scientific evidence to support these uses. This root also has antioxidants that reduce the risk of serious heart disease and diabetes.

NUTRITION: White turmeric is rich in vitamin C, vitamin B6, manganese, iron, potassium, and omega-3 fatty acids.

FOODS & USES: In Indonesia, white turmeric is ground to a powder to make white curry pastes. In India, it's used fresh or pickled. In Thai cuisine, it's used raw and cut in thin strips in certain salads or in Thai chili pastes. It flavors vegetables, bean dishes, and chutneys. Zedoary flowers are in demand. This root is the source of 'Shoti Starch,' which is a substitute for arrowroot and barley. Zedoary is also used in the manufacture of liquors, stomach essences, and bitters. The essential oil produced from the white turmeric roots is used in the production of perfumes and cosmetics.

White tumeric produces beautiflul flowers.

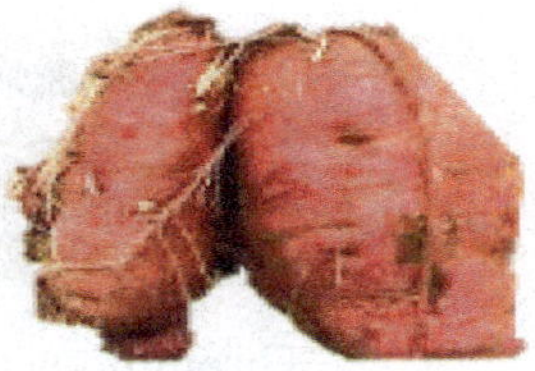

YAMS - WHITE – PURPLE
GREATER YAM – WINGED YAM
UBE - UWALA
Dioscorea alata

This big yam is known throughout Hawaii as purple -'ube', or 'uwala', the greater yam, the winged yam, or 'uhi uala,' the white yam. The Puna area on the Big Island has a great reputation for growing these yams. This specific yam type, known botanically as *Dioscorea alata (L),* is one of the most important staple crops throughout Indo-Asia and the world's most cultivated yam. The yam has been known to man for millenniums and may be the world's earliest cultivated hybrid. Researchers believe it was native to northern India-SE Asia when an Indo-Asiatic farm group intentionally crossed two wild yams 8,000 years ago. It traveled to Africa from Asia around 1,500 BCE. Yams grow untended in the wild, but it's not a wild tuber. You'll find this yam locally, but also online. Keep a close watch when growing yams, so the vines don't grow beyond your boundaries. This is classified as a Category 1 invasive. Young white yams grown in softened soil can grow long, solid, and straight, older roots tend to branch out and can be distorted. With proper cultivation, this yam can grow over 2 meters long and the known record is 81 kilos (178 lbs.). Boiling or roasting of the root is necessary to make it edible. Some varieties of yams and sweet potatoes become purple because of the antioxidant called anthocyanin, that also colors red cabbage, red wine, and purple cauliflower.

HOW TO GROW: Yams have an attractive vine and needs a trellis or a fence for support as it extends. It is sometimes grown as an ornamental. If you or a neighbor has a white yam that flowers, you may get some seeds. Generally, it's best to plant using pieces of the exist-ing vines or roots for propagation. First, find a fresh piece of yam and cut the root into several chunks and let the pieces dry for two days or dip the cut end in wood ash. Fresh-cut chunks could quickly rot when planted. If you want a yam nursery, put pieces in a bucket of dirt and keep it moist. Once they sprout, you can move the baby yams to your garden.

While you're waiting, dig a hole the size of a 5-gallon bucket. (You can grow sizable yams in a bucket on your porch or patio, but harvest before it outgrows the space.). Backfill the hole with rotted manure and compost mixed with the soil. Plant the piece of yam 12cm (5 in.) deep and keep plants a meter apart. This yam prefers full or partial sun and a pH in the range 5.5 - 6.5. It is important to keep the area of the yam vine clean of weeds for the first 3 months. It thrives with regular watering, but do not overwater. Where you plant should be well-drained. Beware of erosion due to the watering; keep the heads of the yams covered at all times so they don't get sunburned. Support, like a trellis, to grow on will boost your plant's growth. When it's climbing, the vines will spread out and expose more leaf surface to the sun. White yam can reach 15 meters in height and it wraps its vine around right to left. Be careful, this yam can become an invasive species.

When the tubers are mature, the vine will yellow and shrink. The root should be ready to harvest in 6 months, normally weighing 5-10 kg. When checking the size of the root, soak the area, carefully take a hand garden trowel and dig away the soil. Try not to gouge the root, as those spots will rot quickly. Carefully pull up the entire plant, using a shovel if necessary. Remove the roots from the vine and brush off the dirt. This potentially invasive plant spreads, so dispose of all parts you will not replant. Keep your white yam roots dry or they may sprout. Remember, these roots are toxic when raw; they must be completely cooked. Wearing protective gloves is a good idea.

MEDICINAL: A paste made from pounding yam can be applied as a treatment for wounds, inflamed hemorrhoids, and skin diseases. Dried powdered yam treats piles. Juice of the tuber is used to kill stomach parasites. It is a laxative and used to treat fever.

NUTRITION: Yam contains per 100grams: 100 calories with tiny amounts of protein and fiber, but has good amounts of vitamins B1, B2, and C. The mineral content is high in potassium, magnesium, calcium, phosphorus, and zinc. As a staple, nutritionally, this yam compares favorably to brown rice and whole wheat. Yams contain phytosterols, alkaloids, tannin, and starch.

FOOD & USES: Several countries are developing processed products such as starch, yam flakes, or powder from surplus supplies. Yams have a mildly sweet, nutty taste, like sweet potatoes. This yam is mainly prepared as a vegetable, comparable to the potato, cut to make French fries and chips, but claimed to be superior to similar potato products. Usually, this yam is peeled, chopped, and boiled like a potato. Older yams can be pounded into a sticky elastic dough called pounded yam. It's usually eaten in soups and stews.

To make yam root flour slice the peeled root into ¼ in thick pieces. Boil for only a few minutes, but not until soft. You want the slices to remain stiff. Dry the slices in the sun and then grind them into flour. Force the dust through a fine sieve to create a uniform texture.

IF ANY YAM TASTES BITTER, DON'T EAT IT.

RECIPES: YAM PUDDING

Ingredients: 500g (1lb.) of white yam peeled, 1 cup of milk, 2 cups water, 2TBS brown sugar or jaggery (coconut sugar), ½ cup grated coconut, ½ TS ground cinnamon, ½ TS ground cardamom, 1TBS soft butter, ¼ cup roasted cashews, almonds, or peanuts (optional).

Method: Chop the yam into small cubes. In a sizable pot, add the milk and water and bring it to a boil. Add the chopped yam. Reduce the heat and simmer for 45 minutes, or until the yam pieces are slightly soft but not mushy. Stir in sugar, ground cinnamon, and cardamom powders. Stir continuously until the liquid in the pan has almost evaporated. Turn off the flame and let the mixture cool. Then add butter and use a blender, hand mixer or whisk until it becomes a light pudding. Then stir in the nuts and top with a sprinkle of ground cinnamon and nutmeg. Serve at room temperature or refrigerate and serve cold.

YAM CAKE

Ingredients: 2 cups of peeled, diced yam, 1 medium onion minced, 1 cup rice flour, 2 TBS cornstarch, 1 TS Chinese 5 spice powder, ½ TS ground black pepper, ½ TS salt. FOR THE TOPPING: spring onions and 2 green chili peppers chopped fine.

Method: Heat a pan over medium heat, and fry the onions until they brown, about 4 minutes. Add the cubed yam to the pan and fry it with the onion until it browns. In a separate bowl, combine the rice flour, cornstarch, and water. Stir until it forms a smooth paste. Make sure there are no lumps in the mixture. Add the flour mixture into the pan slowly and stir until everything forms a thick paste. Add the salt, pepper, and five-spice powder, and mix well. Pour the mixture into a heatproof bowl inside a bigger pot that has water on high heat. Cover and steam the flour-yam mixture for 45 minutes, or until cooked. Let cool, but serve warm, not hot. Before serving, sprinkle the top with the chopped spring onions and sliced chilies. Accompany with spicy chili sauce or chili paste.

DID YOU KNOW? The yam was first domesticated in the highlands of New Guinea from around 10,000 BCE. However, much older remains identified as probably *Dioscorea alata* have been recovered from the *Niah Caves* of Borneo (*Late Pleistocene* period circa 40,000 BCE.)

HAWAIIAN NAMES: pālau, ulehihi, hoʻonohonoho, lehua, piʻa, poni,aha, niʻihau, ʻālela, uhi ʻālela. Bitter yam, hoi; piʻoi, ube, uwuala

I've always felt that having a garden is like having a good and loyal friend. – C. Z. Guest

THE HAWAIIAN HOME GARDEN GUIDE
VEGETABLES

The term vegetable usually means that the fruit, leaf, stem, or root of a plant is edible for humans. This word does not apply to any rule of botany and is subjective. Vegetable is not a scientific or botanical term and is based on regional food and cultural customs. Most vegetables are annual plants, meaning they must be replanted every year.

There is no clear distinction between vegetables and fruits. Most vegetables consist largely of water, making them low in calories and can be eaten either raw or cooked. They are excellent sources of fiber, multiple vitamins including A and C, and minerals such as potassium, calcium, and iron. All the amino acids needed to synthesize protein are available in vegetables. Consuming five or more portions of vegetables a day is recommended for good health. You need to grow your own. Fresh vegetables quickly age and spoil, but their storage life can be extended by preservation methods as dehydration, canning, freezing, fermenting, and pickling. Living in a warm climate can provide fresh veggies all year.

Botanists define most vegetables developed from plant blossoms as fruits. However, at the market everyone knows the difference between a fruit and a veggie. Eggplants, bell peppers, and tomatoes are botanically fruits, as are most grains, and even some spices like black and red pepper. Some vegetables are variable as corn are considered a vegetable only when fresh. When it is dried it becomes a grain.

A vegetable is also defined as a person with a dull or inactive life. If you have a home garden your life will be busy and interesting. Grow healthy veggies and don't be considered one.

PAGE:

Photos courtesy 912crf.com

Garden: One of a vast number of free outdoor restaurants operated by charity-minded amateurs in an effort to provide healthful, balanced meals for insects, birds, and animals.
~ Henry Beard and Roy McKie, Gardener's Dictionary

BROCCOLI - PALAKAĪ
Brassica olerace

Broccoli is a recent addition to Hawaii's farms and home gardens. Broccoli was considered a cool weather crop that would not grow in the tropics. Now, through considerable agri-science manipulation, broccoli is an excellent and profitable crop. There are two types, curding and sprouting. Curding grows white or purple heads similar to cauliflower. Sprouting is the most familiar type with numerous small headed, green shoots.

Broccoli could possibly be the healthiest food you can grow and eat. It contains vitamins C, A, K, B1, B2, B3, and B5. With soluble fiber and many nutrients, broccoli has been researched to provide the human immune system with anti-virus, anti-infection, and anti-cancer capacities. All for only twenty-two calories a serving!

The name, broccoli is derived from the Italian word 'broccolo' for branch, stem, or stalk; and is usually considered an Italian vegetable. The botanical name is Brassica olerace. Broccoli is actually the flower head of the plant. If broccoli is not harvested and left 'to go to seed', the familiar green florets transform into yellow flowers. The florets are healthier in vitamins than the stalk. Broccoli's leaves can also be cooked and contain more vitamin A than the florets.

HOW TO GROW: Good healthy food is the best reason to grow broccoli, only then you can know it has had little chemical sprayed on it. Broccoli and cauliflower are two of the highest sprayed vegetables. If you start from seeds plant in moist potting soil. A variety of broccoli that grows well in the tropics is 'Di Ciccio,' and seeds are available online. To grow broccoli, work plenty of well-rotted chicken manure into the plot. The preferred pH is 6-7. It's best to place your plot in an area that does not get the direct heat of the sun in the late, heated afternoon. Broccoli needs sun, but not heat. Lack of sunlight may produce thin, leggy plants and small heads. Too much heat and the heads will burn and be brown and yellow. If you are earnest about growing broccoli, consider growing it under the shade of a mesh roof. Broccoli plants can grow to 2 feet tall. It is important to plant seedlings at least a foot apart, (2 feet apart will be better and reduce fungus problems.) and 2 feet between rows. A high nitrogen fertilizer may be applied when the plants are 5 weeks old. Like many things in life, growing broccoli takes patience and perseverance.

Broccoli needs constant water to cool the leaves for the heads to fully develop, but be careful they do not become waterlogged. The area requires excellent drainage. They also need space between the plants for air to circulate. The part of broccoli we eat is the stem and its bunch of unopened flower buds. The green buds develop first in one large central head and lat-er in several smaller side shoots. Serious growers recommend cutting the central head leaving two or three inches of stem. This is done after the head is fully developed, but before it spreads out into individual florets. This stimulates the side shoots to develop for later pickings. This permits a continued harvest of broccoli for several weeks. To promote the growth of a second head after the first has been harvested, keep fertilizing and watering. Broccoli is a good veggie for a porch pot.

Broccoli suffers the same pests as cabbage and cauliflower. Aphids and the cabbage worm from the diamond back moth are the main enemies. Fungus is always a potential problem.

MEDICINAL: Broccoli helps fight cancer, especially breast, colon, and lung. It boosts the immune system. It contains antioxidants and a substance called sulforaphane, a powerful cancer fighter and preventative. Broccoli may reduce the risk of heart disease. It is an antioxidant and helps with stress. Researchers have discovered a compound in broccoli that fights the bacteria that causes peptic ulcers much better than modern antibiotic drugs. This same broccoli compound protects against stomach cancer, currently the second most common type of cancer. Broccoli has been found to lower the risk of prostrate and lung cancers.

NUTRITION: A half cup of steamed or boiled broccoli has 22 calories, 2½ grams of both protein and fiber with a tiny bit of fat, and no cholesterol. The same small portion of broccoli has one and a half times the daily amount of vitamin C, plus vitamin A, niacin, thiamin, iron, folate, selenium, phosphorus, potassium, zinc, and magnesium. For those who are lactose intolerant, broccoli is a good source of calcium necessary for controlling high blood pressure, and in prevention of colon cancer.

FOOD & USES: Before preparing, broccoli should be washed quickly and never left to sit in water. To retain the nutrients in broccoli, either steam it, stir-fry it, or boil it in a very small amount of water. Most other preparation methods will cause a nutrient loss of about 25 to 35 percent. Broccoli that has been cooked still has 15 percent more vitamin C than an orange and as much calcium as a cup of milk. If broccoli's smell irritates, add a slice of bread to the pot.

When purchasing, the better broccoli tops will be a darker green than the stems, closed with no yellowing, and should feel crisp. If stored properly in the fridge, unwashed, dry, in a plastic bag, broccoli will keep up to two weeks. However, like most fruits and veggies, it's best fresh for the most nutrients. So, grow it at home.

RECIPES: BROCCOLI PASTA

Ingredients: 3 cups chopped broccoli florets and stems, 1 cup carrots peeled and sliced about ¼ inch thick, 1 garlic clove minced, 1 medium onion chopped small, ½ cup whole milk, ¼ cup grated cheddar cheese, 3 cups boiled noodles (elbows, bowtie, or spaghetti), 2 cups chicken broth, ½ cup boned chicken, (Chicken pieces may also be used.), salt, and spices to taste.

Method: Boil chicken pieces in suitable pot for 10 minutes and retain the remaining broth. Remove chicken bones if you desire. Boil noodles. Combine all ingredients in a suitable baking dish, cover, and bake at 350 degrees for 30 minutes.

CHEESE BROCCOLI

Ingredients: 2 cups chopped broccoli stems and florets, ½ cup sliced peeled carrots, 1 cup cooked rice, 1 garlic clove minced, 1 large onion chopped small, ½ cup water, 1 cup milk, ½ cup grated cheddar cheese, salt, and spices to taste.

Method: In a large frying pan heat the water before adding broccoli, garlic, onion, carrots, and seasonings. Cook for 5 minutes before adding remaining ingredients. Cook for 10 minutes stirring frequently as it will stick and burn.

ALMOND BROCCOLI

Ingredients: 1 head of broccoli - stem and florets sliced a ¼ inch thick, ½ cup almonds, 2 TBS oil, juice from ½ fresh lemon, salt, and spices to your taste.

Method: Steam broccoli pieces for 10 minutes, remove and cool. Heat oil in a frying pan and cook almonds for 3 minutes while adding spices. Remove from heat and add broccoli and lemon juice. Stir and serve.

DID YOU KNOW? China produces the most broccoli, India is second, and the USA is third. Broccoli originated around 6 AD via artificial selection in the Southern Italian Peninsula or Sicily. It was grown only in Italy until a royal marriage in the 1500s moved some seeds to France.

CAULIFLOWER – KĀPIKI PUA

Brassica oleracea var. botrytis

Cauliflower is another of the veggies that makes one wonder how it evolved. It's a bit difficult to grow, but easy to cook in a variety of tasty dishes. Cauliflower is a delicious standard for stir-fry, great crispy raw in salads, creamed, or curried.

Cauliflower's Latin reference is *Brassica oleracea var. botrytis.* It is one of the Aloha State's most popular vegetables. It's part of the cabbage family and believed to have originated in Asia Minor. For 2,500 years, cauliflower has been a part of the diet of the people of Turkey. Early traders brought it to Europe. The Italians became famous for spicy cauliflower salads while the French chefs created creamy cauliflower soups. Asians mix vegetables and curry to make a cauliflower stew.

Cauliflower has a white head, which consists of unformed flower buds that average six inches in diameter. The head is called a 'curd.' The curd forms off a stalk. As the curd is pulled apart, the cauliflower looks like a small tree enclosed in stiff green leaves, which provide protection from the sun. This leaf protection slows or stops the development of chlorophyll, which contributes to the head being white. Raw cauliflower is firm and slightly bitter in flavor.

HOW TO GROW: Cauliflower is very fragile. To offset pests, heat, and drought that may easily damage it, what is grown locally in the islands is often sprayed with too much agri-cultural chemicals. If chemical spray worries you, as it does me, try growing your own. There are hybrid seeds available online that are adapted to the tropics. These cauliflowers will grow even in temperatures of 35C (95F) and produce heads in less than 70 days. The most successful method of starting cauliflower is in seed trays. The optimum soil pH is 6-7. Spray with starter fertilizer and systemic pesticide such as Admire before and after transplanting. Seedlings should be at least four inches tall before transplanting. They should be spaced in the rows at least a foot apart. Cauliflower is best planted in raised beds, so the soil can easily drain. Any interruption in the growing cycle of cauliflower, such as intense heat, drought, or hard rain, may stunt the development of the edible head. Cauliflower needs constant water, and fertilizer high in nitrogen. If cauliflower is kept too wet, the curd or head will turn brown from mildew. Cabbage worms and black rot are cauliflower's most common enemies.

Several types of cauliflower are self-blanching. This means the leaves naturally pull back as the head forms. If this does not happen when two to three inches of the white head shows, tie the outer leaves together over the center of the plant. This protects the head from sunburn and keeps it from turning green giving it a sour flavor. With the proper high nitrogen fertilizer mix (12-12-17-2) and enough water, the head should grow rapidly until it is 6-8 inches in diameter. This will happen one to two weeks after blanching, about 3 months after transplanting. A perfect head of cauliflower has white, firm, tight flowers. Cut the head off at the stem leaving a few leaves attached for protection of the flowers. The flowers will separate and become coarse with a strong taste if you wait too long to harvest. Extremely hot weather will also cause flower separation.

MEDICINAL: Cauliflower contains allicin to improve heart health and reduce the risk of strokes, and selenium, which with vitamin C strengthens the immune system. Cauliflower assists to maintain a healthy cholesterol level. It has folate, a B vitamin needed for cell growth and replication. This white veggie is recommended to women who are pregnant to help properly develop unborn children. is an excellent source of fiber to improve colon health and prevent cancer. It is also a blood and liver detoxifier. Cauliflower with turmeric may prevent or inhibit the spread of established prostate cancer.

NUTRITION: Cauliflower is very nutritional, especially when eaten raw. One cup of raw cauliflower has only 25 calories. It's rich in vitamins K and C. One cup has more than the daily requirement of B6, B5, B3, folate, biotin, magnesium, iron, manganese, and molybdenum. Again, the best cauliflower is organically grown as agricultural chemicals may offset all of cauliflower's nutritional value. Cauliflower and other members of the cabbage family have compounds that activate enzymes, which may disable and eliminate cancer-causing agents.

FOOD & USES: When buying, always seek clean tight white cauliflower heads with no spots. In cauliflower, size has no relation to quality. Uncooked heads are best kept in the fridge stem side down in a paper or plastic bag with a few holes poked in it. This should prevent moisture and mildew from developing and keep up to a week.

RECIPES: TOO HOT CAULIFLOWER

Ingredients: 3 cups steamed cauliflower drained, 1 TBS flour, ¾ cup milk, 1 TBS chopped hot pepper, ½ TS margarine, 1/2 cup shredded cheddar cheese, ¼ cup breadcrumbs, salt, spice to taste.

Method: In a saucepan melt the margarine over a medium flame. Combine flour, spices, and milk in a sealed container and shake vigorously. Add this mixture to the saucepan of margarine stirring constantly. Keep stirring as you add the cheese, then the chopped hot pepper. Place the cauliflower in a baking dish and cover with the mixture. Cover with breadcrumbs. Put in an oven preheated to 350 degrees for 15 minutes. You can also do this by carefully stirring over a low flame on the stove top.

CAULIFLOWER SOUP

Ingredients: 1 nice sized head of cauliflower chopped, 1 TS olive oil, 1 small onion, 1 gar-lic clove, 1 large potato chopped into small pieces, 1 cup chicken broth, 1 cup water, 1 TBS chopped chives, salt, and pepper to taste.

Method: Heat oil in a skillet and cook onion and garlic until soft; add potato, chicken broth, water, and cauliflower. Boil, then simmer covered for 10 minutes until vegetables are soft. Season to taste. Top with a sprinkle of chives. Note: This may also be put into a blender and liquefied.

CHEESY VEGGIE CASSEROLE

Ingredients: ½ pound cheddar cheese, ½ cup margarine, ½ cup chopped of each of carrots, cauliflower, bok choy, sweet green pepper, 1 cup crushed biscuits (crackers), salt, and pepper to taste (Almost any vegetable can be added.)

Method: Cube cheese small and place in saucepan with ¼ cup margarine. Melt over medium heat. Stir constantly or it will stick and burn. Place vegetables in an oven dish and cover with cheese mixture. Stir and sprinkle the top with the crushed crackers mixed with the remaining ¼ cup of melted margarine. Bake uncovered at 350 degrees for 25 minutes.

BRAISED CAULIFLOWER

Ingredients: 1 medium head cauliflower (about 2 pounds), 1 TS sesame oil, salt, pepper and seasonings to taste.

Method: Cut cauliflower into quarters and slice again into ¼ inch thick pieces. Mix slices, oil, and seasonings in a large bowl. Spread on a nonstick baking sheet. Bake 25 minutes at 400 degrees. Turn the cauliflower every ten minutes until cauliflower is browned.

DID YOU KNOW? Cauliflower may yellow when in alkaline water. For whiter cauliflower, add a tablespoon of milk, or lemon juice to the water. Do not cook cauliflower in an aluminum or cast-iron pot. The chemical compounds in cauliflower will react with the aluminum and turn the vegetable yellow. While in an iron pot, it will turn a brown or blue-green color.

Cauliflower is nothing but cabbage with a college education. ~ Mark Twain

CORN- MAIZE - KŪLINA
Zea mays

The Western Hemisphere's primary grain was maize, now called corn in some countries, before the Europeans intruded. Around 8000 BCE, the inhabitants of Mexico probably first grew corn! It's unique from other grains since scientists do not know how the plant evolved. In the Old World, such as Egypt or Greece, no evidence exists of maize in archaeological remains, and no mention of it is made in ancient writings. It is believed to have evolved only in the West. Explorers carried corn to many parts of the world and use it nearly everywhere today, often as a feed grain for animals. Corn is a European word, which means kernel. Botanically it is *Zea mays*.

Corn grows so easily on the Hawaiian Islands that Bayer Corporation began seed production facilities on Oahu, Maui and Molokai. This Hawaiian grown corn seed is shipped to in the U.S. mainland. Try to secure corn seed that says 'sweet' in its name, harvest early before the kernels have a chance to harden. A superstition is that once one ear is picked, all remaining ears will quickly ripen. Fresh corn on the cob can be frozen and stored. First, blanch by boiling for a few minutes and plunged into iced water. Allow to cool before storing in sealing freezer bags and freezing. Frozen ears may last 6 months.

HOW TO GROW: Corn is grown easily in any garden with sufficient light, fertile soil, and enough space. It's especially popular with home gardeners because it tastes better when it's harvested and eaten fresh. Corn is best grown in full sun, in well-drained soil with a pH of 6. Plant 2 seeds, 1 inch deep, every 12 inches. Rows should be about 24 inches apart. To insure proper pollination and formation of the ears, plant at least 4 rows of corn in the garden. For smaller gardens, try planting in a 4 feet by 4 feet block. Corn is a heavy feeder, use rotted chicken manure, or a 12-24-12 mix when the plants are 12 to 18 inches tall. Fertilize corn every eight days until ripe. Each cornstalk should produce at least one large ear. Using this method, you should harvest 16 to 24 ears from a small garden area. It takes 2-3 months from planting until harvest.

Corn regularly requires water. Hot, dry conditions during pollination result in missing kernels, small ears, and poor development of the tips of the ears. Harvest after the kernels are full and milky when pinched. This will occur in 12 to 18 days after the silk first shows. Corn earworms are a constant problem laying eggs on the developing silks or on the leaves near the ear. Tiny caterpillars follow the silks down into the ear, where they feed on the tip. Only one corn earworm will be found per ear because they are cannibalistic, with the largest devouring all others. Once the worm is inside the protective husk covering, there is no effective control. The corn borer can be somewhat controlled by the regular spraying of Malathion or Fastac.

MEDICINAL: In traditional Chinese and Native American medicine, corn silk is used to treat a variety of ailments, including prostate problems, malaria, urinary tract infections (UTIs), and heart disease. Recent research indicates that corn silk may also help reduce blood pressure, cholesterol, blood sugar, and inflammation. Corn silk may be used fresh, but it's often dried before being consumed as a tea or extract.

NUTRITION: Eating corn is good for you. One ear of yellow corn has only 80 calo-ries, 2.5 grams protein, 20 grams carbohydrates, 2 grams dietary fiber, potassium, vitamin A, niacin, and folate. Folate has been found to prevent some birth defects and to reduce the risk of heart disease and stroke.

FOOD & USES: Corn is usually considered a food and most people know about corn meal, cornstarch, or corn-based food additives, like corn syrup. In batteries, cornstarch is often used as an electrical conductor. Cornstarch is a common ingredient used in the production of matchsticks and is a common ingredient in many cosmetic and hygiene items, including deodorant.

It's also energy in ethanol and in hand sanitizers. There are corn-based plastics. Corn syrup is one of the main ingredients in cough drops.

RECIPES: BOILED CORN ON THE COB

Drop washed corn into a pot of rapidly boiling water. Boil for 4 minutes. The cooking time depends on the size and age of the corn. Do not overcook the corn or it will toughen.

CORN PUDDING

Ingredients: 2 cans whole corn or shave the kernels off 4 large ears, 1 cup biscuit /cracker crumbs 2/3 cup flour, 3 cups milk, 4 eggs, quarter cup cooking butter, 2 TBS sugar, a pinch of nutmeg, salt, and pepper to taste.

Method: In a large bowl, mix the corn, cracker crumbs and milk. Beat the eggs well and add to the corn mixture. Add salt, pepper, and nutmeg to your taste; stir well. Pour into a pudding pan or large baking dish and dot with bits of butter. Bake slowly in a 300° oven for about an hour.

SAVORY CORNCAKES

Ingredients: ¾ cup all-purpose flour, ¼ cup yellow cornmeal, ½ TS baking powder, ¼ TS pepper sauce, 1 cup canned creamed corn, 1 egg lightly beaten, ½ cup milk, 1 TBS vegetable oil, 1 large ear of corn - kernels cut off-the cob (2/3 cup), 2/3 cup minced onion

Method: Sift together dry ingredients. Whip the creamed corn, egg, milk, and oil, and stir into the dry mixture. Do not over-mix. Let the batter rest in the refrigerator for an hour or more. Then stir in the corn kernels and the minced onion. Cook the batter in batches, dropping table-spoonfuls onto a greased griddle over medium heat. Cook 2 minutes per side, or until golden. Keep warm in a 200 F. oven while finishing. Makes 24 small corncakes.

CORN PEPPER STIR FRY

Ingredients: 4 cups corn kernels about 5 ears, 1 TBS vegetable oil, 1 TBS cooking butter, 2 sweet bell peppers - diced, 1 small onion - diced, 1 TS ground cumin, salt, and black pepper to taste, 2 TS chopped culantro.

Method: Heat oil in a large frying pan; add corn, peppers, and onion. Once everything is cooked add spices and butter.

CRUSTY CORN CASSEROLE

Ingredients: 1- 16 oz. can cream corn, the kernels sliced off 4 ears of fresh corn, (or one 16 oz can whole corn), 2 beaten eggs, ½ cup sour cream, 2 cups cornmeal, 2 TS baking powder, two TBS brown sugar, 1 stick of butter or margarine, 1 small onion, 1 TBS finely chopped green pepper, ½ small hot pepper (optional).

Method: Preheat oven to 350F. Melt butter in an 8X12 inch glass-baking dish in the oven. Mix all ingredients and pour into the hot buttered pan. Bake 1 hour until a crust forms.

DID YOU KNOW? Evidence has been found for the earliest domestication of corn in Mexico about 9,000 years ago! By volume of production, corn is the third most important food crop of the world behind wheat and rice. Farmers grow corn on every continent except Antarctica. An ear of corn has an average of 800 kernels, arranged in 12 to 16 rows. There is one piece of silk for each kernel. The corncob or ear is part of the corn plant's flower. A pound of corn consists of approximately 1,300 kernels. Each tassel on a corn plant releases as many as 5 million grains of pollen. Corn is an ingredient in more than 3,000 grocery products. One bushel of corn will sweeten more than 400 cans of Coca-Cola, produce 32 pounds of starch, or 2 1/2 gallons of ethanol fuel. The main ingredient in most dry pet food is corn.

When planning for a year, plant corn. When planning for a decade, plant trees. When planning for a lifetime, train, and educate people. ~ Chinese proverb.

EGGPLANT
Solanum melongena

Eggplant ranks among Hawaii's favorite vegetables with the slim Long Purple, and the round Black Beauty popular varieties. Native to India and what is now Pakistan, eggplant was first domesticated over 4000 years ago. In its home region, the eggplant is used in many local dishes and carries a wide range of names. Eggplant is botanically S*olanum melongena.*

Related to the tomato, potato, and the pepper, eggplant is the only member of the deadly Nightshade family to originate in the Eastern Hemisphere. As tomatoes were first believed to be poisonous, eggplant was believed to cause mental illness, and became known as the 'Mad Apple.' The Spanish explorers believed its fruit to be a powerful aphrodisiac. When a variety with egg-shaped white fruit was grown in Germany around 1600, the English gave the name 'eggplant.' The Spanish introduced eggplant to Brazil before 1650 and someone brought seeds to the Hawaiians.

HOW TO GROW: Eggplant is grown in many shapes, sizes, and colors. There are more than 30 varieties of eggplant to choose from. When selected, grown, and prepared prop-erly, anyone will become a true lover of eggplant. 2-3 plants will yield enough for most fam-ilies. Eggplant is best started from transplants spaced 18 to 24 inches apart. Eggplant prefers well-drained soil with a pH of 5.5-7. Pick a sunny location for the best results and space the plants about 2 ft. apart. Water the plants regularly. Use starter fertilizer for transplanting, a high nitrogen fertilizer when the plants are half grown, and again immediately after harvesting the first fruits in 2-3 months.

Verticillium wilt causes yellowing, wilting, and death of the plants. Flea beetles cause tiny holes in the leaves. Damage can be severe on young plants, but can be controlled by ap-plying a suitable insecticide. White flies love the shade of eggplants' broad leaves. Inspect the underside of the leaves regularly. White fly infestation could ruin an entire garden. If you see white specs get the appropriate chemical pesticide and treat the plants without hesitation.

To test if the fruit is ripe, hold it in your palm and gently press it with your thumb. If the flesh presses in and bounces back, it's ready. If the flesh is hard and does not give, the eggplant is too young. Eggplants bruise easily; harvest gently and cut the eggplant keeping the cap and some of the stem attached.

MEDICINAL: An Australian study found eggplant absorbs more fat in cooking than any other vegetable. One serving deep fried absorbed 83 grams of fat in just 70 seconds, four times as much as an equal portion of potatoes and added more than 700 calories. **Cooked, oth-er than frying**, eggplant controls high blood pressure and it can help reduce high cholesterol. Eating eggplant can help cure insomnia.

NUTRITION: 100 grams of cooked eggplant has 27 calories, 1g protein, 6g carbs, 3g fiber (if eaten with the skin), with potassium, manganese, and vitamins C and B-6.

FOOD & USES: Eggplant has an agreeable texture and slightly bitter taste. Cooked eggplant soaks up a lot of oil, so to remedy this, salt and press the air and water out before cook-ing. To peel or not to peel the eggplant, depends on its use in the recipe. Eggplant can be baked, grilled, steamed, or sautéed. It's versatile and works well with tomatoes, onions, garlic, and cheese. My puppy loves raw eggplant. Popular dishes include eggplant Parmigiana. ratatouille. mousakka, stuffed with minced meat, pickled, baked with onions and chopped tomatoes, sliced, grilled, and curried.

RECIPES: EASIEST EGGPLANT

Ingredients: 2 eggplant, 2 TBS vegetable oil, jerk seasoning, cumin, curry powder, hot pepper to taste, ¼ cup grated cheese.

Method: Wash and slice the eggplant making discs, but leaving the skin on. Put in baking pan and coat with oil. Then sprinkle spice of your choice. Bake for 20-30 minutes turning once with a spatula. Sprinkle with cheese and shut oven off. Keep in oven 10 more minutes until cheese melts. Serves four.

EGGPLANT CURRY – you can use different types of eggplant purple, green, or round

Ingredients: 1 large purple or 6 small round eggplants in 2-in. slices, 1 TBS vegetable oil for frying, 1 onion sliced thin, 2 garlic cloves minced, 8 curry leaves, 1 TS each of the following spices, red pepper powder, coriander seed, cumin seeds, turmeric powder, 1 TBS black mustard seeds, 1-2 TBS fresh lime juice, 1 TBS minced coriander leaves, 1 cup thick coconut milk.

Method: In a mortar make a paste from the spice seeds, powders, and garlic. Pour 1 TS of oil, salt, and pepper into a bowl and rub each eggplant slice – both sides - and place in a frying pan and cook over medium flame 3-5 minutes each side, until both sides are beginning to brown. Remove the eggplant and slice each piece in half. Add the remaining oil, spice mix, and onion to the frying pan and brown. Add the curry leaves. Add the eggplant pieces and ½ cup of the coconut milk. Bring to a boil stirring. Reduce heat to a simmer and cook for 5 minutes. Add remaining coconut milk and coriander leaves. Serve with rice or roti.

RATATOUILLE - A great vegetable stew served hot or cold.

Ingredients: 2 TBS olive oil, 2 garlic cloves, crushed, 1 large onion thinly chopped, 1 eggplant cubed, 2 green peppers chopped, 4 large tomatoes chopped, 3 potatoes sliced ¼ inch, 1 TS fresh or dried basil, ½ TS dried oregano, ½ TS dried thyme, 2 TBS chopped fresh parsley (or use 2 TBS Italian seasoning).

Method: In a 4-quart pot heat oil. Add garlic and onions and cook until soft. Stir in eggplant until coated with oil then add peppers. Cover pot and cook over medium heat for 10 minutes, stirring occasionally to keep the vegetables from sticking. Add tomatoes, potatoes, and herbs; mix well. Cover and simmer 20 minutes. The eggplant should be tender, but not mushy. Serve hot or chilled.

BABAGANOUSH: To roast, stab eggplant with a knife and place garlic cloves and spices in the wounds, before carefully turning above an open flame

Ingredients: 1 large eggplant, 2 garlic cloves minced, 1 TBS fresh lemon, and 1 TBS olive oil

Method: With a sturdy fork blend roasted eggplant with all ingredients. Use as a dip or as a sandwich spread. Great on roti, biscuits/crackers, and pita bread.

Agriculture is civilization. ~ E. Emmons

DID YOU KNOW?

55 million tons of eggplant are grown every year 60% are grown in China. Other major producers include India, Egypt, Turkey, and Iran. Eggplants are technically berries, but are regarded as vegetables when cooking. Eggplants can have many colors such as white, green, purple, and purple with white stripes. Eggplants contain 95% water. 50% of the volume is air. The biggest eggplant ever grown weighed 3.06 kg (6 lb. 11 oz.). *Eggplant contains more nicotine than any other edible plant*, with a concentration of 0.01mg/100g and nothing to worry about compared to passive smoking.

There will come a time when only those who know how to plant will be eating.
Chief Oren Lyons

GLOBE ARTICHOKE – FRENCH ARTICHOKE

Cynara scolymus

Artichokes, both the globe and the Jerusalem artichokes are grown throughout the Hawaiian Islands on a limited scale. They are completely different and unique vegetables. The globe artichoke's botanical name is *Cynara scolymus*, while Jerusalem artichokes botanically are *Helianthus tuberosus*. The globe is native to the Mediterranean. Every Hawaiian home garden should grow both types of artichokes for cooking and better health. Globe artichokes have the best antioxidant levels of all vegetables, including anti-inflammatory anti-oxidants.

Globe artichokes are one of the oldest foods and one of the strangest. They are a flow-er bud in their early stage. If they are permitted to mature, they are 7 inches wide and purple. Nothing is easy about the artichoke, but the taste is worth the effort. Young, immature arti-chokes are picked to eat after the hairy, spiny center is removed. The fleshy base of the leaves is also delicious, but must be scraped with your teeth. Besides healthful eating by home growing, there is an enormous market for artichokes. Italy produces over a billion pounds a year, followed by the major world growers of Spain, Egypt, and Argentina.

HOW TO GROW: Globe artichokes are an easy to grow perennial, perfect for the home garden. They prefer full sun, fertile, well-drained soil with a pH of 6.5-7.5, and elevations to 1,000 meters. Of course, seeds are available online. They are drought-resistant, once established. One plant can produce 20 artichokes every season. Your neighbors will be envi-ous. These plants have large leaves and grow up to one meter (3 ft.) tall and wide with sturdy branches with purple flower heads. They wilt and die after flowering. Every 2 years divide all the shoots for new plants.

Start by either finding an existing plant or sprouting 3 seeds in pots. You can grow a globe artichoke as a patio or balcony plant in a big pot. Outdoors, plant the seedlings at least 1 meter (3ft.) apart. Dig a hole big enough that the rootball can fit in and add compost. Plant the leaf crown level with the ground. It's best to pack around the plant's base with mulch to hold in moisture. Weed in the beginning and feed well-rotted manure. To harvest, cut off the flower buds, golf ball size before they open and flower. After harvesting the main head, usually smaller heads appear that can be harvested later.

MEDICINAL: Eating globe artichokes is excellent for your health. They lower 'bad' LDL cholesterol and increase 'good' HDL cholesterol. They are a good source of potassium, which regulates blood pressure, a great source of fiber, reduce the risk of bowel cancers, and combat constipation and diarrhea. Artichokes contain inulin, a type of fiber that acts as a pre-biotic that may help to lower blood sugar levels. Antioxidants including rutin, quercetin, sily-marin, and gallic acid are responsible for these anticancer, anti-inflammatory effects.

NUTRITION: 100 grams of artichoke has only 47 calories, with 4g of protein and 5 of fiber. Globe artichokes are high in folic acid, niacin, vitamins A, C, and K, and rich in calcium, potassium, magnesium, and phosphorus.

FOODS & USES: Never cook artichokes in aluminum pots because they'll turn the bright pots dull gray. To cook, cut offfthe stem, rinse, pry open the petals, and steam the artichoke for half an hour. Scrape the meaty part of the leaves with your teeth and eat the center 'heart.'

DID YOU KNOW? The Greek god, Zeus spotted a beautiful girl named Cynara, while visiting his brother Poseidon. He fell in love, made her a goddess, and took her back to Mount Olympus. Cynara became lonesome and secretly visited her family. When Zeus discovered these trips, he angrily kicked Cynara off Mount Olympus, and turned her into an artichoke. Cynar is an Italian artichoke flavored aperitif liquor.

HOT PEPPER - CHILI – PEPA WELA

Capsicum frutescens

The people of the Aloha State love chili peppers. The Hawaiian type of chili peppers, pepa wela, are small, slim cones 1-2 inches long with a rounded point. These are not sweltering burning hot peppers but instead have a sweet – spicy – with slight heat. Grow your own!

Hot peppers originated in tropical Americas about 5000 years ago. The Incas and Aztecs cultivated the hot pepper, but they mainly used it for medicines. The hot chili pepper is the spicy side of the Capsicum family, which also produces the sweet green bell pepper. The passion for hot peppers can be traced directly to Christopher Columbus. The explorer was searching for spices of the Far East, especially black pepper, when he stumbled upon the potent hot pepper. Five centuries later, hot peppers are grown everywhere the climate permits and are the biggest hit of the modern spice market. By mid-1500 England was growing peppers. A century later, the pepper had won over Europe and the spice paprika was born.

All peppers in Asia and India are called 'chilis' even though their origin from the Western Hemisphere is forgotten. *Chīlli* was the Aztec Indian word meaning hot pepper. The European explorers were overwhelmed with the flavor of these heated spices, so they never forgot their name. It became an English word from the Spanish in the mid-1600s. The reason they are called peppers is because they tasted like dried peppercorns. Black pepper had been imported to Europe from India since Roman times. Chili peppers are associated with Asian/Indian cooking because Portuguese traders brought the chili seeds to the east as part of the spice trade of the 1500s.

Black pepper has nothing in common with hot peppers. Black pepper, one of the world's most common tabletop condiments, is ground from the seeds of a vine grown in Asia. Botanically black pepper is *Piper nigrum* and chilies are *Capsicum frutescens.*

Due to various climates and soils, nature has produced an assortment of hot pepper types. The Western countries grow the Scotch Bonnet/Congo/habanero peppers. Asians/Indians love the tiny hot bird pepper and the slim green and red chilies. The long slender red Cayenne chili is from Guyana, and the jalapeno is from Mexico. These pepper types vary in size, shape, and color. The hottest peppers usually mature to fiery red. Dried peppers are even hotter than fresh. The seeds and membrane are the hottest part of the pepper. All types of hot peppers emit oil that can burn the eyes or skin when handled. The Carolina Reaper is the world's hottest pepper.

HOW TO GROW: Peppers are easy to grow in soil or in patio pots. I suggest starting them in trays from seeds. If you grow several kinds, like I do, label your plants. When they are 4-6 inches tall, transplant to a well-forked bed. Plant about a foot apart and water regularly. Use a fungicide such as Ban Rot in the first weeks and then use a light pesticide as Pestac sparingly. Peppers thrive on very light doses of 20-20-20 fertilizer mix every three weeks. If you are light on the fertilizer, a good pepper tree can produce for almost a year. Pick off the first set of blossoms and that allows the plant to grow stronger and set more peppers during the season. Once the trees start to flower use 12-12-17-2 mix. Water is the biggest enemy to pepper cultivation. The plants must be well drained. Once you have peppers on the trees, reduce the waterings. This will stress the plant and increase the heat of peppers. Harvest when the green fruit changes to yellow or red. Be careful if the peppers are pungent; your hands and eyes may suffer.

MEDICINAL: Pepper is one of the most important plants used as medicine in dif-ferent countries and civilizations. It was used by the Mayas for treating asthma and by the Aztecs for toothaches. If you are a smoker, you will benefit by eating hot peppers every day. Studies have shown that red peppers contain carotenoid, a Beta-cryptoxanthin, which helps to promote healthy lung function. Capsaicin is used as an analgesic in topical ointments, nasal sprays, and dermal patches to relieve pain. Chilies are used as treatments for types of cancers, rheumatism, stiff joints, bronchitis, and chest colds with cough and headache, arthritis, and heart arrhythmias. Peppers improve digestion and can increase your metabolic rate by up to 25%, which makes you less hungry. Hot peppers can make some food safer because they reduce harmful bacteria on food. Low in calories, hot peppers, especially red, contain more vitamin A than carrots and make it easier to stick to a healthy diet since the food has more flavor. Hot chili oil is rubbed into the scalp reputedly as a Caribbean cure for baldness.

NUTRITION: 100 grams of hot peppers has 40 calories, 2g of protein with high amounts of vitamins A, C, B-6, and K, and copper, iron manganese, and magnesium.

FOOD & USES: Hot peppers can be eaten fresh, dried, or pickled, but why does the world have a hot mouth? It seems 'capsaicin,' the active ingredient of hot peppers, fools the body into experiencing pain. Capsaicin causes the brain to produce natural pleasure chemicals called 'endorphins.' These pleasure chemicals remain after the pain of the pepper. The brain remembers the pleasure and forgets the spicy pain. Capsaicin is used in the non-lethal pepper spray, Mace, organic pesticides, and animal deterrents.

HOW TO DRY HOT PEPPERS: If you're drying peppers indoors, especially in a frying pan or wok, the fumes will irritate your eyes. Open your windows and bring in a portable fan or two to keep the air circulating. Always wear gloves when handling hot peppers. Do not scratch your eyes, nose, face, or any other sensitive area of your body after handling. Wash the peppers first. Stove top pan: The fastest and most irritating method. Put the peppers in the pan on high heat and keep stirring until they begin to burn. Natural: String the peppers on some thread and hang them in a dry location. They will take several weeks to completely dry. Oven: Depending on your oven and the size of the peppers, it can take 12 hours to 4 days to completely dry. They should not burn at all. When the dried peppers are cool, store them in glass jars.

RECIPES: MANGO HEAT

Ingredients: 2 mangos, 1-3 hot pepper without seeds (depending on your taste). 1 small green - about to ripen papaya, 1 TS ginger root grated, 1 TS honey, salt, and spice to taste

Method: Slice peeled mangos from the seed, spoon mango flesh into a blender. Slice seeded peeled papaya and add to the blender with pepper and spices. Pour the blended mixture into a small pot and cook over low heat for 10 minutes. Serve with chicken, beef, or fish dishes.

SIMPLE HAWAIIAN PEPPER WATER – preferred island condiment Ingredients: 6-12 chili peppers, 8 oz. water, 12 oz. white vinegar, 3 garlic cloves minced, 1 TS sea salt preferred, ginger piece fingertip size, and ½ TS soy sauce

Method: Combine the ingredients in a saucepan and bring to a boil. Reduce the heat immediately and simmer for 5 minutes. Remove from heat, cool, then transfer to a sterilized bottle.

DID YOU KNOW? Supposedly, the first hot pepper sauce was made from the Tabasco pepper and took its name. If the pepper is too hot, do not drink water because the heat - capsaicin - is an oil and will spread to more parts of your mouth. The Mayans rubbed hot peppers on their gums to stop toothaches. Hot peppers are considered fruits, not vegetables. The Incas believed eyesight was improved by eating hot peppers. The smaller the pepper, the hotter it will be. The most potent peppers are less than 3 inches.

Jealousy is like a hot pepper. Use it mildly, and you add spice to the relationship. Use too much of it and it can burn. -- Ayala Malakh-Pines

KOHLRABI

Brassica oleracea Gongylodes Group

Kohlrabi is a strange-looking vegetable, a relative of the cabbage, but looks like a turnip that's grown above ground. It produces both the green and purple varieties throughout the Hawaiian chain, best in the cooler winter months. It's native to northern Europe and has been cultivated since the late 1400s. The crisp, juicy taste is like a sweet, slightly peppery broccoli.

Kohlrabi's bulb-stem is usually 3 inches (7cm) in diameter with two layers of long, green leaves. Underneath the usually pale green skin is white, dense flesh with a faint cabbage aroma. Often, it's referred to as a root, but technically, it's a stem and eaten for its crunchy and sweet flavor. The bulb is peeled and eaten both raw and cooked, steamed, stir-fried, or even stuffed. The leaves have a delicious mustard/kale taste.

HOW TO GROW: Kohlrabi is a biennial (2 years) plant usually grown as an annual. The waxy leaves merge from the swollen stem. At full maturity, the plant has small clusters of yellow four-petaled flowers. It's easy to grow, and an ideal home garden vegetable, but don't overplant it. Kohlrabi likes hard packed soil, so don't fork too much. Sow kohlrabi seeds ¼ inch deep and they should germinate in a week. When they sprout, space the plants about a foot (30cm) apart. Harvest the plants when the bulbs reach 3 inches (8cm) in diameter. Cut the stem just above the soil line. It's wise to plant kohlrabi far away from broccoli, or cauliflower, as they may crossbreed. The plant matures in 55–60 days after sowing.

MEDICINAL: Consuming kohlrabi will improve your gastrointestinal system, and help your body absorb more nutrients from the food you eat. Rich in potassium, this will keep you alert with plenty of energy. It's rich in vitamin C and B6, which boosts the immune system, protein metabolism, and red blood cell production. Eating this veggie will relieve gout and arthritis. It can be used in poultices directly on the affected areas of the feet, knees, or hands. The leaves and seeds are used as teas to bathe bacteria or fungus infections.

NUTRITION: 100 grams of raw kohlrabi has 27 calories, 2g protein, and 5g fiber. It has tremendous vitamin C, with some vitamins A, K, and B6, potassium, and magnesium

FOOD: Look for firm, smooth-skinned, kohlrabi that are round; elongated tend to be woody. Its leaves and stems are slightly crunchy and cook similarly to greens. Avoid any that have soft spots or yellowing leaves. The leaves should be crisp-looking and intensely green. When raw, kohlrabi is slightly crunchy and mildly spicy, like radishes mixed with turnips. You can toss them in a salad, make a slaw out of grated kohlrabi, or eat them with a drizzle of good oil and a sprinkling of sea salt. Grate kohlrabi into soup, steam it, or shred it for fritters. To roast, steam the bulb for 5 minutes, then roast for another 45 minutes, until the flavor sweetens.

RECIPES: PEPPERY KOHLRABI NOODLES

Ingredients: 3 kohlrabi and 2 carrots peeled and sliced into match-sticks, 4 oz. cooked rice noodles, 2 green chilies chopped, 1 avocado sliced, ½ cup crushed peanuts or cashews, lime slices. Ingredients for dressing: juice and zest of 1 lime, 1 clove garlic minced, 1 TS minced ginger, 2 TBS soy sauce, 2 TBS coconut vinegar, and salt to taste.

Method: Mix dressing and toss with carrot and kohlrabi. Let sit for 1 hour to marinate. Add remaining ingredients and toss. Adjust seasonings to your taste.

NAMES: Botanical: *Brassica oleracea* Gongylodes Group, Kohlrabi, Kohl is German for cabbage and Rabi is Swiss German for turnip because the swollen stem resembles the latter.

OKA - OKRA – OCHRO
Abelmoschus esculentus

The okra plant is a tall growing vegetable from the same family as the hibiscus. This is an easy vegetable to grow, and the blossoms are beautiful big flowers, perfect for a hedge in your Hawaiian home garden.

It is believed that okra originated in Ethiopia before spreading to North Africa and the Middle East and India perhaps a millennium before reaching the Americas. Little is known of its early history and it has never been found growing wild in SE Asia or India. It is logical Arab traders helped spread this vegetable to both Asia and Europe. The name 'okra' itself is shortened from the Nigerian dialect 'okuru.' This green pod is best known as a key ingredient in the thick spicy stew named 'gumbo,' which is Swahili for okra. Types vary from shades of green to white, fat, or slender shapes, with either a ribbed or smooth surface. Green, ribbed pods are common.

Okra are rarely cooked unaccompanied except when it is fried. Usually, a little of it is chopped and added with other vegetables into rice, soups, and stews. Okra alone is generally too 'gooey/slimy. This veggie has a unique flavor and texture, and the juice will thicken any liquid to which it is added. With a taste somewhere between that of eggplant and asparagus, okra mixes well with other vegetables, particularly tomatoes, peppers, and corn. Okra is easily dried to a dull gray on the plant for later use. A pinch of dried okra will thicken any dish similar to corn starch. In some lands, okra is grown just for the seeds. Ripe seeds produce an edible cooking oil and can be roasted and ground as a coffee substitute.

HOW TO GROW: Okra is perfect for a border around a backyard garden, because the 3-6 ft. tall plants produce beautiful blossoms that rival its cousin the hibiscus. Plant the seeds one inch deep and a foot apart. Okra usually grows well in any good garden soil. Six plants produce enough okra for most families. Soak the seeds overnight and they should sprout within 10 days. As with most vegetables make certain the area is well-drained and okra prefers a pH from 6-6.8. Keep the area watered and weeded. The first 4-inch pods should be harvested within 2 months. Since okra grows fast and tall, you can intercrop with a vining bean that will climb the okra.

Pods should be cut (not picked) while they are still immature and tender at two to three inches long. This vegetable must be picked at least every other day. Okra has short hairs that can irritate bare skin, so it's wise to wear gloves and long sleeves when harvesting. A sharp knife will cut the pods and should not harm the plant. When the stem is difficult to cut, the pod is probably too old to use. The large pods rapidly become tough and woody. It's possible to cut the tall, mature okra tees that seem to be finished bearing down to about 1-2 ft. and they will again produce.

MEDICINAL: Okra is recommended for pregnant women because it's rich in folic acid. The mucilage and fiber found in okra helps adjust blood sugar by regulating its absorption in the small intestine where it also helps grow good bacteria (similar to the ones fed by yogurt). This helps the body absorb the vitamin B complex. For adding bounce to your hair, boil horizon-tally sliced okra till the brew becomes slimy. Cool it and add a few drops of fresh lemon juice, then use as the last hair rinse. Your hair will have a youthful spring. Okra is an excellent laxative and treats irritable bowels, heals ulcers, and soothes the gastrointestinal tract. The protein and oil contained in the seeds of okra are a source of first-rate vegetable protein. It is enriched with amino acids like of tryptophan, cystine, and other sulfur amino acids.

NUTRITION: 100 grams of okra has 33 calories, 2g of protein with vitamins A and C, calcium, potassium, and manganese. It has nearly 10% of the recommended levels of vitamin B6 and folic acid. It's a prime source of soluble fiber in the form of gums and pectins. Soluble

An okra tree does not grow taller than its master.~ African Proverb fiber lowers cholesterol and reduces the risk of heart disease. The insoluble fiber keeps the intestinal tract healthy, decreasing the risk of some cancers, especially colon-rectal cancer.

FOOD & USES: Buy only firm, springy pods with no mushy, brown or yellowing spots. Don't wash okra until just before you cook it; moisture will cause the pods to become slimy. Store untrimmed, uncut okra in a paper or plastic bag in the refrigerator crisper for no longer than three or four days or it will turn to mush. Never prepare okra in a cast iron or alu-minum pot, or the vegetable will darken. The discoloration is harmless, but makes the okra look rather unappetizing. To remove some of the stickiness from okra, soak the pods in vinegar for half an hour.

When serving okra as a side dish, cook the whole pods rapidly until crisp-tender, or just tender to minimize the thickening juices. Try the same quick cooking when you are adding okra to any cooked dish in which you want to retain its crisp, fresh quality. Add the vegetable during the last 10 minutes of cooking time. On the other hand, when okra is to be used in a soup, stew, or casserole that requires long cooking, it should be cut up and allowed to thicken with its juices.

RECIPES: OKRA MELEE — This can easily be created from your backyard garden.
Ingredients: 2 cups small okra pods chopped ¼ inch thick, 2 TBS cooking oil (prefer healthwise canola), 1 large onion sliced, two garlic cloves chopped, 2 bay leaves, ½ TS of thyme and basil, 1 sweet bell pepper chopped, 3 large ripe tomatoes chopped, corn kernels cut from 4 ears, ½ cup chicken broth, salt, and pepper to taste.
Method: In a large frying pan, heat oil and add onions, garlic, and spices until onions are limp. Then add bell pepper and continue cooking until onions are clear. Add tomatoes, okra, broth, salt, and pepper. Reduce heat and simmer uncovered for 15 minutes, stirring occasionally. Add corn and cook 5 more minutes longer. Season to taste.

CORNMEAL COOCOO
Ingredients: 1lb. okra chopped small, 1 large onion chopped small, 3 garlic cloves minced, 1 TBS fresh thyme, 2 cups fine cornmeal, 1 TBS oil, 2 cups cold water, 1 TS salt, 2½ TBS butter or margarine, 4 cups of boiling water.
Method: Soak cornmeal in cold water for 5 minutes. In a 2 qt. pot, heat the oil and saut the onion, garlic, and thyme for 2 minutes. Add okra pieces and fry for 1 minute. Carefully add boiling water to okra and continue to cook 10 minutes. Strain out onions and okra. Pour half the liquid into a bowl for later. In the okra water, add soaked corn meal, 1 TBS butter, and salt. Stir and add saved water when it needs it. Stir for 15 minutes and add okra/onions. Use remaining butter to grease a bowl before filling with thick cornmeal. Let cool for 5 minutes. Invert bowl so it drops out.

OKRA SOUP - simple or to your taste
Ingredients: 1lb. young okra pods sliced into ¼ inch rounds, 2 large tomatoes chopped small, 1 cup corn, 1 medium onion chopped, 2 cups water, 1 lemon, 1 hot pepper seeded and minced (optional), pinch of the following to your taste, salt, cumin, black pepper
Method: Place all the ingredients in a saucepan and bring to the boil. Reduce the heat and simmer for about 15 minutes, until the okra is tender.

DID YOU KNOW? Okra is related to both cotton and hibiscus. Every year the world grows 9 million tons of okra. India grows 60%. Because of their nutritional value, scientists are studying how to efficiently harvest okra seeds and turn them into flour. Paper and ropes can be made from the leaves and stems of the plant.

NAMES: okra, *Abelmoschus esculentus*, is also known as lady's fingers, gombo, gumbo, quingombo, okro, okra, bamia, bamie, quiabo.

We should all be eating fruits and vegetables as if our lives depend on it - because they do.~ Michael Greger

PEA EGGPLANT - TURKEY BERRY
Solanum torvum Sw.

The pea eggplant also known as the turkey berry is native to India and its fruits are edible. It's a popular traditional vegetable throughout SE Asia, especially in Thailand, Sri Lanka, and India. Because it is so easy to grow and a hardy plant, the pea eggplant is perfect for the home garden or patio pot. In the West Indies, half-grown, firm berries are boiled and eaten with yams or ackees, or added to soups and stews. Try some in your Hawaiian recipes.

This tasty vegetable plant will blossom within 4 months of planting and will produce fruit all year for at least 3 years, maybe longer. Pea eggplant fruits look like yellow-green berries about 1cm (.4in.) that grow in clusters. They resemble a bunch of green peas that yellow as they ripen and are filled with bitter, flat brown seeds. Its botanical name is *Solanum torvum Sw.*

HOW TO GROW: This is a no-brainer for your home garden because pea eggplant grows like a weed. Plant the seeds in moist, enriched, well-drained soil in full sun. The best pH is 5–6. It can grow in partial shade and can handle periods of drought. Treat the pea eggplant correctly, keep it weeded, watered, and fertilized; it should produce for a few years. You only need a few; space plants a meter apart. This plant will grow to 2000m in elevation. With good care, pea eggplant should reach 1.5 meters (5ft.) tall and beware, some varieties have small thorns. Root crops or pineapples may be planted with pea eggplant.

MEDICINAL: Pea eggplant berry juice can treat almost every ailment from coughs and respiratory-chest ailments, fevers, sore throats, and arthritis. The leaves are dried and powdered to assist diabetic patients in regulating their blood sugar levels. Poultices of the leaves treat wounds and skin diseases such as eczema. The leaves and flowers are boiled into a decoction with honey for fevers and colds. The flowers can be pressed for juice to treat eye problems.

NUTRITION: 100 grams of cooked, seedless pea eggplant has only 40 calories with 2g protein and 4g fiber. It contains vitamins A and C with calcium, iron with trace amounts of manganese, copper, and zinc.

FOOD & USES: The young shoots, leaves, blossoms, and fruits can be eaten as a vegetable. The bitter fruit with its juicy pulp and many small seeds is palatable raw or cooked. The flavor's similar to eggplant and they remain semi-firm after cooking. Like green peas, they add texture and visually enhance a dish. In SE Asia, it's cooked alone or added to stews and soups. Mature sized, but still green and unripe fruit are added whole or sliced to Thai and Indian curries.

To prepare pea eggplant pull apart the cluster and give each a good wash. To soften them, cover the clean berries with water in a frying pan on medium heat for 5 minutes. Remove from heat and let them cool. In bowl, squash the pea eggplant with your fingers or a fork. You'll see the multitude of tiny seeds. Gently rinse until most of the seeds are gone. You'll never get them all.

RECIPES: PEA EGGPLANT AND POTATO CURRY

Ingredients: 1cup pea eggplant washed, boiled, and seeded, 4 potatoes washed and diced, 2 garlic cloves minced, 1 medium onion chopped small, 1TS roasted red chili powder, ½ TS each of turmeric and black pepper powders, ¼ TS black mustard seeds, 12 curry leaves, salt, 2 TBS oil.
Method: In a frying pan or wok on medium heat, add oil and mustard seeds. Stir until mustard seeds pop. Add onion, garlic, and potatoes. Fry until brown. Stir in all spices. Add salt to your taste. Then stir in seeded pea eggplant. Stir until everything is coated. Add curry leaves and stir for 2 minutes. Serve with rice or roti.

DID YOU KNOW? The glycoalkaloid solasodine, which is found in the leaves and fruits of the pea eggplant, is used in India in the manufacture of steroidal sex hormones for oral contraceptive.

ROSELLE – RED SORREL
Hibiscus sabdariffa

Roselle is a type of hibiscus, *Sabdariffa.* These flowers are actually the leaves surrounding the seedpods of the roselle. This bush grows easily on every Hawaiian Island and it's attractive for landscaping. Roselle's a bush with straight reddish stems that grows to eight feet and makes a strong hedge. It can be a great border for a vegetable or flower garden. Fresh or dried, roselle makes a nice addition to flower arrangements. China and Thailand grow the most roselle, but the best quality roselle is grown in the Sudan and Nigeria.

Native to Malaysia, roselle has traveled the world. It is widely cultivated in Asia, Africa, throughout the Caribbean, and Central America. African slaves carried roselle seeds to the Western Hemisphere. By 1700, roselle was cultivated in both Brazil and Jamaica. By the late 1800's roselle was growing in Florida. Roselle has many names such as sour-sour, Florida cranberry, zuring, and red sorrel. This is an attractive and tasty bush for your island home garden.

HOW TO GROW: Roselle will grow in almost any soil, but likes rich sandy soil with a pH of 5.5-6.8. If you direct plant, put the seeds about a foot apart in mounded rows. If you choose to sprout, transplant when they are 4 inches tall. Space the rows about 2 feet apart. Roselle grows easily and needs little fertilizer. Weeding is necessary until the plants reach 2 feet tall. More flowers will be produced if the plants are pruned. A fertilizer mix of 4-6-7 is satisfactory. The root-knot nematode is roselle's main pest, and the mealybug can cause problems. Roselle is usually a 5–6-month crop. It's ready to harvest when the buds are still tender and easily snapped off by hand. It is easier to snap off the flowers in the morning than in the evening. The flowers must be harvested quickly as either the flowers will fall off or get too hard for use.

MEDICINAL: Roselle is an ingredient in many traditional-homeopathic medicine remedies. Drinking the tea will help relieve hypertension, coughs, and hangovers. A paste of the leaves and seeds is used in Egypt as an antibiotic on wounds. Leaves warmed in hot water can be used to draw out boils. The seeds are valued for its mild laxative effect, ability to increase urination, a natural coolant for relief during hot weather, and treats cracks in the feet. They are used to restore physical strength after an illness. The bitter root is used to make an aperitif and a tonic. The plant is also reported to be antiseptic, aphrodisiac, astringent, and a purgative.

NUTRITION: 100 grams of roselle has 49 calories with 1g of protein. It's very nutritional, high in vitamin C, calcium, magnesium, niacin, riboflavin, and iron.

FOOD & USES: Roselle is cultivated only for the dark-red seedpods surrounding the fruit. These are the basis of the popular red non-alcoholic drink. The funnel-shaped flower petals usually are yellow with deep red blotches near the base and grow up to five inches wide. The leaves can also be eaten steamed, as they have a citrus-like taste when added to salads or curries. The seeds can be roasted and eaten, or ground into flour. Throughout Africa, roselle seeds are used to make a coffee substitute or fermented to make a simple meat substitute called 'furundu.' Roselle seeds are a great chicken feed.

Roselle is growing in popularity with food and beverage manufacturers. Pharmaceuti-cal companies are exploring its potential as a natural coloring to replacing some artificial dyes. Roselle is also used in salads, jellies (such as Jamaica's famous rosella jam), sauces, soups, beverages, chutneys, pickles, tarts, puddings, syrups, and wine. Powdered, dried roselle is used to flavor and color commercial herbal teas. Most tasters can't discern roselle drink from cranberry juice in flavor and appearance. Dried roselle will last for months if stored in a cool, dry place. In Africa, roselle is steamed and eaten with ground peanuts. For stewing as sauce or filling.

for tarts or pies, tender pods are cooked with sugar. To prepare a smooth roselle sauce, jam, or chutney; the outer leaves should be passed through a food processor. Another method is after cooking press the mixture through a sieve. Some cooks steam the roselle with a little water until soft before adding the sugar, then boil for 15 minutes. Sweet roselle sauce is a great addition to custards, cakes, or ice cream. Since roselle contains 3% pectin, it's not necessary to add additional to make roselle jelly.

RECIPES: HAWAIIAN CHRISTMAS PUNCH

Ingredients: ½ lb. roselle, 1 cinnamon stick, 10 whole cloves, 1 strip of dried orange peel, 1gallon boiling water, 1 lb. of sugar, 1 TS ground cinnamon, ½ TS ground cloves, 1 cup rum (optional)

Method: Put the dried roselle, orange peel, cinnamon stick, and whole cloves in a gallon jar. Fill with hot water. When cool, cover and leave the mixture steep for 2 days. Then strain off the liquid, add enough sugar to sweeten, add rum, powdered, or ground spices, and let it stand for another 2 days before serving chilled, or with crushed ice. This is a popular drink at Christmas time when roselle is in season.

SIMPLE ROSELLE JUICE

It is not necessary to remove the seedpod to make roselle juice. Wash roselle and place in a large pot, cover with water, and boil for 15 minutes. Strain and add sugar and spices to your taste.

ROSELLE JELLY

Bring 3 cups of roselle juice to a boil and add 2½ cups of sugar. Heat to 200 degrees. Without an intricate thermometer, this will be difficult. Test if the roselle has jellied by spooning a few drops of the roselle into a glass of tap water. If the roselle remains firm and doesn't dissolve, it is jelly. Sterilize empty jelly jars by boiling to avoid contamination. After filling and capping, put the jars in a pot, and boil for another 10 minutes.

SOUR ROSELLE EVERYTHING SOUP

Ingredients: 1 cup roselle leaves, 1 chicken leg or breast, 1 pork chop, ¼ cup shrimp peeled and deveined, 1 fillet of fish salmon or king, the following chopped - 1 onion, 1 bunch culantro, 1 bunch celery, 1 okra, 1ripe tomato, 1 garlic clove minced, salt, and spice to taste

Method: In a suitable pot cook chicken and pork in a quart of water. Remove, debone, and shred the meat, saving the water. In the same water cook the shrimp and the fish. Debone the fish and return to the water. Add the roselle leaves, all vegetables, salt, and spices. Cook 15 minutes. Add shredded chicken and pork. Serve hot.

Roselle pod

FRIED ROSELLE LEAVES

Ingredients: 2 cups 'tender' roselle leaves, 1 cup small shrimp cleaned, 1 cup bamboo shoots (canned) sliced, ½ hot pepper seeded and minced, 1 onion chopped small, 2 garlic cloves minced, 2 TBS canola oil, ½ TS turmeric powder, 1TS chili powder, salt to taste

Method: In a large frying pan heat oil add turmeric powder, hot pepper, onion, and garlic. Add roselle leaves and shrimp with 1 tablespoon of water stirring constantly. Add bamboo shoot slices and cover. Cook at medium heat for 15 minutes or until the roselle leaves become dry and give off some oil. Turn off heat and cover for 10 minutes before serving on a bed or rice or noodles.

A gardener learns more in the mistakes than in the successes. ~ Barbara Borland

SWEET BELL PEPPERS
Capsicum annuum Group

Sweet peppers bring a good price in the markets because they take time and patience to grow. These tasty, sweet cousins of the chili are an easy, plentiful crop for Hawaiian gardeners and can be many bright colors. At home, with daily attention, you can grow magnificent looking and tasting bell peppers. The world loves sweet pepper and grows almost 40 million tons. China grows the most at 19 million tons and Mexico's second at 3.5 million tons.

Columbus sought the well-known spices of the East, especially pungent black pepper. Instead, he discovered sweet and hot (Capsicum) peppers. As vegetables and condiments, both peppers now dominate the world's spices. Scientists believe hot and sweet peppers were first cultivated in Bolivia, South America. Everywhere explorers landed in the American tropics they encountered native Indians growing peppers. By 1600, Spanish, French, and Portuguese explorers had spread peppers through Europe, England, Asia, and India where they became known as chilis.

HOW TO GROW: Sweet peppers are called bell peppers due to their shape. In some countries, the sweet pepper is called 'the mango pepper' or the 'bull nose pepper.' The bell pepper may be colored green, red, yellow, purple, or brown; and be as big as six inches long. Sweet peppers aren't difficult to grow. In fact, all types of peppers thrive in our hot dry season, but our rainy season doesn't mix with most peppers. If you start your plants from seeds, they should sprout in about 2 weeks. They're ready for transplanting when 4 inches tall. To prepare your garden for transplanting, work well-rotted chicken manure into the rows. The best soil pH is 6. Sweet peppers should be spaced at least a foot apart. As seedlings, the biggest pest your peppers will encounter is the mole cricket. A ring of newspaper, half buried around each plant will deter them. Put just a pinch of starter fertilizer in each hole and cover with at least an inch of soil before you plant the seedling. Always avoid getting water onto the pepper plant leaves, as this will encourage disease.

Peppers need a high phosphorus fertilizer like red 12–24-12 once, when the plants are about three weeks old. After they have blossomed use blue 12-12-17-2. Peppers are usually very sturdy plants with few pests, but need water every other day in the dry season. Horned tomato worms, garden snails, or white flies can do severe damage to your plants. Check your leaves. If they are getting eaten, inspect them at night with torchlight to find and destroy the villain. White flies will lay eggs on the underside of the leaves. Mites, in all their many varieties, are mature sweet peppers biggest enemy in . Proper insecticide properly applied can stop white fly and mite infestation if caught early. Most pepper will bear in about 70 days. Do not pull or twist, but cut the fruit so not to damage the stems. With regular water and fertilizer, a pepper can bear for 6 months.

MEDICINAL: Due to vitamin C and beta carotene content, capsaicin, and flavonoids, bell peppers seem to aid in preventing cataracts, prevent blood clot formation, and reduce the risk of heart attacks and strokes. Red bell peppers have significantly higher levels of nutrients than green, and contain lycopene, which helps to protect against cancer and heart disease. Chinese medicine uses bell peppers to treat indigestion, decreased appetite, and swelling. Be careful, these peppers may increase the pains from arthritis.

NUTRITION: 100 grams of raw green pepper is only 31 calories and has three times the daily requirement for vitamin C (four times more than an orange) and the complete require-ment of vitamin A. These two vitamins plus B-6 and folic acid protect against high cholesterol and heart disease. They also contain good amounts of potassium, manganese, and zinc.

FOODS & USES: Bell peppers are among a dozen foods pesticide residue is most frequently detected. Before cutting the pepper, wash it under running water. If the pepper has been waxed, you should also scrub it. Bell peppers are great for stir-fry, salads, pastas, curries, or soups. They're easy to work with, no peeling required. Their great flavor can be enhanced by charring directly on top of a gas stove or under the broiler. Roasted or raw, sweet peppers should be in your garden. To keep sweet peppers fresh, store them unwashed in the fridge. Moist peppers spoil faster. To make your own dried sweet pepper flakes: carve the seeds and white pith from inside the pepper. Then slice the pepper into ¼ in. thick rings. Blanche by putting into boiling water for 4 minutes. Remove and immediately drop into ice water. Let sit for 10 minutes then drain for 1 hour. Put pepper slices on a tray in your oven at 140 F for 1 hour. Turn off heat, open oven door, but let pepper slices sit. Once dried crush with rolling pin or mallet. 1 pepper makes ¼ cup of flakes.

RECIPES: ROASTED PEPPERS

Ingredients: 4 sweet bell peppers preferably of different colors, 1 red onion sliced thin, 2 TBS dried basil or 6 fresh basil leaves, 1 garlic clove minced, 6 TBS olive oil, 2 TS balsamic vinegar (or garlic flavored vinegar), salt, and seasonings to taste

Method: Coat peppers with the olive oil. Place peppers on a cookie sheet in a 300-degree oven until the skin begins to blister. Remove from oven and place in a bowl of ice water. When the peppers are cool, peel and remove seeds. Slice into one-inch strips. Peel and slice onion as thin as possible. Place peppers, onions, garlic, olive oil, vinegar, and seasonings in a bowl and toss. This may be served warm or cold or may be stored in a jar in the fridge.

SWEET PEPPER SAUCE

Ingredients: 3 bell peppers - clean of seeds and membranes - roasted and cut lengthwise into ½ -inch strips, 3 TBS olive oil, 1 red onion sliced, 1 garlic clove minced, 2 nice size tomatoes chopped, salt, and spices to taste.

Method: On low heat in a saucepan sauté the onion and garlic in the olive oil. Add pepper strips and cover. Cook 20 minutes. Add the tomatoes and seasonings and cook for another 5 minutes. Put mixture in a blender and pulse about 25 times, but do not fully blend it. This can be a side dish or a topping for pasta or a sauce marinade. Add beef, chicken, or pork this pepper sauce will become a soup base.

STUFFED PEPPERS

Ingredients: 4 large sweet green or red peppers, 1 lb. minced beef, 1 garlic clove minced, 1 can tomatoes with liquid chopped, ½ cup ketchup, 1 cup cooked rice, salt, and seasoning to taste

Method: Neatly slice the tops off the peppers (save) and remove the seeds and membranes. Place peppers in a saucepan of boiling water for 4 minutes. (Raw peppers may be used.) In a frying pan, brown the minced beef, onions, and seasonings. Add rice and tomatoes to this mixture. Place peppers in a baking dish and fill with the minced beef/rice mix. Mix the ketchup with a ¼ cup of water and pour over the filled peppers. Replace the pepper tops and bake for a half hour at 300 F.

DID YOU KNOW? The sweet pepper is botanically known as *Capsicum annuum* and is the best-known domesticated pepper worldwide. A warning, because of the demand for sweet peppers they are usually pushed and sprayed in cultivation. Bell peppers are very adaptable plants, grown in tropical and temperate climates, as well as very versatile foods. Their cultivation by many cultures is because they have a long shelf life and travel well. Originally, sweet peppers were elongated, wrinkled, and much smaller than today, eventually bred into their current marketable size and shapes. The Dutch have produced attractive deep-purple peppers. Color choices include white, salmon, red, yellow, and chocolate-brown. The heaviest bell pepper grown weighed 738 g (1 lb. 10 oz.), in the UK in 2020. Try your luck to beat that!

To keep the body in good health is a duty...
otherwise we shall not be able to keep our mind strong and clear. ~ Buddha

TAMARILLO
Solanum betaceum

The tamarillo is an attractive shrub native to the Andes Mountain region of South America, the countries of Peru, Chili, and Bolivia. This shrub produces another unique egg-shaped, edible fruit and is perfect for the Hawaiian home garden or as a patio or balcony container plant. These trees are more successful in the higher, cooler elevations. In the hot tropical lowlands, only small fruits develop, but it's still an attractive tree. Its botanical name is *Solanum betaceum.*

A staple of Mexican cuisine, they are eaten raw and cooked in a variety of dishes, particularly salsa verde/green sauce. Seeds were imported to New Zealand in 1890. Only three countries grow tamarillos commercially, New Zealand, Colombia, and Australia.

HOW TO GROW: The tamarillo tree is attractive, with a straight trunk and large leaves that have a musky aroma. Its blossoms are light pink or white in large clusters. One cluster can produce as many as 6 fruits. This tree grows to 5 meters (16ft.) with little attention in deep, fertile, well-drained soils with a pH of 5 to 8.5. Plant seeds in a protected area, a nursery, or big pots. When the sprouts reach 1 meter, they can be transplanted to their permanent location. They should bear fruit within 2 years and be very productive for 6 years, producing 20kg (44lbs.) annually. The life span is about 10 years. The tamarillo tree needs regular water and protection from high winds.

Tamarillo fruit are about 6-10cm (4 long and 4cm (1½ in.) wide and may be yellow, orange, or a deep maroon. Only unripe tamarillos are green. Some varieties have stripes. Red fruits are sour and yellow and orange fruits are sweeter. The ripe flesh is firm with more and larger seeds than a tomato. When you pick them, leave a short piece of the stem attached.

MEDICINAL: Eating a tamarillo or drinking the juice relieves stress and is used to treat severe headaches and migraines. A poultice of warmed/steamed leaves is wrapped around the neck to treat a sore throat. A decoction of tamarillo pulp is a treatment for colds, sore throat, high blood pressure, and liver disease.

NUTRITION: Per 100 grams, tamarillos have 85 calories, and are very high in vitamins A, C, and E with iron, calcium, and magnesium.

FOOD & USES: The flesh of the tamarillo is sweet and tangy, with a complex flavor compared to a combination of tomato, guava, and apricot. The skin and outer flesh can be bitter and best not eaten raw. It's usually eaten by scooping the flesh from a halved fruit. When lightly sugared and chilled, the flesh is used for a breakfast dish spread on toast. They can be halved and sprinkled with sugar and seasoning, then grilled or fried, added to soups, stews, and salads. They can be made into compotes or added to chutneys and curries.

RECIPES: SALSA VERDE

Ingredients: 550g (1¼lb.) tomatillos halved and scooped out, skin discarded, 4 garlic cloves chopped, 1 onion chopped very small, 4 green chilies (more if you desire heat), 12 sprigs of cilantro, 2 TBS oil, salt to taste

Method: Put all ingredients, except cilantro and salt in a suitable pot and cover with water. Bring to a boil and reduce heat and simmer for 10 minutes. Pour into a blender and blend until smooth. Heat 2 TBS oil in a frying pan and add blended salsa. Simmer for 10 minutes. Add salt.

DID YOU KNOW? Before 1967, the fruit was known as the 'tree tomato,' but the new name 'tamarillo' was chosen by the New Zealand Tree Tomato Promotions Council to increase its exotic appeal.

NAMES: tree tomato, tomate serrano, tamamoro, blood fruit in South America

THAI EGGPLANT
Solanum melongena

The small golf ball 'Thai' variety of eggplant is a unique vegetable that is grown throughout the Aloha State. Besides the usual American purple globe variety, I've included two other very different eggplant varieties, the Thai, and the small turkey-berry. These probably are less hybrid-evolved variants of Solanum Melongena, eggplant's botanical name. Before eggplant became purple and elongated, it was egg-shaped and had white, meaty flesh. Thus, the name eggplant. This variety is distinct with its rough green skin and small round shape that has a hollowed interior with tiny seeds. The Thai eggplant is eaten stewed, roasted, or fried.

HOW TO GROW: Thai eggplant is an annual that will grow everywhere up to 2000 meters in altitude. Plant the seeds in a bed or a large pot and they should germinate within 2 weeks. They will be ready for transplanting within 2 months when they're 8cm (3 in.) tall. Expect Thai eggplant to grow to a meter with many branches and large leaves. When arranging in your garden, it's best to keep a meter between plants. Follow the same growing method as with eggplant. This vegetable bush needs regular watering and occasional nitrogen fertilizer, manure, or compost, and they should bear fruit for 3 months. Seeds are available online.

MEDICINAL: All parts of the Thai eggplant, the roots, leaves, and fruits, are used in Ayurvedic-homeopathic treatments. The roast pulp of the Thai eggplant is used to reduce swelling. Once the flesh has cooled smear it around the raised areas. Researchers have found that an infusion made from the fruit has reduced total and LDL cholesterol levels. A tea prepared from the leaves treats asthma, and coughs. The same leaf tea and a juice squeezed from the fruit can be used as a body wash to relieve itching. Scientists have found that Thai eggplant roots contain solanine, a compound that's toxic with fungicide and pesticide properties, one of the plant's natural defenses.

NUTRITION: 100g is 25 calories with potassium,magnesium, and vitamins A & C.

RECIPES: THAI EGGPLANT CURRY

Ingredients: 250g(½lb.) Thai eggplant, stem and seeds removed and quartered, 4 garlic cloves chopped, 1 medium onion chopped, 2 green chilies chopped, 1TS each of black mustard seeds, curry, chili, turmeric powders, and fenugreek seeds, 1 stick of cinnamon, 10 curry leaves, salt, ½ cup of coconut milk, and 2 TBS oil for frying

Method: In a frying pan heat the oil over medium heat and add the garlic, onion, and curry leaves. Stir-fry for 3 minutes and then stir in the mustard seeds and cinnamon stick. Add the Thai eggplant and stir-fry for another 3 minutes. Stir in the remaining ingredients and simmer, stirring for 5 minutes. Reduce heat and pour in the coconut milk. Add salt to your taste. Cover and simmer with frequent stirring for 10 minutes.

THAI EGGPLANT PICKLE - This a good appetizer.

Ingredients: 500g (1lb.) Thai eggplant with stems, seeds removed, quartered, 1 big red onion sliced wide, 12 green chilies sliced long, 1TBS black mustard seeds. All soaked in 3TBS coconut vinegar for half an hour, 1TBS chili powder, 1TS turmeric powder, 1TBS sugar, salt to taste, ½ cup oil,

Method: Combine soaked Thai eggplant with turmeric and salt. Fry in oil for 5 minutes flipping pieces. Remove and cool. Pound mustard seeds into a paste. Combine chili powder and sugar with onion and chili pieces with mustard paste. Combine/coat the fried Thai eggplant quarters.

DID YOU KNOW? There are many types of eggplants: Long, deep purple-globe American. Long thin purple-Chinese eggplant. Purple with white streaks is named fairy tale, Small round purple-Indian, purple oblong-Italian, green thin and long-Japanese, and the round white, the round green, and the round Thai.

TOMATO
Lycopersicon esculentum

Everyone loves fresh home-grown tomatoes, and they grow well on every Hawaiian Island. The entire state produces about 20,000 tons yearly.

Central and South American Indian tribes, such as the Incas and Aztecs cultivated tomatoes since 700 AD. In the 1600s, the early Spanish explorers distributed the tomato to every colony. The Spanish explorers carried the fruit to Europe where tomatoes became popular in Spain, Portugal, and Italy. The French called it 'the apple of love,' while the Germans named it 'the apple of paradise.' Botanically the tomato is *Lycopersicon esculentum.*

HOW TO GROW: Tomatoes are one of the easiest vegetables to grow throughout the islands and perfect for a small backyard garden. Heatmaster, Cherokee Purple, and Calypso varieties are favorites with gardeners up to an elevation of 2,000 meters. The exact days to harvest depends on the variety and it can range from 60 days to more than 100 days. There are 2 basic types of tomatoes. Bush or 'determinate' varieties grow 2 to 3 feet tall. Bush tomatoes provide lots of ripe tomatoes at one short period. Vining or 'indeterminate' varieties grow more leaves, and they produce evenly throughout the season. Stake the vining plants to get the maximum use of space.

If you want to experiment, find a type of tomato you like, get one at the market that is overripe or rotting and bury it an inch deep in your garden. Keep watering and the seeds will sprout. Tomatoes need sun, and soil that doesn't stay moist with a pH from 5.6 to 6.8. Plants should be sprouted from seeds in trays and kept until they are four to six inches tall. Trans-plant to rows separated three feet apart, as tomato vines will spread out. The plants should be spaced 18 inches apart. A 5-10-5 fertilizer should be used sparingly, only once or twice throughout the plants' life. Water twice a week. Too much water or fertilizer will harm toma-toes. Ripe fruits will easily pull off the vine.

To reduce the use of pesticide, plant scented herbs such as basil or dill around the edge of the garden. French marigolds are great also and brighten a garden while chasing pests. Shiny strips of foil flashing in the breeze should keep corn birds from pecking the fruit.

MEDICINAL: A diet containing tomatoes may help reduce cancer, heart disease, and premature aging. Tomatoes contain an antioxidant, lycopene, which reduces the risk of prostate cancer if eaten almost daily. Tomato's color is what makes them so good for your health. The color is lycopene, an incredible antioxidant. Research shows tomatoes have cancer-fighting properties. Antioxidants block the damaging cancer-causing effects of oxygenated free radicals in the cells. The amazing thing about lycopene is that it is about twice as good as other antiox-idants in foods. It has been studied to be effective in preventing breast cancer, lung cancer, and prostate cancer. Lycopene is effective against aging as well. When choosing tomatoes, get the reddest and the ripest because they have the highest amounts of lycopene as well as beta caro-tene, another healthful ingredient. The good news is cooking a tomato with a bit of oil, such as olive oil, helps bring out the effectiveness of the lycopene.

NUTRITION: The British originally believed the tomato to be poisonous, but they're really good for you. Tomatoes contain vitamin C, potassium, folacin, and beta-carotene. A medium tomato has only 25 calories.

FOOD & USES: Tomatoes are used extensively in Spanish, Italian, and Mexican cuisine. Don't store ripe tomatoes in the fridge. Cold temperatures lessen the flavor in tomatoes. More than sixty million tons of tomatoes are produced per year, 16 million tons more than the second most popular fruit, the banana. Apples are the third most popular at 36 million tons, then oranges at 34 million tons, and watermelons with 22 million tons.

RECIPES: EASY TOMATO SAUCE is great with pasta for any Italian favorite as spaghetti, lasagna, or can be used over rice.
Ingredients: 3 cups chopped tomatoes, 4 TBS olive oil, ½ cup chopped onion, 2 garlic cloves sliced as thin as possible, 1 TBS oregano, 3 leaves fresh basil or 1 TBS dried, salt, pepper to taste.
Method: Bring a large frying pan to medium heat and sauté the onion and garlic in the olive oil until the onion is clear, but not browned. Add tomatoes and cook until it becomes thick. Add spices, while reducing the temperature. Stir constantly to avoid burning. Simmer for 15 minutes. To remove excess liquid and thicken the sauce, remove lid from skillet.

EASY OVEN TOMATO CASSEROLE
Ingredients: 3 cups of tomatoes sliced into wedges, 1 cup cheddar cheese shredded, 1 cup breadcrumbs, 1 cup green sweet pepper chopped, 1 onion chopped, 2 garlic cloves chopped, ¼ cup butter or margarine melted, salt, and pepper to taste (corn, okra, long bean may be added)
Method: Combine all ingredients in a large, greased oven casserole dish. Bake at 300 degrees for 45 minutes.

FRIED GREEN TOMATOES is a great side dish for fish or chicken.
Use green tomatoes only after at least one has ripened on that vine.
Ingredients: 5 large green tomatoes sliced ½ inch thick, one cup flour, 1 egg beaten into a cup of milk, 1 cup cornmeal, salt, and pepper to taste
Method: Put the egg and milk mixture in a shallow bowl. Sprinkle the flour on a plate or a piece of waxed paper. When adding spices to the cornmeal, spread it like the flour. Flip each tomato in flour and then dip in the egg-milk mix. Coat each slice with cornmeal mix and place on a plate. Heat oil in a frying pan over medium heat and add the slices. Do not crowd the skillet-Cook until golden brown on each side. Drain on a paper towel and serve hot. Serves six. **SALSA:** a Mexican specialty that can be eaten as a side with chips or as a condiment on sand-wiches. You can chop ingredients either fine or chunky and make it as spicy hot as you prefer. Ingredients: 3 large ripe tomatoes chopped, 1 onion chopped, 1 bunch of chives chopped, 1 or more hot peppers chopped depending on your taste, ¼ cup culantro, juice of 1-2 limes, and a TS of salt.
Method: Toss all chopped ingredients in a bowl, shake the salt, and squeeze the limes over it. Let sit for at least an hour for the flavors to combine.

DID YOU KNOW? The tomato is the world's most popular fruit. A fruit is defined as the edible part of the plant that contains seeds. A vegetable is the edible stems, leaves, and roots of the plant. The US government classified the tomato as a vegetable for trade purposes in 1893. Tomatoes picked green eventually turn red, but will not have good flavor. A vine ripened tomato tastes best. Tomatoes should never be placed to ripen in direct sunlight, as they lose most of their vitamin C. Green tomatoes ripen faster stored in a cool place wrapped loosely in newspapers.

This method may take a week or two. They will last longer stored stem down. Tomato salsa has replaced ketchup as the top selling condiment in the United States. A salsa/ketchup combination is now available. In the late 1600's ketchup began as ke-tsiap, a Chinese sauce of spicy pickled fish. The tomato variety was created in 1800. Heinz started bottling tomato ketchup in 1876.

It's difficult to think anything but pleasant thoughts while eating a home grown tomato. ~ Lewis Grizzard

THE HAWAIIAN HOME GARDEN GUIDE
BEANS – LEGUMES

A bean is an edible seed of a legume plant fit for human consumption. The fruit of legume plant that bears beans in pods. Beans are a seasonal vegetable that requires warm temperatures to grow and mature usually within two months. Some varieties of bean plants can inject nitrogen into the soil as they grow. This decreases the need for additional fertilizer and repairs soils that grew heavy feeder crops as corn and pumpkins. This is why our ancestors, especially the ancient Amerindians, grew the '3 sisters' of corn, squash/pumpkins, and beans together. The beans would feed the soil while climbing up the cornstalks. Natural agriculture in harmony.

Beans are among the oldest food plants. They provided protein 10,000 years ago in the Middle East. Improved by cultivation practices, beans have been feeding people in S.E. Asia since 7000 BCE. But most scientists believe beans originated in South America in 2,000 BCE. Then beans were eaten fresh or dried. Christopher Columbus found green beans growing in fields on the Bahamas Islands.

As the green bean pods mature, they usually yellow and dry up, and the beans inside change from green and harden to their final color. As a vine, bean plants require a support, and their pods ripen gradually. Recent agricultural practices developed the 'bush bean' for commer-cial production that does not require support and the pods all ripen within weeks.

Beans were always an important source of protein throughout the world. Simply dried, they store adequately for months to provide a stable food source. Beans are high in protein, complex carbohydrates, folate, and iron similar to meat in nutrients, but without saturated fats.. Beans also have significant amounts of fiber and soluble fiber.

PAGE:

Courtesy 912crf.com

Any garden demands as much of its maker as he has to give.
But I do not need to tell you, if you are a gardener, no other undertaking will give as great a return for the amount of effort put into it. - Elizabeth Lawrence

BUTTER BEAN – LIMA BEAN – 'ELIMA PĪ

Phaseolus lunatu

Throughout the Hawaiian Islands, the yellow bean pods in the market are known as butter beans because the color resembles butter. These beans are a variety of lima bean, which has larger seed beans. Archaeological evidence from Peru indicates the large-seed bean species were among the first to be cultivated around 6,000 B.C.E. They were domesticated in the South American Andes Mountains. About 1,200 years later, a smaller-seeded variety developed.

During the Spanish conquest of Peru, lima beans were exported with the boxes marked by place of origin, Lima, Peru. That's how the beans got their name, which is incorrectly pronounced. Botanically, lima beans are *P. lunatus* and *P. limensis. I*n native Hawaiian they are *'elima pi.*

ABOUT BEANS: Bush beans develop flowers at the end of branches, while pole or climbing vines produce leaves on the main stem with flowers and fruit. Some varieties of beans have been enhanced to not have 'strings,' called 'stringless.' The term 'snap beans' commonly describes stringless bean pods. Dry beans such as kidney, navy, etc. are snap beans. Edamame are a bush type and usually don't need support. Vining types, sometimes termed 'half-runners,' can be grown without support, but bean production is increased by adding support, while pick-ing time is reduced. Pole beans must have support, a trellis, a fence, or poles to climb.

HOW TO GROW: Butter beans require well-drained soil with a pH of 5.8-6.5, and full sun. Dig a hole about 30cm (1 ft.) in diameter and 15cm (6in.) deep. Mix in rotted compost and manure and soak the soil. Keep the holes 60cm (2ft.) apart. Pole beans need some type of support so put that in before planting. Bean seeds that are bright pink have been treated with a fungicide to increase germination. Plant 4 seeds 2 inches deep per hole and water regularly until they sprout. If butter beans don't get enough water, the flowers will drop prematurely, and you won't get many pods. They should produce blossoms in about a month after planting and ready to harvest in another 30 days. These beans may die and send shoots again the following season. Butter beans should be picked when they are as thick as a pencil.

MEDICINALLY: Butter beans contain both soluble and insoluble fiber. Soluble helps regulate blood sugar levels and lowers cholesterol. Insoluble fiber prevents constipation, irrita-ble bowel syndrome, and digestive disorders.

NUTRITION: 100 grams of butter beans have 115 calories, 8g of protein, and 7g of fiber with vitamins B1 (thiamine), B2 (riboflavin), B3 (niacin), Pantothenic acid 8%, B6, and are very high in B9 (folate) with iron, manganese, magnesium, potassium, zinc, and phosphorus.

FOOD & USES: The pods and shelled seeds, fresh or dried, can be used in soups and stews. Sprouted seeds are used in salads and stir-fry. Steamed young leaves are edible but bitter.

RECIPES: BUTTER BEAN CURRY -

Ingredients: 300g butter beans, chopped 4 garlic cloves, 1 medium onion, chopped, 1 thumb-size piece of ginger, peeled and chopped, ½ TS turmeric powder, 1TBS black mustard seeds, 8 fresh curry leaves, 1 TBS regular curry powder, 1 TS garam masala, 4 cloves, 1 cinnamon stick,1½ cups of coconut milk, 1 dry chili, juice of 1 lime, salt to taste, 2TBS oil, a handful of cilantro

Method: In a frying pan or wok, heat the oil and stir in the mustard seeds until they sputter. Add curry leaves, onion, garlic, turmeric, salt, and gin-ger, and stir until onion is soft. Stir in the curry powder and add the coconut milk. Add the cloves, cinnamon, and the chili. Bring to a quick boil and reduce to simmer. Add butter beans and cook 15 minutes. Add lime juice and garam masala. Garnish with cilantro.

CHICKPEAS – GARBANZO BEANS – ‘O KA PĪKĪ
Cicer arietinum

Protein is essential to human development. Meat, fish, eggs, and nuts are expensive and consume lots of energy to grow and produce and space to store in our households. Beans are a near perfect food. Chickpeas are a great tasty source of protein, 9 grams in every 100. They are one of the easiest beans to consume and because of this they can be made into great recipes from creamy hummus, to falafel, or roasted and sprinkled atop salads. Plus, they're packed with magnesium, phosphorus, and potassium for heart health.

This powerhouse bean is chickpeas in English, but better known throughout the world as garbanzos in the Aloha State and in Central and South American countries. This bean has been grown around the Mediterranean for 8,000 years. The botanical Latin term for chickpea, *Cicer arietinum,* means ‘small ram’ referring to this bean's ram's head shape. These beans are also known as Bengal grams, hommes, hamaz, and Egyptian peas. The Hawaiians call them *‘o ka piki.*

This high-protein legume was probably cultivated first in the Middle East and then traveled to the ancient Egyptians, Greeks, and Romans. In the 1500s Spanish and Portuguese explorers brought this bean to other subtropical regions of the world, and that's undoubtedly how they reached the Western Hemisphere. Today, the main commercial producers of chickpeas are India, Burma, Pakistan, Turkey, Ethiopia, and Mexico.

Chickpeas have been domesticated into many varieties. Types are suited for tropical, sub-tropical, and temperate regions. There are two main types: the Desi, originally in India, and Kabuli, initially from the Mediterranean and Middle Eastern regions and now grown in Canada. The difference between the two main types of chickpeas are ‘Kabuli’ type beans generally have the largest seeds and grow well in cooler regions under irrigation. Desi chickpeas have smaller seeds and yield better in India and other dry climate conditions.

Chickpeas are the cassava of beans. Studies are being completed throughout Hawaii to increase the feasibility and production of chickpeas. It should be a great crop that could restore the soil where pineapples and sugar had previously been cultivated. Chickpeas will grow almost anywhere and with little water. This bean is inexpensive, but try and grow some as an experiment. If you've only ever eaten dried or canned chickpeas, fresh homegrown ones are so much tastier.

HOW TO GROW: The chickpea plant has branches near the ground and will grow to two feet high. Rain usually provides enough water, but chickpeas will thrive with irrigation. In India, they grow chickpea in sugarcane fields. Although usually considered a dry-land crop, chickpea develop well on rice lands. Chickpeas need well-drained soil with full sun. Chickpeas won't produce well in rich soils. The ideal is a relatively poor soil, as high levels of nutrients encourage lush growth, prone to developing mildews and related diseases. Avoid heavy clays or shade. Chickpeas, due to the particular bacteria that live along the plants' roots, can convert nitrogen in the air into the form plants need to flourish. A wide range of soil pH 6-9 will work. The Black Kabuli is a good variety to try.

To grow chickpeas, get raw seeds at the market and wrap in a slightly moist paper towel for a few days until they begin to sprout. Have a nice patch of soil well prepared with few clumps. Plant the sprouting seeds two inches deep, about a foot part. Chickpea requires occasional weeding and slight fertilizing with 12-24-12 and will tolerate long dry stretch-es, but try to regularly water it. It is ready to harvest in 3-5 months depending on the va-riety. These dry pods are more difficult to shell than pigeon peas because they are sticky and cave in rather than split apart. Cows, goats, or sheep will enjoy these plants for forage.

MEDICINAL: Chickpeas provide slow-burning carbohydrates, manganese, and iron needed for a lasting energy supply while its fiber stabilizes blood sugar. Unlike hard to digest meat, chickpeas are low in calories and virtually fat-free. However, chickpea contains 'purine' and individuals with kidney problems or gout may want to avoid these beans. Research has found that a seven-day diet (one meal a day of chickpeas cooked with onions and turmeric powder will drastically reduce your overall cholesterol.

NUTRITION: 100g of chickpeas has 360 calories, and almost 9g protein with plenty of 17g fiber. These beans are loaded with folate and other vitamins with manganese, phosphorus, potassium, and zinc.

FOODS & USES: Chickpea is a very versatile vegetable; consumed as a fresh green vegetable, dried, fried, roasted, or boiled; as a main course, snack food, a sweet, or a condiment. Chickpea is ground into f our; and used for soup, dhal, and to make bread. Chickpeas has a nutty f avor, yet the overall taste is like starchy butter. We usually see beige chickpea, but there are black, green, red, and brown varieties. Eating chickpeas as sprouts will increase the food value.

They should be dry, intact - not cracked, without any insect damage. In an airtight container, chickpeas should keep for a year. Once cooked, it will keep 2-3 days in the fridge. Canned or dry have about the same nutritional value. Like rice, it's best to inspect chickpea before cooking to remove stones, and damaged beans by rinsing them in a strainer.

Chickpea varieties are used in Middle Eastern, Indian, Spanish, Italian, Greek, Asian and North African cooking. Add chickpeas to penne pasta mixed with olive oil, feta cheese, and fresh oregano for a unique tasty lunch, or just add chickpea to simple mixed vegetable soup to enhance its taste, texture, and nutrition.

RECIPES: HUMMUS

Ingredients: 1 lb. well-cooked chickpeas, 2 garlic cloves, ¼ cup fresh lemon juice, ¼ cup water, 1 TS salt, ½ cup sesame tahini spread (optional, 2 TBS olive oil, pepper, and spices to your taste.
Method: put everything in a blender or food processor and blend until smooth. Serve with roti or crackers (biscuits

ROASTED CHICKPEAS

Ingredients: 1 lb. of well-cooked chickpeas, 2 TBS olive oil, 1 TS soy sauce, spices to your taste.
Method: Mix ingredients in a bowl and place on a baking sheet. Bake at 450 degrees for half an hour or until brown and crunchy.

FALAFEL

Ingredients: 1 lb. of cooked chickpeas, 1 large onion chopped fine, 4 garlic cloves minced, 2 TBS chopped parsley, 1 TS coriander, 1 TS cumin, ½ TS salt, 2 TBS flour, spices to taste, and fry oil.
Method: Combine all ingredients in a bowl or food processor, mashing the chickpea. It should become a thick paste which forms into the size of small, slightly flattened, ping pong balls. Fry on high in two inches of oil for a few minutes until golden brown.

DID YOU KNOW? Humans should consume 0.36 grams of protein per pound of body weight. (.36 x150 = 54 = 2 ounces of protein.) The average man should consume 56 grams of protein per day, and the average woman about 46 grams. Save money and have better health – eat beans. 14 million tons are grown worldwide every year. Only more soybeans are grown. Chickpeas are one of the best food sources of vitamin B-9- folate that fights aging. Since the mid-1700s, ground chickpeas have been used as a caffeine-free coffee substitute.

Hard and dry, a chickpea is inedible. Hard and dry, a heart is unlovable. Pre-soak it in dance, music, and art. — Khang Kijarro Nguyen

GREEN BEANS – PINA ŌMA ʻOMA ʻO

Phaseolus vulgaris

Green beans grow easily throughout the Hawaiian chain. I always have a few bean bushes in the garden. It doesn't take many; maybe five plants, to provide a constant supply of the tender beans. Green beans are descended from a bean ancestor that originated in Peru thousands of years ago. Migrating Indian tribes spread this vegetable through South and Central America before the Spanish Conquistadors took them around the world after 1700. Although green beans are not yet a big commercial crop throughout the Hawaiian Islands, more farmers are cultivating them.

There are basically two bean categories: edible pod beans and shell beans. Green beans, Phaseolus vulgaris, also known as snap or string beans, are the most popular edible pod bean. These beans are often called string beans because years ago a fibrous string ran along the seam of the bean. The string was noticeable when you snapped off the ends. Somehow that trademark has been bred out bean of the modern bean varieties. The snapping noise is the reason for its other nickname. Green beans are picked while still immature and the inner bean is just beginning to form. They snap when you break them. The variety may vary in size, but average about four inches in length, deep emerald green, wth a slight point at either end. They contain tiny seeds within their thin pods. There are more than 130 varieties of green beans.

HOW TO GROW: Green beans are easy to grow and worth the effort. They will adapt to almost any loose soil with a pH of 6 or higher. There are two types of green beans. Pole beans grow much like a climbing vine and require a stake or trellis. Bush beans spread up to two feet and do not require structural support. The bush plants produce a lot of beans all at once. The vining pole beans produce beans throughout the growing season. It's best to directly plant into your garden because green beans don't like to be transplanted.

To make a few green bean rows, fork a well-drained area 10 ft. x 4 ft., mixing in some well-rotted chicken manure and a few shovels of limestone. Plant the seeds about 1 inch deep and 4 inches apart while keeping the rows separated by 18 inches. When the sprouts are about 6 inches tall, carefully mold around the base and add about a soda bottle cap of 12-24-12 fertilizer to the base of each bush. Green beans don't need much water; too much water can be an enemy. Water once a week until they flower and start to bear, and then water 3 times a week. Replant every 2 weeks to have a constant supply of these tasty beans. If bugs are eat-ing the leaves, spray once with a light pesticide as Malathion or Pestac. Green beans should bear within 2 months. Beans will nourish the soil by producing much needed nitrogen. Green beans are an excellent crop to plant in a field after a heavy eater like corn, which takes many nutrients from the soil.

MEDICINAL: Eating these delicious beans helps lower high blood pressure and their fiber may also help prevent colon cancer. Green beans reduce the severity of diseases such as asthma, arthritis, and rheumatoid arthritis. Green beans are a good source of riboflavin, which has been shown to help reduce the frequency of migraine attacks. These beans are a very good source of iron. Compared to red meat, green beans provide iron for fewer calories and are totally fat-free. Iron is an integral component of hemoglobin, which transports oxygen from the lungs to all body cells and is also part of key enzyme systems for energy production and metabolism. If you're menstruating, pregnant, or lactating, you need iron!

NUTRITION: Green beans have only 45 calories per 100 grams with 2 g protein and are loaded with many potent nutrients. Green beans are one of nature's vitamin pills: a great source of vitamins C, and B-6, dietary fiber, potassium, folate, iron, magnesium, manganese, thiamin, riboflavin, copper, calcium, phosphorous, protein, omega-3 fatty acids, and niacin.

Please remember, these nutrients are best if you grow them at home using as little chemicals as possible.

FOOD & USES: When purchasing string beans, they should be stiff, and should snap sharply, with a spray of juice from the seam. If soft, they've been around too long and will taste disappointing. Avoid beans with obvious seed bumps pressing up through the pods as they were picked too late and will tend to be tough. I prefer to steam green beans as these beans will continue to cook after you take them out of boiling water. A good rule is to cook green beans as little as possible using the smallest amount of water possible. Fresh green beans may be stored in an airtight container in the refrigerator for about four days. Beans can also be blanched and frozen immediately after harvest for a longer shelf life.

RECIPES: SIMPLE GREEN BEAN AND TUNA SALAD

Ingredients: 1 lb. green beans, tips trimmed and steamed, 2 cans of processed tuna fish crumbled, juice of one lemon, olive oil, salt, and spice to taste.

Method: In a large bowl combine all the ingredients, chill, and serve.

LEMON-WALNUT GREEN BEANS

Ingredients: 3 lbs. green beans, ¼ cup butter or margarine, 6 chives chopped, ½ cup chopped walnuts, (peanuts or almonds can be substituted), 2 TBS chopped fresh or crushed dried rosemary, 3 TBS fresh lemon juice, 1 TBS grated lemon rind.

Method: Steam green beans for 5 minutes. In a large pot with a cover, melt the butter. Add chives and cook until tender. Add green beans, walnuts/peanuts/almonds, rosemary, and lem-on juice; cook, stirring constantly, until thoroughly heated. Sprinkle with lemon rind. Serve immediately.

SIMPLE SKILLET GREEN BEANS

Ingredients: 1 ½ lb. green beans tips removed, 2 garlic cloves minced, 2 TBS olive or virgin coconut oil preferred, ½ TS red pepper flakes or powder, 2 TBS water, ½ TS salt.

Method: In a large skillet or wok, heat the oil over a medium flame. Stir the pepper into the oil. Add green beans and cook, stirring often until the beans begin to brown, about 5 minutes. Stirring constantly, add garlic and salt and cook about 30 seconds. Add water and immediately cover. Cook covered until the beans are bright green and crisp, yet tender - 2 minutes. Serve immediately.

HOT PEPPER GREEN BEANS

Ingredients: 1 lb. green beans-tips removed, 4 garlic cloves minced, 1 medium onion chopped, 1 hot pepper-seeds removed and minced as fine as possible, 3 TBS vinegar, 2 TBS brown sugar, 2 TBS soy sauce, 1½ TS cornstarch, 3 TBS sesame oil, salt, and spices to taste

Method: Steam green beans for 5 minutes. In one bowl combine onion, peppers, garlic, and green beans. In another smaller bowl combine vinegar, soy, cornstarch, and sugar. In a large frying pan heat the oil and add the bean bowl mixture and stir-fry for 2 minutes. Add the re-maining bowl of ingredients and stir-fry until beans are coated.

DID YOU KNOW? Green beans have been cultivated in Mexico for over 7,000 years. The global green bean market is estimated at $32 billion. China produces 75% of the world's total with 22 million tons. Indonesia is second with 950,000 tons and the US is third with 865,000 tons. In home gardens tomatoes, peppers, and green beans are the most commonly grown veggies. The longest green bean was grown in the USA in 1996 grew to a length of 121.9 cm (48.75 in).

Nothing is more the child of art than a garden. ~ Sir Walter Scott

JACK BEAN – AWIKIWIKI
Canavalia ensiformi

The jack or broad bean is a climber that needs to be trained to twine around supports or it can take over your home garden and grows anywhere in Hawaii. It's definitely worth growing for its tasty nutritional benefits. The plant can spread over 10 meters (39ft.. Wrapped on a good support system, the pink-mauve blossoms make it an attractive ornamental plant. The green pods can grow to 36 cm (14in. long with large white seeds.

This species of bean has been grown in Central America since 3,000 B.C.E. Because of its fantastic nutritional content, it's grown everywhere in the tropics and subtropics. Japan and Southeast Asia do large-scale commercial cultivation for its edible seeds and pods.

TAKE NOTE: The beans are mildly toxic. Boiling and discarding the water with a good rinse and then reheating the beans will remove toxicity.

Jack beans *(Canavalia ensiformis)* are white with a black scar, and grow inside, slight-ly curved pods. A similar bean variety known as sword beans (*C. gladiate)* are usually found in straight pods with red beans. The native Hawaiians named it *the awikiwiki* meaning the fast one.

HOW TO GROW: The jack bean is a short-lived perennial, tolerant of dry weather, fast-growing, and will adapt to most soils. This bean is also resistant or immune to most pests. However, if not controlled, the jack will twine around other plants. It works well planted with bananas, cassava, or coffee. It prefers soil rich in organic matter, and a pH in the range 5–6. Jack thrives in the humid, tropical lowlands, but will grow to 1800m in elevation.

Fork mounds a meter in diameter and work in some rotted manure and compost. Plant two pockets 30cm (1ft.) apart. Plant 2 seeds in each 3cm (1in.) deep. They should germi-nate with 3 days. Green pods can be harvested within 3 months, and mature light brown pods for seed are ready after 6 months. Pods reach 30cm (1ft.) long, but are harvested at half that size for eating. Seeds are large, 1cm (1/2 in.) long, and as wide. The pods may have 20 white, smooth seeds.

This plant may help feed the other vegetables in your garden and revitalize your soil. Jack has deep roots, which makes it resistant to droughts. It has a symbiotic relationship with some soil bacteria that create (fix) atmospheric nitrogen. Some nitrogen is consumed by the growing plant, but some can also be used by other nearby plants. At the end of the growing season, leave the roots in the ground to decay and release their nitrogen.

MEDICINAL: In China, the entire plant is pounded and made into poultices to treat boils. Seeds are ground to make a tea for bad coughs and to cleanse the kidneys. In Nigeria, a paste made of the seeds is used as an antibiotic and antiseptic. In Korea and Japan, an extract from the bean is used in soap to treat athlete's feet and acne. Pharmaceutical companies are researching jack as a source for the anti-cancer agents: trigonelline and canavanine.

NUTRITION: 100 grams of fresh green jack has 104 calories with 10g protein, 12g of carbs, 4g of fiber, and 8mg vitamin C. 100 grams of dried mature beans have 350 calories with 20g of protein, 61g of carbs, with lots of potassium, phosphorus, calcium, with traces of copper, niacin, thiamin, and zinc. This bean's protein has good amounts of most essential amino acids.

FOODS & USES: These beans are recommended for cultivation in developing coun-tries as an easy, reliable, and affordable source of abundant protein. Very young seedpods, before the seeds swell, are also cooked like French beans. Young, tender pods are pickled in Japan. Young leaves are cooked like spinach. Jack beans should be boiled, and the water should be drained off to remove any poisonous substances from the beans. Roasted seeds are used to substitute coffee.

RECIPES: It's best to cut off both ends of the beans and pull off the side strings

JACK CURRY

Ingredients: 250g jack beans chopped ¼ in., 3 garlic cloves chopped, 1 medium onion chopped, 6 curry leaves and small piece of pandan leaf if available, ½ TS of each turmeric, chili, and curry powders, 1/3 cup coconut milk, salt to taste, and 2TBS oil.

Method: In a suitable pan, heat the oil and add the garlic, onion, curry, and pandan leaves. Stir in the turmeric power and then the jack pieces. Add salt and stir-fry. Stir in the chili and curry powder. Stir in the coconut milk. Reduce heat and simmer for 4 minutes. Serve with rice or roti.

JACK STIR-FRY

Ingredients: 250g jack bean pods chopped, 1 medium onion chopped, 2 garlic cloves minced, 1 green chili chopped, 6 curry leaves, ½ TS black mustard seeds, ½ TS turmeric powder, 1 TS urad dhal, 1 dry chili whole, ¼ cup grated coconut, salt to taste, 1TBS oil, water as necessary

Method: In a frying pan or wok heat the oil and fry the mustard seeds, add garlic, green chili, and onion as soon as mustard seeds pop. Add the dry chili. Stir-fry for 3 minutes and then add the jack beans, salt and turmeric powder. Sauté for 2 minutes stirring. Add 2 TBS water. Stir for 2 minutes and reduce heat, cover, and simmer for 5 minutes. Stir in grated coconut.

NAMES: Jackkkai – Indian broad bean, jack-bean, coffee bean, wonder-bean, giant stock-bean, horse-bean, horse gram, chickasaw lima bean, Brazilian broad bean, ensiform bean, mole bean, Go-Ta-Ki, overlook bean, pearson bean, watanka

DID YOU KNOW? Beans were probably one of man's irst food when he started walk-ing upright. Beans from thousands of years ago are genetically very close to the beans we eat today. Broad beans, also called fava beans, originally the size of a small fingernail, were grown in Thailand in 7000 BCE; that's before ceramics were developed. Beans were left with the mummies in Egyptian tombs. The legume / bean family *(Fabaceae)* is the third largest family of flowering plants with more than 18,000 different species, and an enormous variety of shapes and sizes. The orchid family *(Orchidaceae)* has the most species with about 20,000 and the sun lower family *(Asteraceae)* is second with about 24,000 species.

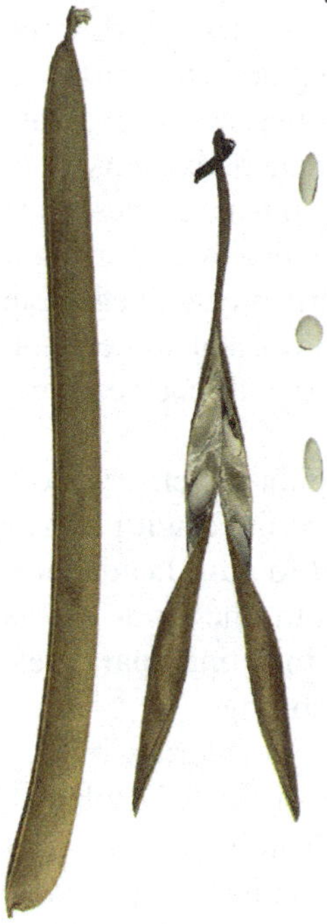

The fruit is technically called a legume or pod that splits open along two seams. Of all the legumes, the peanut is the only variety that develops below the ground. The Central American Liana Entada *(E. gigas*) grows the world's largest bean pods, up to 1.5 m (5 ft,) long. Its enormous vine is like Jack's beanstalk. Costa Ricans call it *'escalera de mono'* or the 'monkey's ladder.' Its big seeds are named "sea hearts" and often washed out to sea to distant lands. China grows the most beans in the world. The USDA recommends adults eat 3 cups of beans every week.

The USA leads the world in producing dried beans; 40% are shipped to 100 di erent countries. North Dakota grows almost a third of the US beans. An average of 8 percent of cooked beans is protein. If you eat beans regularly, you will have less gas from them, and the gas comes from their sugar content. Frozen or dry beans contain less sodium and are better for you than canned. Cooked beans can be frozen for six months. It's best to let them thaw in the refrigerator overnight before reheating. Eating beans lowers the risk of cancer because they're rich in antioxidants. Beans can be made into burgers, cakes, drinks, pies, fudge, mu ins, jewelry, bean-bag chairs, toys, and musical instruments. The largest bean bag was 147.9m³ (5,223 ft. 67.7 in³) was created by Comfort Research in Grand Rapids, Michigan, USA, in September 2017 as an entry into the 2017 ArtPrize international art contest.

Weather means more when you have a garden. There's nothing like listening to a shower and thinking how it is soaking in around your green beans. ~ M. Cox

LABLAB BEANS - HYACINTH BEANS

Dolichos lablab L - Lablab purpureus

The lablab, or hyacinth beans are one of easiest and the most nutritious vegetables you can grow at home. They are known on many islands as seim and grown for their delicious pods and seeds. In warm tropical and subtropical areas, this plant is cultivated both on large-scale commercial farms and in small home gardens as a veggie and ornamental flowering vine. Lablab also has edible leaves and flowers. This delicious bean s not well known in Hawaii. Try it.

These bean vines are fast growing and the bacteria on the roots rejuvenates the soil by injecting nitrogen. This type of vining bean has a symbiotic relationship with certain soil bacteria. The bacteria form root nodules and fix nitrogen from our atmosphere. The bean vine uses some of this nitrogen but other plants growing nearby and the soil can benefit.

Believed to be native to Africa, lablab beans are cultivated around the world. These beans traveled from Africa to India in between 1600 and 1500 BCE. In the 1700s, these beans were grown in Europe and two centuries later lablab beans were introduced to America as an ornamental plant. Lablab beans are used for natural medicine.

HOW TO GROW: This species of bean has many varieties due to extensive breeding in cultivation, but in general, they are annual or short-lived perennial climbing vines. Some types of lablab have white flowers, others maroon-purple, or blue. The main twining stems of these vines can reach 6 meters (20 ft.) in length. Growing on a fence or a trellis for support is recommended. It's best to plant during the last weeks of the rainy season. Pre-soaking the seeds for 2 hours in warm water is recommended. The seeds sprout within a month. Lablab prefers well-drained soil with high organic matter content and a pH between 5.5 and 6. Immature pods are ready to harvest 4 months after planting. The seedpods are 10cm (4in.) long and 15mm (½ in.) wide. Mature pods ripen within 6 months. Lablab is known for its edible seed, but the long taproots, leaves, flowers, and seedpods are edible. The seeds can be black, white, red, or brown depending upon the variety.

The most common varieties of lablab beans are: White Flower-grows beautiful white flowers and thin pods along trellis and fences. It's necessary to harvest young pods every day. Purple Flower is suited for subtropical regions and displays beautiful purple flowers, vines, deep purple pods ideal for a backyard fence or patio. Asia Purple has purple-red flowers on light green stems. This is probably the most popular variety to grow as an ornamental vegetable plant and it produces lots of young edible light green pods. Asia Purple is resistant to heat and drought conditions. Asia White produces flowers 45 days after planting. Young pods and seeds are very tender and delicious. Asia White produces beans for more than 4 months.

MEDICINAL: Lablab leaves are used to accelerate childbirth, regulate menstruation, to treat stomach troubles, and tonsillitis. A decoction of lablab leaves combined with leaves from other plants treat heart problems. The leaves are crushed and sniffed to cure headaches. The flowers treat alcoholism, poisoning, and flatulence. The juice from the pods treats inflamed ears and throats. Dried and roasted fully mature seeds treat intestinal parasites, fevers, cholera, sunstroke, vomiting, diarrhea, alcoholism, and arsenic poisoning.

NUTRITION: 100 grams of raw, immature lablab pods have only 46 calories, but dry mature seeds contain 344 calories with 24g of protein and the same amount of fiber. Dry lablab beans are one of the finest sources of several B-complex vitamins such as thiamin (1.130 mg or 94% of the Daily Value), riboflavin, folates, and niacin. Fresh pods carry vitamin A, a powerful antioxidant, 29% of the Daily Value (DV), 12.6 mg or 21% of DV of vitamin C. Dry beans hold 148% of the DV of copper, 13% of calcium, 64% of iron, 71% of magnesium, 68% of manganese,

26% of potassium, 53% of phosphorus, and 84% of zinc.

Copper helps to maintain a positive outlook, and to focus. Potassium is an essential electrolyte of cell and body fluids, which helps to counter the bad effects of sodium on the heart and blood pressure. Zinc has anti-inflammatory properties which helps to reduce the risk of diseases. Minerals such as selenium, manganese, and zinc assist people having lung disorders.

FOOD & USES: The flowers and immature pods are commonly eaten as vegetables. The mature seed is edible as long as it is thoroughly cooked, boiled twice with the water changed, and a final rinse. They have a mild flavor, very rich in protein, and can be used as a food staple. These beans can also be prepared as 'tofu' or be fermented into 'tempeh' in the same way that soya beans are used in Japan. They can be sprouted and eaten raw like a mung bean sprout. The tender young seedpods and immature seeds can be eaten raw or cooked like green beans. The leaves must be cooked and can be used just like spinach. They can also be dried for later use. The leaves contain up to 28% protein. The blossoms can be eaten raw or cooked in soups and stews. A green dye can be obtained by pounding the leaves.

IMPORTANT NOTE: Dried beans must be boiled twice and rinsed to make them safe for consumption because they contain high concentrations of cyanogenic glucosides. The young, tender green pods are usually what's eaten; in some areas the leaves are eaten the same as spinach.

RECIPES: LABLAB STIR-FRY

Ingredients: 500g (1 lb.) green lablab pods chopped into 1inch pieces, 4 garlic cloves chopped, 1 medium onion diced, 8 medium tomatoes diced, 10 curry leaves, 1 TS of each powder: coriander, cumin, and chili, salt to taste, and 3 TBS coconut oil

Method: In a frying pan or wok coated with the oil fry the garlic, onions, and curry leaves until the onions turn light brown. Add the chili, coriander, and cumin powders. Add the tomatoes and cook for 5 minutes, stirring. Add the salt and bean pieces and stir, simmer covered for 10 minutes on low heat. Remove from the heat, but leave it covered for 10 more minutes. Enjoy with rice.

ROASTED LABLAB BEANS

Ingredients: 500 grams of lablab beans, washed, boiled twice, and rinsed, 6 garlic cloves chopped, 1 large onion chopped, 3 green chilies chopped, 1TS turmeric, 1TS chili powder, 2 cups of water, salt to taste, and 5TBS oil

Method: Soak the lablab beans in water for 8 hours, then boil twice, and rinse before cooking the curry. In a frying pan or wok, heat the oil and fry the onion, garlic, chili, and turmeric until you get that great aroma. Add the lablab beans and stir for 5 minutes. Add 1cup of water and bring to a boil with the pot covered. Add the salt. Enjoy hot with rice or cold on a salad.

LABLAB BEANS AND POTATOES

Ingredients: 250g of green lablab pods washed and chopped, 4 potatoes, peeled and diced, 2 gar-lic cloves chopped, 1 large onion chopped, 2 green chilies chopped, 4 tomatoes chopped, 1TS of the following powders: turmeric, chili, and garam masala, salt, water as necessary, and 2 TBS oil

Method: Heat the oil in a frying pan or wok. Sauté the garlic, onion, and chilies. Add everything and cook with 4 TBS water, adding more if you want a thinner sauce. Serve with rice or roti.

NAMES: hyacinth bean lablab, Australian pea, bonavist, dolichos bean, Egyptian bean, Egyptian kidney bean, Indian bean, jack bean, lubia bean, seim bean

I came to love my rows, my beans, though so many more than I wanted. They attached me to the earth,and so I got strength like Antaeus.

Henry David Thoreau

LONG BEAN – PI ‘A LŌ ‘IHI

Vigna unguiculata ssp. Sesquipedalis

The slender, green long bean is a vegetable favorite on the dinner tables in throughout the islands. These long beans botanical name defines the bean subspecies: *Vigna unguiculata subsp. Sesquipedalis*. There is the purple version of this long bean and all belong to the same family as the cowpea, black-eyed pea, and pigeon pea. Of the many vegetables I’ve researched, this long bean seems to have the most names: Chinese long bean, Peru bean, asparagus bean, string bean, snake bean, snake pea, snap pea, long-podded cowpea, bodi, bora, juro-kusasage-mae (Japanese), dow gauk (Chinese), and sirao. It’s believed to have originated in in Southeast Asia and domesticated to grow long and straight from the cowpea. Beans have been cultivated for at least 7000 years, and there are more than 12,000 species of beans throughout the world. Long bean is an important vegetable in SE Asia, India, the Caribbean, and the Hawaiian Islands.

The long bean has many seems to be more tolerant to the climatic conditions, diseases, and pests of the tropics than regular green beans. Done right, long beans can be an attractive and tasty addition to any home garden. Done without planning, the vines that produce this long bean easily can take over a backyard. Long beans cannot be allowed just to rest on the ground. It needs to be up on a fence or trellis, otherwise it bears poorly.

HOW TO GROW: The long bean is a pencil-thin bean that resembles a green bean although not as crisp and grows to about 2 feet long. It’s usually harvested at 18 inches or less. This bean type is an annual climbing plant with white, yellow, or pale purple flowers and grows quickly in the Caribbean’s tropical climate. It grows so well that the markets always have long beans for sale. It’s best to grow some at home and know there weren‘t many chemicals used.

The long bean prefers a light, well-drained soil with a pH of 5.5 to 6.8, enriched with compost, or rotted chicken manure and grows to maturity in less than 2 months. Being a climbing plant, this bean provides extra work for the gardener. Plant seeds an inch deep, at least 8 inches apart with raised rows separated 3 least three feet. Soak the seeds in water before sowing, for better germination. The distance between rows is necessary because sticks need to be placed every 6-8 ft. along the rows. Some farmers put in the sticks before planting so not to later irritate the roots of young plants. Strings are then pulled between the sticks at 2 or 3 levels of 1, 2, and 4 ft. Then carefully weave the long bean vines between these strings. This bean is perfect for the home gardener who has a chain link fence for it to climb. It can also vine on dead eggplant or hot chili pepper trees. In a small garden plot, it’s best to plant the long bean at the rear, so it does not shade the other vegetables. Care must be taken to ensure the sprawling vines are controlled and do not interfere with other veggies. In the early growth stages use a little starter, 12-24-12 fertilizer mix, and just a pinch of the high nitrogen 12-12-17-2 mix when flowering begins. Long beans require water. If it is an extremely dry season the beans will be short, tough, and stringy.

Fruits grow from open flower to marketable length of 18 to 30 inches in about 9 days. Flowering will occur 5 weeks after sowing. Pick the pods at the tender stage at maximum length, before the seeds mature, or swell. Long beans may grow up to 24 inches long, but it is better to pick them at 12 to 18 inches. Long bean needs only a little fertilizer. It enriches the soil by fixing atmospheric nitrogen in nodules on its roots. With the help of nitrogen-fixing bacteria, the plant makes its own food. It’s wise to rotate planting areas. Mole crickets love the young beans. Aphids are drawn to the pods of this plant. Thrips tend to be a pest early in the season. Spider mites can be a problem, producing a silver-speckled appearance on leaves. Ringing, where the beans turn into a spiral, is one of the most common diseases and this may be also caused by mites. Long bean issusceptible to nematodes and mosaic virus. A chemical spray of an insecticide like Fastac is beneficial. Be careful to spray chemicals at least 2 weeks before harvest.

MEDICINAL: Long beans are reported to have anti-diabetic, blood-purifying, and diuretic properties. Their high fiber is good for digestion. Earaches are supposedly eased by eating Long bean leaves cooked with rice. To stop lactation and wean their children, nursing mothers eat boiled long leaves with potassium alum. (Ugh!)

NUTRITION: These beans are rich in vitamins A and C. 100 grams has 4 grams of protein, 110 mg calcium, 5 mg iron, and calcium.

FOOD & USES: Long beans can be prepared in various ways: stewed with toma-to sauce; boiled and drained, then seasoned with lemon juice and oil; or simmered in butter or oil and garlic. These are great in a salad or in stews. The pale green bean is meatier and sweeter than the dark green bean, which has a less delicate taste. The bean's young leaves can be cooked like spinach and eaten as a vegetable. The seeds can be germinated and used like sprouts. Long bean shoots can be used like asparagus, which is why they are sometimes called asparagus beans.

RECIPES: LONG BEAN STIR FRY

Ingredients: 1 lb. long beans, 1 lb. bok choy, 1 TBS sesame seeds, 2 TS peanut oil, ½ TS sesame oil, 2 TBS soy sauce, salt, and pepper to taste.

Method: Cut off stems from the bok choy, trim ends before slicing into long strips. Blanch the pak choy in hot water, remove and set aside. First steam the long beans until bright green (about 4 minutes), and then combine with bok choy. Heat peanut oil and sesame oil in a hot frying pan or wok. Add the vegetables and pepper to taste. Stir-fry for 2 minutes. Add and soy sauce and sesame seeds. Season with salt and pepper.

SIMPLE CHINESE LONG BEANS

Ingredients: 1 lb. cooked long beans, 3 garlic cloves minced, and ½ cup balsamic vinegar

Method: Mix all ingredients in a bowl and serve at room temperature.

LONG BEAN SALAD

Ingredients: 1 cup long beans cut into 2-inch pieces, 1 cup bean sprouts, 1 sweet bell pepper (prefer red) seeded and sliced into thin strips, 1 medium cucumber sliced thin, 1 red onion, 1 TBS balsamic vinegar.

Dressing: 1/3 cup white vinegar, 1 TBS olive oil, 2 TBS sugar, 1 TBS herbs such as basil and or thyme chopped, 2 garlic cloves minced, salt to taste.

Method: In a large frying pan mix long beans, onion, and vinegar. Cook 2 minutes with con-stant stirring. Take off heat and allow cooling. Mix in the remaining ingredients. Then mix in dressing. Chill for 2 hours before serving.

SAUTÈED LONG BEANS

Ingredients: ½ lb. long beans cut into 1-in. pieces, 1/2 cup chopped onion, ½ cup shrimp, ½ cup pork cut into thin strips (Subsitute chicken, beef, shrimp, or boneless fish chunks.), 2 ripe tomatoes, 2 TBS minced garlic, 2 TBS vegetable oil, salt, and pepper to taste. Method: Heat oil, garlic, and onion. Add tomatoes and cook until soft, stir in pork and shrimp. Simmer stirring occasionally. Add beans and cook until tender. Serve hot.

DID YOU KNOW? Most types of beans have the ability to pump nitrogen from the en-vironment into your soil - making a more nitrogen-rich soil. This is done through nodules on their fine roots. It's a smart move to plant beans where the heavy nitrogen feeding veggies, like corn, were planted the year before.

Beans, beans, the musical food, the more you eat the better you toot.
The more you toot, the better you feel, so eat beans with every meal.

PIGEON PEAS – PI NUNU - PUERTO RICIAN PEA

Cajanus cajan

Known in most countries as red gram or pigeon peas. Hawaiian islanders love them and named them *'pi nunu'* or *'pi pokoliko'* (Puerto Rican pea). These dried peas/beans vary in colors from red to deep purple, brown to black, even to white. Fresh peas are easy to grow and growing your own will help with the national effort to be self-sufficient. Most gram is imported from India, who produces more than 80% of the world's red gram. Some botanists believe these beans originated at least five millenniums ago in Africa; others think it must have been India. They've been found in the tombs of Egyptian pharaohs. World production is about 6 million tons. Apparently, the English fed pigeons these peas on the island of Barbados and the name stuck. This bush bean has been cultivated throughout the Hawaiians since the late 1800s.

This plant is good for the home garden because it'll tolerate bad soil, heat, and drought, and produce a good crop in hot and humid climates. Growing pigeon peas provides organic nitrogen, which improves organic soil structure. This is a great cover crop that will improve and protect soil from erosion and can shade less heat-tolerant crops.

HOW TO GROW: Growing pigeon peas is much easier than picking the crop. If you have a friend or neighbor who is growing peas, ask for some to start your crop. First, select good peas and dry them to save the seeds. Peas are easy to grow in most soils, providing the planting area drains in excessive rains. The best areas to plant are slopes or rises. Soak the seeds for at least a day before planting. Fork the soil and break all dirt clumps until it is fine and soft down at least 10 inches. The bed should be further raised at least 6 inches to provide excellent drainage. Make rows at least 3 feet apart. Plant the seeds 1 inch deep and at least 1 ft. apart. Before the seeds sprout spray the soil with Banrot. When the sprouts are a foot tall, thin to 3 ft. apart.

Water is both the enemy and savior of these peas. Of course, water is necessary for the tree to sprout and grow. Once the pea tree gets big and weighty, it doesn't need much moisture. If torrential rains occur, then the soil loosens, and the tree can collapse. Because of the bushy branches, pea trees don't take well to being staked unless done when the sprouts are small. Roots not only provide stability, but also take available nutrients from the soil.

Moisture also makes insects and fungus thrive. A good spray with a light pesticide like Pestac or Fastac weekly will keep most bugs away. (A light pesticide is one that wears off in about a week when used at the recommended dosage.) Ants seem to love eating pea trees and can cut the trunk to the heart. One type of small black wasp drills the trunk and lays eggs. Fungus is another villain that will ruin a pea crop. There are various types of fungus and it takes a qualified person to define the enemy. Pigeon pea varieties are classified as tree type, tall and dwarf.

The problem with pigeon peas is they take 4-5 months to mature, much longer than other beans. This gives every parasite a chance to attack. Picking the ripe pods is the next phase and very labor intensive. You must select only those pods that have the inside peas almost poking out. The peas should feel hard and firm. By the end of a day's harvest, your neck and back ache, and you are almost cross-eyed from checking the hanging peas. A nice-sized tree can produce ten pounds of peas. Multiply that times two to three hundred trees (the size of a good garden) and decide if you want to grow and sell, or just buy and shell. The trees can be pruned, reduced by a third, and left for a second season, which can bear a heavier crop.

MEDICINAL: A poultice can be made of the young leaves of pigeon peas and applied to sores. Chinese claim powdered leaves help expel kidney stones. Chinese shops also sell dried roots as an anti-toxin, to expel parasite worms, expectorant, sedative, and vulnerary. Salted leaf juice is juice is taken for jaundice. Argentines use a pigeon pea leaf decoction for genital and other skin irritations, especially for females. Leaves are also used for toothache, mouthwash, sore gums, child-delivery, and dysentery. Pigeon pea blossom decoctions are used for bronchitis, coughs, and pneumonia. Scorched seed when added to coffee alleviates headache and vertigo. Fresh seeds are said to help incontinence in males, while immature fruits are believed to treat liver and kidney ailments.

NUTRITION: For vegetarians and everybody, peas are an excellent source of protein. To get the most protein from peas or beans, eat them with unrefined rice or wheat. 100 grams of fresh, green pigeon peas has 140 calories with 7 grams of protein and 5 grams of fiber. They are packed with hard-to-get B complex vitamins with C and K. The fresh peas have plenty of minerals as manganese, magnesium, phosphorus, and zinc. 100 grams of dried peas have 340 calories, 21g protein and 17g of fiber. The dried only have the B vitamins and are a better source of manganese, magnesium, phosphorus, potassium, and zinc.

FOOD & USES: Vitamins are more available if the peas are consumed as sprouts. Sprouts can be cooked in stir fry or added to sandwiches. Tender young beans may also be added still in the pod to stir fry, soups, and stews. Whole dried pigeon peas look like black-eyed peas, with a light tan or beige skin, speckled with small brown spots. They are also sold with the skins removed, in which case, they will be green. And you can also find them split, as is the case with dhal, where they will resemble split mung beans and canned or frozen green pigeon peas.

RECIPES: GOGHANI – a taste of Trinidad

Ingredients: 2 lbs. of very full peas (almost yellow), ¼ lb. salt fish-boiled, cleaned, and mashed, 2 medium onions sliced thin, 5 garlic cloves grated, 2 TBS cooking oil, salt, and spice to taste

Method: In a pot, boil peas till tender with salt and drain. In a frying pan heat oil and fry salt fish, then add peas and all other ingredients. Stir well and eat with rice and or roti.

COCONUT AND PIGEON PEAS

Ingredients: 1lb. of fresh or frozen peas boiled until soft – drained, and rinsed, 1 TBS olive oil, 1 small onion quartered, 1 garlic clove minced, ¼ TS oregano, ½ TS thyme, 1 TS culantro chopped, 1 TS salt, 1 chicken bouillon cube, 3 cups water, 2 cups coconut milk

Method: Heat olive oil in a pot. Add the onion, garlic, oregano, thyme, salt, culantro. Sauté for 1 minute. Add the boiled pigeon peas and cook over medium heat for a minute. Add the chicken bouillon and 3 cups of water. Stir and bring to a boil over medium heat. Mash some of the pigeon peas with a potato masher. Allow to boil about 15 minutes until about half of the water is gone. Add coconut milk and simmer 20 minutes until the mixture reaches a creamy consistency.

PIGEON PEA CURRY

Ingredients: 2½ cups pigeon peas -fresh, frozen, or dry soaked overnight, 2 colocasia/eddos/taro peeled and chopped, or potato, 2 TBS coconut oil, 1 onion diced, 4 garlic cloves minced, 3 chili peppers (size depends on taste), 2 TBS culantro chopped, ½ TS cumin seeds, 2 TBS curry powder, ½ TS fresh ground black pepper, 3 cups water, 1 TS salt.

Method: In a suitable pot, bring the pigeon peas to a boil and cook for half an hour. Heat the coconut oil in a wok or large frying pan and add onion and garlic. Fry for 2 minutes over medium heat. Lower heat and add, culantro, cumin seeds, hot peppers, and black pepper and cook for 3 minutes before adding curry powder. Cook for 4 minutes and colocasia/eddos/taro pieces. Increase heat to medium and stir in pigeon peas. Add water and salt. Bring to a boil stirring and then reduce heat and simmer until tender. Mash taro to thicken. Enjoy on rice.

NAMES: Indian dhal, thuvarai and thuvaram paruppu; in Jamaica: tur or gungo peas; in Africa: the congo pea

PURPLE LONG BEANS
Vigna unguiculata subsp. *Sesquipedalis purpura*

The long bean is one of the few vegetables that I devoted two separate chapters. Purple long beans are a type of cowpea and the seeds are available online. This variety contains anthocyanin, a phytochemical that colors these beans deep purple. These will grow 30 inches, cylindrical, with a grooved, purple skin. These beans taste better harvested when they reach 30-50cm (12-18in.), with about 15 seeds per pod. Softer pods and immature seeds are tender enough to be eaten raw. Longer beans taste better cooked. They have a sweet taste similar to green beans, but a stronger flavor than the usual green yard long beans. These plants and seeds aren't similar to green beans. Grow some of these nutritious purple beans at home. Stun friends and neighbors with purple beans and sell a new bean in the market.

Long purple beans are native to southern Asia and will grow all year throughout the 50th state. Other names are purple-podded asparagus bean, purple snake bean, red noodle beans, and China long purple beans. Botanically known as *Vigna unguiculata* subsp. *Sesquipedalis purpura.* (*Sesquipedalis* in Latin means foot and a half – *purpura* is purple.) This genus of bean has more than 200 species native to the tropics Purple and green long beans will grow where hot temperatures stifle regular green-snap beans.

HOW TO GROW: This plant grows as a vine and should be trellised or provided poles and strings for support. Easy to grow, the purple long bean will do well even in dry areas prone to drought. It's best to soak the seeds for a day before planting. Fork the soil into mounds. These beans do best with a pH between 6.0 and 7.5. Plant two seeds per hole - 2 inches deep. They do not require much water, but a prolonged dry spell with cause less bean production, and those that grow will be tough. Seedlings should be visible within 2 weeks. For a constant supply of beans, plant more every 2 weeks. These beans don't require much fertilizer. Keep the plants off the ground; train the plants to climb the supports. They will grow tall before blossoming. After 60 days, the purple long bean will produce two bean pods per flower. They need to be picked at least every other day.

MEDICINAL: All parts of the purple long beans are used in Ayurvedic-homeopathic traditional medicine. The roots and leaves are boiled to make a paste or poultice to treat gout and skin irritations. Roasted seeds treat insomnia and increase brain power and memory. A decoction of the leaves treats most stomach problems.

NUTRITION: 100 grams of purple long beans have only 47 calories, 3g of protein, with 8g of carbs. These purple beans are extremely high in potassium, calcium, and amino acids with magnesium, phosphorus, iron, and zinc. It's also extremely high in vitamin A, with some C, B1. B2, B3, and B6.

FOOD & USES: Purple long beans can be used much like long green beans. In the market, check the freshness by snapping the beans in half. The crisper the snap, the better. Avoid wilted, discolored beans. It's best to refrigerate and use within four to five days after purchase, before the beans become soft. Purple beans should be boiled or steamed for a very short period and their color won't fade when cooked.

RECIPES: Use the long purple beans the same as green yard long beans.

DID YOU KNOW?

Africa and Asia annually consume 5 million tons of dry long bean seeds and that represents 30% of the total beans grown in the tropics.

WINGED BEAN – PĪNI ĒHEU

Psophocarpus tetragonolobus

The winged bean is a delicious vegetable, that will grow easily almost anywhere in the Hawaiian Islands. It's a superfood because it tastes delicious, a great source of protein, vitamins, and minerals, easy to grow, and restores nutrients to the soil. Like moringa, winged bean can help relieve the world's problems with malnutrition. The winged bean is excellent for nutrition, yet not well known outside of Southeast Asia. It's not globally cultivated and only grown on a small scale.

The pods are described as 'winged' because the beans are long, 6 inches, and about an inch wide. Instead of being round, they are rectangular and boxy, with four fancy curved edges. This legume veggie may have originated in Africa, New Guinea, Mauritius, or Madagascar. No one knows for certain as botanists have never found a wild variety. Like so many vegeta-bles, Arab seafaring traders, brought seeds to many lands. The botanical name is *Psophocarpus tetragonolobus*. Other names include four angle beans, four-cornered beans, Manila bean, dragon bean, or goa bean. These are also labeled asparagus pea because the young pods have a delicate taste similar, like asparagus. This is an easy vining bean to grow, and super nutritious.

HOW TO GROW: Winged bean is perfect for a small home garden. It grows anywhere with hot and humid tropical conditions, with plenty of water. It's best to soak winged bean seeds for a day before planting. I like the more exotic purple winged beans and they seem to take the longest to sprout. These beans will grow in almost any soil with a pH from 5-7, enriched with compost. Loosen the soil in a foot square, plant two seeds about an inch deep, keep the soil moist, and sprouts should appear in a week. Six plants will produce enough for one person. Most vines will fruit within three months. Winged bean plants are usually disease and insect resistant and can bear pods for 3-4 months and longer. They supply a steady source of protein, vitamins, and mineral-rich food year-round.

The winged bean is an attractive climbing vine with delicate white blossoms. The vines may reach four meters and must grow over a support. Weave the vine around a fixed string and as it grows, add horizontal strings. The winged bean provides good-looking landscaping against any wall, but the vines must be kept organized or they will grow in a clump and be less productive. This vine loves the full sun as long as the roots are kept moist. With enough water, the long roots protrude the surface and become thick, good for stir-fries.

Growing winged beans will restore nutrients to your garden soil. These vines grab ni-trogen from the air and move it into the dirt. It needs very little fertilizer. Just plant the seeds, run the strings, eat the beans. Occasionally, add dried cow dung (manure) to the water. The beans should be ready within 3 months after planting. This is a great vegetable because the entire plant is delicious and nutritious. It is also a very sun-sensitive vegetable. The plants don't flower unless the day length is short, less than 12 hours of sunlight. Winter months produce more winged bean. Winged beans should be flexible and shiny green when fresh.

Experiment picking winged beans at different lengths and tenderness. The beans ap-pear and mature in a matter of days. The 2 to 3inch long pods are very tender and can be eaten raw, whole, and best for salads. The longer pods are good to chop for stir-fries, soups, or grilled whole. Older beans should be boiled, no more than 5 minutes. Blanching is great for using bigger beans in salads; bring beans to a boil, drain, and put in ice water. This keeps the beans tender, but crisp. Older beans may darken in color, but taste great with more fiber.

MEDICINAL: Medical studies suggest winged bean may help the body fight infections of micro-organisms. A diet rich in winged bean will help prevent wrinkling of premature aging, reduce severe headaches, help reduce arthritis pain and swelling, boosts and protects the nervous system, provides antioxidants, cleanses the liver but may inhibit blood clotting. They may also produce kidney stones. They use concoctions of winged bean seeds and leaves in homeopathic medicinal treatments for ear infections, gout, and water retention. Leaves may be squeezed, the juice warmed, and dropped into an infected or painful ear.

NUTRITION: 100 grams of these beans when young, immature, contain only about 49 calories. Mature seeds have ten times more, 400 calories. If you're on a diet, eat the immature beans. Protein is high at 6% in all the parts of this plant: the leaves, roots, and beans. There's minimal fat with no cholesterol. These beans are packed with folates, niacin, thiamin, vitamins A and C, potassium, calcium, and copper.

FOOD & USES: The young leaves are excellent steamed or fried. The blossoms are a nice addition to rice. Stir-fried winged bean roots taste like cashews. The seeds can be dried, roasted, ground, and boiled to make an energy-rich hot drink. Raw seeds may also be made into a milk comparable to soy milk. It can be then prepared into tempeh.

RECIPES: WINGED BEAN SIMPLE STIR FRY

Ingredients: 4 garlic cloves of, 4 fresh hot chilies, 500 grams winged bean, 1 TBS fish sauce, 2 TBS sesame or virgin coconut oil, black pepper, and salt to taste.

Method: Use a mortar and pestle to grind the garlic and chilies into a paste. Wash the winged beans, trim edges, and chop into small pieces. Blanche the bean pieces by dropping them into boiling water for two minutes. Then, with a slotted spoon, immediately put the pieces in ice water. Let sit for five minutes. Each type of oil will produce a very different flavor. DO NOT BLEND THE OILS. Heat oil and stir in the garlic and chili paste. Once hot, add winged bean pieces and stir quickly for two minutes. Remove from heat. Add fish sauce. Add salt to your taste.

SIMPLE WINGED BEAN SALAD: Again, the choice of oil used will change the flavor

Ingredients: 400g wing beans, 2 TBS of one of these oils, sesame, virgin coconut, olive, or sunflower, 2 TBS vinegar – again, different vinegars have different flavors, rice vinegar (mirin), apple vinegar, coconut vinegar. Rice + sesame oil + rice vinegar, or olive + sunflower + apple vinegar, or coconut oil + coconut vinegar, one TS white sugar, sesame seeds

Method: Combine the oil-vinegar mixture you choose and stir in sugar. Wash, trim, slice beans lengthwise into very thin, ¼ inch strips, drop into a hot skillet or wok, stir for 2 minutes. Re-move to a bowl and combine with oil mix. Coat with sesame seeds. Add salt and black pepper to taste.

WINGED BEAN WITH CASHEWS

Ingredients: 500g tender green (not brown) winged bean, 25g cashews, 2 garlic cloves, 1 medium onion, 4 dried chilies, 1 TBS oil, salt, and seasoning to taste, 1 TBS sesame oil

Method: Wash and chop wing bean into ½ inch pieces. Drop into boiling water and cook for 3 minutes. Immediately drain and place wing bean pieces in ice water to blanch. Chop cashews, if raw stir for a few minutes in a heated frying pan. Heat oil and add chopped dried chilis, stir, then add chopped onion and garlic. Drain wing bean and combine all, season, and mix with sesame oil.

DID YOU KNOW? All the plant parts, seeds, tender, immature pods, young leaves, flowers, and roots are edible and consumed as food. Flowers are used to color rice and pastries. Winged bean flour can be used as a protein supplement in bread making. Leaves were used in a compound lotion for smallpox in Peninsular Malaysia.

Better beans and bacon in peace than cakes and ale in fear.~ Aesop

THE HAWAIIAN HOME GARDEN GUIDE
LEAFY VEGETABLES

Leafy vegetables are also referred to as greens, or leafy greens. They are plant leaves cooked or eaten raw as a vegetable. Although they come from a variety of plant families, most have nutrition, eating, and cooking methods in common. Leafy vegetables usually mean short-lived herbaceous plants such as lettuce and spinach that grow full cycle in a few months. However, almost one thousand species of plants have edible leaves. Anthropologists believe prehistoric humans consumed five pounds of leaves every day. Leafy vegetables were a major food source. Lettuce is considered among salad greens usually eaten uncooked, added fresh to tossed salads, providing color and great flavor. Leaf vegetables are among the most nutritious of vegetables when compared by fresh weight. They are also among the most productive garden plants relative to nutritional value per square foot of garden space. They are also dollar valuable because they grow rapidly, allowing several crops during a year.

Vegetables are low in calories, low in fat, high in protein relative to calories, high in dietary fiber, iron, calcium, and very high in vitamin C, carotenoids, and folic acid as well as vitamin K. Darker leaves have more vitamins A, C, and calcium. Leafy vegetables are ideal for weight loss and to manage weight on the long-term. Adding more green vegetables to a balanced diet increases the intake of dietary iber which in turn regulates the digestive system and aids in bowel health. The iber cleanses the intestines and removes many dangerous toxins.

Greens contain few carbohydrates, and those are packed in layers of iber, which make them very slow to digest. That is why leafy greens have very little impact on blood glucose. Eating green leafy vegetables may lower the risk for type 2 diabetes. The US Department of Agriculture recommends eating 3 cups of dark green leafy vegetables every week. Green leafy vegetables also serve to maintain eye health, aid in digestion, increase bone strength, and boost the immune system. A Massachusetts research study found that people over 50 who ate spinach, and other dark green, leafy vegetables ive or six times a week had about half the risk of age-related eye / retina problems than those who ate it less than once a month.

This chapter has some common leafy greens and some that may be unfamiliar. I've grown all; they're easy if you have enough water and plant where the afternoon tropical sun doesn't reach them. All are nutritious and delicious. The typical shelf life for most leaf vegetables is one week.

PAGE:

***Courtesy** 912crf.com*

AMARANTH – INDIAN SPINACH – 'ŌPAE INIA
Amaranthus viridis

Amaranth is a tropical leafy food plant that will reach 2 meters (6 ft.). As the plant grows, the leaves are harvested and cooked the same way as spinach. Amaranth originated in Mexico. The European explorers carried it the Far East. The name 'Indian spinach could relate to the Aztecs who grew it for seeds and leaves. It's also called alba or Indian Spinach. Today, amaranth grows in every tropical country worldwide. It's tasty and a good source for traditional medicines.

Its botanical name is *Amaranthus viridis*. Knowing its name is important because there are almost 100 different variations of the *Amaranthus species*. Amaranth is also known as Chinese spinach, but botanically, Chinese spinach in the USA is *Amaranthus tricolor*.

HOW TO GROW: Amaranth is another tropical weed-like vegetable that can be cultivated anywhere. This is a good plant for the home garden. With the proper cultivation, amaranth stands erect and has many leafy branches. It is usually about 45cm (18in.) tall, but can grow to a meter. Like most leafy vegetables, it prefers well-drained fertile soil with a pH from 5-7, in full sun. It can be grown from seeds or cuttings. In a nursery seedbed, planted ½ inch deep, the seeds should germinate within 2 weeks. In another month, they should be 4 inches tall and ready to transplant to their permanent position keeping 46cm (18 in.) between the plants. It cannot grow in the shade and requires regular water. As the leaves grow, snip the young off with a pair of scissors. Don't strip the plant of leaves at one harvest or it will die. Have at least 6 plants.

MEDICINAL: Amaranth leaves are used in poultices (fresh or as dried powder) to treat inflammations, and skin eruptions like boils and abscesses. Juice pressed from the leaf sap is an eyewash to treat infections and conjunctivitis. Juice from the root treats urinary inflammation and constipation. The entire plant is boiled to make a decoction to purify the blood.

NUTRITION: 100 grams of amaranth leaves has 280 calories with 32 grams of protein. The leaves are loaded with potassium, calcium, phosphorus, and iron with vitamins A, B1, B2, B 3, and C. Amaranth seeds are 15% protein and 5% fat. The seeds are edible with a nutty flavor and are nutritious eaten as snacks or added into baked goods. Boiling the seeds in water with coconut milk makes a delicious porridge.

FOOD & USES: Amaranth leaves can be cut at any time. Larger leaves have a stronger flavor, and the smaller leaves are tenderer. This is an excellent, nutritious substitute for spinach. Let the plants flower and wilt to harvest the seeds. Watch carefully for the first few flowers to wilt and brown. Then snip all of the flowers from the plant. Place them in paper bags to dry. Then the flowers must be threshed (beaten) inside a bag to release the seeds. Use water or wind to separate the seeds from their chaff. The amaranth plant is used to create yellow and green dyes. The ash from burning the plants is rich in potash and used to make soap.

RECIPES: COCONUT AMARANTH

Ingredients: 1kg (2lbs.) amaranth leaves washed, 2 cloves of garlic chopped, 1 large onion chopped, 2 tomatoes chopped, 4 cups of water, 2 cups coconut milk, salt to taste, and 1TBS oil
Method: In a large pot, boil the water, add leaves and boil for 10 minutes. Heat the oil in a frying pan and brown the garlic and onions. Add tomatoes and cook only until they soften. Add the drained amaranth leaves. Stir in the coconut milk and cook for 10 minutes. Season to your taste.

NAMES: African/Chinese/Ceylon/Indian spinach, Surinam amaranth, basella, green amaranth, rough pigweed; wild amaranth. *Amaranthus* comprises about 70 species; 40 are in the Americas.

The more you eat greens, the more you get lean. ~ Dr. Joel Fuhman

BOK CHOY - PAK CHOY

Brassica rapa subsp. Chinensis

Bok choy or pak choy is one of Hawaii's favorite, easy to grow, leafy greens. Botanically, *Brassica rapa subsp chinensis* is a member of the cabbage family that forms a small, elongated head with plump white stalks, dark green leaves with a slightly bitter taste. This leafy vegetable has many names. Cantonese Chinese call it bok choi or bok choy; the Mandarin Chi-nese call it pe-tsai. It is also called choi sum, celery cabbage, white cabbage, Chinese cabbage, or Chinese leaves. Bok choy originated in China, but migrated with the Chinese workers sent to S.E. Asia. Europe, Australia, and the Americas to mine gold and construct the railroads in the 1800's. When the job finished, the oriental workers began their own gardens and markets.

HOW TO GROW: Bok choy is easy to grow from either seeds or tray started trans-plants. The best soil pH is between 6 and 7 and definitely not lower than 5. Fork the soil soft, mix in some well-rotted manure (to help retain moisture), and prepare mounded rows. The raised mounds are necessary for drainage, but not higher than six inches so the soil doesn't become too warm. Use about a cup of limestone per row to combat bacteria and fungus. Place plants every six to eight inches. It grows best in direct sun if it's watered once or twice a day. The sun's heat is its biggest enemy causing it to wilt and die. We keep it under a shade cloth for the first 10 days. A high-nitrogen fertilizer mix can be applied lightly once a week. Cabbage worms and flea beetles are the biggest pests. Because the soil is always moist, fungus can be a problem. Bok choy should be ready to pick after a month. It is best to cut it with a knife, and use fresh, but it can last in the fridge for a week. When purchasing, check for leaves with no black or slimy spots.

Because of limited space in Singapore, commercial growers produce bok choy on towers that are stacked at 30 feet high. Singapore bok choy is small, but this green can grow to a foot tall.

MEDICINAL: Bok choy is high in vitamins A, B6, and C, beta-carotene, calcium, and dietary fiber and contains potassium, and iron. It is low in fat, calories, and carbohydrates. The high amount of beta-carotene in bok choy helps to reduce the risk of certain cancers and reduce the risk of cataracts. Bok choy is an excellent source of folic acid. Bok choy is used by traditional Chinese medicine to quench thirst, relieve constipation, promote digestive health, and treat diabetes.

NUTRITION: Bok choy is rich in vitamin C, fiber, and folic acid. All reduce the risk of various types of cancer. Bok choy has more beta-carotene than other cabbages with more potassium and calcium. A perfect food for dieters, one cup of cooked bok choy has only 20 calories, with no fat, but 3 grams of carbs and 3 grams of protein.

FOOD & USES: To prepare bok choy, first rinse thoroughly and shake or pat dry. To keep it fresh, don't wash bok choy until you're ready to use it. Young bok choy has a mild flavor and can be eaten raw while mature stalks are slightly bitter. This bitterness is transformed into a sweet creamy taste by cooking. It can be cooked whole, steamed, or braised. If the vegetable is mature, separate the leaf from the stalk as the stalks should cook longer. After about 2 minutes the stalks soften from the heat, then add the leaves.

Bok choy is a necessary ingredient in many Chinese recipes and almost any stir-fry. The stalks can be shredded and lightly sautéed. It is a great addition to soups or stews.

RECIPES: BOK CHOY SALAD

Ingredients: 1 bunch bok choy chopped into 3-inch strips, 2 TBS vinegar (preferably rice vinegar),

1 TBS soy sauce, 1 garlic clove minced, ½ TS of each sesame oil, canola oil, and dry mustard powder (yellow mustard can be substituted)

Method: Steam bok choy, rinse, and allow cooling. In a jar that seals mix all ingredients and shake well. Coat while tossing bok choy.

GREEN STIR-FRY

Ingredients: 1 bunch bok choy sliced to 1-inch strips, 1 TBS sesame oil, 1 TBS canola oil, 1 cup bean sprouts, 1 cup lablab beans, green beans, or fresh yard long beans cut into 1-inch pieces, 1 small onion chopped small, 2 garlic cloves minced, 1 TS cornstarch, 1 chicken bouillon cube, ½ TS soy sauce, ½ TS sugar, ½ cup water, salt, and spices to taste.

Method: Heat oils in a large skillet; add soy sauce, garlic, onion, and bok choy, sprouts, and lab lab/yard long beans. Stir for 5 minutes. Mix cornstarch, sugar, and chicken bouillon. Pour over vegetables stirring constantly.

BOK CHOY AND CULANTRO

Ingredients: ½ cup tomato sauce mixed with an equal amount of water (ketchup may be substituted), 2 garlic cloves minced, 1 TS minced ginger root, 4 cups bok choy sliced into 1-inch strips, 1 small green sweet pepper chopped, 1/4 cup green onions chopped, 2 TBS fresh culantro, 2 TBS fresh lime juice, 1 TS soy sauce, salt, and pepper to taste

Method: In a large skillet mix half the tomato sauce, the garlic, and ginger over medium heat for 2 minutes. Then add the remaining ingredients. Cook until bok choy is wilted.

BOK CHOY AND CHICKEN

Ingredients: 2 heads of bok choy washed and chopped into 1-inch strips, 1 pound chicken breast sliced into thin strips (approx. ½ by - inch or smaller), ¼ cup soy sauce, ¼ cup white wine, 2 TBS cornstarch, 2 TBS sesame oil, 2 TBS canola oil, 1 chicken bouillon cube dissolved in ¼ cup of water, 1 TBS minced ginger root, 2 TBS each of garlic and chives chopped fine, salt, and spices to taste.

Method: Mix wine, soy sauce, and cornstarch and cover chicken strips. Marinate for at least 2 hours in refrigerator. Over medium heat, heat the oil in a large skillet and brown chicken strips. Add ginger, garlic, chives, and stir for ½ minute. Add bok choy strips and fry for 1 more minute. Pork or beef strips may be substituted.

FRIED BOK CHOY—easy and delicious

Ingredients: 2 bunches of bok choy, ¼ cup toasted almonds, 2 cups oil for deep-frying, 1 TS white sugar, salt, and spices to taste

Method: Separate bok choy leaves and wash thoroughly. It is very important to pat very dry – otherwise it will splatter when fried. Roll individual leaves like a cigar and then slice into shreds. Heat oil and test temperature with one shred. Carefully drop in a large spoonful of shredded bok choy. Fry for only a few seconds, do not let leaves brown. Remove to drain on paper Finish frying and put bok choy in a bowl and stir in sugar and almonds.

DID YOU KNOW?

Bok choy can handle cold better than heat. Like kale, bok choy is a cool-weather crop but if it gets too hot, bok choy vturns stringy and bitter in areas. In Cantonese, bok choy means white vegetable. Bok choy is a symbol of prosperity and good luck. Bok choy is sometimes called a soup spoon because of the shape of its leaves.

Greens are the primary food group that matches human nutritional needs most completely. ~Victoria Boutenko

CABBAGE - KĀPIKI
Brassica maritime

Cabbage is another of Hawaii's favorite vegetables. It is one of the most widely grown vegetables cultivated by both small and medium scale farmers. Cabbage is one of the most popular and nutritious vegetables grown throughout the Hawaiian chain, cultivated mainly for the local market. China produces the most cabbage with 34 million tons.

Cabbage is one of the world's oldest edible greens. It's a hardy vegetable that grows best in fertile soils. It is believed cabbage originated along the shores of the Mediterranean. Three thousand years ago, Homer wrote of the Greek hero Achilles having cabbage.

Cabbage botanically is *Brassica maritime*, from the family known as *cruciferae* or cross bearers. There are many members including cauliflower, bok (pak choy, broccoli, kale, collards, Brussels sprouts, and kohlrabi. All members of this group have succulent, hairless leaves covered with a waxy coating. This waxy coating often gives the leaf surface a greenish-gray, or blue-green color. Green head cabbage is the most familiar throughout the world, but red and purple varieties are available. Seek cabbage seeds specificly to grow in the tropical heat.

HOW TO GROW: Cultivation requires a cool wet growing season and fertile soil. Hot dry weather will stunt the growth and quality of cabbage. The three types of cabbage are: green - the outside leaves are darker green while the inside leaves are smooth and pale green. Savoy cabbage has blue-green-purple crinkly leaves. Red cabbage is usually smaller and denser than heads of green cabbage. The flavor of red cabbage is slightly peppery.

Cabbage may be transplanted or seeded directly in the garden. Transplants are the better option or start your own nursery bed. Cabbage prefers a pH of 6-7. Space transplants 12 to 18 inches apart in the row. Close spacing produces smaller heads. Five days after transplanting use a starter fertilizer. When the plants are half-grown use a high nitrogen fertilizer. Keep the plants weeded. It is necessary to water cabbage throughout the entire growing season to help it survive the intense sun. Gentle spray from a hose will keep a cool head. Mature cabbage heads may split open if hit with a heavy rain after a long hot dry period. It takes about 70-80 days for the heads to grow to harvest-size, about 9 inches or more in width. The most common diseases that attack cabbage are yellow wilt and black rot. Both diseases are transmitted by seeds, transplants, and insects. This can be reduced by using hot water treated seeds. Worms hatched from white or brown butterflies cause extensive damage by eating holes in the leaves. Locally grown cabbage tends to have extreme pesticide residue. It is good practice to discard outer leaves. Grown your own for chemical free.

MEDICINAL: Since cabbage is high in carotene content, regular eating reduces the risk of some cancers. Sulfur in cabbage reduces the growth of tumors, removes toxins, and strengthens our immune systems. Cabbage is rich in vitamins and minerals and reduces the 'bad' cholesterol that hardens arteries. Eating cabbage first will prepare the stomach for a heavy meal and drinks. Crush a raw cabbage leaf in a mortar to use as a poultice on a cut. To fight acne, drink a cupful of the water cabbage is boiled in, or crush a raw leaf and use it as a poultice on oily facial areas for about 20 minutes. A boiled cabbage leaf applied while still very hot to the abdomen will cure a stomachache. Cabbage has a cleansing effect on the stomach and intestinal tract if consumed raw without salt due to its high sulfur and chlorine content.

NUTRITION: All cabbage types are low in calories and excellent sources of minerals and vitamins, especially vitamin C. Cabbage may reduce the risk of some forms of cancer including colon or rectal cancers. Cabbage is also high in beta-carotene, and fiber. 100 grams of cooked chopped cabbage has only 25 calories, 1 gram of protein, and is very high in vitamins C, K, and B-6, with manganese, potassium, calcium, and iron.

FOOD & USES: Cabbage is a great food because it stores well. One medium head (3 pounds) of green cabbage yields 9 cups shredded raw and 7 cups cooked. The upper half of the cabbage head is tenderer and shreds easier than the bottom. Raw cabbage must be eaten within a few days. Cabbage can be stored in the refrigerator for about 2 weeks. Fermented, shredded cabbage is called sauerkraut. Koreans ferment spiced cabbage for kimchee/kimchi.

RECIPES: CABBAGE ROLLS

Ingredients: 1 head of cabbage about 2 pounds - separate leaves and wash thoroughly, 2 onions, 1 bell pepper, 4 garlic cloves - all chopped small, 1 kg. /2 lbs. minced meat (beef or chicken), pepper to taste, 1 cup tomato sauce or ketchup, 2 TBS oil, and 1/2 cup water.

Method: Remove whole leaves from head and wash. Place leaves in pot of boiling water to wilt. This makes the leaves more lexible. Fry the minced meat in a pan until brown. Pour o any excess liquid. Add garlic, onions, and pepper to the meat. Spoon meat onto the wilted leaves. Roll the leaves, tucking in the ends to retain the minced meat mixture. You may need toothpicks to keep the cabbage rolls closed. Place inished rolls into a baking dish and cover with mixture of tomato sauce, or ketchup, and water. Cover and bake for 1 hour at 300 degrees.

SURFERS' STIR-FRIED CABBAGE

Ingredients: Half head green cabbage about 2 ½ lbs. shredded, 1 onion chopped, 1 medium bell pepper chopped, ½ hot chili pepper (optional), 2 garlic cloves chopped, 2 TBS olive oil, juice of ½ a lime, ½ TS turmeric powder, 8 curry leaves, 1 cup water, salt to taste.

Method: Wash cabbage; remove core, and shred cabbage thinly. Heat a large stainless-steel pan with lid over medium-high heat until hot. Immediately add oil, onions, turmeric powder, and bell peppers, and stir for about 1 minute. Add shredded cabbage and stir for another 30 seconds. Add hot pepper and garlic; continue to stir for 15 seconds. Do not allow garlic to brown. Add water, cover, and cook for 10 minutes. Stir occasionally to keep from sticking. Add more water if necessary. When cabbage is done, almost all the liquid will have cooked away.

SLAW WITH BUTTERMILK DRESSING

This slaw can be made using all green cabbage or any combination of green, red, or Savoy. Use 9 cups cabbage, thinly shredded, ½ cup grated carrots, 2 scallions chopped with green tops

Buttermilk Dressing: Use ¼ cup milk, half cup mayonnaise, 2 TBS vinegar, 1 TBS sugar, 2 TS grainy mustard, ¼ TS celery seed. Combine dressing ingredients in a small bowl and refrigerate.

Method: Mix vegetables together in a large bowl. Add dressing, toss using two spoons, refrigerate.

OCEAN MARINA'S APPLE CABBAGE

Ingredients: a 2 lb. head red cabbage, 4 large apples - cored and sliced, 1 cup raisins, 2 TBS vinegar (prefer apple), 2 TBS butter, ½ TS salt, sugar to taste

Method: Slice or grate red cabbage coarsely. Combine cabbage, apples, raisins, vinegar, butter, and salt in a sizable pot. Cover with water. Stir in sugar. Cook until cabbage is tender, stirring occasionally. Add more water if needed. Add small amount of vinegar at a time to increase sour flavor, as desired. Liquid should boil off when done.

DID YOU KNOW? Research shows that one half head of cabbage a day may help to prevent certain types of cancer. The chemical indole may prove to prevent breast cancer. Cabbage odors can be contained if you place a piece of bread on top of the cabbage when cooking in a covered pot. When you need cabbage leaves for cab bage rolls, freeze the whole cabbage first, let thaw, and the leaves will separate easier. To keep red cabbage red, try adding a tablespoon of white vinegar to the cooking water.

CHINESE CABBAGE – PĀKĒ KĀPIKI

Brassica rapa subsp. Pekinensis

Chinese head cabbage will grow everywhere throughout the Hawaiian chain if it's planted to mature during the cool winter months. Places of higher altitude can grow it continually. In native Hawaiian Chinese cabbage is *'pākē kapiki.'* The Chinese cabbage has large, oblong leaves tightly wrapped producing an upright-growing head. It was originally grown in the Yangtze River Delta region. During the 1500s, Li Shizhen, a naturalist, popularized it by bringing attention to its medicinal qualities. Its second name, napa cabbage, is probably derived from the Japanese word *nappa,* meaning leafy green. Other names are Chinese white cabbage, Peking cabbage, or celery cabbage. Chinese cabbage became a staple in Northeastern Chinese cuisine for making 'suan cai,' Chinese sauerkraut. This dish developed into kimchi in Korea.

This leafy vegetable is easy to grow and perfect for container gardening on the patio or balcony in at least 10-inch (25cm) pots. Chinese cabbage is a popular ingredient in Asian cuisines including stir-fries, noodle dishes, dumplings, rolls, and salads. Grow some because it's a perfect vegetable if you're feeding a family; it's nutritious, inexpensive, and filling.

HOW TO GROW: Plant this vegetable where it will not get direct sun for more than 8 hours each day. Plant in well-drained soil mixed with some compost. The opti-mum pH is 6.5-7. Chinese cabbage does not transplant well, so it's best to directly sow the seeds ½ inch deep and 4 inches (10cm) apart. Thin successful seedlings from 12 to 18 inches (30-45cm) apart. Space rows 18 to 30 inches (45-76cm) apart. These plants are sensitive to heat so cover them with a sunshade or if in containers, move them into the shade when the weather warms. Keep the soil evenly moist so that plants grow fast and stay tender. Slow growth can result in plants going to seed. Once a Chinese cabbage bolts, a flower stalk shoots up, that signals the end of leaf growth, and the leaves become bitter. Chinese cabbage can be attacked by flea beetles, aphids, and cabbage worms. The cabbage heads should be ready to harvest in 3 months.

MEDICINAL: Traditional Chinese medicine uses this type of cabbage to break fevers, regulate urination, treat the common cold, constipation, and coughs. When eaten regularly, napa cabbage is exceptionally low in calories, yet can make one feel full and is a good source of antioxidants like carotenes, which reduce LDL (bad) cholesterol. Its fiber content helps improve digestion. Chinese cabbage contains antibacterial and anti-inflammatory properties, such as glucosinolates, that aids people who suffer from digestion problems.

NUTRITION: 100 grams has 12 calories, vitamin C, with potassium, calcium, and magnesium.

FOOD & USES: Raw napa cabbage has a thin, crisp texture and a mild taste and can be eaten raw in salads. Cooked, it softens and gets sweeter, picking up other flavors from the food it's cooked with. Since it softens, it's frequently added to stir-fries and soups during the last stages of cooking.

When selecting napa cabbage, it should feel heavy. Buy heads with firm green leaves that are not wilted or look like they have been eaten by bugs. If you're going to be shredding cabbage for a recipe you can get about 8 cups of shredded cabbage from one medium 2-pound head. It will keep well in the crisper section of the refrigerator for about 3 days. The cabbage has gone bad when you notice brown spots on the leaves. At this stage, it will be bitter and should not be eaten.

Thoroughly rinse napa cabbage before preparing and allow it to drain. Remove the stem, cutting off the bottom inch of the plant. In addition to traditional Chinese recipes, you can also use napa cabbage to line a bamboo steamer. This will help prevent food from sticking to the bottom when cooking.

RECIPES: SUAN CAI - Chinese sauerkraut, Chinese kimchi, pickled Mustard Greens sour and salty, crunchy, and they are an ugly color, but taste good!

Ingredients: 1 head -1kg. of Chinese / napa cabbage, 2 TBS sea salt, 4 hot bird chilies chopped, 1 TBS dried chili flakes, 1 TS black peppercorns, 1 TBS white vinegar, 1 TS cooked rice, a large, sterilized quart/liter jar preferably with a screw tight lid. Make sure the jar is clean by filling it with boiling water.

Method: Combine 1 TBS salt with the chilies, flakes, and peppercorns. Clean the mustard greens very thoroughly. Rinse several times. Pat down the leaves with a paper towel. Rub the leaves with salt and air-dry for 1 hour to draw out the water. Tightly pack the greens into the jar, sprinkling the peppercorns/chili pepper between each leaf. Repeat until you reach the top, making sure to pack in tightly. Fill the jar with water and top with about a tablespoon of white vinegar. This prevents mold. Cover tightly. Seal the jar and store in a cool dark place for 2 weeks or until the leaves turn yellowish. This means they're sour. The pickling water can be reused for more mustard greens. Just pack in the leaves and sprinkle a little more of the salt-pepper mixture and again top with vinegar.

CHINESE CABBAGE STIR-FRY

Ingredients: ½ head of Chinese napa cabbage (225 g or 8oz.), bottom chopped off and the leaves chopped into strips, 122 g/4oz. of oyster mushrooms chopped, 1 medium onion chopped, 3 gar-lic cloves minced, 2 TBS oil, 1 TS oyster sauce

Method: In a wok or frying pan on high flame, add the cooking oil, and stir fry the onion and garlic until it's light brown. Remove half of the garlic and set aside. Add the napa cabbage strips and mushroom pieces and stir fry for about 1 minute. Try not to overcook the cabbage as the white stems should stay crunchy. Stir in the oyster sauce. Top with the fried garlic/onions. Serve immediately.

KAPALAMA CHINESE NAPA CABBAGE SOUP

Ingredients: ¼ head (112g/4oz.) of napa cabbage washed and chopped into strips, 30g/1 oz. of carrot chopped into thin rounds, ½ bunch of cilantro, 112g/4 oz. of noodles - glass/cellophane type preferred, 1 chicken bouillon cube with 2 cups of water, 2 TBS oil, salt to taste

Method: Heat the cooking oil in a wok or frying pan, and add the napa cabbage and carrot. Fry the vegetables for 1 minute. Stir in the stock and add the glass noodles. Return to a boil and simmer until the cabbage is soft and sweet. Stir in salt and add cilantro.

DID YOU KNOW?

The story goes that Empress Dowager Cixi of the Qing Dynasty fell gravely ill and was unable to eat, drink, urinate or move her bowels. She lacked energy, ran a high temperature and suffered respiratory problems. A monk ad-vised that the Empress only consume napa cabbage juice and soup and that is reputed to have saved her life. When the empress regained her health, she praised the napa cabbage as 'The King of All Vegetables.' A crop of Chinese cabbage was grown aboard the International Space Station in 2017. A unique, scientific plant growth device was used and enough was produced included for the crew to dine on, and the re-mainder for scientific study.

The garden is a ground plot for the mind. ~ Thomas Hill, 1577

GOTU KOLA - PENNY WORT
Centella asiatica

Gotu kola or pennywort is an essential veggie green for your Hawaiian home garden. This small perennial leafy green plant is similar on a genetic level to carrots and thrives in tropical wet areas. It's native to China, India, Indonesia, Japan, South Africa, Sri Lanka, and various islands in the South Pacific. Gotu kola is interchangeable with cilantro or parsley as a culinary vegetable, an excellent source of essential vitamins and minerals needed to maintain optimal health. It's perhaps more valuable as a medicinal herb that can benefit your health to reduce your stress level, improve skin conditions, increase long-term memory, and also may help with your sleep disorders. Gotu kola is also known as pennywort, spadeleaf, and tiger herb. Its botanical name is *Centella asiatica.*

HOW TO GROW: This evergreen plant features scalloped, inch-long leaves, like a spade on playing cards, with small pinkish flowers. The plant measures 8 inches tall by about 3 feet wide. It is green all year, in flower from July to August, and the seeds ripen from August to September. Gotu kola is hermaphrodite - has both male and female organs, and is pollinated by insects. The plant is self-fertile. It's perfect for gardens with moist soil and partial shade. Gotu kola is tolerant of most soil types, including clay, sand, and loam, and grows best in a pH of 5.1 to 5.5. Propagate gotu kola by seed planted 1 inch deep spaced 3 inches apart and keep the soil moist; do not allow it to dry out. Transplant the seedlings into different areas of your garden beds or to larger pots. Occasionally, side-dress with a pinch of high nitrogen fertilizer. It will wilt and yellow if not watered daily or the sun is too harsh, but will recover if quickly if soaked with water. Making a sun shade for the hot tropical afternoon sun helps. Wear gloves when working with gotu kola plants, as some people experience skin irritation after touching the leaves.

MEDICINAL: Gotu kola has been used for many centuries as a treatment for respiratory ailments and a variety of other conditions including fatigue, arthritis, memory, stomach problems, asthma, and fever. 10 leaves a day makes the doctor go away. Burmese monks expound that eating gotu kola daily can cure 99 diseases; it's amazing how this leafy green plant can promote health and fitness and slows your aging. Its antioxidant activity is recommended to improve skin conditions, heal wounds, reduces scars, wrinkles, and stretch marks. Prevents hair losses, helps keep your skin glowing. Organic Indian gotu kola supplements are usually of good quality. The rare Himalayan wild-growing varieties are supposedly the best.

Gotu kola herb has been studied by the scientific community. The majority of the claims of its healing and therapeutic properties, including extending a good, quality life, have been documented and validated. You just need to use it raw. For external for acne or boils, directly apply gotu kola juice in a poultice. For coughs and asthma, eat it as a vegetable. For difficulty in passing urine, drink gotu kola juice. The leaves are the most valuable part of the plant. It has remarkable benefits. It has the ability to treat depression, boost blood circulation, and protect the heart. Gotu kola enhances memory and nerve functions, which gives it potential in treating Alzheimer's disease. The positive effect on brain function also makes it an antidepressant and also treats anxiety and stress. It may treat the insomnia that accompanies these ailments. The anti-inflammatory properties of gotu kola may ease the pain of arthritis. In some cases, it can cause headache, nausea, and dizziness. Starting with a low dose and gradually working up to a full dose can help reduce your risk of side effects. It's recommended to take gotu kola for two to six weeks at a time. Be sure to take a two-week break before resuming use. **Don't use gotu kola if you are**: pregnant, breastfeeding, have hepatitis, liver diseases, skin cancer, diabetes, high cholesterol, have surgery within two weeks, or under 18 years old.

Grow your own gotu kola because it is known to absorb heavy metals or toxins in the soil or water in which it was grown. This poses a health risk given the lack of safety testing, be particularly wary of imported traditional Chinese remedies.

Make gotu kola preparations in your kitchen. To make gotu kola tea, dry the leaves in the sun, crush them into a powder and store the powder in airtight jars. To make infused oil, pick the leaves fresh, wash thoroughly, dry in the sun, and add to virgin coconut or olive oil in a glass bottle. Place in the sun until oil darkens. The more leaves used the better, but no more than one handful dried. Gotu kola oil is good to retain your hair with regular scalp massages. Also, make a paste by mixing dried gotu kola leaves with raw aloe vera gel and rubbing it onto your hair. These natural remedies, with drinking the tea, have restored hair. The entire plant is harvested at maturity and used for medicinal purposes, salads, hair care oils, and many other uses.

NUTRITION: 100 grams of fresh gotu kola has 39 calories and provides: calcium-171mg, iron-5.6mg, potassium-391mg, lots of vitamins A and C, and some B2.

FOOD & USES: Gotu Kola is not only used for medicinal purposes but also is a key ingredient in many Indian, Sri Lankan, Indonesian, Malaysian, Vietnamese, and Thai dishes. When purchased, the lily pad-shaped leaves should have a bright green color without any blemishes or discoloration. The stems are edible and similar to that of cilantro. It has a distinctive sweet and bitter flavor with fresh, clean aroma. Gotu Kola is usually bought fresh in a bunch. It will keep in the fridge for only a few days and then turn black. I store in the fridge with the stems in a cup of water. Cover with a plastic bag, and refrigerate. This way the leaves will stay green for up to a week. When using them fresh, it's best to wash the bunch with saltwater. This kills bacteria. It is also used to make Indian curries, Vietnamese vegetable rolls, and a Malaysian salad called '*pegagai.'* Fresh gotu kola can also be juiced and mixed with water and sugar to create the Vietnamese beverage *'nuoc rau ma.'* If chopped or juiced, gotu kola should be used immediately as it will quickly turn black.

RECIPES: KOLA KENDRA – Simple herbal porridge with many health benefits. Ingredients: 1 cup red/brown whole grain rice, 6 cups water, 1 TS fenugreek powder, 5 cloves of garlic peeled and chopped fine, 2 cups gotu kola-cleaned and washed with saltwater and chopped as fine as possible, 1/3 cup coconut milk, and salt (Note: if you want a thicker porridge use less water. The quantity of gotu kola may be adjusted for a stronger taste.) Method: In a suitable pot/wok with a cover on medium heat, combine rice, water, fenugreek powder, and garlic. Bring to a boil and simmer, stirring for ten min-utes, and add coconut milk. Stirring is necessary, so it does not stick and burn. Cook for 2-5 more minutes. Adjust to your taste with salt, sugar, and or pepper.

GOTU KOLA SAMBAL - one of S.E. Asia's most popular dishes Ingredients: 1 bunch gotu kola washed in salt water and cleaned, sliced thin, ½ medium onion sliced thin, 4 green chilies chopped, ¼ cup grated fresh coconut, lime, and salt
Method: in a bowl mix sliced gotu kola, onions, green chili, and fresh coconut. Add lime and salt to taste at the end. Serve while it's still fresh.

DID YOU KNOW? Gotu kola is known in many Asian countries as the longevity herb. The Daoist master and herbalist Li Ching-Yuen was a famous advocate and regularly used the plant. Ancient Chinese medical texts report that he died at the ripe age of 256 years old and attributed his longevity in part to meditation, exercise, and the daily use of gotu kola.

Herbs like goto kola are time-tested remedies with amazing health benefits, but there remains a lack of awareness even among gardeners.

JEWELS of OPAR – PINK BABY-BREATH
Talinum Paniculatum

Jewels of Opar is a unique leafy green native to Brazil and probably imported to the Caribbean by Portuguese-Spanish explorers. Its leaves are delicious in soups and salads, while Ayurvedic-homeopathic medicine uses the entire plant. Botanically, it is *Talinum Paniculatum.*

HOW TO GROW: Jewels of Opar is a perfect addition to any home garden or landscape. This plant makes an attractive border and looks great in a big patio pot. It's an erect perennial, with fleshy lime green leaves. The main stem will grow 2 feet tall, shrubby similar to basil. The stems remain soft and the flowers add another 18 inches. Each plant can have with up to 25 tiny pink flowers that become ruby-orange seed capsules. Jewels of Opar loves a lot of sun. This plant can withstand some drought, but like every green, it grows better with regular watering. The best growing medium is sandy well-drained soil with a pH of 6 to 6.5. Once you have this plant growing, it will reseed itself. Keep it thinned, or it can become invasive.

MEDICINAL: Jewels of Opar is a little-known wonder plant. It's known in the Chinese medicinal practice as Tu-ren-shen, and has been used to tone digestion, moisten the lungs, used topically to treat edema, skin inflammation, cuts, and scrapes, and promote breast milk. The juice soothes sore muscles. A decoction from the roots treats scurvy, arthritis, stomach inflammation, and pneumonia. Jewels of Opar is used extensively as a reproductive tonic. The swollen root is used as a substitute for ginseng. Herbal recipes use Jewels of Opar to increase vitality, treat diabetes, and restore uterine functions postpartum. Researchers are studying if eating or making a poultice of this plant can block neurons transmitting painful impulses making it a natural pain reliever. Other scientists are experimenting with large-scale Jewels of Opar plantings to absorb and destroy contaminants in the soil and groundwater. That's why you should grow your own. This plant absorbs available poisons and heavy metals.

NUTRITION: Every 100 grams of Jewels of Opar has only 65 calories with plenty of calcium (80mg), magnesium (61mg), and loads of potassium (300 mg). It's a significant source of omega-3 fatty acids and antioxidants. It contains such high levels of iron that scientists studying it developed the mantra, 'a leaf a day keeps anemia away.'

FOOD & USES: Young, small tender leaves can be eaten as a raw salad vegetable with a mild flavor and a hint of lemon. The larger, more mature leaves should be cooked. The shoots and leaves may be added to stews and soup as a substitute for spinach. Like spinach, this plant contains some oxalic acid. Anyone troubled by oxalic acid in food should parboil them first. When you pour off the cooking water, most soluble oxalic acid will be flushed away.

RECIPES: SIMPLE CUCUMBER AND JEWELS OF OPAR SALAD

Ingredients: 1 cucumber and 1 carrot peeled and diced, 1 bunch Jewels of Opar washed and sliced, lemon or lime juice, coconut or olive oil, salt, and pepper to taste.

Method: Combine the diced cucumbers (scrape off the seeds to prevent the salad from turning watery), carrot, small tender Jewels of Opar leaves, lemon juice, olive oil, salt, and pepper.

JEWELS OF OPAR CURRY

Ingredients: 1 bunch Jewels of Opar, 1 onion, 3 garlic cloves and 2 green chilies - all chopped, 1 TS turmeric powder, ½ cup coconut milk, juice of two limes, salt to taste, 2TBS oil for frying

Method: In a big pot with a suitable cover cook over medium heat stirring frequently. As you stir flatten/smash the Jewels of Opar. Add coconut milk, reduce heat, simmer, and stir. Add lime juice and combine everything. Enjoy with rice or noodles.

NAMES: Java ginseng, South American ginseng, pink baby-breath, fame flower

KALE

Brassica oleracea var. sabellica

Kale, also known as leaf cabbage, is a vitamin and mineral powerhouse. Because the center leaves don't form a head, it's considered a contemporary version of wild cabbage. Kale comes in many leaf shapes and colors, curly or fat, bluish, and light or dark green leaves. The usual home garden variety has bright green leaves with rough edges. Each variety has a different flavor. Because of its strong taste, kale was grown in many countries mainly as a subsistence food. Growing kale in warmer climates may make it bitter, if it gets too much sun and not enough water. No matter, there are many excellent, tasty kale recipes. This green leafy vegetable became popular as a nutritious health food during the 1990s. Kale contains more calcium than dairy products.

Kale originated 4,000 years ago in the eastern Mediterranean and Middle East. Curly-leaved varieties of cabbage already existed along with flat-leaved varieties in Greece in the 4th century BCE. Over a couple of millenniums, humans bred primitive leaf cabbage into kale, head cabbage, broccoli, cauliflower, Brussel sprouts, and kohlrabi. These vegetable cousins were not accidents of nature, but from humans tinkering and crossbreeding. Our usual kale was developed by the Scottish, the black or dinosaur variety was developed by the Italians. The Russians bred a variety that can withstand very cold temperatures.

HOW TO GROW: As with most greens, kale thrives in well-drained, heavily compost-ed soil with a pH of 5.5-7. It likes full sun in the early day and partial shade in the hot afternoons. Plants should be 6-8 inches apart and kept weeded with regular water. Lacinato, also known Tuscan or dinosaur kale, is reported to be a good edible variety to grow in the hot weather. The Chinese kale variety grows easy. There are also ornamental kales, so closely read your seed pack.

Kale grows best in cooler conditions, doing very well in higher elevation gardens. It should be ready to harvest within 2 months. It can be grown almost anywhere on the Hawaiian Islands. You do not want to dilute the powerful health benefits of your kale by spraying harmful pesticides on it. If you have a pest problem, use the organic hot pepper and garlic or tobacco solutions mentioned earlier. Kale will look good and grow well in porch pots. Harvest is easy. With scissors or garden shears snip off the leaves you need starting from the outside of the plant, they will soon be replaced with new fresh leaves. Always leave a few of the small central leaves attached to encourage growth. Carefully wash each leaf be-fore eating. Properly cared for, your kale leaves will provide healthy meals for months.

MEDICINAL: Kale is high in fiber and sulphur containing nutrients that may reduce the chances of several types of cancers. This green is high in folate, which is good during preg-nancy. Kale has phytonutrients, to reduce inflammation, detox the liver, and protect brain cells from stress. Phytochemicals are chemical compounds produced by plants, generally to help them resist fungi, bacteria, and plant virus infections, and wards off insects and other animals. When humans consume these plants as food, they make us stronger and more immune. Be advised; raw kale contains goitrins that may lower iodine levels and affect the thyroid gland, but they disappear when kale is cooked. Research demonstrates a moderate intake of goitrin-rich vegetables, including kale, is safe for most individuals.

NUTRITION: Kale is one of the best foods packed with vitamins and minerals. 100 grams has 49 calories, 4g of protein. It's got more than twice the daily requirement of vitamins A, K, and C with B-6 and manganese, calcium, copper, and potassium. Researchers compared raw kale with several cooking methods; all resulted in a significant reduction in total antioxidants and minerals, including calcium, potassium, iron, zinc, and magnesium.

***Leafy greens such as romaine lettuce, kale, collards, Swiss chard, and spinach are the most nutrient-dense of all foods.*~** Joel Fuhrman

Steaming it for a few minutes may be the best way to preserve its nutrient levels. Grow your own kale; it's one of the few super foods that's accessible to everyone, everywhere.

FOOD & USES: When you get ready to cook, remove the leaves from the tougher stalks. Look for dark, crisp leaves. You may find kale slightly bitter and there are recipes to manage the bitterness so all can experience nutritional health benefits of kale. Baby greens are not as bitter as mature leaves. Kale is also often added raw to smoothies, juices, and salads.

Soak kale leaves in a large bowl of water until dirt and sand begin to fall to the bottom, about 2 minutes. Lift kale from the bowl without drying the leaves and immediately remove and discard stems. Kale can be used raw or cooked. Young kale leaves add an earthy flavor to raw salad green mixes, and fully mature kale is one of the few leafy greens that doesn't shrink much when it's cooked. It's great sautéed, roasted, stewed, and even baked into Kale chips. Just be careful not to over-cook it, as it can become bitterer.

RECIPES: CRISPY KALE CHIPS

Ingredients: 1 head of kale, washed, dried, stems removed, and chopped into 1½ in. strips – about 3 cups, 2 TBS olive oil, sea salt (preferred), and spices, perhaps ground cumin, to your taste

Method: Preheat oven to 300 F. In a bowl, toss kale strips with olive oil, salt, and spices. Arrange on an oven tray. Bake until crisp – about 20 minutes. Flip at least once at 10 minutes.

EASY KALE STIR-FRY

Ingredients: One bunch kale, washed, stems removed and chopped into 1 in. strips, 2 garlic cloves chopped small, 2 TBS olive oil, 1 TS soy sauce, salt, and spices to taste

Method; Heat olive oil in a wok or frying pan over medium heat. Add garlic and stir until it gets the aroma. Add kale and cover. Cook, stirring, for 5 min-utes until kale is tender. Remove from heat and stir in salt, spices, and soy sauce.

KALE SMOOTHIE

Ingredients: 4 kale leaves, juice of half a lime, 2 large pieces of pineapple, 1 banana, pineapple chunks, fingertip-size piece of ginger, 1 TBS peanuts, piece of a ripe papaya optional

Method: Put all the ingredients into a blender add ¼ cup of water and blend. Add water or iceto the desired consistency.

NORTH SHORE KALE AND COCONUT

Ingredients: one large head of kale, stems removed and chopped into ribbons, 1 shallot sliced thin, 4 garlic cloves chopped, 2 TBS minced lemon-grass, 1 TBS olive oil, 1 TBS fish sauce, 1 TBS soy sauce, 1 cup coconut milk

Method: Heat the oil in a wok or frying pan over a medium heat. Sauté the shallot for 3 minutes. Add the garlic and lemongrass and cook until fragrant, about 1 minute. Add the remaining ingredients, reduce the heat to low, and cover. Simmer for 10 minutes stirring occasionally until the kale is tender. Serve with rice, noodles, or roti.

DID YOU KNOW? The name kale is a Scottish word derived from coles or caulis, terms used by the Greeks and Romans referring to the group of cabbage-like plants. In Scotland, kale is an important part of their traditional diet that in some Scottish dialects the word is synonymous with food. For example, to be 'off one's kale' is to feel too ill to eat. There are dozens of varieties of kale: lacinato, redbor, Gulag Stars, True Siberian, Red Russian, White Russian, Dwarf Blue Vates, Red Nagoya, Chinese kale, Sea Kale, and the six-foot tall Walking Stick Kale.

I love kale. I bake it. I sauté it. I juice it. I massage it. I want other people to join the cult-of- kale. ~ Misha Collns

LETTUCE – LETIU

Lactuca sativa

Lettuce is becoming more popular throughout the Hawaiian chain both in gardens and in salads. It can be a big moneymaking crop if grown properly with cooperation from the weather. Lettuce is a relatively new crop in the tropics since it does not flourish in the brutal sun, or during dry spells. Most gardeners are growing it in raised boxes under partial shade. Hydroponics, growing in water, is another successful method of cultivating lettuce. World production is almost 28 million tons. China grows 16 million tons, the USA 4 million tons, and India 1 million tons.

Researchers believe lettuce, *Lactuca sativa,* was first grown in the Golden Triangle of the Middle East at least five thousand years ago originally cultivated for royalty. Hieroglyphics in tombs of ancient Egyptians depict long leaf lettuce as food to travel with to the world of the afterlife. Romans and Greeks produced lettuce for salads for the nobility. Lettuce is recorded grown in China by 700 AD. In 1520, England's King Henry VIII gave an estate to a farmer who combined lettuce and cherries in a salad.

There are different types of lettuce; loose-heading lettuce is the most common in Hawaii. This type of lettuce, such as bronze miganette, is sweet, has loose leaves bushing out from a center stalk. Romaine grows a longer leafed loose head and is slightly bitter. Loose leaf lettuce is the easiest to grow. Loose-leaf and loose-heading lettuce are more nutritional because they have more leaves exposed to sunlight, which produces more vitamins. All lettuce is sensitive to heat and needs water. Gardeners must be certain to use water from an uncontaminated source.

HOW TO GROW: If you are a salad lover, your garden should have at least two rows of lettuce. Plant a row every two weeks. Work the soil with a fork until it is very loose. To develop properly, lettuce roots need soft soil. Local lettuce producers prefer to create a soil mix of compost and sharp sand. Lettuce can be planted in a partially shaded area, or in the shade of taller crops. In your seedling bed plant ten lettuce seeds per foot a half-inch deep. Space the garden rows at least a foot apart. Once the lettuce has sprouted, thin to eight inches between plants. It is wise to plant an additional row of lettuce every two weeks. That way you will always have a supply of fresh lettuce. Lettuce needs to be weeded regularly. Light watering daily will cause the leaves to grow faster. Too much water may cause a disease. Lightly fertilize with a nitrogen-rich mix once.

Under perfect, controlled conditions, such as in grow boxes, loose-leaf head lettuce can grow to harvest in twenty-one days. Usually, lettuce is ready to harvest in thirty days. For home use, cut every other head in the row so the remaining heads will expand. Because lettuce is a tender vegetable, it is also a very fragile crop. A change in the watering schedule will cause big problems. Too little water causes leaf tips burn, and too much water will cause leaf rot. The rows must drain adequately to reduce this problem. Aphids are a problem that can be cured with the proper insecticide. As lettuce is eaten raw, be very careful what pesticides are used, what strength, and when the poison wears off. Fungus is an enemy during the rainy season.

MEDICINAL: Eating lettuce is good for the nervous system. It is a good food for diabetics or anyone suffering from anemia since it is low in carbohydrates and with high iron content. Lettuce juice or lettuce cooked as soup is a natural remedy for insomnia. Eating lettuce has a tranquilizing effect. Try eating a lettuce salad with a tablespoon of olive oil before going to bed. Sweet dreams. Lettuce tea made from a half cup of lettuce to a cup of boiling water relieves stress and is a good body tonic fighting cold viruses, and works against asthma.

NUTRITION: A friend says lettuce is only crunchy water. He's right because lettuce is 95% water! Lettuce provides vitamins A and C, and minerals such as potassium, iron, and calcium. 100 grams of leaf lettuce has only 8 calories with 1 gram each of fiber, protein, and carbohydrates. The nutritional quality of lettuce depends on growing with minimal chemicals.

FOOD & USES: Lettuce is best stored in the coolest part of the refrigerator. Never keep lettuce close to apples, pears, or bananas. These fruits emit a natural ripening gas that will cause lettuce to quickly decay. Always wash lettuce thoroughly.

RECIPES: Lettuce has few recipes other than salads. However, lettuce soup is tasty.

LETTUCE SOUP – This is an excellent way to use old lettuce, but it may make you sleepy.
Ingredients: 1 head of lettuce chopped in 1-inch pieces, 1 medium onion chopped small, 1 carrot chopped small, 1 garlic clove minced, 1 quart of water, salt, seasonings to taste.
Method: In a large pot combine all ingredients and bring to a boil. Then simmer for 10 minutes. Add seasonings. Serves 4.

CITRUS SALAD DRESSING
Ingredients: Juice of 1 orange and 1 lemon, 2 TBS olive oil, salt, seasonings to taste.
Method: Combine all ingredients and either hand whisk or use a blender. Seal in a suitable container and this dressing will keep for a week in the refrigerator. If you want to warm this dressing, just place the jar in a pot of warm water for 5 minutes.

HONEY MUSTARD DRESSING
Ingredients: 1 TBS Dijon mustard, ¼ cup fresh chives chopped small, 2 TBS clear vinegar, 2 TBS honey, 4 TBS olive oil, salt, and seasonings to taste
Method: In a blender combine all ingredients and pour over your favorite lettuce salad.

BRAISED LETTUCE
Ingredients: ¼ cup olive oil, 1 head lettuce, 2 TBS minced onion, 2 TBS minced garlic, 1 TBS ground coriander, 1 cup water, 1 TBS white vinegar (prefer wine vinegar), salt, and spices to taste
Method: wash lettuce and trim off any undesirable spots, then slice the head in half - across the root. Heat olive oil in a skillet and add the lettuce. Cook until golden on both sides. Add garlic, onions, spices, water, and vinegar. Place skillet in a 350-degree oven for 15 minutes. Remove and let lettuce cool on a plate. Before serving reheat the liquid and pour sauce over the lettuce.

GRILLED LETTUCE
Ingredients: 2 heads of lettuce washed and cut lengthwise, 2 TBS soy sauce, 2 TBS brown sugar, 4TBS white wine, 2 TBS olive oil
Method: Blend soy, wine, sugar, and oil. Brush the lettuce halves with the mixture and place on a grill for 2-3 minutes. Turn the lettuce halves and brush again grilling 3 more minutes. Serve warm.

PEAS AND LETTUCE
Ingredients: 1 head lettuce, 2 TBS chopped chives, 1 TBS butter or margarine, 1 cup boiled pigeon peas, salt, and spice to taste
Method: Wash and cut lettuce into inch strips. In a skillet stir-fry chives and lettuce in butter for a minute, add boiled peas, spices and cover the skillet. Cook for 2 minutes until peas are heated.

DID YOU KNOW? Lettuce is part of the daisy family,*Asteraceae*. There are Egyptian hieroglyphic records of lettuce being grown over 6000 years ago. Ancient Romans gave it the name lactuca, from which the English lettuce is derived. Christopher Columbus introduced lettuce to America during his second voyage to the New World in 1494. In China, lettuce is typically eaten cooked and use of the stem is as important as the use of the leaf.

Lettuce is like conversation; it must be fresh and crisp,
so sparkling that you scarcely notice the bitter in it. - Charles Dudley Warne

MUSTARD CABBAGE
Brassica juncea

Mustard grows easily throughout the Hawaiian chain. The preferred type has wide stalks with the core leaves forming a slight head. You'll love mustard greens, and other tasty dishes that begin with frying mustard seeds. Mustard leaves or mustard cabbage originated in central Asia, in the Himalayas and was first cultivated in northwest India. It spread to western China, Burma, and Iran. Canada and Nepal grow more than 50% of the world's mustard. Bangladesh, Central Africa, China, India, and Japan also commercially grow mustard.

Mustard greens are the leaves of the brown mustard plant, *Brassica juncea* and it's a hybrid of black mustard, *Brassica nigra* and field mustard known as *Brassica rapa.* Its seeds are used to make the French-style brown mustards.

There are several types of mustard greens. Gai Choy known as Chinese mustard, Indian mustard, or head mustard grows from a large bulb. Red mustard has rounded purple leaves. Japanese mustard or mizuna has serrated feathery leaves. American mustard, Southern, or the 'curly' mustard variety is similar to curly kale. For centuries, mustard has been well known in India for its nutritive and medicinal values. The leaves as well as the seeds of this mustard variety are edible, and diverse medicinal uses of its seeds are also well known in many countries.

HOW TO GROW: Mustard cabbages do well in fertile, well-drained soils with a good moisture holding capacity. If drainage is poor, seeds should be planted in raised beds. The pH is best between 5.5 and 7.0. Florida broadleaf, giant red, or mizuna are good varieties to grow in the heated tropics. Plant seeds directly into the soil, spaced about 12 to 15 inches apart. At 3-4 weeks, thin to about 10 inches between plants. Water regularly or the mustard will grow slowly and be tough and bitter with tip burn. Water early in the day so the mustard is dries to prevent fungus. Spray every 2 weeks with a light insecticide like Malathion or Pestac.

Mustard plants are ready in less than 3 months. Plants can be thinned of leaves or harvested for food before they mature. Harvest before the flower stalk begins to appear. (Except for the flowering varieties.) Cut the whole plant at ground level and remove any damaged outer leaves.

MEDICINAL: Korea's kimchi made with mustard leaf is a functional food for nutrition and disease prevention. Mustard leaves, seeds, and mustard oils may prevent cancers. Mustard oil is a rich source of polyunsaturated fatty acids making it heart healthy. Mustard leaves could be an anti-diabesity vegetable (anti-diabetes-obesity). Mustard is one of the many Ayurvedic-homeopathic vegetables that treats patients with various health problems, including dementia and Alzheimer's.

NUTRITION: 100g of mustard leaves has only 34 calories including 1g protein, 2g fiber, with vitamin A, calcium, and iron. Recently, mustard has been clinically cultivated to produce other nutrients including selenium, chromium, iron, and zinc.

FOODS & USES: Seeds of this plant are commonly used in India, Sri Lanka, America, Japan, China and other countries and regions as a traditional pungent spice, a source of edible oil, and protein. Mustard oil is used for massages, and in cosmetics for hair control. The leaves and stems of mustard are edible, but the stems take longer to cook. Discard any tough, woody parts. Mustard greens will lose their bright green or purple color when cooked too long. It's better to blanch or steam mustard greens before stir-frying, sautéing, or pureeing.

Mustard greens have a spicy, horseradish-like taste. Young small, tender leaves will be milder. Bigger, older leaves can be bitter. Raw mustard greens lightly dressed with lemon juice, olive oil, and salt make for a refreshing, peppery salad. Stir-fry curly mustard in sesame oil with garlic and dried red chilies. Mustard greens are delicious in a soup or s stew with white beans or with miso. Make pesto using blanched curly mustard greens instead of basil. Serve blanched mustard greens the Japanese way, seasoned simply with soy sauce. Fermented mustard greens are a popular, nutritional side dish throughout Asia known as dua cai chua in Vietnam, gundruk in Nepal, and kimchi in Korea.

RECIPES: SARSON KA SAGG – INDIAN MUSTARD GREENS - a North Indian soup made with fresh mustard greens and other green leafy vegetables. *Sarson* translates to mustard greens, and *saag* translates to creamed spiced greens.

Ingredients: 2 bunches of mustard greens -1 lb., 1 bunch spinach– ½ lb., 4 small red chili peppers, all previous washed, hard stems removed, and chopped, 5 garlic cloves minced, ¼ TS turmeric powder, ¼ cup corn meal or corn flour, 1 ½ TBS dried fenugreek leaves, salt to taste, 2 cups water

Tempering: 1/4 cup neutral oil, 3TBS ghee, butter, or margarine, 1 small onion or 2 scallions chopped small, 3 TBS ginger minced, 2 TBS cilantro or culantro.

Method: In a wok or big pot, add 2 cups of water and heat over medium. Add the mustard greens, spinach, green chili peppers, garlic, with turmeric. Reduce heat to low and simmer an hour occasionally stirring. Remove from heat, uncover, and stir the saag, adding the minced garlic. Using a whisk, blend into a rough purée. Remove any missed stems. Return to medium heat and stir in the corn flour and cook for 5 minutes to remove the floury taste. Then stir in the dried fenugreek leaves. Add salt to your taste. Reduce the heat to low, add ½ cup water. Stir occasionally; add more water if necessary, to thin it.

Tempering: Heat a medium saucepan over medium heat. Add the oil, ghee, and onion. Sauté until the onion is golden – about 5 minutes. Stir in minced ginger and sauté for another min-ute. Stir this oil mixture into the mustard soup. Garnish with cilantro. Serve with roti, tortillas, noodles, or rice.

GOOD OLD MUSTARD GREENS

Ingredients: 1lb. mustard greens - washed and chopped, 1 large onion sliced thin, 2 garlic cloves minced, 3TBS chicken broth, 1 TBS olive or canola oil, ½ TS dark sesame oil, salt, and pepper to taste

Method: In a pot over low heat, sauté the onions in the oil until they brown about 5 minutes. Add minced garlic and stir for another minute. Add mustard greens and broth. Cook, continually stirring until mustard just begins to wilt. Toss the greens with sesame oil and season with salt and pepper.

DID YOU KNOW? Mustard could have been humans' first condiment. Egyptian pharaohs stocked their tombs with mustard seeds to accompany them after death. Romans were the first to grind the spicy seeds into a spreadable paste and combine it with a flavor ful liquid—usually, wine or vinegar. French monks, who mixed the ground seeds with 'must,' or unfermented wine, inspired the word mustard, from the Latin *mustum ardens* which translates to 'burning wine.' The Greeks used a poultice of mustard seeds to draw the poi son from scorpion stings and used mustard paste to ease pains and aches. In modern times, mustard helps to lose weight, relieve asth ma, grow hair, strengthen the immune system, treat skin problems, and lower cholesterol. Dijon mustard gets its unique taste from using the acidic juice of unripe grapes rather than plain vinegar.

NAMES: *Brassica juncea,* is also known as Indian mustard, Chinese mustard, oriental mustard, leaf mustard, or mustard greens.

The best fertilizer is the gardener's shadow. ~ Unknown

SALAD GREENS

Salad greens are wise choices for nutritional plants for the home garden. Varieties of greens provide many fresh salad choices and can be added to soups and stir-fries. These various greens can be valuable to hotels and restaurants. All these greens will grow anywhere in the Aloha State.

Man has eaten greens for thousands of years. Most greens started cultivation around the Middle East and the Mediterranean areas. The Egyptians, Greeks, and Romans bred the plants we enjoy in today's salads. Grow a few of each type and see what tastes best.

HOW TO GROW: All leafy salad greens are grown in the same manner as lettuce or spinach. Depending on your space and how many you choose to grow, they may be started by seeds in porch pots or directly into your garden. All leafy greens do not do well in extremely hot weather and require regular watering. The soil should be 5.5 - 6.5, with plenty of compost worked in along with some well-rotted manure. Plant 6-10 inches apart. Most leafy veggies like full sun in the morning and at least partial shade in the afternoon. All enjoy elevations higher than sea level.

As the season heats up, mulching around the plants is a good procedure to keep the roots cool. There are varieties of every green that botanists have developed to grow in the tropics. Use a high nitrogen fertilizer sparingly when the leaves reach 2-3 inches. Most leafy greens can have the outer leaves cut with scissors for eating and the plant will regenerate and keep producing nice tender greens every week. Carefully wash each leaf before eating. Properly cared for, your greens will provide a sustainable amount of healthy leaves for months or longer. Inspect your plants daily for insect damage. You do not want to dilute the powerful health benefits your greens by spraying harmful pesticides all over it. Use organic pest chasers like pepper and garlic water or tobacco water. Time your crops and have these greens ready to harvest just as the heat begins.

CHICORY: Chicory leaves (*Chichorium intybus)* are eaten similar to celery. The buds and roots are boiled and used as a spice. 100g is 30 calories with 1g protein and vitamins A, C, and K. It has calcium, magnesium, and phosphorus._

ENDIVE is an annual plant grown for salad leaves like lettuce and botanically *Chicho-rium endivia*. The green leaf is oblong, curled, and fringed. Heads are ready to harvest in 3 months. Endive has 17 calories per 100grams 1g protein, with vitamins A and C, potassium, and calcium.

ESCAROLE - (C*hichorium endivia latifolium*) is another type of green, less bitter, with broad, flat leaves. Harvest in 3 months when the leaves are 6 inches. It has 69 calories in 100 grams with vitamins a and C with potassium and iron.

RADICCHIO is also known as red chicory: (*Cichorium intybus var. foliosum)*. Heads are ready when they feel firm in about 2 months. Harvest quickly before it gets bitter. Radicchio will last in the fridge 3 weeks in a perforated bag. 100 grams has 23 calories with some vitamin C, potassium, and iron.

ROMAINE LETTUCE is known as cos lettuce (*Lactuca sativa L. var. longifolia*) and is more tolerant of the heat than other greens. This is the usual lettuce served in Caesar salads. It's usually ready in 2 months when it's 6-8 inches tall. Per 100 grams it has 17 calories, 1g protein, vitamins B-6 and C with potassium and iron.

Leafy vegetables are good inexpensive nutrition and will treat a variety of ailments as abdominal and bladder pains, rheumatism, and coughs. They are best when you grow your own.

SESSILE JOYWEED
Alternanthera sessilis

Sessile joyweed is found throughout the Hawaiian Islands as an ornamental ground cover. It is an aquatic plant grown in ponds and marshes and is cultivated worldwide in tropical and subtropical countries. It's a perennial loaded with manganese and vitamins A and C. Eating the bitter-sweet leaves, stems, and flowers is nutritious. Easy growing, the purple base stems with the attractive, shiny white flowers, make it perfect for the home garden as an ornamental hedging plant because as it grows longer, roots will form at the nodes. It can be invasive.

As an aquatic plant, sessile joyweed is native to Central and South America and was spread everywhere by the Spanish and Portuguese explorers. For a green source of vitamin C, they kept it growing in dark moist areas of their ships.

HOW TO GROW: Use seeds if you can find them. I bought some mature sessile joy-weed at the market and put the stems into a glass of water near a window where it would get partial sunlight. Roots formed in less than a week. I transplanted to a grow box that gets only partial sun, and kept the soil moist. As it lays prostrate, and grows longer it will sprout many branches with nice, fleshy leaves. This is a tasty green, sort of a miniature spinach that lays flat. Snip off what you need for dinner. Sessile joyweed grows like a weed. Just keep it moist and not too much direct hot sun. Designate a growing area because if you are successful, it can take over. The white blossoms are 2.5mm long and are excellent for bees.

MEDICINAL: Traditional homeopathic medicine uses sessile joyweed as a diuretic and laxative. When it's made into a broth or tea, it has cooling properties. It has been used to treat difficult urination, skin diseases especially with burning sensations, diarrhea, and indigestion, poor breast lactation with new mothers, enlarged spleen, fever, and hemorrhoids. A paste made from the leaves helps wounds heal quicker. The plant is rich in vitamin A and beta carotene and is believed to be beneficial for improving vision. This plant is an ingredient in hair oils and eye cosmetics.

NUTRITION: 100 grams of sessile joyweed has only 75 calories packed with magnesium and potassium. It is also a good source of vitamins A & C and Beta carotene.

RECIPES: SESSILE JOYWEED AND COCONUT

Ingredients: 300g washed sessile joyweed, 1 small onion, 2 green chilis – all chopped, 100g fresh coconut grated, juice of 1 lime, ¼ TS red chili powder, salt to your taste, and 1 TBS oil for frying

Method: Inspect the green and select the less mature stems and leaves. Thoroughly rinse the sessile joyweed leaves and then chop in manageable sections and then chop as small as possible. In a saucepan with the oil, combine sessile joyweed leaves, chopped onion, grated coconut, chilies and/or chili powder, and salt. Heat, stirring over medium heat for 5 minutes and ready.

SESSILE JOYWEED STIR-FRY

Ingredients: 2 cups washed and chopped sessile joyweed, 1 large onion chopped small, 2 garlic cloves chopped, ½ TS each of cumin seeds, mustard seeds, turmeric, 1 TBS each chickpeas, black gram, 3 dry red chilies chopped, ¼ cup grated coconut, 10 curry leaves, salt to taste, 1 TBS oil

Method: Heat oil in a wok or large pan and roast cumin, mustard seeds, and both beans. Then add chili pieces, onion, and curry leaves. Fry until onion browns. Stir in turmeric. Add sessile joyweed leaf and cover. Cook on medium for 5 minutes. Remove from heat and add coconut. Enjoy.

DID YOU KNOW? Other names for sessile joyweed are, water amaranth, and dwarf copperleaf spinach. In Hindi, it is gurdid, koypa, ponnanganni, matsyaakshi, and gudari saag

My salad days, when I was green in judgment. ~ William Shakespeare

SPINACH - SPINAKA
Spinacia oleracea

Spinach grows well throughout the Aloha State, but needs regular water and if it's commercially produced it requires irrigation. The botanical name of the leafy vegetable universally named spinach is *Spinacia oleracea* and it belongs to the same family as beetroot. It was first cultivated in southwestern Asia over two thousand years ago. Arab traders carried spinach to Persia, now Iran. Irrigation was necessary to grow this green leafy vegetable in a hot dry climate. Centuries later, Arab traders spread spinach to Europe and to China where the name still translates as 'Persian green.' The Italians take some credit for civilizing spinach. When an Italian countess, Catherine de Medici, (who was from Florence, Italy) married the King of France, she came with her own cooks who prepared spinach 'her way.' Since then, spinach dishes have been referred to as 'a la Florentine.'

HOW TO GROW: Spinach grows in most climates where there is sufficient water, especially in sandy soil. The best pH is 6-8. Spinach is difficult to transplant, so it's best to plant directly into the soil. Fork a row about ten inches deep and 6 inches wide to make a 4 -inch mound. Plant the seeds a half inch deep. Thin the sprouts to 6 inches apart. Keep watered so the soil remains damp and spinach should mature in 3-4weeks. Spray with a mild insecticide such as Pestac or Malathion every 2 weeks. Fertilize with a high nitrogen mixture every 10 days. Harvest when leaves reach the desired size. The whole plant can be pulled or cut at the base. Single leaves may be picked from plants one layer at a time, giving inner layers more time to develop. Don't wait too long to harvest because the bigger leaves will become bitter. Indian summer is a good variety to grow that will withstand the heat of tropical summers. Always wash leafy greens well before using then in the kitchen.

MEDICINAL: In the cartoons, Popeye gets extra strength by consuming spinach. Eat more spinach to reduce your risk of age-related health problems as muscular degeneration, cancer, heart disease, and neural tube defects. Lutein and zeaxantin are two carotenoids supplied by spinach that help keep your eyes healthy. Carotenoids and the antioxidant vitamins C and E in spinach are also believed to reduce the risk of cancer, heart disease, stroke, and cataracts. And the healthy dose of potassium and calcium found in spinach can help regulate your blood pressure. Spinach and other leafy greens also provide folic acid, which is known to reduce the risk of heart disease. Spinach may even improve your memory.

NUTRITION: One cup of fresh spinach has only 40 calories and has twice the daily requirement of vitamin K. This vitamin is essential to keep human bones healthy. Spinach is also a great source of vitamins A, C, magnesium, and folate. Eating this green prevents cholesterol from blocking arteries causing heart attacks or strokes and reduces high blood pressure. Spinach may be a rival with carrots for benefiting eyesight by keeping the eye muscles strong and reducing the incidence of cataracts. One cup of boiled spinach provides a third of the daily requirement of iron. Iron is necessary for bone growth.

RECIPES: GARLIC SPINACH

Ingredients: 1½ pound spinach-cleaned and trimmed, 10 cloves garlic minced, 2 TBS olive oil, 1 TBS, butter, 2TBS fresh lemon juice, salt, and spice to taste

Method: Mix minced garlic and olive oil. In a large skillet, melt the butter and heat this mixture adding spinach until it is just beginning to wilt (four to five minutes). Place spinach in a bowl and mix with salt, spices, and lemon juice.

CURRY SPINACH

Ingredients: 1-pound fresh spinach, 6 garlic cloves minced, ¼ cup tomato paste (or ketchup), ½ TS turmeric, 1 TS coriander powder, ½ TS cumin powder, ¼ cup water, salt, and spices to taste

Method: In a large skillet heat oil add garlic and tomato paste, stir over medium heat for 2 minutes. Add spinach and seasonings, stir well, cover. Add water and reduce heat to low. Cook for 5 minutes. Serve warm.

AFTER THE LAST WAVE SPINACH PIE

Ingredients: 2 pounds fresh spinach (or bok choy), 2 garlic cloves minced, 1 medium onion chopped, 2 TBS olive oil, 4 eggs, one TS lemon juice, ½ pound feta cheese, ½ pound grated cheddar, ¾ cup whole milk, salt, and spice to taste, one pie crust

Method: In a large skillet cook onion and garlic in oil until just browning. Add spinach until it wilts. Blend in all ingredients except cheddar cheese. Pour into pie shell. Bake at 350 degrees for half an hour. Cover with grated cheddar and return to oven for 5 minutes.

SPINACH ROCKEFELLER

Ingredients: 2 pounds fresh spinach, ¾ cup breadcrumbs, 2 TBS olive oil, ½ cup butter or margarine, ½ cup grated Parmesan (cheddar may be substituted) 2 eggs beaten, 2 cloves of garlic minced, 2 small onions minced, 1 small hot pepper seeded and minced, 1 large firm tomato sliced, salt, and spice to taste

Method: In a large frying pan sauté spinach in oil until it wilts, add breadcrumbs, butter, garlic, quarter cup cheese, onions, pepper, and spices. Stir to keep the mixture from sticking. Cook for half an hour before removing from heat. Arrange tomato slices in a large baking dish. Spoon spinach mixture on each slice. Cover with remaining cheese and bake for five minutes at 350 degrees

SPINACH POTATO TORT

Ingredients: 2 pounds fresh spinach, ½ cup breadcrumbs, 2 large potatoes, 2 TBS butter or margarine, 1 TBS olive oil, 1 medium onion chopped, 2 garlic cloves minced, 2 eggs, ½ cup grated Parmesan cheese (cheddar may be substituted), 2 sweet peppers-cored and sliced, 2 packages sliced turkey breast, ½ cup grated Mozzarella cheese (or cheddar cheese), salt, and spice to taste

Method: In a pot, boil potatoes until tender. In a large skillet sauté onions and garlic. Add spinach until wilted. Drain all excess liquid before adding beaten eggs, cheese, and ¼ cup bread crumbs. Grease a large baking dish with butter and dust with the remaining breadcrumbs. Slice the potatoes and cover the baking dish's bottom. Spread half the spinach mixture evenly over the slices. Then spread half of the peppers, turkey, and cheese. Repeat again beginning with potato slices, spinach, peppers, and turkey. Top with a layer of potatoes brushed with olive oil and seasonings. Bake at 400 degrees for half an hour.

DID YOU KNOW?

Spinach has an undeserved reputation for being high in iron. In 1870, Dr. E von Wolf measured the iron content of spinach, but placed the decimal point in the wrong position so the iron content of spinach was overstated by ten times. Sixty-seven years later the mistake was discovered, again by German chemists. Iron and calcium in vegetables are not usually fully absorbed by the human body. Spinach contains a chemical called oxalic acid, which binds with iron and calcium and further reduces the absorption of these minerals. To improve iron absorption, spinach should be eaten with vitamin C-rich foods such as orange juice, tomatoes, or citrus fruit.

The answer isn't another pill. The answer is spinach. ~ Bill Maher

SUNSET HIBISCUS – TONGAN SPINACH

Abelmoschus manihot

Living among the Hawaiian Islands there are a great variety of nutritious greens to enjoy at the dinner table. Sunset hibiscus, known as bele, or edible hibiscus, is a towering, shrubby perennial native to China that spread through India, Indonesia, to northern Australia. This is one of the easiest greens to grow. This plant produces many young shoot tips and succulent young leaves that are supremely nutritious, eaten fresh or added to soups and curries. The leaves can be as large as a slice of bread and on sandwiches, one leaf can be used instead of lettuce. Botanically, this leafy green vegetable is *Abelmoschus manihot*.

HOW TO GROW: In English, sunset hibiscus' name is sunset musk mallow and is a large perennial herb that grows up to 6 ft. tall and can spread 3 ft. wide and are good for a container crop. This plant will grow in the lowlands up to 2000m above sea level and thrives in the rainy season. Edible hibiscus must be propagated by cuttings. The same as planting cassava, watch which way the shoot is pointing and bury at least 2 nodes. Sunset hibiscus grows throughout the year providing a constant supply of highly nutritious shoots and leaves. Once it produces and you've trimmed it, leave the primary stem in the field and it'll grow a new shoot. It's best to use a 12-inch stem cutting or branch that isn't too young and not too old. Plant maintenance is generally done earlier when the stem is newly planted. Cuttings will begin to produce shoots in 2-3 weeks. They only require weeding and watering if it's a dry spell. As you trim the leaves, more young leaves grow. Keep the plant pruned to facilitate picking the shoots. At 3 months, harvesting should begin of sprouts in the low altitude gardens and about 6 months in the highlands. Young shoots are cooked or sold in the local market that day. Storage for more than 3 days will cause sunset hibiscus' leaves to wither and turn bitter.

MEDICINAL: For centuries in China, sunset hibiscus treats chronic kidney disease inflammations, oral ulcers, and burns. In the Pacific Islands and Indonesia, it treats kidney pain, lowers high cholesterol, eases childbirth, increases breast milk, and protects against osteoporosis. Scientific researchers have identified more than 120 phytochemical ingredients from the flowers, seeds, stems, and leaves, including: flavonoids, amino acids, nucleosides, polysaccharides, organic acids, steroids, and volatile oils. This plant acts as an anti-diabetic and treats kidney functions, an antioxidant, anti-inflammatory, analgesic, antidepressant, antiviral, antitumor, and has positive effects on cognitive functions, and reduces bone loss, both due to aging.

Cooked leaves, shoots are used throughout Papua New Guinea to treat colds, sore throats, stomach aches, and diarrhea. It's a good baby-food, because the young leaves are almost fibreless and are easy to puree after boiling.

Sunset hibiscus good health tea is prepared with a ¼ cup of fresh leaves chopped fine and steeped in 1 cup of boiling water. This tea treats sore throats, nausea, and mouth ulcers. Rubbed on the skin it reduces inflammation, cleans and helps to heal wounds, insect bites, and rashes.

NUTRITION: Sunset hibiscus is an extremely nutritious vegetable and contains 12% protein. Per 100g of edible leaves and stem tips has 150 calories contain, 4g protein and 1g fiber. It is a rich source of vitamins A and C, calcium, and iron. It contains the essential amino acids isoleucine, leucine, lysine, methionine, phenylalanine, threonine, valine, and histidine, which cannot be made by the body and must be obtained from foods.

FOOD & USES: Sunset hibiscus is usually steamed or boiled sometimes in coconut milk. Like pandan, it's used to wrap fish or chicken. It can be stir-fried and used to substitute for spinach.

The produce manager is more important to my children's health than the pediatrician.

~ Meryl Streep

TROPICAL LETTUCE - INDIAN LETTUCE (#2)
Lactuca indica

Lettuce is one of the most difficult greens to grow in the tropics for salads. There is a type of native, tropical lettuce similar to the dandelion. Known as 'tropical lettuce,' its botanical name is *Lactuca indica*. This plant is believed to have originated in China, Taiwan, and southern Japan. Chinese immigrants probably carried it through SE Asia. Now, it's been nat-uralized in North America, Europe, Africa, and Asia. It has been introduced either on purpose or accidentally to parts of South America and Australia. It's known as wild lettuce or one of the many milkweeds. The plant is harvested wild or cultivated for food and medicine, especially in Malaysia, Indonesia, and the Philippines. This plant is well suited for the Hawaiian climate.

HOW TO GROW: Tropical lettuce is started by cuttings or by seeds that are available online. This green is a perennial; when you cut it back, it will re-sprout. It has a long taproot and when cultivated with some organic fertilizer and regular water it can reach a meter tall and spread a meter wide with fleshy green blade leaves. The flowers resemble a wild daisy with white to pale yellow petals and bright yellow centers. Tropical lettuce will grow almost anywhere. It wants full sun in a well-drained sandy loam soil with a pH of 5-6, but can't be waterlogged. This plant is good for a window box or porch pot. Tropical lettuce likes a little high nitrogen fertilizer. The leaves can be trimmed repeatedly as they grow. Immature leaves will be sweeter than full-grown leaves. This plant matures in 2 months.

MEDICINAL: In traditional medicine, the leaves are boiled and steeped to make a tonic for stomach-digestive ailments and used for a detoxification regimen. The milky white sap contains 'lactucarium,' used for its anodyne, antispasmodic, digestive, diuretic, hypnotic/narcotic, and sedative properties. It's taken internally to treat of insomnia, anxiety, neuroses, hyperactivity in children, dry coughs, whooping cough, and rheumatic pains. Lactucarium is weak in young plants and stronger when the plant flowers. Commercially, the plants' heads are cut and the juice is scraped several times a day. **CAUTION: Never use this plant without the supervision of a skilled practitioner. Normal doses cause drowsiness, an excess will cause restlessness. An overdose can cause death by cardiac paralysis.** The sap is applied externally to remove warts.

NUTRITION: Tropical lettuce has 2% protein with some vitamin C, beta-carotene, riboflavin, calcium, and iron. The leaves also contain 6 antioxidative phenolic compounds.

FOODS & USES: Tropical lettuce is eaten raw, boiled, or steamed and used for a wrapping when grilling or frying meats, vegetables, fish, or deep-fried tofu similar to pandan. When used as a wrap, it is often dipped in a peanut sauce, black bean sauce, or a mixture of soya bean, lime juice, garlic, and green chilies. It can also be cooked and stirred into soups or steamed and mixed with rice, meat, and vegetables. This wild lettuce has a nice flavor when stir-fried. Eaten fresh, the leaves have a slight bitter after taste that can be hidden with various salad dressings. It will keep for a couple of days when stored in the refrigerator, but it is recommended to use immediately to preserve quality. It's cultivated to feed rabbits, pigs, poultry and even fish in Asia. The leaves can feed silkworms as a substitute for mulberry. Juice extracted from the whole plant is used as an ingredient in commercial cosmetic skin conditioners.

DID YOU KNOW? *Lactuca* comes from the Latin *actuca* or *lactucae* referring to the milky white sap produced when the plant is cut. *Indica* means Indian in origin used loosely to include Asia.

WATER SPINACH - MORNING GLORY
Ipomoea aquatica

Water spinach is a delicious leafy green grown in a moist medium for its tender shoots. Its botanical name is *Ipomoea aquatica*. Water spinach's origin is disputed. Many botanists believe it is native to China while others believe it originated in India. There is a case that the Chinese explorer, Zheng He, may have distributed this plant during his travels through SE Asia, Sri Lanka, India, and Africa. Other names are water morning glory, Chinese watercress, kang-kong in SE Asia, Chinese convolvulus, or swamp cabbage. This green is very nutritious. It's wise to grow this nutritious plant because many grow it in polluted water increasing the absorption of heavy metals. Buying from a market, wash water spinach extremely well.

HOW TO GROW: Moist clay soils rich in organic compost are the best to grow water spinach with a pH range from 5 to 7. Partial shade increases production along with a nitrogen-phosphorus-potassium fertilizer twice a month. Water spinach can grow in raised beds separated by irrigation ditches. Plant the seeds directly into the beds or create a nursery and transplant seedlings when they reach 4 inches. Keep 6 inches between the plants and water heavily every day. The crop is ready for harvest in two months. Pull the whole plant.

The wet method is a more accepted way to grow water spinach. This method uses lat ields surrounded by raised banks, like old rice paddies. The seedlings must be grown in a nursery bed. When the seedlings are six weeks cut foot long pieces to transplant into the soft mud-marsh paddy. Each 12-inch transplant should have at least 6 leaf nodes. Plant these into the mud-marsh 15 inches apart. Flood with an inch of water. The wet method provides several harvests; the irst in about a month. Cut the upper part of the main shoot at the waterline. This will cause the water spinach to sprout sideways and upwards, and you'll pick the fresh vertical branches weekly.

Water spinach blossoms are white with a purple center, trumpet-shaped, and an inch wide. Keep some lowers for seeds for planting. It naturally grows in waterways, and requires little care. This nutritious green is used in Indonesian, SE Asian, Filipino, and Chinese cuisines.

MEDICINAL: The young shoots are a mild laxative and used for fever. The leaves are crushed as a poultice on sores and boils. Ringworm is treated using a paste from the blossoms. The plant has shown it possibly regulates glucose uptake and reduces blood glucose levels.

NUTRITION: 100g of water spinach has only 19 calories with 2g of iber, and 2.6g of protein. This green is loaded with vitamins A, B1 (thiamin), B2 (ribo lavin), B9 (folate), with some B3 (niacin). Water spinach has calcium, iron, magnesium, phosphorus, potassium, and some zinc.

FOOD & USES: The leaves and young shoots may be cooked or eaten raw. Fresh, water spin-ach tips are added to salads, older leaves are cooked, stir-fried, steamed, or boiled with oil, onions, garlic, vinegar, and soy sauce for a few minutes and eaten in various dishes. It's cooked with fish and a sweet and spicy sauce.

RECIPES: EASY WATER SPINACH STIR-FRY

Ingredients: 500 grams of water spinach washed and chopped to 2 inches (5cm), 1 large onion chopped, 5 cloves of garlic chopped, 1 TBS fresh ginger chopped small, two green chilies chopped or 2TS chili paste, 1 TBS soy sauce, salt to taste, 1 TBS lemon or lime juice, 1 TBS oil

Method: In a large frying pan or wok coat with oil. Over medium heat fry garlic, ginger, and onion until just beginning to brown. Then add chilies, soy sauce and salt. Reduce heat and stir in chopped water spinach. Stir. Cover for two minutes. Mix in lemon or lime juice. Serve with noodles or rice. Chicken, pork, prawn, shrimp, or ish pieces may be added.

THE HAWAIIAN HOME GARDEN GUIDE

NUTS

The term nut is applied to many seeds that are not botanically true nuts. All nuts are seeds, but not all seeds are nuts. Nuts are both the seed and the fruit and cannot be separated. Seeds come from fruit, and can be removed from the fruit, like almonds, cashews, walnuts, and pistachios, which were once inside fruit. Any large, oily kernel found within a shell and used in food may be considered a nut. While a wide variety of dried seeds and fruits are called nuts, only a certain number are considered truly nuts.

Because nuts generally have high oil content, they are a highly prized food and energy source. Many seeds are edible by humans and used in cooking, eaten raw, sprouted, or roasted as a snack food, or pressed for edible oils and cosmetics. Nuts and seeds are also a significant source of nutrition for wildlife.

Nuts are cholesterol-free, except if processed with an oil. Unless salt is added to nuts, they naturally contain just a trace of sodium. Nuts can be used in many ways. Whole, flaked, ground nuts, and nut butters are widely available. Nuts can be added to sweet dishes, cakes, and biscuits. Nut butters can be added to soups and stews to thicken them. Nuts, both tree nuts and peanuts, are among the most common food allergies.

Consumption of various nuts such as almonds and walnuts can lower serum LDL or 'bad' cholesterol. One study found that people who eat nuts live two to three years longer than those who do not. However, this may be because people who eat nuts tend to eat less junk food and be more conscious of their weight. Nuts may be enjoyed simply, eaten out of hand. You don't have to get out your cookbook to enjoy nuts. The essence of eating healthy with nuts is not to overindulge.

PAGE:

For women, eating just two handfuls of nuts a week may extend their lives as much as by jogging four hours a week. ~ Michael Greger

ALMONDS

Prunus amygdalus

Almonds easily grow above 2500 ft. elevation with cool and dry growing conditions. The true almond tree is rare throughout the Hawaiian Islands, so grow one. Throughout the world, almonds are the most widely grown and eaten tree nut, cultivated for thousands of years before they acquired their original name, 'the Greek nut.' Botanists consider almonds fruit, and part of the plum family originally native to the area of western Asia and China. Primitive man prized these tasty nuts as a food staple since it stored well. According to anthropologists, early nomads created a trail mix of ground almonds with chopped dates, and used sesame oil to roll it into little balls. By 4,000 B.C.E., humans learned to cultivate almonds and they were grown by almost every ancient civilization. Handfuls of almonds were found in King Tut's tomb, to nourish him on his journey to eternity. The United States (57%), Spain (10%), and Iran (5%) are the top producers.

HOW TO GROW: To grow almonds, first decide if you have enough space for a sizeable tree, but they can also be grown in pots and sizable containers. Search for a sprouted nut. The almond is extremely adaptable and is grown to 3500 meters above sea level. Almost every island has slopes perfect for this nut. Because it has a deep root system, the almond can withstand severe drought and poor soils, but the roots will seek a constant water source like your septic tank or soak-away. The best pH is 7-8.5. Always put ant bait close to young trees. The big ants will eat all the new leaves, and kill, or stunt the new tree. To gain the maximum yield from your almond tree spray every other month with a soluble 20-20-20 fertilizer mixed with a light pesticide. Monthly broadcast a cup of 12-12-17-2 blue fertilizer around the base of the tree. In the dry season, water the tree soaking the ground every other week. To contain the size of the tree you can cut (top) the center stem of the tree after three years. It is best to top trees on the full moon. This will make the lower branches grow out, but it should not get any taller. An almond tree can take as long as 5-12 years to produce almonds, but a mature almond tree can typically produce fruit for 25 years. Almond trees technically produce drupes, a type of fruit with a fuzzy layer called a hull, and a hard shell that contains the almond. The drupes grow from pollinated blossoms and in 4 months split open, permitting the almond to dry. On almond plantations, mechanical tree shakers are used to get ripe almonds to drop. The nuts dry in the sun for 10 days. Almonds are mechanically hulled in large farms. It takes 1.1 gallons of water to grow one almond and to grow 500g (1lb.) takes 1,900 gallons. All nut trees consume a lot of water to produce.

MEDICINAL: Almonds are loaded with flavonoids in the nut's skin, and are anti-oxidants linked to improving breathing. Scientists believe these chemicals may even prevent respiratory diseases such as asthma, emphysema, and chronic bronchitis. Almonds are good eating nuts, high in monounsaturated fatty acids, and lower your cholesterol level, reducing the risk of heart disease. Almonds contain significant amounts of magnesium, vitamin E, fiber, and potassium, all of which are beneficial to a healthy heart and for strong bones. **NOTE:** Almonds contain 25.2 mg of cyanide per kg, but most disappears when processed. **DO NOT EAT RAW ALMONDS!**

NUTRITION: 100grams has almost 600 calories with 450g from fat, 20g of protein and loads of potassium and iron. One ounce of almonds contains about 10% of the recommended daily allowance of calcium, a great non-dairy source for vegetarians. Almonds are rich in vitamin E, with just a handful (30g, about 20 nuts) providing 85% of the Recommended Daily Intake (RDI).

FOOD & USES: These nuts are ground to make almond butter, almond flour, almond crackers, and almond milk, or roasted, salted, and seasoned to be enjoyed straight from the can or bag. The hulls are rich in nutrients and used as livestock feed. The shells are crushed and can be burned for heat. Almonds are sold raw or natural, roasted, or dry-roasted. Almonds are roasted (deep-fried) in oil usually (bad) highly saturated coconut oil. The process adds about ten calories per ounce of nuts, or a little more than a gram of fat (mostly saturated fat, if coconut oil is used). Dry-roasted almonds are not cooked in oil and better for your health, but they may be salted or contain other ingredients, such as corn syrup, sugar, starch, MSG, and preservatives. I've found that the best way to prepare raw almonds is to toast them in a dry skillet over low heat, stirring frequently, until golden and fragrant, about 5 minutes. Remember, remove the almonds immediately from the skillet or they're likely to scorch and have a burnt taste. You can also toast almonds in a baking pan in a 350°F oven for 10 minutes. Slivered and sliced almonds will take less time than whole almonds. Blanched almonds are briefly heated in boiling water and then shocked in cold water to stop the cooking, which kills enzymes that would cause them to deteriorate. Almonds are a tasty addition to any curried dish and can be mixed with chilled, cooked rice and raisins to make an easy and tasty salad. To make delicious almond-flavored 'milk' place one-cup of fresh roasted almonds with 4 cups of water in a jar. Tightly screw on the lid and refrigerate for 1-2 days, but no longer as it may ferment. Blend until smooth.

RECIPES: ALMOND ONE POT NUTRITIOUS STIR-FRY

Ingredients: ½ cup blanched almonds, 6 cups assorted vegetables (green beans, carrots, onions, cabbage, cauliflower, etc.) chopped small and thin, 1 TBS fresh ginger root minced, 2 garlic cloves chopped, 1/3 cup water, 2 TBS cornstarch, 1 TS sesame oil, 3 TBS soy sauce, 3 cups cooked rice (prefer brown), 3 TBS cooking oil, salt, and spices to taste

Method: In a large skillet, wok, or frying pan heat half of the oil – 1½ TBS - on medium heat. Add almonds and garlic; cook for five minutes. Carefully remove almonds and garlic with a spoon and set them aside. Add the rest of the oil, increase the heat to high, and then add ginger and vegetables. Stir-fry for five minutes and reduce heat to medium. In a bowl, mix the corn-starch, soy sauce, and water and pour over vegetables. Toss for 2 minutes over medium heat. Mix in seasonings, sesame oil, almonds, and rice.

COCONUT AND ALMOND RICE

Ingredients: 28g (1oz.) crushed almonds, 2 cups low fat milk, 2 cups uncooked rice, ½ cup coconut milk, ¼ TS salt, and seasonings to your taste

Method: Put almonds in a plastic bag and crush with a spoon. In a medium pot bring milk and salt to a boil before adding rice. Reduce heat, cover, and cook till rice is done. Add coconut milk, remove cover and simmer for five minutes. Stir in almonds with a fork to fluff the rice.

DID YOU KNOW? Almonds are stone fruits related to cherries, plums, and peaches. California produces almost a billion pounds of almonds yearly. The almond industry brings in an astounding $11 billion dollars annually. It takes more than 1.2 million beehives to pollinate California's almond crop (over 550,000 acres). Chocolate manufacturers currently use 40% of the world's almonds. The botanical name for almonds is *prunus amygdalus*. **Never eat raw almonds**. There are two types of almonds, sweet and bitter. The bitter are toxic with high amounts of cyanide. Sweet have no poisons and are safe to eat. The primitives who first ate almonds in ancient times must have roasted them to remove the poison. There are also two types of shells. The US type has a soft shell, and Spanish almonds are hard. Indian almonds are the most sought after.

ARECA NUT – BETEL NUT

Dypsis lutescens

Areca nut or betel nut is the seed of the areca palm cultivated in Sri Lanka, India, Japan, China, the Philippines, Africa, and East Indies. This tree is grown sparsely throughout the Hawaiian chain, usually by immigrants from Micronesia. Chewing betel nut in public parks and many public places is illegal in Hawaii. It is commonly referred to as betel nut, from the leaves that wrap into a preparation known as 'paan.' Areca palm seeds are mixed with betel leaves, (that's how they became betel nut), and chewed as a stimulant throughout India, Asia, and the South Pacific. The exotic-looking areca palm has dwarf varieties that only grow to 1.5m while the regular varieties grow 10 to 20m tall with feather fronds. These trees will make your garden or patio royally tropical. Grow the areca palm for decoration, not for consumption. Its botanical name is Dypsis lutescens. The betel leaf, *Piper betle,* is a relative of black pepper, *Piper nigrum* and another natural intoxicant, kava, *Piper methysticum.*

Areca palms are native to the Philippines or Malaysia. Archaeologists discovered remains in Thailand, Indonesia, and the Philippines to support the nut and leaf have been chewed for more than 4 millenniums. Today, chewing areca/betel nut is a popular culture in many Asian countries, similar to smoking cigarettes, with the same health hazards.

Even though it resembles a round, brown nut in appearance, the areca nut is considered a fruit. It's not eaten raw or cooked. Commercially, its available fresh, dried, and/or cured. It contains arecoline, an alkaloid stimulant similar to nicotine, and coupled with tannins when chewed it has a smooth, sharp taste. No one knows how, when, or where the areca nut and the betel leaf were combined to produce the enjoyable energy boost coupled with an overall positive feeling.

Areca nuts are similar to miniature coconuts, round or oblong. The outer husk and inner nut mature from a soft, easy-to-pull apart, green, to firm coarse yellow-orange husk securing a hard, light brown seed/nut. World production of areca nuts exceeds 1.4 million tons, and India grows more than half.

HOW TO GROW: The easiest way to grow an areca nut palm, other than a nursery, is to find a tree. Carefully remove some offshoots/suckers growing at its base. Keep the roots intact. It's best to first replant in a large pot that can be moved to give partial shade and be easily watered while watching the young tree grow and harden to the tropical elements. Decide where to plant it and, if planting several in a line or as a border they need a 10 meter (30ft.) spacing. Check with nurseries; there are dwarf varieties. This tree likes the cooler morning sun and should avoid the harsh afternoon. It likes well-drained soil with the optimum pH 6.1-6.5 and will thrive up to 1000 meters above sea level. To transplant, dig a hole the size of a five-gallon bucket and fill with rotted compost and manure. This is not a drought tolerant tree and requires regular water. It should bear fruit within 5 years and may have multiple annual harvests. At that time, the trees should be heavily fertilized with 10kg (22lbs.) of rotted manure. If you are storing the harvested seeds, consider many pests and insects also enjoy their taste. Unripe or ripe nuts are shelled, boiled, and sun-dried, then crushed with lime and catechu, a red dye made from boiled pieces of areca palm wood.

MEDICINAL: Chewing areca nut tastes good and is stimulating, BUT there is sufficient evidence that areca nut, with or without tobacco, can cause cancer. Since chances of getting cancer are high, areca nut should only be consumed infrequently in small doses. Pregnant women should avoid areca nut to avoid fetus harm. Chewing may cause low blood pressure and shortness of breath. In Ayurveda, the nut is considered an aphrodisiac and improves appetite. Areca palm leaves produce negative ions, just like air purifying machines.

BLACK WALNUT - WEST INDIAN WALNUT
Juglans nigra - Juglans jamaicensis

Early American settlers found the native black walnut tree (*Juglans nigra)* growing from Florida to Canada. If you have some property in the highlands of one of the Hawaiian Islands, this would be a great tree to have growing for nothing more than shade, and perhaps some nuts in the future. This tree can live a century. If you have an acre or more, commit to a black walnut grove for a future where beautifully grained, dark brown lumber will be extremely expensive. Your trees will help scrub the atmosphere to clean the air. In nature, these trees are individuals found scattered in a forest. Four generations can carve their initials in the same spot.

Plant the West Indian black walnut (*Juglans jamaicensis*) to help save a species of tree. This walnut in is found in tropical Cuba, the Dominican Republic and Haiti, and Puerto Rico. Contrary to its biological name, one is not native to Jamaica. Per Wikipedia: In Puerto Rico there are less than 20 trees of this species remaining. It is also rare in Cuba and Hispaniola. This tree is listed as an endangered species by the US. I've found no record of this tree in Ha-waii, but since it survives in the Caribbean, it should grow in the Aloha State.

EVERY ISLAND NATION SHOULD HAVE AT LEAST TWO WEST INDIAN WALNUTS IN THEIR BOTANICAL GARDENS. LET'S SAVE THIS SPECIES.

HOW TO GROW: This will be a tall tree that craves sunlight and its huge root system makes it storm resistant. Plant it at least 30 feet from your house in rich, deep, well-drained, but often moist soils with a pH of 5.5 to 7.5. This should be deep soil, free of rocks with a deeper water table that doesn't get waterlogged even after storms. Black walnut trees do best in warm, not hot, and even mild climates. Hawaii is perfect with frequent rains. A black walnut is a fast grower and will reach the maximum height of up to 80 feet in 30–35 years.

In a big pot or a designated seeding area – where you won't accidentally hit with the weedwacker or a lawn mower - plant a dozen or more walnuts 3 inches deep. If using a pot, be certain that it can accept a taproot a foot long before you transplant. If you plant several out-doors to create a private walnut grove, space the seeds 20 feet apart. To increase your efforts, plant two or more nuts in each spot, a foot apart. Circle each seed with a clear plastic water bottle with the bottom and top cut off. Bury half of the bottle around the walnut/seed. Place a marker so you can watch the progress. Keep a record of the planting date. It's best to plant the seeds in October. Don't let the soil dry out, but don't keep it overly wet. Usually, walnut seeds need to go through a cold period and then sprout in the spring. Walnut seedlings grow fast. Watch them closely because they can grow a meter/yard tall in the first year, and even more the second. Leaves will sprout when the daytime temperatures reach 70 F. If they are not pollinated by neighboring trees, they may set by self-fertilizing.

The quantity of nuts won't be the same every year and peak production will take ap-proximately two decades. Black walnut is susceptible to a disease called 'a thousand cankers.'

Black walnut trees are antisocial. They can be allelopathic, meaning its roots can re-lease a chemical known as the 'juglone' substance that harms other plants. This chemical provides this tree variety a competitive advantage to have its own space. Black walnut initially requires weeding, thinning the higher branches, and pruning to create trees that will produce the optimum amount of lumber for your grandsons or great-grandsons.

Harvesting begins when the nuts fall from the trees. A green fleshy outer hull protects the walnut inside. They look like tennis balls. It turns dark brown as it ages. Inside the husk, is

God gives the nuts, but he does not crack them. ~ Franz Kafka the walnut encased in its shell. Inside the shell is the nutmeat. Be warned, wear rubber gloves when cleaning the husks from black walnuts. The oil stains and is very difficult to remove. Trya hammer to smash the dry husk and then work what remains on the nut between your hands. A tan oil will seep out of the mess of skins. Use a wire brush when washing the bare nuts in a bucket of water. If any nut floats toss it away. Spread the clean nuts in a cool dark area for at least 2 weeks before cracking them open to get the meat. If unshelled the nuts can store for 10 months if kept cool and dry. You can freeze the meat.

MEDICINAL: Black walnut hulls contain juglone, a chemical that is antibacterial, antiviral, anti-parasitic, and a fungicide. Black walnut is reportedly used to treat ringworm, yeast, and candida infections. A skin wash is made from 3 black walnut husks boiled in a gal-lon of water. Powdered hulls can be mixed with talc for personal use. Try only a tiny bit first to see if there is any allergic reaction. Walnut oil is rich with nutrients and antioxidants and may improve your memory and concentration.

NUTRITION: 100 grams has a whopping 900 calories. They are high in fat and shockingly without any protein, minerals or vitamins. Eating walnuts will provide serotonin, important for a healthy emotional balance and fights depression. Most walnut oils contain an omega-3 fatty acid called alpha-linolenic acid. If included in your daily diet, it lowers the risk of heart disease, cancer, and fights the inflammation of rheumatoid arthritis. You only need to eat 2 nuts daily – 4 half nut meats.

FOODS & USES: China grows half of the world's black walnuts, picking over a million tons yearly. The US is second with 600,000 tons. Commercially, black walnut is an im-portant tree because the wood is a deep brown color and easily worked. Walnuts are cultivated for their distinctive and desirable taste. The outer husks of black walnuts were used to make ink, medicine, and as a dye for leather and cloth. Missouri, Ohio, and Kentucky have 35% of the black walnut trees in the US. One pound of black walnuts sells for more than $10 wholesale and $15 retail. The price for quality black walnut lumber $2.00 a board foot for sawlogs (used to make lumber) and $3.00 - 10.00 aboard foot for veneer logs (used to make fine face veneer for furniture or architectural paneling Walnut wood is traditionally used for gun stocks, furniture, flooring, paddles, and coffins. Walnut shells are often used as an abrasive in sand blasting or other circumstances where a medium hardness grit is required.

Black walnuts are tasty in cookies and cakes, and a crunchy addition to salads.

RECIPES: BLACK WALNUT COOKIES

Ingredients: 2 cups all-purpose flour, 1 cup brown sugar, 1 egg, 1 cup black walnuts chopped, ½ cup / 4 oz. soft butter, ½ TS each of: pure vanilla extract, baking soda, TS salt, cinnamon and nutmeg to taste and ½ cup granulated sugar for dipping

Method: In a bowl mix dry ingredients, flour, baking soda. In another bowl combine brown sugar butter, egg, water, and vanilla. Combine the dry with the creamed and add the nuts. Whisk the mixture and then ladle tablespoons of the mixture onto a greased cookie sheet. Com-bine cinnamon, nutmeg, and granulated sugar and sprinkle on top of cookie dough. Bake in an oven preheated to 375 for 12 minutes.

DID YOU KNOW? The world's largest and the US national champion black walnut tree lives on Sauvie Island, Oregon. At 5ft. from the ground, it is 8 ft. 7 in (2.62 m) in diame-ter and 112 ft. (34 m) tall, and its crown spreads of 144 feet (44 m).

Time spent amongst trees is never wasted time. ~ K. Mayer.

BRAZIL NUTS

Bertholletia excels

Researching the Brazil nut provided insight to the importance of wilderness to this tree. It's believed Portuguese explorers brought trees to the Hawaiian Islands in the early 1700s, but my research didn't find anything definite. The climatic conditions are similar to the Caribbean where they grow on every island, but the complex natural fertilizing system maybe missing. For avid naturalist gardeners, the unique Brazil nut tree can be started from natural seeds purchased online. The chances of ever getting a nut are slim because of the complex natural fertilization. Once you find raw seeds for sale, buy several. File some of the shells to make sprouting easier. Soak them in the shell overnight. Once sprouted, keep transplanting to larger pots. There are dwarf varieties for sale; try to find one for your patio. This tree requires tropical heat and plenty of water.

Big, reaching 150 feet and up to 6 feet in diameter, a Brazil nut tree isn't exactly perfect for every backyard because of its size and long root system. Planting a Brazil tree could be-come a time capsule as they can live for more than 500 years. (Some trees in remote Brazilian highlands are estimated to have existed for 1000 years!)

Brazil nuts aren't considered a true nut, but instead the seed of the *Bertholletia excels* species. They're called Brazil nuts because they're from the Amazon non-flooding rain forest. Amazingly these trees do not take to easy cultivation. Brazil nuts come almost entirely from collectors in the wild. There are very few Brazil nut tree plantations. In fact, more Brazil nuts come from Bolivia than Brazil. More than 20,000 tons of these nuts are harvested each year. Bolivia collects about 50%, Brazil 40%, and Peru 10%.

HOW TO GROW: The Brazil nut produces better in the wild than cultivated because they depend on the complexity of nature more than other trees. Their pale white blossoms have very sweet nectar, but it is protected by a cover flap on the blossom. This must be pollinated by large-bodied bees with long tongues. All these circumstances and necessities are only available in the wilderness. Small, long-tongued male bees use the scent of the sweet nectar to attract large bodied, long-tongued orchid bees, which eventually pollinate the Brazil nut blossoms. This is Mother Nature at her best. Without the blossoms, the bees do not mate. A lack of bees means the trees won't get pollinated and no Brazil nuts. The pods take more than a year to ma-ture after pollination. This 'natural' system is a bit too complex for many cultivators.

The Brazil nut tree is grown from one of the seeds inside the nuts. The huge pod drops and is often chewed open by an animal, like large black rats, to eat some of the rich nuts. Some of the remaining nut/seeds germinate in the shade, and sprout. New trees never develop until a big tree falls and makes space for the necessary exposure to sunlight. This wait could be a decade. Once the conditions are right and they mature, after 10 years, the trees begin to produce fruit.

MEDICINAL: Brazil nuts are an excellent source of selenium, a vital mineral and antioxidant that may help prevent heart disease. Two Brazil nuts can provide your entire daily intake of selenium. Brazil nuts are particularly healthy because selenium makes their protein content 'complete'. This means that, unlike most plant proteins from beans, etc., the proteins contained in Brazil nuts have all the necessary amino acids for optimal growth in humans just like meat and fish. Brazil nuts also contain small amounts of radioactive radium. This is not because of elevated levels of radium in the soil, but due to the incredibly developed roots of the tree. The radium is not dangerous.

NUTRITION: Eight Brazil nuts equal a whopping 180 calories with 18 grams of fat, 4 grams of protein, 3 grams of carbohydrates, and very significant amounts of magnesium and thiamine.

FOOD & USES: Brazil nut oil is also used to lubricate clocks, an ingredient in artists' paints, and in the cosmetics industry. Cutting any Brazil nut trees is prohibited by law in Brazil, Bolivia, and Peru. Because of the length of time these trees survive and their unique station in nature, Brazil nut trees should be protected everywhere. Illegal cutting of timber and slash and burn clearing of land is a continual threat.

There are high amounts of fat in Brazil nuts. The fats are unsaturated, and healthy when eaten in moderation. Eight medium Brazil nuts count as one serving, an ounce. Since these nuts are relatively high in fat, you shouldn't eat them more than three times per week. Buy and crack open a pod of fresh Brazil nuts and roast them yourself on a baking tray at 350 for 5 minutes. Once you have the roasted nuts you must try some recipes.

RECIPES: BRAZIL NUT BREAD PUDDING

Ingredients: 6 slices toasted bread buttered, cut into inch-long strips, ½ cup Brazil nuts sliced or chopped small (crushed almonds may be substituted), 2 large eggs, ¼ cup sugar, pinch of salt, 1 TS vanilla essence, 2 cups whole milk

Method: Arrange bread pieces in layers in a buttered baking dish, sprinkling each layer with Brazil nuts. Beat eggs slightly, add sugar, salt, vanilla, and milk. Mix well and pour over bread. Sprinkle top with Brazil nuts and bake in a 325-degree oven for one hour.

ROASTED BROCCOLI WITH BRAZIL NUT PESTO

Ingredients: ¼ cup Brazil nuts chopped, ½ cup chopped parsley, 2 TBS water, 1 large garlic clove minced, ½ cup fresh basil chopped, ½ TS lemon zest, ¼ cup olive oil, 4 TBS Parmesan cheese grated, pinch of salt, 2 heads of broccoli cut into 4 inch-long florets, spices to your taste

Method: Mix the parsley, Brazil nuts, water, basil, garlic and lemon zest in a food processor and pulse to a coarse paste. Add the ¾ of the olive oil, salt, and the Parmesan and process to a slightly smooth paste. In an oven preheated to 450 degrees toss the broccoli with the remaining olive oil and spread in an even layer on rimmed baking sheets. Roast broccoli until the broccoli is browned and crisp-tender. Put the broccoli on a serving dish, pour the pesto on top, mix, and serve.

QUICK BRAZIL NUT SOUP

Ingredients: 1 cup Brazil nuts chopped into big pieces, 1 large onion chopped, 4 cups water, ¾ TS salt, 8 medium tomatoes diced, vegetable/canola oil for frying.

Method: Sauté onion until tender. In medium pot add water, onion, and salt. Bring to a boil, reduce to a simmer, before adding tomatoes and Brazil nut pieces. Simmer 10 minutes and serve.

DID YOU KNOW? Brazil nuts are so rich in oil they will burn like a candle when lit. The oil is often used in shampoos, soaps, hair conditioners. These nuts are also called para nut, walnut, butternut, cream nut, or castanea. Brazil is named after a tree, but not this nut tree.

People who will not sustain trees will soon live in a world that cannot sustain people. ~ Bryce Nelson

CANDLENUT – KUKUI TREE

Aleurites moluccanus

The Hawaiian name is *kui-kui* or *kukui* and this evergreen is the perfect tree for a home garden on any island. The kukui/candlenut is a beautiful tropical rainforest tree that can grow to almost 100 feet, with a wide, 25 ft. spread of overhanging branches adorned with small white owers. One of the original canoe plants, to the native Hawaiians, the kukui nut/candlenut represents protection, peace, knowledge, and light. Strung together, kukui nuts are worn as leis or bracelets. The kukui is Hawaii's state tree and the blossom is Molokai's island regional ower.

The kukui/candlenut originated in the Indonesian islands and have been dated in archaeological sites to 10,000 BCE. Before the original Hawaiians learned to press the oil from the kukui nuts, they burned the nuts for light. Utilizing nature at its best, nuts were laced onto palm leaves' main stems and burned. Each provided a quarter-hour flame and became a clock

HOW TO GROW: The kukui/candlenut tree grows better in full sun, but can toler-ate partial sun, with regular rain, up to about a 3000 ft. elevation. This tree can handle most conditions, plenty of water or drought, and any type of soil if it drains. It prefers a pH of 5-8, high humidity. Sow the hard seeds in a sunny spot in moist potting soil and cover with a glass to contain the heat and moisture to speed germination, which may take 3-4 months. The traditional method is to soak the seeds overnight. Then press the seeds into soft earth and cover with a bucket full of dried leaves. Burn the pile of leaves, then quickly and carefully uncover the warm seeds. Toss them into a tub of ice water. This should crack the shells and help speed germination. You can also try with a hammer. Also plant some uncracked soaked seeds. This tree will grow in a pot on your lanai.

Kukui/candlenut trees make good windbreaks, privacy screens, and property markers. Because of the wide dense canopy, they're great shade trees and the spreading roots hold the soil. Fruiting can be continuous with blossoms and fruits at all stages of ripeness on a tree. They should bear within 7 years. A good-size tree usually bears twice a year and should produce 60kg of nuts. The hard-shell nut is 1½ -2½ in. with a ¾ in. white, greasy kernel, the source of the oil. It tastes like a Brazil nut, slightly toxic raw, but edible cooked or roasted.

In areas/ plantations where many of the kukui/candlenut trees are cultivated, the nut/seeds fall and are carried away by animals. This tree is considered invasive in several coun-tries: American Samoa, Ecuador, the Dominican Republic, the Cook Islands, and Brazil.

MEDICINAL: Almost all parts of the kukui/candlenut tree are useful for traditional medicine, leaves, nuts bark, roots, sap, and owers. The light-colored nut oil is practically odorless, easily absorbed, and great as a massage oil. It's used in soaps, candles, and lotions. People use it to stimulate hair growth. Internally, it's a laxative, treats insomnia, and improves digestion. Externally kukui/candlenut oil helps muscle-joint aches, heals dry skin, and soothes insect bites.

NUTRITION: 100 grams has 470 calories with 49 g of fat, and 7g of protein. These nuts contain good amounts of potassium, phosphorus, calcium, and magnesium, vitamins B-1 and B-3.

FOOD & USES: This tree is used for everything from lighting, dyes, food, to tattoo ink. The nut oil was used as a varnish and to preserve ropes and fish nets. The trunks were hollowed to make small canoes. The wood is not hard and not resistant to insects or decay. Kukui/candlenuts are best ground and added to recipes. They can be sliced, roasted in a skillet, and added to curries and sauces, sprinkled on dishes 'Inamona' is a paste made from roasted kukui/ candlenuts mixed with sea salt use with the Hawaiian dish, poke – raw fish.

CASHEWS

Anacardium occidentale L.

Cashews are one of God's practical jokes and one of nature's strangest creations. It is an egg-shaped, brilliant yellow, orange, or red fruit atop a brown wrinkly attached nut. It looks as if it were attached as an afterthought. Related to the mango, this soft, juicy fruit is incredibly sweet, but the raw nut is severely bitter. Over the centuries, cashew nuts have been experimented with on all the Hawaiian Islands. In the 1930s, a few hundred trees were tried on Oahu. Recently, small cashew orchards have been planted on Oahu, The Big Island, Maui, and Kauai.

Portuguese explorers discovered the trees, which they called 'caju' in northeastern Brazil. Around 1560, the Portuguese transported the original cashews to Africa's east coast. Africans cultivated the trees for the nuts, which they sold back to the Portuguese. The nuts were shipped to India to be shelled and roasted. India soon took over cashew cultivation and now is the world's largest exporter of these delicious nuts with Brazil second, and Africa third. Together, they produce 200,000,000 pounds a year.

The cashew fruit or 'apple' is a false fruit that develops from the blossom. The nut is the cashew seed. You will never see cashews sold in the shell. The nut is surrounded by a double shell that contains a caustic resin, which is a potent skin irritant, also found in poison ivy. The delicious apple is seldom sold since they are too fragile to market and spoil within a day. The apples contain tannin and begin to ferment immediately. East Indians prepare the fruit into liquor they call 'fenny.' Juice of the cashew fruit will badly stain clothes.

HOW TO GROW: Cashew makes a beautiful garden tree usually growing to less than 20 feet tall. Find a tree vendor or visit with an owner of a cashew estate. You can try to sprout a tree from a raw nut, good luck. You'll need a mature unshelled nut. These seeds are viable for up to 4 months. Dry the fresh seed in the sun for 3 days and soak in water overnight before sowing in good quality potting soil. If you're lucky, the seeds will germinate anywhere from 4 days to 3 weeks. Cashew trees need at least 6 hours of direct sunlight. Cashew prefers poor sandy well-drained soil with a pH of 5-6.5. With regular watering and monthly doses of high nitrogen fertilizer mix, trees bought from nurseries should bear in five years. Cashews have a good local market, however, getting the nuts are not easy. First, pick and twist the nut from the fruit. Try to avoid staining your clothes from the fruit juice. Unshelled cashew nuts will keep up to 2 years. Do not attempt to break the shell before roasting; cashew shell contains caustic oil that can burn your skin.

MEDICINAL: Cashews are rich in two categories of antioxidants that may help reduce inflammation, strengthen the immune system, and protect against diseases. Cashew oil is rich in vitamin E and good for treating skin problems, as removing scars and corns. The powdered nut can be mixed with water into a paste and applied to wounds for quick healing. The leaves, bark and cashew apple are antiseptic, stop diarrhea, reduce fever, lower blood sugar, blood pressure and body temperature. Cashew tree products have long been alleged to be effective anti-inflammatory agents, counter high blood sugar and prevent insulin resistance among diabetics. Cashew seed extract stimulates blood sugar absorption by muscle cells.

NUTRITION: One ounce of cashew nuts has 150 calories with 12 grams of fat, no cholesterol, and 9 grams of carbs. Cashews have less total fat than peanuts or almonds, and more than half of their fat is unsaturated fatty acids. This unsaturated fatty acid is oleic acid, the same healthy monounsaturated fat found in olive oil. Oleic acid is good for the heart, even in diabetics. If you are health conscious, dry-roasted cashews have a lower fat content. Iron, copper, zinc, and magnesium are found in cashews.

FOOD & USES: Count 15 cashews as a handful. For recipes, 1 pound of shelled cashews equals 3 cups. The fruit can be prepared into a marmalade jelly or canned whole. To roast cashews at home, place nuts in a preheated oven at 175 degrees for 20 minutes. At room temperature, shelled, unroasted cashew nuts spoil quickly since they contain much oil. Refrigerated nuts can remain for half a year. Large-commercial cashew operations use the nutshells to make varnish, insecticide, paint, and even rocket lubricant. For this primary reason, cashews are never sold in the shell. Cashew Nutshell Liquid (CNSL) has been processed like petroleum since the 1930s. The solids are used to make automobile brake shoes, while the liquid is in resins and epoxy coatings.

It's easy to prepare a cashew version of peanut butter by just placing a pound of roasted nuts in a food processor until you get the desired consistency of crunchy or creamy style. Cashews are a great addition for tasty variations of vegetable-rice, salads, stir-fries, pastas, or steamed vegetable dishes. Always add the cashew nut after the cooking process is complete, as heat will cause the nuts to soften and dissolve. This softening is why other nuts are preferred to use for baking recipes rather than cashews.

RECIPES: SPIRITED PINE-NUT SOUP

Ingredients: 250g (½ lb. cashew nuts chopped, 28g (1oz. salted butter or margarine, 1 red Spanish onion minced, 1 cup pineapple pieces without juice, 2 cloves of garlic chopped, 4 red potatoes peeled and diced, 2 cups vegetable bouillon, ¼ cup dark rum (optional, 2 TBS corn-starch with 4 TBS water, 2 cups water, salt, and spices to taste

Method: In a large skillet, fry onion and garlic until soft. Add cashews, pineapple, rum, water, bouillon, and spices. Bring to a boil. Simmer on very low heat for thirty minutes. Stir corn-starch into the four TBS of water until smooth. Add to soup stirring constantly until the mixture thickens.

CASHEW RICE SPECIAL

Ingredients: ¼ cup chopped cashews, 1 ½ cups brown rice, 3 cups water, ¼ cup raisins, 4 cloves, 1 TS cinnamon, 2 TBS soft butter or margarine, 2 bay leaves, salt, and other spices to taste

Method: Always rinse rice before cooking until the water is clear. In a suitable pot combine rice, water, cinnamon, cloves, bay leaf, and salt. Cover pot and bring to a boil, then cook on low for twenty minutes. Remove from heat, but keep covered for fifteen minutes. With a fork blend in butter, cashews, and raisins.

NUTTY TOMATO SAUCE

Ingredients: 4 pounds diced plum tomatoes, 1 large onion chopped, 1 cup water, 5 garlic cloves minced, 1 TS each fresh or dried marjoram, rosemary, and oregano leaves, 1 cup cashews chopped as small as possible, salt, and spice to taste

Method: Bring water to a boil in a large skillet. Add all ingredients. Cook on high, but stir fre-quently so it does not stick. After 20 minutes, pour or spoon off the excess liquid. Stir cashew pieces into the sauce. Serve over noodles, rice, or vegetables.

DID YOU KNOW?

Cashews are one of the best-tasting nuts on earth and now rank #1 among nut crops in the world with 4.1 billion pounds. Also known as the blister nut because it is related to poison ivy. The nut can be found in industrial products such as paints and brake liners. Botanically cashew is *Anacardium occidentale.*

No tree to make shade?
Blame not the sun, but yourself. ~ Chinese proverb

CHINESE CHESTNUTS
Castanea mollissima

Chestnuts are grown sparsely throughout the Hawaiian chain, but you can plant one for an ornamental. The European or Spanish chestnut is believed native to Asia Minor. Chestnuts were one of the earliest tree crops to be domesticated and were even mentioned in Chinese poetry more than 5000 years ago. The Japanese chestnut (kuri) was cultivated before rice. Malabar chestnut trees are more common, but they aren't related to the Chinese or American varieties. The ancient Greeks introduced chestnut cultivation to the Mediterranean region about 3.000 years ago. The colonists of North America discovered the countryside was a continuous forest of chestnut trees. The American chestnut (*Castanea dentata*) was the most common tree from Maine to Georgia and west to the Mississippi. Then in 1904, the chestnut blight disaster hit. Supposedly, it spread from some imported Asian chestnut trees planted on Long Island, New York, and in less than 40 years, it killed almost every American chestnut.

Today, Asia is saving the American chestnut production. Now the four main species are European, Chinese, Japanese, and American chestnuts. A well-maintained tree is a family heir-loom with a lifespan of 200 to 800 years, depending on the species. Europe and Asia have a well-developed chestnut industry. South Korea and China produce more than 40% of the world's chestnuts. The United States imports more than 6,000 tons of chestnuts worth $20 million from Italy, China, and S. Korea. Chestnut farms are long range goals with great profit potential. Every nation with highlands should invest and intercrop until heavy nut production begins.

Now gardeners can plant Chinese chestnut trees, which are blight resistant, and are the best acclimated for the Aloha State. Chinese chestnut trees may sound exotic, but the species is an emerging tree crop in North America. Many gardeners are growing Chinese chestnuts for the nutritious, low-fat nuts, but the tree itself is attractive enough to be an ornamental. The differences are slight: American chestnut leaves are narrower producing smaller nuts. But the Yankee trees stand straighter while the Chinese spread wider. All create a momentary mess in the yard. Dropping chestnuts present a possible litter prob-lem, and end walking barefoot on the lawn. Roasted chestnuts can be edible treat. Chestnuts are considered 'a grain that grows on a tree,' with similar nutrition to brown rice.

HOW TO GROW: Chestnuts are considered self-sterile; at least two trees are required to facilitate cross-pollination and generally produce more nuts. Chestnuts can grow at 1000 meters of elevation in full sun and well-draining fertile soil with a pH 5.5-6.0. It's recommended to avoid recently cleared land to help resist the root rot. Space these trees 50 ft. apart and away from buildings, walkways, and roads. Chestnut trees have reddish-brown or grey bark that is smooth in young trees but becomes rough and furrowed in old trees. You'll want to protect young trees from squirrels, raccoons, deer, and other hungry pests. Once ma-ture, the trees can reach up to forty feet tall and wide. These trees usually bear within 5 years.

You can try sowing fresh seeds in a pot. They are slow to germinate. Plant grafted trees in moderately acidic, well-drained soil where they'll get lots of sun. Chinese chestnut, is native to China and possibly Korea, noted for its resistance to chestnut blight. It's a medium-sized, low-branched, deciduous tree that grows 40'- 60' tall with an open, rounded crown. Dunstan, Lucky 13, and Carpenter are recommended Chinese/American hybrids. Blossoms are beautiful, but only for a few days and are followed by edible chestnuts en-cased in spiny, 2-3in. pods. Each pod contains multiple nuts. Regular water is necessary for the tree to mature and produce nuts. If the trees suffer in a drought, the nuts will be smaller and fewer. Heavy salty air will also reduce production, but the great cool shade remains.

MEDICINAL: Eating chestnuts regularly helps improve digestion, manage diabetes, nourishes the spleen, kidneys, and stomach, lowers blood pressure, improves circulation, and soothes inflammation. A poultice made from the leaves reduces inflammation. A tea made from the bark reportedly will help break a fever.

NUTRITION: The chestnut is a highly nutritious food. 100 grams of chestnuts has 200 calories, significant amounts of vitamins B-6 and C, but only a trace of fat compared to peanuts or cashews. The excellent protein of chestnuts has amino-acid content like that of an egg. They are the only nuts that contain vitamin C.

FOOD & USES: Although Chinese chestnuts are sometimes sold in farmers' markets, Spanish chestnuts are more often sold commercially. First, remove the outer shell and inner bitter tasting skin to obtain the edible kernel. Chestnuts can be eaten raw, but they're also eaten candied, boiled, steamed, deep-fried, grilled, or roasted. They can be dried and milled into flour, which can then be used to prepare breads, cakes, pastas, or used as thickener. Boil the in-shell nuts whole for 30 minutes and then cut in half and using a teaspoon, scoop out the soft kernel flesh inside. You can a roast in-shell chestnuts in the oven, over hot embers, or in the microwave, **BUT FIRST YOU MUST PIERCE THE SHELL** to prevent nut explosions. They can be used to stuff vegetables and poultry. They are available fresh, dried, ground, or canned (whole or in puree). Chestnut is of the same family as oak, and likewise, its wood contains many tannins. This renders the wood very durable, and it naturally resists the outdoor elements. It also corrodes iron slowly, although copper, brass, or stainless metals are not affected. Usually, there are 30 nuts to a pound.

When you buy chestnuts, pick ones that are hard, glossy, and round. Avoid any with little holes or scratches by insects. Choose ones that feel heavy. It has been awhile since they were harvested. Cook them ASAP.

RECIPES: CHESTNUT HUMMUS

Ingredients: ½ lb. cooked chestnuts, juice of 1 lemon, 1 garlic clove, 4 TBS olive oil

Method: In a blender, blend all ingredients adding a few tablespoons of the water the chestnuts boiled in as needed for preferred consistency. Serve with veggie pieces, roti, or biscuits.

CHESTNUT RICE

Ingredients: 2lbs. boiled fresh chestnuts, 3 cups short grain rice, 4 cups water, 2 TBS soy sauce, 2 TBS rice vinegar, salt and sugar to taste, 4 kale or spinach leaves chopped

Method: Combine all ingredients in a rice cooker except kale and nuts. Place them on top of the rice mixture. Cook. Stir the rice and serve.

DID YOU KNOW? The name chestnut is from early English 'chesten nut,' from the Old French word 'chastain.' Roman soldiers ate chestnut porridge before going into battle. Chestnuts should not be confused with: horse chestnuts, which are not related and mildly poisonous

to humans, or water chestnuts also unrelated, but with a similar taste from a water plant. Alexander the Great and the Romans planted chestnut trees across Europe while on their various campaigns. A Greek army is said to have survived their retreat from Asia Minor in 401–399 BC thanks to eating chestnuts. The Hundred-Horse Chestnut is the largest and oldest known chestnut tree in the world. Located on the eastern slope of Mount Etna in Sicily it is generally believed 2-4,000 years old. It had a circumference of 57.9m (190 ft.) when it was measured in 1780. Above-ground the tree has since split into multiple large trunks, but below-ground these trunks still share the same roots.

You are like a chestnut burr, prickly outside, but silky-soft within, and a sweet kernel, if one can only get at it. Love will make you show your heart some day, and then the rough burr will fall off. — Louisa May Alcott

COCONUT - NIU

Cocos nucifera

There's a Hawaiian legends that tells of the Tahitian chief, Pōka'ī, who brought the first coconut trees to Hawaii. Coconut trees thrive in Hawaii's climate on all the islands. Some of the ancient groves still exist on the Big Island and Molokai. In west Maui, there is one of the few remaining farms devoted to coconuts. The Hawaiian word for coconut is *'niu'* pronounced nee-oo.

The coconut's origin over 3,000 years ago is a mystery. Recent DNA testing by biologists at Washington University in St. Louis discovered two possible origins of the coconut, SE Asia including the Philippines, Malaysia, and Indonesia. The other is India and Sri Lanka. Indonesia produces 17 million tons, the Philippines 15 million ons, and India 14.6 million tons. In ancient Sanskrit tablets, the coconut is named 'kalpa vriksha,' which means 'the tree that provides everything necessary for life.' Spanish explorers gave name, 'coco', meaning monkey face, because the 3 dark spots at the dry nuts' base look like a wide-eyed face. Botanically coconut is referenced as *Cocos nucifera.*

The coconut is one of nature's most useful trees. What would shade our beaches; where would we tie our hammocks, and how would we make curries without coconuts? Ocean currents have distributed floating nuts throughout the world. The nuts will only grow between 28 degrees north or south latitude, the closer to the equator the better. All the hundreds of varieties of coconuts belong to the same species. Trees can grow to 80 feet or bear as short as 6 feet. The nuts can be red, yellow, orange, green, brown, or double.

HOW TO GROW: To grow a tree, simply find the type of nuts you enjoy. Scan the area to find any nuts, which may have already sprouted. If none are growing, select mature nuts that still have the husk on them, and you hear water slosh when you shake it. Remember, you need at least two trees to string a hammock. Soak them in water for 2-3 days. I find the easiest way to sprout a nut is to place it in a drain. Be watchful that a heavy rain doesn't wash them away. OR place the soaked nut in a container filled with potting soil. Nuts kept damp usually sprout a stalk in 3-4 months. Be careful when removing the sprouted coconut that you don't hurt the new roots. Plant the coconut point side down and leave one-third of the coconut above the soil and should bear 5-6 years after planting. It's recommended to plant dwarf coconut palms at least 20 feet apart, from fence lines, or structures. Certain trees can bloom 10 times a year, so nuts will be continuously available. It takes almost a year for a nut to mature on the palm. Yearly a tree usually will produce 40 or more nuts. Fresh nuts can be stored to 3 months.

MEDICINAL: Grated coconut treats stomach burning due to acidity, treat colic, ulcers, and improve strength. Coconut oil rubbed into the scalp increases hair growth, treats dandruff, and headaches. Coconut water is a diuretic and used to soften skin. A mix of coconut water and grated coconut treats hiccups. Coconut water is sterile and intravenously used as a substitute for glucose.

NUTRITION: Coconut meat has 354 calories for 100grams with 3g of protein and 33g of fat. The meat has plenty of potassium, iron, and magnesium. Coconut water per 100 grams has 19 calories 1g of protein with potassium but is loaded with sodium. A coconut will provide ½ to 1 cup of water. Coconut milk has 230 calories to 100g with 2 g of protein and 24 of fat with potassium and magnesium. 100 grams of coconut oil is 100 grams of fat with 860 calories and no noticeable vitamins or minerals. One coconut should yield 3 cups grated and 1 cup of liquid.

FOOD & USES: Coconuts may be harvested at different ages for different purposes. A young, six-month-old nut has jelly inside, as the white meat has not formed. It's also used in the cosmetic industry for moisturizers. Mature nuts are processed for their oil, a saturated fat, used for cooking, religious ceremonies, or soap. Pressing the meat of mature nuts makes coconut milk. Grating the mature meat also produces the world's most familiar type of coconut slivers used in cooking, baking, and candy. Coconut flesh can be dried, grated, powdered, flaked, toasted, frozen, or reconstituted. Processed coconut has less than 3% moisture, but is 68% oil. The coconut palm leaves can be woven into hats or stripped to make simple brooms. The husks can be stripped into a fiber called 'coir' and woven into mats, fishnets, paint brushes, and rope.

The almost clear liquid in a coconut is not coconut milk, but coconut water. To make true coconut milk boil equal parts of shredded coconut and water, simmer, and strain it. 'Coco Loco' or other varieties of sweetened cream of coconut used in pina coladas are not canned coconut milk.

RECIPES: POL SAMBOL – try an easy spicy coconut condiment specialty

Ingredients: 100g (¼ lb. grated coconut, 6 dried red chilies/hot peppers (more to your taste, 1 medium onion chopped small, ½ fresh lime, ½ TS sugar, salt to taste

Method: In a mortar, grind the red chilies and salt to a paste. Add the grated coconut and grind again. Grind in the sugar and then the onion. Remove to a bowl and squeeze the lime juice.

EASY COCONUT PIE

Ingredients: 4 eggs beaten, 1 cup sugar, ½ cup bakers' flour, 4 TBS melted butter or margarine, 2 cups grated coconut, 1 TBS vanilla extract

Method: Mix all ingredients together, preferably in the above order. Pour into a greased baking dish or pie pan. Bake at 350 degrees for 40 minutes.

COCONUT RICE PUDDING

Ingredients: ½ cup uncooked rice, 1 cup grated coconut, 1 cinnamon stick (or 1 TBS ground cinnamon, 2 cups coconut milk, ½ cup sugar (prefer brown

Method: In a medium saucepot place rice and cinnamon stick, cover with water, and bring to a boil. Reduce heat and simmer until water is absorbed. Remove cinnamon stick before adding coconut milk vanilla and sugar. Simmer until rice is creamy. Stir in on ½ cup grated coconut. Remove from heat and allow to cool. Cover with ½ cup grated coconut. Sliced mango or pineapple pieces may be added before serving.

COCONUT ROTI – SKILLET BREAD

Ingredients: 1 cup all-purpose flour, ½ cup grated coconut, 1 TBS chili flakes (optional, salt, ¾ cup of water, and oil for greasing the pan

Method: In a bowl, combine flour, shredded unsweetened coconut, salt, chili flakes, and ½ cup water and mix. If the mixture feels dry, add the remaining water, and knead into a ball. Separate into 4 small balls. Grease your hands and on a floured surface stretch or roll each ball out to the size of a small plate. Place in a warm frying pan and cook for 5 minutes or until you smell it burn and flip. Cover and cook for 3 more minutes. Do the same for the rest.

DID YOU KNOW? More than 20 billion coconuts are produced each year. A third of the world depends on coconut for food or their economy. Coconut oil was the leading vegetable oil until soybean oil. Falling coconuts kill 150 people every year - 10 times the people killed by sharks. Islands in the Indian Ocean used discs carved from coconuts as currency until 1910. Sri Lanka is the world's largest producer of coconut arrack liquor, more than 45 million liters annually from the sap of unopened flowers of the coconut palm

He who plants a coconut tree plants food and drink, vessels, and clothing, a home for himself, and a heritage for his children. ~ South Seas saying

KOLA NUT - COLA NUT – BISSY NUT

Cola nitida

Kola nut will grow in the Hawaiian highlands. I've seen photos of kola trees on Maui. This tree is believed to be native to Africa and the slave trade introduced it throughout the Caribbean. Kola nuts have been used as a natural stimulant for millenniums.

Throughout Western Africa, a small piece of nut is chewed before each meal to pro-mote digestion, but also suppresses hunger. Europe got its first taste of the kola nut in the 1500s, delivered by Portuguese traders. Nigeria exports more than half of the kola nuts the world consumes. Africa, especially the Sub-Sahara, consumes the majority. Plates of these nuts are offered to visitors in African villages as a symbol of hospitality and kindness. Almost everyone in Nigeria and adjoining African countries enjoys artificial energy the kola nut provides, including truck drivers, students, and menial laborers.

The kola nut is about the same size and appearance as the nutmeg. Every nut has 2-4% caffeine. Botanically known as, *C. acuminata* and *C. nitida* are the two commercial kola nut varieties. *Cola nitida,* is closely related to*Theobroma cacao*, the cocoa tree.

HOW TO GROW: The kola tree is an evergreen tree that loves the hot, humid, trop-ical lowlands and will grow to 800 meters. Growing a kola nut tree in your home garden is a family project because it will bear after a dozen years, may outlive two generations, and reach a height of 20 meters (66ft). A newlywed father may plant a tree and his great-grandchildren will play in is shade and enjoy the fruits. First, find some kola nuts for seeds. This tree prefers a rich, well-drained soil and a position in full sun with a pH in the range 5–6. Start by planting a few nuts in containers. Big white kola nuts, about 15g, germinate the best. Red nuts have a lower rate of success. These trees can also be grown from cuttings and that'll be quicker and more certain because it may take 6-12 months for the nuts to germinate, if they germinate. Once they've sprouted transplant them outdoors, at least 10 meters from houses and fence lines. Young trees require some shade and regular water. Plant them in the partial shade of another tree or bush with a shorter lifespan like cocoa. As the kola tree grows it will shade the cocoa.

Kola nut trees' growth is slow, with trees reaching only 3m (10ft.) in 4 years. They start to bear nuts when they've reached a dozen years, and you should get a dozen kilos of nuts. With good care, they'll continue to bear for 90 years. 300 nuts is a good annual yield per tree.

A properly maintained kola nut tree can bear continuously. They have attractive yellow blossoms edged with purple that produce big fleshy fruits up to 12cm (5in.) long and 7cm (3in.) wide, bigger than your hand. The pods look like clam shells and will split open as they ripen, showing the seeds packed similar to cacao. Inside the fruit, about a dozen round or square seeds develop in a white seed-shell. The seeds can be white to maroon. Initially, kola nuts are bitter and astringent, but mellow and gain flavor with age. The nut's aroma is sweet and rose-like. The tree's sapwood is pinkish-white and the heartwood is dull yellow.

MEDICINAL: In herbal medicine, the kola nut is usually soaked in alcohol to make tinctures to help 'drive' other herbs into the blood and may help increase oxygen levels in the blood. They may be chopped and boiled, drunk as a tea, although bitter, and the powder can be taken in a capsule. For thousands of years, kola nuts have been used in folk medicine as an aphrodisiac and an appetite suppressant, and to treat morning sickness, migraine headache, and indigestion. It is used in Jamaica to treat food poisoning. They counteract hunger and thirst, treat morning sickness and nausea in pregnant women. It has also been ground into a paste and applied directly onto the skin to treat wounds and inflammation. Other kola nut uses include fighting infection and clearing chest colds. Grate or coarsely grind the kola nuts and place in alcohol for 24 hours. Strain and discard the nuts and you have a crude extract with a strong smusky flavor. This should be used in small amounts.

Kola nuts are mainly used as a stimulant to stay awake, withstand fatigue, promote better concentration, and help people lose weight by diminishing appetite. In addition to caffeine, the nuts contain theobromine, the ingredient that is found in chocolate and reputed to bring a sense of well-being. This may explain the mild euphoria and optimism from chewing the nuts.

NOTE: Chewing the kola nut while drinking alcohol is dangerous. The nuts' caffeine tricks you into not realizing how alcohol-impaired (drunk) you really are. If you suffer from hypertension stay away from kola nuts as they will increase blood pressure.

NUTRITION: The kola nut contains Vitamins B1, C, and E. They are very high in potassium, magnesium, and calcium with iron, zinc, phosphorus, and micronutrients. Kola nuts are 2-4% caffeine and theobromine, with healthy tannins, alkaloids, saponins, and flavonoids. Theobromine is also found in green tea and chocolate.

FOOD & USES: The first taste of kola nut is bitter, but it sweetens after mixing with saliva. Frequent chewing of the kola nut can also lead to stained teeth. The nut can be boiled to extract the caffeine. Kola nut is used by the pharmaceutical industries, and demand for the nut is also increasing for produce soft drink sodas, tonic wines, essential oils, and scented candles and many other goods. The kola pod husk is used in manufacturing poultry feeds, liquid de-tergent, and organic fertilizer. The nuts are a source of varying shades of burnt orange dye for cloth. However, *kola nut* is not a durable dye and will fade with excess washing and exposure to the sun.

RECIPES: KOLA CITRUS HERBAL TEA – good for whatever ails you

Ingredients: 2TBS chopped kola nut, zest of: 2 oranges, 1 lemon, 1 lime, 1TBS fresh ginger, 1TS vanilla extract, ½ TS nutmeg ½ TS cinnamon

Method: Soften the kola nut pieces by simmering for 30 minutes alone in a liter of water. Then add the remaining ingredients and simmer for 20 minutes. Strain and cool overnight in the re-frigerator. Use within a few days.

DID YOU KNOW? A Portuguese explorer visiting Africa in 1587, observed that many people he'd encountered used the kola nut to relieve thirst and improve the taste of water. Kola nut, commonly called 'bissy' in Jamaica, came to the island on a slave ship in 1680. In the 1880s, a pharmacist in Georgia, John Pemberton, took caffeine extracted from kola nuts and cocaine-containing extracts from coca leaves and mixed them with sugar, other flavorings, and carbonated water to invent Coca-Cola, the first cola soft drink. As of 2016, the cola recipe no longer contained kola nut extract. Kola nut was used as the main ingredient in cola drinks for many years but is now mainly used as a f avoring agent. In 2007, the UK supermarket chain Tesco introduced the American Premium Cola that uses kola nuts, spices, and vanilla. The Igbo, a tribe in southern Nigeria, consider the kola nut tree to be the first tree and fruit of the earth. They consider the nut to be a symbol of hospitality, kindness, and brotherhood. The nut is also thought to improve the flavor of any food and to counteract the effects of drinking tainted water.

The kola nut stays longer in the mouths of those who value it. ~ An African proverb.

MACADAMIA NUT

Macadamia integrifolia

The macadamia nut, or Queensland nut, is native to Australia where it was consumed for thousands of years by the Aborigines. This is a fast-growing, medium-sized tree with thick, dark green foliage which grew in the rain forests. The first plantation was established in the 1880s. It wasn't until the development of successful grafting techniques and the introduction of mechanical processing that commercial production became feasible.

This is a perfect tree for your home garden in Hawaii. Because macadamias are difficult to grow, slow to bear, and specific in natural locations, production has not kept pace with increased demand, thus rendering the product costly. Grow your own! Previously, most commercial production was in their native Australia and in Hawaii. However, given the successes of Hawaii, many other tropical countries are devoting large tracts of land to macadamia plantations. It is time Hawaii look beyond tourism and get a larger, steady revenue from macadamia orchards. South Africa is now the largest producer of macadamia nuts.

HOW TO GROW: Two types of macadamias are grown commercially: the smooth-shelled *Macadamia integrifolia* and the rough-shelled *M. tetraphylla.* Grafted seedlings are recommended and do best in the low elevations up to 2500 ft., perfect for the Caribbean. Macadamia trees grow 60 ft. high and 40 ft. wide with clusters of scented pink or white flowers. Bunches of up to 20 nuts follow. This tree requires rich well-drained soil with a pH of 5.0-6.5, and 70-150 in. of rain annually. Plant with full to partial sun in a protected area. Since the branches of can be brittle, high winds are a problem. The varieties, Kau and Pahala, are more wind resistant. Mac nut trees are picky about fertilizer and soils for the optimum harvests.

Macadamia trees bear within 5 years and reach full potential at 12 -15 years. Because of this slow development, beehives and honey production near the orchard can generate early revenue. Intercropping macadamia nuts with faster bearing crops like bananas or coffee is a good financial option.

MEDICINAL: Macadamia nuts are recognized for their energetic and antioxidant properties. Consuming 5 to 20 nuts per day would help lower cholesterol levels and heart risks.

NUTRITION: 100g of Mac nuts has 73g of fat, 7g of protein, and are high in calcium, iron magnesium, phosphorus, and vitamin B.

FOOD & USES: The highest quality macadamia kernels contain at least 72% oil. Kernels with less than 72% oil are usually immature and harder, and they become over brown when roasted. These nuts are eaten either plain, cooked, spiced, sugared, grilled as an aperitif, caramelized in ice cream or in tasty desserts, roasted and salted or used in pastries and candies. There's even macadamia liquor. Macadamia oil has a buttery flavor, great for roasting, baking, and deep-frying. Macadamia shells make activated charcoal for use in carbon filters for water and air purification systems. Construction particle board made of mac shells is water resistant.

DID YOU KNOW? The Aborigines mixed mac nut oilwith their body paints and clay to increase durability. Pure mac nut oil was used to soften their skin and mixed with other herbal extracts to treat ailments. The first Europeans to eat these nuts were Walter Hill who had asked a young helper to crack some nuts to plant. The helper ate some and said they tasted excellent and didn't get sick. Hill ate some and became 'the founder' of the mac nut industry. Hill gave the scien-tific name, *Macadamia intergrifolia,* to honor Dr. John MacAdam, a noted Aussie scientist.

PEANUTS – PINANI

Arachis hypogaea

A bottle of roasted ground nuts doesn't last long in our household. Peanuts are not really nuts, but a type of bean that forms underground. These tasty 'nuts' have a variety of names including ground nuts, ground peas, monkey nuts, and goobers. Peanuts originated on the slopes of the Andes Mountains in South America at least 3500 years ago. Throughout the Hawaiian Island chain everyone loves peanuts, and on every island they're grown in small gardens. The local Hawaiian word for peanuts is *'pīnani.'* Botanically, they are *Arachis hypogaea.*

Spanish and Portuguese explorers brought the peanut to Europe, Africa, and Asia, including the Caribbean. Peanuts were used as a basic food aboard early sailing ships because of their high nutritional value and low cost. Virginia, in the United States, harvested the first crop for sale in 1840. Roasted peanuts began at Barnum's Circus in 1870. Peanut butter was created in 1890 as a paste for elderly patients who couldn't chew, but really, the ancient Incas created peanut butter and flavored it with ground cocoa beans. The Aztecs used mashed peanut paste to cure fevers. Today, over half of the peanuts grown in the USA are used to make peanut butter.

HOW TO GROW: The peanut is not a 'usual' plant since it has blossoms above the ground, yet the fruit (the peanut) grows beneath the soil like a potato. Each peanut is actually a seed. There are two basic types of peanut plants, runners and bush. The runner spreads out like a watermelon with nuts growing from the main branches. The bush type is similar to a typical bean plant growing more than two feet tall. Once the small yellow flowers are self-pollinated, the bud of the nut begins to grow below the surface. One plant can produce 40 nuts.

Peanuts are a 'long crop' taking about 5-6 months to mature. They are great for the soil as they enrich it with the nitrogen they produce while forming the nuts. Raw nuts (not salted, baked, or otherwise processed) are easily found at most supermarkets. Soak the raw nut in a damp cloth for 2 days. Work the soil at least a foot deep with a fork and pull up into a mound 6 inches high and wide. Plant 2 nuts together about 2 inches deep, and space plantings about 6 inches. Sprouts should appear in 8 days. Blossoms should appear a month and a half later. Use a 12–12–17-2 fertilizer mix every 2 weeks for 6 weeks then switch to 7–11-27 once a month. Water regularly. Using a fork, carefully harvest the plants. Each plant should produce 20-40 nuts. Determining when to harvest is a problem. Most farmers pull a bush and check. If the pod breaks from the root easily, they're ripe. If the nuts are pale pink and watery, the plant needs more time to mature. Raw nuts are 40% water so they must be dried to prevent spoiling.

MEDICINAL: If you eat one ounce of peanuts every day the unsaturated fats in peanuts will reduce the risk of heart disease by 25%. Rich in folate and niacin (vitamin B3) will increase the HDL, good cholesterol, by as much as 30%. Protein and dietary fiber makes up 25% of each peanut. The most unique property of peanut butter is its high content in resveratrol, a substance that's been shown to have strong anti-cancer properties. Peanut allergy is one of the most common food allergies found worldwide. In the United Kingdom, peanut allergy is present in between 0.4 and 0.6% of the whole population.

NUTRITION: Peanuts are a very 'fatty food.' Every ounce of raw peanuts provides 7 grams of protein, 7 grams of fat, and 3 grams of fiber. A cup has over 600 calories because they are over 75% oil. However, it's healthy oil; monounsaturated, not polyunsaturated. Research-ers in Great Britain discovered that women who ate 5 ounces of peanuts weekly reduced heart attacks by a third. Peanuts' nutritional value includes protein, fiber, calcium, potassium, iron, copper, and vitamins E and B complex especially B-3.

FOOD & USES: Very little, if any of the peanut plant is wasted. Peanut oil is used for lubrication, cosmetics, paint, soap, and lamp oil. It is the 6th most important oil seed in the world. Cattle farmers use the peanut as a food source to fatten their livestock. Peanut shells are compressed into artificial fireplace logs, kitty litter, and particleboard for home construction. The skins are processed into paper. Peanuts are a big money crop throughout the world today bringing in over 4 billion US a year. The peanut market expanded with George Washington Carver in 1903. Carver developed over 300 uses for the peanut including soap, cheese, medicine, grease, ink, and bleach. It only takes 720 peanuts to produce a one-pound jar, and 26 million tons of peanuts are harvested every year!

Peanuts can be eaten raw. Raw nuts may be added to soups, salads, or stir-fry. Placing them in a baking pan in a 350-degree oven for half an hour will roast nuts.

RECIPES: PEANUT STEW

Ingredients: 2 medium onions chopped, 5 medium tomatoes chopped, 6 garlic cloves minced, 1 cup okra, 3 cups potatoes cubed, 2 TBS olive oil, 5 cups water, ½ cup tomato sauce (or ketchup, 1 TBS ground cumin, 1 TBS chili powder, 1 hot pepper (remove before serving, 2 cups spinach or bok choy chopped, ¼ cup fresh mint leaves chopped, ¾ cup chunky peanut butter, ½ cup chopped roasted peanuts. Salt and spices to taste

Method: In a large stockpot, mix onions, garlic, potatoes, tomatoes, and olive oil. Saute' about 5 minutes then simmer. Add water, tomato sauce, spices, and mint leaves. Continue to simmer for 10 minutes. Add okra, spinach, or bok choy, and peanut butter. Cook for 5 minutes until stew begins to thicken. Serve bowls garnished with chopped nuts.

PEANUT SPICY SAUCE

Ingredients: 1 cup chunky peanut butter, 3 garlic cloves minced, 1 large hot pepper chopped fine, 2 TBS fresh ginger peeled and chopped, 3 TBS brown sugar, 4 cups coconut milk, 3 TBS soy sauce, 1 TBS molasses, 1 TBS fresh lime juice

Method: In a suitable pot over low heat, whisk together all ingredients until well combined.

BOILED PEANUT BREAD

Ingredients: ½ pound crushed peanuts, 1 pound cassava peeled and grated, ½ pound pumpkin peeled and grated, 3 TBS milk powder, sugar, spices, and salt to taste

Method: In a large bowl combine all ingredients and divide into 6 equal portions. Place each portion on a piece of greased aluminum foil. (Traditionally banana leaves were used as packets.) Boil packets in a suitable pot for 1 hour.

DID YOU KNOW?

Some African tribes believed peanuts had souls, so they cast gold replicas of the ground-nuts. Peanut butter was originally made for toothless people. The average European eats less than a tablespoon of peanut butter a year. The furthest thrown a peanut has ever been thrown was 37.9 meters. In the United States yearly exports is between 200,000 and 250,000 metric tons of peanuts. Peanut products are used to treat malnutrition in emergencies and even saves lives.

No man in the world has more courage than the man who can stop after eating one peanut. ~ Channing Pollock

PILI NUTS

Canarium ovatum

Pili nuts are good trees for the home garden to generate conversation and healthy nuts. Called the world's hardest nut to crack, pili nuts taste like cashews and almonds, but remain unknown outside of their native Philippine Islands, but can grow anywhere with a tropical climate. Centuries ago, indigenous peoples found pili trees growing, ate the ripe nuts, and the boiled encasing pulp. Now, pili nuts are considered a healthy snack and a valuable commercial export.

The present limited cultivation cannot fill the demands of the international market for the pili nut and its by-products. Countries haven't yet developed efficient cultivation or harvesting technologies. Until recently, there was almost no commercial planting of this crop; fruits were collected from wild trees. Pili nuts were introduced to Hawaii during the 1940s. Within the last 50 years, cultivation has expanded to the Philippines, Hawaii, and Central America.

HOW TO GROW: To hold the pili tree captive, old native Filipino pili nutters claimed the pili could only sprout under a mother tree and could not be domesticated. That has been discounted. Pili trees can be grown from fresh seeds, but that's risky and germination takes 1-3 months. It can reach 6 ft. in 2 years. You need a male and female tree. The pili is not self-fertile. It's recommended to cleft or wedge graft material from identified female trees that produce desirable nuts. One male pili tree can pollinate twenty female trees. The ideal soil is deep, rich loam with a pH of 4.5 to 7.0, with full to partial sun. Pili trees can grow on a straight trunk to 60 ft. with dense, rounded foliage, good for shade. Its brown bark flakes in large pieces. Small creamy-white blossoms are found on both male and female trees that produce an oval fruit 1½ - 3 in. long. The trees begin to bear within 6 years and reach optimum production in 10 years. The young nuts are green and ripen to a dark purple to black with a thin yellow pulp encasing a thick-shelled triangle nut holding one buttery white kernel.

One tree usually produces 35kg of nuts. The fruits are husked, the exterior pulp removed, and dried for 3 days so the interior nut kernel loosens from the shell. The hard shells have to be cracked by hand using a long heavy blade. Machines smash and waste the kernels. Shelling a maximum of 100kg a day is one of the aspects that makes pili nuts expensive.

MEDICINAL: Resin from the pili nut tree has been used as an ingredient in ointments and because of its pasty consistency, antiseptic properties, and quick drying, it's used as a natural plaster to heal wounds. Raw pili nuts are used as a laxative.

NUTRITION: One ounce (28g) of pili nut has 200 calories with 3g of protein and contains magnesium, potassium, and calcium. They're rich in omega fatty acids, phosphorus, and have high levels of protein and all eight essential ami-no acids. Pili nuts are also rich in vitamin E. This promotes a healthy heart and lowers cholesterol. They are also processed into pili nut oil, which is said to be better than olive oil.

FOOD & USES: Annually, the Philippines exports about 5,000 tons of pili nuts. The pili nut meat is the main food source, eaten raw, sugar-coated, salted, or roasted. Pili nut powder is used in pastries and ice cream. The kernel and the pulp can be cold pressed for excellent cooking oil. When boiled, the nuts' pulp is eaten as a vegetable that compares in taste and texture to avocados. It's combined in salads after blanching in hot water for 3 minutes. Also, oil can be extracted from the nuts' outer pulp and used for cooking or as a substitute for cotton-seed oil in the manufacture of soap. Fresh sap gathered from slashed shallow cuts on the trees' trunks is also used for igniting fires. The stony shells are excellent media for orchids and anthuriums. The hard shells are formed into bricks and used as fuel. The resin is an ingredient in paints, varnish, and sealants.

KANAMI-HAOLE – TROPICAL ALMOND

Terminalia catappa L

Kanami-haole is Hawaii's name for the tropical almond, native to S.E. Asia and brought to the island chain before 1800. This is another good productive, shade tree for the home, but it's big, growing to 40 meters. In ancient times, Chinese or European merchants probably brought some nuts, or the dried tropical almond nuts floated, easily crossing the oceans, and took root along coastlines. Blessed, a few nuts grew, and now the tropical almond is a popular local nut.

Propagated by seeds, the tropical almond tree is extremely hardy against drought, wind, and salt. This is a fast-growing, beautiful, and unique tropical tree. Unless you want a very tall tree, cut the center stalk after 3-4 years to stunt its growth. Cut it during a full moon and the branches will spread. Plant it at least 6 meters from your house and far from your septic tank. The tropical almond grows near beaches, river mouths, and on coastal plains up to 300 meters of elevation. The greenish-white blossoms are small, arranged in dainty clusters along a stem. As they mature, the leaves change from green to yellow, to a brilliant orange-red. The color changes are why this is a great landscape ornamental. This tree requires cleanup; it drops its leaves twice a year. It begins to bear when the leaves drop. The fruits leave a black stain if left in place for more than a day. From blossoms to harvest takes 3 months. The edible fruits are green and turn red when ripe.

MEDICINAL: The tropical almond tree is a natural pharmacy. Parts treat dysentery, leprosy, coughs, jaundice, indigestion, headaches, colic, pain and numbness, fever, diarrhea, sores, skin diseases, diabetes, etc. An infusion of the leaves treats jaundice. The leaves are used to treat indigestion. Externally, the leaves may be rubbed on breasts when feeding a child to cure pain or can be heated, applied to sore parts of the body, or used in a poultice for swollen rheumatic joints.

NUTRITION: 100 grams has 600 calories and 20 grams of protein. The oil content is 61g. The seeds are a rich source of zinc, magnesium, iron, and calcium.

FOOD & USES: These nuts may be enjoyed fresh, raw, and without seasoning. They are sometimes curried or shredded into a mix with coconut, onions, hot peppers, and spices. The seeds contain about 50% oil, which unlike other oils, is harmless to heart patients. The oil can be used in cooking and for soap. The fruits yield a black dye. The wood is moderately hard and used for cabinet work, construction, boat, bridges, floors, paneling, boxes, and crates.

RECIPES: TROPICAL ALMOND CONFECTIONARY

Ingredients: ½ cup each of the following: tropical almond (AKA - Indian almonds) sliced in half and roasted, oil, ghee, and corn flour mixed with ½ cup water and 1 TS red food color, 3 cups sugar, 3 cups water, 1 TBS ground cardamom

Method: In a small pan on low heat, melt the oil and ghee and set aside. In a pot or wok on high heat, combine the sugar and 3 cups of water. Bring to a boil, stirring and add the corn flour mixture. Reduce heat to medium and cook for 5 minutes. Add a spoonful of the oil-ghee mix every 2 minutes, constantly stirring. In 10 minutes, all the oil should be added. Cook for a total of 30 minutes after adding corn flour mix. Stir constantly for the final 10 minutes. There will be some splatter as the oil is released. Carefully spoon out the floating oil. It should be about half a cup. At 25 minutes, stir in the tropical almonds. Remove from heat and transfer to a glass bowl. Cool before serving.

NAMES: Almond, West Indian almond, false kamani, Sri Lanka almond, Malay almond, Barbados almond, *Hawaii:* kanami-haole, *Thailand:* tau khon, *China:* lan ren shu,

THE HAWAIIAN HOME GARDEN GUIDE
BULB & STEM VEGETABLES

Bulb vegetables are eaten for the aroma and flavor they add to simple food. Bulbs can be considered storage organs. At the bottom of a bulb is a thin, flat disc called the basal plate. Fine roots grow from this plate. The body of a bulb is made up of layers of fleshy scales, which can be considered a type of modified leaves. It is here these plants store food, mostly sugars. Bulbs are usually biennials grown as annuals. In the center of the bulb is the bud for the next year's plant. Leaves arise from underground stems with long sheathing bases, or straws. Most of the edible bulbs are of the *Allium* family – onions, chives garlic, shallots, and leeks.

These seasoning bulbs are easy to grow and integral in what is called 'a kitchen garden' to give a fresh taste to your dishes. All are easy to grow in containers and can be done in small pots on your window ledges, no matter what season.

If not able to grow your own and are forced to buy, always select firm bulb vegetables that are unblemished, absent of mold or dark spots. The green part of the chive, spring onion, and leek should be firm, bright, and shiny, not withered. Bulb vegetables can be stored for long periods and are not only delicious, but very good for you. Some bulb vegetables, like garlic, chives, and onions, are also well known for their medicinal qualities.

Celery is the most common stem vegetable. It is very tasty, used mostly for flavoring along with the bulb vegetables. In most vegetables, stems support the entire plant and have buds, leaves, flowers, and fruits. They are also a vital connection between the leaves and roots to transfer water and mineral nutrients from roots upward, and organic compounds and some mineral nutrients in any direction within the plant. Bulb plant stems of chives, garlic, onion, and leeks, are usually called straws. They are also used for flavoring.

PAGE:

Drawings Courtesy of R. Douglas

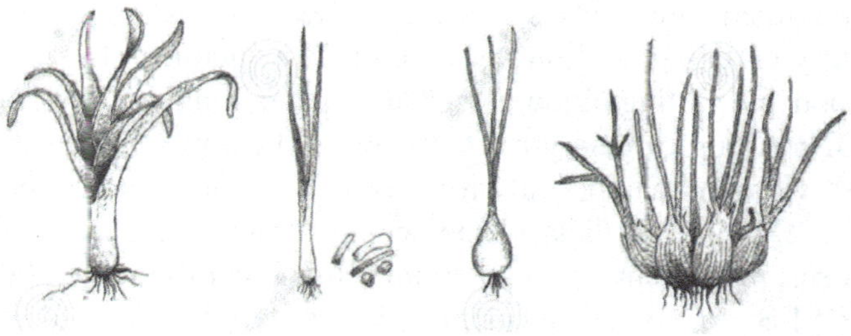

I like gardening — it's a place where I find myself when I need to lose myself. ~
Alice Sebold

CELERY
Apion graveolens

I'd be lost cooking without celery. Part of our garden is dedicated to kitchen herbs with chives, culantro, sage, thyme, parsley, and of course celery. Celery is a kitchen requirement along with carrots, onions, and potatoes. Its distinctive flavor makes it a great addition to many cooked dishes. Hawaii grows more than 500 tons yearly and imports about double that amount. Most of the celery is grown in the Waimea area of The Big Island. Most islands grow a local variety, smaller and stronger tasting than the imported type with larger stalks. Leafy celery stalks grow to a foot tall joined at the base. It belongs to the same family as carrots. Most people consume the stalks, but the leaves, roots and seeds can also be used as a food and seasoning, as well as a natural medicinal remedy.

Celery originated around the Mediterranean Sea and North Africa more than three thousand years ago. The Greeks at the time of Troy considered celery a religious plant and used celery and parsley without any difference. The Romans gave it a name meaning 'strong smelling' and used celery as an herb for cooking, but also considered it a superstitious bad omen. The Chinese recognized that celery could reduce blood pressure. Modern varieties of celery have varying degrees of bitterness. Eating celery did not become popular until the 1700s.

HOW TO GROW: Celery grows on every Hawaiian Island. Its best garden location is facing the wind at less than 200 meters of elevation. Celery is not easy to grow under the tropical sun, because it needs plenty of water with some shade. It also takes lots of patience because it's a 4-to-5-month crop. After deep forking a bed, raise it about six inches. (A box is best.) It's best to purchase transplants or sprout seeds indoors. Seeds are available, but are tiny and newly sprouted seedlings are very fragile to the sun. Transplants should be 3-4 inches tall and somewhat hardened to the sun before moving to their specific spot in your garden. Place the shoots 6 inches apart. Water at least twice a day and use a high-nitrogen fertilizer mix sparingly every 2 weeks. First spray with a fungicide such as Ban Rot and then spray every 2 weeks with a pesticide such as Fastac or Pestac. Do not use strong pesticides as the poison will not wear off by harvest. Researchers have listed celery as one of twenty vegetables usually sold with residual garden chemicals. Celery is best grown at home, so you know what you're eating!

MEDICINAL: Celery is a heart-friendly vegetable. It contains compounds that relax the arterial muscles, which regulate blood pressure. Juice from 4 stalks can cause about a 15% reduction in blood pressure. Celery consumption ights arthritis, rheumatism, and gout. Its anti-in ammatory properties reduces swelling and pain around the joints. Celery also assists in removing uric acid crystals that build around joints. This 'stalk/stem' vegetable contains vitamin C, which strengthens the immune system, so eat celery to relieve a cold. Eating celery daily may reduce artery-clogging cholesterol. It contains a chemical that lowers levels of stress hormones in your blood, permitting blood vessels to expand, reducing pressure.

NUTRITION: A cup of celery contains only 20 calories. The leaves are valuable as they contain the most vitamins. Supposedly it takes more calories to eat and digest celery than there is in the celery. It is high in dietary iber with vitamins A, B2, B5, B6, C, K, calcium, magnesium, phosphorus, magnesium, and potassium. It is low in saturated fat and cholesterol.

FOOD & USES: Always wash celery and keep it in the fridge. If there's space, store it upright in the fridge in a glass of water. Since celery has high water content, room temperature will cause it to wilt. Chopped celery can be added to tuna or chicken salad, and the leaves can be used in salads, stir-fry, and soups. If you're a juice lover, mix some celery with carrots in a blender.

RECIPES: BROILED OR GRILLED CELERY – unique side dish

Ingredients: 10 stalks of celery - better to use the big, imported stalks, 1 small onion chopped, 2 garlic cloves minced. ¾ quarter cup grated cheddar cheese, salt, and spice to taste.

Method: Cut celery stalks four inches long. Put in a baking dish under the broiler or on foil on the grill or coal pot. Once both sides are slightly brown, turn celery so that the indented part is up. Fill with onions and garlic. Cover with cheese. Return to cooking until the cheese melts.

BAKED CELERY

Ingredients: 12 celery stalks with leaves, 2 TBS butter, ¼ cup grated cheddar cheese, salt, and spices to taste

Method: Wash celery and trim the root end. Put into boiling water for 10 minutes. (If you prefer more crunch- don't boil.) Chop celery into 2-inch pieces and place in a greased baking dish. Cover with grated cheddar cheese, spices (add a topping of breadcrumbs or chopped nuts). Bake at 200 degrees for 15 minutes.

CELERY-BANANA SOUP

Ingredients: ¼ cup celery chopped small, 3 bananas green to just ripe any type - peeled and chunked, 3 cups boiling water, 1 TBS butter or margarine, 1 cup condensed milk, salt, and seasoning to taste

Method: In a pot with a cover combine bananas, celery, water, and seasonings. Boil for 10 minutes. Add milk and butter. Simmer covered stirring infrequently for 10 minutes. Serve warm.

CELERY STIR-FRY

Ingredients: 2 cups celery chopped small, 1 cup grated coconut, 1 medium onion chopped small, 2 garlic cloves, 1 TS chopped fresh ginger, 2 green chilies ground into paste, ½ TS fresh ground black pepper, 3 TBS lime juice, 2 curry leaves, salt, and spices to taste

Method: Combine all ingredients in a suitable pan or wok. On medium heat for 5 minutes, stir until celery wilts and coconuts begin to brown. Continue to stir for 5 more minutes, but reduce heat to low. Remove from heat and add lime juice, salt, and your choice of spices.

SWEET AND SOUR CELERY

Ingredients: 1 bunch of imported celery-leaves removed and chopped into 2-inch pieces, 2 cups water, 3 TBS lemon juice, ½ hot pepper-seeded and minced (optional), ¼ cup vegetable oil, 1 TBS brown sugar, 1 TS salt,

Method: In suitable deep skillet combine all ingredients except celery and bring to a boil while stirring. Add celery, cover, and simmer about 20 minutes. Remove from heat and keep covered until celery cools. Chill and serve as an appetizer.

DID YOU KNOW? King Tut's tomb contained a shroud adorned with garlands of wild celery, olive leaves, willow, lotus petals, and cornflowers. Ancient Romans considered celery an aphrodisiac. Scientists now know celery contains a pheromone, androsterone, that attracts. Casanova, the famous Italian lover ate celery to keep up his stamina. One ounce of celery seeds will grow an acre of celery. Celery's botanical name is *Apion graveolens,* a member of the *Umbelliferae* family with carrots, parsley, dill, cilantro, caraway, cumin, and the poisonous hemlock. The world's heaviest celery, 42 kg, was grown Gary Meeks in Malvern UK in 2018.

Using the recommended Oriental dosages to lower mild cases of high blood pressure, one would eat about four ounces of celery (about a cup of chopped celery) daily. You should begin to see results after only a week or two.

CHIVES

Allium schoenoprasum

Chives are the smallest member of the onion family and prized for their flavor. They are also one of the rare vegetables believed to be native to the northern hemispheres of both America and Euro-Asia. Chives are grown almost anywhere and everywhere for their delicate flavor.

Fresh chives are available in local markets all year, but it is easy to grow your own and a perfect addition to any Hawaiian kitchen herb garden. Chives can be grown in clumps or sin-gly. The dark green, hollow, thin leaves or straws reach 20 inches. Since no large underground bulb forms like regular onions, the leaves are the flavor. Chives are perfect for the kitchen garden because it is a perennial. It will grow in most places on every island and it just keeps growing.

HOW TO GROW: Chives can be grown from seed, but it takes weeks for the tiny seeds to sprout, and it is easy for them to be damaged by the severe sun. It is quicker and eas-ier to start with seedlings. Chives want moist, well-drained, very loose soil with a pH of 6-7. Chives grow best in full sun, yet will tolerate partial shade, and most soil types. Plant chives along edges of your garden to keep the other plants free from insects. Its flowers attract bees, important for pollination. Chives' pale purple flowers are star-shaped with six petals.

The leaves/straws can be harvested after the chives are six inches tall. When you are cook-ing, simply cut or use scissors to snip the leaves two inches above the ground. The plant will continually regenerate leaves, allowing for a continuous harvest. Do not cut off all the leaves of one plant clump at one time. This will permit the plant to rejuvenate and permit that same clump of plants to be cut over and over.

MEDICINAL: Consuming chives can lower blood pressure, but it takes more than the few usually in a meal. The usual amount will improve digestion and stimulate appetite. Chives can reduce fluid retention. Researchers discovered eating a lot of chives may help reduce the risk of prostate cancer, by as much as 50 percent. The Romans believed chives reduced the pain of sunburn, and a sore throat.

NUTRITION: Chives are high in vitamins A and C, potassium, and calcium. 100 grams of chives has 30 calories.

FOOD & USES: It was believed that bunches of dried chives hung around a house would ward off disease and evil. Chives are usually used fresh and are a common addition to baked potatoes, cream soups, eggs, and many other dishes.

RECIPES: QUICK GOURMET COCONUT CRAB SOUP

Ingredients: 3 TBS fresh chopped chives, 1 stalk celery chopped small, 1 small onion and 1 carrot chopped small, 2 cups coconut milk, 2 cans cream of asparagus soup (cream of tomato or celery can be substituted), ¾ cup water, 1 TBS curry powder, 1 TBS soy sauce, 1 TS fresh lemon juice, ½ pound fresh crab meat, salt, and pepper to taste

Method: Combine coconut milk, soup, one TBS chives, onion, celery, carrot, water, curry powder, soy sauce, and lemon juice in a large pot. Simmer, constantly stirring. Add crab and simmer until crab is cooked through. Add salt and spice to your taste and sprinkle with the remaining chives before serving with rice or roti.

HIGH-LIFE MASHED POTATOES

Ingredients: 2 TBS chives chopped, 5 pounds potatoes-peeled, cut into chunks and boiled, ¾ cup heavy cream, ¼ pound of butter, 1 cup sour cream, 1 cup grated cheddar cheese, 3 slices cooked bacon - crumbled, salt, spices, and pepper to taste

Method: In a two-quart pot, warm the cream and butter until hot, yet not boiling. Remove from heat.

Add sour cream, salt, spices, pepper, and the cooked potatoes. Mash until smooth. Add cheddar cheese, bacon, and chives. Serve warm.

BRITS' COTTAGE PIE - It's a shepherds' pie only if made with minced lamb.

Ingredients: 4 TBS chopped chives, 1-pound minced beef, 1 small can mushroom pieces, ½ cup beef broth, 1 medium onion chopped, 1 carrot chopped small, 2 garlic cloves minced, salt, spice, and pepper to taste, 1 TBS English (Worcestershire Sauce, 1 TBS flour, ½ cup heavy cream, 2 cups mashed potatoes, 2 cups cheddar cheese grated

Method: Divide mashed potatoes in half. Layer 1 cup of potatoes on the bottom of a 10-inch-square, 3-inch-deep, greased baking dish. (Almost any size including a pie dish will work. In a large heavy skillet sauté minced beef, mushrooms, onion, carrot, garlic, salt, and pepper until it is almost dry. Stir constantly to break the beef into small pieces. Add the English (Worcestershire Sauce and flour and stir for about a minute. Pour in beef broth. Then stir in the heavy cream. Simmer until everything thickens. Pour over the potatoes in the baking dish. Combine the remaining mashed potatoes with chives, 1 cup of grated cheese, and spices. Spread evenly on top of minced beef layer. Sprinkle 1 cup of grated cheese on top. Bake for 45 minutes at 375 degrees, until cheese has melted and is slightly burnt.

PAPAYA-PINEAPPLE SALSA

Ingredients: 2 bunch chives chopped small, 1 cup ripe papaya seeded and chopped, 1 cup fresh pineapple chopped, ¼ cup red onion chopped small, 1 hot pepper seeded and minced, 1 garlic clove minced, 1 bunch of culantro chopped small, 2 TBS lime juice, 1 TBS vinegar, pinch of salt

Method: Combine all ingredients in a suitable bowl and cover. Chill before serving.

HAWAIIAN CRAB PUFFS

Ingredients: 2 TBS chopped chives, 1 cup crabmeat, picked clean-fresh or canned, ½ cup shredded cheddar cheese, 1 TBS English/Worcestershire sauce, 1 TS dry mustard, 1 TS lemon juice, 1 TBS dill weed, ½ cup butter, 1 cup beer (optional), ¼ TS pepper, 1 cup all-purpose flour, 4 large eggs, 1 TS baking powder, ½ TS salt.

Method: Combine crabmeat, cheese, chives, English sauce, dry mustard, lemon juice, and dill in a bowl. In a large pan over medium heat melt butter, add beer, salt, and pepper. Remove from heat and stir in flour. Return to heat and add eggs one at a time. Add baking powder and crab. Whip until a nice dough forms. Put spoonfuls on a baking sheet and bake at 400 for half an hour or until golden brown.

DID YOU KNOW?

Chives are an herb with a mild onion flavor that grow in clumps like grass. They are referred to only in the plural because they grow in clumps rather than as individual plants. Only the leaves, also known as scapes or straws, have the flavor. Fresh chives can be stored in the fridge in a plastic bag for up to a week. Do not wash until ready to use them, as excessive moisture will promote decay. Chives botanical name is *Allium schoenoprasum* derived from the Greek meaning reed-like leek. Chives are also the smallest and only species of *Allium* (onions) native to both the New and Old World. Its English name, chive, is from the French word cive, which was derived from cepa, the Latin word for onion.

Gardening simply does not allow one to be mentally old, because too many hopes and dreams are yet to be realized. ~ Allan Armitage

GARLIC - KALIKA
Allium sativum

Among Hawaiian cooks, garlic, known as *kālika,* is indispensable in the local cuisine. Garlic is believed to have originated in the hot, dry Kirgiz Desert region of Siberia in Russia. In this region, the summers are dry, hot, and arid. That climate made garlic a very tough plant that can flourish in any soil and climate with little care. Garlic was cultivated by almost every civilization in history including at the gardens of Babylon. The Egyptians worshiped the garlic herb as a god. They also used garlic as currency. A slave could be bought for fifteen pounds of this bulb. Garlic was spread to Europe by returning crusaders after the Holy Wars. Warrior Vikings carried supplies of garlic in their boats. By 1000 C.E., garlic was grown in virtually the entire known world and was universally recognized as a valuable plant.

Although Hawaii has an excellent climate to cultivate garlic, most must be imported. Most garlic comes from China that grows more than 21 billion tons a year. Soft neck varieties like Creole are reported to grow well. Plant in the winter months and harvest in the spring.

HOW TO GROW: Growing garlic is easy. First, take two or three heads of nice fresh garlic and wrap them in a damp paper towel or paper bag. Place in the vegetable drawer of your refrigerator. The cloves inside the head will sprout. In your garden, fork a row ten inches deep and break the clods. The ultimate pH for garlic is 6-7. Plant the individual sprouted cloves every three inches with the sprouted side pointing to the surface. With enough water, light, and potassium-rich fertilizer, each clove will become a head. While growing, carefully snip a few of the garlic's green stalks (not all) and use in dishes. With some patience, islanders can grow their own garlic.

MEDICINAL: Garlic is valued for its taste, but also its medicinal benefits. No one knows how much garlic must be ingested to improve health, but experts feel the best results come from using raw garlic. Louis Pasteur discovered garlic kills bacteria. It was used as an antiseptic in World War II when sulfa drugs were scarce. Scientists have found that when raw garlic is cut or crushed, it creates a compound that kills at least twenty-three types of bacteria. Yet when garlic is heated it forms a different compound that reduces blood pressure and cholesterol. Garlic contains vitamins A, B, and C and stimulates the immune system. It may reduce the risk of stomach cancer and be a treatment for AIDS. Sixty kinds of fungi are also killed by garlic, including athlete's foot and vaginitis. If you fear garlic breath will end a romantic mo-ment, eat parsley or fennel seeds.

NUTRITION: No one eats a massive amount of garlic. Most prepared dishes have 3 garlic cloves. One teaspoon of garlic has 4 calories, slight amounts of protein and fiber, with lots of potassium, calcium, magnesium, and vitamin C.

FOOD & USES: The paper-like skin can be saved in a jar with water to sprinkle as a natural insecticide in your garden. Peeled garlic cloves can be soaked in vinegar for two days. When the vinegar is drained off, you have garlic-flavored vinegar. Cover the same garlic cloves with cooking oil and refrigerate creating garlic-flavored oil. Both the vinegar and oil should be used within three months. Whole heads of garlic may be baked as a spread for bread or crackers/biscuits. Remove as much outer skin as possible, but leave the head intact. Expose the cloves by cutting the top of the head. After placing the heads in a covered dish, pour olive oil on the heads and add your seasoning and pepper. Bake in the oven at 350 degrees for an hour. The cloves should be soft enough to squeeze out. To make a roasted garlic salad dressing, roast 6 crushed cloves with 4 tablespoons olive oil. Put in a 300-degree oven until the oil crackles. Mix with vinegar and spices of your choice to make a flavorful vinaigrette. To make garlic salt put three pressed cloves in a half cup of salt in a sealed jar. Let stand in the fridge for a few days. Use the garlic in cooking and the salt for flavoring. Smashing 6 cloves and whipping them into a quarter pound of butter or margarine makes potent garlic butter for tasty garlic bread or a base for cooking anything.

RECIPES: BRUSCHETTA – GARLIC BREAD

Ingredients: 1 loaf of French bread or local butter bread cut into 1-inch-thick slices, 2 TBS olive oil, 4 cloves garlic minced, 3 ripe tomatoes diced, 10 fresh basil leaves (or 2 TBS Italian seasoning), salt, and spice to taste

Method: On a cookie sheet or an oven pan, grill bread on both sides, either under the broiler or in a skillet. Mix all other ingredients and coat one side of the bread. Return to broiler for 2 minutes. For variety, add cheese or chopped hot peppers.

GARLIC ROAST PEPPERS

Ingredients: 6 sweet peppers (prefer red), 1 tomato chopped, 3 large onions, 6 cloves garlic chopped, 2 TBS olive oil, 2 TBS fresh or dried thyme, and salt and pepper to taste

Method: Slice onions into long strips, mix with tomato and garlic. Slice the top off the bell peppers and brush the insides with olive oil. Fill peppers with onion-garlic-tomato mixture and sprinkle each with thyme. Place in a covered oven dish and bake at 350 for 40 minutes. Uncover and place under broiler for 3 minutes or until edges of peppers just begin to brown.

GARLIC SHRIMP

Ingredients: ¼ pound of sweet butter or margarine, 4 garlic cloves minced, 2 pounds cleaned medium or large shrimp, 2 TBS olive oil, 1 TBS basil, ¼ cup white wine, salt and spice to taste, the juice of a fresh lemon or lime

Method: In a large frying pan, melt butter while adding garlic, basil, olive oil, and wine. Add shrimp and sauté until they turn slightly pink. Squeeze the lemon over the skillet. Serve with rice or pasta.

GARLIC COLD REMEDY

Crush a few garlic cloves. Put them in a bowl and cover with olive oil. Let sit for at least half an hour. Rub the oil on your feet and wear socks to bed. Sleep well as the garlic enters your system and clears your lungs of any cold.

HEALTH NOTES: Eating garlic fights heart disease, cancer, and colds, lowers blood cholesterol levels while re-ducing plaque in the arteries. Garlic is also known to be an aphrodisiac. Garlic has more ger-manium than any other herb. Germanium is an anti-cancer agent. Eating garlic may make you less allergic. Eating a clove of garlic every morning will fight asthma. During the World Wars, British medics used garlic to treat wounds when the supply of sulfur drugs ran out. Throughout history, garlic has been used as a strong antiseptic and as a strong antibacterial, antifungal, antiviral and anti-parasitic herb. These properties have been verified in countless studies.

DID YOU KNOW? April 19 is National Garlic Day in the USA. Garlic is sometimes called the 'stinking rose.' The smell of garlic can be removed by running your hands under cold water while rubbing a stainless-steel object. According to Christian mythology, after Satan vanished from the Garden of Eden, garlic grew from his left footprint and onion grew from his right. The botanical reference is *Allium sativum.* In 1985, the largest head of gar-lic, 1.19 kg (2 lb. 10 oz.), was grown by Robert Kirkpatrick in California, USA.

In the kingdom of spices, garlic is the king ~ Michael Bassey Johnson

LEEKS
Allium ampeloprasum

Leeks can grow everywhere in Hawaiian Islands when the temperatures are cooler. In the low elevations, they should be planted in August and September and harvested in the winter. In islands' cooler higher elevations, leeks can be planted throughout the year. Leeks are of the lily family, the same as the onion, but sweeter than the standard onion. Leeks have been around for 4 millenniums. When the Israelites fled Egypt, leeks are mentioned in the Bible as one of the foods the Israelites missed. The Romans brought leeks to Great Britain.

Although big leeks don't taste the best, there is a yearly competition in England to grow the largest leeks and the winners can range from 4 to 5 inches in diameter. Most leeks are used in soups. Better tasting leeks are slim, not fat, with clean white bulbs, and firm, tightly rolled dark green tops that are definitely not be yellowed or wilted. The base should be at least a half-inch in diameter. The younger the leek, the more delicate the flavor and texture. If the leek is limp at all, don't waste your money. Leeks can be an onion substitute in recipes, yet onions will never replace the unique flavor in leek recipes. Depending on freshness, leeks can keep in the fridge for up to a week. Leeks must be wrapped tightly to store in a fridge or other foods will begin to have a 'leeky' taste. Uncooked leeks can be sliced thin and added to salads.

HOW TO GROW: Leeks take a lot of patience, but you can grow many leeks in a small space. Start them from seed in a seedling tray and transplant them after three months (when they about chive size) into the ground. The leek growing routine is to fertilize once about four months after transplanting with 12-24-12, mold, and pull up dirt around the base, every month to make the white onion-like base grow longer. The higher you mold leeks by pulling the soil up, the more you'll reduce the chances of a tough green outer casing, giving you a tenderer white stem. They take about 6 months to mature, but remain good in the bed for at least another year. Leeks grow to about 2 feet in length and 1-2 inches in diameter, and when cut they have a mild onion aroma without tears.

MEDICINAL: Leeks have most of the same medicinal properties as garlic but to a lesser degree. Leeks are a laxative and a diuretic. Poultices of warm leeks are used to draw boils, reduce inflammation of bruises and gout. A syrup made from leeks is a good treatment for sore throat, laryngitis, and an inflamed respiratory tract. The water that leeks are boiled in makes a good rinse for both the skin and hair. It will reduce the itch from insect bites. Eating leeks reduces the risk of prostate and colon cancer.

NUTRITION: 100grams of leeks has 29 calories with 2g of protein. They are loaded with calcium, iron, and vitamin C with trace amounts of carotene, thiamine, and niacin.

FOOD & USES: Before cooking, always wash leeks and peel off the outer layers of skin. For many, leeks are more digestible than onions. Add chopped leeks to salads, egg dishes, and soups for a new flavor. Potatoes are tasty when cooked with leeks. Remember, don't add other seasonings as garlic, or strong onions that will overwhelm or hide the leek's subtle flavor.

RECIPES: POTATO-LEEK CAKES

Ingredients: 1 large potato peeled and grated, 1kg (2lbs.) leeks sliced thin, 4 eggs beaten, ½ cup breadcrumbs, ¼ cup grated cheese, 1 TS salt, ½ cup cooking oil (canola)

Method: Combine everything and chill in the fridge for at least an hour. Shape into small hand-sized cakes. Fry in hot oil until brown on both sides. If cakes do not hold together add one-third cup of flour. Serve with chutney or sour cream.

CURRIED LEEKS

Ingredients: 1kg (2lbs.) leeks cleaned and trimmed, 2 TBS curry powder, 1 bay leaf, 4TBS cooking oil (olive or canola), ¼ cup sultana yellow raisins, 1 apple peeled and sliced thin, ½ cup white wine (non-alcoholic may be substituted), salt, and spice to taste

Method: Make a layer of cleaned leeks in a frying pan. Cover with curry powder, bay leaf, oil, raisins, and apple. Add wine (grape juice) and boil. Cover and cook for about 20 minutes. Un-cover and simmer until no liquid remains. Serve hot with rice.

WHITE CURRY WITH LEEKS

Ingredients: 1kg leeks peeled and chopped, 4 garlic cloves minced, 2 green chilies chopped, 10 curry leaves, 1 TS of each of the following spices: curry powder, ground cumin, coriander, fennel, fenugreek, 4TBS coconut oil, 3cups coconut milk, salt to taste

Method: Heat the oil in a large pot or wok on high and sauté the spices, curry leaves, and garlic stirring for 5 minutes. Add coconut milk and bring to a boil. Add leeks, chilies. And salt. Cook 20 minutes. Serve with rice.

LEEK EGGS

Ingredients: 4 small leeks, ¼ cup butter, 1 clove of garlic minced, ¼ cup grated cheese, 4 eggs beaten, ¼ cup condensed milk, 2 TBS flour, 1 TBS baking powder, salt, and seasonings to taste Method: Mix all ingredients in a greased baking dish or bread pan. Bake at 350 for half an hour. Let cool before serving so it holds its shape.

LEEK POTATO SOUP

Ingredients: 2 to 4 cleaned leeks, 4 medium potatoes peeled and diced, 1 TBS olive oil, 2 cups water, 4 cubes vegetable bouillon, 1 cup milk, 2 TBS chopped parsley, salt, and seasonings to taste

Method: Sauté potatoes and leeks in olive oil for 5 minutes. Add bouillon, water, and seasonings, bring to a boil, and then simmer for half an hour. While stirring, add milk and simmer. Serve warm.

LEEK CHEESE AND BACON TART

Ingredients: 1 TBS olive oil, 3 leeks chopped, 2 cloves of garlic minced, 1TS fresh or dried rosemary, 1 pack ready-rolled puff pastry, ¾ cup cream cheese with garlic and herbs4 strips of fried bacon crumbled, ½ cup cheddar

Method: Preheat oven to 200C/400F. Saute' leeks and garlic for 3 minutes. Unroll the pastry onto a baking sheet. Cover evenly with cream cheese leaving 1in space around the edges. Spread leeks, bacon and grated cheddar. Pull the edges of pastry dough to double as a crust. Bake for 20 minutes.

DID YOU KNOW? The word leek comes from the Anglo-Saxon name for the plant, leac. The roman Emperor Nero ate loads of leeks because he believed it improved his voice. In the 6th century, St. David, the patron Saint of Wales, directed sol-diers going into battle to wear leeks tied to their hats so they could be identified from their enemy, the Saxons. They beat the Saxons and the leek became the national emblem of Wales. The world's longest leek (measured 136.0 cm (4 ft. 5.5 in.) in 2021. The world's heaviest leek weighed 10.7 kg (23 lb. 9 oz.) in 2018. Both were grown in the UK.

The highest reward for man's toil in the garden is not what he gets for it, but what he becomes by it. ~ John Ruskin

ONION - ANIANI

Allium cepa

The onion is an important part of everyone's diet in the Aloha State. It'll take some patient, concentrated effort to become self-sufficient in onions. Try short-day onion seeds; the Texas Early Grano onion and the Bermuda white should work. For a red variety, Red Creole is tolerant of heat. It's recommended to plant late in the year, October or November.

Onions have been known since the Bronze Age. With garlic, onion is mentioned in the oldest part of the Bible, desired by the Israelites after leaving Egypt for the Promised Land. Onions were found in the Egyptian tomb of King Tut. Columbus carried onions to the Western Hemisphere and their popularity spread among Amerindians. Wild onions grow on almost every continent.

HOW TO GROW: Onions will grow in most soils if it is deeply forked, with some shade from the heated afternoons when the plants are young. Later, onions need full sun and a long 11–12-hour day to develop a big bulb. The perfect pH is 6-7 and forking softens the soil so the onion bulb can swell and develop. Build up the row about six inches higher than the garden and soak the soil. With your finger create an inch-deep groove in the center of the mounded row. Carefully, sprinkle the tiny onion seed in that groove sparingly. Onion seeds are smaller than a pencil point. The best way to plant is to work the seeds through your fingers, dropping as few as possible. Beware of 'dumping' too many seeds in one part of the row. Once the seeds hit the dirt, they are almost invisible. Cover the groove and a few days later lightly sprinkle the rows with water. Do not water heavily as the seeds will wash out. Within ten days sprouts should show. Thin the sprouts to two inches apart. Weeds will kill the onions so gently pull any unwelcome visitors to the row. Using a hoe, pull dirt up onto the onions. Water lightly three times a week, more if it is extremely hot. Heat will burn out the stems. Use a 12–24–12 fertiliz-er sparingly every other week, then after six weeks switch to 12–12–17–2. Too much fertilizer will hurt the onions. Onions mature in 3-4 months. The last month reduce watering so the bulbs harden. Once the green stem begins to wither and turn brown, the onions are ready for harvest. Pull the onions and dry in a shaded area. Bright sun may discolor the onions. Onions may need at least a week to dry completely.

MEDICINAL: Eating onions gives some protection against heart disease and colon cancer. They may also reduce the frequency and strength of asthma attacks. It seems the more pungent onions, especially yellow, are better for you as they have more antioxidants. Their po-tent anti-inflammatory properties may also help reduce high blood pressure and protect against blood clots. A recent study shows quercetin, a compound found in onions, helps lower cho-lesterol. Eating onions may help prevent inflammation and hardening of arteries, beneficial to people with high cholesterol.

NUTRITION: 100 grams of raw onions has 40 calories with 1 gram of protein. Onions are a good source of vitamins B-6 and C, chromium, manganese, potassium, and phosphorus.

FOOD & USES: Onions may be eaten raw, broiled, boiled, baked, creamed, steamed, fried, French (deep) fried in oil, or pickled. Onions can be chopped and dried in the oven. Use the lowest setting and remove when thoroughly dry, yet not brown. Store at room temperature in airtight containers. Onions also can be frozen. Chop and place on a tray in the freezer. When frozen, remove and place in freezer containers or bags, and seal. This way they freeze evenly. This allows you to chop many onions at one time and then remove the amount you need. Frozen onions should be used for cooking only. Whole frozen onions can be baked.

RECIPES: TWO-DAY ONION SOUP

Ingredients: 1kg (2 lbs.) peeled sweet onions, 1/2 cup butter or ghee, 2 TBS paprika, a bay leaf, ¾ cup flour, 3 quarts beef bouillon, 1 cup white wine, 2 TBS browning sauce, salt, and pepper to taste, ½ pound grated Swiss or cheddar cheese. French bread.

Method: Slice onion a ¼ inch thick, place in a large soup pot and sauté slowly in melted butter. Add spices. Cook for 10 minutes longer before adding bouillon and wine and browning sauce. Simmer for 2 hours. Refrigerate overnight. Pour into oven-proof bowls topped with a slice of French bread and grated cheese. Place in a preheated 350 F oven for 20-30 minutes. Serves 6.

ONION SAMBAL – an easy to make, delightful taste of South East Asia

Ingredients: 500g (1lb.) onions chopped small, 10 curry leaves, I tomato diced, 1 green chili, 1 green cardamom pod, 1 cinnamon stick or ½ TS cinnamon powder, 2 cloves, 1TS brown sugar, 1 TBS tamarind paste, 4TBS oil, salt to taste

Method: In a pot or wok heat oil over a medium flame. Once hot, add the onions and fry until they brown. Add the remaining ingredients. Stir frequently for 10 minutes. This stores well in the fridge for a week as a condiment for anything.

ONION RINGS

Ingredients: 4 large white or yellow onions, 1 cup flour, 1 cup beer, 4 cups frying oil, 3 TBS sugar, salt, and pepper to taste.

Method: Blend the flour and beer thoroughly in a large bowl. Then the batter must sit covered at room temperature for at least 3 hours. The batter can be adjusted to thick or thin by adding more flour or beer. When ready to use stir in the sugar and seasonings. Slice the onions at least a ¼ inch thick and separate into rings. Heat oil in a deep pot. The oil must be hot enough that when a battered piece of onion is dropped into it, it immediately sizzles. Dip the onion rings in the batter and drop into the oil. Do not crowd the pot, as the rings will stick together. The onion rings will rise to the top of the oil when they are cooked. Fry them until golden brown. This can also be done as a 'Blooming Onion.' Slice a whole onion almost all the way through in both directions, leaving about a ¼ inch to hold it together. Put the entire onion in the batter and fry. Then slice and eat.

CARMELIZED ONIONS

Ingredients: 2 TBS oil olive or canola preferred, 3 large onions (red or white), sliced, pinch of salt, 2 TBS brown sugar, 2TBS balsamic vinegar

Method: Heat the oil in a large skillet or wok. Add onions with the salt and cook slowly for half an hour until they become soft and golden. Stir to prevent sticking and burning. Add a splash of water to the pan and mix well. Add sugar and vinegar for a sweet flavor. Cook 5 minutes until the sugar dissolves and vinegar has evaporated.

DID YOU KNOW? The Hawaiian word for onions is 'aniani.' Some people shy away from onions because they make you cry. When you cut into an onion, it releases a sulfur compound into the air. When it contacts with water, it is converted to sulfuric acid that stings your eyes. To stop the tears, before chopping chill peeled onions in the refrigerator. To get the onion smell off your hands, rub with lemon juice or vinegar. China grows the most onions with 29 million tons and India is second with 20 million tons a year. U.S. onion consumption increased 50% in the last 20 years. The ancient Egyptianshe worshiped the onion They believed that its spherical shape and concentric rings symbolized eternity. The world's heaviest onion was 8.5 kg (18 lb. 11.84 oz.) grown by Tony Glover in the UK in 2014.

An onion can make people cry,
but there's never been a vegetable that can make them laugh. ~ Will Rogers

RHUBARB

Rheum rhabarbarum

Rhubarb is mainly grown in the cooler highlands of the Hawaiian Islands usually at altitudes above 1,000 meters. Eaten raw, the stalks are crisp, similar to celery, with a strong, tart taste, but it's usually cooked, stir-fried, curried, or sweetened into desserts and pies. Rhubarb's color can be reddish-maroon to pink, or pale green.

Rhubarb's botanical name is *Rheum rhabarbarum*. A wild variety known as false rhubarb is *Rheum rhaponticum*, and probably an ancestor of today's cultivated rhubarb.

The exact origin of rhubarb is unknown. Chinese have used the rhubarb's root for medicinal purposes for thousands of years. They named it rhubarb 'the great yellow' (*dà huáng*). Marco Polo saw it cultivated in the northwestern mountains of China. Rhubarb was imported along the Silk Road, and reached Europe in the 14th century. It was much more expensive than opium and spices like cinnamon, and saffron. Originally, the roots were used for medicine. Later, the fleshy edible stalks were cooked and eaten.

WARNING!Do not eat the leaves of the rhubarb plant because they are poisonous.

HOW TO GROW: Rhubarb is a perennial vegetable that grows easily, but needs cool weather to really produce. Choose a permanent place in your garden or yard to cultivate rhubarb because this is a long-term, at least a 2-year project. First, ind someone who has the base of the rhubarb plant, known as crowns or roots. Rhubarb wants well-drained, fertile soil with a pH of 6 to 6.8, and preferably in full sunlight. Remember, this is a long-term project. Dig large deep holes, the size of a 5-gallon bucket; ill with a mixture of soil, compost, and rotted manure. Space rhubarb about 4 feet apart and plant the roots 2 inches deep. It's best to cover the rhubarb crowns with a layer of straw/cut grass and manure. That will keep the crowns moist and provide nutrients. Water regularly. If you keep the plants weeded, insects and diseases should not be a problem.

Permit your rhubarb plants to become established; don't harvest any stalks during the irst growing season. Harvest the stalks when they reach 18 inches long. Discard the leaves! Always let at least 2 stalks remain on each plant to ensure continued production. That way you can harvest for up to 20 years without replacing your plants. After the third year, the harvest period should be every 8 to 10 weeks. If the stalks become thin, stop harvesting; this means the plants need organic fertilizer, manure. Dig up and split the rhubarb crowns/roots every 3 to 4 years.

MEDICINAL: The *Rheum* species contains at least 60 different types and many hybrids. The type baked into pies differs from the medicinal rhubarbs, which are generally consid-ered inedible. For thousands of years, traditional Chinese medicine has used rhubarb roots as a laxative. It was one of the irst Chinese medicines imported to the West. Rhubarb also appears in medieval Arabic and European prescriptions. Today, scientists are exploring the various rhubarb species to treat ailments as dermatitis, pancreatic cancer, and diabetes. Rhubarb, stewed or juiced, is a tonic to aid the blood and the digestive system. Studies show that rhubarb helps lower your bad cholesterol levels as well as your total cholesterol.

NUTRITION: 100 grams of rhubarb has 116 calories, and is very high in potassium, and vitamins C and K1. Rhubarb is rich in antioxidants, particularly anthocyanins, which causes it to be red, and proanthocyanidins. These antioxidants have anti-bacterial, anti-inflammatory, and anti-cancer properties.

FOOD & USES: Boil rhubarb and enjoy it as a soup (with some sweetener.)

Rhubarb is delicious in preserves, puddings, and pies. Rhubarb is also used as a favor-ng agent. For cooking, the stalks are often cut into small pieces and stewed / boiled with a small amount of water. Rhubarb stalks have a high concentration of water, so little is needed. Stewed rhubarb, like applesauce, is usually eaten cold with added sugar, cinnamon, nutmeg, or ginger. Pectin, or sugar with pectin, can be added to the same stewed mixture to make jams. Rhubarb can make a fruity drink like lemonade and rhubarb compote.

Rhubarb should be processed and stored in glass or stainless steel containers, which are unaffected by its high concentration of oxalic acid content. Because of the same oxalic acid (the compound that makes the leaves and roots toxic to eat), rhubarb extracts are used to clean metal, tan leather, and repel insects. A natural pesticide is made by boiling the leaves in water for 20 minutes. This simple mixture will kill insects and fight fungus in your home garden.

WARNING!!! Keep all rhubarb roots away from children and pets.

RECIPES: RHUBARB POTATO CURRY:

Ingredients: 500g rhubarb washed, peeled, and chopped into 1 in. pieces, 2 large potatoes peeled and chopped, 3 garlic cloves, one large onion, 4 green chilies - all chopped, 10 curry leaves. 1TS each of curry, turmeric, black and red pepper powders, 1TS each of black mustard and fenugreek seeds, 1 stick of cinnamon, salt to taste, oil for frying, 1 cup water, and ½ cup thick coconut milk

Method: In a large pot or wok on medium heat, heat the oil and fry the garlic, onion, curry leaves, potatoes, and cook for 5 minutes. Stir in all the spices and salt. If you want, add some sugar. Add the chili pieces and fry for 2-3 minutes. Stir in the coconut milk. Cook for 1 minute, reduce the heat to low, and add the rhubarb pieces. Cook for two minutes. You may add extra water. Enjoy.

EXOTIC RHUBARB ACHAR:

Ingredients: 3 stalks of rhubarb washed, skinned, and chopped into ½ inch pieces, 4 garlic cloves and 2 green chilies chopped, 1½TS chili powder, ½ TS each of the following: black mustard seeds, fenugreek seeds, turmeric powder, 1TBS jaggery (coconut sugar) or brown sugar, 10 curry leaves, ¼ cup oil-gingerly preferred

Method: Heat 2 tablespoons of oil in a pan and add mustard seed. When they pop, add the fenugreek seeds, curry leaves. Fry, stirring for 1 minute. Add chopped rhubarb, green chili garlic cloves, chili powder, turmeric powder, salt and jaggery/brown sugar, and the rest of the oil. Once the rhubarb is soft, switch off the flame. Store in a glass jar.

RHUBARB JUICE:

Ingredients: 2 lbs. rhubarb chopped into 1-inch pieces, 8 cups of water

Method: Put rhubarb pieces into a pot with the water. Boil over high heat, then reduce the heat and simmer for 15 minutes. Strain through a cheesecloth. After 20 minutes, the juice will separate, leaving a yellowish sludge at the bottom. Decant the clear pink liquid into a clean bottle or pitcher, leaving the sludge behind. Chill for a few hours and serve over ice.

DID YOU KNOW? In Finland, Norway, Canada, Iceland, and Sweden, a sweet treat for children is a tender stick of rhubarb, dipped in sugar. In Chile, Chilean rhubarb, which is only very distant-ly related, is sold on the street with salt or dried chili pepper. Humans have been poisoned after ingesting the leaves. That was a particular problem during World War I when the leaves were mistakenly recommended as a food source in Britain. The toxic rhubarb leaves are used in flavoring extracts after the oxalic acid is removed by treatment with calcium carbonate. The world's heaviest rhubarb was 2.67 kg (5 lb. 14 oz.), grown by E. Stone (UK) of the UK in 1985.

SHALLOTS

Allium ascalonicum

Maybe you thought shallots were only smaller onions. In groceries, you'll find them near the onions and garlic. Buy some shallots and try growing them in a pot or garden. While delicate enough to eat raw, shallots can be diced, minced, sliced, or roasted. The small red bulbs offer a wonderful, but not overwhelming, soft, sweet onion flavor. Like mild garlic with a hint of onion, they are especially good used in fish and vegetable recipes. The shallot is small; their pink-purple skin is papery with pale purple and white flesh. Shallots should grow on every island with a little love and lots of patience. Plant October to December.

Shallots are used for both their nutritional and aromatic properties. They are used in Middle Eastern, Indian, Asian, Chinese, European, and Mediterranean cuisines. Botanists believe shallots originated in Central or Southeast Asia before Arab traders carried them to India and the Mediterranean. The ancient Greeks named shallots after finding them in a Palestinian port now known as Ashkelon in Israel. Ancient Egyptians used them as medicinal remedies.

HOW TO GROW: Shallots are usually grown from sets or bulbs, and they are planted the same as garlic cloves. Separate each bulb and plant an inch deep, 4 - 6 inches apart with the pointed end facing up in rich, well-drained soil, with a pH 5-7. They grow to 1-2 feet tall; 6-12-inch spread. Shallots require full sun, but in the beginning partial shade assists in surviving the afternoon heat. A removable sunshade is a good idea. Their bulbs grow in clusters, like garlic. When planting, keep the tops of the bulbs a little above ground. Later as they develop, keep molding soil around the shallots as water will tend to wash it away. It's recommended not to fertilize with animal manure. Instead use a 12-12-17-2 mixture.

Shallots require regular water, but be sure the soil isn't always moist. You can cut some of the green tops to use as green onions. Leave a portion of the stems intact to feed the bulbs. Shallots are ready to harvest in 3-6 months. Shallots will sprout again if you leave them in the ground at the end of the season. When they are ripe, as with onions, shallot tops start to yellow and fall. Shake off excess soil and let them cure in a dry, shady spot for a couple of weeks. You can store shallots for up to 4 months if kept cool. There are basically three types of shallots. Some consider the French gray shallot, also known as griselle, with gray skin, as only 'true' shallot. It still grows wild in Central to Southwest Asia. Pink or 'Jersey' shallots with a reddish color are the most common and as delicious as the French grays. Echalion shallots are a cross of a regular shallot and an onion. Also known as 'banana' shallots, they are larger like an onion with the mild flavor of a shallot.

MEDICINAL: Shallots contain a vegetable pigment, quercetin –a flavonoid – that strengthens small blood vessels and protects against heart disease and diabetes. Researchers also believe quercetin reduces bad cholesterol. All allium vegetables are recognized for their ability to kill and inhibit cancer cells, which diminishes the risk of cancer. The most important benefits of shallots are: as a high source of antioxidants that improve heart health, in cancer and diabetes prevention, as an anti-inflammatory, and as an antimicrobial, that might also help fight obesity, and helps prevent or treat allergies. They have been used as a remedy for sore throat, infections, and bloating.

NUTRITION: They are slightly high in calories – 100 grams of shallots contain 72 calories, with 2.5 grams of protein and 17 grams of carbs. Shallots are full of fiber, vitamins, minerals, and antioxidants. They are high in vitamin A, pyridoxine, and potassium with calcium, phosphorus, iron, and copper. Shallots are an excellent source of selenium a trace element that slows aging, fights dementia, and keeps skin and hair healthy.

FOOD & USES: Shallots have a mild onion/garlic flavor and can be used in any recipe calling for onions, especially where you want a milder taste. They differ from green onions where the white and green parts are often used differently in cooking. Shallot tops and roots both have a stronger flavor than green onions.

WARNING!!! Like other members of the *Allium* family, all **parts of the shallot contain a toxin that affects red blood cells in cats and dogs, causing anemia and sometimes death.**

Remove the papery skin and root end. Slice the shallot in half length-wise. Place the cut side down, and slice long ways, stopping just before you reach the root end. This will keep the layers together and allow for easier slicing. Use the tip of the knife to slice. Once you peel, slice or chop a shallot, the raw shallot should be placed in an air-tight container and can stay fresh in your fridge for about 5 days.

They are tasty sautéed in butter and added to recipes. They are great for sautéing, stir-frying, and can also be braised or roasted in chunks or whole. When buying, pick the firm bulbs with bright, taut skin. Avoid those with damage or bruises. Small shallots have a mild taste and larger ones taste stronger. 'Melting' is a term used with shallots because when they are cooked, they soften, and their lavor combines beautifully into dishes. They can be diced, sliced into rings, and fried, roasted whole either in their skins or peeled, or pickled. Raw shallots also make a great addition to salad dressings, and if you find them fresh, their green tops can be used as an aromatic seasoning or garnish, similar to spring onions.

RECIPES: MILKY RICE WITH CHILI SAMBAL - exotic taste of Sri Lanka
Ingredients: 1 cup white rice, 2 cups water, 1 cup coconut milk, pinch of salt

VEGETARIAN CHILI SAMBAL: Ingredients: 1 cup peeled shallots, 4 dried red chilies 3 TBS red chili powder, 2 TBS fresh lemon juice, a pinch of salt
Method: Cook rice until it's soft and most of the water is gone. Add the coconut milk and cook for 5 minutes until the rice becomes milky. In a frying pan quickly sauté the shallots and red chilies, only a few seconds. In a mortar, pound it into a paste. Add lemon juice. Serve atop the cooled rice.

PICKLED SHALLOTS

Ingredients: 1kg shallots, 1 cup coarse sea salt or kosher saltPickling vinegar:1 TBS black peppercorns, 1 TBS coriander seeds, 1 TBS yellow mustard seeds, 10 cloves, a few pieces of mace blades, one dried red chili chopped (optional), 2 bay leaves, 3 cups white wine vinegar, plus 3½ TBS of wine vinegar, ½ cup light brown soft sugar
Method: The day before pickling, put the shallots in a large bowl and cover with boiling water. Cool, then drain and peel the skins and cut off the root ends. In a large bowl, dissolve the coarse sea salt with 2 ½ cups boiling water. Add 5 cups of cold water, then add the shallots. Cover and soak overnight. The next day, drain and rinse the shallots several times. Pack the shallots into sterile jars. For the pickling vinegar, put all the whole spices in a medium saucepan and roast over a low heat until they give off the aromatic smell. Add the dried chili last. Then add the bay, pour in all the vinegar and sugar, let it dis-solve, and simmer. Pour the hot vinegar over the shallots and seal while hot. Ready to eat in a month if you can wait.

DID YOU KNOW? Ancient Egyptians considered shallots sacred. The shallot was carried to France the by the survivors of last Crusade. Botanists have trace shallot's origin to the village of Ascalom in Israel. Bigger shallots have less flavor.

NAMES: Shallot, French shallot, gray shallot, Spanish garlic. They are called *kanda* in India, *ham* in the Philippines, and *brambang* in Thailand.

Growing food was the first activity that gave us enough prosperity to stay in one place, form complex social groups, tell our stories, and build our cities. ~ Barbara Kingsolver

THE HAWAIIAN HOME GARDEN GUIDE

VINES

Vines refer to any climbing or trailing plant. They may be dense, airy, bushy, shiny, colorful, attractive, vertical, and productive. A vine is basically a long stem that uses energy, seeking sunlight to produce fruit. Vine vegetables are considered 'small space' vegetables because they need just a bit of soil and lots of vertical growing room to climb. One of the best ways to make efficientuse of the area in a home garden is to use vertical space. You'll get more production per square foot by using vines trained to climb instead of allowing them to sprawl. Keeping vines off the ground can be healthier, fighting soil bacteria, fungus, and dampness that could damage the plant before it produces.

Fences can support climbing fruits and vegetables. All that's needed is to plant some vining vegetables at the base of the fence, then water, and fertilize as needed. Nature will take care of the rest. Allow vines as cucumbers, squash, melons, and long bean that usually sprawl on the ground to climb. Building supports for climbing vegetable plants adds character to your garden. A tall tripod made of bamboo is a beautiful and easy support for vining plants. Push bamboo posts deeply into the ground and wrap at the top with garden twine or wire. Wrap wire around the legs of the tripod to add additional support to the structure and provide additional surface area, like rungs on a ladder, for the cucumbers or long bean to cling on. Sow seeds directly around the base of each pole, encouraging them to climb as soon as they emerge.

Tie trailing cucumbers, squash, and pumpkins to a sturdy frame or tripod. The vine will be able to support the fruit, in most cases. If you have a very heavy squash or pumpkin developing, a sling made from tying the sleeves of an old shirt to the fence, tripod, or trellis, and using the trunk of the shirt will make it extra secure.

The world's longest gourd was 3.954 meters (12 ft. 11 in.) and was grown in Serbia in 2019. The heaviest gourd weighed 213.41kg (470½ lbs.) was grown by Steve Connolly in the USA in 2020.

PAGE:

I have no hostility to nature, but a child's love to it. I expand and live in the warm day like corn and melons. ~ Ralph Waldo Emerson

ASH MELON – WINTER MELON – WAX GOURD

Benincasa hispida

Ash melon is a delicious, nutritious, unique tropical melon, that's similar to a pale watermelon (round or oblong). When it's young, it's covered in a fuzzy coating of fine hairs, and a foggy gray color. That gives it the name, ash melon. Young ash melon has thick white flesh that tastes sweet. As it matures, the taste becomes bland. The fruit goes bald and a waxy coating develops that gives it another name, wax gourd. Botanists believe it originated in Japan or Java. The botanical name is *Benincasa hispida.* Ash melon is a member of the cucurbits family along with cucumbers, melons, squash, and pumpkins.

HOW TO GROW: Ash melon is an annual climbing vine that can extend to 6 meters (20ft.) in length. Although scarce, this will grow anywhere throughout the Hawaiian chain. Get seeds online. It's comparable in taste to chayote/christophene, especially in stir-fries. Mature fruits can vary in weight to 50 kg. Unless supported by a strong trellis, these vines will grow over other ground plants. Properly supported, it's an attractive vine with yellow flowers and broad leaves.

This vine likes full sun in compost-rich, well-drained soil. It requires regular watering to produce big fruits. The best pH is 5.6 to 6.8 and is reasonably drought tolerant. It can be grown up to 1,500 meters in elevation. Plant the seeds 3cm (1in.) deep and a meter (39in.) apart. The seeds should germinate within 20 days. It's best to train the vines to climb and keep them off the ground. Plants take 5 months to have mature fruits, but immature fruits can be eaten after 3 months. The fruits can weigh from 5 - 50 kilos with a mild flavor. (Stronger in immature fruits with a juicier texture). Due to the natural wax coating, ash melon will store for several months, and in the right conditions, as long as a year.

MEDICINAL: Ash melon has been used as a food and medicine for thousands of years in Asia and especially in China. All parts of the fruit are used medicinally. The skin or rind is diuretic and treats urinary problems. The skin is burned and then applied to relieve the pain of deep wounds. The seeds treat inflammations, bruises, and coughs. Seed oil is used to expel parasites. The fruit's flesh is considered an aphrodisiac and treats lung diseases such as asthma. Research has found these fruits contain anti-cancer aromatic compounds. The fruit's juice treats nervous disorders. It is believed, daily consumption of ash melon sharpens your focus, and increases your brainpower, and energy without the nervous jitters of caffeine from drinking coffee. Eating ash melon will strengthen your overall immune system.

NUTRITION: 100 grams of ash melon has 13 calories and is a valuable source of phosphorus, calcium, and magnesium. It has good amounts of vitamins B1, B2, C, and niacin. The seeds are rich in protein and oil.

FOOD & USES: These gourds should have smooth skin when buying without bruises, cuts, or indentations. They can be the same size, shape, and color of a watermelon, but with a distinctive white, ash-coated surface. The white residue is harmless, but sticky when wet. It's best to wash the gourd before cutting. Expect the inside to be white, crisp, and not soggy with an even texture. They are eaten as vegetables in curries, as flavoring for soups, pickles, and pre-serves. The wax coating is used to make candles. Blossoms and young leaves can be steamed and eaten as a vegetable like spinach or added to soups. The seeds can be pressed for oil.

RECIPES: ASH MELON ENERGIZER

Blend 3 cups peeled and seeded ash melon with the juice of 2 limes. Add salt and fresh mint to your taste. Serve on ice.

ASH MELON SOUP

Ingredients: 250g (½ lb.) ash melon skinned, seeded, and chopped into 1-inch chunks, 2 cups dried or fresh mushroom (shiitake preferred), 2 cups chicken or vegetable broth depending on your inclination and taste, a 1-inch cube of fresh ginger peeled and sliced, salt, and pepper to taste. 1 green/spring onion and or 2 sprigs of cilantro chopped as a garnish per bowl.

Method: If using dried mushrooms place in a bowl of water, cover, and rehydrate for 30 minutes. Squeeze out excess liquid. In a wok or suitable pot, bring water to a boil, add ash melon and simmer for 20 minutes, until tender. Add mushrooms, chicken or vegetable broth, ginger, salt, and pepper. On low heat, simmer for 30 minutes. Serve hot garnished with spring onions and or cilantro.

CANDIED ASH MELON

Ingredients: 500g (1lb.) ash melon skinned, seeded, and chopped into sticks 5cm (2 in.) long and 1cm (½ in.) wide, 1TS baking soda (NOT baking powder), 500g (1lb.) sugar (brown or white) to your taste.

Method: Put ash melon chunks into a pot of water with 1 TS of baking soda and bring to a boil for 1 minute. Drain ash melon in a strainer or colander. On a medium flame, heat the sugar with ½ liter (2 cups) of water until dissolved. Let it boil for two minutes and reduce heat to low. Carefully stir in the drained ash melon pieces into the sugar syrup. Simmer for 10 minutes but do not permit it to boil. Remove from heat and keep everything in the same pot cooling for a full day. Then carefully spoon out the ash melon pieces and in a suitable area let drain undisturbed on waxed paper. Store in an airtight container. Ground nutmeg and cinnamon, black or red pepper powder may be sprinkled before these candied pieces become solid.

CURRIED ASH MELON

Ingredients: 2 cups ash melon peeled, seeded, and chopped into chunks, 2-inch piece of ginger peeled and sliced, 1 large tomato chopped, 4 green chilies chopped, 5g tamarind paste, 1TS turmeric powder, salt, and pepper to taste, 6 curry leaves, 3 sprigs cilantro, 1TBS coconut oil, 1TBS black mustard seeds

Method: In a suitable pan or wok, on medium flame, heat the oil and add ash melon, tomato, ginger, chilies, cilantro, and tamarind paste until vegetables soften, stirring for 5 minutes. Stir in the turmeric powder, salt, and pepper. Remove from heat and let vegetables cool. Then put the mixture in a blender until creamy. In the same, now empty pan, on medium flame, add 1TS oil, mustard seeds, and curry leaves. Heat until mustard seeds sputter. Remove from heat and combine with creamy ash melon mixture. Serve with rice or roti.

ASH MELON AND WATERMELON SMOOTHIE

Peel, seed, and chunk equal parts of ash melon and watermelon. Put in a blender with plain yogurt (or water). Add honey and blend. Serve on ice.

DID YOU KNOW? Medical and pharmaceutical companies are interested in ash melon. An initial research study in 2000 published in *Fitoterapia*, demonstrated that this melon's juice helps patients withdraw from opioid addiction. Research in 2001, showed that consuming ash melon treated stomach ulcers. An Indian study in 2003 found that ash melon was a strong anti-depressant. A 2005 study found that oil extracted from ash melon seeds reduced the blood supply to cancerous tumors. It is recommended to drink one glass of ash melon juice every day for a week and you'll feel more energized, alert, and focused.

NAMES: ash gourd, ash pumpkin, Chinese watermelon, preserving melon, white gourd, tallow gourd.

Food tastes better when you eat it with your family. ~ Everyone

BITTER MELON – 'AWA 'AWA MELON

Momordica charantia

Bitter melon or *'awa 'awa melon,* is a mystery vegetable. It's not attractive, not sweet, but easy to grow and one of the healthiest foods you could ever 'learn' to eat. This bumpy-skinned fruit of a vine is the strangest and bitterest member of the melon family. Bitter melon is considered a 'strange' food because it's seldom sweet. People seem to love a bitter-hot taste. Bitter melon fruit is for all those who love a truly bitter taste. A mature fruit should first be sliced and salted, then squeezed to remove the bitter juice before cooking. However, small young fruit are almost sweet. The mystery is why so few people grow it, and so few people know how good it is for you. Try to enjoy the bitter taste for all its healthful virtues! This vine will grow anywhere in the islands.

Bitter melon is a tropical fruit, not a vegetable. This fruit has a distinct looking exterior with warts and an oblong shape. When cut in half it shows a relatively thin layer of flesh surrounding a central seed-pith-filled cavity with large flat seeds. Seeds and pith appear white in unripe fruits, ripening to red. The flesh is crunchy and watery similar to cucumber or chayote/christophene. Bitter melon is usually eaten before it fully ripens. It can be eaten when ripe and turns yellowish, but it becomes bitterer as it ripens. The fully ripe fruit turns orange and mushy, and is too bitter to eat. It uniquely splits into segments that curl to expose seeds covered in bright red pulp.

HOW TO GROW: Bitter melon is easy to grow from seeds. It is best planted along a fence or anywhere the vine can climb. Because bitter melon seeds are scarce, first visit the market and search for an overripe, yellowish fruit. Set the fruit out until it softens and then remove the seeds to dry. The best pH for bitter melon is 5.5 – 6.7. Dig several small holes along a fence line. Plant about four seeds per hole. Water regularly and in a few days, bright green sprouts will appear. As the vine grows, carefully weave it onto the fence. Bitter melon is a natural climber. Spray occasionally with a mild pesticide and water-soluble fertilizer. In a few weeks, yellow blossoms will appear. Water every other day and once a month sparingly use a high nitrogen fertilizer mix at the roots. Birds will be the biggest pests to your bitter melon. This vine easily grows out of control.

MEDICINAL: The bitter juice can be helpful to diabetics. A tea of the leaves and blossoms provides natural relief for high blood pressure. In 1999, a Bangladeshi clinical trial was conducted to examine the effect of bitter melon on patients with Type 2 Diabetes. The researchers recorded the patients' sugar levels both without food intake for 12-24 hours and after taking 75g of glucose. They then administered a bitter melon pulp suspension to diabetic patients and 86 out of the 100 responded to the vegetable intake, showing a significant 15% reduction in fasting and post-meal serum glucose levels.

Eating it will even improve your sleep. The high beta-carotene and other properties in bitter melon make it one of the finest vegetable-fruits that help alleviate eye problems and improve eyesight. Bitter melon roots are used to treat eye-related diseases. Bitter melon juice may be beneficial in the treatment of a hangover for its alcohol detoxification properties. It also helps cleanse, repair, and nourish liver problems due to alcohol consumption. This bitter juice can also help to build your immune system and increase your body's resistance to infection. Take two ounces of fresh bitter melon juice and mix with a cup of honey diluted in water. Drink daily to improve asthma, bronchitis, and laryngitis.

Regular consumption of bitter melon juice has also been known to improve psoriasis condition and other fungal infections like ringworm and athlete's feet. The fruit is a coolant aids digestion, a laxative, increases appetite, cures gas pains, blood diseases, anemia, urinary discharges, asthma, ulcers, and bronchitis.

NUTRITION: 100 grams of bitter melon has only 43 calories with 1g of protein. It's an extremely good source of potassium, calcium, magnesium, vitamins A, B-6, and C. It is rich in iron, copper, and manganese, and contains twice the beta-carotene of broccoli, twice the calcium of spinach, and twice the potassium of a banana. Regular consumption of bitter gourd juice has been proven to improve energy and stamina levels.

FOOD & USES: Bitter melon can be added to soups, stews, and stir fry. This fruit is eaten green, curried, or sliced thin for salads or stir-fried with onions and garlic. It gets bitterer as it ripens. When it turns orange and mushy, toss the fruit, but save the seeds to plant.

This fruit can also be pickled. The traditional East Indian way to prepare bitter melon to cook is to peel the skin off and cut it into thin slices. Salt the pieces and set out where it can get full sun for a few hours to reduce its bitterness. After a few hours, squeeze out the excess salty, bitter water by hand and rinse with water a few times. Now it is ready for any recipe.

RECIPES: BITTER MELON STUFFED WITH CURRIED SHRIMP

Ingredients: 4 large bitter melons, 2 onions, 4 cloves of garlic, 1 TS cumin, 500g (1 lb.) of cleaned and deveined medium shrimp chopped, 1 cup coconut milk, 3 TBS curry masala, 2 TBS canola oil, salt to taste.

Method: Slice bitter melon along one side so the other side's skin acts as a hinge. Remove all seeds creating a small pocket. Drop bitter melon in boiling water for ive minutes. Drain on a clean kitchen towel and allow it to cool. Put all ingredients except shrimp in a blender to produce a ine consistency. Heat oil in a large frying pan and add blended ingredients for 3 minutes. Add chopped shrimp. Stuff the bitter melon with the shrimp-spice mixture and place it in a covered baking dish. Pour in the coconut milk. Bake for 30 minutes at 350-degrees.

FRIED BITTER MELON

Ingredients: 1 large bitter melon, 4 onions chopped, 1 large tomato chopped, 4 garlic cloves minced, 2 TBS canola oil, salt, and hot pepper to taste.

Method: Remove seeds and then slice the bitter melon into half-inch thick pieces. Salt and let sit for half an hour. Then squeeze and rinse the pieces. In a large skillet, heat the oil, add on-ions, garlic, pepper, then add the bitter melon pieces. Cover and add 2 TBS water. Cook for 15 minutes.

BITTER MELON SALAD

Ingredients: 1 whole bitter melon, cleaned and sliced, 1 medium onion, sliced thin, 1 medium tomato sliced, 1 TS olive oil, ½ cup vinegar (balsamic vinegar preferred), a pinch of sugar to balance the tang of the vinegar, salt, and pepper to taste

Method: Combine vinegar, olive oil, salt, and pepper in a bowl. Taste the mixture and add a pinch more of sugar if desired. Arrange the onion, tomato, and bitter melon on a shallow salad server. Cover with the mixture and serve without stirring.

DID YOU KNOW? Bitter melon's botanical name is *Momordica charantia*. It is also known as bitter melon, balsam pear, bitter cucumber, bitter gourd, karolla, African cucumber, balsam pear, bitter apple, bitter gourd, bitter pear melon, cindeam-or, carilla plant, *Africa*: concombre, *Korea*: karela, *China:* ku gua, kuguzai, *Philippines*: margose, wild cucumber, and ampalaya.

Grow what you love. The love will keep it growing. ~ Emilie Barnes

BOTTLE GOURD – 'UMEKE 'UMEKE

Lagenaria siceraria

The bottle gourd grows everywhere throughout Hawaii and is a great, nutritious vegetable for your home garden. Its Hawaiian name is 'umeke 'umeke. There are a few varieties that are differently shaped. Round types are known as calabash gourds. They are not the same as the calabash that grows on a tree. There are two varieties, cultivated and sweet, or wild and bitter. The sweet variety are used for cooking and the bitter variety is favored for medicines.

Bottle gourd is usually considered a squash, but it's an edible gourd, either straight or with a curved neck, that can be cooked into a variety of dishes. Originally, it was cultivated in tropical Africa, then Asia, and finally reached the Western hemisphere. These gourds were used by humans as far back as 8,000 years ago in the form of musical instruments, containers, nat-ural canteens, and fishing floats. Bottle gourd is one of the most widely distributed vegetables in our world.

HOW TO GROW: Bottle gourd is a fast-growing annual with hairy stems, long forked branches, and a pleasant fragrance. The plants may be grown easily from seed, but require a long hot growing season. First, fork the soil and mix in compost and manure that will help retain moisture. Soak the soil before planting 3 seeds to a mound about 4 inches high and 30cm (1ft.) in diameter. Bottle gourd prefers a pH from 6.5 to 7.5 and will grow to 2500 meters in elevation. This vine will adapt and grow almost anywhere with proper, regular care. Make the mounds beneath strong supports, as a trellis, and keep the mounds 2 meters apart. Once they sprout, within a week, thin to only two plants per mound. In two more weeks, the vines will appear. The white blossoms and dense green leaves properly climbing a support is a nice-look-ing ornamental. Blossoms should appear within 2 months and this vine should produce for 6 months. A well-cared for plant should produce 10 fruits. Once on the vine, the fruits take 2-3 months to mature.

MEDICINAL: Bottle gourd is considered a very healthy vegetable. Ayurvedic medicine utilizes every part of the plant. The practitioners want to use the wild, bitter variety. Eating sweet cultivated vegetable reduces bad cholesterol levels. The juice stabilizes blood sugar level and maintains blood pressure for diabetics. Morning is the best time to drink a small glass on an empty stomach, which will help lower cholesterol levels and maintain a healthy heart. Make the juice fresh every day; it does not store well. Peel and try it. If it tastes bitter, dump it. It's advised to drink the juice alone, not blended with other vegetables. You can add ginger or fresh mint for flavor. The juice treats urinary tract infections and helps regulate the bowels. An oil pressed from the leaves and roots is used externally to treat skin diseases. A leaf paste soothes insect bites.

NUTRITION: This vegetable is 90% water and a rich source of vitamins and minerals. 100 grams contains 14 calories and provides daily needs of 13% of vitamin C and 7.36% of zinc with 174 mg of potassium, 13 mg of magnesium, and 15 mg of phosphorus.

FOOD & USES: Bottle gourd in the market should be firm, slender, feel light in weight, and be a light green. Brown spots mean it's not good. The young fruits are edible and are usually cooked as a vegetable. The flesh is white, firm, with a nice texture and a mild taste. The seeds and skin of young bottle gourds are edible, but as it matures, they lose some of their tenderness. Young shoots and leaves are cooked, and the seeds can be used in soups. Flesh of young fruits are used for icing on cakes.

Before cooking, use a vegetable peeler, and then cut the peeled bottle gourd in half lengthwise. Remove the seeds and spongy interior. Wash and save the seeds to add to soups or stir-fry. For a stronger taste, bake or slow roast the gourd before chopping it. Bottle gourd can then be pureed and added to soups, an Indian curry, or chutney. Some stuff the hollowed gourd with spices, meat, and rice and roast or bake. Bottle gourd pairs well with eggplant, onion, toma-toes, and peppers and tasty seasoned with a garlic, ginger, and fennel.

For millenniums, people have turned bottle gourds into tools and utensils. Mature gourds are made into water bottles, dippers, spoons, and many other utensils and containers. They also make good birdhouses, ornaments, and musical instruments. Designs lightly scratched into the skin of developing fruit will develop into scars that remain intact in the mature fruits. Wash bottle gourds with soapy water, allow them to dry, and then wipe the gourds with a mild solution of chlorine bleach ensures that the surface is clean. Place the gourds in a well-ventilated area away from direct sunlight for a week. The skin will harden and darken. After a week, the outside of the gourd should be dry, but it takes months for the interior to cure.

Move the gourds to a dry, dark area where they can remain for at least six months and suspend each with twine. This permits a good air low under and around the gourd. It is best to check on them every few days. Discard any that begin to decay, shrivel, or become soft. A gourd is ready for use when it feels light, and you can hear seeds rattling. Then, you can carve, paint, wax, shellac, or decorate it any way you wish. Use a small bit on a drill to enlarge a hole; do not use large bits or you will break the gourd. The gourd will be full of seeds and soft ibrous material. Use a long, curved wand, like a straightened metal coat hanger, to break up this ma-terial and pull it out of the gourd. Once cleaned, put a few stones into the gourd and shake it to loosen additional material.

RECIPES: FRIED BOTTLE GOURD AND SPICY YOGURT

Ingredients: 20 - ½ in. slices of peeled, seeded bottle gourd, 1in. cinnamon stick or ½TS pow-der, 2 black cardamom or 1TS cardamom powder, 1TS ginger powder, 1TS fennel powder, 1cup water, 1cup plain yogurt, 3 TBS oil, salt and pepper to taste, mint, or cilantro as a garnish
Method: In a frying pan or wok, heat 2TBS of the oil and brown the bottle gourd slices. Re-move the slices onto a paper. In the pan, heat the cinnamon and cardamom until you smell them. Remove from heat and carefully add the water to the hot pan. Bring to a boil and add the fried bottle gourd slices, ginger and fennel powders, and salt. Simmer and stir in the yogurt. Simmer, stirring (whisk) for 5 minutes as it thickens. Serve with rice.

CURRIED BOTTLE GOURD

Ingredients: 1 bottle gourd, peeled, seeded, and chopped into ½ in. pieces, all the following chopped: 2 garlic cloves, 1 large onion, 1 large tomato, and 4 green chilies, 1TBS oil, 1TBS cumin seeds, 1TS Turmeric powder, 1TS black pepper powder, ¼ cup water, salt to taste, and cilantro
Method: In a frying pan or wok, heat the oil and add cumin seeds. Carefully stir until they pop. Stir in the onion, chilies, and garlic. Sauté for 3 minutes before adding the bottle gourd and tomato pieces, turmeric and black pepper powders, and salt. Carefully stir in the water and use a spatula to scrape the pot to loosen everything. Bring to a boil, reduce the heat and simmer for ive minutes uncovered. Garnish with cilantro.

DID YOU KNOW? The fruits of some cultivated bottle gourd may be more than 1 meter (about 3 feet) long. Botanists believe bottle gourd gourds originated in Africa. However, archaeological evidence shows it was in Peru around 12000 B.C.E., in Thailand about 8000 B.C.E., and in Zambia around 2000 B.C.E. It may have spread so far because mature fruits have hard, dry skin that is waterproof. A mature dried bottle gourd can float over the oceans a year and the seeds remain fertile.

NAMES: milk gourd, white-flowered gourd, calabash gourd, run-ning or climbing vine. In India, it's known as lauki or doodhi, and in Chinese: hulu or huzi.

BUTTERNUT SQUASH

Cucurbita moschata

Butternut squash is a Hawaiian favorite and grows abundantly on all the islands. This type of 'winter' squash originated in either Central America or northern South America and are more tolerant of hot, humid weather with greater resistance to disease and insects. This squash is an aggressive grower and prefers moderately moist soil with full access to sunlight. Butternut squash is considered a winter squash, but is a warm-season annual vegetable vine that trails along the ground or climbs. It has yellow fruit-bearing lowers. Summer squash have thin skins and winter are hard and thick to store longer. Grow some and bake, boil, or grill them.

HOW TO GROW: Butternut squash is another easy to grow vine. Unlike its fellow other squashes, it is eaten after it reaches the mature fruit stage when the rind has become thick and hard. Work the soil with a fork, adding rich compost; make a mound 46cm (18in.) high. This allows the soil to heat around the seeds and roots. They like full sun and a pH of 6-6.5. Plant 5-6 seeds per hill about 10 cm (4in.) apart and 2.5cm (1in.) deep. Keep the soil moist, but not soggy. In about 10 days, the seeds will sprout. When they are about 15cm (6in.) high, thin out the weakest, leaving three plants per hill. Trim the shoots back when the main vine is 60cm (2 ft.) long. This will concentrate the plant's energy to produce lowers and fruit rather than leaves and vines. Water plants regularly and start feeding plants weekly when they begin to lower. Butternut squashes are hungry plants. Remove any leaves covering the young squashes so they ripen more fully and consider lifting the fruits off the ground onto bricks or straw, to ripen. The squash will be ready in less than 4 months. It's best to harvest the fruits when they are small when the skin is hard (can't be punctured with the thumbnail) and uniformly tan and have more lavor. When harvesting, leave a 1-inch stem on each fruit. Use a clean knife to snip the stem.

MEDICINAL: The seed is eaten fresh or roasted for the relief of abdominal cramps and distension due to intestinal parasites. About 800 peeled seeds are reported to make a safe and effective treatment for tapeworm. They are ground into a fine flour, then made into an emulsion with water and eaten. Then take a purge to expel the tapeworms or other parasites from the body. The boiled root of butternut squash supposedly increases mothers' breast milk.

NUTRITIONAL: 100g of butternut has 45 calories and 1g of protein with potassium.

FOOD & USES: These squash have a mild lavor and are easy to cook. They can be grilled, boiled, fried, baked, added to pasta, and used in soups and salads.

RECIPES: BUTTERNUT COCONUT CURRY

Ingredients: 1500g (1lb.) 1 medium-sized butternut squash peeled & cubed, 3 garlic cloves chopped, 1/2 medium red onion chopped small, 3/4 cup water,1 TS turmeric powder, 1 TS ground black mustard, 1/2 TS each of the following: cinnamon powder, ground cumin, red chili powder, ground black pepper, salt, 1 cup coconut milk

Method: In a medium pan or wok on medium heat, add the cubed butternut squash, chopped onion, garlic, and all the spices, salt, and water. Cook the butternut squash for about 5 minutes. The butternut squash should be half-cooked. Add the coconut milk and stir for 3 minutes while cooking, so the curry sauce doesn't curdle or stick. Serve warm with rice or roti.

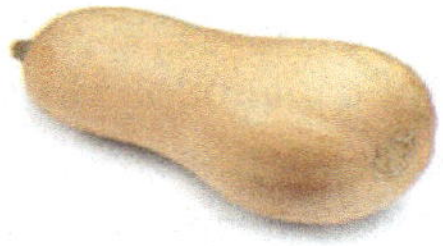

DID YOU KNOW? Butternut squash is an elongated pear shape with pale tan peel and bright orange flesh. The name 'butternut' comes from its buttery flesh and nutty flavor. Butternut squash is the sweetest winter squash. Its flesh is thick and moist and hides few seeds; its skin is comparatively thin and easy to peel. The heaviest weighed 25.17 kg.

CANTALOUPE – MUSKMELON

Cucumis melo

Cantaloupe is always a treat whenever I find them in the market. These sweet orange melons may cost a bit more, but we owe ourselves a sweet taste. Cantaloupes are often recognized as the muskmelon and a member of the same vine family of pumpkin, cucumber, watermelon, and squash. All cantaloupes are muskmelons, but not all musk-melons are cantaloupes. Honey dews, casaba, and Persians are other types of muskmelons. Muskmelon, also known as *Cucumis melo,* is a species of melon that belongs to the gourd family. Muskmelon has ribbed, tan skin and a sweet, musky flavor and aroma. 'Cantaloupe' refers to two varieties of muskmelon: North American cantaloupe *(C. melo var. reticulatus)* and the European cantaloupe *(C. melo var. cantalupensis).*

Typically, cantaloupe is round with a mesh-looking gray skin, sweet-tasty or-ange flesh with a unique pleasant smell. It is believed cantaloupes originated in Persia, be-tween Iran and Turkey, thousands of years ago. This melon was well known by the Greeks and Romans 2000 years before Christ. On his second voyage in 1494, Columbus brought seeds for cultivation on Hispaniola and before 1650 it was a cash crop for Brazil. The modern, true variety of cantaloupe was first cultivated in Cantaloupe, Italy, around 1700 AD, thus the name. These delicious melons will grow anywhere in Hawaii's fertile soil.

HOW TO GROW: Cantaloupes should be planted in mounds about six inches high to permit drainage. They love the full sun and rich moist soil. The best pH is 6-6.8. Work the soil with a fork and place some rotted manure at the base before building the mound. Plant four to six seeds per mound about four inches apart and a half-inch deep with about six feet between mounds. Insects like the cucumber beetle, and the vine borer love cantaloupe. Use various recommended insecticides. Fertilize these melon plants regularly and give water once or twice a week. The main killer of this type of muskmelon is powdery mildew fungus. It is wise to give your plants a good drenching of an appropriate fungicide every two weeks until they blossom. Cantaloupes will not cross-pollinate with watermelon, cucumber, pumpkin, or squash. Keep the vines from tangling each other. It is time to harvest when the stem end of the fruit dries out. That end of the cantaloupe should be soft when you press it with your finger. It is overripe if the entire melon is soft. Do not purchase any cantaloupe that still has part of the stem attached as it was picked before it was fully ripe. The skin, underneath the mesh, should be a yellow or cream color when ripe.

MEDICINAL: Cantaloupe is rich in antioxidants that can help prevent cancer and heart diseases. Potassium in cantaloupe helps excrete sodium, reducing high blood pressure, especially in those with salt-sensitive hypertension. Cantaloupe juice with potassium can reduce muscular cramps. A special compound in this muskmelon relieves the nerves, calms anxieties, and helps against insomnia. When going through a stressful period, drink this melon's juice regularly. The potassium content helps to rebalance and normalize the heartbeat. This in turn sends oxygen to the brain and regulates the body's water balance. The natural nutrients and minerals found in cantaloupe juice provide a unique combination to help the body recover from nicotine withdrawal when trying to quit smoking. Smoking also quickly depletes vitamin A, but cantaloupe juice can help replace it with its beta-carotene.

NUTRITION: A tasty cantaloupe will not spoil your diet as a cup of cantaloupe has only 50 calories. Eating half of a cantaloupe will provide the daily requirement of vitamins A and C, folic acid, and potassium. Cantaloupe has no fat or cholesterol and provides the neces-sary fiber. One serving of a quarter of a medium melon provides more than 400 % of your daily vitamin A, and it also provides nearly 100 percent of your daily vitamin C! Cantaloupe has high levels of beta-carotene, folic acid, potassium, and dietary fiber. It is also one of the very few fruits that has a high level of vitamin B complex, B1 thiamine, B3 niacin, B5 pantothenic acid, and B6 pyridoxine. Cantaloupe has no fat or cholesterol and provides fiber in the diet.

FOOD & USES: Some interesting ways to use cantaloupes are to make a refreshing drink by mixing some sparkling soda water with fresh-squeezed cantaloupe juice or pureed flesh. Another is to blend cantaloupe with mango slices and add lemon juice to prepare a unique cold soup. Cut a cantaloupe in half and remove the seeds before filling the cavity with ice cream or vanilla yogurt or fruit salad.

RECIPES: CANTALOUPE SORBET

Ingredients: 1 cantaloupe - peeled, seeded, and chunked, ¼ cup fresh orange juice, ½ cup sugar - optional, ½ TS salt

Method: Start with 2 cups of cantaloupe in a blender with the remaining ingredients. Blend until everything is smooth and sugar has dissolved. Pour into shallow dishes with covers and freeze. This will be smoother if you remove from the freezer after 3 hours, blend again, and refreeze.

CANTALOUPE PIE

Pie Ingredients: 1 medium cantaloupe - peeled, seeded, and cubed, ½ cup sugar, 3 TBS cornstarch, 3 eggs separated - 3 yolks for pie and the whites for the topping, 4 TBS butter or margarine, 1 TBS vanilla extract, a pinch of salt, 1 store-bought pie shell - graham cracker preferred Meringue ingredients: 3 egg whites, ¼ cup sugar, ½ TS vanilla extract

Method: Pie – blend cantaloupe slices until smooth, which should make at least 2 cups of puree. Pour into a large mixing bowl and add sugar, egg yolks, vanilla, and melted butter. Use a mixer or hand whisk. Pour mixture into pie shell and bake for 45 minutes at 350 degrees. To top with meringue beat egg whites until soft peaks form. Then slowly add sugar then vanilla. If this is too time-consuming, top pie with whipped cream or Dream Whip.

CANTALOUPE CHICKEN

Ingredients: ½ cantaloupe - peeled seeded, and diced, 1 sweet mango - peeled, seeded and chopped, 2 cooked chicken breasts steamed, grilled, or baked (6 thighs may be used) remove bones from chicken, 1/4 cup sour cream, 1 bunch chives chopped, jumbo pasta tube shells like manicotti or cannelloni, 1 TBS fresh lime juice, 1 TBS mustard – Dijon type preferred, salt, and spice to taste

Method: Marinade chicken to your taste before cooking, cool, remove bones, and chunk. Cook pasta, rinse, and cool. Combine diced cantaloupe, mango, sour cream, chives, and spices; then stuff the shells. Serve chilled.

DID YOU KNOW? True 'cantaloupe' *(Cucumis cantalupensis)* from Cantaloupe, Italy, is a hard-shelled melon not grown much outside the Mediterranean countries. It is believed the first 'true' cantaloupe grew in the Pope's garden. Muskmelons with soft rinds and netted surface markings are popular worldwide. In some parts of the world, it is also known as the rock melon, for its rough skin looks like a rock. Leaving an uncut cantaloupe at room temperature for two or three days will make the fruit softer, juicier, and sweeter. The world's biggest cantaloupe was grown in Kentucky USA in 2019 and weighed 67 pounds 1.8 ounces (30.47 kilograms) with a circumference of 136.625 inches (347 cm).

When the world wearies and society fails to satisfy, there is always the garden. ~
Minnie Aumonier

CHAYOTE - CHRISTOPHENE

Sechium edule

We eat a lot of crunchy chayote in stir-fry. My friend calls it the 'vine mushroom' since it acquires the taste of whatever it's cooked with. The flesh is bland, tasting like a cross of a potato and a cucumber. Chayote is a pear-shaped member of the squash family, originated in Central America and cultivated by the Mayan and Aztec Amerindians. Chayote is now cultivated in the world's tropics from Australia and Madagascar to China and Algeria. It has many names, Chayote in Spanish, custard marrow to the British, and vegetable pear or mirliton to the US. The flavor is similar to a zucchini summer squash, but chayote has only a single seed. There are 4 basic varieties, smooth or prickly, green, or white. All types grow well in the soil and climate of the Aloha State.

HOW TO GROW: Chayote grows as an attractive vine, but it needs a lot of attention. This vine loves the sun, but also needs plenty of water and humidity, and a fence or a trellis. The easiest method to grow this vegetable is to locate a farmer and beg a plant. Failing that, select two chayote at the market. Ask the vendor if they have any that are over-ripe and budding. If not, set the chayote in a warm window, but not in direct sun. In a few days it will start to shrivel and wrinkle, and soon sprout a bud. Plant the seed bud upwards in a clay pot with sandy soil. Grow chayote in loose, well-drained, but loamy soil rich in organic matter. This vine prefers a soil pH of 6.0 to 6.8. Lightly fertilize with 12-24-12. Once the plant catches move it outdoors where the vine can climb. Provide it with some shade with a banana leaf or a board. Do not fully cover it. Water regularly and use 12-12-17-2 mix when it begins to blossom. This veggie takes 4-5 months to bear. Chayote tends to produce better the second season. Although chayote is self-pollinating, it seems to like having brothers or sisters around. You'll probably get more fruit if you plant a second vine on a close fence.

Chayote requires a lot of water. It especially likes to be cooled down in the heat of the day with a light spray. I have successfully grown it in the extreme heat of the dry season by spraying it with a hose every afternoon. The chayote was on a fence overhanging my grow box kitchen garden so both benefited from the water. The chayote also provided some shade for the ground level plants. This vine loves to grow across an old ishing net strung between trees.

MEDICINAL: Chayote is used in Ayurvedic-homeopathic medicine. Infusions of the leaves are used to dissolve kidney stones, cure other kidney diseases, and treat arteriosclerosis and hypertension. Infusions made from the fruit are used to relieve urine retention. Slices of raw chayote rubbed on the skin will treat acne. A slice of chayote under a plaster will draw the poisons from a boil.

NUTRITION: One cup of chayote has only 25 calories with almost no fat or carbohydrates. It has some iber and Vitamin C, B-9 (folate), and K with potassium and trace calcium. However, it is a source of sodium (salt).

FOOD & USES: Chayote is very versatile and can be eaten raw, grated, or sliced, boiled, and mashed, fried – especially good in stir fry, or baked. Chayote takes on the taste of the spices used with it. Raw chayote may be added to salads or salsas, most often marinated with lemon or lime juice. In Latin America, chayote is used like pumpkin and makes it into a sweet pie. In the Philippines, it's used to make chop suey, stir-fries, and soups. The vines are utilized on some islands to make sturdy, woven ropes, hats, and baskets.

RECIPES: BAKED CHAYOTE

Ingredients: 4 chayotes, halved and seeded, 2 TBS olive oil, 1 bunch culantro chopped, salt, and spice to taste

Method: Wash, but do not peel the chayote halves. Place in baking dish on the cut side. Brush with olive oil or melted butter and sprinkle with the culantro, salt and spices. Bake at 350 degrees for 40 minutes.

CHAYOTE SOUP

Ingredients: 2 chayotes peeled, seeded, and cubed, 1 large onion (red preferred) chopped, 4 large ripe tomatoes chopped, 2 garlic cloves sliced thin, 1 bunch culantro or cilantro chopped, 4 TBS olive oil, half cup water, ½ hot pepper seeded and minced (optional)

Method: In a large skillet heat the oil before adding the garlic and onion. Then add tomatoes, culantro/cilantro, salt, spices, and water. Simmer for half an hour. Top with grated cheese, and or breadcrumbs.

CHAYOTE ONION QUICHE

Ingredients: 3 chayotes- peeled, seeded, and cubed, 1 large onion, 1 medium red sweet pepper sliced into rings, 1 firm tomato chopped, ¼ cup butter, 2 eggs beaten, ¼ cup milk, ½ cup grated cheddar cheese, 1 unbaked pie shell, salt, and spice to taste.

Method: Sauté onions and chayote in butter until cooked but still firm. Mix in the tomato. Add half of the cheese, salt, and spices and pour into the unbaked pie shell. Mix the eggs with the milk and pour into shell. Cover with remaining cheese and pepper rings. Bake at 350 degrees for 45 minutes, or until the eggs are cooked. This can be changed into an omelet by omitting the pie shell.

CHAYOTE SWEET PEPPER SALAD

Ingredients: 2 chayotes peeled, seeded, and sliced thin, 1 large sweet pepper (preferably red) - cored, seeded and sliced into match sticks, 1 TS olive oil, 2 limes for juice, salt and spices to taste

Method: In a bowl, mix the chayote and sweet pepper pieces with the oil, lime juice and seasonings. Let stand for at least 20 minutes before serving.

CHAYOTE CASSEROLE

Ingredients: 2 cups chayote- peeled, seeded, and cubed, 250 g (½ lb.) minced beef or chicken, 1 medium onion chopped, 2 garlic cloves minced, ½ sweet bell pepper chopped, ¼ cup tomato sauce or ketchup, 2 TBS butter or margarine, 2 TBS canola oil, 1 bunch cilantro chopped, ¼ cup breadcrumbs TBS, salt, and spice to taste

Method: In a frying pan brown the onion and garlic with the minced meat in the oil then adding the sweet pepper and chayote pieces. Mix in tomato sauce, cilantro, salt and spices before dumping into a casserole dish greased with butter. Cover with breadcrumbs before baking at 350 degrees for 45 minutes.

DID YOU KNOW? Chayote's botanical name is *Sechium edule,* a member of the *Cucurbitaceae* family, a subtropical member of the squash/gourd family. It is a pear-shaped fruit, has a single seed and a taste simi-lar to zucchini. It is also known in *English*: christophene, Madeira marrow, merliton, vegetable pear, custard marrow, in the *French West Indies*: chouchoute, bricne. Other names include vegetable pear, cho-cho, soussous, chuchu, choko, pipinella, mango squash, and huisquil, sayote, tayota, choko, chocho, chow-chow, fence grown squash, and alligator pear. In Brazil the name is chuchu or xuxu, which is also a term of affection the same as sweetie. For all these exotic names this fruit of a vine is simply eaten as a vegetable. The young root tubers are also eaten.

Gardening is the purest of all human pleasures. ~ Francis Bacon

CUCUMBER - KUKAMA
Cucumis sativus

Every Hawaiian Island can grow lots of cucumbers. This vine fruit is always a favorite in the local markets. Cukes grow easily all year in various soils and excellent climate. Every home garden should have at least two cuke vines.

The phrase 'cool as a cucumber' is an apt one. Growing in a field on a hot summer day, the interior flesh of a 'cuke' is many degrees cooler than the outside air temperature. Cool and moist due to their high-water content, cucumbers belong to the same family as pumpkins, zuc-chini, watermelons, and other squashes. The botanical name is *Cucumis sativus*.

Cucumbers are one of the oldest cultivated vegetables, farmed since 8,000 B.C.E. and probably native to India. Cave excavations have revealed that the cucumber has been grown as a food source for over 3000 years. Aristotle praised the healing effects of cured cucumbers 8 centuries before Christ. Cucumbers spread to China about 200 B.C.E. and showed up in Europe in Roman times. One Roman emperor is reported to have eaten fresh cucumbers every day of the year, grown by artificial methods in the off-season. Columbus brought cucumbers to the New World on one of his voyages, and the vegetable soon spread to English and Spanish colonies, and to the Native Americans. Cucumbers come in a variety of sizes, some up to 2 feet long. Pickles are cucumbers that have been cured in a brine or vinegar solution.

HOW TO GROW: Cucumbers grow and produce with little care and require attention about twice a week. Cucumbers grow best on slightly acid soils or pH 5.8 to 6.5. Lime should be applied if the soil test shows a pH 5.5 or less. Rows should be 3 to 4 feet apart. Plant seeds 1/2 to 1 inch deep. When sprouts appear, thin the seedlings to one plant every 12 inches in the row or to three plants every 36 inches in the hill or mound system. If the vines climb, it'll save garden space.

Depending on variety, and time of year planted, cucumbers usually take 40 to 55 days from sprouting to picking. Pick cucumbers while they are still tender, crisp, and green. Remove large fruits from the vine so that new fruits are encouraged to grow. Cucumber plants have shallow roots and require ample soil moisture at all stages of growth. When fruit begins setting and maturing, the plants need water. For best yields, incorporate compost or well-rotted manure before planting. You should side-dress plants with nitrogen fertilizer when they begin to vine. Cucumber beetles should be controlled from the time young seedlings emerge from the soil. In small gardens, the vines may be trained on a trellis or fence. Do not handle, harvest, or work with the plants when they are wet. Keep weeds a foot or more from each plant's roots. Do not use herbicides/weed killer chemicals. Watch for aphids, leaf miners, beetles, and fruit worms. If insects become a severe problem, spray, or dust with an approved insecticide, but always wait until after 10 A.M. to spray so pollinating bees are not killed.

MEDICINAL: Ayurvedic medicine uses cucumber to treat difficulties in urination, excessive thirst, headache, insomnia, and holds good promise in liver cancer treatment. Ex-ternally it conditions skin and relieves pain. Boiling the leaves combined with cumin powder treats throat infections. For diabetics, cucumber stimulates insulin production. Cucumbers are mild laxatives.

NUTRITION: 100 grams has only 34 calories with a half gram of protein with trace amounts of almost every nutrient. They have a high content of vitamin K. Scientific research has shown cucumbers contain significant amounts of phyto-nutrients which have a wide array of human health benefits. Fiber and vitamin A are also lost by peeling.

FOOD & USES: A pickle is a cucumber that has been soaked in a brine, vinegar, or other solution. Last year in the United States, over 5 million pounds of pickles were consumed; 9

pounds per person per year! Cucumbers brought from their native India helped begin a tradition of pickling in the Tigris Valley over 4,000 years ago. Ancient sources not only refer to the nutritional bene its of pickles, but they have long been considered a beauty aid. Cleopatra attributed her good looks to a hearty diet of pickles. Julius Caesar fed pickles to his troops be-lieving they lent physical and spiritual strength.

Christopher Columbus brought pickles to the New World. The great navigator grew cucumbers for the purpose of pickling on the island of Haiti. Before Amerigo Vespucci set out to explore the New World, (Amerigo is who the Americas are really named after.) he was a pickle peddler in Seville, Spain. Since food spoilage and the lack of healthy meals were such concerns on long voyages, he loaded up barrels of pickled vegetables onto explorer ships. Hundreds of sailors were spared the ravages of scurvy because of Vespucci's understanding of the nutritional benefits of pickles. The French explorer Cartier found cucumbers growing in Canada in 1535. In the seventeenth century, Dutch fine food fanciers cultivated pickles as one of their prized delicacies. The area, now New York City, was home to the largest concentration of commercial picklers.

RECIPES: REFRIGERATED DILL CHIPS The secret to the crisp texture is the sugar; so do not reduce the sugar in the recipe.

Ingredients: 2 quarts of cucumbers sliced an eighth to a quarter-inch thick, 1 medium onion again sliced thin, 1 TBS salt, 1½ cups of sugar, and ½ cup white distilled vinegar.

Method: First, mix the cucumbers, onion, and salt in a large bowl, cover tightly, and permit it to rest for at least two hours at room temperature. Then drain any water from the mixture. Completely dissolve the sugar in the vinegar and pour over the cucumbers. Pack into tight sealing containers or use zip-lock bags. Immediately put into the freezer. Pickles will be ready to eat in one week and will keep in the freezer for at least a year.

BAKED CUCUMBER AU GRATIN

Ingredients: 2 cucumbers, 2 cups grated cheese, 4 TBS butter, salt, and pepper to taste.

Method: First peel the cucumbers & cut them into 3-inch pieces. Slice each piece in half and remove the seeds. Cook the cucumber in boiling salted water for 10 minutes, then drain and pat dry. Put a layer of cucumber slices in the base of a buttered ovenproof dish. Sprinkle with a third of the cheese, and season with salt & pepper. Repeat these layers, Finishing with cheese. Dot the top with butter. Bake cucumber gratin in the center of a preheated oven at 400 deg. for 30 minutes.

EASY CUCUMBER SOUP

Ingredients: 6 cups chicken broth, 3 large cucumbers - peeled, quartered, and seeded, 250g (½ lb.) fresh mushrooms, 1 bunch chives chopped, and fresh parsley

Method: Using a large saucepan, bring chicken broth to a boil. Slice cucumbers into thin slices about a quarter-inch thick. Wash mushrooms and cut into slices the same as the cukes. Cook cucumbers and mushrooms in broth for 10 minutes or until tender. Before serving, add chives to the soup. Use salt and pepper to your taste and then garnish with parsley.

DID YOU KNOW? The World's longest cucumber is 107cm (42 in.) and grown in Wales in 2011. The heaviest cucumber is 12.9 kg (23 lb. 7 oz.) and was grown in the UK in 2015. Rub a freshly cut cu cumber over the shoe, its chemicals will provide a quick and durable shine that not only looks great, but also repels water. To avoid a hang over, eat a few cucumber slices before going to bed, wake up refreshed and headache free. A fast and easy way to remove cellulite, rub a slice or two of cucumbers along your problem area for a few minutes, the phytochemicals in the cucumber cause the collagen in your skin to tighten, firming up the outer layer and reducing the visibility of cellulite.

LUFFAH – LOOFA
` *Luffah acutangula*

This exotic-looking vegetable sponge will grow anywhere in Hawaii. Luffah, also spelled loofah, and known as the vegetable sponge. It is a vining gourd grown mostly for its useful fibrous skeleton. Young fruits can be cooked the same as squash, used as a vegetable in stews, or as a replacement for cucumbers. The luffah gourd plant, botanically *Luffah acutangula* or angle gourd, is a fast-growing annual tropical or subtropical climbing vine that can reach 30 feet. The mature gourd can be used as an organic bath or kitchen sponge. Smooth luffah, *L. cylindirca,* is larger and more cylindrical. A relative, *Luffah aegyptica* is superior for sponge purposes. Home grown vegetable sponges make excellent gifts.

Angled luffah originated in southern Asia or India and has been cultivated since ancient times, but has now become naturalized worldwide. Young fruit are called Chinese okra, because of the okra-like shape and external ridges. This versatile gourd is used in many Asian recipes because of its slightly bitter-sweet juiciness, and the somewhat spongy texture soaks in spicy flavors. Mature fruits become very bitter.

HOW TO GROW: Luffah is a lowland tropical plant that can be grown up to elevations of 500 meters. It's best to scrape the hard shell of the luffah seed coat with a fingernail file and soak them for at least a day before planting. Plant 5 seeds in each container or hole. Seedlings should sprout in a week and then grow rapidly. Luffah produces best when grown in well-drained soils with the pH in the range 5.5–7 and getting full sun. To create the proper vine gourd growing area, put in some strong upright stakes at least 6 feet tall about 12 feet apart. Dig deep holes, 1 foot in diameter at the base of these stakes. Fill the holes with rotted manure, or compost mixed with the original soil. This will create a mound. If using a fence for a trellis, dig several mounds about a foot apart. These vines grow upward and will wrap around anything. Thin or transplant sprouts keeping two plants in a mound spaced 6 feet apart. Stretch strong nylon cord between the stake tops so the vine can run. Keep plants moist, but don't overwater.

After the five-petal, yellow blossoms appear, the loofah fruits develop. Plants can produce fruit within 60 days. Pick young gourds for eating two weeks after the fruit's set. The baby gourds are edible when only 10-18cm (4-8in.) long and look like okra pods or small zucchinis with smooth but crunchy flesh filled with tiny soft seeds. They can change from tender to inedible in a week. They quickly grow into club-like fruits, termed pepos, which can be 75cm (30in.) in length and 12cm (5in.) in diameter. Fruits are a tapered cylinder-shape, tawny in color, and bitter, with 8–10 ribs. These gourds' insides have three fibrous chambers with hard black oval seeds.

Expect to pick twice a week throughout the three-month season. It's best to remove with a knife as not to damage the vines. Each plant may produce 15 luffahs. When the fruit feels lighter with dry dark skin, they're ready to harvest. For sponge production, the fruits are left for two months on the vines till turning brown. Shake the seeds out of completely dried fruits.

MEDICINAL: Luffah gourd is a natural detoxifier, purifies the blood, strengthens the immune system, and helps to cool down a fever. The seeds are emetic and purgative, eaten to expel intestinal worms. Juice from the leaves is applied to skin problems such as eczema, and used as an eyewash to treat conjunctivitis. The gourd's fibers can be boiled in water, to make a strong luffah tea, and then used as medicine. This tea prevents and treats colds. It's also used for sinus problems. Some people use it as a wash to soften their skin. Women use luffah tea to regulate menstrual periods and nursing mothers use it to increase milk flow. A poultice of warmed luffah fibers treats arthritis, muscle, and chest pains. Luffah seed oil also treats skin diseases. The whole luffah sponge may be rubbed against the body as a gentle exfoliant on the skin and increases blood circulation. Luffah charcoal, prepared by heating luffah fibers in a closed container, is applied directly to the skin for shingles in the face and eye region. Charan-tin, a saponin, and peptide, an amino acid, in this vegetable help to lower blood sugar levels as well as urine sugar levels. The high fiber content helps digestion and the excretory system.

NUTRITION: Luffah gourd is a very low-calorie vegetable. Per 100 grams, it has only 20 calories and rich in anti-oxidants. It has vitamins A and C riboflavin, niacin, and essential amino acids with moderate levels of calcium, magnesium, potassium, iron, and phosphorus.

FOOD & USES: Sponges are prepared by soaking mature fruit in water until the skin and seeds are washed away. Sun dry and pound the fiber interior on a hard surface. Use a full round section in your shower and slice the fiber cylinder in half for a kitchen scrubber. Mature fruit is dried and its fibrous insides are used as a skin brush for bathing. They can also be used as a natural scrubber for kitchen wares, pots and pans. When making fruit drinks and wines, luffah makes a good filter. The main vine is used as a temporary tying rope for firewood and crops. The fibers also work as packaging material, filter material, or as a textured craft material for decorations. The entire mature plant, especially the seeds, can be boiled and used as an insecticide.

When growing on the vine, young loofah fruits, no larger than 6-inches, may be cooked like summer squash and used in stews and soups. In the tropics, luffah gourd is available all year in the markets. Choose immature, green, firm gourds with a healthy stem. Avoid oversize, mature, as well as limp, soft fruits with surface cuts, cracks, and bruises. Fresh luffah gourd does not last long. Stored in the refrigerator, it should be used within two or three days or it'll wilt and become limp. Scrape the ridges and slice the fruit. Peeled luffah should be cooked because its skin contains foul-smelling compounds which make it unappetizing raw. Like in squashes and gourds, luffah should not be overcooked. Luffah has a mild, delicate flavor and soft texture and goes well in lentil curry and mixed-vegetable stews.

RECIPES: LUFFAH GOURD AND EGG STIR-FRY

Ingredients: 500g (1lb.) luffah gourd, peeled and sliced into 1-inch pieces, 2 cloves of garlic chopped, 2 large eggs beaten, 1 TBS fish sauce, chili pepper, salt, and 1 TBS oil

Method: In a wok or frying pan over a medium flame, heat the oil and stir-fry the garlic until brown. Stir in the luffah pieces with the fish sauce. Cook until soft, about 1 minute, then add the eggs. Stir 1 minute, until eggs are set. Season with salt and chili pepper. Serve with rice or noodles.

LUFFAH GOURD CURRY

Ingredients: 3 cups of luffah gourd chopped into 2-inch cubes, 2 cloves of garlic, 3 medium tomatoes, 2 green chilies, 1 inch of ginger, 1big onion – all chopped, 1TS of these spices: black mustard seeds, cumin in powder and seeds, fennel seeds, coriander powder, chili powder, and curry masala, salt to taste, ½ cup water, and 1TBS oil

Method: In a suitable pan or wok, heat the oil and add the mustard, cumin, and fennel seeds. When the mustard seeds pop, stir in the garlic and ginger. Add the remaining chopped vegetables and fry until the onion becomes clear. Stir in all the spices and the chopped tomato.When it starts to liquefy, add the luffah gourd, salt, and water. Cook until the curry sauce thickens. Enjoy with rice or roti.

DID YOU KNOW? It has various names: luffah squash, Chinese okra, sponge gourd, angled luffah, dishcloth gourd, ridged gourd, sponge gourd, vegetable gourd, strainer vine, ribbed loofah. The longest gourd in the world was a 'luffah' grown in China in 2008 which measured 4.55m (15ft.). Prior to WWII, most of the luffah grown in the USA was used as filter material in ships' boilers.

MALABAR SPINACH

Basella alba

Malabar spinach is an edible,perennial vine found in tropical Asia and Africa, where it's eaten as a leafy green vegetable. It grows so easily on the islands; it became Hawaiian spinach. It's not a true spinach, but the leaves bear a resemblance. It's native to the Indian subcontinent, Southeast Asia, and New Guinea. Spinach (botanically *Spinacia oleracea*) originated about 2000 years ago in the Middle East and traders introduced it to India and ancient China via Nepal.

Hawaii's great soil and climate produces over a hundred plants that provide edible green leaves. There are 2 types of Malabar spinach: 1) dark green almost round or heart-shaped leaves with white lowers - *Basella alba; 2)* reddish-purple leaves and stems - *Basella alba Rubra*. Grow some vining healthy greens from a porch pot.

HOW TO GROW: Malabar spinach is a fast-growing, soft-stemmed vine that can grow to 10 meters (33 ft.) when supported by a trellis or strong lines. Its thick, heart-shaped leaves have a mild flavor, and an excellent hot weather substitute for spinach that may be grown as a vegetable and as an ornamental. It grows best in sandy loam soils rich in organic matter with a pH ranging from 6.5 to 6.8 up to 500 meter elevations. Grown by seeds, it germinates within three weeks, and pre-soaking the seeds a day in warm water shortens the sprouting time. Place individual seedlings into your garden about half a meter apart. Stem cuttings, about 20cm (8 in.) apart will easily root and 2-3 plants should feed a family. This vine will take a lot of space and needs support to keep it from crushing the other veggies in your garden. Growing in partial shade increases the size of the leaves. Malabar spinach prefers a hot humid climate with lots of sun and needs constant moisture to prevent blossoming, which makes the leaves bitter. The vines will eventually have white or red spikes of blossoms that will become loads of berry-like black or red fruits.

MEDICINAL: Traditional medicine uses almost all parts of the plant. Malabar spinach roots are astringent and cooked to treat diarrhea. A paste of the root is applied to swellings. The leaves and stems are cooked and eaten as a laxative and a paste of the leaves is applied externally to treat boils. Its juice is a safe laxative for pregnant women, and a decoction has been used to ease labor. Studies report extracts from the berry-like fruit have potential as cancer treatments.

NUTRITION: 100g of Malabar spinach has only 19 calories. It's high in vitamins A and C, potassium, and magnesium, with some calcium and iron. Malabar spinach has a high oxalate content, but half of regular spinach, and is reduced by steaming. Oxalate may cause kidney stones.

FOOD & USES: Malabar spinach may be eaten cooked or raw, and the taste differs considerably. The small berries are used as a rouge, a dye for coloring foods, and also as an ink for official seals. A leaf infusion of the leaves is prepared as a tea substitute.

RECIPES: MALABAR SPINACH AND COCONUT STIR-FRY Ingredients: about 3 meters (8 ft.) of Malabar spinach stem with leaves to feed two people-washed and sliced into strips, 4 cloves of garlic chopped, 1 medium onion chopped, 2 green chilies chopped, ½ cup grated coconut, 1TS 1turmeric powder, juice of two limes, salt to taste, and 1 TBS oil for frying Method: In a wok or frying pan, heat the oil and brown the garlic and onion over low heat. Add grated coconut and stir for 3 minutes. Add turmeric and chili pieces, and then the Malabar spinach. Stir for a few minutes, but as soon as the spinach wilts, remove from the heat. Stir in lime juice and enjoy.

The greatest gift of the garden is the restoration of the five senses. – Hanna Rion

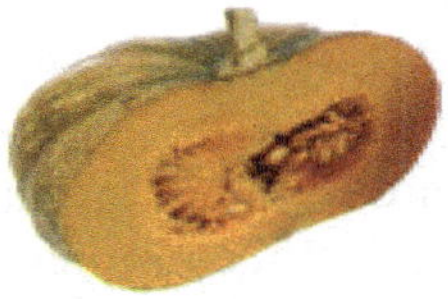

PUMPKIN – PAUKENA

Cucurbita mixta

Pumpkin is Hawaii's most popular member of the winter squash family. These are large and small varieties of hardback melons usually with orange or yellow flesh. Varieties of these pumpkins grow easily on every island. There's no record of which group of early explorers probably brought pumpkin seeds to the Hawaiian Islands. The local name is 'paukena.' These are not for carving. People love to eat pumpkin and they are grown on all continents except Antarctica. They are even grown in Alaska and Siberia. China grows 8 million tons and India 5 million tons.

Pumpkins originated in Central America. They were among the first crops grown for human consumption in the Western Hemisphere. Seeds from related plants have been found in Mexico, dating back over 7000 years. Native Amerindians used pumpkins centuries before the Europeans arrived. Pumpkins soon became a staple in the explorers' diets. They returned to Europe, where they became a new foodstuff. Early settlers used pumpkins for stews, soups, and desserts. In addition to cooking, the New World Europeans also dried the shells and cut strips to weave into floor mats. The settlers learned from the natives to make pumpkin pie by filling a hollowed-out shell with milk, honey, and spices; then baking it.

HOW TO GROW: Pumpkins should be planted on mounds about a foot high to in-crease drainage from the roots. Because pumpkins vine, they require a minimum of fifteen square feet per seed mound. Plant three to four seeds per mound, one inch deep. Allow 5 to 6 feet between hills, spaced in rows six feet apart. When the young plants are well established, thin each hill to the best two or three plants and keep free from weeds by hoeing and shallow cultivation. Irrigate if an extended dry period occurs in early summer. Pumpkins will tolerate short periods of hot, dry weather. A long dry period will cause small fruits, however, a hard rain after a dry period will cause the fruit to split open. A long, wet period will cause the fruit to rot.

Bees are necessary for pollinating squash and pumpkins and may be killed by insecticides. When insecticides are used, they should be applied only in late afternoon or early evening when the blossoms have closed for the day and bees are no longer visiting the blossoms. As new blossoms open each day and bees land only inside the open blossoms, these pollinating insects should be safe from contact with any potentially deadly sprays. Cut pumpkins from the vines carefully, using pruning shears or a sharp knife, and leave 3 to 4 inches of stem attached. Snapping the stems from the vines results in many broken or missing 'handles.' Pumpkins without stems usually do not keep well. Wear gloves when harvesting fruit because many varieties have sharp prickles on their stems. A powdery appearance on the fruit is the best stage for harvesting.

MEDICINAL: They are also high in fiber. Eating pumpkin seeds helps men avoid prostate cancer. They were once recommended as a cure for freckles and a cure for snakebites. Pumpkin has alpha-carotene and beta-carotene, which are potent antioxidants, and the body converts them to vitamin A. Vitamin A promotes healthy vision and ensures proper immune system function. The beta-carotene in pumpkin may also reverse skin damage caused by the sun and act as an anti-inflammatory for sore muscles and joints. In scientific tests, pumpkin has been shown to reduce blood glucose levels, improve glucose tolerance, and increase the amount of insulin the body produces.

NUTRITION: 100grams of cooked pumpkin has 26 calories, 1g protein with some potassium, magnesium, and calcium. They are rich in vitamins A and C. The orange-flesh is a dead giveaway that pumpkin is a source of beta-carotene, which is a powerful antioxidant.

FOODS & USES: Pumpkins are used to make pumpkin butter, pies, custard, bread, cookies, and soup.

RECIPES: PUMPKIN SEEDS - A GREAT HEALTHY SNACK

Ingredients: pumpkin seeds, salt, and spice to taste.

Method: Wash seeds. In a pot of water dissolve 2 TBS salt and spices – hot pepper, curry, or cumin - whatever is your taste. Bring to a boil and add seeds. Simmer for fifteen minutes. Drain, add more salt/spice if desired, and bake in a 350 oven for 10 minutes. Eat them as a great snack. (Use less salt or add sugar).

CURRIED PUMPKIN

Ingredients: 250g (½ lb.) pumpkin chopped into cubes, 2 garlic cloves minced, 1 medium onion chopped small, 2 green chilies chopped small, ½ TS each: fenugreek seeds, Madras curry powder, turmeric powder, mustard seeds, 10 curry leaves, ½ cup of coconut milk, and 1 TBS oil

Method: In a skillet or wok, heat the oil in a pan over a medium flame, add garlic, onions, chilies, and curry leaves. Sauté for a few seconds and then add pumpkin cubes. Stir for a minute. Add the remaining ingredients. Cook until the curry is thick, and the pumpkin is soft.

PUMPKIN ALL-NATURAL FACE MASK - exfoliates and soothes

Ingredients: ¼ cup pureed pumpkin, 1 egg, 1 TBS honey, and 1 TBS milk.

Method: Mix, then apply it to your face, wait for 20 minutes or so and wash it off with warm water

PUMPKIN BREAD

Ingredients: 2 cups cooked pumpkin mashed, ½ cup canola oil, 2 cups sugar (can be less if you desire), 1 egg, 1 TBS cinnamon, 1 TS salt, 2 TS baking soda (not powder), 2½ cups bakers' flour, (Optional – ½ cup raisins, peanuts, or coconut)

Method: Mix all ingredients until the batter is smooth. Fill greased bread pans. Bake at 350 degrees for 90 minutes. Watch after an hour as small baking pans will cook faster.

PUMPKIN PANCAKES (FRITTERS)

Ingredients: 1 cup bakers' flour, 1 cup boiled-mashed pumpkin, 1 TBS sugar, 1 cup milk, 2 eggs, 2 TBS baking soda, 2 TBS oil, 2 TS cinnamon, salt to taste

Method: Combine all dry ingredients in a good-sized bowl. In another bowl, mix wet ingredients. Then mix dry and wet together. An electric mixer will save effort. Spoon batter on a greased griddle or skillet and cook both sides until golden brown. Top with stewed pumpkin.

DID YOU KNOW ? The world's largest pumpkin was 2,624.6-lb from Belgium in 2016. The U.S. grew 1.1 billion pounds of pumpkin production in 2017, and 99% of all those pumpkins were sold for decorations. Americans don't eat, but rather carve faces in pumpkins for All Hallows Eve (Halloween). The tradition originally started with the carving of turnips by Irish immigrants. They found pumpkins easier to carve for the ancient holiday. Pumpkins are about 90% water. Pumpkins belong to the *Cucurbitaceae* family and most tropical versions are *Cucurbita mixta.* A French explorer in 1584 first called them 'gros melons,' which was translated into English as 'pompions.' Its name is from the medieval French word 'pompom', meaning 'cooked by the sun.' It was not until the 17th century that they were first referred to as pumpkins. The word 'pumpkin' appeared for the first time in the fairy tale Cinderella.

There are lots of people in this world who spend so much time watching their health that they haven't the time to enjoy it. ~ Josh Billings

SNAKE GOURD
Trichosanthes cucumerina

Snake gourd is fast-growing vine member of the gourd family. It is a relative of cucumbers, pumpkins, and tomatoes. This is an attractive, conversation starter ornamental for your Hawaiian home garden. The weird-shaped, edible, long, slender fruits are often curved like a snake. These unique gourds can reach more than six feet long and 4 inches wide. It's believed to be native to India and is now grown everywhere in the tropics and subtropics. It's eaten immature as a vegetable, much like the summer squash, and it can also be a beautiful ornamental vine on a fence or trellis. Snake gourds' white blossoms are delicate and beautiful, and the long fruits are eye-catching. The blossoms are unique, with a strong fragrance and only open at night, with long fine hairs between the petals. These hairs stay furled during daylight, but uncurl for a lacy display, but for only one night. Seeds are available online.

Snake gourds are eaten when immature and have a bland taste that requires seasonings. As they mature, their flesh becomes tougher and bitterer, and their rind turns dark red as it hardens. Scientists are interested in this gourd because it's a great food for vegetarians, it's a highly productive crop, and it's a cooling agent of natural medicine.

HOW TO GROW: The snake gourd is a fantastic home garden plant that grows fast, and produces a lot over several months. To grow straight fruits, rather than curved, use a fence or a trellis for support so the fruits grow straight down. It is best to soak the seeds overnight. Dig a hole about a foot deep and wide; mix compost with this soil and refill the pit to create a mound at the base of your fence or supports. Plant the seeds directly into the mounds or germinate them in a nursery and then transplant. Snake gourd prefers well-drained soil with a pH of 6-6.7. Plant more seeds than needed because the germination rate is only about 60%. When the fruit is small, you can tie a weight to the flower end of the gourd to make a straighter fruit. For eating, harvest snake gourds when young, before two months after planting.

MEDICINAL: Snake gourd is a very nutritious vegetable and a natural antibiotic, cough medicine, and laxative. This **vegetable** is a tonic for the heart, helps stimulate the production of body fluids, and relieves dehydration. Snake gourd is best known for a cooling effect in the body. Besides being a heart-friendly vegetable, snake gourd reduces body acids.

NUTRITION: 100 grams of snake gourd flesh contains only 20 calories. It is rich in dietary fiber with vitamins A, B6, and C. This vegetable has high levels of zinc and manganese with other minerals like magnesium, calcium, and phos-phorus. This fruit also contains many medicinal compounds.

FOOD: Choose a snake gourd that is firm to the touch and is light green; avoid any gourd that looks like they have wrinkles on them. It has soft, bland flesh. Some types, when immature, have an unpleasant odor and a slightly bitter taste, but both disappear in cooking. As they mature and turn yellow to red, the gourd becomes too bitter to eat, but it contains the seeds packed in a red spongy pulp that's used in Africa as a substitute for tomatoes. The shoots and leaves are also eaten as greens. Snake gourd seeds are edible and great for your health. To prepare snake gourd, wash, using a sharp knife and without pushing too hard, scrape off the white layer leaving a slippery green skin ready for your recipes. It's best to rub some salt on the peeled snake gourd to remove any naturally occurring wax.

Enjoy the season of mists and mellow fruitfulness,
Close bosom-friend of the maturing sun. ~ John Keats

RECIPES: FRIED SNAKE GOURD – ASIAN STYLE

Ingredients: 500g snake gourd washed and skinned, 4 garlic cloves chopped, 1 large onion chopped, 4 green chilies chopped, 10 curry leaves, 1TS turmeric powder, 1TS red chili powder, ½ cup grated coconut, and 2TBS oil.

Method: Slice the gourd in half and scrape out the spongy center seeds. Slice it thin into semi circles. Put on a plate and sprinkle with salt. Let sit for 15 minutes and then squeeze out any extra moisture. In a wok or frying pan, heat the oil and cook the garlic, curry leaves, chilies, and onions until they brown. Stir in the chili pieces, turmeric, and chili powders, and fry for a few minutes before adding the gourd pieces. Stir-fry everything until the snake gourd pieces are soft, about 10 minutes. Then add the grated coconut. Cook, constantly stirring for 5 minutes. Enjoy with rice.

SNAKE GOURD CURRY

Ingredients: 500g of washed and skinned snake gourd, 1 large onion chopped, 4 green chilies chopped, 6 curry leaves, 1TS curry powder, 1TS turmeric powder, ¾ cup thick coconut milk or curd, salt to your taste, 2TBS oil

Method: Chop the snake gourd into small chunks. In a wok or frying pan, add the snake gourd pieces, and then everything except the coconut milk. Add the oil last and then stir to combine all the ingredients. Cook covered, stirring occasionally on medium for 5 minutes. Mix in the coconut milk. Continue to simmer covered on low heat 5 minutes. Add salt.

FRIED SNAKE GOURD RINGS

Ingredients: 2 cups snake gourd sliced into ½ inch circle pieces with the center seeds re-moved so they're rings, ¼ cup rice flour, ¼ cup gram flour, 2 cloves of garlic minced, 1 green chili chopped tiny, 1 large egg, ½ TS turmeric powder, ½ TS chili powder, salt to taste, oil for frying

Method: In a mixing bowl, combine all the ingredients except snake gourd and oil. Add a little water, 2 TBS at a time, until you have a thick batter and add 1 TBS of oil. Dip the snake gourd rings into the batter and coat. In a wok or frying pan, heat the oil for deep frying. Carefully place the snake gourd rings individually into the oil and fry on both sides until brown. Drain on a paper towel or newspaper. Serve hot with a sea-soned sauce.

SNAKE GOURD RIATA

Ingredients: 250g snake gourd peeled and diced, 2 cups of yogurt, 4 green chilies chopped, 1 medium onion chopped, 2TBS grated coconut, 1 TBS black gram/urad dahl, 1TS black mustard seeds, salt to taste, and 2TBS oil

Method: In a suitable pan, heat 1TBS oil on medium and fry the mustard seeds and black gram for ½ minute. Add green chilies and onion, sauté for 2 minutes, then add the snake gourd. Cook for 2 minutes and then stir in the grated coconut. Turn off the heat and allow to cool for 10 minutes. Stir in the yogurt and add salt.

DID YOU KNOW? Seeds were sent to Europe from China in about 1720, and the snake gourd became well known throughout Europe for the oddness of its shape, not for food. A century later, Thomas Jefferson grew it in America. Fully mature snake gourds can become hard enough to be carved into didgeridoos, a wind instrument. The fruits can easily reach 1.5m (5ft.) in length and in 2017 a gardener in New York grew the longest snake gourd ever at 2.63m (8.62 ft.). **NAMES** Snake gourd, dummalla, the serpent foud, chichinga, padwal, potakaaya, pathola, pudalankaai, dhunduli. The botanical name is *Trichosanthes cucumerina.* Tricosanthes means hairy flower.

Happiness is a vine that takes root and grows within the heart, never outside it.~ Khalil Gibran

SPAGHETTI SQUASH
Cucurbita pepo

Vegetable spaghetti or noodle squash, is an oblong, pale yellow variety of winter squash that'll grow almost anywhere throughout the islands. Named after its shredded lesh that resembles spaghetti noodles, it's often eaten like string pasta. This is a unique vegetable, and easy to grow in your home garden. Spaghetti squash's origin is confused. One source says it originated in China, created in the early 1900s by an agricultural research facility. Another source says it was a popular vegetable in the countryside of northern Manchuria. In 1934, a Japanese seed company improved this squash, and 2 years later the Burpee Seed Co. introduced it to the US. Israel developed a hybrid variety, 'Orangetti,' in 1986 and released seeds in 1990. Surprisingly, none of the creating botanists gave this squash a matching unique Latin name. It remains *Cucurbita pepo* L.

HOW TO GROW: Spaghetti squash are a winter squash. Most are cylinder-shaped fruit about a foot (30cm) long with round ends with a hard ivory/yellow skin at maturity. The spaghetti requires well-drained soil, reinforced with compost and some rotting manure with a pH of 6-6.8. It's best to fork and loosen the soil and work it into a 2 ft. diameter mound. Soak the mound and plant 6 seeds about 2 in. deep. Keep the mounds about 6ft. (2 meters) apart. Do not plant different types of squash in the same vicinity as there's the possibility of cross pollination.

As the seedlings grow, thin to 3 per mound. Water the roots not the leaves. Overwatering will cause root rot. With all squash, mulching is good; shredded newspapers or leaves (banana works well) will help the soil retain moisture and ight weeds. Give a high nitrogen fertilizer sparingly at the roots every 3 weeks. Do not get fertilizer on the leaves as it will burn them. The leaves' color are the best indicator of the plant's health. Watch out for lea beetles. A weekly spray of the hot pepper onion garlic natural insecticide will help.

Spaghetti squash matures approximately 3 months after seeding. It's best to harvest with a sharp knife or garden sheers. Leave an inch of the vine attached to the ripe squash. Mature spaghetti squash will be pale to bright yellow with a similar colored lesh. Spaghetti squash can be stored at 50-60 F for 6 months or at room temperature for several weeks

MEDICINAL: Native Americans used squash to treat intestinal parasites and urinary problems. Research has found eating squash has blood cleansing properties. It's used as a general tonic during rehabilitation after an illness. The best way of absorbing the nutrients and healthful properties of all squash is to enjoy it as a lightly cooked soup.

NUTRITION: Spaghetti squash has 50 calories per 100 grams. This squash is high in fiber, an excellent source of folic acid, potassium, magnesium and small amounts of vitamins A and B6. Our friend Dick recommends eating squash when they are young and tender.

FOOD & USES: When cooked, the texture of the spaghetti squash is tender with a slight crunch and a mild flavor. The flesh is thick, moist, with stringy pulp and flat, cream-colored seeds. Spaghetti squash can be baked, boiled, steamed, or even microwaved. It can be served with or without sauce as a substitute for pasta. Cooking the whole squash is an easier method, but takes more time. If baking a whole squash, pierce with a sharp knife to avoid a very messy explosion. Bake for one hour at 375 F and let stand 5 minutes. Remove seeds and pulp. Dry and eat the seeds. Gardener Dick recommends eating young and tender squash.

You can boil the squash whole – again after puncturing all the way through – for 30 minutes. Then test with a fork. I prefer to cut the squash in half, using a serrated knife, and scoop out the seeds. Once cooked, and cooled, use a fork to scrape out the spaghetti strands. Add your favorite topping: tomato marinara, butter, or cheese and spiced breadcrumbs.

SPINE GOURD – HEDGEHOG GOURD
Momordica dioica

Spine gourd or hedgehog gourd is Hawaii's name for an egg-shaped fruit. Its yellow-green fruit have a thick layer of soft spines covering a juicy pulp filled with tiny seeds. It's a relative of the cucumber and bitter gourd. It's eaten and enjoyed like squash: stuffed, curried, or fried. This is an attractive, very different, and healthy vegetable for the home garden,

The botanical name for the spine gourd is *Momordica dioica,* a vining plant grown for its fruits. It is often confused with *Leucas Zeylanica,* which is a weed sometimes eaten for its bitter leaves.

HOW TO GROW: Spine gourd can be considered a perennial crop, but it will produce better if replanted every season into soil enhanced with compost and manure. The best pH is 5.5-7. It's best to grow on a support such as a trellis to keep it off the ground for better fruit production. Start with seeds and once the plants are two feet long, train them to climb the support. At 6 weeks, treat with a fertilizer containing equal parts nitrogen, phosphorus, and potassium. After another month, sparingly use nitrogen-rich fertilizer every two weeks. Regular watering will increase fruit production. Reduce fertilizer and water once the fruit is dark green. Seeds are available online.

This vine is usually cultivated during the rainy season. The fruits should be ripe when harvested within 3 months of planting. Spine gourds change from light green to yellow if overripe.

MEDICINAL: Eating spine gourd is good for urinary problems. A tea made from the boiled roots is used for diarrhea and lowering blood pressure. It's a traditional herbal remedy for diabetes mellitus as it reduces blood sugar levels in diabetic patients because it has a high content of plant insulin. Anything high in fiber and water is a smart choice for a diabetic's diet.

Spine gourd with carotenoids, like **l**utein, preven**ts** some eye diseases, cardiovascular diseases, and even **cancers.**

NUTRITION: It has only 17 calories per 100 grams with 3g of protein, 33mg of calcium, 42mg phosphorus, and is rich in vitamins A and B-6 folate.

FOOD: For nutritional benefits, spine gourd should be cooked and eaten garden fresh, or used when you buy it. It will last only 3 days in the fridge. Vegetables high in water content should be cooked in covered pans with water added so their nutrients don't evaporate while cooking.

RECIPES: SPINE GOURD STIR FRY

Ingredients: 500g of spine gourd quartered, 2 medium onions sliced, ½ TS turmeric powder, 1TS red chili powder, ½ TS garam masala powder, ½ TS sugar, 2 TBS oil, 10 curry leaves, salt to taste

Method: Coat a frying pan or wok with 2 TS of the oil and sauté the spine gourd with the sugar for 10 minutes. In another saucepan, heat the remaining oil and brown the onions. Add curry leaves, chili, turmeric, and garam powders and stir until combined. Add spine gourd and fry on medium for 10 minutes. Enjoy.

SPINE GOURD FRITTERS

Ingredients: 500g spine gourd washed and sliced in half lengthwise, 2 TBS red chili powder, 2TS coriander powder, 2-3TBS tamarind juice, salt to taste, ½ cup rice flour, oil to fry

Method: Make a thick paste of the tamarind juice, chili, and coriander powders with salt. Coat the spine gourd slices. In a wok, heat the oil until it smokes. Drench spine gourd spiced slices in rice flour and carefully fry 1 minute on each side until golden brown.

SQUASH

Cucurbita

The word 'squash' applies to dozens of vining vegetables that usually have orange or yellow blossoms and orange or yellow flesh. Squash is part of the gourd family; *Cucurbita* is a variety of vines that grow nutritious and tasty foods. It's believed all squash are native to the western hemisphere, originating millenniums ago in the Andes Mountains. Squash was a main-stay of the Amerindians, grown as one of the 'three sisters' of native agriculture along with beans (*Phaseolus vulgaris*) and maize (*Zea maize*). The three crops were grown together, with the maize providing support for the climbing beans, and shade for the squash. The vines provide ground cover to limit weeds and keep the soil moist. Beans increased nitrogen in the soil for all 3 crops.

Gourds, pumpkins, and squash are often lumped into the same category, because they come from the same plant family. Pumpkins may be a type of squash, gourds are not. Since squash is usually cooked (although varieties as zucchini can be eaten raw in salads) it's assumed to be a vegetable, but squash is botanically considered a fruit because it has seeds and blossoms.

Squash is easy to grow. Most types require well-worked soil enriched with compost and rotted manure with a pH of 5.5-6.8 that will drain after heavy rains. They only require 2 inches of water a week. Overwatering and water on the leaves will cause problems.

WHY EAT SQUASH? They are grown primarily for their seeds and fleshy fruits. The seeds are sources of oils and proteins. The seeds are eaten raw, dried, or milled into a butter or flour. Squash seeds contain approximately 50% oil, which is extremely nutritional and non-saturated. Squash seeds are more than a third protein. Cultivated worldwide, squash, pumpkins, and gourds are an important food source. China and India grow 65% of the world's squash crop.

There are two families of squash: summer and winter. Winter squash – (*Cucurbita maxima)* flesh has less moisture and a hard stiff skin. When the soft flesh is re-moved what remains is a hard shell-like skin. That is why they store better and longer. A few winter squash types are: cushaw, hubbard, butternut, acorn, delicata, and spaghetti.

Summer squashes (*C. pepo)* have soft, shiny skins. When picking, while the skins are soft and tender, test the skin with your thumbnail to determine if the squash is ready. The plants are essentially similar towinter squash, both have large leaves. The vine stems may be long and trailing or shorter, more upright and bushy.

Summer squashes come in all shapes and sizes. Young, immature squash is eaten with tiny seeds, and nutrient-rich tender flesh. Small patty pan squash can be cooked whole. Larger zucchini loses flavor and are better for baking or soups. You can roast squash, stir-fry, grill, bake, and even puree it into soup. There are hundreds of ways to cook squash. If picking or buying from a market, plan to use summer squash within a week. Squash can be grated and frozen for about 9 months.

In the tropics, squash is grown all year, but they grow better in the dry sea-son. During the rainy season, it's best to teach the vines to climb a fence or a trellis to protect against fungus. Varieties of squash take between 50 and 100 days to mature.

They're named **WINTER SQUASH** because they store well and are eaten during cold winters

ACORN: Shaped like oak tree acorns, yellow and deep green, these vines were perhaps one of the first squash grown by the Amerindians. Acorn squash can grow to 1-3 pounds.

BANANA SQUASH: Cylinder-shaped, they are pinkish-orange and can grow up to 3 ft. with a thick skin. Only eat the flesh.

BUTTERCUP SQUASH (not butternut): These resemble a candy peanut butter cup. Sweet, buttercup squash has orange flesh and green skin, excellent to puree for soup or delicious sliced and roasted.

BUTTERNUT SQUASH: These have a light beige/orangish skin with deep orange flesh.

CUSHSAW SQUASH: This a green or yellow crookneck squash that loves hot and humid areas because it was developed in the Caribbean.

DELICATA SQUASH: A cylinder-shaped squash with an edible skin that has thin green stripes through its ridges. Its best flavor comes out when it's sliced and roasted.

HUBBARD SQUASH: This squash has a tough skin, but a sweet flavor and grow to 15 lbs.

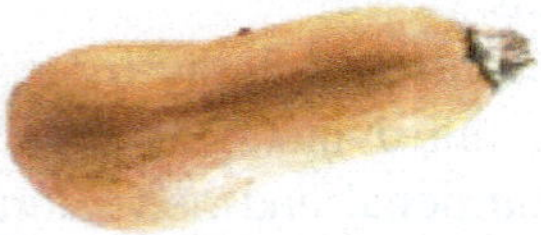

HONEYNUT SQUASH: This is a mini-butternut squash with a richer, sweeter flavor.

KABOCHA SQUASH: Known as Japanese pumpkin, it's a smaller variety with a flavor similar to butternut squash. Its edible skin has green tones good for vegetable tempura.

RED KURI SQUASH: This squash is onion-shaped and deep red that's great roasted or grilled in soups or casseroles.

SPAGHETTI SQUASH: Yellow when mature, oval-shaped spaghetti squash has stringy, noodle-like flesh.

SWEET DUMPLING SQUASH: Weighing less than a pound it's a small mild-flavored pumpkin.

TURBAN SQUASH: This mild-flavored squash is red or green with white stripes and looks like two small pumpkins growing together

They're named **SUMMER SQUASH** because they're best eaten fresh during the warm months.

COUSA SQUASH: A thin-skinned, light-colored, shorter, stouter zucchini with a mild sweet flavor.

CROOKNECK SQUASH: Bright yellow slender, curved necks, with swelled bodies provide the name. These can have smooth, bumpy, or rough skins. Best harvest small and immature or they can be watery. Crooknecks can be grilled, stir-fried, or used in tasty soups.

YELLOW STRAIGHTNECK: These squash are straight versions of the bent crooknecks and easier to store

PATTY PAN or SCALLOPED SQUASH: These squash are named because they resemble a scalloped cake or pie pans. These are best picked young and tender. The mature squash must be peeled before frying or roasting. They, and most squash, can be made into griddle cakes.

TATUMA SQUASH: This pale green, firm squash is from Mexico and also named calabacita.

TROMBONCINO SQUASH: This is named after the brass horn the horn and can grow to 15-20 lbs. with few seeds. Great for stir-fry.

ZEPHYR SQUASH: This squash has a yellow top and a green bottom with a thicker skin than a zucchini. Zephyrs have a nutty flavor.

ZUCCHINI SQUASH: A dark green, straight cylinder squash first cultivated in Italy. There are gray, yellow, striped, and round versions. All are great steamed, fried, or grilled.

ROUND ZUCCHINI SQUASH: These are grapefruit-shaped zucchinis with the same taste. Also named 8-ball the grapefruit-sized round zucchini can be stuffed or carved and used as a bowl.

RECIPE: SQUASH PANCAKES: All squash can be peeled, seeded, and grated

Ingredients: 3 cups, grate one medium onion, chop a few chili peppers –optional. Mix with 1 cup all-purpose flour, one egg, 1 TS sugar, and a pinch of salt.

Method: Let the mixture sit for 10 minutes. Heat 1-2 TBS oil in a skillet. Then drop in spoonfuls of the squash batter. Brown on one side, press and flip with a spatula. Brown. Then cool and drain oil on a paper towel. Enjoy.

DID YOU KNOW?

Squash comes from the Narragansett Amerindian word 'askutasquash.' This roughly translates to 'eaten raw or uncooked.' Summer and winter squash are related to melons like honeydew and watermelon. A single cup of butternut and most other squash with orange or yellow flesh provides more than the daily requirement of vitamin A. In 2019, Ukraine was the number 3 squash and pumpkin producer in the world after China who produced 8.8 million metric tons of squash. India was second. The entire squash plant is edible; leaves, tendrils, shoots, stems, flowers, seeds, and fruit can be eaten. Squash are commonly made into candies in Latin America

You can't control everything in your life,

but you can control what you put in your body. ~ Unknown

WATERMELON – ʻĀMELEKA

Citrullus lanatus

Āmeleka, Hawaiian for watermelon, is loved in the islands. Watermelons are 91% wa-ter and 6% sugar, and are eaten as a dessert, but they are not just sugary sweets. They have high amounts of vitamin C and are low in fat and sodium, which makes them healthy food. Hawaii grows over a half-million tons of watermelon yearly. Annual consumption is 16 lbs. per person.

Watermelons are 91% water and 6% sugar, and eaten as a dessert, but they are not just sugary sweets. They have high amounts of vitamin C and are low in fat and sodium, which makes them healthy food. Every island home garden should have a watermelon patch. They grow easily. This is a good money crop for small farmers to sell in the local markets.

Some anthropologists believe watermelon originated in the Kalahari Desert of Southern Africa 5,000 years ago. Other researchers claim it originated from a species that grew wild in the Nile valley. It was first cultivated in Africa and it spread to the Mediterranean, the Near East, and India. Egyptian hieroglyphics depict the earliest watermelon harvest on tomb walls dating back 4000 years. Watermelons were left in tombs as food to nourish the dearly departed in the afterlife. By the 7th century, it reached India and from there China grew its first water-melons in the 10th century. Capt. Cook is credited with bringing watermelon seeds to Hawaii and probably some came from Mexico. China is the largest producer of watermelons, growing 80 million tons a year.

Watermelon is a great treat your garden can produce and it is nutritionally good for you. Related to cucumbers, pumpkins, and squash, there are over 1,200 varieties of watermelon.

HOW TO GROW: Watermelons, *Citrullus lanatus,* require good sandy soil, a pH of 6-6.8, and should be planted at least four feet apart in rows six feet apart, in beds raised six to twelve inches to allow for drainage. Watering should be done by soaking, not sprinkling, which damages the leaves. The vine of the watermelon plant branches in many directions, with numerous large leaves. The watermelon flower is not very showy and must be pollinated by honeybees, to produce fruit. Use a 20-10-10 fertilizer mix when they begin to vine, and 12-12-17-2 when flowering. It takes the watermelon plant about 3 months to become full-grown, and it is ready for harvest when the part of the rind touching the ground changes from white to pale yellow.

MEDICINAL: Chew and eat watermelon seeds because they contain 'cucurbocitrin' to aid in lowering blood pressure and improving kidney function. The seeds yield a yellow oil rich in linoleic, oleic, and stearic acids. The sweet watermelon surprisingly has only half the sugar content (5%) of an apple. It tastes sweeter because sugar is its main taste-producing agent. Watermelons are ideal for health as they do not contain any fat or cholesterol, are high in fiber content and vitamins A and C, and a good source of potassium. Eating watermelon will reduce inflammation that contributes to asthma, atherosclerosis, diabetes, colon cancer, and arthritis.

NUTRITION: Watermelon is practically a multi-vitamin unto itself. 100 grams has 30 calories, with 1g protein, 20 grams of carbohydrates, phosphorus, iron, calcium, potassium, vitamins C and A. Watermelon is also high in disease-fighting beta-carotene. Lycopene and beta-carotene work with plant chemicals, which are not found in vitamin/mineral supplements. Watermelon is the leader in lycopene among fresh fruits and vegetables. Watermelon contains a high concentration of lycopene and regular consumption may reduce the risks of prostate cancer.

FOODS & USES: Watermelon is grown in over 96 countries worldwide. Watermelons are very fragile and cannot be harvested by machines. Instead, they are carefully tossed by

Watermelon — it's a good fruit. You eat, you drink, you wash your face. ~ Enrico Caruso workers on a relay that runs between the fields and the truck. Every part of the watermelon, including the seeds and the rind, are edible.

RECIPES: WATERMELON SLUSH — All you need is a blender.

Ingredients: 5 cups watermelon with the seeds removed, 1 cup sugar syrup optional - boil ½ half cup water with 1 cup sugar for 1 minute and cool, 2 TBS fresh lime juice

Method: Put watermelon into a blender. Pour into a suitable freezer dish. Stir in lime juice and syrup. Freeze solid. Makes 4 servings. Covered this will keep in your freezer for 3 months.

FRIED WATERMELON — For those who don't count calories!

Ingredients: 3 cups of watermelon - seeded and cut from the rind. The red flesh should be cut into squares or circles about an inch thick. 1 cup flour, ¼ cup cornstarch, 2 egg whites beaten, and 3 cups oil for frying, powdered sugar.

Method: Coat watermelon shapes with flour. Mix egg whites and cornstarch with just enough water to make a thick batter. Heat the oil in a deep saucepan. Coat the watermelon pieces with the egg cornstarch batter and put them into oil. Fry pieces until light brown. Remove and drain the melon pieces on a paper towel. Sprinkle with powdered sugar.

THREE-DAY PICKLED WATERMELON RIND

Ingredients: 1 kg (2lb.) watermelon rind cleaned of seeds and red flesh, 2 TBS pickling spice, 2 cups brown sugar, 4 cups distilled white vinegar, 5 cups water, 1 TBS allspice, 1 TBS whole cloves, 2 TBS cinnamon

Method: Cut rind into 1-inch cubes. Mix pickling spice, cinnamon, cloves, allspice, and sugar with 4 cups water and 1 cup vinegar. Boil for 5 minutes. Then cover and soak rind pieces for a day. Drain the liquid mixture from the rind and reheat it. Then pour it over the rind again and let stand for another day and cover with 2 cups water mixed with one cup vinegar. Let stand overnight. Then boil all together for 5 minutes and allow to cool. Put into clean sealing bottles and refrigerate. If you want spicy pickles, add half of a hot pepper minced to the initial mix.

WATERMELON COLADA

Ingredients: 4 cups watermelon - seeded and cubed, ½ cup water, 2 TBS brown sugar, 1 lime sliced, 2 TBS fresh mint, ice

Method: Combine watermelon, ice, and sugar in a blender and blend to the consistency you desire - smooth or chunky. Put a lime slice and some mint leaves into every glass and squash with a spoon to release the flavors. Cover with the watermelon liquid and enjoy.

TROPICAL SALAD

Ingredients: 2 cups watermelon seeded and chopped into ½ -inch cubes, 2 ripe mangos chopped into ½ - inch pieces, two ripe avocados chopped into ½ - inch cubes, 1 small red onion chopped, 1 lime sliced thin, 2 TBS vinegar, ½ cup fresh orange juice, 1 TBS cilantro minced, 1 TBS fresh lime juice, 2 TS olive oil, 1 TBS grated orange zest, ¼ TS salt, and black pepper to taste

Method: Pour orange juice into a large bowl. Stir in the olive oil and orange zest, season with pinch of salt, and black pepper. Toss the watermelon, and mango in the dressing. Stir in the onions and cilantro. Sprinkle avocado cubes with lime juice, and season.

DID YOU KNOW? By weight, watermelon is the most consumed melon in the world. Last year, the United States grew and consumed over 2 billion kilos of watermelon. The leading commercial growers of watermelon include Russia, China, Turkey, Iran, and the United States. The heaviest watermelon weighs 159 kg (350.5 lb.) and was grown by Chris Kent in the USA.

When one has tasted watermelon, he knows what the angels eat. ~ Mark Twain

ZUCCHINI – COURGETTE

Cucurbita pepo var cylindrica

Squash are separated into summer and winter types. The long season, odd-shaped, hard-skinned squash that store well are usually referred to as winter squash. Smaller, short season types, which are eaten before the skin and seeds thicken, are the summer squashes. They have a mild nutty taste like corn. Most hybrids grow well and can be planted all year throughout Hawaii.

The most common summer squash is zucchini squash that usually has smooth, thin dark green skin and creamy white flesh. Crooknecks with yellow skin and white flesh and white saucer-shaped patty pan or scalloped squash are other common varieties of summer squash, but less available in the Hawaiian Islands. Grow your own! Botanically summer squash is *Cucurbita pepo.*

Modern day squash developed from the wild squash, which originated 10,000 years ago in an area of southern Mexico. Squash developed varieties and a better taste as it spread throughout the Western Hemisphere. The word 'squash' comes from the Narragansett Amerindian word askutasquash, which means, 'eaten raw or uncooked.' Christopher Columbus brought squash back to Europe, and Portuguese and Spanish explorers took squash to the rest of the world. Presently China and Japan are the largest squash growers on the planet.

Zucchini is a member of the cucumber and melon family. Although zucchini has been grown in Central America for thousands of years, modern zucchini was developed in Italy.

HOW TO GROW: Squash will grow in almost any well-drained garden. One or two plants should be enough to supply any family. Be certain you have fresh, first-generation seeds as cross-pollination of last year's crop may give you a surprise and produce pumpkins. Since squash has shallow roots, prepare six-inch-high mounds about three feet apart by forking the soil. The preferred pH is 6-7, and they like a place in full sun, but protected from strong winds. Place a shovel of rotted manure in the base. Plant four seeds about one inch below the surface. Use a fungicide every two weeks for the first six weeks to prevent powdery mildew. Just as the squash begins to vine, carefully pull dirt, and mold the roots. Cucumber beetles and whiteflies are common pests that must be treated with chemicals. Water regularly and use a high-nitrogen fertilizer mix every other week. They should be ready in 6-8 weeks.

Summer squash grows rapidly and should be ready to pick in ten days after flowering. This squash should be picked when they are immature and tender; you can eat everything, even the skin. It is best to use a knife to remove the fruit to not damage the vine.

MEDICINAL: Cooked zucchini is easy to digest, making it a treatment for constipation and acid reflux. Zucchini treats colds, aches, and various health conditions. Zucchini is also rich in antioxidants. Zucchini contains a lot of water and fiber which promote healthy digestion and may help lower blood sugar levels in people with type 2 diabetes. The fiber, potassium, and carotenoids in zucchini help to lower blood pressure.

NUTRITION: Per 100 grams, zucchini has only 17 calories with 1g of protein. Zucchini is an excellent source for vitamins A, B-6 folate, and C, potassium, calcium, folate, and magnesium. A zucchini has more potassium than a banana.

FOOD & USES: Squash male blossoms can be eaten raw or cooked by dipping in a batter and frying. If you don't want to fry the male blossoms, they are a delicious addition to any salad. Summer squash can be prepared on the barbecue grill, steamed, boiled, sautéed, fried, or stir-fried. Squash combines excellently with onions, tomatoes, and okra in vegetable stews. Handle summer squash delicately as small cuts in the skin can easily cause decay. Zucchini will keep about a week in the fridge. It can be frozen, but the flesh will get soft.

Spices that go well with summer squash are marjoram, cumin seeds, parsley, dill, rosemary, and savory. Use them sparingly otherwise, they will hide the real flavor of the squash. Shredded or diced summer squash can be added to salads or sandwiches. They can also be steamed, boiled, baked, fried, and stuffed. Because squash are mostly water, little is neces-sary to steam them in a covered pot.

RECIPES: BAKED ZUCCHINI

Ingredients: 2 medium zucchinis (or other squash) washed and sliced, 1 medium onion chopped, 2 carrots shredded, 1 sweet pepper cored and chopped, 2 medium tomatoes chopped, 2 leaves or 1 TBS of sweet basil (or processed Italian seasoning), ½ hot pepper seeded and minced, 2 garlic cloves of minced, ½ cup cheddar cheese shredded (optional), salt, and spices to taste

Method: Mix all the vegetables and spices and pour into a casserole dish. Cover with cheese and bake at 350 degrees for 45 minutes. This same recipe can be used by splitting the zucchini or other squash lengthwise and coring the centers of the seeds. Pour the vegetable mix into the cavity and cover with cheese. Depending on the squash size you may have to increase cooking time to 1 hour.

COOL ZUCCHINI SOUP

Ingredients: 4 medium zucchinis sliced a ¼ inch thick, 1 large onion chopped, 3 cups water, half-ripe avocado diced, 2TBS lemon juice, ½ hot pepper seeded and minced, 4 garlic cloves minced, 1 bunch of cilantro, ½ TS English/Worcestershire sauce, salt, and spices to taste

Method: Place zucchini pieces, chopped onions, garlic, hot pepper, water, and salt into a large – 6-quart pot and bring to a boil. Then simmer for 8 minutes. Cool before pouring into a blender. Blend until smooth. Add English sauce and the lemon juice and blend again. Chill for 3 hours before serving. Garnish each bowl with a few pieces of avocado and sprigs of cilantro.

ZUCCHINI AND BANANA BREAD

Ingredients: 2 cups grated zucchini, 2 cups mashed bananas, 3 eggs beaten, 1 cup brown sugar, 4 cups bakers' flour, 1 TS cinnamon, 1 cup vegetable oil (canola preferred), 1 TS vanilla extract, 1 cup grated coconut

Method: Combine eggs, sugar, oil, and vanilla in a large bowl. After this is mixed add zucchini, then slowly add the flour. Next add cinnamon, coconut, and bananas. After thoroughly mixing everything into a smooth batter, spoon into two greased bread pans and bake at 350 for at least 50 minutes. Cool, but zucchini bread is best served warm.

SURFER'S DELIGHT - ZUCCHINI AND LENTILS

Ingredients: 1 medium zucchini chopped, 1/2 cup lentils, two cups of water, ½ TS turmeric, three TBS butter or ghee, 2 garlic cloves minced, ½ hot pepper seeded and minced, 1TS ginger minced, ½ TS curry masala, salt, and other spices to taste

Method: In a 4- quart pot place the washed lentils in the water with the turmeric and bring to a boil. Simmer for half an hour until the lentils are tender. In a large skillet heat the butter and brown the onion with the garlic. Add the pepper, ginger, and zucchini and cook for 5 minutes. Add the cooked lentils to the skillet stirring in the curry masala. Cook for 15 more minutes.

DID YOU KNOW? The word zucchini comes from 'zuc-ca' the Italian word for squash. Zucchini, like all squash, has its ancestry in the Americas. However, the varieties of squash typically called 'zucchini' were developed in northern Italy in the second half of the 19th century. The potassium and magnesium content of sum-mer squash helps to reduce high blood pressure and the risk of heart attack and stroke. The world's largest zucchini on record was 69 1/2 inches long, and weighed 65 lbs. Bernard Lavery of the UK.

When I pass a flowering zucchini plant in a garden, my heart skips a beat.

~Gwyneth Paltrow

THE HAWAIIAN HOME GARDEN GUIDE
GRASSES

For lack of a better term, these plants are lumped together as 'grasses.' Rice is a domesticated grass that produces a grain that feeds half the world's population. Sugar cane is a type of perennial grass that satisfies most of the world's sweet tooth. Tobacco and marijuana are not food, spices, or herbs, but just agricultural plants. Marijuana, rice, and sugar cane originated in Asia; while the world has the Amerindian to thank for tobacco. All are extremely profitable to produce. A third of the world's population uses tobacco every day, while 5% enjoy marijuana. Legal marijuana now makes $10 billion a year. 3.5 billion people depend on rice daily.

All of the 'grasses' in this classification are easy to grow. Once they begin to grow, they dominate and grow like weeds. One tobacco plant can produce enough seeds to plant an acre. Rice, the whole grain brown variety, is extremely nutritious, while sugar cane, tobacco, and marijuana consumption may lead to health hazards. According to my Net research, the world produces one and three quarter billion tons of sugar annually and consumes 175,000 tons every minute! From the American Heart Association: it recommends 9 teaspoons or less of added sugar for men a day and 6 teaspoons or less for women. There are about 4 grams of sugar in a teaspoon. At 36 grams or 1.25 ounce/28 gr. a day for a year consumes 31 pounds for men and 20 for women. But the reality of sugar consumption is much higher, almost 80 grams a day or 60 pounds annually.

Tobacco companies produce five and a half trillion cigarettes every year. That's almost a thousand cigarettes for every man, woman, and child in the world. Asia, Australia, and the Far East are the largest consumers. One and a half billion cigarettes are smoked daily!

Brown rice, especially when accompanied with beans, is the perfect food. To create white rice, the germ and the inner husk or bran is removed. Then the grain is polished, usually with glucose or talc. Brown rice has 349% more fiber, 203% more Vitamin E, 185% more B6, and 219% more magnesium than refined white rice. With 19% more protein, brown rice is a more balanced food. White rice does include 21% more thiamin, B1, which is added in the enrichment process. Brown rice has a low Glycemic Index, 55 compared to white rice's 70, or even more with additional processing, such as parboiling, which posts at 87. For reference, a donut is 76. Diabetes develops later in life because of consuming too many foods with a high Glycemic Index.

Many whole foods are high in complex carbohydrates and remain healthy nutritious foods. The nutrients and fiber have been removed from refined or simple carbohydrates. Research has linked a diet of mostly refined carbohydrates to an increased high risk of obesity, heart disease, and type 2 diabetes. Eat whole grains and live longer.

PAGE:

Photo by author

Gardening is medicine that does not need a prescription.
And with no limit on dosage. - Author unknown

HORSE PURSLANE

Trianthema portulacastrum L

Horse purslane is a tasty, easy-to-grow green, a common weed, that's native to South Africa. The English name is horse purslane and can grow anywhere in Hawaii. It's perfect for a garden or as an attractive ground cover. There are 3 varieties of horse purslane: green, golden, and a large-leaved golden variety. It can also be grown in pots. Horse purslane leaves contain and absorb water, so it's most often curried with coconut milk or combined with dhal/lentils, and when prepared as a stir-fry mixture, it's cooked over high heat. Its botanical name is *Trianthema portulacastrum L.* It has various traditional uses against diseases and some bioactive compounds.

HOW TO GROW: Horse purslane is a perennial herb in the marigold family, with an expanding root system. The plant usually lays prostrate and branches out. It's a perfect edible ground cover with numerous small blossoms, but can easily become invasive. The best time to plant is during the rainy season, with good humidity and temperature conditions. Horse purslane seeds are small, kidney-shaped, and dark brown. Plant the seeds 1cm (½ inch) deep and more than half of the fresh seeds should germinate within a week. Seedlings start erect, but soon the weight of their leaves on thin stems will pull them over. A plant can grow to 2 meters (7 feet) long. This is a quick crop; the green leafy plant is ready to be harvested within 45 days after emerging from the soil. Its leaves are edged with a purple vein and the stems are red-dish-purple. The single blossoms are pale green outside and deep rosy pink inside. The small fruit is a capsule with usually 2 seeds. Horse purslane has a long, fibrous taproot system. The way it grows laying down with many branches, helps it quickly cover open cultivated fields. It can become invasive.

MEDICINAL: Horse purslane is useful in treating all types of inflammations and bladder problems. In large doses, the roots can cause miscarriages and are used to relieve obstructions of the liver, as a laxative, and to relieve asthma. The leaves are diuretic. They are used in the treatment for swelling of the feet, jaundice, and bladder irritations. The thickness of the leaves makes them good for wound-dressings or poultices. Inhaling the juice of the horse purslane leaf is believed to relieve migraines. A decoction of the herb treats intestinal parasites and rheumatism. It's also a possible antidote to alcoholic poisoning.

NUTRITION: 100 grams of horse purslane has only 21 calories, but is packed with vitamins A, C, & B complex with high levels of potassium, calcium, magnesium, iron, and phosphorus. It's one of the richest green plant sources of omega-3 fatty acids. Since it contains no cholesterol, horse purslane provides the beneficial omega-3 fatty acids without the cholesterol of fish oils. It lowers the cholesterol and triglyceride levels, and raises the beneficial high lipoprotein.

FOOD: Horse purslane leaves have a slightly salty flavor and tasty in salads and stir-fries.

RECIPES: HORSE PURSLANE WITH COCONUT

Ingredients: 1 bunch horse purslane leaves washed and chopped into 2-inch pieces, 2 garlic cloves chopped, 1 medium red onion chopped, 2 green chilies chopped, 1TS black peper, ½ cup coconut milk, ½ cup grated coconut, 2 TBS oil for frying, and salt to taste

Method: Make a thin paste combining the grated coconut, garlic, and black pepper. In a wok or big frying pan, on low heat, add oil and fry the onion and chilies. Add the horse purs-lane leaves. Stir-fry for 3 minutes and then add the coconut paste. Simmer for 5 minutes. Add coconut milk and salt. Stir for another 5 minutes. Enjoy with rice or noodles.

NAMES: Desert horse purslane, black pigweed, and giant pigweed.

MARIJUANA – GANJA - PAKALOLO

Cannabis sativa – Cannabis indica

Marijuana, cannabis, ganja, weed is one of the world's biggest cash crops. In 2020, 51 countries have legalized the use of marijuana for medical purposes, pain relief. It is illegal to use, possess, cultivate, transfer, or trade in most countries. Canada and Uruguay are the only two countries where commercial cannabis production and sale is legal nationwide. In 2001, Portugal became the first European country to abolish all criminal penalties. Hawaii legalized cannabis for medical treatment in 2000. At the time of this writing, there are 8 medical dispensaries spread over Oahu, Maui, Kauai, and the Big Island. Today, an ounce of medical marijuana in the USA averages over $300. In 2020, it became legal to possess 3 grams of marijuana for recreational purposes.

The Hawaiian word, 'pakalolo' means numbing tobacco. According to Maui Grown Therapies the f rst seeds may have entered on a whaling ship. Strains like *Maui Wowie*, became famous in the 1970s. *https://mauigrown.com/the-history-of-cannabis-in-hawaii/*

Marijuana is the innocent hemp plant used around the world for tens of thousands of years. Around 7000 - 8000 BCE, the f rst fabric is believed to have been woven by prim-itive man from dried hemp weed.Ma, a Chinese name for hemp, *Cannabis sativa,* predates written history and has been used to describe medical marijuanasince at least 2700 BCE.

Cannabis is indigenous to Central and South Asia. Evidence of the inhalation of cannabis smoke can be found as far back as 3000 BCE. It is also known to have been used by the ancient Hindus of India and Nepal thousands of years ago. That was the origin of *Cannabis indica*. The herb was called ganjika in Sanskrit, shortened to ganja in the modern Indic language. Hemp quickly adapted to thrive in the pre-latrine soils around man's early settlements, which quickly led to its domestication.

Another shocker is that Europeans brought marijuana/hemp to the western hemisphere for their own purposes. It was first grown in Chile - South America in 1545, and North America in Virginia in 1606. From the end of the Civil War until 1912, virtually all hemp in the US was produced in Kentucky. Until the beginning of the 19th century, hemp was the leading cordage fiber and rivaled flax as the chief textile fiber of vegetable origin. Hemp was described as 'the king of fiber-bearing plants,' the standard by which all other fibers are measured. As 'hemp', cannabis sativa is valued for how easy it grows, and the many uses. Hemp can be made into textiles, rope, or oils for fuel or to create food products high in potassium. The waste product from manufacturing these items can be used to make paper.

HOW TO GROW: I won't tell you how to grow it except it grows easily like its name, a weed. It is dangerous and illegal to get or possess the seeds. However, marijuana seeds are a huge business worldwide going for f ve to twenty-f ve US dollars each from the internet! If that shocks you then get ready. An eighth of an ounce - three and a half grams - of good tasting, illegal hemp in New York or Los Angeles can go for f fty US or more, and a pound is upwards of f ve thousand dollars! Is there any wonder why ganja is now the world's biggest cash crop? Many analysts contend the market value of ganja produced in the United States alone exceeds $35 billion - far more than the crop value of such staples as corn, soybeans, and hay. The value of the yearly marijuana crop of the world is beyond a trillion dollars. California grows more than one-third of the US national harvest, worth an estimated $14 billion, exceed-ing the value of that state's grapes, vegetables, and hay combined. Marijuana is the largest revenue-producing crop in Alabama, California, Colorado, Hawaii, Kentucky, Maine, Rhode Island, Tennessee, Virginia, and West Virginia, and one of the top f ve cash crops in 29 other states.

MEDICINAL: Marijuana has shown positive effects on cancer, AIDS, and glaucoma. Its use is effective on AIDS patients from its ability to increase a person's appetite as well as relieving nausea allowing a patient to regain weight. Marijuana reportedly helps glaucoma patients by reducing ocular pressure, which can cause damage to the eye. It is the most effective treatment for chronic nausea. It is not physically addictive, but is psychologically addictive. Warnings: coughing, asthma, upper respiratory problems, difficulty with short-term memory loss, racing heart, agitation, laziness, confusion, paranoia, and possible psychological dependence can occur with use.

Smoking marijuana is bad for your health. The carcinogenic by-products of combusted plant matter inhaled into the lungs should be avoided whenever possible. It is currently held that there are no known carcinogens in marijuana when not smoked. Therefore, this drug should be taken orally and absorbed into the body through the digestive system using capsules, medicinal teas, and various edibles.

Supposedly Christopher Columbus brought cannabis sativa to the Americas on his first voyage. In 1619, the Virginia Assembly passed legislation requiring every farmer to grow hemp. Hemp could be exchanged as legal tender or money in Pennsylvania, Virginia, and Maryland. The US Declaration of Independence was written on hemp paper!

Cannabis was criminalized in South Africa in 1911, 1913 in Jamaica, and during the 1920s in the United Kingdom and New Zealand. Canada criminalized marijuana in the Opium and Drug Act of 1923, and the United States outlawed it with the Marihuana Tax Act of 1937. This was after the Mexican Revolution of 1910, when Mexican immigrants flooded into the U.S., introducing to American culture the recreational use of ganja.

Although only two countries in the world have fully legalized the drug for personal use, more than 51 countries have decriminalized its use and/or its cultivation in limited quanti-ties. Medicinal use of the plant is already legal in a growing list of countries, including Austra-lia, Belgium, Canada, the Netherlands, Israel, and 33 states of the USA. Dealing and trafficking are still illegal. Medical marijuana centers now appear like syndicated coffee shops on streets in American cities.

The legality of cannabis for medical and recreational use varies from country to country, in terms of its possession, distribution, and cultivation, and how it can be consumed, and what medical conditions it can be used for. These policies in most countries are regulated by three United Nations treaties: the 1961 Single Convention on Narcotic Drugs, the 1971 Convention on Psychotropic Substances, and the 1988 Convention Against Illicit Traffic in Narcotic Drugs and Psychotropic Substances. Cannabis is classified as a Schedule I drug under the Single Convention treaty, meaning that signatories can allow medical use but that it is consid-ered to be an addictive drug with a serious risk of abuse. (Wikipedia.)

The marijuana 'high' comes from tetrahydrocannabinol (THC), concentrated in the flowers. The minimum amount of THC required to have a perceptible effect is about ten micrograms per kilo of body weight. Aside from a subjective change in perception - most notably enhanced mood, effects include increased heart rate, lowered blood pressure, impairment of short-term memory, working memory, coordination, and concentration.

There are basically two types of ganja. *Cannabis indica* differs from *cannabis sativa* in that it is a shorter plant with broader leaves with less of a euphoric high. Sativa is found practically everywhere around the world, excluding Antarctica. Indica is chosen for its medicinal properties, in particular the anti-epileptic, anti-inflammatory, and stimulant properties.

As the worldwide economy withers and taxes climb, illegal/contraband products like marijuana that relieve stress while making an enormous profit are more desirable.

I don't think pot is more dangerous than alcohol. ~ Barack Obama

Everyone knows that you can get a nice safe high in an Amsterdam coffee shop. In the Netherlands, the term coffee shop has come to mean a place where hashish and marijuana are available. In 1998, they began the ACD or Amsterdam Coffeeshop Directory. It lists cool places to smoke. Above all, the ACD is about unashamed cannabis tourism.

Marijuana is less toxic than many of the drugs that physicians prescribe every day. ~ Dr. Joycelyn Elders

DID YOU KNOW? Uruguay was the first country to legalize mari-juana in 2013, but it took until 2017 for sales to begin. 1.250 packages were sold in the first 6 hours. Historically canvas used for paintings were made of cannabis/hemp. Hops, necessary to brew beer, and cannabis are of the same family. Before the Internet, there was ARPANET in 1969 used by students to share information. Supposedly, in 1971, the first online sale ever was MIT students selling cannabis to Stanford's students. Bob Marley was buried with his Bible, his guitar, and bud of marijuana. Bhang" is an Indian milkshake whose main ingredient is marijua-na. According to the U.N., 158.8 million people around the world use marijuana, which is over 3.8% of the world's population.

NAMES: *Thai:* Kạychā. *French:* Chanvre, *Spanish:* Cáñamo, *Italian:* canapa, *Dutch:* Hennep, *Hawaiian:* Pakalolo, *Russian:* konoplya, *German:* Hanf, *Hindi:* Charas, Swedish: *Hampa*

Photos and graphs *courtesy: 912crf.com*

Marijuana (cannabis) has been used medicinally for centuries. It has beenshown to be effective treating a wide range of symptoms and conditions. It is irre-sponsible not to provide the best care we can as a medical community, care that could involve marijuana. We have been terribly and systematically misled for nearly 70 years in the United States, and I apologize for my own role in that.

~ Dr. Sanjay Gupta - Neurosurgeon

PANDAN – SCREWPINE LEAF

Pandanus amaryllifolius

Commonly known as pandan, the screwpine is a broadleaf plant with fragrant leaves that grows throughout Southeast Asia. Most Asian households grow the most fragrant variety and use pandan in many ways for its great fresh aroma. It grows easy and is great for any garden and chef, especially for BBQ, in the islands. This plant is available in nurseries and online in Hawaii. It grows easy and is great for any garden or landscape, but the plants rarely flower.

This plant is thought to have spread from Malaysia throughout the Asian tropics from Micronesia to Australia and Africa. Pandan has been cultivated for over a thousand years. It is also known as the screwpine leaf, the name bestowed by European seafarers. It has a unique appearance, like a fan with long, dark green leaves that are ribbed and sharp-pointed.

The scientific name is *Pandanus amaryllifolius,* and the family consists of over 750 species that are widely distributed from Hawaii to Equatorial Africa. Pandan or screwpine leaf should not be confused with the bigger common screw pine tree, *Pandanus utilis*.

Pandan is easy to grow, and easy to cook with, but doesn't have any aroma until its leaves are cut and crushed. Fragrant, fresh pandan leaves provide natural food flavoring and coloring, makes a good insect repellent, air freshener, food wrapping, and are woven into tokens of affection. The leaves have to be bruised, heated, or boiled to release their complex flavor, a rosy, almond, and milky sweet, vanilla-like flavor. Dried leaves are flavorless and only used for coloring.

HOW TO GROW: Pandan is a sterile plant, meaning it has no seeds and can only be grown from suckers or cuttings. I found good cuttings in the market and placed them in jars filled with water. I changed the water every day for two weeks the roots sprouted. Then I transplanted them to my kitchen spice garden. Pandan is attractive for functional landscaping.

These plants require sun for 6 hours. Pandan does not tolerate hot direct sunlight. Mine get morning light. These plants need regular water, but be careful, it's easy to overwater. If the leaves get limp and yellow, too much water. It grows very slowly, so give it a year to fill out, and should eventually grow to two meters tall. Use a high nitrogen fertilizer like 20-10-10. It's best to have several plants to harvest leaves from. You should let it grow for at least six months, around a half meter tall, and limit cutting to only a quarter of the leaves. The lower leaves will always be the best smelling. It is best to use a pair of kitchen scissors to remove the necessary leaves, so you don't damage the plant by pulling. Get the older, mature leaves at the bottom first, so the top younger leaves will help the plant keep growing. Don't stunt the growth of this plant.

MEDICINAL: Pandan leaves have many uses in traditional medicine. Extracts may relieve fevers, gas, and indigestion. They help diabetics keep blood sugar levels steady after eating. The roots and leaves make a tea for relief from anxiety, help with sleep, relieve headaches, will reduce uric acid levels to ease the painful symptoms of gout and reduce blood pres-sure. Pandan leaf extract applied to the scalp and skin relieves dryness and dandruff. It's good to treat sunburn.

To give your dog a good smell and fight fleas and ticks, scrub with pandan leaves and bathe them with the tea. Pandan leaves will treat pain and soothe inflammation because they contain tannin and essential oils. Soak leaves in coconut oil for several days and then rub the oil on the sore areas. A balm can be made for treating arthritis.

NUTRITION: 100g of pandan leaves has 24 calories, 6g carbohydrates, no protein or fat.

FOOD & USES: Pandan is used to wrap meats or seafood before they are fried, steamed, and even barbecued. The leaves transfer a sweet, rose flavor to these dishes. Bruised leaves or their extract is used to flavor rice and puddings. Australians use whole leaves to wrap chicken and meats for grilling. Pandan is known as '*The vanilla of Southeast Asian cooking.*'

Using pandan is simple. Just cut and tie one or two leaves in a knot and add them to a pot of rice, soups, stews, and curries. The aroma and flavor of the leaves will be absorbed into the dish.
Don't eat pandan leaves!

Juice is basically what you get when you blend at least ten clean pandan leaves with water. Rough the leaves by twisting them. Extract is the sediment that separates to the bottom of pandan juice after you leave it for a day. The dark, thick sediment is basically the essence. In India, it's named 'Kewra.' Carefully pour off the top and drink it or mix with tea. Don't freeze the extract, but you can freeze and store whole unwashed leaves in a sealed plastic bag for a month in the freezer. In the fridge, leaves will lose flavor after three days, while the juice and extract are good for about five days. The extract contains little water and has a strong, almost bitter taste before being added to a recipe, so adjust how much you use.

PANDAN BALM FOR SORE MUSCLES On a low flame, gently heat 2 cups of coconut oil. Cut 6 or more mature pandan leaves into one-inch pieces. Carefully stir pieces into coconut oil. Cook on low, constantly stirring. Once you smell the pandan, turn off, and let cool. Store in a sealed bottle away from the sun. Rub the cool pandan balm onto any sore muscle.

RECIPES: EASY PANDAN RICE: Wash rice, put in cooker with required water, add the pandan leaves, and cook as usual.

PANDAN WATER: Wash three leaves and cut them into small pieces. Put in a liter pot of water and bring to a boil. Use more leaves for a darker color and more flavor. (Lemongrass and ginger make great additions.) Simmer for fifteen minutes. Let cool and enjoy.

PANDAN SEAFOOD CAKES

Ingredients: 250 grams fish filet, or peeled prawns or a combination chopped very small, 30 grams bacon ends (minced pork), 2 bulbs lemongrass, 1 medium red onion peeled, 6 cloves of garlic, 1 green chili, 1 sweet green or red pepper, 2 eggs, 1 cup all-purpose flour or breadcrumbs, 1 TS of each, turmeric, curry, and red chili pepper pow-ders, 1 TBS soy sauce, sugar, salt, pepper, and spices to your taste. Oil or margarine for frying. 10 pandan leaves and wooden toothpicks for wrapping.
Method: In a blender, combine all ingredients and make a stiff paste. If not stiff enough, add flour and or breadcrumbs by the spoon. Too much will make these cakes crumble. Spoon to form cakes that are the width of big pandan leaves. Wrap each cake with a piece of pandan and hold together with wooden toothpicks. Wet hands will help shape the cakes. Heat a large nonstick skillet or wok with oil. Carefully place the seafood cakes in the frying pan with a spatula. Cook on one side and flip and flip again before removing. The seafood cakes should be golden brown.

SPICY PANDAN CHICKEN

Ingredients: 1 lb. of chicken breast or any part without bone. Cut into many 2-inch pieces, 1 TBS oil for frying, 1 TBS roasted curry powder, 1TS chili powder, 1 TBS soy sauce, salt and pepper to your taste, enough wide mature pandan leaves to wrap the chicken pieces, wooden toothpicks to secure the leaves

Method: With a sharp knife make cuts on every chicken piece halfway on both sides in opposite directions. In a bowl first rub soy sauce into every piece and then combine the curry and chili powders with salt and pepper and rub that into the cuts on the chicken. Wrap with the pandan leaves and hold together with wooden toothpicks. Heat oil in a wok or deep skillet and fry for about 5 minutes flipping often. Unwrap and enjoy.

DID YOU KNOW? Crushed pandan leaves make a great air freshener for a car or bathroom. A folded pandan leaf under a pillow will help bring a good night's sleep. In India, pandan leaves are sacred to the Hindu Gods, Shiva and Ganesh.

RICE – LAIKI

Oryza sativa L

Rice is Hawaii's favorite belly-filler. Yearly, every person consumes 45kg, 100lbs. The first immigrant plantation workers on the Islands were the Chinese, and they wanted rice. During the mid-1800s rice paddies were built along Waikiki. Once upon a time, 10,000 acres of the islands were planted in long grain rice. Sugar was the only bigger crop. Japanese immigrants wanted short grain rice imported from California. By WWII, Hawaii's rice industry was fading. On Kauai, families still have small rice fields. Various nationalities seem to have their own brand of starches. Those from North America and Europe love wheat. Central and South Americans love corn, and Asians love rice. The world's rice producers are China, India, and Indonesia.

Rice is a domesticated version of a type of grass. Just like wheat, barley, or rye, its seeds, or grains of rice, are consumed as a major food staple. Rice is a source of carbohydrates and has fed more people longer than any other grain; for at least ten thousand years! The first cultivation is believed to have been from wild Asian rice from the area around the Yangtze River. Chinese legends refer to rice as a gift of survival given to man from the animals after the great flood. The 'great flood' destroyed all plants, and everyone was starving. A dog appeared to the people with rice seeds tangled in its tail. They planted the seeds, the rice grew, and mankind survived.

Rice cultivation moved to India. About 3500 years ago rice came to Africa in the Niger River delta. Around 10 AD, Southern Europe began farming rice. Portuguese explorers brought rice to Brazil, and Spain spread it to Central and South America. Rice is second only to corn in world cultivation. More than half of the world's population depends on rice as a staple food. Rice cultivation is usually identified with the flooded rice plains or paddies of Southeast Asia. Rice has developed a unique versatility, enabling it to grow even in the desert conditions of Saudi Arabia. Asian countries, including India, consume an average of a half-pound a day per person, or 180 pounds a year. Americans only eat about twenty-five pounds of rice a year. Rice increased in cost since April 2008. Rice-producing countries like China and India reduced exports to feed their own people. China produces 180 million tons a year and eats only 135 million tons, India grows 135 million tons and eats 85 million tons, the USA grows only 8 million and eats only 3.5 million tons. More corn is eaten worldwide than rice.

Long-grain varieties such as basmati and jasmine are suited for hot, humid climates. Medium and short-grain rice was developed for temperate and mountainous regions. Non-mechanized rice cultivation has previously been ideal for countries with high rainfall and cheap labor. Modern technology makes rice cultivation less labor-intensive, but still requires irrigation.

Most rice lovers have tried a few of the many types such as arborio, basmati, Bhutanese red rice, black forbidden, black japonica, calrose, carnaroli, glutinous, jasmine, kalijira, koshihikari, poha, shahi, vialone nano, and others. Many of these types also have versions that are aromatic, brown, wild, converted, instant, and bleached white. Any variety of rice with different spices, nuts, vegetables, beans, meats, or seafood makes a belly filling, nutritious meal. Brown rice retains its outer brown shell, which contains the proteins and minerals, and is much more nutritious than refined white, which is mainly just carbohydrates. Compared to corn and wheat, rice is lower in fat and protein. Agro scientists have developed strains of 'miracle rice' with more protein. It is still wise to compliment rice with soy, animal, or fish protein. Rice cannot be used to make bread because it does not usually contain gluten.

HOW TO GROW: Grow rice at home as an experiment in a clean, solid plastic bucket with no cracks or holes. Purchase long-grain brown rice, unrefined, and chemical-free. White rice will not work because it has been processed. Put about ten inches of dirt or potting medium in the bucket. Fill with water to about two inches higher than the soil. Then throw in a handful of brown rice. It will sink into the soil. Keep the bucket in the sun and the water level about two inches above dirt until the rice sprouts. When your plants (look like lawn grass have grown to six inches, add more water to about four inches deep. Add no more water. In the coming weeks, evaporation will lower the water in the bucket. The bucket should not have any standing water when the rice plants are ready to harvest. If you keep the bucket warm, it usually takes four months for rice to mature. The green stalks will dry and whither to gold when they are ready. Cut your stalks with scissors. Wrap in a brown paper bag and put in a warm breezy place for about three weeks. Roast the rice in your oven at about 150 degrees for an hour. Husk it by rubbing it between your hands. Every bucket should grow enough for a meal, a small meal.

Traditional rice farming is hard work, planting seedlings by hand in wet fields. Modern techniques include laser leveling the fields before shallow plowing and flooding to five inches deep. Now, fast airplanes sow the seeds. The seedlings get nutrients from the water, which also repels weeds. It usually takes four months for the green seedlings to mature to three-foot-tall golden-brown stalks topped with grains of rice. Rice is usually harvested about a month after the rice plants have flowered. The fields are drained, and the plants are cut halfway up the stalk. Brown rice is sheathed in the bran, which is removed by rubbing the grains against each other to produce white rice. The rice must be dried and milled. One acre can yield 8,000 pounds of rice.

MEDICINAL: Brown rice is naturally better for you compared to white rice because it has more fiber. This is better than white rice fortified with vitamins. If you experience diarrhea eat a hand full of uncooked rice. As you digest it, the rice will scrape your intestines and the runs will be over. A bush medicine remedy is to boil rice in excess water and strain. Mix with hot pepper and apply externally for gout. The water rice boils in has cosmetic and health benefits. Still warm, this water can be drunk as a tea for energy. Once it cools, it can be used to bathe with, and it will soften the skin. As a hair rinse it makes healthier and softer hair.

NUTRITION: Natural brown rice has about a calorie a gram and is low in saturated fat, cholesterol, and sodium. It is also a good source of selenium and manganese.

Botanically rice is *Oryza sativa L.* of the *Gramineae* family. There are three main rice types. *Indica* has long slender grains, *japonica* is a shorter, plumper medium-grain rice, and *javanica* is long and thick. Each type of rice has long, medium, and short-grained varieties.

FOOD & USES: Cooking rice is easy. First, wash and rinse the rice before cooking. Look for stones and impurities. For every cup of rice add two cups of water. Choose a pot with a lid that seals. Bring the rice and water to a boil uncovered. Once it boils, reduce heat to a simmer and cover. You might want to put a piece of foil under the lid to make a good seal. Do not open for half an hour unless you smell it burning. Add spices and enjoy.

DID YOU KNOW? Long grain rice is light and fluffy because the grains do not stick together. The taste of white, long grain rice is subtle. Brown rice has a somewhat nutty flavor. Medium grain rice is shorter than long grain and a little plumper. It's stickier than long, but not as sticky as short grain. It's used for creamy dishes, like risotto or paella, and perhaps even certain desserts. Short grain rice is almost round and is featured in many Oriental and Caribbean specialty foods. It is especially popular for Japanese sushi as it sticks together quite easily. 'Converted' or 'parboiled' rice is steamed under pressure before it's milled. This makes the grain harder, preventing overcooking. It also helps maintain several important vitamins and minerals and provides a slightly different flavor. Instant rice or pre-cooked is preboiled, dehydrated, and packaged. It only needs to boil a few minutes before it's ready to eat.

SICKLE SENNA
Senna tora

Leaves from sickle senna or sickle pod, which comes from a shrub that grows in the Aloha State, is an excellent nutritional source of vitamins and minerals. It is believed to be native to northern India and S.E. Asia. In many countries, this plant is used for food and traditional medicines, and can be located growing wild. It's commercially cultivated on a small scale for its seeds, which are used as a dye. Its common English name is sickle senna. Botanically it is known as *Senna tora* or *Cassia tora.* Classified as an annual, many consider it a weed because it can be invasive

Sickle senna is a sprawling, bushy shrub, but it's known to have an unpleasant smell. It is known as stinking cassia. With proper cultivation, it can grow to 2 meters and have attractive yellow flowers. It is an easy to grow perennial that will live for 2 years, produces many seeds, and can become invasive.

HOW TO GROW: The best way to start a sickle senna plant is to sandpaper or care-fully nick the hard coat on several seeds. Then soak them overnight in water. Plant the seeds in a pot with a combination of sand and compost. Thin to only 2 or 3 plants. Sickle senna does best in well-drained soil of a mixture of compost and manure with a pH of 5-7 in full sun. With regular water, and it should be beautiful. Young leaves and immature beans/pods are consumed as vegetables.

MEDICINAL: Use the leaves externally treat skin problems as infections, sores, and insect bites. Boiled in tea or eaten, they purge parasites. The seeds are eaten, combined with a leaf decoction, to treat conjunctivitis. A poultice applied with a paste made from ground roots, mixed with lemon juice is a treatment for ringworm and itchy skin. A boiled decoction of the seed pods treats fevers. A salad made with young leaves treats constipation.

NUTRITION: Research has shown that raw seeds from the sickle senna plant contain 15% protein, with trace amounts of iron, zinc, manganese, and cobalt. Researchers found that when sickle senna seeds were fed to rats, there was a decrease in blood cholesterol. The leaves contain 6% protein with good amounts of calcium, copper, magnesium, potassium, zinc and iron. Studies show that high consumption of roasted seeds may cause stomach problems, but were okay when used only as a supplement in roti, bread, chutneys, stir-fries, etc.

FOOD & USES: The seeds may be boiled or fried in the pods similar to beans. as a side-dish. Roasted seeds are drunk as a coffee substitute. Tender young leaves are steamed or boiled and fermented leaves are used as a condiment. Soaking sickle senna plant seeds, leaves, and roots produces dyes of various colors. The stems are woven into floor mats and fences.

RECIPES: EASY SICKLE SENNA LEAVES STIR-FRY

Ingredients: 500g (1lb.) of sickle senna leaves washed and dried, 3 garlic cloves chopped, 1 medium onion chopped, 2 green chilies chopped, 1TS black mustard seeds, 1 TS chickpea dhal/split lentils, 4 TBS chopped nuts-cashews, almonds, or peanuts, salt to taste, 2 TBS oil-
Method: In a frying pan or wok, heat the oil and add the mustard seeds and chickpea dhal. Stir until they sizzle. Stir in the garlic, onion, and chilies. Stir in the leaves and cook for 5 minutes. Remove from heat and stir in the salt and chopped nuts.

DID YOU KNOW? Sickle senna seeds contain gums that are commercially extracted from wild plants for food manufacturing processes. Sickle senna is often confused with Chinese senna or sicklepod, *Senna obtusifolia.*

NAMES: sickle senna, sickle wild sensitive-plant, sickle pod, tora, coffee pod, foetid cassia, chakunda, stinking cassia, java bean, low ssnna, peanut weed

SUGARCANE - KO

Saccharum officinarum

Botanists believe sugarcane was one of the canoe plants brought to Hawaii from Polynesia. Captain Cook traded iron for sugar. The first sugar mill was on Lanai in 1802. By the 1840's sugar was becoming king. All islands have stopped producing sugar. Maui closed the last mill in 2016. The sap from the thick stalk stores energy as sucrose. From this juice, sugar is extracted by evaporating the water. Most people don't realize that sugar cane is a type of perennial grass. Sugar cane is native to southern Asia. Agronomists believe a specific type of cane was developed in India while other types could have developed independently in Indone-sia. Archaeologists believe the people of New Guinea were the first to domesticate sugar cane around 8000 B.C.E. Crystallized sugar was reported 5000 years ago in India. Sugar cane's botanical name is *Saccharum officinarum*.

Around 700 AD, Arab merchants carried sugar to most of the countries surrounding the Mediterranean Sea. By the tenth century, there wasn't a village from Syria to Spain that didn't grow sugar cane. On the second voyage, Christopher Columbus brought sugar cane to the Caribbean. The word sugar comes from the Sanskrit word *sharkara,* which means material in a grandular form. In Arabic, it is *sakkar*; Turkish is *sheker*; Italian is *zucchero*; and Yoruba speakers in Nigeria call it *suga*.

HOW TO GROW: Growing sugar cane is easy; all you need is the necessary space and some cane. The latter might be hard to find, as it seems to be disappearing from our culture. A line of planted cane makes a sweet hedge line, but must be constantly trimmed or it can take over. Cane likes a slightly acidic soil around pH 6.5, but will grow in soils from 5 to 8.5. Fork your area so the dirt is loose.

Next, check out the stalks of cane you are about to plant. Near every joint in the stem is a little pointed leaf, which will take root. Chop the stalk two inches after this joint and before the next joint. Stick the cane in the loose dirt so the joint is covered. If you keep the soil moist, a shoot should develop in about two weeks. In about six months you should have sizeable cane. Keep it trimmed and enjoy the trimmings.

MEDICINAL: Cane juice can be purchased from several roadside vendors. It is a natural high energy drink that is a great mix for fresh fruit juices. Sugarcane juice helps strengthen your liver and is thus suggested as a remedy for jaundice and works as a laxative. Too much sugar can increase the overall risk of heart disease. Sugar changes the muscle protein of the heart as well as the pumping mechanics of the heart. A 2009 study found that glucose consumption accelerated the aging of cells in the body. A 2012 study found that excess sugar consumption was tied to deficiencies in memory and overall cognitive processing. Chemical manufacturers use sugar to grow penicillin.

NUTRITION: Fresh sugarcane juice is a good source of riboflavin, calcium, and magnesium. Cane juice provides glucose and renews energy. One tablespoon of cane juice has 45 calories. One tablespoon of white granulated sugar has 46 calories, and brown has 34.

FOOD & USES: Sugar cane is eaten raw, crushed to make cane juice, and of course processed into white rum. Cane first is burned in the field to remove the dead leaves or trash. After cutting, juice is extracted by crushing or mashing the stalks. These mashed stalks are termed bagasse. The juice is boiled into a concentrate and the sugar crystallizes. Sugar processing was done through to the 1800s with a boiling house. At the top of each furnace were up to seven large metal basins called coppers. Each copper was smaller and much hotter than the preceding one.

The cane juice was placed in the first copper kettle and heated with some limestone added to remove impurities. The juice was skimmed, then channeled to the other copper kettles.

The result of this first boiling and removal of the sugar crystals produces first molas-ses, which has the highest sugar content since only a small percentage of the sugar has been re-moved. A second boiling creates second molasses with a slightly bitter taste. Blackstrap molas-ses is produced after a third boiling of the original cane juice. Unlike refined sugars with zero health benefits, blackstrap molasses contains significant amounts of vitamins and minerals, especially calcium, magnesium, potassium, and iron. Today mills located in sugar plantations extract raw sugar from sugarcane. Raw sugar is a yellowish tan. Refineries purify raw sugar by bubbling sulfur dioxide through the cane juice. This bleaches all impurities transparent, creat-ing refined white sugar, almost pure sucrose.

Today Brazil, India, and China are the world's largest producers of sugar. Brazil pro-duces about thirty tons of raw sugar cane per acre. Every ton separates into 1500 pounds of juice, which is 300 pounds of sucrose, 1,200 pounds of water, and 500 pounds of bagasse/crushed cane. Beyond a sweetener, one acre of sugar cane can produce 420 gallons of ethanol fuel.

Products made from sugar cane include molasses and rum. Cane sugar syrup was the traditional sweetener in soft drinks for many years, but was replaced by less expensive, and less sweet, corn syrup. Rock candy is enjoyed by people everywhere.

RECIPES: SUGARCANE ROTI - Cane juice is substituted for water.

Ingredients: 2 cups rice flour, (or 1 cup rice flour and 1 cup baker's flour), 1 cup cane juice, 1 TBS butter or ghee, 1 TS salt

Method: In an appropriate heavy frying pan, heat the sugarcane juice until it boils. Stir in butter, flours, and salt. Stir until all the flours are fully combined with the cane juice. Remove from the heat and allow to cool. Then work the dough with your hands until there are no lumps. Form dough into balls and roll on wax paper into flat roti. Roast as any roti and serve. These roti will be stiffer than usual roti because of the cane juice.

CANED RICE

Ingredients: 4 cups cane juice, 2 cups white rice (Basmati preferred)

Method: In a 2-quart pot heat the cane juice. After washing the rice, combine it with the cane juice. Cook over low heat stirring constantly until mixture becomes smooth. When the mixture is very thick, remove from heat and refrigerate. Serve with chopped mint.

SUGAR CANE SEAFOOD

Ingredients: 2 pounds peeled deveined medium to large shrimp (2 pounds of fish fillets can be substituted), 2 1-foot lengths of mature sugar cane, 1 bunch of chives chopped, ½ cup water, 1 TS fresh mint chopped, 1 TS salt

Method: Wash sugar cane and carefully cut lengthwise pieces that will fit into a baking dish, the smaller the pieces, the better. Line the dish with the cane pieces. Lay the washed shrimp on top of the cane, pour in the water. Sprinkle shrimp with the salt and chives. Cover tightly and bake at 300 for 30 minutes. Uncover and bake for five minutes. Serve with rice. Garnish with mint. Use the remaining liquid as a sauce.

DID YOU KNOW? Sugar is a carbohydrate that occurs naturally in fruits and vegetables. Sugar is most highly concentrated in sugar beets and sugar cane. Lemons contain more sugar than strawberries. In the late 1500s, a teaspoon of sugar (4.5 grams) cost the equivalent of $5 in London. A can of Coke has 39 grams of sugar and a can of Pepsi has 41 grams. That is about 7 teaspoons or 13 cubes of sugar per can. (28 grams is an ounce!) Sugar hardens asphalt. It slows the setting of ready-mixed concrete and glue,

Sugarcane is always sweet, people only sometimes so. ~ Myanmaran proverb

TOBACCO – PAKA
Nicotiana rustica

Tobacco is a true Western Hemisphere - Caribbean plant. It is believed tobacco began growing in the Americas about 6,000 B.C. The first attempt to grow tobacco commercially in Hawaii was in 1903 on the Big Island at Hamakua. It's estimated 10% of the population are daily smokers. According to health experts, tobacco use within Hawaii is a leading cause of death. Hawaii became the first state to raise the minimum age for smoking to 21. Its local name is paka. A small group on Kauai are growing tobacco and making boutique cigars.

Early Amerindians used it in religious and medicinal practices. Tobacco was believed to be a cure-all; to dress wounds, as well as a pain killer. Chewing tobacco was believed to relieve the pain of a toothache. The Spanish word '*tabaco*' originated in the Arawak - Taino language of the Caribbean. In Taino, it referred to a roll of tobacco leaves, like a cigar. It may also have resulted from Tobago - a kind of Y-shaped pipe that covered both nostrils used to sniff tobacco smoke. On Columbus' first landing at San Salvador, natives brought fruit, wooden spears, and certain dried leaves that had a distinct fragrance. Each item seemed prized by the natives. Columbus accepted the gifts. The fruit was eaten, but the pungent 'dried leaves' were discarded.

When Columbus landed in Cuba, natives wrapped dried tobacco leaves in palm or maize leaves and lit one end. The natives commenced 'drinking' the smoke from the other end. A sailor, Rodrigo de Jerez, became the first confirmed European smoker. He returned to Spain with the habit, but the smoke billowing from his mouth and nose frightened his neighbors. He was imprisoned for seven years. When he was released, smoking was a Spanish craze.

A few decades ago, tobacco was king, the most popular of all herbs, botanically *Nicotiana rustica*. This plant has lately acquired an unsavory reputation. Indoors it smells horrible, and when overused, it has been proven to cause many deadly diseases. The Caribes and Arawaks used tobacco at infrequent ceremonies, and certainly not twenty times a day. The major reason for tobacco's growing popularity in Europe was its supposed healing properties. It was believed tobacco could cure almost anything, from bad breath to cancer. In 1571, A Spanish doctor named Nicolas Monardes wrote a book about the history of medicinal plants of the new world where he claimed tobacco could cure 36 health problems. In 1588, A Virginian named Thomas Harriet promoted smoking as a viable way to get one's daily dose of tobacco. Unfortunately, he died of nose cancer because it was popular then to exhale smoke out through the nose.

During the 1600s tobacco was so popular that it was frequently used as money, literally 'as good as gold'! It wasn't until the 1900s that the cigarette became the major tobacco product. By 1901 three billion cigarettes and six billion cigars were sold. In 1902, the British Phillip Morris built a New York headquarters to market its cigarettes, including a now famous Marlboro brand. 840 packs of cigarettes are sold every second in the U.S. That's more than one million every hour!

HOW TO GROW: Most tobacco is grown commercially today, but it is attractive and can be grown in the backyard garden as an ornamental, or for personal consumption. Tobacco is a member of the nightshade family that includes tomatoes, peppers, and eggplants. Growing it is very similar to growing tomatoes. Getting the seed or a starter plant is the first hurdle. We got ours from a friend who has it growing wild. Even though family members were smokers they did not grow their own. Seeds of various varieties can be bought inexpensively from the Internet. Get ready, the seeds are miniscule. It is said a teaspoon of tobacco seeds will grow six acres of tobacco.

The second hurdle is to plant the seeds. Tobacco seeds are not much larger than a pin prick and care should be taken when sowing seed as to not sow too thickly. An easy way to spread the seed evenly is to mix it two parts sand to one-part seeds. The best method is to start the seeds in cups or trays. Use moistened commercial potting soil mix. Tobacco plants feed heavily on nitrogen and potash, but sprouted seeds can be burnt by nutrients. Commercial pot-ting soil mix has no nutrients. DO NOT COVER the seeds with any soil as they need sunlight for germination. If covered too deeply the seed won't germinate at all. Tobacco seed has three requirements for germination: light, moisture, and a temperature of at least 65 degrees F.

Tobacco seeds should germinate in about a week to ten days. Although these seeds need to be kept moist, don't hit them with a heavy spray that can wash them into the soil. It is best to use a hand spray bottle adjusted to a fine mist. Your plants should never dry completely. Cover the starter containers with sheets of white paper so the seedlings don't get burned by the hot sun.

As a start try about six to ten plants. Tobacco seedlings should be hardened before you transplant into your field or garden. A week in partial sun should be adequate, but two weeks is better. The seedlings are ready to transplant into your garden or bigger containers when they are about six inches tall. Once repotted, water with a starter fertilizer. The tobacco plant needs full sun to grow broad leaves. If you plan to grow tobacco in your garden plot space the tobacco plants two feet apart in the row, and space rows three feet apart. Keep transplants from drying out by transplanting either on a cloudy day, or in the evening. Water the plants thoroughly once transplanted. During dry weather water each evening till plants become established. Tobacco also requires less water to produce in the island's arid areas.

Mold the base of each plant and keep the plot weeded. Tobacco roots grow quickly with thousands of small hair-like feeder roots close to the soil surface. Care should be taken when hand weeding to not damage the roots. Fertilize each plant with a pinch of 12-12-17-2 bearing salt. Molding will give support to these plants, which can grow taller than six feet. To-bacco likes full sun. Plants grown in partial shade will produce slender leaves. When growing tobacco for harvesting, remove both flowers and suckers from your plants to enable the plant grow large leaves. Your tobacco will be ready to harvest about 3 months after germination.

Around the time flower-heads start to form (if you haven't chopped them off) the plants are full grown, and the bottom leaves will be ready to pick. Pick them when they show signs of yellowing. Cut a slit near the butt end of the center rib of each leaf, feed a thin wire or cord through these slits to hang them. The leaves should be hung about an inch apart, some where dry, out of the way and preferably warm - like garage rafters. You can make acceptable smoke by drying the leaves, slicing them thin. Put that in your pipe and smoke it!

MEDICINAL: Insect bites are relieved after applying a paste powdered tobacco leaves and water. Tobacco when eaten, if your stomach can handle it, will supposedly relieve gas. A hair rinse or body wash of tobacco leaves boiled in water will clean the hair while killing lice.

NUTRITION: Cigarette smokers have lower vitamin C, and carotenes, including b-carotene levels than nonsmokers. Epidemiological evidence indicates that a relationship exists between the incidence of cancer, heart attacks, and cataracts, and serum b-carotene levels. Vitamin E is the body's most effective antioxidant. Smokers have been found to have lower levels of plasma vitamin E than nonsmokers. Antioxidant nutrients seem to have protective roles against cancer, heart disease, cataract formation, cognitive (mind) dysfunction, and perhaps some other diseases. Cigarette smoking promotes oxidation.

RECIPES: THE BEST RECIPE FOR GOOD – DON'T SMOKE

NATURAL GARDEN PESTICIDE: Nicotine is a powerful insect poison. Boil 1 TBS tobacco per quart of water. Let sit for a day. Transfer to a spray bottle.

THE HAWAIIAN HOME GARDEN GUIDE
TREES

A tree is defined as a perennial plant with an elongated stem / trunk, supporting its branches and leaves. Trees reduce erosion and help moderate the climate. They scrub carbon dioxide out of the atmosphere, storing carbon in their bodies. A single tree can absorb as much carbon annually as a car makes while driving 26,000 miles. If 20 million trees were planted, the earth, and its population will be provided with 260 million more tons of oxygen. Those same 20 million trees will also remove the 10 million tons of CO2. But, sadly, an estimated 16 billion trees are lost every year because of forest management and deforestation. Forests provide living quarters for many animals and plants. Tropical rainforests are filled with a great variety of earth's life forms. Trees provide shade and shelter, construction timber, cooking and heating fuel, and fruit for food plus they provide many other uses.

Our planet Earth has more than 80,000 edible plant species. Over 1,400 trees and plants are native to the Hawaiian Islands, but less than 25% of the total island land remains in natural forests. The high forests are the source of Hawaii's few major rivers. Hawaii has more than a third of the state, 1.5 million acres, in forest. Even bustling Oahu is 36% forest. The largest area of remaining native dry forest is on the Big Island of Hawaii. Kauai also has a large dry tropical forest. These dry forests only cover 2% of the land mass. Slash and burn affected the lowlands on all the islands.

Trees are the longest living organisms on Earth, and never die of old age. The oldest tree in the Hawaiian chain is on Maui: Okilani, an ancient Ohi'a tree. It was discovered 4 decades ago, and is estimated to be over 4,000-years-old. It's not only the oldest known living thing on Hawaii, but one of the world's oldest trees. The oldest recorded planted tree may be the banyan tree in Lahaina, the former capital of Hawaii. The Smith family planted it in the 1870s.

California holds the record for the oldest living trees. Some of the state's bristlecone pines and giant sequoias are 4,000-5,000 years old. Methuselah, an estimated 4,852-year-old ancient Bristlecone Pine, is one of the oldest living trees in the world.

Trees give us most of our fruit, even spices like nutmeg, cinnamon, cocoa, cloves, and our favorite beverages of coffee and tea. The trees of this section are unique. The cotton tree remains essential to a few of the Hawaiian Islands. Noni bears fruit, although not sweet and with an unusual smell. Nevertheless, Tahitian Noni surpasses a billion dollars in sales every year dealing with seventy countries. The annatto/'lipstick' tree is cultivated in the tropics, the Caribbean, Central and South America, India, Sri Lanka, and Indonesia for the dye that the seeds yield. Years ago, islanders utilized a lot of annatto seeds to color and flavor foods and dye fabric. Calabash is an ancient tree native to the tropical Western Hemisphere. European explorers brought it to Asia as a functional tree to produce hard-shelled gourds.

Moringa is considered by many countries as a miracle plant that can solve many of their nutritional and medical problems. Coffee and tea are undoubtedly the most important, most transported, and most valuable of any producing trees. Yearly, 3 million tons of tea are produced worldwide. A few well-spaced coffee trees and tea bushes in the back yard and you are on your way to caffeine self-sufficiency.

Coffee and tea are big business worldwide. Independent coffee shops in the USA do more than $12 billion in sales every year. After petroleum, coffee is the world's most traded commodity, yet most of the world's coffee is grown by small-scale coffee farming families in over 50 countries, all of them located in the tropics. Brazil produces 30% of total coffee and Colombia ranks second. More than 7 million tons of green coffee beans are hand-picked every year. For every pound of gourmet coffee sold, small-coffee farmers earn an average of just 1/8 of a US dollar! Americans consume 400 million cups of coffee per day and 4 million espresso coffees are sold daily in Italy. Coffee, cocoa, and spices could provide a national income alternative to fossil fuels.

Courtesy R Douglas

PAGE:

The best time to plant a tree was 20 years ago. The next best time is now. ~ Chinese proverb

NONI

CALABASH

COFFEE

It is not so much for its beauty that the forest makes a claim upon men's hearts, as for that subtle something, that quality of air that emanation from old trees, that so wonderfully changes and renews a weary spirit.

~ Robert Louis Stevenson

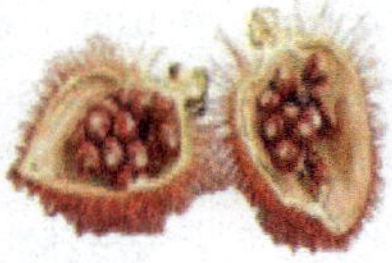

ANATTO – 'ALAEA – LIPSTICK PLANT

Bixa orellana

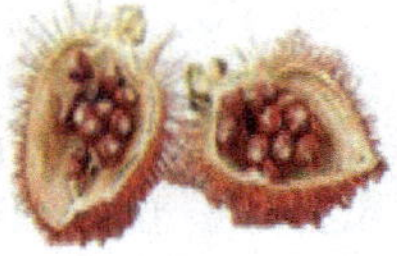

Years ago, islanders utilized a lot of annatto to color and flavor foods. This plant is cultivated in warm regions of the world, such as the Caribbean, Central and South America, India, Sri Lanka, and Indonesia for the dye that the seeds yield. As with many traditions, now it's seldom seen unless at vending stands along less traveled roads. The Hawaiian name is *'alaea*. The usual Eastern Caribbean name is roucou or oucou. The most common name is annatto. It comes from the fruit of the achiote, botanically the *Bixa orellana* tree. Annatto is native to Central and South America. This is a perfect home garden plant for an aspiring chef.

After the explorers encountered Amerindians as the Mayans, Incas, and Aztecs colored in brilliant red, annatto seeds and pulp were imported to Europe in the 1500s. Commercial cultivation began in India two centuries later. The dye was supposedly used by the Amerindians as an antidote to poison caused by eating the wild / bitter version of cassava that had not been properly processed to remove the natural cyanide poison. World production of annatto seeds is estimated to be 11,000 tons with Brazil the largest grower. Peru exports over 4,000 tons a year.

HOW TO GROW: Our tree started from a cutting, but annatto can be grown from seeds. It will grow almost anywhere if the soil is well-drained. Start this shrub in a sizeable container and keep the growing medium slightly damp until new leaves appear. This plant will grow in all conditions from a moderately wet to arid climate and prefers well-drained soil with a pH of 5.5-6. It needs the usual starter fertilizer until it begins to show pods. Then switch to a bearing fertilizer. Transplant when it's too big for your container. These shrubs make nice ornamental bushes at the entrance to driveways, sidewalks, or paths.

This is usually a short, attractive evergreen shrub, but it can grow to over twenty feet with shiny heart-shaped leaves, sometimes with red veins. The three-inch pale pink blossoms bear a strange, hairy fruit that is heart-shaped with red prickly spines. It may be yellow, red, or maroon. Red is the most common variety. When ripe, the pod splits in half to reveal about fifty seeds encased in a red pulp.

MEDICINAL: As an herbal remedy a mixture of annatto pulp and seeds boiled in oil makes a salve that helps heal small cuts and burns while preventing scarring and blistering. A decoction of the leaves and pulp relieves stomach disorders like indigestion and will help relieve asthma. Annatto leaves in a bath is refreshing. Leaves heated in oil will reduce the pain of a headache when pressed to the forehead. People take annatto for diabetes, diarrhea, fevers, fluid retention, heartburn, malaria, and hepatitis. It's used as an antioxidant, bowel cleanser, and to repel insects.

NUTRITION: Annatto seeds are a food rich in fiber and a protein source: 11.50% protein, 6.74% moisture, 42.19% total carbohydrates and 28.45% fiber. The seeds contain the minerals potassium, manganese, calcium, copper, iron, and magnesium. They also contain high levels of the amino acids: lysine, isoleucine, and leucine. More research is required, but annatto could become an easy to grow, low environmental impact protein food for developing countries.

FOOD & USES: The dye obtained from the pulp of the seed is called bixin and used worldwide as a red-orange dye for coloring rice, cheeses, soft drinks, oil, butter, and soup. The dye is also used in some regions to dye textiles. In the past few years, banana growing countries prepared a banana version of ketchup using annatto for the red color. It was difficult to discern the banana variety from the traditional tomato ketchup. We used the annatto to color and flavor special dishes. The Amerindians used it as the original war paint and to give a startlingly attractive color to their bodies while protecting from the sun and insects. Amerindians supposedly used the seeds as an aphrodisiac. Annatto seeds 'supposedly' are given to bulls to make them more aggressive for bullfighters.

RECIPES: ANNATTO SEASONING

Pick at least a gallon of annatto pods, cut, and scrape out the seeds. From personal experience wear rubber gloves and do this procedure on a piece of plastic to ease clean up. Do this carefully with a teaspoon because it really stains! Place everything in a clean bucket and cover with clean fresh water. Allow to soak for at least a day. Strain the water into a next bucket. Rub the seeds between the palms of your hands to remove any remaining pulp. I recommend wearing latex gloves and not getting too splashy. Strain again and repeat until the seeds fail to give off any color. In a cast iron or stainless pot (It may stain aluminum.) bring the strained annatto liquid to a boil and simmer for 10 minutes. Add salt and any herbs you desire for your personal taste. Cover and let sit for a few hours. Allow to cool before pouring in bottles. If you were neat – as few home cooks are - cleanup is easy. If not use diluted bleach to clean surfaces and hands. I suggest keeping this in the refrigerator.

ANNATTO OIL

Ingredients: ½ cup cooking oil, 2 TBS annatto seeds

Method: In an appropriate frying pan heat the oil over medium heat for 3 minutes. Add the seeds while continuously stirring. Heat for 1 minute. Allow to cool. Strain or not depending on the intended use. Refrigerate.

ANNATTO RICE

Ingredients: 2 cups cooked rice, 1 lb. minced pork or chicken, 1½ TBS annatto seeds, 1 hot chili seeded and minced, 3 garlic cloves minced, 1 cup diced tomato, 2 bay leaves, 3 TBL oil, salt, spices, and black pepper to taste

Method: In a frying pan or wok, heat the oil and annatto seeds over medium heat until they pop. Lower the temperature and cook until the oil yellows, about 5 minutes. Remove from heat and carefully spoon out the annatto seeds and discard. Return the oil to medium heat and sauté the hot pepper and garlic for half a minute. Then stir in the minced meat, tomato, and bay leaves. Stir in the cooked rice and season. Remove bay leaves before serving.

DID YOU KNOW?

The Carib Indians used annatto seeds for body paint. In the 1600s, Europeans used the seeds in chocolate recipes and the English used them to add color to cheeses. The seeds have a slight peppery taste and lemon aroma. They are a preferred organic coloring agent. They have been called the poor man's saffron because of the color they make. These seeds produce a basting oil used in Asian chicken and a coloring for Chinese pork. The seeds are usually ground into a fine powder and steeped in water or oil to obtain the color and flavor.

NAMES: *Hawaii:* 'alaea, *Sri Lanka:* rata kaha, *English:* lipstick plant, colorau, *Mexico:* k'u-zub, and pumacua. *Brazil:* annato and urucu, *Germany:* urucum, *Dominica and the French West Indies*: roucou, oucou, *Colombia:* achiot, and *Amerindian:* arnotto.

Our bodies are our gardens – our wills are our gardeners.

~ William Shakespeare

CALABASH – 'AMIA - BOLLE

Crescentia cujete

The calabash tree is native to the Caribbean, South, Central, and North America. European explorers brought it to Asia as a functional tree producing hard-shelled gourds. The Hawaiian name is la ʻamia. It's scarce, but available online. It's a unique ornamental addition to any Hawaiian home garden especially if you are interested in handcrafting the unique fruit. Today, the calabash isn't used functionally, instead, this gourd is more decorative.

Native Amerindians planted useful calabash trees throughout the Caribbean, Mexico, Central and South America centuries before Columbus. It is uncertain where the tree originated. This tree was priceless to those Indians. Fibers from the bark were twisted into twine and woven into ropes. The hard branches and trunk made tools and tool handles. The wood has elasticity and can be shaved into strips like the apple tree and woven into baskets. The gourd-like fruit was indispensable for cups and water containers, made into musical instruments, and containers.

A Benedictine monk in the 1500s reported the Caribbean Taíno Indians (in Puerto Rico) cut eye holes in large calabashes to pull over their heads. This was camouflage when they hunted river and lake birds. The Indians wore the calabash and waded into the water. The floating calabash didn't frighten the birds, and they could easily grab the birds' legs from underwater. This tree's wood was used later by Europeans for cattle yokes, tool handles, wooden wagon wheels, and ribs for building boats.

HOW TO GROW: The calabash is now considered a unique tropical tree that produces gourds or 'calabashes.' It is a great backyard tree because it doesn't get really big, usually about 30 feet, and can be pruned to a desirable height and shape. The fruit can be made into artful gifts. Botanists classify the calabash as *Crescentia cujete.* This tree can be grown from the small flat seeds of dried fruit, cuttings, or root suckers. Calabash can take most soils from fine to course if it is well-drained. The best pH is 6.6-7.5. The calabash can survive blasts of salty sea air. Most calabash develop at least two trunks with a few strong branches that seem to twist with a natural sag. For some strange reason, orchids love to nest in a calabash tree.

Water the tree during the dry season and fertilize with a cup of 12-12-17-2 bearing fertilizer every two months for superior calabash development. The calabash will bloom and bear at the same time. The large light green bell-shaped blossoms hang directly on the large branches and trunk. The branches are long and spread outward with almost no secondary branches. After 4-5 years, they'll produce fruit that can grow to a foot or more in diameter. The blossoms usually bloom at night. The only downside to this tree is that the blossoms are pollinated by bats. The fruit matures in about seven months.

MEDICINAL: The fruits are poisonous. Properly prepared, the fruit treats colds, diar-rhea, pneumonia, and intestinal irregularities. It is also used for relief from menstrual pains and to ease childbirth and procure an abortion. The leaves can be used in the treatment of dysentery, colds, lung diseases, toothache, wounds, and headaches. The bark is used to clean wounds. A tea made from this tree's leaves is used to treat high blood pressure.

FOOD & USES: Young fruit is occasionally pickled, but the pulp is poisonous. The seeds are poisonous if consumed raw. Cooked seeds are used to make a beverage. The leaves are cooked and used in soups. A syrup and a popular confection called 'carabobo' are made from the seed.

The Taíno Amerindians gave the musical world two percussion instruments made from the calabash, the maracas, and the güiro. Maracas are still made from small, dried calabash filled with a few pebbles or hard seeds. The traditional güiro (also called the fish is made from a hollowed-out calabash. The outside of the calabash is carved with grooves. A set of flexible wood sticks, a pua, are banded together and either scraped or rubbed up, down, or across the ridges, with a fast or slow tempo. That creates fine harmonious clicks that blend into the infamous 'cha-cha' sound. The calabash güiro is not only recognized as a Caribbean percussion instrument. It is used in musical pieces ranging from Igor Stravinsky's ultra-classical *The Rite of Spring*; to the famous rock and roll band REM. Listen for it in the classic music of The Drifters' *Under the Boardwalk.*

Today, calabash art has bypassed the basic utensil. With a bit of carving, a splash of paint, and a coat of varnish, the calabash can be crafted into a unique container, purse, or wall art.

CALABASH ART

Ingredients: 1 calabash, 1 of bucket water, a rough scrub brush, a Scotch scrub pad, sand-paper, paint (to your taste and varnish (optional

Method: First, the freshly picked calabash is prepared by cutting (and keeping an access and removing the seedy pulp. Then soak the shell in water and scrub with a rough brush and then finish with a green Scotch pad. Put it somewhere in partial shade, but breezy, to dry. Once dried the calabash will appear carved from wood. Rub both the exterior and interior with sandpaper until smooth. Create whatever you want: a güiro, or maracas, a vase for potted plants, or a bowl to hold your keys. Paint as the spirit moves you. Enjoy for years.

DID YOU KNOW? The unusual calabash blossoms develop from buds that literally grow out of the main trunk and limbs, a condition termed cauliflory. Like many other large-flowered *Cauliflorous* species, calabash trees are commonly pollinated by small bats when the pollen on the upper side of the blossom is placed on the head and shoulders of the bat. In Central America, the seeds of the calabash are toasted and ground with other ingredients including rice, cinnamon, and allspice to make the drink 'horchata.'

Güiro

Calabash Blossom

Maracas

DID YOU ALSO KNOW? The Mayan name for calabash was luch. The flowers of the calabash tree bloom in the evening and emit a slight aroma. Bats pollinate the tree when they blossom at night. The flowers close and wither away by noon time the next day.

Do not confuse the calabash tree, *Crescentia cujete,* with the calabash or bottle gourd, *Lagenaria siceraria.*

NAMES: Krabasi, kalebas, huingo, Spanish: jicara

COFFEE - KOPE
Coffea

Coffee is the popular beverage prepared from the roasted seeds of the coffee bush. Known as cafe, java, mud, Joe, or Nescafe; the drink contains caffeine, the most prevalent stimulant in the world. Usually served hot, also enjoyed chilled, one cup of coffee contains 80 to 120mg of the stimulant caffeine, depending on the method of preparation. Coffee is the world's sixth-largest agricultural export in terms of value, behind wheat, corn, soybeans, palm oil, and sugar. Coffee is also second only to oil as the world's most heavily traded commodity.

HAWAIIAN COFFEE

Coffee in Hawaii started more than a century after plantings began in the Caribbean. In 1817, Don Francisco de Paula Marin planted the first coffee trees, but they didn't thrive. A second attempt was successful by the governor of Oahu. In 1825, Chief Boki, planted Brazilian coffee seedlings. These Brazilian trees were the official great grandparents of the Hawaiian coffee industry. Their seedlings were carried to all the outer islands. A French variety of coffee was introduced to the Big Island three years later and became the god parents to the 'Kona' variety.

A sugar grower, Herman Weidemann, had the foresight to try an Arabica strain, Typica, from Guatemala. That planting in 1892, was better adapted to the Big Island's Kona area. Typica is the same strain that made famous Jamaica's Blue Mountain brand. Today, coffee is grown on 5 islands: Kauai - the largest coffee farm, Oahu and Molokai - only one coffee plantation each, Maui with many independent small farms, and the Big island has what's termed 'the Kona Belt,' 30 miles long and 2 miles wide stretch of coffee farms. Today, coffee farms occupy almost 7,000 acres, producing 6.4 million pounds at a value of over $50 million. The Hawaiian word is '*kope*.'

Hawaii's climate is almost perfect to grow your own coffee. These trees are an attrac-tive addition to any Hawaiian home landscape whether covered in snow white blossoms or bearing red coffee cherries. Seedlings are available on every island. As with every aspect of the home garden, with time, practice, and patience you can grow and process your own tasty cup of coffee.

HISTORY

Coffee is one of Africa's gifts to the world. Supposedly, the bush was first discovered by accident in the Ethiopian highlands. There is more than one legend of coffee's origin. A thousand years before Christ, two tribes, the Oromos and Bongas, were at war in the Ethiopia region of Kefa (cafe). The Oromos warriors ate a stimulant made from crushed coffee beans rolled in animal fat. However, the Bongas won and enslaved the Oromos marching them to the Congo Highlands of Harrar to be sold. The slaves who ate the harsh tasting Robusta version coffee planted a few seeds. Their effort took roots in the highlands' rich soil and transformed into the more flavorful Arabica, which is today's most common type of coffee.

Another origin story begins twelve hundred years ago with a shepherd named Kaldi. Supposedly Kaldi watched his goats prancing among the coffee bushes eating red berries containing the beans. The shepherd ate a few berries and proceeded to romp with his flock. Kaldi related this experience to his Imam. After the holy man watched the shepherd and the goats' bizarre energy, he picked some berries to share with the herder. That night the holy man and his monks were extremely energetic for the teachings. As this legend reports, when they finally slept, the prophet Mohammed revealed to the Imam the berries enhanced wakefulness and wakefulness promoted prayer, and prayer was better than sleep.

That Imam and his monastery became famous throughout Arabia for the spirited praying of its brethren, known as dervishes, who chewed on the berries during sermons. Even today, Suni's start the whirling dervish ceremonies after drinking a brew made from coffee beans. Coffee drinking spread with Islam through northern Africa and the eastern Mediterranean into India.

In the 10th century, an Arabian physician wrote the earliest comments about coffee. By chewing the leaves or berries, coffee use became a religious event. Around 1000 A.D., the irst coffee was brewed from the leaves of the coffee plant. A legend of Yemen, famous for Mocha coffee, tells of an exiled sheik who brewed bitter coffee from green berries. The energy from caffeine probably aided in his surviving his exile.

The method of preparing coffee still in use today, first boiling and then roasting the beans, started a thousand years ago in Arabia. However, by parching or boiling coffee export beans (a roasted bean can't grow) production remained only in Arabia until 1500. The coffee trees belonged to the King. It was punishable by death to possess either a living tree or a living coffee seed, which is usually misnamed as a bean.

The popularity of the drink spread as methods of preparing co ee were continually re-ined. For three centuries, coffee grounds were consumed with the hot water. Then, patient settling left the grounds at the bottom of the cup. The first public coffeehouse opened around 1475 in Constantinople. In the home, coffee was only consumed after elaborate ceremonies. Coffee became such an integral part of Arab life that Turkish men had to provide the beverage to their wives or face divorce. When it irst appeared in Africa and Yemen, coffee was a religious intoxicant. This usage in religious rites among the Suni branch of Islam led to it being put on trial in Mecca for being a 'heretic' substance as wine. The caffeine effects of coffee was forbidden among orthodox and conservative Imams in Mecca in 1510 and Cairo in 1530.

Italian merchants from Venice traded coffee with other spices throughout the Mediterranean. The first coffee imported into Europe was known as '*qhaweh*', or 'Arabia wine'. An Indian pilgrim smuggled fertile seeds strapped to his belly from Mecca and brought coffee trees to Italy in the 1500s. The Dutch founded the first coffee estate in Indonesia in 1696. Around 1715, the Dutch gifted several coffee trees to the French Royal Botanical Garden. From this garden sprung the Caribbean brand of coffee. The first coffee tree in the Western Hemisphere was brought from France to the Island of Martinique in the 1720s. Gabriel Mathieu de Clieu, a naval officer, raided some coffee sprouts from the Royal Botanical Garden at night. He sailed for Martinique, where those sprouts fathered at least 18 million coffee trees during the next half-century throughout suitable French colonies. Brazil lured their coffee industry with romance. In 1727, a Brazilian colonel sweet-talked the wife of the governor of French Guiana for seedlings. Coffee has become an important part of many cultures throughout the world today as both a stimulant and a cure for digestive disorders. It was found to stimulate conversa-tions. During the mid-1600s coffee gained acceptance in New York. Within a quarter-century, coffeehouses became popular throughout all of New England. Before 1600, the favorite drink at breakfast was beer!

The coffee houses in France, England, and the American colonies proved to aid the spread of political opinions. In 1675, King Charles II tried to close London coffee houses to stop the discussion of liberal ideas. The desire for coffee made the king bow; eleven days later, the coffee houses reopened. The French opened the irst coffee cart 325 years ago at the St. Germaine Fair in Paris. This was the beginning of door-to-door delivery of coffee heated by charcoal urns in Paris. During the 1700s, some coffeehouses served as barbershops and casinos. In 1773, the Boston Tea Party beginning the American Revolution was planned in the Green Dragon coffee house. The Merchants' coffee house in New York was the site of US Government headquarters just after the beginning of the American Revolution. The New York Stock Exchange started as a coffee house. Lloyd's Coffeehouse became Lloyd's of

London. The Baltic Coffeehouse became the London Shipping Exchange. The Jerusalem Cafe evolved into the East India Company. Until recently, the runners at the British Stock Exchange were called waiters because it began as a coffee house. A cup of coffee cost a penny in the first English coffeehouses. Coffeehouse life was addictive for those wanting to discuss ideas. Due to the penny charge, coffee houses were called 'penny universities.'

Coffee was accidentally decaffeinated in 1903. Seawater damaged a shipment and removed most of the caffeine from the beans. Using this method, German coffee importer Ludwig Roselius began extracting caffeine from coffee beans. His creation was 'Sanka,' French for *sans caffeine* or 'without caffeine'.

HOW TO GROW: It takes nearly a year for a cherry to mature after irst lowering, and a tree about 5 years of growth to reach full fruit production. While coffee plants can live up to 100 years, they are generally the most productive between the ages of 7 and 20. Proper care can maintain and even increase their output over the years, depending on the variety. The average coffee tree produces 10 pounds of coffee cherries per year or 2 pounds of green beans. All commercially grown coffee is from a region of the world called the Coffee Belt. The trees grow best in rich soil, with mild temperatures, frequent rain, and shaded sun. It's easy to grow co ee at home either in soil or large containers. Since coffee is a variety of tropical evergreen, trees can be used to landscape as shrubs. The trees begin full yield when five or six years old and continue to harvest for another decade. Coffee plants may live for 60 years. Trees are kept trimmed to six feet to get the best yield and to make picking easier. If not pruned, a coffee tree can reach 40 feet.

A coffee tree starts with a fresh-picked coffee cherry that should germinate in three months. Soak the cherry in water for a day before planting lat side down about 1½ inch deep in fast-draining potting soil. The preferred soil pH is 4.9-5.6, should be moist, never soggy, but watered daily.

After about a year and the tree is a few feet tall fertilize with 10-10 -10 every month. The plants will grow to about 10 feet if given ample root room, but a mature root system can extend six feet. Simple pinching of branches produces a bushier plant. The coffee cherries turn from yellow to orange and then bright red, about six months after blossoming. While in bloom, the coffee tree is covered with 30,000 white lowers that develop into fruit after a day. Trees need trimming every 10 years and then cannot be harvested for the next 2.

Each ripe red cherry has two beans. Coffee cherries do not ripen at the same time. In fact, it will have blossoms and berries together in various stages of ripening. Only the ripe berries can be picked and should be washed with water and fermented in a small container until the pulp falls off. There should be only one day between harvesting and processing. Immerse cherries in water and skim off the hollow floaters. The sinkers go through a pulping machine, then dry for 1-2 days. Wash again with fresh water. Spread out the harvested cherries to dry for 3 weeks in the hot sun on a concrete slab. Turn periodically until the outer shell of the cherries turns brown and the beans rattle around inside.

To make homegrown coffee you need to be self-taught on roasting and that can be done in your oven. The smell of this process may fill your home as you burn the chaff. Your beans should be placed in something perforated, a steel strainer or veggie steamer, so all sides get evenly heated at 250 degrees F. for about seven minutes. Increase the oven temperature to 450 degrees. Within ten minutes the beans should crackle. Then stir them so they roast consistently. About every two minutes check them until they have achieved a color slightly lighter than what you desire. As the beans cool, they will continue to roast. High altitude beans seem bitterer and need longer roasting. Mill, brew, drink, and savor the lavor of your personal plantation coffee!

MEDICINAL: Coffee is a purely medicinal substance. Medicinal substances do not nourish, but alter the healthy condition of the body. Coffee can impair the desired effect of homeopathic medication. People drink coffee to relieve mental and physical fatigue and to increase mental alertness. Coffee is also used to prevent Parkinson's disease, gallstones, type 2 diabetes, gastrointestinal cancer, lung cancer, and breast cancer. Other uses include treatment of headaches, low blood pressure, obesity, and attention deficit-hyperactivity disorder (ADHD). Rectally, coffee is used as an enema to treat cancer, and drinking more than 3 cups of coffee daily may significantly reduce the risk of rectal cancer. Drinking caffeinated beverages like coffee seems to increase blood pressure in elderly people who experience dizziness after meals. Caffeine belongs to a group of drugs called xanthines, central nervous system stimulants. The effects of caffeine are physical, not psychological. Physicians regard one cup of coffee with 150 milligrams of caffeine as a therapeutic dose. Caffeine invigorates, quickens, and clarifies thought, and expands idea association.

NUTRITION: Black coffee contains no significant amounts of the macronutrients, fat, carbohydrate, or protein, and therefore contains only 1 or 2 kcal per 100ml with some micronutrients, notably potassium, magnesium, and niacin.

FOOD & USES: Twenty-five years ago, most American coffee was made from percolators. This is a bad way to make coffee since the coffee continually boils, but sounds and smells nice. 90 years ago, the drip coffeemaker was created when a German housewife, Melitta Bentz used a coffee filter of blotting paper from her son's notebook. The introduction of the Melita version with cone-shaped drip units improved the taste.

The coffee table replaced the tea table in most American living rooms in the 1920s. Coffee breaks began during World War II when employers reasoned coffee increased the pro-ductivity of their workers.

COFFEE CONSUMPTION

Plain black coffee is a natural beverage with no added sugar or preservatives and does not contain any calories. It is the world's second most popular drink after water. North America and Europe drink coffee an average of about one cup of coffee to three glasses of water. Last year, coffee drinking in the USA was 22.1 gallons per person and is the world's most popular beverage with over 400 billion cups consumed each year, or 1.4 billion cups every day. More than half of the US adult population daily drinks some type of coffee beverage, averaging about three cups.

By 1960 coffee had become the beverage of choice for teenagers, who went to modern coffee houses to hear the new folk music. In 1965, Nescafe introduced freeze-dried soluble coffee, and two years later invented a way to capture more aroma and flavor from every single coffee bean. Today, every corner of the world treats itself to a cup of Nescafe.

Coffee is grown commercially in over 70 countries throughout the world. In Brazil, over 5 million people work in the coffee trade. Most of those are involved with the cultivation and harvesting of more than 3 billion coffee plants. Only 20% of all harvested coffee beans are high-quality premium beans. The quality of a cup of coffee does not only depend on the bean blend, but also on the ratio of the amount of water and coffee used when brewed.

Nine out of ten coffee drinkers have coffee at breakfast. Work is the second most prominent location. Americans drink an estimated 350 million cups of coffee a day. Coffee is second only to oil as the world's most valuable commodity.

Coffee travelled a long journey since the end of the first millennium when Arabs soaked green coffee beans in cold water to make the first coffee. James Mason patented the first American coffee percolator in 1865. A German housewife introduced cone-shaped drip units that improved coffee's quality. Ironically, it was George Washington, not the first American president, but a Belgian living in Guatemala who created the first instant coffee in 1906 before he immigrated to the United States. In 1930, the Brazilian government approached the Swiss company Nestlé to produce a quality cup of naturally flavorful coffee by only adding water. Nestlé experimented for seven years before creating Nescafé in 1938. During World War II, American forces launched Nescafé in Europe, because it was included in their food rations. Coffee was first called 'a cup of Joe' during WWII when American servicemen (G.I. Joes) became big coffee drinkers.

The number of cups of gourmet coffee beverages consumed per drinker per day is 2.5 cups. The average American adult consumes 10.2 pounds of coffee beans per year. The largest gourmet coffee purveyor, Starbucks, had net earnings of $920 million, compared to $494 million in fiscal year 2005. Starbucks opened in 1971 in Seattle, named after the first mate in *Moby Dick.* The chain now has 30 blends and at its height in 2007 opened three new stores every day. Gourmet coffee lovers will spend as much as 45 hours a year waiting in line. The average person who buys gourmet coffee spends the equivalent value of a round-trip plane ticket to Los Angeles every year.

COFFEE PRODUCERS

Coffee is one of the world's most important commodities, with seven million tons produced every year. A good coffee tree can produce two pounds of raw coffee per year. One acre of coffee trees will produce up to five tons of coffee cherries, which reduces to one ton of beans after hulling and milling. Coffee increases in volume by 20% during roasting. Daily, Americans consume 300 million cups of coffee. Four thousand coffee beans are necessary to produce one pound of roasted coffee, or fifty beans for each cup. Coffee grows only in fifty-six tropical countries around the world. Annually it is worth 45 billion USD to the growers.

Worldwide approximately 20 million people work in the coffee industry. Small farmers grow most of the world's coffee and rarely share in the huge profits. The mostly third-world governments, facing huge international debts, must sell their coffee crop regardless of price. These coffee producers compete against each other, often resulting in an over-supply, which drives prices down. The world's largest coffee producer is Brazil with over four billion coffee trees. Colombia comes in second with nearly two-thirds of Brazil's production. Recently, Vietnam has flooded the market with large quantities of green beans.

A few giant manufacturers and retailers control the international coffee market, which wildly fluctuates. When the price of coffee rises, estates plant more trees. When these trees start producing, it causes a glut and the price falls. Then, new bushes are not planted, there is a shortage and the price rises. Small producers often must sell when their harvest is ready with little or no choice of who buys. In the UK where 90% of the coffee sold as 'instant' - just two multinational food companies, Nestlé and Kraft, account for 70% of all retail coffee sales.

Coffee is of the botanical family *Rubiaceae*, with more than sixty different varieties. Only two have any economic significance, Arabica, and Robusta. Three-quarters of world production is of the milder, higher-quality Arabica produced largely in Latin America. Arabica has been known from prehistoric times and is believed to originate in southwest Ethiopia, specifically from Kaffa; thus its name. There are 900 different flavors of Arabica. While more susceptible to disease, it is considered to taste better than Robusta. It takes a coffee tree three to four years to produce berries. The Arabica species self-pollinates, whereas the Robusta uses cross-pollination.

Robusta, with 40% more caffeine, probably originated in Uganda and is grown where Arabica will not thrive. Many commercial coffee blends use as Robusta as an inexpen-sive substitute for Arabica. Robusta tastes bitter with little flavor compared to Arabica. Robusta beans are used in some espresso blends to provide the foamy head while reducing the ingredient cost.

I have measured out my life with coffee spoons. ~ T.S. Eliot

Another variety, Liberica from Western Africa, has no great importance in coffee trade. Two more bean types are Mocha from Yemen, and Java from Indonesia. The modern coffee trade is much more specific about the origin, labeling coffees by country, region, and sometimes even the estate where the beans were grown.

Coffee means subsistence to more than 100 million people. Each year some 7 million tons of green beans are produced worldwide. Most of which is hand-picked, dried, sorted, (andsome aged) before coffee is transformed into the familiar roasted coffee. Roasting coaxes golden flavor from a bland bean. All coffee is roasted before being consumed either by the supplier or roasted at your home. It takes heat to turn coffee's carbohydrates and fats into aromatic oils, burn off moisture and carbon dioxide, and alternately break down and build up acids, unlocking the characteristic coffee flavor.

Among the numerous luxuries of the table…coffee may be considered as one of the most valuable. It excites cheerfulness without intoxication; and the pleasing flow of spirits which it occasions…is never followed by sadness, languor, or debility. ~ Ben Franklin

DID YOU KNOW?

Coffee as a beverage has taken about 1,000 years to evolve to its current presence in specialty coffee shops. In the 1400s the first coffee shop opened in Constantinople, where the Turks thought the drink was an aphrodisiac. The first coffee house in Europe opened in Venice in 1683, while coffee was available in Europe as early as 1608, mostly for the rich. In the year 1763, there were over 200 coffee shops in Venice. Italy now has over 200,000 coffee bars. The first commercial espresso machine was manufactured in Italy in 1906. Southern Europe enjoys the flavor of dark-roasted coffee. A coffee tree can be harvested during its fifth year and many times throughout every year. Coffee cherries usually contain two beans. Cherries with three beans is considered a sign of good luck. Brazil is responsible for 30 to 40 % of total world output. Approximately 2,200 ships only haul coffee beans each year. In December 2001, Brazil produced a scented postage stamp to promote its coffee - the smell should last between 3 and 5 years.

NAMES: Cafe, Café, Espresso, Expresso, Java, Mocha

We want to do a lot of stuff; we're not in great shape. We didn't get a good night's sleep. We're a little depressed. Coffee solves all these problems in one delightful little cup. ~ Jerry Seinfeld

COTTON – PULUPULU

Gossypium barbadense

Cotton is a shrub related to the hibiscus and okra. It's undeniably one of the world's most important plants and one of the oldest that has remained basically unchanged for thousands of years. Fiber from the plant is spun nto yard or thread and becomes cloth that remains soft and breatheable. Animal skins are limited in quantity and not porous. Cotton is the most widely used natural fiber cloth in clothing today. Cotton is also a mystery because it's native to tropical and subtropical regions around the world, including the Americas, Africa, Egypt, and India. Cotton was independently domesticated in the Eastern and Western Hemispheres. Circa 800 A.D, Arab merchants brought cotton cloth to Europe. In 1492, Columbus discovered cotton growing in the Bahama Islands. By 1500, cotton was generally known throughout the world.

The areas that have been found to have the most varieties of wild cotton are Mexico, followed by Australia and Africa. Cotton fabric dates to 5,000 BCE in the Indus Valley and even earlier in Peru. India started growing cotton in 3,000 BCE. The invention of the cotton gin lowered production costs, increased its worldwide cultivation and use. World production is estimated at 25 million tons, equaling 110 million bales, each year. Cotton cultivation occupies 2.5% of the world's fertile land. India is the world's largest producer and exporter of cotton. The United States is the second largest exporter followed by China, Brazil, and Pakistan.

In Hawaii, the cotton plant, *Gossypium tomentosum,* is known as *pulupulu* (or huluhu-lu-hair). Due to the excellent weather, it's a perennial shrub that can produce for 3-5 year. Hawaiian cotton easily grows on the hot, dry, windy coastal plains and in the dry forests, usually within the sheltered areas of the islands. Native/naturalized Hawaiian cotton can live with the salt air. Waterlogged soil from high rainfall kills it. The Hawaiian variety was crossed with other cotton varieties, and the hybrids are immune to the insect pests that destroy cotton crops. The cotton gin (slang for engine) was invented in 1793 and could do the work 10 times faster than by hand. It made possible processing large quantities of cotton fiber for the textile industry. Within a decade, the value of the U.S. cotton crop rose from $150,000 to more than $8 million. Why would you grow cotton? Perhaps to understand how our with this ancient valuable plant, or to learn the ancient art of a spinning machine and make yarn for your loom. A simpler use, cotton plants have flowers resembling hibiscus and seed pods that can be used in dried flower arrangements. All can create an art form.

HOW TO GROW: Cotton grows in warm climates and cannot tolerate frost. It has a 5-month growing season that cannot be interrupted by cold weather. It requires soil enriched with plenty of compost and a pH of 5.8-8. Plant 3 cotton seeds an inch deep, in full sun and keep moist. Space plants 2 ft. apart. They should sprout within a week. Water regularly and after 30 days use some high nitrogen with potassium fertilizer. Cotton is a heavy feeder and drinker. Because of cotton being an ancient crop, it attracts many pests. Do not over water and plant the cotton trees far apart to get good air circulation. Interplant garlic, onions, and a variety of aromatic herbs like mint to ward off insects. There are many genetically improved seed varieties that fight insects.

At about 4-5 weeks, the plants will begin branching. Cotton takes lots of tender lov-ing care. At 8 weeks the attractive white blossoms will appear. After they have been pollinated, they will turn pink. (Cotton is self-pollinating.) At this point the plants produce a boll, the seed pod of the plant, which becomes the 'cotton ball.' The bolls have sharp spiny edges, which can poke you at harvest time. Periodic pruning will keep the shrub full and stops the lower branches from spreading. The seeds are attached to wispy fibers. Nature evolved the cotton plant when the boll opens the seeds can fly on the wind to grow in new places.

There should be cotton flowers after 2 months and another 50 days before the seed pods, called bolls, form. That's what has the cotton. Harvest after bolls split open, during a period of dry weather. Cut cotton on long stems and dry in bunches, like herbs. Each cotton tree can have up to 100 pods/bolls. Wear gloves to harvest. You can twist the pods off, but use shears unless you have a huge cotton plantation. Then it would be picked by machines. All the cotton pods on a plant won't be ready to harvest at the same time. Take only the bolls that are ready; leave the others for another day. Spread the cotton to dry in a cool, dark area. Then the seeds must be separated. Egyptian fine is the cotton that is famously used in sheets and fabrics. It produces a white long, fine fiber. It has a longer growing period and matures in 155 days. Cotton will grow on your patio in large pots used for trees. Keep the pots dark colored to keep the roots warm. It can be an indoor houseplant, but for it to flower, you'll need grow lights.

MEDICINAL: Cotton leaves are used in a tea to treat respiratory problems as asth-ma, coughs, and throat infection. The same tea can be used as a skin wash to treat acne, in-sect bites, and inflammation. The leaves are used as a natural bandage to heal wounds.

NUTRITION: Cottonseed is full of protein but poisonous to humans and most animals. The U.S. Department of Agriculture recently approved a genetically engineered cotton that has seeds that are safe to eat.

FOODS & USES: Cotton seeds are a valuable by-product. The lint is removed from the seeds by a process similar to ginning. Some linter makes candle wicks, string, cotton balls. It's used in cellulose products like rayon, photographic film, and cellophane. The cotton seed first must have the toxic compound Gossypol removed before it should be processed into oil or ingested. The cleaned seeds are crushed, and the kernel is separated from the hull and squeezed to obtain cottonseed oil. It's used for cooking, shortening, soaps, and cosmetics. A refining residue called soap stock makes linoleum, oilcloth, waterproofing materials, and paint bas-es. The seed hulls are used for fertilizer, plastics, and paper. A liquid made from the hulls, called furfural, is used in the chemical industry. The remaining mash is used for livestock feed.

DID YOU KNOW? Cotton has been spun, woven, and dyed since prehistoric times. It clothed the people of ancient India, Egypt, and China. The earliest evidence of cotton is preserved in copper beads and carbon dated to 5500 BCE found in Pakistan. Cotton bolls discovered in a cave near Tehaucan, Mexico, have been carbon dated to as early as 5500 BCE. In Peru, cotton cultivation has been dated to 4200 BCE. Spanish who explored Mexico and Peru in the early 1500s found the people growing cotton and wearing clothing made of it. The Greeks and the Arabs were ignorant of cotton until Alexander the Great attacked. Eli Whitney claimed a patent on the cotton gin in 1793, though a machinist named Noah Homes built the first cotton gin 2 years earlier. Cotton comes naturally in a variety of colors. White cotton was the standard; commercial brown cotton was grown for personal use by slaves and poor whites. Pink, green, blue, and yellow varieties have all been popular. Today, people often bleach cotton to achieve a uniform color. January 15, 2019, China's Chang'e 4 spaceship grew cotton from seeds on the Moon's far side. Cotton is usually measured in bales, 0.48 cubic meters (17 cubic feet) and weigh 226.8 kilograms (500 pounds).

There are few sights more pleasant to the eye than a wide cotton field when it is inbloom. It presents an appearance of purity, like an immaculate expanse of light, new-fallen snow.

~Solomon Northup

THE CURRY TREE
Murraya koenigii

The curry tree is essential in Indian, and SE Asian cooking. These trees are available throughout the Hawaiian Islands. The botanical name is *Murraya koenigii.* These leaves are a mainstay in curries and many natural medicines. Usually called curry leaves, they translate as the leaves of the sweet neem tree that's native to India, Sri Lanka, the Andaman Islands, and Bangladesh.

This tree is very hearty and immigrants, who could not leave the tasty leaves behind, carried its seeds to many distant lands. It is now cultivated worldwide in countries with a tropical climate or indoors. Curry leaves are not the same as cur-ry powder. They are often added to this popular spice mixture. These leaves are used in cooking to add flavor to rice dishes, curries, and dahls/dried lentils or peas.

The earliest mention of the curry leaf is in the first century AD among ancient Southern Indian manuscripts. These same documents indicate the word curry is from the dialect of south India. *Kari* translates as spiced sauces. These leaves provide a unique fla-vor and are best used fresh, usually fried with the chopped onion in the first stage of any curry. Raw leaves smell so good and have a flavor that tastes like a combination of citrus, anise, and lemongrass. Cooked, the leaves have a mild slightly sharp taste with the aro-ma of cashews. Curry leaves are valued for their flavor, but eating them provides sever-al health benefits. All aspiring chefs should grow your own tree. Ask around on the social pages and you'll find someone with a curry tree. These trees grow easily in the Caribbean.

HOW TO GROW: A mature curry leaf tree grows to five meters with a brown to a dark green trunk with dots on the bark. It is deciduous, growing clusters of small fragrant, self-pol-linating, white blossoms in summer. These produce shiny deep purple berries, 40 to 80 in a cluster, less than a half-inch long when ripe. The berries aren't used in cooking. The berries are edible, but taste 'medicinal.' Find some fresh seeds to plant; dried won't germinate. Stem cut-tings can also be used. The soil must remain moist, but not wet in a mostly sunny location. For the best results, fertile, humus-rich, moisture-retentive but well-drained, light soil with a pH of 5.5 - 6.5. When your plant is still young, don't apply strong fertilizers. Young curry trees are not heavy feeders; once a month give it a light soluble fertilizer. To increase foliage growth, apply 3 teaspoons of iron sulfate monthly. Give it 1 teaspoon of Epsom salt (magnesium sulfate) in 1 li-ter of water every 3 months. A natural fertilizer can be made by diluting yogurt. Curry trees are perfect for a pot on the patio or balcony. If you see leaves disappearing inspect for caterpillars.

MEDICINAL: Since ancient times, curry leaves have been used in Ayurvedic-ho-meopathic medicines. They are used to assist diabetics to help reduce high blood sugar lev-els and are believed to be anti-inflammatory and antibacterial. A tonic of boiled leaves is a stimulant for the digestive system and also for healthy hair and skin. Curry leaf oil may cure acne pimples, athlete's foot, ringworm, itches, boils, and help to heal burns. The oil is used to strengthen bones and fight osteoporosis, calcium deficiencies, and after chemo-therapy. Eating dishes with curry leaves may reduce high cholesterol. Significantly, stud-ies show curry leaves contain substances that may protect against nervous system condi-tions like Alzheimer's disease. Studies with curry leaves grown in Malaysia found they had powerful anticancer effects against aggressive types of breast, colon, and cervical cancers.

Curry leaves have an unusual and pleasant scent that makes it a prime ingredient in many soaps, body lotions, air fresheners, perfumes, bath, and massage oils. It is wide-ly used in aromatherapy at spas and health clinics for face steams and hair treatments.

NUTRITION: A hundred grams of curry leaves provide 100 calo-ries. They're rich in carbohydrates, proteins, and fiber with other minerals such as calcium, phosphorus, iron, magnesium, and copper. They're rich in vita-mins A, B, C, and B2, nicotinic acid, antioxidants, amino acids, and flavonoids.

FOOD & USES: Although most commonly used in southern and western Indian, and Sri Lankan curries, leaves from the curry tree can be used in many other dishes to add flavor. In Cambodia, they toast the leaves in an open flame or roast them until crispy and then crush into a sour soup dish called Maju Krueng. Curry leaves are best when boiled, steamed, or sautéed. They are usually stripped from a fresh stem, fried in hot oil with other spices, onions, garlic, and ginger, mustard seeds, and chili peppers for a base when making a dish or as a savuce to flavor a dish. The leaves are edible and do not need to be removed before eating. They add a unique, uplifting flavor to stews, curries, soups, rice and bean (dahl) dishes. They accent the flavors of lentils, yogurt, coconut milk, oyster sauce, pea shoots, eggplant, pork, and fish. They will keep up to two weeks in the refrigerator when stored fresh in a sealed container and up to six weeks in the freezer. However, for the best medicinal properties, fresh leaves must be used.

RECIPES: EXOTIC - BANANA FLOWER STIR FRY – that big purple ball that dangles below the banana bunch that's often tossed and wasted can be a delicious dish.
Ingredients: 1 banana flower, ½ cup orange lentils, 2 cloves garlic, 1 medium on-ion sliced thin, 3 green chilies or 1 hot pepper chopped small, ½ cup grated coconut, ½ TS turmeric powder, 30 curry leaves, 2 TBS coconut oil, 2 TBS water, salt to taste.
Method: Rinse your hands with cooking oil to prevent the banana from staining them. Be care-ful with your clothes. Remove petals one at a time. Put each floret in a separate bowl until you find the white center. Cube the white portion of the flower and put into a bowl of water. Cook the banana asap to keep from darkening. Remove black stamen from florets and the cup-shaped part of the florets. Chop remaining portion and put in a bowl of water. Rinse both the florets and white cubes twice, drain, pat dry and place on a plate. Mix with 1 TBS of oil so they do not blacken. Pressure cook red lentils until one whistle. Do not overcook as you want it crisp, not mushy. Drain and keep excess water. In a frying pan heat oil, add onion, garlic, curry leaves, and chili peppers. Fry, stirring, until onion is clear. Stir in ¼ TS turmeric power, add banana flower pieces and salt. Stir and add 2 TBS water. Cover on low heat until the flower is almost cooked. Stir frequently. In another bowl, combine ¼ TS turmeric powder and the grated coco-nut. Add lentils and coconut mix to the banana. Cook, stirring for 5 minutes. Serve with rice.

CURRY LEAVES-CASHEW-COCONUT RICE - easy and tastes great
Ingredients: ½ cup cashews chopped, ¼ cup coconut flakes (fresh best), 1 cup rice-Basmati preferred, 5 TBS vegetable oil – canola preferred, 1 TBS yellow mustard seeds, 2 cardamom pods or ¼ TS ground cardamom, 12 curry leaves, 1 medium onion sliced thin, salt to taste
Method: In a cast iron skillet or similar on low heat, separately toast the cashew nuts and coconut flakes until light brown. Put in separate bowls so the flavors do not bleed/combine. Wash rice and cook as usual in a sauce pan. Heat the oil on medium in a skillet suitable to later hold the rice. Add mustard seeds, cardamom, and curry leaves stirring until mustard seeds pop, 2-3 minutes. Add onion and cook until lightly browned. Remove from heat and scrape all loose. In a bowl, fluff the rice, add a pinch of salt, and stir in the spices and oil with two-thirds of the cashews and coconut. Use remaining cashews and coconut as a garnish.

DID YOU KNOW? The US and British governments banned importing fresh curry leaves because of concerns about the spread of citrus greening disease

NAMES: sweet neem, *Hindi:* karipatta, *Caribbean.* karrapouli, *French:* carripoule

MAUBY –MABI - MAVI - SOLDIERWOOD
Colubrina ellipticain - Colubrina arborescens

What's in a name? With plants it can be confusing between the scientific Latin terminology and a multitude of local names. Mauby/soldierwood drink, like roselle/sorrel, was always a special holiday drink at our home. The drink is made from the dried bark of a small tree native to the northern Caribbean, south Florida, and Central America. West Indians call the tree mauby, but during my research, I found this tree has many botanical names, but all are of the *Colubrina* family. Some botanists say the mauby/mabi/mavi tree is *Colubrina ellipticain*, or *Colubrina reclinata.* All are members of the buckthorn family. In Florida, it's known as soldierwood or nakedwood. I've seen the mauby drinks – Mauby Fizz – in Hawaiian markets, but never the tree or bark but have references that it grows. With the climate, rainfall, and soil similar, grow your own.

Most people don't know the tree, but enjoy the refreshing drink. Some claim it as an aphrodisiac, others say it helps for arthritis, but everyone knows it is a great coolant on a hot day.

The mauby/soldierwood is a hardwood and a nice, functional backyard tree. The tree will be shrouded in attractive purplish green leaves at the height of the rainy season. That's when the bark should be harvested. During the dry season it will shed most of its leaves as the sap drains to the roots. This tree not only drops its leaves, as it matures it will have a multitude of small berries that have three seeds. I would not recommend this tree if you have a swimming pool.

HOW TO GROW: Find someone with a tree and search under it for a seedling or search the social pages online. This is another virtually effortless tree since it only needs direct sunlight a few hours a day, shelter, to be staked when hard winds blow, and well-drained soil with a pH of 6-8. Once or twice a month during the dry season, I recommend giving it a five gallon bucket of water. As the wet season ends, broadcast two cups of starter (12-24-12) fer-tilizer around the roots. Small green blossoms appear usually in July and the berries come on from September to March. Then switch to a bearing fertilizer such as 12-12-17-2. Molding with mulch after broadcasting the fertilizer will help it to feed and not wash away. Molding will give support to the young roots and help the soil conserve nutrients and water. This tree tolerates both dry weather and some salt spray.

This tree has few enemies except for fungus which can be protected with occasional spraying of appropriate fungicides. Some mauby trees are evergreen, but you won't know until after planting. All the mauby I've made and drank has only been prepared from the bark with some other herbs. My research found sources that also use the leaves and berries.

MEDICINAL: Unsweetened and tangy, which is very seldom, the refreshing bark decoction is used as a bitter tonic for diabetes – lowering blood sugar, hypertension - lowering blood pressure and cholesterol, and stomach disorders. In a small study published in the *West Indian Medical Journal* (2005 Jan; Vol 54(1):3-8.), 'a mauby syrup and coconut water mixture reduced blood pressure in more than 50% of hypertensive subjects. The research results showed significant decreases in either the mean systolic, the mean diastolic pressure, or both.›

Drinking mauby, bathing in mauby water, and a poultice of mauby leaves are reported to relieve arthritic pains. A tea made from the leaves and the wood is used as a remedy for rheumatism while the extract is used in antiseptic baths. A decoction made from the boiled wood, mixed with milk, is used to build up the blood, especially after childbirth. A bark tea, combined with anise, nutmeg, mace and sugar, is considered a diuretic. The tea is also considered an aphrodisiac.

FOODS & USES: The shiny seeds of this and related species have been made into necklaces and similar ornaments. The heartwood is yellowish brown; the sapwood is whitish or light brown. The wood is hard, heavy, strong, and durable. It is used chiefly for posts, and formerly for piling because of its resistance to decay in water. It is also employed in general carpentry and construction where sufficiently large pieces are available.

The bark is steeped in water to make a cooling, fermented drink known as 'mabi champan.' People love mauby! Trinidad locally produces more than one and a half million gallons a year of mauby concentrate. Due to the lack of local bark, it's imported from Haiti and the Dominican Republic. Mauby can be bought as a pre-made syrup and mixed with water or soda. To the mauby novice, the first taste is sweet, but changes to a bitter aftertaste. It may provide a dose of the trots (diarrhea). Many find it an acquired taste. The bark of the mauby tree is usually sold by local farmers but it can also be purchased online. Grow your own tree and you'll enjoy this great healthy drink.

RECIPES: MAUBY ISLAND TOTAL REFRESHER- enough for a party or for gifts
Ingredients: six pieces of mauby bark, one gallon of fresh water, four sprigs of marjoram, one bunch anise (optional), one stem of rosemary, a half cup of peeled ginger root sliced thin, one stick of cinnamon, one TS ground nutmeg, a dozen cloves, two TBS Angostura bitters, three cups brown sugar (more or less to taste), one TS dry yeast
Method: Boil the mauby bark alone in one quart of water and let sit covered for at least an hour. Then add everything except the yeast. Bring up the heat and simmer for an hour before letting it cool preferably overnight. Add the one TS of yeast to a tablespoon of water and combine with batch. Then carefully bottle filling each to the neck. Let sit in a shaded place again overnight. Chill and enjoy. Using white or brown sugar will contribute to the darkness of this drink. The bitters will buffer the slightly bitter aftertaste.

Mauby is sweet and bitter, made as a concentrate or a fizzy drink, and is taken to aid digestion. ~ Eat the Weeds

NAMES: In the *Dominican Republic* and *Puerto Rico:* maví (or mabí), in *Haiti* and *Martinique:* mabi, and maubi in the *Virgin Islands* and *Dutch Caribbean Islands*. In the *Florida Keys:* soldierwood or nakedwood.

Trees are sanctuaries. Whoever knows how to speak to them, whoever knows how to listen to them, can learn the truth. ~ Herman Hesse

MORINGA – MIRACLE PLANT
Moringa oleifera

Many countries consider moringa the miracle plant that can solve many nutritional and medical problems. In Hawaii, moringa grows in almost any area, and is considered drought resistant. Native to India, this tree has been an integral part of traditional natural medicines for centuries. This tree's leaves and seed pods are a food source packed with protein, vitamins, and minerals. The National Academy of Sciences rates moringa as 'possibly the planet's most valuable undeveloped plant.' Botanically, it's *Moringa oleifera.* In Hawaii, its marunggay or kalamungay.

Having a moringa tree will provide lots of healthy food that can be prepared in many different ways. The leaves are similar to spinach, the drumstick pods can be cooked like green beans, and even the roots taste like spicy horseradish.

HOW TO GROW: Moringa trees are so easy to grow some countries consider them to be an invasive species. This tree prefers a sunny spot, in sandy-loamy soil as long as it is well-drained to an elevation of 2000 meters. Moringa will produce the best leaves and pods in fertilized soil with a pH of 6.3-7 with regular watering of at least 80cm (32in.). The recommended method of planting this tree to guarantee a good root system is to dig a hole 30cm (12in.) wide and deep. Refill with a combination of the soil with the addition of manure and/or compost. The seeds have a high rate of germination and can be planted directly into this hole. Seedlings can also be sprouted from 1m long cuttings, 4cm in diameter, perfect if you are harvesting branches for the leaves.

Trees from cuttings may produce in one year. In the second year, these trees produce 300 drumstick pods. Properly fertilized and watered, the crop size will increase every year. A mature tree can produce more than 1,000 pods. A moringa tree looks great with its full sagging branches and can grow to 12m (40ft.) with light gray bark. The pale-yellow blossoms smell good and should appear within 6 months. For ease of picking, trees are annually topped to keep them at 3m (10ft.). The dark green seed pods resemble drumsticks, 40cm (15in.), hanging from the branches. The dark brown seeds have natural wings; the reason moringa is listed an invasive species.

MEDICINAL: Almost all parts, the bark, sap, roots, leaves, seeds, and flowers, of the moringa tree are eaten or used in Ayurvedic-homeopathic traditional medicine. It is used regular-ly to make a nutritional tonic. Moringa reduces inflammation, is an antioxidant, an aphrodisiac, and boosts the immune system. Eating leaves is prescribed for anemia and believed to increase women's milk production. Juice pressed from the leaves treats diabetes, reduces cholesterol, and stabilizes blood pressure. Moringa blossoms are used to treat inflammations, muscle diseases, tumors and enlargement of the spleen. The blossoms are made into an infusion and used as an eyewash. **WARNING!** Do not use during pregnancy, moringa blossoms can cause a miscarriage.

Moringa roots treat the cardiovascular system, stimulate appetite, female reproductive issues, arthritis, and are used externally as a natural antiseptic. These roots contain concentrat-ed phytochemical compounds that are found throughout the rest of the plant, and can provide therapeutic benefits for many conditions. Modern scientific studies of the roots show they con-tain elements to treat ovarian cancer. **BE CAREFUL** before using moringa roots, because of the high concentration of spirochin, an alkaloid in the root. In small doses, 35mg, it can accel-erate the heartbeat, but at larger doses of 350mg or more, parts of the nervous system may be paralyzed. Studies also show the seeds relieve the pain of rheumatism and gout. Moringa seeds have a high iron content and treat anemia and a natural antibiotic that treats fungal infections. Oil pressed from moringa seed protects the hair and skin cells from damage because it con-tains hydrating and detoxifying elements. It also has some healing properties for skin allergies, irritations, wounds, blemishes, and stretch marks. Naturally occurring isothiocyanates are the main anti-inflammatory and anti-carcinogenic compounds in moringa leaves, pods and seeds.

NUTRITION: Eating moringa is the same as taking a natural vitamin and mineral supplement. 100 grams of cooked moringa has 64 calories with 9g of protein, 2g fiber, 1.5 g fat, and is an excellent source of vitamins A, thiamine-B1, riboflavin-B2, vitamin C, pyridoxine-B6, niacin-B3, and folates. Moringa is also high in calcium, magnesium, manganese, phosphorus, potassium, zinc, and iron. The leaves are the most nutritious part of the plant; one cup of fresh, chopped leaves (21g) contains 7 times more vitamin C than oranges and 15 times more potassium than bananas with calcium, protein, iron, and amino acids. The bulk of the nutrition is found in the young leaves; nutrients are lost with age. The best, most nutritious moringa leaves are harvested while the trees are saplings.

Drumsticks or immature seed pods are high in vitamin C, potassium, magnesium and manganese, all of which may be reduced by the preparation method. The seeds, cooked in recipes similar to peas or roasted like nuts, have lots of vitamin C and lesser amounts of B vitamins.

FOOD: Young drumstick fruits are cooked as a vegetable, often chopped and curried or used in dahls (dried lentils and peas) and lentil soups with the flavor of sweet green beans or asparagus. Drumsticks smashed to a pulp are fried or mixed with curried vegetables. The outer skin of the seed pod can be tough, so drumsticks are often chewed, spitting out the skin. Or the flesh and tender seeds are sucked leaving the skin tube. Shredded roots make a spicy sauce similar to horseradish sauce as a condiment and can also serve medical purposes.

Finely chopped, tender young moringa leaves are added to soups, vegetable dishes, and salads often instead of, or combined, with coriander. Moringa leaves can be preserved by freezing or drying powdered with variable preservation of their nutrients. Powdered moringa leaves are added to soups, sauces, and smoothies. Powdered leaves taste great when used to enrich yogurt, cheese, pastries. Moringa leaf powder is an effective handwashing soap with antiseptic properties.

Moringa blossoms make a delicious tea and are tasty, mixed with salad greens or fried in butter as a snack. Some say they taste like mushrooms. These lowers are considered a delicacy in many places. Moringa lowers can be dried in the shade, or in an oven under low temperature for tea. Let the lowers steep in hot water for 5 minutes for the best flavor. Moringa tea is nutritional; people drink very strong moringa lower tea at the first sign of the cold to boost their immune system, or for a sore throat. Be careful, they have a laxative effect. They contain a good amount of both calcium and potassium.

The seeds are added to sauces or eaten as a fried snack. Ground into a powder, moringa seed is used to increase the protein, iron, and calcium content of wheat lour baked goods. From the seeds comes an oil with a light, pleasant taste that is good for cooking. This oil is also used for body massage and in aroma therapy. Moringa oil contains 4 times more colla-gen than carrot oil. It's used in soaps, shampoos, perfumes, and other skin care products.

After the oil is cold-pressed, the remaining pulp is dried, powdered, formed into a cake, and then used to purify water. It successfully removes dirt particles and harmful bacteria from muddy river water and makes it potable. This method is safer and healthier than using aluminum sulfate and other chemicals, such as chlorine, for water purification. Proteins in moringa seeds causes particles to clump together, making filtration easier.

RECIPES: MORINGA LEAF STIR-FRY

Ingredients: 100g young moringa leaves washed and shredded, 1 medium onion chopped, 3 green chilies chopped, ½ cup grated coconut, 1/2 TS of each: turmeric, black mustard seeds,red chili powder, salt to taste, and 1TBS oil

Method: Combine salt and turmeric powder with the shredded leaves and set aside. In a frying pan or wok, heat the oil and add mustard seeds until they splutter. Add the onions, green chilies, and sauté it until the onion browns. Add the leaves, stir, and cook covered on low heat for 4 minutes.

Remove the lid and cook, stirring constantly for another 5 minutes. Add the grated coconut stirring for 2 minutes. Add salt if necessary.

MORINGA VEGETABLE BROTH

Ingredients: 2 cups moringa leaves washed and dried, 2 green chilies split lengthwise, 1 medium onion chopped, 1 small potato cubed, 1 eggplant cubed, 1 sprig of curry leaves, 1TS turmeric powder, ½ cup coconut milk, 2 cups water, juice from 2 lemons or limes, and salt to taste

Method: In a suitable covered pot, combine the potatoes, eggplant, water, and salt and cook until potatoes are not quite soft, 7 minutes. Add curry leaves, turmeric, and coconut milk and bring to a boil. Remove lid and cook for 2 minutes. Stir in the moringa leaves and cook for only 1 minute. Remove from heat. Before you serve this dish, stir in lemon or lime juice.

MORINGA SOUP

Ingredients: 250g (1/2lb.) moringa drumstick pods cut to 3in pieces, 1 medium onion chopped, 500g taro root peeled and cubed, 2 green chilies chopped, 1 lime quartered, 6 curry leaves, 1TS fennel seeds, ½TS cumin seeds, ½TS turmeric pow-der, salt to taste, fresh ground black pepper, 4 cups of water, and 1TS coconut oil

Method: In a suitable pot, boil the moringa drumstick pieces until soft. Remove from heat. When cool, transfer to a mortar and grind the drumstick pieces until you can remove the skins. Combine the skinless drumstick mush with all ingredients except the lime wedges and black pepper Bing to a boil on medium heat and then reduce the heat and simmer for 20 min-utes. When serving, sift some black pepper and squeeze a piece of lime over each soup bowl..

MORINGA SMOOTHIE

Ingredients: 1cup almond, soy or regular milk, (or ice cream), 1/2 cup fresh moringa leaves, or 1TBS moringa leaf powder, 1 banana chopped. Add mango pieces, blueberries, strawberries, whatever your taste.

Method: Fill a bender with the milk or ice cream, moringa leaves, and fruit pieces. Blend until creamy and enjoy immediately.

DRUMSTICK SHRIMP CURRY

Ingredients: 1 cup moringa drumsticks parboiled, peeled (Boil moringa enough so the skin comes off easily.), 500g shrimp or prawns, peeled and cleaned, 2 large onions chopped small, 2 tomatoes chopped small, 1-inch piece of ginger and 4 garlic cloves minced, 4 green chilies chopped, 1TS chili powder, 1TS coriander powder, ½ TS cumin powder, ½ TS turmeric powder, 1TS curry powder, 1 stick of cinnamon, 3 cloves, ¼ cup grated coconut, salt to taste, and 3 TBS oil

Method: In a frying pan or wok, heat 3TBS oil. Add the chopped onions, garlic, ginger, and green chilies and fry until onion browns. Add all the powders and mix well. Add the tomatoes and parboiled drumstick pieces and cook for 4 minutes. Add the prawns and stir until it's well coated with the spice mixture. Cook for 2 minutes and then add the coconut and salt with 1 cup of water. Cook covered for 5 minutes until you get your desired curry consistency. Enjoy with rice.

DID YOU KNOW? Moringa may possess anti-fertility qualities and isn't recommended for pregnant women. Chemicals in the bark may make the uterus contract and lead to a miscarriage. Moringa leaves are dried at low temperatures for better preservation of the vital nutrients. Then they are ground into deep green powder.

NAMES: The miracle tree, benzolive tree, the ben oil tree, or the horseradish tree

Although few people have ever heard of it today, Moringa could soon become one of the world's most valuable plants, at least in humanitarian terms.

~ Noel Vietmeyer

NONI

Morinda citrifolia

Noni is a strange plant that grows easily in the Hawaiian chain. The fruit looks like a pineapple that has been shaved. I first encountered noni when a friend into natural health foods offered the fruit to me. Noni has a varied reputation of being a cure all, or a fake. Like aloe vera, it is a strange plant, is very bitter, but noni also smells, perhaps stinks.

Botanically, noni is *Morinda citrifolia* and is known by a variety of names like nono, nonu, great morinda, Indian mulberry, nunaakai in India, dog dumpling in Barbados, mengkudu in Malaysia, beach mulberry, and cheese fruit. The name noni is derived from Hawaiian where it was one of the original canoe plants. This weird fruit comes from a tree related to the coffee family; native to areas ranging from Southeast Asia to Australia, including Polynesia.

Historically, the noni plant was known in Sanskrit as 'ach' and attributed special prop-erties by ancient physicians. It has been known and used in northwest India in Ayurveda, the practice of holistic medicine, for thousands of years. East Indians use powdered extracts from roots, leaves, and the fruit of the plant as a sedative and for other medicinal purposes.

HOW TO GROW: First, find someone that has noni and will give you a branch to plant. Noni will grow almost anywhere in almost any condition, but, as with almost everything, the better the soil, the better it will grow. It will thrive in full sun, or partial shade, and in a vari-ety of soil conditions. Noni grows best in a pH of 5-6.5. The best approach is to start your plant in a pot. Transplant once the plant seems hardy and is about a foot tall. First fork the soil and break all clumps. Keep the area weeded to reduce competition for water and sun. Fertilize after three months with 12-24-12. Use a bearing fertilizer as 12-12-17-2 when noni begins sprouting fruits in a year and a half. Water regularly during dry spells. Every month this tree should have about 15 pounds of fruit. Keep it trimmed to the size of a shrub because it can reach 15 feet.

Not an overly attractive plant, it doesn't take up much space in the backyard. Noni has large, dark green, deeply veined leaves, and should continuously have flowers and fruits. Noni is a multiple fruit, like a pineapple; bunches of simple fruits, each from its own flower with a single pistil. The simple fruits grow together to form a multiple fruit. Noni begins to stink as it ripens. This smell has gained it the names 'cheese fruit' or even 'vomit fruit.' It is oval-shaped and reaches four inches. This fruit is at first green, then yellow, and almost white as it ripens.

Noni was virtually unknown outside the Pacific Islands until the 1990s. Hawaiian noni was first marketed as a health supplement in powdered capsule form and later as noni juice. Recently organic noni fruit has become popular. Today noni is big business. Tahitian Noni International is the largest noni product manufacturer that markets and sells the reputed health benefits in the form of noni juice, noni cosmetics, and pet care products. Three hundred other companies sell noni health supplements, which is a multi-billion-dollar industry. Noni is one of the chief agricultural exports of Polynesia. Some noni juice has water added despite claims of 100% pure noni juice. Manufacturers also sweeten this juice to improve the taste. Lack of regulation means you must carefully research all noni products.

MEDICINAL: Traditionally, the leaves of the noni tree were used topically for healing wounds. In herbal Polynesian medicine, noni has been used for many health conditions, such as constipation, diarrhea, skin inflammation, infection, and mouth sores. Today noni is adverted to cure everything from arthritis, atherosclerosis, bladder infections, boils, bowel conditions, burns, cancer, chronic fatigue syndrome, circulatory weakness, colds, cold sores, diabetes, drug addiction, eye inflammation, fever, fractures, gastric ulcers, gingivitis, headaches, heart disease, hypertension, improved digestion, immune weakness, indigestion, kidney dis-ease, malaria, menstrual disorders, respiratory disorders, ringworm, sinusitis, sprains, strokes, thrush, and wounds. There's no real evidence, that noni is effective for any of these conditions.

Studies evaluating the effects of noni suggest it may have anti-cancer, pain-relieving, and immune system-enhancing effects. However, these studies used extremely high doses that would be difficult to obtain from taking the juice. More importantly, there's no reliable evidence about the safety, or effectiveness of noni for any health condition in humans. Noni juice is high in potassium and should be avoided by people with kidney disease.

NUTRITION: 100 grams of noni fruit contains 95.67% water, 15.3 calories, 0.43 g protein with lots of vitamin C, and some calcium. 100 grams of noni juice has 47 calories with less than a gram of protein and vitamin C, biotin, and folate.

FOOD & USES: Noni has many seeds. Despite its strong smell and bitter taste, the fruit is categorized as a starvation food to be eaten if nothing else is available. However, it is a staple in some Pacific islands, eaten raw or cooked. Southeast Asians and Australian Aborigines consume the fruit raw with salt or cook it with curry. The seeds are edible when roasted.

Noni fruit are ripe when yellow to pale white. Once picked, they can be placed in glass jars and kept lightly capped until they soften and release their juices; or ripen the fruit until they split by setting in the sun. Some recommend keeping noni in jars 6 weeks until fermentation begins. Noni should be crushed over a bucket that has a sieve incorporated, so the juice can be strained. The juice should be stored in a cool place or refrigerated. Noni often has a bitter taste. Mix it with water, sugar, or additional fruit juices. Pure noni juice has only ten calories per cup.

RECIPES:

CRANBERRY NONI: 2 ounces noni juice, ½ cup cranberry juice, ¼ cup orange juice, fill glass with club soda

GRAPE NONI: 2 ounces noni juice, ½ cup grape juice, 1 TS fresh lemon or lime juice, fill the glass with lemon-lime soda.

NONI FRUIT SALAD:

Ingredients: ¼ cup noni juice, grapefruit slices, crushed pineapple, bananas, cherries, orange pieces, grapes, mangoes, and papaya pieces. Combine, chill, and serve.

CURRIED NONI:

Ingredients, 3 small green noni fruit - sliced thin, 1 TBS sesame oil, 1 sweet pepper, cauliflower, broccoli, and carrot - all chopped, 1 cup coconut milk, 1 TBS curry powder

Method: Heat oil and add curry powder for one minute, add noni and veggies. Add coconut milk and bring to a boil. Reduce heat and simmer for 10 minutes. Serve with pasta or rice.

DID YOU KNOW? Noni trees were originally brought to Hawaii by ancient Polynesians as one of the original 27 'canoe plants.' There are 6 varieties of noni trees that produce fruit. However, there is only one type that is medicinal. The fruit of the noni tree resembles a young breadfruit. The aroma is described as disgusting and the flavor matches. Noni juice products have become popular worldwide. Several countries have banned importation and many others have banned noni advertisements because of fantastic unproven health claims.

A healthy attitude is contagious,
but don't wait to catch it from others. Be a carrier. ~ Tom Stoppard

TEA - KĪ
Camellia sinensis

In your home garden, you can grow true tea – from the *Camellia sinensis* bush. Anywhere throughout the Hawaiian Islands the tea bush, known as 'kī,' will grow. Like everything, some places are better than others. From planting to serving, your first cup of tea will take 3 years.

Tea is an aromatic beverage prepared by pouring hot water over cured or fresh tea leaves. Tea is an evergreen tree believed native to the tropical and subtropical areas of Southeast Asia. Originally, all variations were medicinal treatments. It's mentioned in a Chinese medical text circa 200AD. During the Tang Dynasty in the second half of the first millennium, tea drinking became a social habit of all classes. In the 1500s, Portuguese explorers brought the beverage, then named *chá,* to Europe. The Dutch became the first tea merchants. A century later, the British began Indian tea plantations. The tea bag was born in 1907 by an American tea merchant, Thomas Sullivan, who used them to brew samples of his teas. At first, the tea bags were small silk drawstring bags.

Today, tea is planted throughout SE Asia, from Sri Lanka, India, China, to Japan. In the equatorial tropics, it can grow up to 2,000m in elevation. In 2019, 6.5 million tons were produced with the big producers - China: 2.4 million tons, India: 900,000, and Sri Lanka grew 340,000 tons.

Tea probably is only second to water as the most consumed daily beverage because of the ease of preparation and its caffeine content. Tea has many tastes, aromas, and colors. Sri Lanka black tea, Chinese green, and Darjeeling, all varied flavors, sweet, nutty, and floral, to satisfy a world of tea drinkers. Green tea is made from leaves steamed and dried, while black tea leaves are withered, rolled, fermented, and dried.

Leaf size differentiates tea plants. Tea bushes have two notable varieties: Chinese and Japanese teas use *Camellia sinensis* var. *sinensis,* which have the smallest leaves. Indian teas use the big-leaf plant: *Camellia sinensis* var. *assamica.* Darjeeling tea is considered a hybrid between Chinese small-leaf tea and Assam-type large-leaf tea. Another mid-sized tea leaf is known as the Cambodian-type: *C. assamica* subsp. *Lasiocaly* and was originally classified as a variation of Indian/Assam teas.

The tea business was first attempted in the late 1800s, in Hawaii, but other, more profitable, less finicky crops of sugar cane and pineapples took over. Tea resurfaced at the begin-ning of this millennium as pineapple and sugar faded into bad memories. Tea, like coffee, has serious global competition for taste, and branding is necessary. Kona is known for the perfect coffee growing conditions. Hilo and several elevated areas on the Big Island are perfect for teas. China and Japan have been growing teas for thousands of years, while India and Sri Lanka for more than a century. It'll take patience and practice to grow and process the perfect brands of Hawaiian teas.

HOW TO GROW: If you have the space, and prune the tea trees to keep them small, you can grow different trees for different flavors. Uniformly trimmed tea bushes make an excellent boundary hedge. Tea plants are started from seeds and/or cuttings. Tea flowers are self-sterile and require cross-pollination by insects to produce seeds. Germination takes 1-2 months. At first, seedlings need to be shaded. Tea produces best in a well-drained soil with a pH of 4.5-6.0 and annually they need 50 inches of rain/water. At six months to a year, they can be transplanted to their permanent place. Consider how large you will let each tea bush spread. It will 2-3 years before a new plant is ready to pick and maybe a dozen years for a plant to make seeds.

High elevations to 1,500 m (4,900 ft.) are often where the best quality tea is grown. Tea grows slower and develops a better taste. Bushes are trimmed to waist height to make picking easier. The shorter the bushes the faster new leaves sprout. Tender, new leaves raise the quality of the tea. During the right conditions, a short tea bush can grow new leaves every 1-2 weeks. Carefully hand pick the new, 2-3 leaves sprouts. 10 kg of green shoots produce 2.5 kg dried tea.

Growing and processing/drying tea has a steep learning curve. But the tea is your own home grown. Spread one layer of fresh tea leaves on metal trays and dry in the sun for a full day until the leaves begin to wilt. Everything is a variant of the local temperature and humidity. Some 'black' teas are made by fermenting/aging a ball of the wilted leaves in a very humid place for 4-6 hours or until the color changes to a yellowish orange with a pleasant aroma. To produce green tea, fresh picked leaves are steamed or heated to prevent oxidation and preserve the natural essence. Store your finished tea in sealed air-tight containers.

MEDICINAL: The Chinese regard tea as a cure-all, but there are hundreds, if not thousands of varieties of tea. The usual beverage is considered an antitoxic, diuretic, expectorant, and stimulant. Teabags can be used as poultices onto baggy or tired eyes, compressed onto the temples to ease a headache, or rubbed on sunburned areas to relieve the pain. **BEWARE:** there is evidence linking the tannin in tea to esophageal cancer where tea is regularly drank. High consumption may also cause unpleasant nerve and digestive problems.

NUTRITION: Black and green teas have very few calories id drank plain. They both contain good amounts of vitamins B-2, C, D, and K with traces of calcium, magnesium, iron, and zinc. However, the noted health benefits of green tea are from its micro-nutrients and antioxidants.

FOOD & USES: For 3,000 years, dried, cured tea leaves have been used to brew a stimulant/caffeine beverage. *Chasei* is a tea extract; *Teu-cha* powdered ceremonial tea. Tea extract flavors are used in some alcohols, frozen dairy desserts, candy, baked goods, gelatins, and puddings. Refined oil pressed from the tea seeds is different from cottonseed, corn, or sesame oils in that it is a non-drying oil. Tea is a potential source of food colorings. Steam distillation of black tea yields an essential oil.

BREWING TEA: A good cup of tea begins with a tablespoon of your favorite leaves. Steep in 180-210 F hot water for 3-5 minutes. Steeping a tea longer will not make it stronger only more bitter. Add more tea leaves or bags for a stronger cup.

USE TEA IN YOUR RECIPES: Instead of regular, plain water in soups, breads, cakes, or cookies, use strong brewed tea or milk tea. Before grilling or frying, rub meat, chicken, or fish with green or black teas. Green tea leaves can be used as an herbal spice in making tasty, different steamed rice or vegetable dishes. On the barbecue, use tea as a dry rub or to smoke a new flavor into your dinner. Milk tea can flavor smoothies, puddings, and custards. Make tea ice cubes.

DID YOU KNOW? The most drank beverage worldwide is known as: te, cha, and chai; and in English as tea, cha or char, and chai. The Portuguese, who traded with Macao tea merchants is the 1590s picked up the Cantonese pronunciation of *cha*. In the 1600s, the Dutch gave us the word tea from the Chinese *tê* or the Malay *teh*. The spiced tea term *chai* is from a northern Chinese dialect variation of *cha* that the Persians added *yi*. In the 1900s both English and Hindi added *chai*.

NAMES: Tea, assam, black tea, Orange Pekoe, ch'a, Darjeeling, dust, green tea, gunpowder, hyson, iced tea, imperial, Keemum, Lapsung Souchong, Oolong, orange pekoe, souchong, silver tip, sen-cha, twanky,

We are like tea, we don't know our own strength until we're in hot water. ~ Sister Busche

VEGETABLE HUMMINGBIRD TREE – 'OHEA
Sesbania grandiflora

What a name! The vegetable hummingbird tree has been naturalized throughout the Hawaiian Islands. It's a short ornamental tree with tender leaves, green fruit, and flowers that are eaten alone as a vegetable salad or mixed into curries. Hummingbird leaves are prepared stir-fried or mixed with coconut into a delicious sambal. These leaves are not seasonal and always available. This could be one of your staple greens. It probably originated in SE Asia, within Malaysia or Indonesia. Many Asian countries claim this nutritious tree. In its native Asian countries, it grows on raised beds between rice paddies, along roadsides, and in backyards. The botanical is *Sesbania grandiflora.* Please grow a few because unfortunately, the vegetable hummingbird is listed as endangered in the Hawaiian chain. Locals named it *'ohai.*

More people need to learn how good this tree tastes and how good it is health-wise. This tree is so revered that bulk shipments of leaves are exported from India and Sri Lanka to cold weather countries as England and Canada. Leaves are dried and exported as nutritional teas.

HOW TO GROW: The hummingbird tree is perfect for your yard, a fast-growing perennial evergreen. It's small, to 10 m (30ft.), with few branches. All it desires is full sun and protection from strong winds. That means shade for you. It grows easily from seeds if they are soaked for a day in hot water. You can carefully scratch the hard seed shell to increase germination. Every bean/fruit has many seeds. This tree will grow from sea level to 800m (2500 ft.). It prefers a pH of 5.5 - 8.5, moist soil, and tolerates floods and long droughts.

The edible blossoms are white or pink that make long, thin pods 50–60 cm (21 inches) long with 20 or more seeds. The hummingbird tree is considered a legume tree because its long pods release their seeds by splitting open along two seams. The desired leaves can be 30 cm (10 inches). The bark is light gray wrapping soft, white wood.

MEDICINAL: Every part of the hummingbird tree is used for traditional-homeopathic medicine. The juice of the leaves and flowers treat fevers, headaches, nose, sinus, and throat ailments. The bark and leaves are useful in poultices for inflammation and wounds. Eating the flowers with the stamen removed and the leaves relieve constipation. The juice of the bark mixed with honey treats chronic intestinal disorders. Hummingbird tree will treat intestinal parasites/worms. A few drops of juice extracted from the blossoms may brighten your vision.

NUTRITION: This tree is a nutritional powerhouse! The leaves and blossoms of the hummingbird tree are high in calcium, potassium, iron. Per 100 grams of leaves are 321 calories, 36.3g protein, 7.5g fat, 47.1g carbohydrate, and 9.2g of fiber. Nutritionally, 100 grams has 1684mg calcium, 258mg phosphorus, 2,005mg potassium K, with loads of vitamins A and C. The blossoms per 100 grams have 345 calories, 14.5g protein, 3.6g fat, 77.3g carbohydrate, 10.9g fiber, 145mg calcium, 290mg phosphorus, 5.4mg iron, 1,400mg potassium, and almost as much vitamin and C as the leaves. Oil pressed from the seeds is also nutritious and contains 12.3% palmitic, 5.2% stearic, 26.2% oleic, and 53.4% linoleic acids. Five-day-old sprouts from hummingbird seeds are rich in vitamin C at 166mg per 100 grams.

FOOD: The tender leaves, green fruit, and flowers are eaten alone as a vegetable or mixed into curries or salads. Often there are butterfly caterpillars or their eggs on these leaves, so check both sides of the leaves when washing. The large blossoms may be dipped in batter and fried in butter. Remember, the nutritional content depends on the type of oil used and the temperature to fry. So, use butter, sunflower, virgin coconut oil, or canola oil on low heat. Tender pods can be steamed and eaten the same as string beans. Young leaves chopped fine are also steamed, boiled, or fried. White blossoms are eaten raw as salad, but the red blossoms are said to be bitter.

Hummingbird tree blossoms are a key ingredient in anti-aging skin creams. The blossoms contain many nutrients and tannins that can smooth wrinkles and increase the production of collagen.

This tree's inner bark can serve as fiber. Although the white, soft wood, is not durable, it can be used for cork. The wood is like bamboo and used in Asian construction. A gum called katura is red when it first drips from the tree and turns black as it dries. This sap is astringent, slightly soluble in water, and when applied to fishing lines, it makes them waterproof. A bark extract is toxic to cockroaches.

RECIPES:

SHREDDED HUMMINGBIRD TREE STIR-FRY- A great green dish.

Ingredients: One bunch hummingbird tree leaves-100g, washed and shredded-sliced into thin strips, 1 medium onion chopped, 4 cloves of garlic chopped, 5 green chilies chopped, 1 cup grated fresh coconut, 1TS turmeric powder, 1 TS red chili powder, ½ TS ground black pepper, 1 TBS lime juice, and salt to your taste.

Method: In a frying pan or wok, combine all ingredients over low heat and quickly stir fry for 3 minutes. Shut off the burner but leave ingredients and let the heated pan continue to cook them.

HUMMINGBIRD TREE LEAF CURRY WITH EGG

Ingredients: 100g hummingbird tree leaves washed and chopped, 1 cup of thick coconut milk, 2 eggs beaten, 3 cloves of garlic chopped, 3 green chilies chopped, 1 large onion chopped, 10 curry leaves, 1 TBS each of roasted curry powder, roasted chili powder, and turmeric powder, 1 TS fenugreek powder, and salt to taste

Method: In a large frying pan or wok, over medium heat, put the hummingbird tree leaves, onions, garlic, chilies, and curry leaves. Fry for 2 minutes and then add coconut milk. Stir in all the spices. Keep continually stirring for 5 minutes. Add the eggs and cook for 3 more minutes. Enjoy.

CHICKEN NOODLES WITH HUMMINGBIRD TREE LEAVES

Ingredients: 250g minced chicken, 1 bunch 100g hummingbird tree leaves washed and shredded, 1 large onion chopped, 4 cloves of garlic chopped, 4 green chilies chopped, 2 carrots peeled and chopped small, 10 curry leaves, 100g cashews, almonds, or peanuts chopped, 1 TBS curry powder, 1TBS sugar, 1 TBS chili powder, 1 TS fenugreek, 2 TBS coconut oil, 1 large pack of rice or wheat noodles, and salt to taste

Method: In a pot boil the noodles per the package directions. In a wok or frying pan, heat 2 TBS of the oil and fry the sugar until it smells, stir in the curry powder and fenugreek. Add the garlic, onions, chilies, and carrots, and fry for 3 minutes. Add the minced chicken and fry until cooked. Stir in the hummingbird tree leaves and fry for 2 minutes. Add nut pieces and add salt to taste. Stir-fry for 2 minutes and add the drained noodles. Enjoy.

DID YOU KNOW? In India, hummingbird tree blossoms are sacred to the Hindu god Shiva.

NAMES: Agati, agathi, English: scarlet wistaria tree, vegetable hummingbird, West Indian pea, Fr ench: agati à grandes fleurs, colibri végétal, fagotier, fleur papillon, gros mourongue, pois valette, pois valier, sesbanie à larges fleurs; Spanish: baculo, báculo, cresta de gallo, gallito, pico de flamenco, sesbania agata, zapaton blanco; Indonesian: turi

Keeping your body healthy is an expression of gratitude to the whole cosmos- the trees, the clouds, everything.

~ Thich Nhat Hanh

APPENDIX

All material in this book has been researched and compiled from reliable sources. Everything between its covers is provided for your information only, and should not be considered as medical advice or instruction. No action or inaction should be taken based solely on the contents of the information. Readers should consult appropriate health professionals on any matter relating to their health and well-being.

DO NOT consider any of this information to be medical advice or instruction. No action or inaction should be taken based solely on these articles. Readers, please consult health professionals on any matter pertaining to their health and well-being.

THE HAWAIIAN HOME GARDEN GUIDE will not take any responsibility for any adverse effects from the use of plants. Always seek advice from a professional before using a plant medicinally. The use and dosage of herbs and foods vary with circumstances such as climate (whether it is hot or cold), the person's health and age, and the manner the dose is administered. Herbs and foodstuffs should be used for medical purposes only with proper medical advice. The medicinal plants must be grown without chemical fertilizers or pesticides.

PAGE:

Imagine all the food mankind has produced over the past 8,000 years. Now consider that we need to produce that same amount again – but in just the next 40 years if we are to feed our growing and hungry world.

~ Paul Polman

THE HAWAIIAN HOME GARDEN GUIDE
CALORIE & NUTRITION GUIDE

If your garden effort has limited space, finances, and energy, search through the nutrition and remedies sections and learn the best selections to grow at your home.

PLEASE TAKE NOTE: All material in this book has been researched and compiled from reliable sources and is provided for your information only. No action or inaction should be taken based solely on the contents of the information. Readers should consult appropriate health professionals on any matter relating to their health and well-being. This information is not be considered as medical advice or instruction. DO NOT consider any of this information to be medical advice or instruction. No action or inaction should be taken based solely on these articles.

THE HAWAIIAN HOME GARDEN GUIDE will not take any responsibility for any adverse effects from the use of plants. Always seek advice from a professional before using a plant medicinally. The use and dosage of herbs and foods vary with circumstances such as climate (whether it is hot or cold), the person's health and age, and the manner the dose is administered. Herbs and foodstuffs should be used for medical purposes only with proper medical advice. The medicinal plants must be grown without chemical fertilizers or pesticides.

Courtesy 912crf.com

He that takes medicine and neglects diet wastes the skills of the physician.
~ Chinese proverb

A **ACEROLA CHERRY / WEST INDIAN CHERRY:** One hundred grams has 32 calories high in vitamin C, potassium, and iron.

ACKEE: When properly prepared, the ackee is delicious, rich in vitamin A, zinc, iron, potassium, and calcium. A good-sized fruit usually has 150 calories. Ackee provides enough protein that it can be the center of a meal.

AIR POTATO: Air potato has about 100 calories per 100 grams. It is 7.5% protein, and very rich in potassium, niacin, vitamin C, magnesium, and calcium, but it's also high in sodium.

ALLSPICE: A teaspoon of ground allspice has only six calories with some vitamin C and calcium.

ALMONDS: 100 grams has almost 600 calories with 450g from fat, 20g of protein and loads of potassium and iron yet amazingly no cholesterol. One ounce of almonds contains about 10% of the recommended daily allowance of calcium, a great non-dairy source for vegetarians. Almonds are rich in vitamin E, with just a handful (30g, about 20 nuts) providing 85% of the Recommended Daily Intake (RDI).

ALOE VERA: Per cup, aloe has 36 calories. Aloe vera contains a good amount of vitamin C: 9.1 g for every 1 cup of aloe vera juice. Aloe vera also contains essential vitamins like vitamin A - beta carotene, E, B12, B-9/ folic acid, and choline. The plant also contains calcium, chromium, copper, iron, selenium, magnesium, manganese, potassium, sodium, and zinc.

AMARANTH: 100 grams of amaranth leaves has 280 calories with 32 grams of protein. The leaves are loaded with potassium, calcium, phosphorus, and iron with vitamins A, B1, B2, B 3, and C. Amaranth seeds are 15% protein and 5% fat. The seeds are edible with a nutty flavor and are nutritious eaten as snacks or added into baked goods. Boiling the seeds in water with coconut milk makes a delicious porridge.

ANNATTO / ROUCOU: Annatto seeds are a food rich in fiber and a protein source: 11.50% protein, 6.74% moisture, 42.19% total carbohydrates and 28.45% fiber. The seeds contain the minerals potassium, manganese, calcium, copper, iron, and magnesium. They also contain high levels of the amino acids: lysine, isoleucine, and leucine. More research is required, but annatto could become an easy to grow, low environmental impact protein food for developing countries.

APPLES: 100 grams of raw apple has 52 calories with significant levels of vitamin C and potassium.

ARAMBELLA / POMMECYTHERE: Arambella fruit have 160 calories for 100 grams because 10% is sugar, 85% is water, which is why it is great for juice. The fruit is good source of vitamin C. potassium, carbohydrates and fiber.

ARROWROOT: Arrowroot has 65 calories for 100 grams with some protein, potassium, iron, vitamin B-6, and magnesium.

ASH MELON: 100 grams of ash melon has 13 calories and is a valuable source of phosphorus, calcium, and magnesium. It has good amounts of vitamins B1, B2, C, and niacin. The seeds are rich in protein and oil.

AVOCADO: One cup of avocado has 236 calories, but is also rich in vitamins K, B-6, C, and copper. They contain oleic acid, which helps lower cholesterol, potassium to lower blood pressure, and folate to lower the risk of heart attacks. Avocado contains a high percentage of lutein, an important nutrient for healthy eyes.

B **BAEL FRUIT - GOLDEN APPLE:** 100 grams of bael fruit has only 140 calories with 60mg vitamin C, 55mg of vitamin A, very rich in potassium, and calcium. They also provide niacin, riboflavin, and thiamine.

BALATA: A hundred grams of balatas, about a big handful, have sixty-two calories with two and a half grams of protein and ten grams of fat.

BANANAS: Bananas not only taste good, but also are a healthy quick energy food packed with vitamins with 2 grams of protein and 4 grams of fiber. Bananas are rich in potassium, vitamins B complex, A, and C. One banana is about 99.5% fat free with about 90 calories comprised of 75 percent water, 20% starch and 1% sugar. Compared to apples, bananas have less water, fifty %more food energy, four times the protein, half the fat, twice the carbohydrate, almost three times the phosphorus, nearly five times the vitamin A and iron, at least twice the other vitamins and minerals.

BASIL: Basil has very few calories and is a good source of vitamins A, C, and K, magnesium, and potassium.

BAY LEAVES: Since a recipe might only call for 1-2 bay leaves, their nutritional value is negligible. 100 grams of bay leaves – probably enough for two years – has 315 calories with good percentages of vitamins B-6 and C, potassium, iron, calcium, and magnesium.

BEET ROOT: Beets are rich in folic acid, which fights heart disease and anemia. Beets are high in fiber. Fiber is good for the intestines and helps regulate blood sugar and cholesterol. One cup of fresh cooked beets has only seventy-five calories of which one and a half grams are fiber and the same amount of protein with eight grams of carbohydrates. Beets also contains vitamin A, potassium, and phosphorus.

BILIN – BILIMBI: 100 grams of bilimbi has 30 calories with vitamins A and C, cal-cium, phosphorus, and potassium.

BITTER MELON / CARAILLI: 100 grams of bitter melon has only 43 calories with 1g of protein. It's an extremely good source of potassium, calcium, magnesium, vitamins A, B-6, and C. It is rich in iron, copper, and manganese, and contains twice the beta-carotene of broccoli, twice the calcium of spinach, and twice the potassium of a banana. Regular consumption of bitter gourd juice has been proven to improve energy and stamina levels.

BLACK PEPPER: Black pepper, especially fresh ground, is almost a small vitamin pill. One teaspoon of black pepper contains 6 calories with manganese, copper, calcium, vitamins C and K, iron, phosphorus, potassium, and selenium.

BLACK WALNUT: 100 grams has a whopping 900 calories. They are high in fat and shockingly without any protein, minerals or vitamins. Eating walnuts will provide serotonin, important for a healthy emotional balance and fights depression. Most of walnut oils contain an omega-3 fatty acid called alpha-linolenic acid. This 'good fat' can reduce your risk of developing heart disease. If included in your daily diet, it lowers the risk of heart disease, cancer, and fights the inflammation of rheumatoid arthritis. You only need to eat 2 nuts daily – 4 half nut meats.

BOK CHOY: Bok choy is rich in vitamin C, fiber, and folic acid. All reduce the risk of various types of cancer. Bok choy has more beta-carotene than other cabbages with more potassium and calcium. A perfect food for dieters, one cup of cooked bok choy has only 20 calories, with no fat, but 3 grams of carbs and 3 grams of protein.

BOTTLE GOURD / LAUKI: This vegetable is 90% water and a rich source of vitamins and minerals. 100 grams contains 14 calories and provides the daily needs of 13% of vitamin C and 7.36% of zinc with 174 mg of potassium, 13 mg of magnesium, and 15 mg of phosphorus.

BRAZIL CHERRY / GRUMICHAMA: 100 grams has 60 calories with vitamins A and C with calcium, iron, and phosphorus.

BRAZIL NUTS: Eight Brazil nuts equal a whopping hundred and eighty calories with 18 grams of fat and thiamine. Brazil nuts are an excellent source of selenium, a vital mineral and antioxidant that may help prevent heart disease. Two Brazil nuts can provide your entire daily intake of selenium. Brazil nuts are particularly healthy because selenium makes their protein content 'complete.' This means that, unlike most plant proteins from beans etc., proteins contained in Brazil nuts have all the necessary amino acids for optimal growth in humans just as meat or fish.

BREADFRUIT: Nutritionally a breadfruit has about a hundred calories of which two grams are protein with less than one gram of fat. It has twenty-five grams of carbs. Breadfruit is a good source of vitamin B, and there is more vitamin C in the riper fruits. It contains calcium, phosphorus, and iron.

BREADNUT / CHATAIGNE: 100 grams of breadnuts has 360 calories, of which 13g protein, 6g fat, 76g carbs, and 14g fiber, with vitamins C, B6, calcium, phosphorus, potassium, iron, thiamin, folate, and niacin. In 100 grams, there are good quantities of four valuable amino acids: 3g of methionine, 2.6g of leucine, 2.4g of isoleucine, and 3.1g of serine.

BROCCOLI: A half cup of steamed or boiled broccoli has twenty-two calories, two and a half grams of both protein and fiber with a tiny bit of fat, and no cholesterol. The same small portion of broccoli has one and a half times the daily amount of vitamin C, plus vitamin A, niacin, thiamin, iron, folate, selenium, phosphorus, potassium, zinc, and magnesium.

BUTTER BEANS: 100 grams of butter beans have 115 calories with 8g of protein and 7g of fiber with vitamins B1 (thiamine), B2 (riboflavin), B3 (niacin), Pantothenic acid 8%, B6, and are very high in B9 (folate). They contain iron, manganese, magnesium, potassium, zinc, and phosphorus.

BUTTERNUT SQUASH: 100 grams of butternut squash has 45 calories and 1 gram of protein with loads of potassium.

C **CABBAGE:** All cabbage types are low in calories and excellent sources of minerals and vitamins; especially C. Cabbage may reduce the risk of some forms of cancer including colon or rectal cancers. Cabbage is also high in beta-carotene, vitamin C, and fiber. A half-cup of cooked cabbage has 16 calories, 3 grams fiber, 3 grams carbohydrates, and 18 milligrams of vitamin C.

CAIMATE / STAR APPLE: 100 grams of caimate has 60 calories with calcium and phosphorous.

CALAMUS ROOT - SWEET FLAG: Calamus root is a vegetable multi-vitamin and mineral. 100 grams has only 120 calories, but 15g are protein, 6g of fiber, a very high content of vitamin E with vitamins C and A. This root has very high amounts of potassium, sodium, magnesium, and calcium, with lesser amounts of phosphorus, copper, and zinc.

CANDLENUT: 100 grams has 470 calories with 49 g of fat, and 7g of protein. These nuts contain good amounts of potassium, phosphorus, calcium, and magnesium, vitamins B-1 and B-3.

CANISTEL: 100 grams of canistel has 140 calories, with 35 grams of carbs, 1 gram of protein and plenty of calcium, phosphorus, iron, niacin, carotene, and vitamin C.

CANTALOUPE: A tasty cantaloupe will not spoil your diet as a cup has only 50 calories. Eating a half a cantaloupe will provide the daily requirement of vitamins A and C, folic acid, and potassium. Cantaloupe has no fat or cholesterol and provides necessary fiber.

CAPE GOOSEBERRY: 140 grams of cape gooseberry has 74 calories with 2g of protein, and has high quantities of vitamins A, B1, B3, and C with good amounts of iron, phosphorus, and calcium.

CARDAMOM: Nutritionally per 100 grams cardamom has 300 calories, 70 grams of carbohydrates, 10 grams of protein, and 7 grams of fat, with 12 grams of fiber. It is high in calcium, iron, magnesium, phosphorus, and potassium. It also has vitamins C, B, thiamin, niacin, and folate.

CARROTS: A half-cup serving of cooked carrots contains four times the recommended daily intake of vitamin A, with 35 calories, 1 gram protein, 8 grams of carbohydrates, 2grams fiber.

CASHEW: One ounce of cashew nuts has 150 calories with 12 grams of fat, no cho-

Laugh a lot. It burns a lot of calories.~ Jessica Simpson lesterol, and 9 grams of carbs. Cashews have less total fat than peanuts or almonds, and more than half of their fat is unsaturated fatty acids. This unsaturated fatty acid is oleic acid, the same healthy monounsaturated fat found in olive oil. Oleic acid is good for the heart, even in diabetics. If you are health conscious, dry-roasted cashews have a lower fat content.

CASSAVA: Cassava is such an important food because it is high in starch, which produces energy. A half pound of cassava has 215 calories, of which a quarter is carbohydrates, with one gram of fat, and three grams of protein. It also has Vitamin C.

CAULIFLOWER: One cup of raw cauliflower has only 25 calories. It is very rich in vitamins K, and C. One cup has more than the daily requirement of vitamins B6, B5, B3, folate, biotin, magnesium, iron, manganese, and molybdenum.

CAYENNE PEPPER: 2 teaspoons of cayenne pepper has about 10 calories. It is a good source of vitamins A and C, has the complete B complexes, and is very rich in calcium, manganese, and potassium.

CELERY: A cup of celery contains only 20 calories and has a good dose of vitamin C and B-6, folic acid, and fiber. Celery also has minerals such as manganese and calcium. The leaves are valuable as they contain the most vitamins.

CHENETS/GENIPS: Chenets have about 60 calories per quarter pound with some calcium, phosphorus, and iron.

CHAYOTE / CHRISTOPHENE: One cup of chayote has only 25 calories with almost no fat or carbohydrates. It has some fiber and vitamin C. However it is a source of sodium (salt).

CHIVES: One tablespoon of chives has only 1 calorie. Chives are high in vitamins A and C, potassium, and calcium.

CHESTNUTS: The chestnut is a highly nutritious food. 100 grams of chestnuts has 200 calories, significant content of vitamins B-6, C only a trace of fat compared to peanuts or cashews. The excellent protein of chestnuts has amino-acid content like that of an egg. They are the only nuts that contain vitamin C.

CHICK PEAS / GARBANZO BEANS / CHANNA: 100g of chickpeas has 360 calories, and almost 9g protein with plenty of 17g fiber. These beans are loaded with folate and other vitamins with manganese, phosphorus, potassium, and zinc.

CHOCOLATE: Cocoa powder is actually good for you! It has nearly twice the antioxidants of red wine, and up to three times the antioxidants found in green tea. Cocoa also contains magnesium, iron, chromium, vitamin C, zinc, and other minerals at 12 calories per tablespoon.

CILANTRO / CORIANDER: 4 grams has only 1 calorie with some vitamins A and C.

CINNAMON: One tablespoon of cinnamon has 17 calories with only 1 calorie from fat and 5 from carbohydrates. It is a source of manganese, calcium, and iron.

CLOVES: 2 grams or 1 teaspoon of ground cloves contains 6 calories with 1 gram of fiber, lots of magnesium (55% of the daily value) and some vitamin K.

COCONUT: 80 grams or 3 ounces of coconut has 280 calories, with 220 are from fat. It has some dietary fiber, minimal protein with some iron, calcium, and vitamin C. One cup of coconut water has 45 calories with 4 from fat.

COCO PLUM / ICACOS PLUM / FAT PORK: These fruits are so rich that 100 grams has 50 calories with some vitamin C. calcium, iron, and phosphorus. The chemical features of the coco plum/coco plum include flavonoids, terpenoids, steroids, and tannins. Scientific medical interest in flavonoids has increased because many of them exhibit anti-inflammatory, antiviral, antibacterial, as well as anticancer abilities.

COFFEE: Black coffee contains no significant amounts of the macronutrients, fat, carbohydrate, or protein and therefore contains only 1-2 kcal per 100ml with some micronutrients, notably potassium, magnesium and niacin.

CORN: One ear yellow corn has only 80 calories, 2.5 grams protein, 20 grams carbohydrates, 2 grams dietary fiber, potassium, vitamin A, niacin, and folate.

CREPE GINGER: This plant is a good source of nutrients and natural antioxidants. 100 grams of crepe ginger leaves has 19% Protein and 12% fiber with plenty of vitamins C, E, and A. This plant contains lots of iron with calcium, magnesium, and potassium.

CUCUMBER: A four-inch cucumber has 20 calories, 1 gram fiber, 1 gram carbohydrates, with calcium, vitamins A, and C.

CULANTRO: Other than taste, culantro has slight food value. A quarter cup of leaves has only 4 calories with virtually no fat, fiber, cholesterol, or carbohydrates. The plant is report-edly rich in calcium, iron, carotene, and riboflavin. Leaves are an excellent source of vitamins A, B 1, B 2, and C.

CUMIN: 6 grams of cumin has about 20 calories of which half are from fat. It also has some iron and calcium

CURRY TREE / KARRAPOULI: 100 grams of curry leaves provide 100 calories. They're rich in carbohydrates, proteins, and fiber with other minerals such as calcium, phosphorus, iron, magnesium, and copper. They're rich in Vitamin A, B, C, and B2, nicotinic acid, antioxidants, amino acids, and flavonoids.

CUSTARD APPLE: Custard apples are a well-balanced food having protein, fiber, minerals, vitamins, energy, and little fat. They are an excellent source of vitamin C, a good source of dietary fiber, vitamin B6, magnesium, potassium, with some B2, and complex carbohydrate. 100 grams of custard apple flesh has 100 calories. A small custard apple will weigh around 250 grams and will provide your entire daily intake of vitamin C.

To eat is a necessity, but to eat intelligently is an art. ~ La Rochefoucauld

D **DILL:** One gram of dill has no calories It is rich in vitamin C and one tablespoon contains more calcium than in a third cup of milk.

DRAGONFRUIT / PITAYA: Dragon fruit are rich in vitamins C, E, B1, B2, B3, phosphorus, fiber, magnesium, iron, niacin, potassium, and calcium. There are only 60 calo-ries per 100 grams. Like everything, consume dragon fruit in moderation because of fructose levels. The tiny seeds are a good source of omega-6 and omega-3 fatty acids. The fruits' phos-phorus promotes anti-aging properties and lycopene fights cancer. Eating it helps to promote the growth of two types of healthy bacteria in your stomach: lactic acid bacteria and bifido bacterial.

DURIAN: Durian is nutritious, but fattening at 150 calories per 100 grams. Diabetics beware; durian has a high sugar content. Eat small portions: per quarter pound, it has 150 cal-ories, with 3 grams of protein and 35 grams of carbohydrates, and 6 grams of fat. It is high in vitamin C, vitamin B-6, thiamine, potassium, and manganese.

E **EDDOES / TARO**: Eddoes provide a good dietary fiber at 110 calories per adult serv-ing with no cholesterol, but two grams of protein.

EGGPLANT: Eating eggplant is good for you. It can reduce high cholesterol. One cup of cooked eggplant has 27 calories, 1 gram of protein, 6 grams of carbohydrates (two grams of fiber if eaten with the skin, phosphorus, potassium, and folate.

ELEPHANT APPLE / CHALTA: Chalta or wood apple is a source of slight protein, calcium, and phosphorus. One apple has about a hundred calories.

F **FENNEL:** 3 grams of fennel have about 3 calories. Fennel contains manganese, calcium, potassium, magnesium, phosphorus, and vitamin C.

FENUGREEK: The fenugreek plant is quite nutritious, high in proteins, ascorbic acid, niacin, and potassium. 100 grams (3.5oz.) of fenugreek seed has 323 calories, 40g of protein, 58g of fiber 25 g with vitamins B2-riboflavin, B3-niacin, B-6, B6-folate (B9), and some vitamin C. It's packed with calcium, has loads of iron, magnesium, manganese, phosphorus, zinc, and some potassium. These strange little seeds contain antioxidants, powerful phytonutrients, including choline.

FINGER ROOT: 100g of finger root has 80 calories, 1.8g protein, with vitamins B-6 and C, potassium, magnesium, and iron.

G **GALANGAL:** 100 grams of galangal contains 45 calories and is also a source of sodium, iron, vitamins A and C.

GARLIC: 1 cup of garlic, almost a quarter pound, has only two 200 calories with plenty of calcium, vitamin C, and iron.

GIANT GRANADILLA / BARBADINE: Giant granadilla is a good source of vitamins B-6 and C, with magnesium, calcium, and potassium. 100 grams has only 60 calories with 2g of protein and 10g of fiber.

GINGER: 2 grams of ginger have 8 calories, only 30 milligrams of carbs, 4 milligrams of fiber, and just 3milligrams of protein, with no noticeable vitamins or minerals.

GOTU KOLA: 100 grams of fresh gotu kola has 39 calories and provides: calcium-171mg, iron-5.6mg, potassium-391mg, lots of vitamins A and C, and some B2.

GOVERNOR PLUM / CERISE PLUM / *Flacourtia indica*: 100 grams of governor plum fruit has only 94 calories and contains many minerals. It is an excellent source for iron, potassium, calcium, niacin, and vitamin C.

GOVENOR PLUM / MARTINIQUE PLUM / *Flacourtia inermis*: Governor fruit has all the vitamins C, A, B, thiamine, riboflavin, and niacin which boost immunity. 100 g has 87 calories. The fruit is rich in vitamins C, carotene, and minerals like calcium, magnesium, and iron.

GRAPEFRUIT: One grapefruit has 80 calories with one gram of protein, 18 grams of carbohydrates, and 3 grams of fiber. It is an excellent source for vitamins A and C, most B vitamins with folic acid, pectin, calcium, potassium, and magnesium.

GREEN BEANS: Green beans, have only 45 calories per cup, yet are loaded with many potent nutrients. They are an great source of vitamins A, C, and K. green beans are a good source of dietary fiber, potassium, folate, iron, magnesium, manganese, thiamin, riboflavin, copper, calcium, phosphorous, protein, omega-3 fatty acids, and niacin.

GUAVAS: Guavas are high in vitamins A and C, phosphorus, and niacin. Some guavas have four times the vitamin C of an orange. A quarter pound of guavas has only 60 calories.

GUINEA ARROWROOT / TOPPEE TAMBU: Guinea arrowroot is more than a starchy, belly filler. Per 100 grams these roots have 60 calories, 9 grams of carbs, a half gram of protein, with calcium, phosphorus, and some vitamin C.

H **HAUSA POTATO / CHINESE POTATO:** Per 100grams, hausa potatoes have 40 calories, 1g protein, 1g fat, high in calcium and iron with lesser amounts of potassium, phos-phorus, manganese, and magnesium.

HOG PLUM: A quarter pound of hog plums has about 40 calories with a lot of carotene, calcium, and thiamine.

HORSERADISH: Horseradish has only 2 calories per teaspoon. High in fiber, vitamin C, folate, pantothenic acid, magnesium, potassium, manganese, riboflavin, vitamin B6, phosphorus and copper.

HORSE PURSLANE: 100 grams of horse purslane has only 21 calories, but is packed with vitamins A, C, B complex with high levels of potassium, calcium, magnesium, iron, and phosphorus. It's one of the richest green plant sources of omega-3 fatty acids. Since it contains no cholesterol, horse purslane provides the beneficial omega-3 fatty acids without the cholesterol of fish oils. It lowers the cholesterol and triglyceride levels, and raises the beneficial high lipoprotein.

HOT PEPPER / CHILI: 100 grams of hot peppers has 40 calories, 2g of protein with high amounts of vitamins A, C, B-6, and K, and copper, iron manganese, and magnesium.

I **INDIAN GOOSEBERRY:** Indian gooseberry is a good source of calcium, phosphorus, tryptophan, lysine, methionine, and a spectacular source of vitamin C. Enriched with iron, carotene, chromium, and fiber a long w ith antibacterial properties. 150 grams has only 66 calories. It's excellent for maintaining he of mind, body, and behavior (*vata/pitta/kapha*) and treat the underlying cause of many health problems.

INDIAN JUJUBE / DUNKS: 100 grams, about 3 Indian jujubes, has only 80 calo-ries with 1 gram of protein and 10 grams of fiber. Dried fruits have more calories than fresh ones. They are high in vitamin C and potassium. The fruit is often eaten to reduce anxiety. Research studies indicate the jujube may improve memory and help protect brain cells from damage by nerve-destroying compounds. Studies with mice even suggest that extracts of jujube seed may treat dementia caused by Alzheimer's. The seeds themselves are not usually consumed.

J **JABONG FRUIT:** A cup of jabong fruit/pomelo has seventy calories and is high in potassium.

JACK BEANS: 100 grams of fresh green avara has 104 calories with 10g protein, 12g of carbs, 4g of fiber, and 8mg vitamin C. 100 grams of dried mature beans have 350 calories with 20g of protein, 61g of carbs, with lots of potassium, phosphorus, calcium, with traces of copper, niacin, thiamin, and zinc. This bean's protein has good amounts of most essential amino acids.

JACKFRUIT / KATAHAR: One cup of raw jackfruit has about 150 calories, with 4 grams of fat and 2 grams of protein. It provides 10% of the daily requirement of vitamin A and 20% of vitamin C with calcium, iron, zinc, and phosphorus.

JALAPENO PEPPER - One of these hot spicy peppers has only 20 calories with plenty of vitamins A and C.

JAVA PLUM – INDIAN BLACKBERRY: Per 100 grams, java plum has 250 calories. These fruits are extremely high in potassium, and has plenty of calcium, magnesium, phosphorus, with vitamins C and A. Fruits have some of the highest levels of natural folic acid and recommended for pregnant women. They also are an excellent source of vitamins B3 and B6. They have no cholesterol, but some sodium.

JERUSALEM ARTICHOKES / SUNCHOKES: 150 grams of raw Jerusalem artichoke has 110 calories, 3g of protein, and is an extreme source of potassium, phosphorus, iron, magnesium, and important trace minerals of zinc and copper plus vitamins A, b1, B2, and B3. This root is an excellent source of vitamin B9, folate, and beta-carotene.

JEWELS of OPAR: Every 100 grams of Jewels of Opar has only 65 calories with plenty of calcium (80mg), magnesium (61mg), and loads of potassium (300 mg). It's a significant source of omega-3 fatty acids and antioxidants. It contains such high levels of iron that scientists studying it developed the mantra, 'a leaf a day keeps anemia away.

K **KALE:** Kale is one of the best foods packed with vitamins and minerals. 100 grams has 49 calories, 4g of protein. It's got more than twice the daily requirement of vitamins A, K, and C with B-6 and manganese, calcium, copper, and potassium. Researchers compared raw kale with several cooking methods; all resulted in a significant reduction in total antioxidants and minerals, including calcium, potassium, iron, zinc, and magnesium. Steaming it for a few minutes may be the best way to preserve its nutrient levels. Grow your own kale; it's one of the few superfoods that's accessible to everyone, everywhere.

KALO –BIG TARO – DASHEEN: 250 calories to a cup. It has low fat, no cholesterol, one gram of fiber, but 60 grams of carbohydrates with healthy Palmitic, oleic, and linoleic acids and adequate levels of the other essential amino acids. It is low in protein, high in potassium, but contains moderate quantities of calcium, phosphorus, and magnesium.

KOHLRABI: 100 grams of raw kohlrabi has 27 calories, 2g protein, and 5g fiber. It has tremendous vitamin C, with some vitamins A, K, and B6, potassium, and magnesium

KOLA NUT / COLA NUT / BISSY NUT: The kola nut contains Vitamins B1, C, and E. They are very high in potassium, magnesium, and calcium with iron, zinc, phosphorus, and micronutrients. Kola nuts are 2-4% caffeine and theobromine, with healthy tannins, alkaloids, saponins, and flavonoids. Theobromine is also found in green tea and chocolate.

KUMQUAT: 100 grams have 3.8 grams of protein, a great source of vitamin A with calcium and potassium.

L **LABLAB BEANS / HYACINTH BEANS / SEIM:** 100 grams of raw, immature lablab pods have only 46 calories, but dry mature seeds contain 344 calories with 24g of protein and the same amount of fiber. Dry lablab beans are one of the finest sources of several B-complex vitamins such as thiamin (1.130 mg or 94% of the Daily Value), riboflavin, folates, and niacin. Fresh pods carry vitamin A, a powerful antioxidant, 29% of the Daily Value (DV), 12.6 mg or 21% of DV of vitamin C. Dry beans hold 148% of the DV of copper, 13% of calcium, 64% of iron, 71% of magnesium, 68% of manganese, 26% of potassium, 53% of phosphorus, and 84% of zinc.

Copper helps to maintain a positive outlook, and to focus. Potassium is an essential electrolyte of cell and body fluids, which helps to counter the bad effects of sodium on heart and blood pressure. Zinc has anti-inflammatory properties which helps to reduce the risk of diseases. Minerals such as selenium, manganese, and zinc assist people having lung disorders.

LANGSAT / DUKU: 100 grams of langsat/duku has only 60 calories with good amounts of calcium, phosphorus, vitamin A, thiamine, and riboflavin. The high content of riboflavin helps reduce the pains of migraines. White langsat skin brightening body scrub is formulated from the extract of langsat fruit.

LEEKS: Leeks are a good source of fiber, leeks folic acid, calcium, iron, potassium, manganese, vitamins B6 and C. For many, leeks are easier to digest than onions. A cup of raw leeks has 60 calories, but a half-cup of boiled leeks has only sixteen calories.

LEHUA MAOLI - MALANGA – YAUTIA – TANNIA: 100grams of malanga has 105 calories with 2 grams of protein. It contains high amounts of potassium, phosphorus, calcium, and magnesium with trace amounts of copper, zinc, manganese, and selenium. It has decent amounts of vitamins A, C, B9 (folates), and beta carotene.

LEMON: A lemon has only 25 calories with one gram of protein, eight grams of carbs with calcium, iron, potassium, and phosphorous. One hundred ml of lemon juice has 50 ml of vitamin C (ascorbic acid).

LEMONGRASS: This herb has a concentration of folate and is a source of synthetic vitamin A. Lemongrass oil contains a high percentage of citral, used in perfumes and ionone. Through distillation they are converted into synthetic vitamin A. Lemongrass is a good source of iron, zinc, and especially manganese. One cup has only 63 calories from 3 grams of fat and 2 grams of protein (if you swallow). Lemongrass is high in Omega 3 and 6 fatty acids.

LETTUCE: Lettuce provides vitamins A and C, and minerals such as potassium, iron,and calcium. A cup of lettuce is ten calories with one-gram each of fiber, protein, and carbohydrates.

LIME: One lime has only twenty calories with absolutely no fat, sugar, or cholester-ol. Citrate of lime and citric acid are also derived from this fruit. One lime contains a third of the daily requirement of vitamin C as ascorbic acid. Limes also have some fiber and potassium.

LONG BEAN / CHINESE YARD LONG BEAN / BODI: These beans are rich in vitamins A and C. 100 grams has 4 grams of protein, 110 mg calcium, 5 mg iron, and calcium.

LOQUAT / JAPANESE PLUM: 100 grams of loquat have 47 calories, 1½ g protein with vitamin A, calcium, magnesium, phosphorus, and potassium.

LOTUS ROOT / LOTUS YAM: 100 grams of lotus root has74 calories, 0-fat, 40 mg sodium, 556 mg potassium with 17.4g carbohydrates, 4.8g dietary fiber, 0-sugars, and 2.5g pro-tein. This vegetable is very nutritious and contains large concentrations of vitamins and miner-als including Vitamin B, Vitamin C, iron, potassium, copper, thiamin, and zinc, among others.

LUFFAH / LOOFA: Luffah gourd is a very low-calorie vegetable. Per 100 grams, it has only 20 calories and rich in antioxidants. It has vitamins A and C riboflavin, niacin, and essential amino acids with moderate levels of calcium, magnesium, potassium, iron, and phos-phorus.

LYCHEE / LICHI: This fruit is 70% water and a great source of hydration for the skin and has a powerful compound 'oligonol' to protect the skin from UV light and free radicals that can damage the skin cells. Lychee is perfect for dieters, 100 sweet grams has only 66 calories, and because it has only a tiny amount of saturated fats and no cholesterol. It is rich in vitamins C and A with many B vitamins, potassium, thiamin, niacin, folate, and copper.

M **MACADAMIA NUT:** 100g of Mac nuts has 73g of fat, 7g of protein, and are high in calcium, iron magnesium, phosphorus, and vitamin B.

MACE: 5 grams of mace has 25 calories. It's high in calcium, phosphorus, and magnesium.

MALABAR SPINACH: 100 grams of Malabar spinach has only 19 calories. It's high in vitamins A and C, potassium, and magnesium, with some calcium and iron. Malabar spinach has a high in oxalate content, but less than half of regular spinach, and is reduced by steaming. Oxalate may cause kidney stones.

MAMEY SAPOTE / MAMEY APPLE: A hundred grams of mamey apple pulp has only 45 calories with some calcium, phosphorus, iron, vitamins A & B, and thiamine.

MANDARIN ORANGE: Per 100 grams, mandarin oranges have 47 calories with vi-tamin C, calcium, magnesium, potassium, phosphorus, and iron.

MANGOS: Mangos not only make you feel good, they are great health wise because they contain plenty of fiber, vitamins A and C, and potassium. One large mango has about 100 calories with no cholesterol, and half of the necessary daily fiber.

MANGOSTEEN: Mangosteen have 60 calories per 100 grams of flesh with calcium, phosphorus, and iron.

MARJORAM: 1TBS has five calories with some vitamin A and C, calcium, and iron.

MAUBY: No significant nutrients.

MINT: Two leaves of mint have hardly any calories with some calcium and potassium.

MORINGA / MIRACLE PLANT: Eating moringa is the same as taking a natural vitamin and mineral supplement. 100 grams of cooked moringa has 64 calories with 9g of protein, 2g fiber, 1.5 g fat, and is an excellent source of vitamins A, thiamine-B1, riboflavin-B2, vitamin C, pyridoxine-B6, niacin-B3, and folates. Moringa is also high in calcium, magnesium, manganese, phosphorus, potassium, zinc, and iron. The leaves are the most nutritious part of the plant; one cup of fresh, chopped leaves (21g) contains 7 times more vitamin C than oranges and 15 times more potassium than bananas with calcium, protein, iron, and amino acids. The bulk of the nutrition is found in the young leaves; nutrients are lost with age. The best moringa leaves are harvested while the trees are saplings.

Drumsticks or immature seed pods are high in vitamin C,potassium, magnesium, and maganese, all of which may be reduced by the preparation method. The seeds, cooked in recipes similar to peas or roasted like nuts, have lots of vitamin C and lesser amounts of B vitamins.

MUSTARD: A cup of cooked mustard greens has only 20 calories with vitamins K, A, E, and C, potassium, calcium, and iron.

MYSORE RASPBERRY: 100 grams of mysore raspberries has only 28 calories with 1 gram of protein, vitamins B-6, C, and E, and manganese.

N **NAMNAM:** Per 100 grams: vitamin A 150–500 I.U., vitamin C 25 mg, and extremely high in flavonoids.

NONI: 100 grams of noni fruit contains 95.67% water, 15.3 calories, 0.43 g protein with lots of vitamin C, and some calcium. 100 grams of juice has 47 calories with less than a gram of protein and vitamin C, biotin, and folate.

NUTMEG: Nutritionally nutmeg is high in calories, 12 calories a teaspoon, two thirds of which come from fat. It has no vitamins and only some iron and calcium.

O **OKRA / LADY FINGERS:** A half-cup of cooked ochro has only 25 calories; 2 grams of fiber, 1.5 grams protein, vitamins A and C, calcium, potassium, and manganese. Nearly 10% of the recommended levels of vitamin B6 and folic acid are also present.

ONIONS: Chopped raw, mature onions have 60 calories per cup, 2 grams fiber, and one-gram protein, with six grams carbohydrates. Onions are a good source of vitamin C, chromium, manganese, potassium, and phosphorus.

ORANGES: Each orange has about 60 calories packed with vitamins C and B-1, fiber, and folate. One orange contains about 50mg. vitamin C; or two-thirds of our daily need.

OREGANO: Oregano's pungent taste is reduced by cooking, so it is best added to any dish just shortly before serving it. Oregano has 5 calories per tablespoon with some protein. It is high in vitamins A, C, E, and K, and minerals as calcium, iron, magnesium, phosphorus, and potassium.

P **PAK/BOK CHOY:** A perfect food for dieters, one cup of cooked bok choy has only 20 calories, with no fat, but 3 grams of carbs and 3 grams of protein.

PAPAYA: A cup of papaya has 120 calories with good doses of vitamins A, E, and C, folic acid, potassium, and fiber.

PAPRIKA: Paprika has only 6 calories per TS with a good bit of calcium and potassium.

PARSLEY: One ounce of fresh parsley has only ten calories with two thirds of the daily requirement of vitamin C while rich in minerals as calcium and potassium. Parsley contains antioxidants such as vitamin C, beta-carotene, and folic acid. These antioxidants fight many diseases such as arthritis, asthma, diabetes, colon cancer, heart attacks, and strokes.

PARSNIPS: Per 100 grams, parsnips have 75 calories, 36mg of calcium, 29mg of magnesium, 375mg or potassium, 17mg of vitamin C, and 22.5mg of vitamin K.

PASSION FRUIT: A half-cup of juice has fifty calories with one gram of protein and eleven grams carbohydrates. It is a good source of vitamins A and C.

PEA EGGPLANT / TURKEY BERRY: 100 grams of cooked, seedless pea eggplant has only 40 calories with 2g protein and 4g fiber. It contains vitamins A and C with calcium, iron with trace amounts of manganese, copper, and zinc.

PEACHES: 100 grams of a juicy peach has only 40 calories with considerable amounts of vitamins A and C and calcium, phosphorus, and potassium.

PEANUTS: Every ounce of raw peanuts provides seven grams of protein, seven grams of fat, and 3 grams of fiber. A cup has over 600 calories, because they are over 75% oil. However, it is healthy oil, monounsaturated, not polyunsaturated.

PEACH PALM / PEWA: Pewa fruit contain carotene, calcium, phosphorus, ascorbic acid, vitamin A, and nicotinic acid. Fruit should be boiled within 2 to 4 days of picking.

PIGEON PEAS: For vegetarians and everybody, peas are an excellent source of protein. To get the most protein from peas or beans, eat them with unrefined rice or wheat. 100 grams of fresh, green pigeon peas has 140 calories with 7 grams of protein and 5 grams of fiber. They are packed with hard-to-get B complex vitamins with C and K. Fresh peas have plen-ty of minerals as manganese, magnesium, phosphorus, and zinc. 100 grams of dried peas have 340 calories, 21g protein and 17g of fiber. The dried only have the B vitamins and are a better source of manganese, magnesium, phosphorus, potassium, and zinc.

PILI NUTS: One ounce (28g) of pili nuts has 200 calories with 3g of protein and con-tain magnesium, potassium, and calcium. They're rich in omega fatty acids, phosphorus, and have high levels of protein and all eight essential amino acids. Pili nuts are also rich in vitamin E. This promotes a health heart and lowers cholesterol. They are also processed into pili nut oil, which is said to be better than olive oil.

PIMENTO SEASONING PEPPERS: 4 pimento peppers have about 10calories with little food value, but add so much great taste.

PINEAPPLE: Two slices of a regular-sized pineapple should be about a hundred grams and has 60 calories with no fat or cholesterol. Pineapple is a good source of vitamin C.

PLANTAINS: Plantains are belly fillers, not for the dieters, as they are mostly carbohydrate, approximately 40 grams per half, with180 calories, very high in potassium approximately 500 milligrams per serving. Plantains are a very healthy fruit to eat. This is because they do not contain any cholesterol or sodium, and are low in fat. They also contain traces of calcium, iron, and potassium, and are high in vitamin A.

POMELO/PUMMELO/JABONG FRUIT: A cup of pomelo has seventy calories and is high in potassium.

POMEGRANATE: Pomegranate A hundred 100 grams of pomegranate has seventy calories with plenty of potassium, vitamins C and B5

POTATO: Per 100 grams, potatoes contain 80 calories, 2g protein, and good amounts of calcium, phosphorus, iron, potassium, and magnesium, with some vitamins A, B-1, B-2, and C.

PUMPKIN: One cup of cooked pumpkin has 24 calories, one gram of protein. Pumpkin is a good source of thiamin, niacin, vitamins A, C, E, B6, folate, pantothenic acid, iron, magnesium, phosphorus, riboflavin, potassium, copper, and manganese.

PURPLE YAM: Purple yam contains per 100grams 100 calories with tiny amounts of protein and fiber, but has good amounts of vitamins B1, B2, and C. The mineral content is high in potassium, magnesium, calcium, phosphorus, and zinc. As a staple, nutritionally, this yam compares favorably to brown rice and whole wheat. These roots contain phytosterols, alkaloids, tannin, and are a source of starch.

R **RADISH / DAIKON / MORAI:** Daikon also known as white or long radish is consid-ered a 'winter radish' while the red 'globe' is a summer type. All radishes are very flavorful and low in calories. A half cup has only 12 calories with a good balance of fiber, potassium, and folate.

RED BANANAS: Every red banana has about 115 calories with 400 mg potassium and 15% of your daily requirements for vitamin C and B6, with one gram of protein. Red bananas have more vitamin C than the usual yellow types. The redder the fruit the more nutritious. Eat at least one banana a day and it comes in truly germ proof packaging because its thick peel is an excellent protection against bacteria and other contamination.

RICE: White, long-grain, regular, not enriched, cooked, without salt 205 calories—one cup brown medium grain 218. Natural brown rice has about a calorie a gram and is low in saturated fat, cholesterol, and sodium. It is also a good source of selenium, and manganese.

ROSEMARY: Rosemary has about 3 calories per TS with some calcium and iron.

RHUBARB: 100 g of rhubarb has 116 calories, and is high in potassium, vitamins C and K1. Rhubarb is rich in antioxidants, particularly anthocyanins, which makes it red, and proanthocyanidins. These antioxidants have anti-bacterial, anti-inflammatory, and anti-cancer properties.

ROSELLE / RED SORREL: 100 grams of roselle has 49 calories with 1g of protein. It's very nutritional, high in vitamin C, calcium, magnesium, niacin, riboflavin, and iron.

RUTABAGA: Per 100 grams of rutabaga, there are only 38 calories, with lots of vita-min C, with potassium, calcium, and magnesium.

S

SAFFRON: One teaspoon has 4 calories with some potassium.

SAGE: Sage contains six calories in two grams with some calcium and iron. Sage is a good source of vitamins A and C.

SAPODILLA: A raw sapodilla has about 120 calories with calcium, potassium, vitamins A, C, and folate.

SCREWPINE LEAF / PANDAN: 100g of pandan leaves has 24 calories, 6g carbohydrates, no protein or fat.

SESSILE JOYWEED: 100 grams of sessile joyweed has only 75 calories packed with magnesium and potassium. It is also a good source of vitamins A & C and Beta carotene.

SHALLOTS: They are slightly high in calories – 100 grams of shallots contain 72 calories, with 2.5 grams of protein and 17 grams of carbs. Shallots are full of fiber, vitamins, minerals, and antioxidants. They are high in vitamin A, pyridoxine, and potassium with calcium, phosphorus, iron, and copper. Shallots are an excellent source of selenium a trace element that slows aging, fights dementia, and keeps skin and hair healthy.

SICKLE SENNA: Research has shown that raw seeds from the sickle senna plant contain 15% protein, with trace amounts of iron, zinc, manganese, and cobalt. Researchers found that when sickle senna seeds were fed to rats, there was a decrease in blood cholesterol. The leaves contain 6% protein with good amounts of calcium, copper, magnesium, potassium, zinc and iron. Studies show that high consumption of roasted seeds may cause stomach prob-lems, but were okay when used only as a supplement in roti, bread, chutneys, stir-fries etc.

SNAKE GOURD: 100 grams of snake gourd flesh contains only 20 calories. It is rich in dietary fiber with vitamins A, B6, and C. This vegetable has high levels of zinc and manganese with other minerals like magnesium, calcium, and phosphorus. This fruit also contains many medicinal compounds.

SOURSOP: A 100 grams of soursop has 60 calories with good amounts of calcium, phosphorus and amino acids. The fruit contains significant amounts of vitamins C, B1, and B2.

SPAGHETTI SQUASH: Spaghetti squash has 50 calories per 100 grams. This squash is high in fiber, an excellent source of folic acid, potassium, magnesium and small amounts of vitamins A and B6.

SPINACH: One cup of fresh spinach has only 40 calories and over twice the daily requirement of vitamin K. This vitamin is essential to keep human bones healthy. Spinach is also a great source for vitamins A, C, magnesium, and folate.

SPINE GOURD / HEDGEHOG GOURD: It has only 17 calories per 100 grams with 3g of protein, 33mg of calcium, 42mg phosphorus, and is rich in vitamin A and B-6 folate.

STAR APPLE / CAIMATE: Per 100 grams, Star apples have 62 calories, more than 2 grams of protein, with vitamins C, B1, B2, and B3. This fruit is a good source of calcium, iron, and phosphorus.

STARFRUIT / CARAMBOLA / FIVE FINGER: One nice star fruit has about 20 calories and is high in carbohydrates, calcium, phosphorus, vitamins A, B, and C, plus amino acids. Starfruit is a good source of dietary fiber, and can help to reduce blood pressure and cholesterol.

STRAWBERRY / QUEEN OF BERRIES: Per 100grams of strawberries, there are only 33 calories. They are high in vitamin C with trace amounts vitamins E, Band 1, B2, B3, B5, B6, B9 (folic acid), and K. The mineral content is high in manganese and potassium, with trace amounts of calcium, iron, magnesium, phosphorus, and potassium.

SUGAR APPLE: A medium-sized fruit is about 80 calories with a little protein, but one gram of fat. They contain good amounts of phosphorous and calcium.

SUGAR CANE: Sugar cane juice provides glucose and renews energy. One tablespoon of cane juice has 45 calories. One tablespoon white granulated sugar has 46 calories, and brown has 34.

SUMMER SQUASH: Summer squash such as zucchini are low in calories with only 16 to a cup. Zucchini and crookneck squash are an excellent source for vitamins A and C, and minerals such as potassium, calcium, folate, and magnesium.

SWEET PEPPER: One cup of raw sweet green pepper has only 24 calories and has three times the daily requirement for vitamin C (four times an orange), and the complete requirement of vitamin A. These two vitamins plus B-6 and folic acid protect against high cholesterol and heart disease. Peppers are rich in vitamins B6, B1, folic acid, manganese, and potassium.

SUNSET HIBISCUS - SUNSET MUSK MALLOW: Sunset hibiscus is an extremely nutritious vegetable and contains 12% protein. Per 100g of edible leaves and stem tips has 150 calories and contains, 4g protein and 1g fiber. It is a rich source of vitamins vitamin A and C, calcium, and iron. It contains the essential amino acids isoleucine, leucine, lysine, methionine, phenylalanine, threonine, valine, and histidine, which cannot be made by the body and must be obtained from foods.

SURINAM CHERRY: 100 grams of Surinam cherries has 50 calories with vitamins A and C, calcium, and phosphorus.

SWEET POTATOES: The sweet potato is very nutritious. A half-cup of cooked sweet potato supplies 2 grams of protein, four grams of fiber, vitamins A and C, calcium, beta-carotene, manganese, and folic acid. One ounce steamed sweet potato has 24 calories.

T **TAMARILLO:** Per 100 grams, tamarillos have 85 calories, and are very high in vitamins A, C, and E with iron, calcium, and magnesium.

TAMARIND: Tamarind pulp has only fifty calories in two ounces with some protein and fiber. Tamarind is high in calcium, phosphorus, and potassium.

TANGELOS: Tangelos have about 100 calories per fruit with plenty of potassium, and of course vitamin C.

TANGERINE: One tangerine, approximately 90 grams has 50 calories with 1g protein, loads of vitamin C and potassium with traces of vitamin B-6 and magnesium.

TANNIA: One cup of cooked tannia has about a hundred calories, mostly water and carbohydrates with a little protein, and virtually no fat. Tannia is a great source of calcium, phosphorous, and iron These roots also supply vitamin A, thiamin, and riboflavin.

TARO - KALO – EDDO: 100 grams of taro root has 112 calories, 2 grams of protein with no cholesterol. The roots contain vitamins A, C, B-6, and E, folate, niacin, potassium, and calcium. The leaves are rich in vitamin C, phosphorous, calcium, iron, riboflavin, thiamine, and niacin.

TARRAGON: One teaspoon of ground tarragon has 14 calories with plenty of potassium. A half ounce of fresh tarragon equals a third of a cup. One tablespoon of fresh tarragon equals one teaspoon of dried.

TEA: Black and green teas have very few calories if drank plain. They both contain good amounts of vitamins B-2, C, D, and K with traces of calcium, magnesium, iron, and zinc. However, the noted health benefits of green tea are from its micro-nutrients and antioxidants.

THAI EGGPLANT: 100g is 25 calories with potassium, calcium, magnesium, vitamins A & C.

THYME: One gram of thyme has three calories with calcium, vitamin A, and iron.

TONKA BEANS: There is no accurate calorie counter for tonka beans, but they are a quarter fat with a larger percentage of starch.

TOMATOES: Tomatoes contain vitamin C, potassium, folacin, and beta-carotene. A diet containing tomatoes may help reduce cancer, heart disease, and premature aging. A medium tomato has only 25 calories.

TROPICAL ALMOND / WEST INDIAN ALMOND: 100 grams has 600 calories and 20 grams of protein. The oil content is 61g. The seeds are a rich source of zinc, magnesium, iron, and calcium.

TROPICAL LETTUCE / INDIAN LETTUCE: Tropical lettuce has 2 percent protein with some vitamin C, beta-carotene, riboflavin, calcium, and iron. The leaves also contain 6 anti-oxidative phenolic compounds.

TURMERIC: This root has 8 calories per teaspoon with manganese, iron, vitamin B-6 and potassium.

TURNIP: Turnips greens are very nutritious; 100 grams of boiled leaves have only 20 calories, 1g protein, loaded with vitamin K, and vitamins A, C, folate, and lutein. The root per 100 grams has 28 calories with vitamins C and B6, folic acid, calcium, and potassium.

W **WATER APPLE /WAX JAMBU / ROSE APPLE:** Water apples are good for most pregnancies with vitamin A and iron. According to early writings, a water apple salad is a ceremonial dish for new mothers. These bland fruits help keep expectant mothers hydrated during morning sickness. The leaves have an analgesic property that can help relieve post-birth pain. One hundred grams or one fruit has twenty-five calories, with some thiamine, riboflavin, niacin, vitamins A and C, slight calcium, magnesium, phosphorus, iron, and no cholesterol. This fruit is 90% water.

WATER HYSSOP / BABY TEARS: 100 grams of water hyssop leaves cooked has only 38 calories and contains 2g protein, 1g fiber with high quantities of vitamin C, calcium, phosphorus, and iron.

WATERMELON: Watermelon is practically a multi-vitamin itself. Watermelon contains about 10% of the daily requirement of potassium, which helps regulate heart functions and normalize blood pressure. One wedge, or a quarter of a small melon, has 90 calories, 2 grams protein, 20 grams of carbohydrates, 1.5 grams of fiber, potassium, and vitamins A and C.

WATER SPINACH: Per 100g of water spinach has only 19 calories with 2g of fiber, and 2.6g of protein. It's loaded with vitamins A, B1 (thiamin), B2 (riboflavin, B9 (folate), with some B3 (niacin). Water spinach has calcium, iron, magnesium, phosphorus, potassium, and some zinc.

WEST INDIAN PEA / VEGETABLE HUMMINGBIRD TREE: This tree is a nutritional powerhouse! The leaves and blossoms of the hummingbird tree are high in calcium, potassium, iron. Per 100 grams of leaves are 321 calories, 36.3g protein, 7.5g fat, 47.1g carbo-hydrate, and 9.2g of fiber. Nutritionally, 100grams has 1684mg calcium, 258mg phosphorus, 2,005mg potassium K, with loads of vitamins A and C. The blossoms per 100 grams have 345 calories, 14.5g protein, 3.6g fat, 77.3g carbohydrate, 10.9g fiber, 145mg calcium, 290mg phosphorus, 5.4mg iron, 1,400mg potassium, and almost as much vitamin and C as the leaves. Oil pressed from the seeds is also nutritious and contains 12.3% palmitic, 5.2% stearic, 26.2% oleic, and 53.4% linoleic acids. Five-day-old sprouts from humming bird seeds are rich in vi-tamin C at 166mg per 100 grams.

WHITE TURMERIC / ZEDOARY: White turmeric is rich in vitamin C, vitamin B6, manganese, iron, potassium, and omega-3 fatty acids.

WHITE YAM / GREATER YAM: White yam contains per 100grams, 100 calories with tiny amounts of protein and fiber, but has good amounts of vitamins B1, B2, and C. The mineral content is high in potassium, magnesium, calcium, phosphorus, and zinc. As a staple, nutritionally, this yam compares favorably to brown rice and whole wheat. These roots contain phytosterols, alkaloids, tannin, and are a source of starch.

WINGED BEANS: 100 grams of these beans when young, immature, contain only about 49 calories. Mature seeds have ten times more, 400 calories. If you're on a diet, eat the immature beans. Protein is high at 6% in all the parts of this plant: the leaves, roots, and beans. There's minimal fat with no cholesterol. These beans are packed with folates, niacin, thiamin, vitamins A and C, potassium, calcium, and copper.

Y

YAM: Yam flesh is poisonous raw, but cooking makes it safe and edible. One cup of cooked yam contains 150 calories with five grams of fiber, vitamins C and B6, potassium, and manganese.

Z

ZUCCHINI / COURGETTE: Per 100 grams, zucchini has only 17 calories with 1g of protein. Zucchini is an excellent source for vitamins A, B-6 folate, and C, potassium, calcium, folate, and magnesium. A zucchini has more potassium than banana.

Courtesy: R Douglas

The doctor of the future will no longer treat the human frame with drugs, but rather will cure and prevent disease with nutrition. ~ Thomas Edison

THE HAWAIIAN HOME GARDEN GUIDE
FOOD REMEDIES

If your garden space is limited, and want to try different natural teas, decoctions, salves, etc. read below and choose what works for you.

There are two great medicines: Diet and Self-Control. ~ Max Bircher

PLEASE TAKE NOTE: All material in this book has been researched and compiled from reliable sources and is provided for your information only. No action or inaction should be taken based solely on the contents of the information. Readers should consult appropriate health professionals on any matter relating to their health and well-being. DO NOT consider any of this information to be medical advice or instruction. No action or inaction should be taken based solely on these articles.

THE HAWAIIAN HOME GARDEN GUIDE will not take any responsibility for any adverse effects from the use of plants. Always seek advice from a professional before using a plant medicinally. The use and dosage of herbs and foods vary with circumstances such as climate (whether it is hot or cold), the person's health and age, and the manner the dose is administered. Herbs and foodstuffs should be used for medical purposes only with proper medical advice. The medicinal plants must be grown with-out chemical fertilizers or pesticides.

The way you think, the way you behave, the way you eat, can influence your life by 30 to 50 years. ~ Deepak Chopra

A **ACEROLA / WEST INDIAN CHERRY**: Historically, acerola cherry has been used to treat dysentery and fever. Research suggests that this cherry possesses anti-inflammatory and astringent properties. It's rich in vitamin C and plus over 150 phytonutrients have been found in the acerola cherry. This fruit also contains potassium, magnesium, and other minerals.

ACKEE: In traditional-homeopathic medicine ackee's ripe arils are combined with sugar and cinnamon to treat fever and dysentery. The new leaves are crushed and applied to the forehead to ease a severe headache or migraine.

AIR POTATO: The air potato variety with white spots on its large, round leaves and also on the fruits are not eaten, but used in traditional medicine. The juice of the air potato is used externally to sanitize wounds, usually as a poultice, to treat cuts, sores, boils, and inflammations. Juice squeezed from the roots expels parasites and when combined with palm oil, it's rubbed to soothe rheumatism. Fruits are used to treat boils and fevers. The leaves are steamed and applied for conjunctivitis/pinkeye. Air potato con-tains the steroid diosgenin, which produces synthetic commercial steroidal hormones for birth control pills and steroids used to boost men's testosterone and used in bodybuilding.

ALLSPICE: In the past, allspice was used to treat indigestion and gas. It was eaten as treatment for stomachaches, vomiting, diarrhea, fever, and colds.

ALMONDS: Almonds are loaded with flavonoids in the nut's skin, and are antioxidants linked to improving breathing. Scientists believe these chemicals may even prevent respiratory diseases such as asthma, emphysema, and chronic bronchitis. Almonds are also a significant source of magnesium and potassium, which contributes to strong bones. Almonds are good eating nuts, high in monounsaturated fatty acids, and lower your cholesterol level, reducing the risk of heart disease. Almonds contain significant amounts of magnesium, vitamin E, fiber, and potassium, all of which are beneficial to a healthy heart.

ALOE VERA: For centuries, aloe vera has been used for medicinal purposes. The secret is its concentrations of nutrients and vital substances that include water, vitamins A, B, C, and E, more than 20 minerals, and both fatty and amino acids. It seems scientists didn't believe aloe actually cured burns, so it was tested. Aloe is now proven to heal irst and second degree burns, but it has not been proven to protect from sunburn. Drinking aloe juice improves blood glucose levels for diabetics, and helps patients with liver disease. Aloe juice may reduce symptoms and in lammation of stomach ulcers. It will also reduce gum disease and dental plaque. Aloe vera extracts are antibiotic and ight fungus. Aloe Vera is used to treat skin ailments as insect bites, acne, sunburns, rashes, scars, blemishes, sores, eczema, and psoriasis. Aloe simply patted on the skin once a day is sufficient for the desired results. It is drank to treat blood pressure, hypoglycemia, arthritis, ulcers, constipation, poor appetite, digestive disorders, diarrhea, and hemorrhoids.

AMARANTH / INDIAN SPINACH: Amaranth leaves are used in poultices (fresh or as dried powder) to treat inflammations, and skin eruptions like boils and abscesses. Juice pressed from the leaf sap is an eyewash to treat infections and conjunctivitis. Juice from the root treats urinary inflammation and constipation. The entire plant is boiled to make a blood purifier decoction

ANNATTO / ROUCOU: As an herbal remedy a mixture of annatto pulp and seeds boiled in oil makes a salve that helps heal small cuts and burns while preventing scarring and blistering. A decoction of the leaves and pulp relieves stomach disorders like indigestion and will help relieve asthma. Annatto leaves in a bath is refreshing. Leaves heated in oil will reduce the pain of a headache when pressed to the forehead. People take annatto for diabetes, diarrhea, fevers, fluid retention, heartburn, malaria, and hepatitis. It's used as an antioxidant, bowel cleanser, and to repel insects. Annatto pulp and seeds boiled in oil makes a salve that helps heal small cuts and burns, preventing scarring, and blistering. A decoction of the leaves and pulp relieves stomach disorders like indigestion, and will help asthma. Annatto leaves in a bath will be refreshing. Leaves heated in oil will reduce a headache when pressed to the forehead.

APPLES: An apple a day may keep the doctor away and so will a tablespoon of good raw apple cider vinegar. Apple cider vinegar with its sour taste and strong smell is mostly apple juice with yeast added. That ferments the juice into alcohol and then into acetic acid. Raw apple cider vinegar (not plain cider vinegar) is known to be a good source of acetic and ascorbic acid (vitamin C), mineral salts, amino acids, and other key components of good nutrition, but it is also a well-loved folk remedy thought to ease digestion, fight obesity and diabetes, wash toxins from the body, kill lice, and reverse aging. Add 1 to 2 tablespoons to water or tea. One researcher found vinegar improves blood sugar and insulin levels in people with type 2 diabetes. Apples also contain some chemicals that seem to be anti-bacterial, anti-inflammatory, and kill cancer cells. Apple peel contains ursolic acid assumed to build muscle and increase metabolism.

ARAMBELLA / POMMECYTHERE: Arambella leaves smell great, but are slightly sour and are used for lavoring, particularly curries. Indonesians make a dish with steamed leaves, salt ish, and rice. Tree sap can be used for medicinal poultices. A tea made from the bark is a supposed remedy for diarrhea. The fruit is good source of vitamin C, and the juice can be used for a remedy for diabetes, heart ailments, and urinary problems.

ARROWROOT: Arrowroot is easily digested, and can be used as food for babies and people recovering from illnesses or medical treatments, especially heart ailments. It helps urinary problems and eaten daily, it lowers cholesterol. It's used to treat intestinal disorders, diarrhea, and irritable bowel syndrome. The root's iber helps push foods through your system efficiently while its nutrients are absorbed, preventing constipation, and helping control blood sugar and diabetes.

ASH MELON: Ash melon has been used as a food and medicine for thousands of years in Asia and especially in China. All parts of the fruit are used medicinally. The skin or rind is diuretic and treats urinary problems. The skin is burned and then applied to re-lieve the pain of deep wounds. The seeds treat in lammations, bruises, and coughs. Seed oil is used to expel parasites. The fruit's lesh is considered an aphrodisiac and treats lung diseases such as asthma. Research has found these fruits contain anti-cancer aromatic compounds. The fruit's juice treats nervous disorders. It is believed, daily consumption of ash melon sharp-ens your focus, increases your brainpower, and energy without the nervous jitters of caf-feine from drinking coffee. Eating ash melon will strengthen your overall immune system.

AVOCADO: Avocados are excellent for moisturizing and nourishing mature skin. This fruit helps prevent wrinkles when used as a mask. Avocados contain at least 14 minerals with vitamins A and E essential for skincare as we age. Gently scrub your face before applying the avocado mask. Combine half a good-sized avocado with one egg yolk and a tablespoon of honey and blend. Apply mixture immediately to your face, neck, shoulders, and hands. Relax in a cool, shaded place with the mask for half an hour. Rinse with warm water and apply a moisturizer.

B **BAEL FRUIT - GOLDEN APPLE:** The dried pulp of the bael fruit is astrin-gent. It reduces irritation in the digestive tract and is an excellent remedy in cases of diar-rhea. For hemorrhoid treatment, prepare a decoction of unripe fruit with fennel and ginger. As a sexual stimulant, prepare a decoction of only the unripe fruit. Ripe bael fruit is a laxative and relieves in lammation. It supports the healthy function of the stomach. Bael sherbet (fruits boiled in water) can improve your digestion. Don't indulge and eat too much; it can reduce your breathing rate, heart beats, and cause drowsiness.

Bael fruit has plenty of vitamin C to boost immunity and provide relief from the common cold. It's also effective in reducing high blood pressure and the risks of cardio-vascular diseases. The tree's sap, Feronia gum, reduces elevated blood glucose levels as an essential Ayurvedic treatment for diabetes. Research is needed to discover if prolong use could be a cure. Boil bael leaves in water and then wash your body to heal wounds and reduce body odors. However, eating leaves or drinking the tea may cause abortion and sterility in women.

BALATA: Legends report the sap is so creamy it could be substituted for cow milk.

However, over indulging the sap will constipate you.

BANANA: The New England Journal of Medicine reported a banana a day can decrease the risk of death from strokes by as much as 40% in certain cases. Bananas have many medicinal uses. Because they are very high in iron they are great for treating anemia. The high potassium and low salt help reduce blood pressure. The high fiber can overcome constipation. Bananas naturally contain tryptophan, which the body converts into serotonin – known to improve your mood and generally make you feel happier, fighting depression. The banana neutralizes burning stomach acid and reduces irritation by coating the lining of the stomach. Bananas keep blood sugar levels up and help to avoid morning sickness. Banana skin rubbed on a mosquito bite will reduce swelling and irritation. B vitamins in bananas can calm jittery nerves while potassium helps normalize the heartbeat. If you're trying to quit smoking, bananas help the body recover from the effects of nicotine withdrawal and are a great snack to assist in weight loss. Supposedly unsightly warts are removed by covering it with a banana skin, yellow side out.

BASIL: A basil leaf will relieve the pain of a mouth ulcer and basil tea will soothe sore gums. Basil tea is also a good soothing remedy for arthritis or rheumatism. This herb improves blood circulation by fighting bad cholesterol, while reducing the chance of irregular heartbeats.

BAY LEAVES: Put bay leaves in a damp washcloth to alleviate the pain of head-aches, especially migraines. Bathing in a bath with bay leaves can treat skin rashes and the pains from sore muscles or arthritis. These same leaves hung in a wardrobe can protect your clothes from moths. Bay oil (if you can ind it) increases blood circulation and is reported to prevent baldness. Bay leaf helps the body process insulin more e iciently, which lowers blood sugar levels. It has also been used to treat stomach ulcers. Bay leaf has anti-inflammatory, anti-oxidant, antifungal and anti-bacterial properties. Bay also treats rheumatism, and colic.

BEET ROOT: Beet juice is considered as one of the best vegetable juices. It's a rich source of natural sugar. It contains sodium, potassium, phosphorus, calcium, sulfur, chlorine, iodine, iron, copper, vitamins B1, B2, C, and P. This juice has easily digest-ible carbohydrates, but the calorie content is low. Red beet juice is high in iron, and it re-generates and reactivates the red blood cells, supplies fresh oxygen to the body. It is useful in treating anemia. Beet juice is a good treatment for jaundice, hepatitis, and nausea. Add a teaspoonful of lime juice to increase its medicinal value. Fresh beet juice mixed with a tablespoonful of honey taken every morning before breakfast will help heal a gastric ulcer. Beet juice combined with carrot and cucumber juices is one of the finest cleansers for kidneys and gall bladder. Beet juice acts as a natural tonic for dry tired skin. Blend one beet root with half cup of cabbage, set some juice aside. Add a tablespoon of mayonnaise or olive oil with beets and cabbage, apply to your face. Mix beet juice with water and freeze the ice cubes. Use a cube to cleanse and tone your skin morning and evening.

BILIN / BILIMBI: In the Philippines, the leaves are applied as a paste or as a poultice on insect bites, and to reduce the swelling and pain of rheumatism. An infusion made from the leaves or blossoms treats coughs and is taken as a tonic after childbirth. A decoction made from the leaves treats hemorrhoids. Bilimbi juice treats hypertension and regulates blood sugar. Consumed in excess, bilimbi's oxalate acid can lead to a risk of kidney stones and even kidney failure.

BITTER MELON / CARAILLI: The bitter juice can be helpful to diabetics. A tea of the leaves and blossoms provides natural relief for high blood pressure. In 1999, a Bangladeshi clinical trial was conducted to examine the effect of bitter melon on patients with Type 2 Diabetes. The researchers recorded the patients' sugar levels both without food in-take for 12-24 hours and after taking 75g of glucose. They then administered a bitter melon pulp suspension to diabetic patients and 86 out of the 100 responded to the vegetable intake, showing a significant 15% reduction in fasting and post-meal serum glucose levels.

Eating it will even improve your sleep. The high beta-carotene and other proper-ties in bitter melon make it one of the finest vegetable-fruits that help alleviate eye problems and improves eyesight. Bitter melon roots are used to treat eye-related diseases. Bitter melon juice may be beneficial in the treatment of a hangover for its alcohol detoxification properties. It also helps cleanse, repair, and nourish liver problems due to alcohol consumption. This bitter juice can also help to build your immune system and increase your body's resistance to infection. Take two ounces of fresh bitter melon juice and mix with a cup of honey diluted in water. Drink daily to improve asthma, bronchitis, and laryngitis.

Regular consumption of this bitter juice has also been known to improve psoriasis condition and other fungal infections like ringworm and athlete's feet. The fruit is a coolant, aids digestion, a laxative, increases appetite, cures gas pains, blood diseases, anemia, urinary discharges, asthma, ulcers, and bronchitis.

BLACK PEPPER: Black pepper is considered good for the digestive system and fights bacteria. Pepper soothes nausea and increases body temperature to fight fevers and chills. Its spicy hot flavor makes the nose and throat produce (water) a lubricating secretion and assists any-one who needs to cough up and clear their lungs. Pepper was also used as an ointment to relieve skin afflictions and hives. However, coarse ground black pepper does irritate the intestines.

BLACK WALNUT: Black walnut hulls contain juglone, a chemical that is an-tibacterial, antiviral, anti-parasitic, and a fungicide. Black walnut is reportedly used to treat ringworm, yeast, and candida infections. A skin wash is made from 3 black walnut husks boiled in a gallon of water. Powdered hulls can be mixed with talc for per-sonal use. Try only a tiny bit irst to see if there is any allergic reaction. Walnut oil is rich with nutrients and antioxidants and may improve your memory and concentration.

BOK CHOY / PAK CHOY: Bok choy is high in vitamins A, B6 and C, beta-carotene, calcium, and dietary iber and contains potassium, and iron. It is low in fat, calories, and carbohydrates. The high amount of beta-carotene in bok choy helps to reduce the risk of certain cancers and reduce the risk of cataracts. Bok choy is an excellent source of folic acid. Bok choy is used by traditional Chinese medi-cine to quench thirst, relieve constipation, promote digestive health, and treat diabetes.

BOTTLE GOURD / LAUKI: Bottle gourd is considered a very healthy vegetable. Ayurvedic medicine utilizes every part of the plant. The practitioners want to use the wild, bitter variety. Eating this sweet cultivated vegetable reduces bad cholesterol levels. The juice stabilizes blood sugar level and maintains blood pressure for diabetics. Morning is the best time to drink a small glass on an empty stomach, which will help lower cholesterol levels and maintain a healthy heart. Make the juice fresh every day; it does not store well. If it tastes bitter, dump it. It's advised to drink the juice alone, not blended with other vegetables. You can add ginger or fresh mint for lavor. The juice treats urinary tract infections and helps regulate the bowels. An oil pressed from the leaves and roots is used externally to treat skin diseases. A leaf paste soothes insect bites.

BRAZIL CHERRY / GRUMICHAMA: In Brazil, an infusion of 10 g of leaves or bark in 1½ cups of water treats rheumatism.

BRAZIL NUTS: Brazil nuts health benefits are known to the world, due to the fact that the nuts are the richest source of the mineral selenium. An ounce of Brazil nuts contains 544 mcg. Selenium found in Brazil nuts is also effective in reduc-ing the risk of breast cancer, prostate cancer, and other cancers. It also helps to maintain the proper functioning of the thyroid and is good for building the immune system.

BREADFRUIT: Breadfruit roots, leaves, and its latex sap are used in natural-homeopathic medicine. Breadfruit root and leaves are chewed for arthritis, asthma, back pain, diabetes, fever, gout, high blood pressure, liver disease, and toothaches. The latex sap is eaten for diarrhea and stomach pain. It's reported that research has shown breadfruit leaf extract inhibits the enzyme that assists the development of prostate cancer. These leaves also kill cancer-causing cells.

BREADNUT/CHATAIGNE: A tea of yellowing leaves to lower blood pressure, treat respiratory problems and to fight diabetes. A decoction of the bark used to heal wounds and treat dysentery. Crushed leaves are used for thrush and juice from stems of leaves for ear infections

BROCCOLI: Researchers discovered a compound in broccoli that fights the bacteria that cause peptic ulcers much better than modern antibiotic drugs. This same broccoli compound protects against stomach cancer, currently the second most common type of cancer. Broccoli has been found to lower the risk of prostate and lung cancers.

BUTTER BEANS: Butter beans contain both soluble and insoluble fiber. Soluble helps regulate blood sugar levels and lowers cholesterol. Insoluble fiber prevents constipation, irritable bowel syndrome, and digestive disorders

BUTTERNUT SQUASH: The seed is eaten fresh or roasted for the relief of abdominal cramps and distension due to intestinal parasites. About 800 peeled seeds are reported to make a safe and effective treatment for tapeworm. They're ground into a lour, then made into an emulsion with water and eaten. It is then necessary to take a purge to expel the tapeworms or other parasites from the body. The boiled root of butternut squash supposedly increases mothers' breast milk.

C **CALABASH:** The fruits are poisonous. Properly prepared, the fruit treats colds, diarrhea, pneumonia, and intestinal irregularities. It is also used for relief from menstrual pains and to ease childbirth and procure an abortion. The leaves can be used in the treat-ment of dysentery, colds, lung diseases, toothache, wounds, and headaches. The bark is used to clean wounds. A tea made from this tree's leaves is used to treat high blood pressure.

CABBAGE: Since cabbage is high in carotene content, regular eating reduces the risk of some cancers. Sulfur in cabbage reduces the growth of tumors, removes toxins, and strengthens our immune systems. Cabbage is rich in vitamins and minerals and reduces the 'bad' cholesterol that hardens arteries Eating cabbage irst will prepare the stomach for a heavy meal and drinks. Crush a raw cabbage leaf in a mortar and use it as a poultice on a cut. To fight acne, drink a cupful of the water cabbage is boiled in, or crush a raw leaf and use it as a poultice on oily facial areas for about 20 minutes. A boiled cabbage leaf applied still very hot to the abdomen treats a stomach ache.

CAIMATE / STAR APPLE: Caimate tree bark is rich in tannin used to tan leather and is thought to fight cancer. A drink made from the boiled bark is used as a stimulant tonic. Cooked caimate is used to reduce a fever. The leaves are grated and applied to cuts to reduce infection. The leaves are boiled and the resulting liquid drank to combat hypoglycemia.

CALAMUS ROOT - SWEET FLAG: The calamus root is used in traditional Ayurve-dic-homeopathic medicine to revitalize the brain, improve the memory, and nervous system. This root simultaneously boosts energy, and you feel focused and alert. It's also a remedy for digestive disorders like gas, nausea, and loss of appetite. Chewing the root will kill the pain of a toothache and is supposed to break the addiction to tobacco. For two millenniums, calamus root has been used as an aphrodisiac throughout the Middle East, India, China, and S.E. Asia.

CANDLENUT: Almost all parts of the candlenut tree are useful for traditional medicine, leaves, nuts bark, roots, sap, and lowers. The light-colored nut oil is practically odorless, easily absorbed, and great as a massage oil. It's used in soaps, candles, and lotions. People use it to stimulate hair growth. Internally, it's a laxative, treats insomnia,, and improves digestion. Externally, candlenut oil helps muscle-joint aches, heals dry skin, soothes insect bites.

CANISTEL: In Mexico, a decoction of the astringent bark is taken to lower a fever and applied on skin eruptions in Cuba. A preparation of the seeds is used as a remedy for ulcers.

CANTALOUPE: Cantaloupe is rich in antioxidants that can help prevent can-cer and heart diseases. Potassium in cantaloupe helps excrete sodium, reducing high blood pressure, especially in those with salt-sensitive hypertension. Cantaloupe juice with po-tassium can reduce muscular cramps. A special compound in this muskmelon relieves the nerves, calms anxieties, and helps against insomnia. When going through a stressful period, drink this melon's juice regularly. The potassium content helps to rebalance and normalize the heartbeat. This in turn sends oxygen to the brain and regulates the body's water balance. The natural nutrients and minerals found in cantaloupe juice provide a unique combination to help the body recover from nicotine withdrawal when trying to quit smoking. Smoking also quickly depletes vitamin A, but cantaloupe juice can help replace it with its beta-carotene.

CAPE GOOSEBERRY: In Ayurvedic-homeopathic traditional medicine, cape gooseberry treats ailments from malaria to toothaches and rheumatism. It is considered a diuretic and relaxant. Infusions treat indigestion, fever, and jaundice. Poultices of steamed leaves treat inflammations and sore muscles.

CARDAMOM: A medicinal, perhaps aphrodisiac, drink can be made by steeping seeds in hot water, or you can just suck on a cardamom seed. This is good for the throat, respi-ratory tract, teeth, breath, and stomach. Green cardamom is used to treat tooth and gum infec-tions (the same as cloves), treat throat problems, lung congestion, and also digestive disorders.

CARROTS: Eating carrots may protect against strokes, and heart disease. Carrots are rich in beta-carotene, which the human body converts to vitamin A. Beta-carotene fights some forms of cancer, especially lung cancer. Carrot is rich in alkaline elements that cleanse and rejuvenate the blood. Carrots also supply the human body with vitamin C, calcium, phosphorus, sodium, sulfur, chlorine, and contain traces of iodine. The mineral contents in carrots lie very close to the skin and should not be peeled or scrapped off. A good washing is always necessary. Carrot juice strengthens the eyes and is a great treatment when rubbed on dry skin. Carrot juice drank daily may cure colic, colitis, and peptic ulcer. Raw carrots are good for fertility. Eaten after a meal, a raw carrot destroys germs in the mouth, prevents bleeding of the gums and tooth decay. Carrots speed up digestion. Carrots eaten regularly prevent gastric ulcer and other digestive disorders. Carrot soup is a natural remedy for diarrhea. Carrots fight all body parasites including intestinal worms.

CASHEWS: Cashews are rich in two categories of antioxidants that may help re-duce inflammation, strengthen the immune system, and protect against diseases. Cashew oil is rich in vitamin E and good for treating skin problems, removing scars and corns. The leaves, bark and cashew apple are antiseptic, stop diarrhea, reduce fever, lower blood sugar, blood pressure and body temperature. Cashew tree products have long been alleged to be effective anti-inflammatory agents, counter high blood sugar and prevent insulin resistance among diabetics. Cashew seed extract stimulates blood sugar absorption by muscle cells.

CASSAVA: Cassava leaves and roots have been a folk remedy for tumors and cancers, which may be due to the B17 content, also known as laetrile. Vitamin B17 is also found in some seeds like apricots, peaches, and apples. B17 stimulates hemoglobin red blood cell count. The bitter cassava variety is used to treat diarrhea and malaria. The leaves are used to treat hypertension, headache, and pain. Cubans commonly use cassava to treat irritable bowel syndrome, the paste eaten in excess during treatment.

CAULIFLOWER: Cauliflower has compounds that activate enzymes that may dis-able and eliminate cancer-causing agents. Cauliflower also helps the liver neutralize poisons.

CAYENNE PEPPER: Cayenne is a food spice, but it also has been used as a miracle herb for the digestive and circulatory system. Peppers like cayenne have reputation for causing stomach problems including ulcers. That is unfounded because hot peppers actually may help prevent problems by killing bacteria you may have eaten, and stimulates the stomach to secrete a natural protective coating that prevents ulcers. Cayenne also helps the body secrete hydrochloric acid necessary for digestion. Once you have good digestion all the other organs of the body get the proper nutrients. Consumed as a tea or in a capsule, cayenne pepper lowers the effects of asthma, and clears congestion. Rubbed on the skin, mixed with a skin cream as a paste, will reduce the pain of arthritis or stop an itch. Studies demonstrate if you consume a lot of 'pepper' your chances of having a heart attack or stroke are lowered. It is a good source of vitamins A and C, has the complete B complexes, and is very rich in calcium and potassium. Gargling with cayenne pepper tea will help relieve a sore throat.

CAYENNE OINTMENT: Good for aches and sore muscles Ingredients: 1 cayenne pepper chopped very fine, ½ cup vegetable oil, 2 TBS natural beeswax grated. (Get this from a beekeeper or a hair dresser who waxes). Heat oil on medium heat in a small sauce pan. Add minced pepper. Cook without boiling for 5 minutes. Remove from heat and pour through a strainer to remove pepper and seeds. Add beeswax and reheat until wax melts. Pour into suitable container or small jar and cool. Try on your next muscle ache.

CAYENNE JUMP UP JUICE- more spike than coffee without caffeine Ingredients: 2 TBS apple cider vinegar, 1/8 teaspoon (a pinch) of baking soda, ¼ TS cayenne pepper powder, 1 glass of hot water First, turn your head away as you add the cayenne powder to the vin-egar and water so you don't inhale as the vinegar (acid) bubbles with the baking soda (alkaline). Sip slowly.

CELERY: Celery is a heart-friendly vegetable. It contains compounds that can relax the artery muscles, which regulate blood pressure. It also causes blood ves-sels to constrict. Juice from 4 stalks can cause blood pressure to reduce 15%.

CEYLON GOOSEBERRY / TROPICAL APRICOT: The Ceylon gooseberry is packed with antioxidants (*that fight cancer and heart disease),* polyphenols, *(help digestion and metabolism, plus fights diabetes)*, and anthocyanins *(that fight viruses, cancers, and inflammation)*. In a 2007 study published in the *Journal of Agricultural and Food Chemistry,* Ceylon gooseberries have 10 anthocyanins and 26 carotenoids. In anthocyanin-rich juice, like gooseberry, scientists found that the compounds reduced the risk of heart attacks. The fruits' skin must be utilized because all anthocyanins had higher concentrations in the skin than in the pulp.

CHAYOTE / CHRISTOPHENE: A tea made from chayote leaves is a bush treatment for hypertension and is reported to dissolve kidney stones.

CHESTNUTS: Eating chestnuts regularly helps improve digestion, manage diabetes, nourishes the spleen, kidneys, and stomach, lowers blood pressure, improves circulation, and soothes inflammation**.** A poultice made from the leaves reduces inflammation.

CHICK PEAS / GARBANZO BEANS / CHANNA: Unlike hard to digest meat, chick peas are low in calories and virtually fat-free. However, chick peas contain 'purine' and individuals with kidney problems or gout may want to avoid these beans. Research has found that a seven day diet (one meal a day) of chick peas cooked with onions and turmeric powder will drastically reduce your overall cholesterol.

CHOCOLATE: In moderation, chocolate actually reduces blood pressure. Men over fifty who eat a small bit of dark unsweetened chocolate every day live longer! Chocolate contains many health benefits from flavonoids, which act as antioxidants that protect the body from aging caused by free radicals, which can cause damage that leads to heart disease. Dark chocolate contains a large number of antioxidants. Studies show consuming a small bar of dark chocolate everyday can reduce blood pressure in individuals with high blood pressure. Dark chocolate has also been shown to reduce LDL cholesterol (bad cholesterol) up to 10 percent. Chocolate tastes good and stimulates endorphin production, which gives a feeling of pleasure. It also contains caffeine and other substances which are stimulants.

CILANTRO / CORIANDER: See coriander.

CINNAMON: Medicinally cinnamon is considered a mild tranquilizer and relieves nausea and gas. It is an antibiotic that fights some fungi and bacteria better than over the counter medication. One-half teaspoon of cinnamon each day may reduce blood sugar and cholesterol in Type II diabetes sufferers. A tea of cinnamon and ginger fights a cold or flu, and indigestion.

CLOVES: Cloves have many health benefits, including keeping blood sugar regulated and fighting bacteria. Clove oil is an important natural antibacterial drug. It's used in dentistry, pharmaceuticals, and aromatherapy. It's also used as an analgesic; clove oil is recommended for inhalation to treat sore throat, colds, coughs, and any breathing problems. One test-tube study found that clove extract helped stop the growth of tumors and promoted cell death in cancer cells. Clove oil boosts concentration. In addition, it revitalizes, energizes, and serves as an aphrodisiac. It's also a natural food preservative due to its anti-bacterial and antifungal effects. Studies show that the beneficial compounds in cloves could help promote liver health. The compound eugenol may be especially beneficial for the liver. Another study researched the effects of clove extract 'nigericin' and was found to increase the uptake of sugar from the blood into cells, increase the secretion of insulin, and improve the production of insulin both on human muscle cells and in mice with diabetes.

COCONUT: Since coconut water is sterile, it has been used intravenously as a substitute for glucose. Coconut water kills bacteria that cause ulcers, throat infections, urinary tract infections, gum disease, cavities, pneumonia, and other diseases. Kills fungi and yeasts that cause ringworm, athlete's foot, thrush, diaper rash, and other infections. Expels or kills tapeworms, lice, and other parasites. Helps reduce health risks associated with diabetes. Improves calcium and magnesium absorption and supports the development of strong bones and teeth. Improves digestion and bowel function. Relieves pain and irritation caused by hemorrhoids. Is heart healthy; improves cholesterol ratio reducing risk of heart disease. Helps prevent liver disease, protects against kidney disease, bladder infections, and dissolves kidney stones.

COCO PLUM / ICACOS PLUM / FAT PORK: A tea made from coco plum leaves is reported to help control type II diabetes. A combination of fruits, bark and leaves boiled in water will produce a tea that will fight severe diarrhea. A tea made only from the bark may help kidney ailments. The teas can be used externally as washes to treat skin problems.

COFFEE: Coffee is one of the most heavily researched commodities in the world today and the general conclusion is coffee drinking is perfectly safe. There is no conclusive evidence to suggest a moderate amount of coffee is bad for you. Research has shown caffeine consumption may have a small effect on blood pressure, however, scientists do not consider coffee drinking to be an important risk factor for hypertension. The key risk factors are known to be a low potassium intake, high sodium (salt) intake, slack lifestyle, and obesity. However, if you drink three to four cups - at a sitting - several times a day, coffee can slow pulse rate, raise blood pressure, contract blood vessels under the skin, and dilate blood vessels of the kidneys, muscles, skin, and heart. Finally, caffeine in large a mounts makes the heart contract harder while it's pumping. Caffeine is a stimulant. It increases alertness and concentration, intensifies muscle responses, quickens heartbeat, and elevates mood. Its effects derive from the fact that its molecular structure is similar to that of adenosine, a natural chemical by-product of normal cell activity. Adenosine is a regular chemical that keeps nerve cell activity within safe limits. When caffeine molecules hook up to sites in the brain where adenosine molecules normally dock, nerve cells continue to fire indiscriminately, producing a jingly feeling sometimes associated with drinking coffee, tea, and other caffeine products. A remarkable remedy for rheumatism and gout is said to be; a pint of hot, strong, black coffee, seasoned with a teaspoonful of pure black pepper, thoroughly mixed before drinking, and the preparation taken just before retiring

CORIANDER: Recent studies have supported its use as a stomach soother for both adults and colicky babies. Coriander contains an antioxidant that helps prevent animal fats from turning rancid, killing meat-spoiling bacteria and fungi. These same substances in cilantro also prevent infection in wounds. Coriander has been shown to improve tummy troubles of all kinds, from indigestion to flatulence to diarrhea. Weak coriander tea may be given to children under age two for colic. It's safe for infants and may relieve their pain and help you get some much-needed sleep. Cilantro and coriander contain substances that kill certan bacteria and fungi, thereby preventing infections from developing in wounds. Sprinkle some coriander seed on minor cuts and scrapes after thoroughly washing the injured area with soap and water. Intriguing new studies suggest that coriander has anti-inflammatory effects. Since the pain of arthritis is caused by inflammation coriander oil may help

CORN: Eating corn is good for you. Corn is high in fiber, niacin, folate and some vitamin A. Folate has been found to prevent some birth defects and reduce the risk of heart disease and stroke. Fiber keeps the intestinal track running smoothly

COTTON: Cotton leaves are used in a tea to treat respiratory problems as asthma, coughs, and throat infection. The same tea can be used as a skin wash to treat acne, insect bites, and inflammation. The leaves are used as a natural bandage to heal wounds.

CREPE GINGER: Crepe ginger's most common therapeutic use is for the prevention and treatment of diabetes and to maintain general good health. In contrast with most medicinal plants, crepe ginger tastes good and the medicinal benefits can be enjoyed raw, as a salad. Preparation is simple for medicinal purposes. All parts of the plants are used; leaves can stimulate appetite, treat skin issues, and as a bath for patients with a high fever. Juice pressed from leaves is used internally for eye and ear infections. The roots are edible and treat intestinal parasites, constipation, and have anti-oxidant properties. Juice from the root is a headache remedy. As a blood tonic, studies have shown that a diet with crepe ginger regularly reduces LDL-cholesterol and increases plasma insulin, tissue glycogen, HDL-cholesterol, and serum protein. Powdered crepe ginger leaves are available in some areas.

CUCUMBERS: Cucumbers are mild laxatives. Cucumbers are a good source of B vitamins and carbohydrates and can provide an energy burst that can last for hours. To avoid a hangover eat a few cucumber slices before going to bed, wake up refreshed, and headache free. Cucumbers contain enough sugar, B vitamins, and electrolytes to replenish essential nutrients the body lost, keeping everything in equilibrium. A fast and easy way to remove cellulite, rub a slice or two of cucumbers along your problem area for a few minutes, the phytochemicals in the cucumber cause the collagen in your skin to tighten, firming up the outer layer and reducing the visibility of cellulite. Works great on wrinkles

CULANTRO: Culantro and coriander are regarded as a drug. It is thought to be an aphrodisiac. If a large quantity is eaten, it acts as a narcotic. The seeds are a common remedy for gas pains and are chewed to ease the pains of birth labor.

CUMIN: Cumin stimulates your appetite while helping the stom-ach to relieve gas. It will reduce nausea during pregnancy. Cumin could be called the 'breast spice' because it supposedly increases both lactation and size.

CURRY TREE: Since ancient times, curry leaves have been used in Ayurvedic-ho-meopathic medicines. They are used to assist diabetics to reduce high blood sugar levels and are believed to be anti-inflammatory and antibacterial. A tonic of boiled leaves is a stimulant for the digestive system and also for healthy hair and skin. Curry leaf oil may cure acne pimples, athlete's foot, ringworm, itches, boils, and help to heal burns. The oil strengthens bones and fight osteoporosis, calcium deficiencies, and after chemo-therapy. Eating dishes with curry leaves may reduce high cholesterol. Significantly, studies show curry leaves contain substances that may protect against nervous system conditions like Alzheimer's disease. Studies with curry leaves grown in Malaysia found anticancer effects against aggressive breast, colon, and cervical cancers.

CUSTARD APPLE: Custard apples are a well-balanced food having protein, fiber, minerals, vitamins, energy, and little fat. They are an excellent source of vitamin C, a good source of dietary fiber, vitamin B6, magnesium, potassium, with some B2, and complex carbohydrate. 100 grams of custard apple flesh has 100 calories. A small custard apple will weigh around 250 grams and will provide your entire daily intake of vitamin C.

He who has so little knowledge of human nature as to seek happiness by chang-ing anything but his own disposition will waste his life in fruitless efforts.
~ Samuel Johnson

D **DASHEEN / KALO / TARO:** The leaf juice of dasheen is considered a coagulant, stimulant, and increases blood circulation. It's a useful treatment for internal hemorrhages, earaches and discharges, and swollen lymph nodes. The root's juice is a pain killer and a laxative. Eating dasheen stimulates the appetite and helps cure constant thirst.

DILL: Dill weed contains carvone, which has a calming effect and aids digestion by relieving intestinal gas. High in calci1 TBS of seeds equals a quarter cup of milk. Dill is believed to increase lactation in nursing mothers and is used in a weak tea for babies to ease colic, encourage sleep, and get rid of hiccups. Crushed dill seeds, mixed with water, is used to strengthen fingernails. Chewed dill seeds can cure bad breath.

DRAGONFRUIT / PITHAYA: Dragon fruit stems and flowers are used to treat diabe-tes, as a diuretic, and promote healing wounds. The fruit lowers cholesterol, boots immunity, and helps digestion. In Taiwan, it's used by diabetics as a substitute for rice to add fiber to their diet.

DURIAN: The leaves, roots and fruits are used in homeopathic medicine to treat fever, jaundice, and skin eruptions. Durian quickly produces energy. Durian builds muscle tissue and improves blood pressure. It may slow or stop the spread of cancer cells, lower cholesterol, and has antibacterial properties. The downside of durian is to never combine it with alcohol. Research demonstrates that durian prevents alcohol from dissipating and increases alcohol levels in the blood causing nausea and vomiting.

E **EDDOES / LITTLE or HAIRY TARO:** Eddoes are very high in starch, and are a good source of dietary fiber. Oxalic acid may be present in the corm and especially in the leaf, and these foods should be eaten with milk or other foods rich in calcium to remove the oxalate in the digestive tract. Absorbing a large quantity of the oxalate ion into the blood stream poses health risks, especially for people with kidney disorders, gout, or rheumatoid arthritis. Calcium in the body reacts with the oxalate to form calcium oxalate, which is highly insoluble and is suspected to cause kidney stones. Food allergies most frequently afflict children, with cow's milk being the most common allergenic food for infants, followed by eggs, peanuts, tree nuts, and soybeans. Poi, eddoes mashed to liquid, is considered a substitute for soy milk in infants allergic to both soy and cow's milk. In addition, the easy digestibility and other character-istics of poi might make it a nutritional supplement for weight gain in patients with conditions

EGGPLANT: Eating eggplant is good for you. It can reduce high cholesterol. One cup of cooked eggplant has 27 calories, 1 gram of protein, 6 grams of carbohydrates (two grams of fiber if eaten with the skin, phosphorus, potassium, and folate. An ingredient in common egg-plant has been shown to cure cancer. The eggplant extract is a phytochemical called solasodine glycoside, or BEC5. The types of cancer treated by eggplant are both invasive and non-invasive nonmelanoma skin cancers. In every case the cancers went into remission and did not return. Australians have been curing their skin cancers using these phytochemicals for decades.BEC5 acts by killing cancer cells without harming any other healthy cells in the human body. BEC5 can also be used to treat actinic keratose, the precursor to cancer, as well as age or sunspots on the skin. Skin cancer is now reported to be the most common illness in men over the age of 50.

ELEPHANT APPLE / CHALTA: The fruit is much used in India as a liver and cardiac tonic, and, when unripe, as an astringent means of halting diarrhea and dysentery and effective treatment for hiccough, sore throat and diseases of the gums. The pulp is used in poultices on bites and stings of venomous insects, as is the powdered rind. Juice of young leaves is mixed with milk and sugar candy and given as a remedy for biliousness and intestinal troubles of children. The powdered gum, mixed with honey, is given to overcome dysentery and diarrhea in children.

Oil derived from the crushed leaves is applied to relieve an itch, and the leaf decoction is given to children as an aid to digestion. Leaves, bark, roots, and fruit pulp are all used against snakebite.

F **FENNEL:** Fennel is used as an eyewash and was thought to increase breast milk. Fennel is believed to curb eating and great for dieters. It will reduce gas and stomach cramps. In medieval times this herb was hung over doors to ward off evil spirits. It is reputed to stimulate strength and courage, and increase the eater's life span. Fennel is in mouth fresheners, toothpastes, desserts, antacids. It ights anemia, indigestion, latulence, constipation, colic, diarrhea, respiratory disorders, menstrual disorders, and eye care. With carrot juice, fennel is a very good treatment for night blindness, or to strengthen the optic nerve. Add beet juice to make a remedy for anemia resulting from menstruation. Fennel juice assists convalescence. The French use it for migraine and dizziness. Boiling fennel leaves and inhaling the steam can relieve asthma and bronchitis. Fennel is used after cancer radiation and chemotherapy treatments to help rebuild the digestive system. Ground fennel seed tea is believed to be good for snake bites, insect bites, or food poisoning. It increases the low o f urine.

FENUGREEK: For thousands of years, fenugreek has been used in alternative Ayurvedic-homeopathic medicine to treat skin conditions and many other diseases. The plant treats bronchitis, fevers, sore throats, wounds, swollen glands, skin irritations, diabetes, ulcers, and some cancers, notably colon cancer. Fenugreek has been used to increase breast milk. One 14-day study in 77 new mothers found that consuming fenugreek herbal tea increased breast milk production, which helped the babies gain more weight. These seeds are also considered an aphrodisiac; men use fenugreek supplements to boost testosterone.

Fenugreek appears to slow the absorption of sugars in the stomach and stimulate insulin production. Both of these effects lower blood sugar in people with diabetes. The seeds have been used as an oral insulin substitute, and seed extracts have been reported to lower blood glucose levels in laboratory animals. In one study of type 1 diabetics, researchers added 50 grams of fenu-greek seed powder to the participants' lunches and dinners for 10 days. A 54% improvement was found in 24-hour urinary blood sugar, and also reduced in the total and LDL cholesterol levels.

As a poultice, a paste of fenugreek seeds and-or leaves, are wrapped in cloth, warmed, and applied directly to the skin to treat local pain and swelling, aching muscles, gout, wounds, and eczema. The downside of using fenugreek is that your perspiration and urine may smell like maple syrup. The same odor may happen with breast milk and/or a breastfed baby may begin to smell like maple syrup. If you are allergic to pea-nuts or chickpeas, fenugreek is in the same family and may cause an allergic reaction.

FINGER ROOT: Finger root has shown to possess anti-allergic, antibacterial, anticancer, anti-inflammatory, antioxidant, and anti-ulcer activities and heals wounds. Several research studies have shown finger root to have aphrodisiac qualities, increasing the quantity and quality of sperm. Finger root is an ingredient in a popular Indonesian herbal medicine known as Jamu, used by Indonesian women post childbirth to strengthen the uterus.

G **GALANGAL:** It can add flavor, antioxidants, and anti-inflammatory compounds to your dishes and provide many health benefits. These include protecting you from infections and potentially from certain forms of cancer. Test-tube studies suggest that the active compound in galangal root, known as galangin, may kill cancer cells or prevent them from spreading Emerging evidence suggests galangal root may boost male fertility

GARLIC: Garlic is valued as a spice, but also for its medicinal benefits. No one knows how much garlic must be ingested to improve health, but experts feel the best results come from using raw garlic. Louis Pasteur discovered garlic kills bacteria. It was used as an antiseptic in World War II when sulfa drugs were scarce. Scientists have found that when raw garlic is cut or crushed it creates a compound, which kills at least 23 types of bacteria.

Yet when garlic is heated it forms a different compound that reduces blood pressure and cholesterol. Garlic contains vitamins A, B, and C and stimulates the immune system. It may reduce the risk of stomach cancer and be a treatment for AIDS. Sixty kinds of fungi are also killed by garlic, including athlete's foot and vaginitis.

GIANT GRANADILLA / BARBADINE: The fruit is valued in the tropics an appetite enhancer. In Brazil, the flesh is used as a tranquilizer to relieve nervous headache, asthma, diarrhea, dysentery, neurasthenia, and insomnia. The seeds are reported to have a narcotic effect in large doses. The leaves boiled in water are used for bathing skin eruptions. Poultices made from the leaves are applied to liver ailments. .

GINGER: Ginger is an excellent natural remedy for nausea, motion sickness, morn-ing sickness, and general stomach upset due to its carminative effect that helps break up and expel intestinal gas. Ginger tea has been recommended to alleviate nausea in chemotherapy patients primarily because its natural properties do not interact in a negative way with other medications. It is a safe remedy for morning sickness, since it will not harm the fetus. Some studies show ginger may also help prevent certain forms of cancer. Ginger is used medicinally because it stimulates and strengthens the stomach, breaks colds and coughs, diarrhea, rheumatism, and especially for nausea. The root can be boiled and pounded into a paste applied to the forehead to ease headaches or made into a poultice to sooth arthritis. To make ginger tea, slice some ginger root, put it in a tea ball and place in a teapot. In spite of it being a natural remedy, it's important that any medicinal use of ginger be discussed with a physician, as it must be taken in moderation to avoid gastric irritation.

GOTU KOLA: Gotu kola has been used for many centuries as a treatment for respi-ratory ailments and a variety of other conditions including fatigue, arthritis, memory, stom-ach problems, asthma, and fever. 10 leaves a day makes the doctor away. Burmese monks expound that eating gotu kola daily can cure 99 diseases; it's amazing how this leafy green plant can promote health andfitness and slows your aging. Its antioxidant activity is recommended to improve skin conditions, heal wounds, reduces scars, wrinkles, and stretch marks. Prevents hair loss and helps keep your skin glowing. Organic Indian gotu kola supplements are usually of good quality. The rare Himalayan wild-growing varieties are supposedly the best.

Gotu Kola herb has been studied by the scientific community. The majority of the claims of its healing and therapeutic properties, including extending a good, quality life, have been documented and validated. You just need to use it raw. For external use acne or boils, directly apply gotu kola juice in a poultice. For coughs and asthma, eat it as a vegetable. For difficulty in passing urine, drink gotu kola juice. The leaves are the most valuable part of the plant. It has remarkable bene its. It has the ability to treat depression, boost blood circulation, and protect the heart. Gotu kola enhances memory and nerve functions, which gives it potential in treating Alzheimer's disease. The positive effect on brain function also makes it an antidepressant and treats anxiety and stress. It may treat the insomnia that accompanies these ailments. The anti-inflammatory properties of gotu kola may ease the pain of arthritis. In some cases, it can cause headache, nausea, and dizziness. Starting with a low dose and gradually working up to a full dose can help reduce your risk of side effects. It's recommended to take gotu kola for two to six weeks at a time. Be sure to take a two-week break before resuming use. Don't use gotu kola if you are pregnant, breastfeeding, have hepatitis or other liver diseases, skin cancer, diabetes, high cholesterol, have surgery within two weeks, or are under 18 years of age. Grow your own gotu kola because it is known to absorb heavy metals or tox-ins in the soil or water in which it was grown. This poses a health risk given the lack of safety testing, be particularly wary of imported traditional Chinese remedies. Make gotu kola preparations in your kitchen. To make gotu kola tea, dry the leaves in the sun, crush them into a powder and store the powder in airtight jars. To make infused oil, pick the the leaves fresh, wash thoroughly, dry in the sun and add to virgin coconut or olive oil in a glass bottle. Place in the sun until oil darkens. The more leaves used the better, but no more than one handful dried. Gotu kola oil is good to retain your hair with regular scalp massages. Also, make a paste by mixing dried gotu kola leaves with raw aloe vera gel and rubbing it onto your hair. These natural remedies, with drinking the tea, have restored hair. The entire plant is harvested at maturity and used for medicinal purposes, salads, hair care oils, and in many more uses.

GOVERNOR PLUM – CERISE PLUM – *Flacourtia indica:* The decoction from the bark is used to relieve arthritis pains. Most parts of the plant are used for cough, pneumonia, and bacterial throat infection. It has also been used for diarrhea. In 2010, The American Eurasian

Journal of Scientific Research published a study that demonstrated governor plum leaves contain potent antioxidants, which may slow down the signs of aging and reduce stress. A juice pounded from the leaves treats asthma and bronchitis. Governor plum fruit stimulates the appetite, fights jaundice and enlarged spleens. The tiny seeds are ground into a paste and used to ease rheumatic pain. A decoction of the root in combination with the leaf sap treats malaria and relieves body aches. The bark is used for rheumatic pain and as a gargle for hoarseness. In 2011, *The International Journal of Drug Development and Research* published a study demonstrating compounds in the leaves contain significant antimicrobial and antibacterial qualities useful to treat infectious diseases and inflammation. To treat diabetes boil governor plum with water and strain. NOTE: Eating governor plum 72 hours before a carcinoid tumor exam may result in a false positive. This is because of the fruit's high level of serotonin. Doctors check for cancerous tumors by measuring serotonin levels in urine. If higher than normal, doctors give the prognosis of cancer.

GOVERNOR PLUM / MARTINIQUE PLUM / ***Flacourtia inermis*****:** The fruits have great nutritional value, but the governor is also known for Ayurvedic-homeopathic treat-ments. The fruits treat jaundice and spleen problems. The juice and a leaf decoction are used to treat diarrhea, dysentery, and piles. The fruits, bark, leaves, and roots are used in concoc-tions for certain health conditions, especially for arthritis, some bacterial infections, nausea, sore throats, and to relieve chest congestion. The leaves are dried and powdered to relieve bronchitis and asthma. The bark and leaves are used to lessen the pain of toothaches and fight bleeding gums. A poultice of the roots treats skin sores and boils. In Malaysia, juice squeezed from the roots is used to fight herpes. In South India, they grind the seeds to a powder and combine with turmeric root and rub women after childbirth to lessen the pain.

GRAPEFRUIT: Grapefruit lowers some cholesterol, helps digestion, and reduces gas. It treats water retention, urinary, liver, kidney, and gall bladder problems. Grapefruit rubbed directly on the skin fights pimples and greasiness. The leaves have antibiotic properties.

GREEN BEANS: Eating these delicious beans helps lower high blood pressure, and their iber may also help prevent colon cancer. Green beans reduce the severity of diseases such as asthma, arthritis, and rheumatoid arthritis. These beans are a good source of riboflavin, which has been shown to help reduce the frequency of migraine attacks. Green beans are a very good source of iron. In comparison to red meat green beans provide iron for a lot less calories and are totally fat free. Iron is an integral component of hemoglobin, which transports oxygen from the lungs to all body cells, and is part of key enzyme systems for energy production and metabolism.

GUAVA: The guava tree leaves are also a natural astringent that are used to stop diarrhea. They can be pounded into a poultice for wounds, boils, and aches. In fact, guava leaves can be chewed to relieve a toothache. Amazon Indians use a tea of the leaves as a remedy for sore throats, nausea, and to regulate menstrual periods. Tender leaves are chewed for bleed-ing gums and bad breath. If chewed before drinking alcohol, it is said to prevent hangovers. A poultice of guava blossoms is reported to relieve sun strain, conjunctivitis, or eye injuries. The leaves are also used in a broth as a diuretic and in the treatment of cystitis.

H **HAUSA POTATO / CHINESE POTATO:** Hausa potato leaves are used in Ayurvedic-homeopathic medicine to treat dysentery, and bloody urine. The juice from the leaves is used for disorders. The roots treat hemorrhoids, nausea, throat infections, and are used to kill parasites. A paste of the root treats skin problems as burns, wounds, sores, and insect bites.

HORSERADISH: Horseradish is a gastric stimulant that helps digest rich foods. It is richer in vitamin C than an orange, and works as an antiseptic. It is valued for its medicinal properties to relieve respiratory congestion. A poultice reduce aches from arthritis or rheumatism.

HORSE PURSLANE: Horse purslane is useful in treating all types of inflammations and bladder problems. In large doses, the roots can cause miscarriages and are used to relieve obstructions of the liver, as a laxative, and to relieve asthma. The leaves are diuretic. They are used in the treatment for swelling of the feet, jaundice, and bladder irritations. The thickness of the leaves makes them good for wound-dressings or poultices. Inhaling the juice of the horse purslane leaf is believed to relieve migraines. A decoction of the herb treats intestinal parasites and rheumatism. It's also a possible antidote to alcoholic poisoning.

HOT PEPPER / CHILI: Pepper is one of the most important plants used as medicine in different countries and civilizations. It was used by the Mayas for treating asthma and by the Aztecs for toothaches. If you are a smoker, you will benefit by eating hot peppers every day. Studies have shown that red peppers contain carotenoid, a Beta-cryptoxanthin, that helps to promote healthy lung function. Capsaicin is used as an analgesic in topical ointments, nasal sprays, and dermal patches to relieve pain. Chilies are used as treatments for types of cancers, rheumatism, stiff joints, bronchitis, and chest colds with cough and headache, arthritis, and heart arrhythmias. Peppers improve digestion and can increase your metabolic rate by up to 25%, which makes you less hungry. Hot peppers can make some food safer because they reduce harmful bacteria on food. Low in calories, hot peppers, especially red, contain more vitamin A than carrots and make it easier to stick to a healthy diet since the food has more flavor. Hot chili oil is rubbed into the scalp reputedly is a Caribbean cure for baldness.

I **INDIAN GOOSEBERRY:** Indian gooseberry has been used in Ayurvedic medicine for thousands of years. Today people still use the fruit of the tree to make medicine. A plant with high tannin content, particularly its fruits, bark, and leaves, this fruit is a known traditional medicine to treat a wide range of conditions like fever, constipation, cough, and asthma. It's most commonly used to reduce total cholesterol levels, including the fatty acids called triglycerides, without affect-ing levels of the 'good cholesterol' or high-density lipoprotein (HDL. It also aids in strengthening immunity, detoxing the liver, treating diabetes, diarrhea, nausea, and cancer, but there's no good scientific evidence to support these uses. However, in test-tube studies, Indian gooseberry extract has been shown to inhibit the growth of cervical and ovarian cancer cells. Ayurvedic formulations containing Indian gooseberry have been linked to liver damage, but it's not clear if taking Indian gooseberry alone would have this effect. The Chinese use this fruit to treat throat inflammation.

INDIAN JUJUBE / DUNKS: Jujubes have been used by Chinese and Indians for many millenniums, there are many medicinal uses. This fruit has more vitamins A and C than apples. In fact, Indian jujubes have twice the vitamin C than citrus and are used as a tea to cure sore throats. These fruits are used to calm digestive and intestinal ailments. A tea of the bark will help fight diarrhea. It is believed that a diet of jujubes will cure baldness. Beware and don't eat too many raw fruits as they have a laxative quality. This fruit improves stamina and strength, stimulates the immune system, helps liver functions, is sedating, and serves as an all-purpose tonic. Raw jujubes are wrapped or poulticed over cuts and ulcers. They are used against blood circulation ailments and fevers. Combined with salt and hot peppers they ease indigestion and gas. Dried ripe fruit is a mild laxative. Jujube seeds are sedative and are taken, sometimes with whole milk, to halt nausea,

Not everybody is intelligent, but every body is intelligent. ~ Mokokoma Mokhonoana vomiting, and abdominal pains especially during pregnancy. Pulverized roasted seeds combined with cooking oil are rubbed to relieve rheumatic areas. Leaves are applied as poultices to assist liver ailments, asthma, and fevers. A decoction of the bitter, astringent bark will ight diarrhea and dysentery, and relieve gingivitis. A paste made from the boiled bark is applied on sores. Jujube root decoction will make a good purge. The powdered root is dusted on wounds. Juice of the root bark is said to alleviate gout and rheumatism. An infusion of the flowers serves as an eye lotion.

J **JACK BEAN:** In China, the entire plant is pounded and made into poultices to treat boils. Seeds are ground to make a tea for bad coughs and also to cleanse the kidneys. In Nigeria, a paste made of the seeds is used as an antibiotic and antiseptic. In Korea and Japan, an extract from the bean is used in soap to treat athlete's feet and acne. Pharmaceutical compa-nies are researching avara as a source for the anti-cancer agents: trigonelline and canavanine.

JACKFRUIT/ KATAHAR: A one-cup serving of raw jackfruit has about 150 calories, with 4 grams of fat and 2 grams of protein. It provides one-tenth the daily requirement of vitamin A and 20% of vitamin C with calcium, iron, zinc, and phosphorus. The benefits of the antioxidant seeds fights cancer, hypertension, aging, and ulcers. Powdered seeds relieves indigestion.

JALAPENO PEPPERS: Jalapeños contain a substance called capsaicin that has shown to have anti-cancer effects. However, the amount needed to achieve this effect is rela-tively high, up to eight habanero peppers per week (roughly equivalent to 24 jalapeños). Cap-saicin in jalapenos not only causes the tongue to burn, it also drives prostate cancer cells to kill themselves. These peppers are naturally high in vitamins A and C, and also bioflavinoids, helps strengthen blood vessels, and makes them more elastic to adjust to blood pressure fluctuations. All hot peppers make us sweat causing fuid loss, temporarily reducing overall blood volume.

JAVA PLUM / INDIAN BLACKBERRY: The Java plum is respected for its medicinal contributions. The fruit juice can be applied to the skin as an astringent. It will stimulate the appetite, relieve gas, prevent scurvy, and cause the body to excrete water. It is also a treatment for diarrhea, indigestion, asthma, sore throat, and skin diseases as ring-worm. Extracts from the seeds are used to lower blood pressure. The fruit is loaded with vitamin C and iron, increases hemoglobin. It is astringent to fight acne. 55 mg of potassium per 100g keeps your heart healthy. Seeds, bark, and leaves are used to treat dysentery and reduce glucose in the blood and excess sugar in the urine of diabetic patients. **NOTE:** Because it lowers blood sugar, one should not consume java plum two weeks before and af-ter surgery. It's not recommended to eat this fruit on an empty stomach or with dairy products. And like every fruit, eat too much and you'll get sick with fever and body aches.

JERUSALEM ARTICHOKES / SUNCHOKES: The Jerusalem artichoke is used in traditional medicine to treat diabetes, heart disease, and rheumatism. The presence of insulin stimulates the growth of bifido bacterium to fight harmful bacteria in the intestines reducing concentrations of possible cancer-causing enzymes, prevents constipation, and boosts the immune system. Potassium-rich Jerusalem artichokes improve bone health and reduce the risk of osteoporosis.

JEWELS of OPAR: Jewels of Opar is a little-known wonder plant. It's known in the Chinese medicinal practice as Turen-shen, and has been used to tone digestion, moisten the lungs, used topically to treat edema, skin inflammation, cuts, and scrapes, and promote breast milk. The juice soothes sore muscles. A decoction from the roots treats scurvy, arthritis, stomach inflammation, and pneumonia. Jewels of Opar is used extensively as a reproductive tonic. The swollen root is used as a substitute for ginseng. Herbal recipes use Jewels of Opar to increase vitality, treat diabetes, and restore uterine functions postpartum. Researchers are studying if eating or making a poultice of this plant can block neurons transmitting painful impulses making it a natural painful impulses making it a natural pain reliever. Other scientists are experimenting with large-scale Jewels of Opar plantings to absorb and destroy contaminants in the soil and groundwater. That's why you should grow your own. This plant absorbs available poisons and heavy metals.

K **KALE:** Kale is high in fiber and sulphur, containing nutrients that may reduce the chances of several types of cancers. This green is high in folate, which is good during pregnancy. Kale has phytonutrients, to reduce inflammation, detox the liver, and protect brain cells from stress. Phytochemicals are chemical compounds produced by plants, generally to help them resist fungi, bacteria, and plant virus infections, and wards off insects and other animals. When humans consume these plants as food, they make us stronger and more immune. Be advised; raw kale contains goitrins that may lower iodine levels and af-fect the thyroid gland, but they disappear when kale is cooked. Research demonstrates a moderate intake of goitrin-rich vegetables, including kale, is safe for most individuals.

KALO / TARO: Since ancient times, this root has treated various ailments as asth-ma, arthritis, diarrhea, and skin problems. The corm juice is a laxative, and pain reliever. The leaves of the plant are anti-diabetic, anti-inflammatory, will expel intestinal parasites/worms. A decoction of the leaves is drunk to promote menstruation. The leaf juice treats conjunctivi-tis. In New Guinea, the leaves are heated over a fire and wrapped in a poultice to draw boils

KOHLRABI: Consuming kohlrabi will improve your gastrointestinal system, and help your body absorb more nutrients from the food you eat. Rich in potassium, this will keep you alert with plenty of energy. It's rich in vitamin C and B6, which boosts the immune system, protein metabolism, and red blood cell production. Eating this veggie will relieve gout and arthritis. It can be used in poultices directly on the affected areas of the feet, knees, or hands. The leaves and seeds are used as teas to bathe bacteria or fungus infections.

KOLA NUT / COLA NUT / BISSY NUT: In herbal medicine, the kola nut is usually soaked in alcohol to make tinctures to help 'drive' other herbs into the blood and may help increase oxygen levels in the blood. They may be chopped and boiled, drunk as a tea, although bitter, and the powder can be taken in a capsule. For thousands of years, kola nuts have been used in folk medicine as an aphrodisiac and an appetite suppressant, and to treat morning sickness, migraine headache, and indigestion. It is used in Jamaica to treat food poisoning. They counteract hunger and thirst, treat morning sickness and nausea in pregnant women. It has also been ground into a paste and applied directly onto the skin to treat wounds and inflammation. Other kola nut uses include fighting infection and clearing chest colds. Grate or coarsely grind the kola nuts and place in alcohol for 24 hours. Strain and discard the nuts and you have a crude extract with a strong musky flavor. This should be used in small amounts.

Kola nuts are mainly used as a stimulant to stay awake, withstand fatigue, promote bet-ter concentration, and help people lose weight by diminishing appetite. In addition to caffeine, the nuts contain theobromine, the ingredient that is found in chocolate and reputed to bring a sense of well-being. This may explain the mild euphoria and optimism from chewing the nuts.

KUMQUAT: Kumquats are high in fiber and aid digestion. The skin contains several useful oils that fight cancer. The juice reduces the chances of kidney stones. Because they contain pectin, vitamin C and oils, they are antibacterial and anti-inflammatory. Eating kumquats is good for treating respiratory ailments.

L **LABLAB BEANS / HYACINTH BEANS / SEIM:** Lablab leaves are used to accel-erate childbirth, regulate menstruation, to treat stomach troubles, and tonsillitis. A decoction of lablab leaves combined with leaves from other plants treat heart problems. The leaves are crushed and sniffed to cure headaches. The flowers treat alcoholism, poisoning, and flatulence. The juice from the pods treats inflamed ears and throats. Dried and roasted fully mature seeds treat intestinal parasites, fevers, cholera, sunstroke, vomiting, diarrhea, alcoholism, and arsenic poisoning.

LANGSAT / DUKU: Langsat/ duku bark can be boiled and drank for diarrhea and ma-laria. A paste made from the seeds is used to reduce fevers. A poultice of the bark is used to fight the poison of a scorpion sting. Dried peels are burned to repel mosquitos and smells good enough to double as incense. Juice pressed from the leaves is used as eye drops for inflammation. **Langsat should not be consumed by diabetics.** The sweet juicy flesh contains sucrose, frutose, and glucose.

LEEKS: Eating leeks helps the bowels, fights arthritis, and reduces the risk of prostate and colon cancer. Eating leeks is also reputed to keep your voice from becoming hoarse.

LEHUA MAOLI - MALANGA – YAUTIA – TANNIA: This root is used in Brazilian traditional medicine to prevent osteoporosis. Eating malanga regularly will reduce hypertension and rheumatism. Used as a poultice it will make boils form a head. The juice is used for skin care. Research is being done if this root inhibits the proliferation of leukemia cells.

LEMONS: Lemons, rich in vitamin C, strengthen the body's immune system as an antioxidant, protecting cells from damage. For a cough, sore throat, or cold, make a syrup by blending one tablespoon lemon juice with two tablespoons honey. You can also try lemon juice mixed with salt and ginger as a tonic for a cold. The aroma of lemon oil has been tested to reduce stress in aromatherapy. Lemon juice mixed with water will aid digestion, and pat an insect bite with raw lemon to stop the itch and reduce swelling. Lemons help cleanse the body through per-spiration and as a natural diuretic. Lemon juice is believed to actually cleanse the liver of toxins.

LEMONGRASS: Ever listen to an advert for modern medicine? The list of negative side effects is usually longer than the positive. The future of our planet is to 'go green.' Lemongrass – citronella oil - is being used in cleaning products, soaps, and shampoos. Us-ing natural lemongrass presents little risk of reaction when compared to the chemicals used in commercial products. Lemongrass will safely moisturize the skin and treat damaged hair.

Before patent medicines, people of various cultures relied on lemongrass tea or oil. It was used for a variety of ailments, from respiratory issues to antifungal. As an antioxidant, lemongrass is essential in Ayurveda. It is commonly used to treat colds, congestion, nausea, indigestion irregular menstruation, diarrhea, and sore muscles and joints. India also used the oil to treat fungal infections of the skin. The essential oil of the plant is used in aromatherapy.

The Chinese use this herb in similar ways, while Cubans use the oil today to reduce blood pressure. Studies indicate the lemongrass variety *Cymbopogon citratus* possesses anti-amoebic, antibacterial, antifungal and anti-inflammatory properties.

Lemongrass tea is known as 'fever tea.' Chop fresh or dried leaves and use 1 teaspoon of lemongrass per cup of boiling water. Boil more leaves and steep over-night for a stronger, concentrated liquid infusion. Studies show drinking lemon-grass tea infusions for 30 days can increase red blood cells/hemoglobin concentration.

Some people apply lemongrass as its essential oil directly on the skin at specific painful areas. Rub it onto the temples and behind the ears for headaches. Rub around the navel for stomach and abdominal pains. Anywhere a muscle aches apply as you would a menthol balm for a sprain. Smelling the essential oil of lemongrass is used as aromatherapy for relief from muscle pain, headaches, nausea, and anxiety.

Chewing the lemongrass blades can prevent mouth/oral infections by stopping the growth of bacteria that can cause cavities in the mouth and sore gums. Always try a small amount to see if you have a unique, adverse reaction. It's not recommended to use lemongrass if you're pregnant.

LETTUCE: Eating lettuce is good for the nervous system. Since it is low in carbo-hydrates it is a good food for diabetics and with high iron content it is good for anyone suf-fering from anemia. Lettuce juice or lettuce cooked as soup is a natural remedy for insomnia. Eating lettuce has a tranquilizing effect. Try eating a lettuce salad with a TBS olive oil before going to bed for sweet dreams. Lettuce tea made from a half cup of lettuce to a cup of boiling water relieves stress and is a good body tonic fighting cold viruses, and against asthma.

LIME: Ancients used the lime for medicinal purposes. During the Mid-dle Ages fragrant limes were used to ward moths from hanging clothes just as moth-balls do today. Sailors loved the lime since it prevented the weakening disease of scurvy.

LONG BEAN / CHINESE YARD LONG BEAN / BODI: Yard long beans are reported to be anti-diabetic, blood-purifying, and diuretic properties. Their high fiber is good for digestion. Earaches are supposedly eased by eating yard long bean leaves cooked with rice. To stop lactation and wean their children, nursing mothers eat boiled yard long leaves with potassium alum.

LaOQUAT / JAPANESE PLUM: Loquats have been used in traditional medicine for thousands of years to treat diabetes, arthritis, rheumatism, and nausea. Powdered loquat leaves treat diarrhea, depression, and even help with a hangover. The leaves are used in Japan to make biwa cha, a beverage which treats skin conditions and bronchitis or other respiratory illnesses.

LOTUS ROOT / LOTUS YAM: Lotus root increases blood pressure, main-tains proper enzyme activity, can uplift your mood, and possibly prevents some cancers. The combination of copper and iron serves to improve blood circulation in the body and in-crease energy. The potassium in lotus root also serves to help regulate blood pressure. It re-laxes the blood vessels and improves blood flow. A large amount of dietary fiber helps to improve the digestive process and reduce constipation. It's also a low-calorie food that helps you to maintain and lose weight while still getting some protein and nutrients in your diet. Vitamin B contains pyridoxine, which helps to improve your mood and relax your mind. It even helps to reduce headaches and serves to lower stress. It's another reason the lotus flower is associated with peace and tranquility. Pyridoxine also helps to lessen the risk of a heart attack and improves heart health overall. Ayurvedic-homeopathic medicine uses lotus root, the flowers, stamens, rootstock, seeds, and leaves to treat bleeding piles, snake bites, diarrhea, coughs, and skin diseases. Powdered lotus root is sold as an herbal medicine.

LUFFAH / LOOFA: Luffah gourd is a natural detoxifier, purifies the blood, strengthens the immune system, and helps to cool down a fever. The seeds are emetic and purgative, eaten to expel intestinal worms. Juice from the leaves is applied to skin problems such as eczema, and used as an eyewash to treat conjunctivitis. The gourd's fibers can be boiled in water, to make a strong luffah tea, and then used as medicine. This tea prevents and treats colds. It's also used for sinus problems. Some people use it as a wash to soften their skin. Women use luffah tea to regulate menstrual periods and nursing mothers use it to increase milk flow. A poultice of warmed luffah fibers treats arthritis, muscle, and chest pains. Luffah seed oil also treats skin diseases. The whole luffah sponge may be rubbed against the body as a gentle exfoliant on the skin and increases blood circulation. Luffah charcoal, prepared by heating luffah fibers in a closed container, is applied directly to the skin for shingles in the face and eye region. Charantin, a saponin, and peptide, an amino acid, in this vegetable help to lower blood sugar levels as well as urine sugar levels. The high fiber content helps digestion and the excretory system.

LYCHEE / LICHI: These fruits are anti-hyperglycemic and can lower blood sugar levels in diabetics. Consult with your doctor before eating many to avoid side ef-fects. These fruits are loaded with vitamin C, which builds the immune system, protects against colds, infections, and fights inflammation. Tea is sometimes made from lychee husks/skins to treat smallpox and diarrhea. The seeds are ground in India to treat stomach ailments. Sore throats are treated with a decoction of bark, root, and blossoms.

OF ALL THE HOME REMEDIES, A GOOD WIFE IS BEST.

~ Unanimous

M **MACADAMIA NUT:** Macadamia nuts are recognized for their energetic and antioxidant properties. Consuming 5 to 20 nuts per day would help lower cholesterol levels and heart risks.

MACE: Mace and nutmeg are very similar in culinary and medicinal properties.Both spices are efficient in treating digestive and stomach problems, relieve intestinal gas and flatulence. It can reduce vomiting, nausea, and general stomach uneasiness.

MALABAR SPINACH: Traditional medicine uses almost all parts of the plant. Malabar spinach roots are astringent and cooked to treat diarrhea. A paste of the root is applied to swellings. The leaves and stems are cooked and eaten as a laxative and a paste of the leaves is applied externally to treat boils and sores. The flowers are used as a poi-son antidote. Its juice is a safe laxative for pregnant women, and a decoction has been used to ease labor. Studies report Malabar spinach functions as a central nervous sys-tem depressant. Extracts from the berry-like fruit have potential as cancer treatments.

MALAY ROSE APPLE / POMMERAC: The Malay rose apple has many medicinal uses. An extract of the bark is used as an astringent to fight infections. The bark is pounded into a mix with sea salt, filtered through coconut husks, and poured into deep wounds. The root is used to soothe itches, and is effective against dysentery, and as a diuretic. Brazilians used the rose apple as a remedy for diabetes and constipation. The juice of crushed leaves can be used as a skin treatment, and can be steeped into baths.

MAMEY SAPOTE / MAMEY APPLE: The oil pressed from the seeds is used as a skin ointment and as a hair dressing believed to stop hair loss. The oil is also reputed to be diuretic and is used as a sedative in eye and ear ailments. The pulverized seed shell is mixed with wine as a reputed a remedy for coronary ailments, kidney stones and rheumatism. The seed residue, after oil extraction, is used for poultices on skin problems like boils. Drinking tea made from the bark and leaves treats arteriosclerosis and hypertension.

MANDARIN ORANGE: For hundreds of years, Chinese traditional medicine has used mature mandarin orange peel to improve digestion, relieve intestinal gas and bloating, and resolve phlegm. Immature mandarin orange peel treats liver and stomach problems.

MANGOS: Mango contains a stomach-soothing enzyme. The skin of the unripe fruit is an astringent, and a stimulant tonic. The bark is also astringent and will dry a runny nose. Mango pickles preserved in an oil and salt solution are used throughout India. However, these pickles (achar), if extremely sour, spicy, and oily, are not good for health and should be specially avoided by those suffering from arthritis, rheumatism, sinusitis, sore throat and hyperacidity. The ripe mango is anti-scorbutic, diuretic, laxative, invigorating, fattening, and astringent. It tones up the heart muscle, improves complexion, and stimulates appetite. The fruit is beneficial in liver disorders, loss of weight, and other physical disturbances. The unripe mango protects men from the adverse effects of hot, scorching winds. A drink, prepared from the unripe mango by cooking it in hot ashes and mixing the pith with sugar and water, is an effective remedy for heat exhaustion and heat stroke. Eating raw mango with salt quenches thirst, and prevents the excessive loss of sodium chloride and iron due to excessive sweating.

MARIJUANA: Marijuana has shown positive effects on cancer, AIDS, and glaucoma. Itis effective on AIDS patients from its ability to increase a person's appetite as well as relieving nausea allowing a patient to regain weight. Marijuana reportedly helps glaucoma patients by reducing ocular pressure, which can cause dam-age to the eye. It is the most effective treatment for chronic nausea. It is not physically addictive, but is psychologically addictive. Warnings: coughing, asthma, upper respi-ratory problems, difficulty with short-term memory loss, racing heart, agitation, laziness, confusion, paranoia, and possible psychological dependence can occur with use.

Smoking marijuana is bad for your health. The carcinogenic by-products of combusted plant matter inhaled into the lungs should be avoided whenev-er possible. It is currently held that there are no known carcinogens in marijuana when not smoked. Therefore, this drug should be taken orally and absorbed into the body through the digestive system by means of capsules, medicinal teas, or various edibles.

MARJORAM: A teabrewed from marjoram leaves may help with indigestion, headache, or stress. Externally dried leaves flowers may be applied as poultices to reduce the pain of rheumatism. Marjoram is considered the most fragrant essential oil among all herbs used in aromatherapy. It is also a warming and soothing massage oil for muscle aches. It fights asthma, headaches, and soothes digestion. Marjoram is used to loosen phlegm, and is a decongestant used to fight bronchitis, and sinus headaches. It is useful as a tonic for the nervous system. Marjoram may be more calming than oregano, to soothe the nerves, reduce tension and stress. One component in marjoram is flavonoids, which relieve insomnia, tension headaches, and migraines

MAUBY: Unsweetened and tangy, which is very seldom, the refreshing bark decoction is used as a bitter tonic for diabetes – lowering blood sugar, hypertension- lowering blood pressure, and cholesterol, and combating stomach disorders. According to the University of the West Indies, mauby combined with coconut milk may lower blood pressure.

MINT: Peppermint is one of the oldest and best tasting home remedies for indigestion, headache, and rheumatism. After dinner mints are used to improve bad breath, ease gas, and aid digestion. Peppermint lessens the amount of time food spends in the stomach by stimulating the gastric lining to produce enzymes which aid digestion. It relaxes muscles, lessens stress, has antiviral and bactericidal qualities, clears congestion related to colds and allergies, and eases intestinal cramping. Regular use of mint is very beneficial for asthma patients, as it is a good relaxant, and gives relief in congestion. But, over-dosage may irritate as well. Mint juice is an excellent skin cleanser. It soothes skin, cures infections, itching, and is also good for pimples.

MORINGA / MIRACLE PLANT: Almost all parts, the bark, sap, roots, leaves, seeds and flowers, of the moringa tree are eaten or used in Ayurvedic-homeopathic traditional medicine. It is used regularly to make a nutritional tonic. Moringa reduces inflammation, is an antioxidant, an aphrodisiac, and boosts the immune system. Eating leaves is prescribed for anemia and believed to increases women's milk production. Juice pressed from the leaves treats diabetes, reduces cholesterol, and stabilizes blood pressure. Moringa blossoms treat inflammations, muscle diseases, tumors and enlargement of the spleen and are made into an infusion and used as eyewash.

WARNING! Do not use during pregnancy, moringa blossoms can cause a miscarriage. Moringa roots treat the cardiovascular system, stimulate appetite, female reproductive issues, arthritis, and are used externally as a natural antiseptic. Modern scientific studies of the roots show they contain elements to treat ovarian cancer. **BE CAREFUL** before using moringa roots, because of the high concentration of spirochin, an alkaloid in the root. In small doses, 35mg, it can accelerate the heartbeat, but at larger doses of 350mg or more, parts of the nervous system may be paralyzed. Studies also show the seeds relieve the pain of rheumatism and gout. Moringa seeds have a high iron content and treat anemia and a natural antibiotic that treats fungal infections. Oil pressed from moringa seed protects the hair and skin cells from damage because it contains hydrating and detoxifying elements. It also has some healing properties for skin allergies, irritations, wounds, blemishes, and stretch marks. Naturally occurring isothiocyanates are the main anti-inflammatory and anti-carcinogenic compounds in moringa leaves, pods and seeds.

MUSTARD: Mustard was always important in medicine. Mustard oil can be so irritable that it can burn the skin. It is used diluted as a liniment for sore muscles. Powdered mustard is used in mustard plasters to fight bad coughs and colds. Various peoples have used mustard for snake bites, bruises, stiff neck, rheumatism, and respi-ratory troubles. It is delightful in bath water or as a foot bath to ease aches and pains.

MYSORE RASPBERRY: Mysore raspberries treats skin rashes, insect bites and stings, gout, and constipation. It also helps to lower the chances of cancer and heart disease.

N

NAMNAM: This is another one of those obscure fruits that boasts a long list of traditional medicinal uses. This fruit is not readily available; you either own a tree, or know someone who owns a tree. The fruits are reported to have useful medicinal properties and are used in various folk medicinal preparations. In some places in India, namnam fruit are sold in the market only for medicinal purposes. It's used to treat loss of appetite. The seeds yield an oil used in India to treat skin diseases. In Indonesia, namnam is used to treat diarrhea.

Little research has been done to discover the true medicinal value of this fruit. Namnam fruit contains vitamin C, vitamin A, and antioxidants to in-crease natural immunity. The leaves can be prepared as a tea, which also increas-es natural immunity, helps the flow of urine, and treats kidney stones. Though stud-ies have thus far been inconclusive, namnam has great potential in fighting leukemia.

NONI: Traditionally, the leaves of the noni tree were used topically for healing wounds. In herbal Polynesian medicine, noni has been used for many health conditions, such as constipation, diarrhea, skin inflammation, infection, and mouth sores. Today noni is advertised to cure everything from arthritis, atherosclerosis, bladder infections, boils, bowel conditions, burns, cancer, chronic fatigue syndrome, circulatory weakness, colds, cold sores, diabetes, drug addiction, eye inflammation, fever, fractures, gastric ulcers, gingivitis, headaches, heart disease, hypertension, improved digestion, immune weakness, indigestion, kidney disease, malaria, menstrual cramps, menstrual disorders, respiratory disorders, ringworm, sinusitis, sprains, strokes, thrush, and wounds. There is no real evidence, however, that noni is effective for any of these conditions.

Studies evaluating the effects of noni suggest it may have anti-cancer, pain-reliev-ing, and immune system-enhancing effects. However, these studies used extremely high doses that would be difficult to obtain from taking the juice. More importantly, there's no reliable evidence about the safety, or effectiveness of noni for any health condition in hu-mans. Noni juice is high in potassium and should be avoided by people with kidney disease.

NUTMEG: Nutmeg was so important because of the purported medicinal proper-ties of its seeds. It is an astringent and stimulant, as well as an aphrodisiac. At the height of its value in Europe, nutmeg was carried to demonstrate wealth. Ground nutmeg is also smoked in India and used for asthma, fever, and heart disease. Nutmeg oil is used in perfume, toothpastes, and cough medicines, and in pharmaceutical drugs. The oil is used externally for rheumatism, and gargled for bad breath and toothaches. Nutmeg oil contains myristicine, which somewhat controls stomach gases, and if consumed in large doses causes hallucina-tions. Nutmeg oil stimulates the brain reducing mental exhaustion and stress. It improves the quality of your dreams, making them more intense and colorful. It is a good remedy for anxiety as well as depression. Nutmeg oil treats muscular and joint pain and it is an excel-lent sedative. Nutmeg's relaxing aroma comforts the body, and increases blood circulation.

O **OKRA / OKA:** Okra has many valuable nutrients. It is a prime source of sol-uble fiber in the form of gums and pectins. Soluble fiber lowers cholesterol and reduc-es the risk of heart disease. The other half is insoluble fiber, which keeps the intestinal tract healthy, decreasing the risk of some forms of cancer, especially colon-rectal cancer.

ONIONS: Eating onions gives some protection against heart disease and co-lon cancer. Onions may also reduce the frequency and strength of asthma attacks. Onions in soups are especially good to help ward off flues. It seems the more pun-gent onions, especially yellow, are better for you as they have more antioxidants.

ORANGES: Each orange has about sixty calories packed with vitamins C and B-1, fiber, and folate. One orange contains about 50mg. vitamin C; or two-thirds of our dai-ly need. Vitamin C is necessary to fight colds and infections, helps reduce asthma attacks, and rheumatoid arthritis while decreasing the chance of colon cancer. It also helps heal wounds and broken bones. Pregnant women need more vitamin C because it helps form collagen, a protein that helps form bones, and makes the expectant mother's body easily absorb iron.

Remember how our grannies had dried orange peels hanging for use in flavor-ing tea? Seems they knew more then, than we do today. Orange peels contain compounds known as 'polymethoxylates (PMF', which has been tested to lower cholesterol significantly – as much as some prescription drugs! Just dry the peel and add to tea or boiling water.

OREGANO: The ancient Greeks applied poultices of oregano leaves to treat sores, and muscle aches. Chinese herbal doctors use oregano to lower fevers, fight vomiting, diarrhea, jaundice, and itchy skin. Europeans use this herb for improved digestion, and to soothe coughs. Germans produce oregano-based cough syrups. Oregano fights the common bacterial disease known as giardia amoeba, common throughout the world. It can cause serious illness and oregano proved to be more effective as treatment than a prescription drug. Oregano is very good for you, helping with digestion by increasing the flow of bile. Studies have shown oregano fights viruses, fungus, bacteria. The French oregano soaps may have been the first to really be antibacterial. This is a great herb to drink as a tea whenever you have a cold or fever.

P

PAK CHOY / BOK CHOY: Bok Choy is rich in vitamin C, ffiber, and folic acid. All reduce the risk of various types of cancer. Bok choy has more beta-carotene than other cabbages, and more potassium and calcium.

PAPAYAS: Eating papayas helps prevent diabetic heart disease. Nutrients in papayas prevent cholesterol from clogging arteries reducing the possibility of strokes. Papaya may lower cholesterol, prevent colon cancer, and reduce inflammation caused by rheumatoid arthritis. Researchers believe papaya may be great for your sense of sight. The high vitamin A (300% of the daily need) content of papaya will help smokers prevent emphysema, or lung inflammation from second hand smoke.

PAPRIKA: Red peppers, from which paprika is ground, have much more vitamin C than oranges. The high heat of commercial processing destroys much of the vitamin C in paprika so look for the sun dried variety. Paprika is a good source for beta-carotene, and is considered both a stimulant, and a blood pressure regulator. It fights bacteria, and aids digestion. Try substituting at least part of the salt you use on cooked food with paprika, and you will be pleasantly surprised. As an antibacterial agent and stimulant, paprika can help normalize blood pressure, improve circulation, and increase the production of saliva and stomach acids to aid digestion.

PARSLEY: Parsley is a natural diuretic and can be used by women to alleviate irregular menstrual cycles. Parsley eaten raw or as tea may ease the bloating that occurs during the time of the month. Chewing parsley will help with bad breath from food odors such as garlic or onions. A tea made from parsley will stimulate as much as a cup of coffee, but watch out as parsley has been reported to be an aphrodisiac. A poultice of its leaves can be used for insect bites or stings

PARSNIPS: Parsnips were used in herbal medicine and considered an aph-rodisiac. Parsnip tea is an excellent diuretic and helps to cleanse by stimulating the production of urine. Parsnips are rich in potassium to keep the heart healthy. Eating parsnip also helps to make your bones stronger, lowers the chances of developing diabetes, helps to reduce cholesterol, and prevents the onset of depression.

PASSION FRUIT: In Madeira, the juice of passion fruits is used to stimulate appetites and as treatment for gastric cancer. In Puerto Rico, where the fruit is known as 'parcha,' it is widely believed to lower blood pressure. Fresh passion fruit is high in beta carotene, potassium, vitamin C, dietary fiber, and polyphenols. They are antioxidant and anti-inflammatory effects plant compounds that they may reduce your risk of chronic inflammation and heart disease. Passion fruit is usually safe to eat and good for you, but some people are allergic to it. This is more likely if you're allergic to latex. Passion fruit seeds are packed with piceatannol, another polyphenol that may improve insulin sensitivity in obese men and potentially reduce the risk of type 2 diabetes. **PEA**

EGGPLANT / TURKEY BERRY: Pea eggplant berry juice can treat almost every ailment from coughs and respiratory-chest ailments, fevers, sore throats, and arthritis.

dried and powdered to assist diabetic patients in reg-ulating their blood sugar levels. Poultices of the leaves treat wounds and skin diseases such as eczema. The leaves and flowers are boiled into a decoction with honey for fevers and colds. Juice from pressed flowers treats eye problems.

PEACHES: Chinese traditional medicine uses the peach to aid digestion, promote circula-tion and reduce fatigue. A tea of peach leaves treat fever, headache, skin problems, ulcers, and malaria. A tea made from the bark and dried leaves treats coughs. Root bark is used for dropsy and jaundice.

PEACH PALM / PEWA: Pewa is recommended as a remedy for head-ache and a bellyache. The oil from the seeds is used as a rub to ease rheumatic pains.

PEANUTS: Researchers in Great Britain discovered women who ate five ounces of peanuts weekly reduced heart attacks by a third. If you eat one ounce of peanuts every day the unsaturated fats in peanut butter will reduce the risk of heart disease by 25%. Rich in folate and niacin (vitamin B3) will increase the HDL, good cholesterol, by as much as 30%. Twenty-five percent of peanuts consists of proteins and dietary fiber. The most unique property of peanut butter is its high content in resveratrol, a substance that's been shown to have very strong an-ti-cancer properties. As many as one per cent of the world's population are allergic to peanuts. This allergy accounts for over three-fourths of all deaths related to food allergies each year.

PIGEON PEAS/RED GRAM: A poultice can be made of the young leaves of pi-geon peas and applied to sores. Chinese claim powdered leaves help expel kidney stones. Chinese shops also sell dried roots as an anti-toxin, to expel parasite worms, expecto-ant, sedative, and vulnerary. Salted leaf juice is taken for jaundice. Argentines use a pigeon pea leaf decoction for genital and other skin irritations, especially for females. Leaves are also used for toothache, mouthwash, sore gums, child-delivery, and dysentery. Pigeon pea blossom decoctions are used for bronchitis, coughs, and pneumonia. Scorched seed when added to coffee alleviates headache and vertigo. Fresh seeds are said to help incon-tinence in males, while immature fruits are believed to treat liver and kidney ailments.

PILI NUTS: Resin from the pili nut tree has been used as an ingredient in ointments and because of its pasty consistency, antiseptic properties, and quick drying, it's used as a natural plaster to heal wounds. Raw pili nuts are used as a laxative.

PINEAPPLE: Pineapple is another fruit that not only tastes good, but also is good for you. Pineapple juice is close to our stomach juices. Consumed moderately, pineapple aids digestion. It has plenty of fiber and helps the body relieve fluids especially mucus from nasal passages. Never consume an unripe pineapple,as it can be poisonous causing throat irritations and diarrhea.

PLANTAIN: Young plantain leaves are a survival food raw in salad or cooked. They are very rich in vitamin B1 and riboflavin. The herb has a long history of use as an alternative medicine dating back to ancient times. A traditional healing salve can be made by placing one large whole plantain chopped (skin too) in a large non-metallic pan (ceramic or Corning ware), with one cup vegetable oil or lard; cover and simmer on low heat till all is mushy and green. Strain while hot. Once cool put in a container that seals tight. It is good for burns, insect bites, rashes, all sores, and used as a night cream for wrinkles.

POMMELO: This fruit fights atherosclerosis and helps regulate blood pressure. Pommelo is great for dieting because it quenches your appetite and has only thirty-five calories per hundred grams. Limonoids found in pummelo are being studied for their cancer -properties. Eating a pummelo is invigorating and increases stamina and lifts your spirit.

POMMERAC: See Malay rose apple.

POTATO: Potatoes are reported to be a general remedy, a laxative, a sedative, increase urine flow, and increase breast milk. It's a traditional remedy, externally applied boiled or raw, for burns, corns, and warts Eating potatoes may worsen arthritis. Potato sprouts are considered toxic.

PUMPKIN: Eating pumpkin seeds helps men avoid prostate cancer. They were once recommended as a cure for freckles and a cure for snakebites. The orange-flesh are is a giveaway

A wise man ought to realize that health is his most valuable possession.~ Hippocrates are is a dead giveaway that pumpkin is a source of beta-carotene, which is a powerful antioxidant. Beta-carotene is converted to vitamin A in the body. Vitamin A is essential for healthy skin, vision, bone development, and many other functions.

PURPLE YAM: A paste made from pounding purple yam can be applied as a treatment for wounds, inf amed hemorrhoids, and skin diseases. Dried powdered yam treats piles. Juice of the tuber is used to kill stomach parasites. It is a laxative and used to treat fever.

R **RADISH:** Radishes and their green tops are an excellent source of vitamin C. The leaves contain six times the vitamin C content of their root, and are also a good source of calcium. Red radishes provide the trace mineral molybdenum, and a good source of potassium. Long white radish/daikon provide a very good source of copper. Historically radishes used as a medicinal food for liver disorders. They contain a variety of sulfur-based chemicals to increase the f ow of bile, which helps to maintain a healthy gallbladder and liver, and improve digestion. Daikon aids in digestion of fatty fried foods. A half cup of daikon daily will prevent kidney or gallstones.

RED BANANAS: Bananas are said to contain everything a human body needs including all eight amino acids, which our bodies can't self-produce. Bananas are a good source of fiber and potassium. The fruit is a mild laxative, good for cardiovascular health, protects against strokes and ulcers. It reduces water retention and is preferred for anemic patients because it's rich in iron. Red bananas have more vitamin C than the usual yellow types. The redder the fruit the more nutritious. Eat at least one banana a day. It comes in a truly germ-proof package because its thick peel is an excellent protection against bacteria and other contamination. The flower of this plant is used to treat ulcers, dysentery, and bronchitis and cooked f owers are good food for diabetics.

RICE: Brown rice is naturally better for you compared to white rice because it has more fiber. This is better than white rice fortified with vitamins. If you experience diarrhea eat a hand full of uncooked rice. As you digest it, the rice will scrape your intestines and the runs will be over. A bush medicine remedy is to boil rice in excess water and strain. Mix with hot pepper and apply externally for gout. The water rice boils in has cosmetic and health benefits. Still warm, this water can be drunk as a tea for energy. Once it cools, it can be used to bathe with, and it will soften the skin. As a hair rinse it makes healthier and softer hair.

ROSEMARY: Rosemary stimulates the circulatory and nervous systems while it also calms indigestion. It will help fight severe headaches. Rosemary's essential oil will ease muscle strains and aches. It is a usual ingredient in shampoos and hair conditioners that fights dandruff and (if you believe it) hair loss. Boil ten sprigs fresh rosemary in two cups water for thirty minutes. Cool and store in a tightly sealed container. Refrigerate. Once a week combine one half cup of this rosemary water with a cup of warm water and work through hair after shampooing. It is thought that adding rosemary to your diet may reduce the chances of having a stroke, or contracting Alzheimer's. Rosemary is a good herb to use as a mouthwash because it soothes sore throats and freshens breath. Chinese combine it with borax to treat baldness. Rub your pets with powdered rosemary leaves as a natural f ea and tick repellent due to its antimicrobial properties.

RHUBARB: The *Rheum* species contains at least 60 different types and many hybrids. The type baked into pies differs from the medicinal rhubarbs, which are generally considered inedible. For thousands of years, traditional Chinese medicine has used rhubarb roots as a laxative. It was one of the first Chinese medicines imported to the West. Rhubarb also appears in medieval Arabic and European prescriptions. Today, scientists are exploring the various rhubarb species to treat ailments as dermatitis, pancreatic cancer, and diabetes. Rhubarb, stewed or juiced, is a tonic to aid the blood and the digestive system. Studies show that rhubarb helps lower your bad cholesterol levels as well as your total cholesterol.

ROSELLE / RED SORREL: Sorrel has many bush medicine remedies. Drink-ing the tea will relieve hypertension, coughs, and is a cure for hangovers. A paste of the leaves and seeds is used in Egypt as an antibiotic on wounds. Leaves that have been warmed in hot water can be used on the skin to draw out boils or ulcers.

ROUCOU: See Annatto

RUTABAGA: Rutabagas have folic acid and antioxidants to fight inflammation, prevent premature aging, and reduce cancer risks. They are a natural diuretic and treats chronic constipation. A cold remedy is to peel and blend a rutabaga into a paste and mix with honey. Use a teaspoon 3 times a day. The oil from the seeds can be mixed with camphor as a remedy for arthritis and rheumatism.

S **SAFFRON:** Large dosages of saffron can be fatal. It is considered an excellent stom-ach tonic and helps digestion and increases appetite. It is also relieves tension and fights depression. It is a fact since antiquity, crocus was attributed to have aphrodisiac properties. Rubbing a salve made from saffron into achy joints is an old folk remedy for gout. Because of saffron's high price, it is unlikely that you will find it ready-made in health food stores. You can make your own, however, by blending a few threads of saffron into petroleum jelly. Spread a thin layer of the salve on the affected areas in the morning and evening. Use the salve until the joint pain abates. Saffron alleviates fatigue and exhaustion, primarily because it works to strengthen the heart and nervous system. It aids digestion, by increasing appetite and gastric juice production. When added to some homeopathic preparations, it also relieves nosebleeds. Saffron milk is a tasty, soothing drink that can be helpful in relieving cardiac problems. To make it, bring one cup of milk just to boil, add a pinch of saffron. Reduce the heat and simmer the mixture for two minutes. Sweeten with honey to taste, and drink it once a day.

SAGE: Medicinally smoking sage helps relieve asthma, and sage tea will help fight a fever, or a sore throat. To hide gray hair put one-half cup of dried sage in two cups of water and simmer for half an hour. Cover and allow the sage water to steep overnight. Over a basin, pour the rinse on your hair at least ten times reusing the same sage water. On the last rinsing, leave the sage water dry on the hair before rinsing with fresh water. Repeat every week un-til your hair is the desired shade. Then rinse every month to retain your hair color. For sore throats, try mixing a sage tea with apple cider vinegar, and salt for gargling. Sage is reported to have moisture-drying properties, and can be used as an antiperspirant. It can also be used as a compress on cuts and wounds. Clinical studies have also shown that it can lower blood sugar in cases of diabetes. Pour boiling water over two teaspoons of sage leaves. Let mixture steep for ten minutes. Strain, then drink one to two cups per day to relieve symptoms of coughs, tonsillitis, and respiratory infections. Because sage has a powerful antiseptic effect, combine sage with your toothpaste. This will help remove plaque, and will strengthen bleeding gums.

SAPODILLA: Medicinally young sapodilla fruits can be boiled and eaten as a bush remedy for diarrhea. Tea made from the yellowed leaves is a reputed remedy for coughs and colds. Eating crushed seeds are claimed to expel bladder and kidney stones, while the liquid essence of the crushed seeds is used throughout the Yucatan as a tranquilizer. The seeds can be crushed into a paste to sooth insect bites. Chicle is used in Central America as a primitive dental filling.

SESSILE JOYWEED: Traditional homeopathic medicine uses sessile joyweed as a di-uretic and laxative. When it's made into a broth or tea, it has cooling properties. It has been used to treat difficult urination, skin diseases especially with burning sensations, diarrhea, and indigestion, poor breast lactation on new mothers, enlarged spleen, fever, and hemorrhoids. A paste made from the leaves helps wounds heal quicker. The plant is rich in vitamin A and beta carotene and is believed to be beneficial for improving vision. This plant is an ingredient in hair oils and eye cosmetics.

SCREWPINE LEAF / PANDAN: Pandan leaf extracts may relieve fevers, gas, and indigestion. They help diabet-ics keep blood sugar levels steady after eating. The roots and leaves make a tea for re-lief from anxiety, help with sleep, relieve headaches, will reduce uric acid levels to ease the painful symptoms of gout, and reduce blood pressure. Pandan leaf extract applied to the scalp and skin relieves dryness and dandruff. It's good to treat sunburn.

To give your dog a good smell and fight fleas and ticks, scrub with pandan leaves and bathe them with the tea. Pandan leaves will treat pain and soothe inflamma-tion because they contain tannin and essential oils. Soak leaves in coconut oil for sev-eral days and then run the oil on the sore areas. A balm can be made for treating arthritis.

SHALLOTS: Shallots contain a vegetable pigment, quercetin – a flavonoid – that strengthens small blood vessels and protects against heart disease and di-abetes. Researchers also believe quercetin reduces bad cholesterol. All allium vegetables are recognized for their ability to kill and inhibit cancer cells, which diminishes the risk of cancer. The most important benefits of shallots are: as a high source of an-tioxidants that improve heart health, in cancer and diabetes prevention, as an anti-inflammatory, and as an antimicrobial, that might also help fight obesity, and helps prevent or treat allergies. They have been used as a remedy for sore throat, infections, and bloating.

SICKLE SENNA: Use the leaves externally treat skin problems as infections, sores, and insect bites. Boiled in tea or eaten, they purge parasites. The seeds are eaten, combined with a leaf decoction, to treat conjunctivitis. A poultice applied with a paste made from ground roots, mixed with lemon juice is a treatment for ringworm and itchy skin. A boiled decoction of the seed pods treats fevers. A salad made with young leaves treats constipation.

SNAKE GOURD: Snake gourd is a very nutritious vegetable and a natural antibiotic, cough medicine, and laxative. This vegetable is a tonic for the heart, helps stimulate the production of body fluids, and relieves dehydration. Snake gourd is best known for a cooling effect in the body. Besides being a heart-friendly vegetable, snake gourd reduces body acids.

SORREL: See roselle.

SOURSOP: Ripe fruit juice is said to be diuretic, and a remedy for blood in the urine and urethritis. Taken when fasting, it is believed to relieve liver ailments, and leprosy. Pulver-ized immature fruits, which are very astringent, are decocted as a dysentery remedy. To draw out chiggers and speed healing, apply the flesh of a soursop as a poultice unchanged for three days. In the Materia Medica of British Guiana, it is recommended to break soursop leaves in water, "squeeze a couple of limes therein, get a drunken man and rub his head well with the leaves and water and give him a little of the water to drink, and he gets as sober as a judge in no time." This sobering or tranquilizing formula may not have been widely tested, but soursop leaves are regarded throughout the West Indies as having sedative or properties. In the Dutch Antilles, the leaves are put into one's pillowslip or strewn on the bed to promote a good night's sleep. An infusion of the leaves is commonly taken internally for the same purpose. It is taken as an analgesic and antispasmodic in Esmeraldas Province, Ecuador. In Africa, it is given to children with fever and they are also bathed lightly with it. A decoction of the young shoots or leaves is regarded in the West Indies as a remedy for gall bladder trouble, as well as coughs, catarrh, diarrhea, dysentery and indigestion. It is said to 'cool the blood', and to be able to stop vomiting, and aid delivery in childbirth. The decoction is also employed in wet compresses on inflammations and swollen feet. The chewed leaves, mixed with saliva, are applied to incisions after surgery, causing the incision to disappear without leaving a scar. Mashed leaves are used as a poultice to alleviate eczema and other skin afflictions and rheumatism, and the sap of young leaves is put on skin eruptions. Research suggests a connection may exist between eating soursop and atypical forms of Parkinson's disease because of the high content of annonacin present in soursop. What a shame because annonacin is also supposed to cure cancer!

SPAGHETTI SQUASH: Native Americans used squash to treat intestinal para-sites and urinary problems. Research has found eating squash has blood-cleansing properties. It's used as a general tonic during rehabilitation after an illness. The best way of absorbing the nutrients and healthful properties of all squash is to enjoy it as a lightly cooked soup.

SPINACH: Spinach protects against heart disease, arthritis, and types of cancer. Eating this green prevents cholesterol from blocking arteries causing heart attacks or strokes, and reduces high blood pressure. Spinach may be a rival with carrots for benefiting eyesight by keeping the eye muscles strong and reducing the incidence of cataracts.

SPINE GOURD / HEDGEHOG GOURD: Eating spine gourd is good for urinary problems. A tea made from the boiled roots is used for diarrhea and lower-ing blood pressure. It's a traditional herbal remedy for diabetes mellitus as it reduc-es blood sugar levels in diabetic patients because it has a high content of plant in-sulin. Anything high in fiber and water is a smart choice for a diabetic's diet.

Spine gourd with carotenoids, like lutein, prevents some eye diseases, cardiovascular diseases, and even **cancers.**

STARFRUIT / CARMBOLA / FIVE FINGERS: Star fruit juice will clean brass and silver. It will also remove rust stains from white clothes. East Indians use this five-finger fruit widely for many medicinal purposes. The juice will help reduce a fever and quench the associated thirst. Boiled fruit will relieve diarrhea or a hangover. A salve made by continuously boiling the fruit to almost nothing is good for eye infections. Eating the ripe fruit is said to reduce hemorrhoids. A poultice of crushed leaves will fight ringworm. The powdered seeds are said to have a sedative effect, and are useful in fighting children's colic. In bush medicine, the starfruit tree is a virtual pharmacy. In India, ripe starfruit is used to halt hemorrhages and to re-lieve bleeding hemorrhoids. Starfruit is used by Brazilians to fight kidney and bladder problems. It is also used to treat eczema. A decoction made from the fruit will overcome severe nausea and vomiting. Starfruit tree leaves can be plastered on the tem-ples to soothe headaches. Beware hydrocyanic acid has been detected in the leaves, stems, and roots. Powdered seeds serve as a sedative in cases of asthma and colic.

STRAWBERRY / QUEEN OF BERRIES: The entire strawberry plant was used to treat depressive illnesses. The leaves and roots were brewed into a tea. Strawberries help to regu-late and stabilize blood sugar. In 2011, doctors discovered eating 35 strawberries a day reduced the complications of diabetes, such as kidney disease and neuropathy. Strawberries are an excellent source of folic acid, an essential for pregnant women to protect against severe birth defects. Romans used strawberries to treat fainting spells, fevers, throat infections, gout, and bad breath.

SUGAR APPLE: A tea made of sugar apple leaves is considered an excellent tonic, especially for colds and diarrhea. A bath in the leaves relieves severe arthritis and rheumatism.

SUGAR CANE JUICE: Sugarcane juice is believed to have many medicinal properties. It supposedly strengthens the stomach, kidneys, heart, eyes, brain, and sex organs. Cane juice is beneficial to drink to break fevers. Sugarcane juice has been found to be very beneficial for preventing as well as treating sore throat, cold and flu. Being alkaline in nature, sugarcane juice helps the body in fighting against cancer, especially prostate and breast cancer. Sugarcane provides glucose to the body, which is stored as glyco-gen and burned by the muscles, whenever they require energy. Therefore, it is considered to be one of the best sources of energy. If you have been exposed to heat and physical activity for too long, drink sugarcane juice. It will help rehydrate the body quickly. Mixing sugarcane juice with lime juice, ginger juice, and coconut water will give good results for enlarged prostate gland. Sugarcane juice is said to speed up the recovery process after jaundice, especially when mixed with lime juice. The juice sucked from the sugarcane can prove highly valuable in case of weak teeth due to lack of proper exercise resulting from excessive use of soft foods. It gives a form of exercise to the teeth and makes them strong.

The art of medicine consists in amusing the patient while nature cures the disease.

Voltaire

It also keeps the teeth clean and increases their life. Sugarcane juice is a fattening food. It is thus an effective remedy for thinness. Rapid gain in weight can be achieved by its regular use. The dew which collects on the long leaves of sugarcane is useful in several eye disorders. When instilled in the eyes, it is an effective medicine in defective vision, cataract, conjunctivitis, burning of the eyes, and eye-strain after excessive reading.

SUNSET HIBISCUS / SUNSET MUSK MALLOW: For centuries in China, sunset hibiscus treats chronic kidney disease inflammations, oral ulcers, and burns. In the Pacific Islands and Indonesia, it treats kidney pain, lowers high cholesterol, eases childbirth, increases breast milk, and protects against osteoporosis. Scientific researchers have identified more than 120 phytochemical ingredients from the flowers, seeds, stems, and leaves, including: flavonoids, amino acids, nucleosides, polysaccharides, organic acids, steroids, and volatile oils. This plant acts as an anti-diabetic and treats kidney functions, an antioxidant, anti-inflammatory, analgesic, antidepressant, antiviral, anti-tumor, and has positive effects on cognitive functions, and reduces bone loss, both due to aging. Cooked leaves, shoots are used throughout Papua New Guinea to treat colds, sore throats, stomach aches, and diarrhea. It's a good baby-food, because the young leaves are almost fibreless and are easy to puree after boiling. Sunset hibiscus good health tea: ¼ cup fresh leaves chopped fine and steeped in 1 cup of boiling water treats sore throats, nausea, and mouth ulcers. Rubbed on the skin it reduces inflammation, cleans, and helps to heal wounds, insect bites, and rashes.

SURINAM CHERRY: A leaf infusion treats fevers and increases appetite. A leaf decoction is a cold remedy. The leaves contain citronella and are used in South American homes spread over the floors to repel flies.

SWEET POTATO: Sweet potato is believed by Chinese medicine to be supplement-ing and warming to the stomach. However, patients suffering from indigestion, or heat-dampness should not eat too much sweet potatoes as it can cause swelling of the stomach and abdominal pain. The leaves of sweet potato are bitter in taste, and are anti-diabetic. They are helpful in lowering blood sugar. These roots are extremely useful in treating asthma. Sweet potato contains phytoestrogens helpful for women's reproductive systems. Sweet potato on your regular diet will fight colon, kidney, prostate, and intestinal cancer. They help prevent acid stomach and ulcers. Use the water the potatoes are boiled in for muscle and joint aches and arthritis.

T

TAMARILLO: Eating a tamarillo or drinking the juice relieves stress and is used to treat severe headaches and migraines. A poultice of warmed/steamed leaves is wrapped around the neck to treat a sore throat. A decoction of tamarillo pulp is a treatment for colds, sore throat, high blood pressure, and liver disease.

TAMARIND: Tamarind pulp can be applied directly on inflammations, and used as a rinse for a sore throat. Pets infested with fleas or ticks can be washed, then rinsed with strong tamarind water, and let dry on them. Boiled tamarind leaves and flowers can be used as poultices for sprains and arthritis. A tea from the tree's bark makes an excel-lent tonic. Hard tamarind heartwood makes the best hoe handles, and mortars and pestles.

TANGELOS: The tincture of the peel is used as a pleasant bitter tonic to treat coughs and choking.

TANGERINE: Tangerine peel/skin are chewed to treat asthma, indigestion, clogged arteries. They possibly help prevent cancer especially lung, colon, and rectal cancer. The essence of tangerine peel helps repel the side-effects of chemotherapy.

TANNIA: As a kitchen remedy, tannia may be used to regulate energy, support diges-tion, and disperse congestion.

TARRAGON: Tarragon increases appetite, improves circulation of blood, and helps in the proper distribution of nutrients, oxygen, hormones and enzymes throughout the body. It stimulates the brain, nervous system, digestive system, circulatory system, and the endocrinal system. This in turn stimulates the whole metabolic system and as a result, growth and immunity are stimulated.

TEA: The Chinese regard tea as a cure-all, but there are hundreds if not thou-sands of varieties of tea. The usual beverage is considered an antitoxic, diuretic, expectorant, and stimulant. Teabags can be used as poultice onto baggy or tired eyes, compressed onto the temples to ease a headache, or rubbed on sunburned areas to relieve the pain. **BEWARE:** there is evidence linking the tannin in tea to esophageal cancer where tea is regularly drunk. High consumption may also cause unpleasant nerve and digestive problems.

THAI EGGPLANT: All parts of the Thai eggplant, the roots, leaves, and fruits, are used in Ayurvedic-homeopathic treatments. The roast pulp of the Thai eggplant is used to reduce swelling. Once the flesh has cooled smear it around the raised areas. Researchers have found that an infusion made from the fruit has reduced total and LDL cholesterol levels. A tea prepared from the leaves treats asthma, and coughs. The same leaf tea and a juice squeezed from the fruit can be used as a body wash to relieve itching. Scientists have found that Thai eggplant roots contain solanine, a compound that's toxic with fungicidal and pesticidal properties, one of the plant's natural defenses.

THYME: Thyme's best use medicinally is as an antiseptic, but it also has expectorant, anti-spasmodic, and deodorant properties. It aids in digestion, and as such, is excellent when combined with fatty meats that often cause gastrointestinal problems such as duck, lamb, and pork. Herbalists use thyme in infusions, extracts, teas, compresses, bath preparations, and gargles. Recent studies indicate that thyme strengthens the immune system. Nightmares were treated with thyme tea. Thyme oil was used during World War I to treat infection and to help relieve pain. Small amounts of this herb are sedative, but larger amounts are stimulant. Thyme is used against hookworm, roundworms, and threadworms. Thyme also warms and stimulates the lungs, expels mucus, and relieves congestion such as asthma. It also helps deter bacterial, fungal, and viral infections. Thyme has always been used as a poultice for wounds, insect bites and stings. It is a good wash for sore eyes, and a hair rinse to fight dandruff.

TOBACCO: Insect bites and dog bites are relieved after applying a paste of powdered to-bacco leaves and water. Tobacco when eaten, if your stomach can handle it, will supposedly relieve gas. A hair rinse or body wash of tobacco leaves boiled in water will clean the hair while killing lice.

TOMATO: The British originally believed the tomato to be poisonous, but this veg-etable is really good for you. Tomatoes contain an anti-oxidant, lycopene that reduces the risk of prostate cancer if eaten almost daily. Tomato is excellent for purifying and rejuvenating your skin, removing acne scars, and healing sunburn. Rub tomato slices directly onto your clean skin, focus on pimply areas. Tomatoes contain vitamin C which has healing powers, and an acid which eliminates dead skin and opens pores, making skin soft and radiant. Tomatoes are rich in vitamin A that prevents overproduction of sebum that causes acne. To remove dark circles under your eyes make a paste using one tablespoon of tomato juice with a half teaspoon of lemon juice, a pinch of turmeric powder and a pinch of gram flour. Keep the paste under your eyes for 10 minutes. Dark circles will disappear after several treatments.

TONKA BEANS: The bark is prepared as a decoction to bathe fever patients. The seeds are fermented in rum and used for snakebites, cuts, contusions, coughs, rheumatism, and as a shampoo. The seed oil is dropped into the ears for earaches and ear infections. They are used in local supernatural potions to bolster courage, attract good luck with money, and in finding love. In fact it is known as the love wishing bean. Supposedly if you wish and throw seven tonka beans into a river, your wish will come true.

TROPICAL ALMOND / WEST INDIAN ALMOND: The tropical almond tree is a natural pharmacy. Parts treat dysentery, leprosy, coughs, jaundice, indigestion, head-aches, colic, pain and numbness, fever, diarrhea, sores, skin diseases, diabetes, etc. An infusion of the leaves is used to treat jaundice. The leaves are used to treat indigestion. Externally, the ternally, the leaves may be rubbed on breasts when feeding a child to cure pain or can be heated, applied to sore parts of the body, or a poultice for swollen rheumatic joints.

TROPICAL LETTUCE / INDIAN LETTUCE: In traditional medicine, the leaves are boiled and steeped to make a tonic for stomach-digestive ailments and used for a detoxification regimen. The milky white sap contains 'lactucarium,' used for its anodyne, antispasmodic, digestive, diuretic, hypnotic, narcotic and sedative properties. It's taken internally to treat of insomnia, anxiety, neuroses, hyperactivity in children, dry coughs, whooping cough, and rheumatic pains. Lactucarium is weak in young plants and stronger when the plant flowers. Commercially, the plants' heads are cut and the juice is scraped several times a day. CAUTION: Never use this plant without the supervision of a skilled practitioner. Normal doses cause drowsiness, excess causes restlessness, and overdoses can cause death through cardiac paralysis. The sap is applied externally to remove warts.

TURMERIC: Occasionally shredded and used fresh, turmeric is more often dried and powdered for use. The roots are boiled for hours, dried for days or weeks, and then ground into powder. Turmeric removes accumulation of cholesterol in the liver and promotes a healthy circulatory system. It is also a great stomach tonic for gas, or indigestion. East Indian women with lovely, velvety skin often attribute it to consuming turmeric. It contains manganese, vitamin B6, and iron with about four calories to a gram. To make a tea from turmeric, pour 8 ounces of boiling water over a half-teaspoon and let sit covered for five minutes, then strain, if necessary. Drink two or three cups daily, as desired. Mix turmeric powder with cooking oil to make a thick paste. Put on the skin over wounds, bites, bruises, etc. Cover with bandage and leave on several hours. It washes off and the color disappears quickly. For a toothache, paint this your face over the toothache.

TURNIP: Turnips have been used to treat rheumatoid arthritis, edemas, headaches, jaundice, hepatitis, and sore throats. Turnip is rich in glycosylates and isothiocyanates that have anti-tumor properties.

W **WATER APPLE /WAX JAMBU / ROSE APPLE:** Medicinally the astringent blossoms are boiled and used to treat fevers and diarrhea. A decoction of the astringent bark is a local application on thrush. It boosts good HDL cholesterol, increases metabolism, and prevents constipation. Because it is mostly water with some potassium, it helps pre-vent muscle cramps. The leaves of water apple can be used as a tea as a possible supple-ment for type II diabetes patients. This fruit is being studied for numerous pharmacological antidiabetic and antioxidant properties. Experiments with various parts of the tree are being conducted for anti-inflammation, wound healing, antibacterial, and anticancer properties.

WATER HYSSOP / BABY TEARS: *Bacopa monnieri* has been used in Ayurvedic/traditional medicine treatments for centuries. Water hyssop supposedly provides a cornucopia of treatments. But there are very few scientific studies to back up the Ayurvedic-homeopathic natural medicinal claims. Taken regularly, this herb purifies the blood, strengthens the immune system, and may improve cognition and memory. It increases sex drive, cures impotence, is an antidepressant, lowers cholesterol, treats arthritis, insomnia, mental fatigue, inflammations, asthma, and rheumatism. A poultice of the boiled plant is placed on the chest to treat bronchitis and other coughs. Burns are treated with this plant's juice as a wash. A tea of water hyssop leaves (fresh or dry) is relaxing and relieves anxiety and stress. Take these supplements only after consulting a doctor or a health expert. DON'T use it while taking birth control pills or during estrogen replacement therapy.

WATERMELON: Watermelon is high in disease fighting beta-carotene. Research also suggests that red-pigmented foods provide this protection. Lycopene and beta-carotene work with plant chemicals, which are not found in vitamin/mineral supplements. Watermelon is the leader in lycopene among fresh fruits and vegetables. Watermelon contains such high concentrations of lycopene that may help reduce the risks of prostate cancer. Watermelon seeds contain 'cucurbocitrin' to aid in lowering blood pressure and improve kidney function. The sweet watermelon surprisingly has only half the sugar content of an apple. It tastes sweeter because the sugar is its main taste-producing agent.

WATER SPINACH: The young shoots are a mild laxative and treat diabetes and fever. The leaves are crushed as a poultice on sores and boils. Ringworm is treated using a paste made from the blossoms. The roots treat opium or arsenic poisoning, and after drinking contami-nated water. The plant possibly regulates glucose uptake in the liver and reduces blood glucose levels.

WEST INDIAN PEA / VEGETABLE HUMMINGBIRD TREE: Every part of the hummingbird tree is used for traditional-homeopathic medicine. The juice of the leaves and flowers treat fevers, headaches, and nose, sinus, and throat ailments. The bark and leaves are useful in poultices for inflammation and wounds. Eating the flowers with the stamen removed and the leaves relieve constipation. The juice of the bark mixed with honey treats chronic intestinal disorders. Hummingbird tree will treat intestinal para-sites/worms. A few drops of juice extracted from the blossoms may brighten your vision.

WHITE TURMERIC / ZEDOARY: Various parts of the white turmeric plant are used in Ayurveda for the treatment of different ailments. Traditional medicine claims it treats cancer, ulcers, diarrhea, dysentery, and even toothaches. Preparation methods be-gin with careful washing with lots of water to remove most of the exterior protein and wa-ter-soluble nutrients. The rinsing is supposed to remove a poison that is yet to be identi-fied. Zedoary is used for stomach pain, loss of appetite, indigestion, as a mosquito repellent, and many other conditions, but there is no good scientific evidence to support these uses. This root also has antioxidants that reduce the risk of serious heart disease and diabetes.

WHITE YAM / GREATER YAM: A paste made from pounding white yam can be ap-plied as a treatment for wounds, inflamed hemorrhoids, and skin diseases. Dried powdered yam treats piles. Juice of the tuber is used to kill stomach parasites. It is a laxative and treats fever.

WINGED BEANS: Medical studies suggest winged bean may help the body fight infections of micro-organisms. A diet rich in winged bean will help prevent wrinkling of premature aging, reduce severe headaches, help reduce arthritis pain and swelling, boosts and protects the nervous system, provides antioxidants, cleanses the liver but may inhibit blood clotting. They may also produce kidney stones. They use concoctions of winged bean seeds and leaves in homeopathic medicinal treatments for ear infections, gout, and water reten-tion. Leaves may be squeezed, the juice warmed, and dropped into an infected or painful ear.

Y **YAM**: Yam flesh is poisonous raw, but cooking make sit safe and edible. Wild Asian yam has traditionally been used in herbal medicine to treat organ system function, especial-ly the kidneys and the female endocrine system. Diosgenin, a natural occurring steroid, in yam makes it an herbal remedy for arthritis, asthma, eczema, carbuncles, diarrhea, menstrual disorders, and certain inflammatory conditions and may help reduce the risk of osteoporosis. Yam extracts are used as a natural alternative to hormonal replacement in women who have reached the age of menopause. Yams' vitamin B6 has been used as a natural herbal supplement for premenstrual syndrome (PMS) in women, especially with the accompanied depression.

Z **ZUCCHINI / COURGETTE:** Cooked zucchini is easy to digest, making it a treatment for constipation and acid reflux. Zucchini treats colds, aches, and various health conditions. Zucchini is also rich in antioxidants. Zucchini contains a lot of water and fiber which promote healthy digestion and may help lower blood sugar levels in people with type 2 diabetes. The fiber, potassium, and carotenoids in zucchini help to lower blood pressure.

Homeopathy cures a greater percentage of cases than any other method of treatment. Homeopathy is the latest and refined method of treating patients economically and non-violently. ~ Mahatma Gandih

THE HAWAIIAN HOME GARDEN GUIDE NATURAL REMEDIES

The home remedies mentioned in the following text are meant to be informative and of historical value. These natural remedies have been derived from various sources. The author states these remedies have not been tested or veri ied. Any natural medicine guide does not replace the advice of your doctor. Never use any home remedy or other self-treatment without being advised to do so by a physician. Please be careful combining anything with prescription drugs. Always store any herbal preparations and prescription drugs away from children and pets. No action or inaction should be taken based solely on the contents of the information. Readers should consult appropriate health professionals on any matter re-lating to their health and well-being

DO NOT consider any of this information to be medical advice or instruction.

No action or inaction should be taken based solely on these articles. Readers, if any of the following conditions are new and persist, please consult health professionals on any matter pertaining to their health and well-being.

THE HAWAIIAN HOME GARDEN GUIDE WILL NOT TAKE ANY RESPONSIBILITY for any adverse effects from the use of plants. Always seek advice from a professional before us-ing a plant medicinally. The use and dosage of herbs and foods vary with circumstances such as climate (whether it is hot or cold), the person's health and age, and the manner the dose is administered. Herbs and foodstuffs should be used for medical purposes only with proper medical advice. The medicinal plants must be grown without chemical fertilizers or pesticides.

ACNE: cabbage, mint, grapefruit, soursop leaves,

Apply affected acne areas with a paste made fom pounding dried orange peel with a tablespoon of pure (boiled) water. Apply cucumber leaves or grated pieces of cucumber to acne areas. Rub the affected acne area with a fresh cut clove of garlic. Drink at least a quart of pure water every day to keep your skin free of impurities that cause acne and pimples. Grapefruit can be rubbed directly on the skin to alleviate pimples and greasiness. Mint juice is an excellent skin cleanser. The sap of young soursop leaves is put on skin eruptions.

ANEMIA: bananas, beets, manioc/cassava

APHRODISIAC/SEXUAL IMPOTENCE: carrot, garlic, ginger, onion, nutmeg, parsley, saffron

Sex is a basic instinct like hunger. Sexual activity, however, demands complete concentration and relaxation. It cannot be performed in haste and tension. Many persons, therefore, su er from sexual dysfunctions. The most common male sexual dysfunction is impotence. The main problem of secondary impotence is the apprehension created by an earlier failure, which generates a good deal of anxiety for the next time. Impotence takes three forms. There is primary impotence when the man's erectile dysfunction is there from the very beginning of sexual activity and he simply cannot have an erection. Secondary impotence is the most common and this implies that the man can normally attain an erection, yet fails on one or more occasions in between normal activity. The third form is advancing age. Garlic is one of the most remarkable home remedies found beneficial in the treatment of sexual impotence. It is a natural and harmless aphrodisiac. According to an eminent sexologist of the United States, garlic has a pronounced aphrodisiac effect. It is a tonic for loss of sexual power due to any cause, and for sexual debility and impotency resulting from sexual overindulgence and nervous exhaustion. Two to three cloves of raw garlic should be chewed daily. Onion is another important aphrodisiac food, second only to garlic. It increases libido and strengthens the reproductive organs. The white variety of onion is, however, more useful for this purpose. A carrot can eaten with half of a boiled egg, dipped in a tablespoon of honey, once daily for a month or two. This recipe increases sexual stamina. The juice extracted from ginger is a valuable aphrodisiac, and beneficial in the treatment of sexual weakness. For better results, half a teaspoon of ginger juice should be taken with a half-boiled egg and honey, once daily at night, for a month.

ARTHRITIS: apples, basil leaf, ginger, green beans, horseradish, leeks, spinach, sugar apple leaves, tamarind blossoms, yam

Drinking raw vegetable juices is beneficial, especially a mixture of carrot, beet, and cucumber. Green salad with lemon juice, cooked vegetables like pumpkin, spices like ginger, coriander, and turmeric all help to keep body joints painless. Apples, oranges, grapes, and papayas should be eaten frequently. Tomatoes, potatoes, and some peppers can hinder body joint actions. If these cause a problem they need to be excluded from your diet. Allergic foods may cause joint pain as a symptom. Identify any problem-causing foods and stop eating them. Consume less tea, sugar, yogurt, chocolate, or fried foods. Change your eating habits for better joint health. Raw potato juice may be a successful treatment for arthritis. Slice a potato thin, without peeling. Place the slices in a bowl filled with cold water overnight. Drink the water the following morning on an empty stomach. Fresh juice can also be extracted from potatoes. A medium-sized potato juiced and diluted with a cup of water should be drunk first thing in the morning. Extract a cup of juice from any fresh green leafy vegetable, mixed with equal proportions of carrot, celery, and red beet juice. This regimen should dissolve any deposits around the joints. Every day drink a cup of fresh pineapple juice and it will help reduce swelling and inflammation. A diet of only eight or nine bananas daily for three or four days is advised in the treatment of arthritis. The juice of one lime diluted with water taken first thing in the morning is a remedy for arthritis. The natural iodine in sea water may relieve arthritis pain. If sea bathing is not possible, the patient should relax for thirty minutes every night in a tub of warm water, in which a cup of common salt has been mixed. Blend a few threads of saffron into petroleum jelly and spread a thin layer of the salve on the affected areas in the morning and evening. Use the salve until the joint pain abates. A bath in sugar apple leaves relieves severe arthritis. Boiled tamarind leaves and flowers as poultices treat sprains and arthritis.

ASTHMA: almonds, carrot, coffee, Indian jujube, fennel, green beans, lemon, lettuce, marjoram, mint, nutmeg, onions, oranges, spinach, starfruit, garlic, basil leaves, honey yam.

Various remedies include: Add fifteen drops of fresh garlic juice in warm water and drink for asthma relief. Combine a quarter cup of onion juice with one tablespoon water and a pinch of black pepper and drink. Consume three cups of a combination of fresh carrot and spinach juice daily. Add 30-40 leaves of basil to a liter of water, strain the leaves, and drink the water throughout the day as an effective aid for asthma. For quick relief from asthma breathe the vapors from a jar of fresh honey. At every meal drink a glass of water with the juice of one lemon. Lettuce tea made from a half cup of lettuce to a cup of boiling water relieves stress and is a good body tonic fighting colds, viruses, and asthma. Ground nutmeg is also smoked in India for asthma. A couple of cups of strong, regular black coffee will also have a beneficial effect on asthma. Powdered starfruit seeds serve as a sedative in cases of asthma.

BAD BREATH: cumin, fennel, mint, nutmeg, rosemary

BALDNESS/HAIR: bay leaf oil, black pepper, lime seeds, onions, rosemary, sage, thyme

Various remedies include: Massage one tablespoon on the bald patches daily of one cup mustard oil boiled with four tablespoon henna leaf. Apply to bald patches daily a mixture of ground lime seeds and black pepper corns in equal proportions. Scrub the bald area with onions until it becomes red. Then apply honey. Rub scalp daily with one teaspoon of vegetable oil in which raw mangoes have been preserved for over a year. The Chinese combine rosemary and borax to treat baldness. To make a rosemary hair rinse boil ten sprigs fresh rosemary in two cups water for thirty minutes. Cool and store in a tightly sealed container. Refrigerate. Once a week combine one-half cup of this rosemary water with a cup of warm water and work through your hair after shampooing. You will be amazed at the luster. To hide gray hair put one-half cup of dried sage in two cups of water and simmer for half an hour. Cover and allow the sage water to steep overnight. Over a basin, rinse hair at least ten times reusing the same sage water. On the last rinsing, allow the sage water to dry on the hair before rinsing with fresh water. Repeat every week until your hair is the desired shade. Then rinse every month to retain your hair color. Thyme is a good hair rinse to fight dandruff.

BLOATING: cumin, ginger

The most effective home remedies for bloating are herbal teas. You can brew chamomile, ginger, peppermint, or basil tea. Another natural remedy is to boil one tea spoon of ground cinnamon with water, add some honey.

BLOOD CLEANSERS: aloe vera, carrots,

BLOOD PRESSURE/ HYPERTENSION (lower)) beetroot, bitter melon leaf tea, bananas, celery, breadnut leaves, chocolate, chayote, garlic, green beans, jackfruit, mauby bark, nutmeg, onion, papaya, paprika, roselle, soursop, sour tamarind leaves, spinach, watermelon seeds

Reducing intake of salt/sodium, while increasing potassium and magnesium may lower blood pressure. Consume bananas, melons, grapefruit, oranges, cabbage, cauliflower, and other fresh vegetables and fruits rich in potassium. Magnesium is present in nuts, rice, wheat germ, beans, soy, and also in bananas. Supplementing with oral calcium will also help. Celery assists the dilation of the muscles that regulate blood pressure. Celery juice, carrot juice, and water may lower your BP. Garlic helps to lower cholesterol and increase circulation of blood. Drink a glass of raw beetroot juice twice a day for at least a week. Prepare a bottle of equal proportions onion juice and honey, and drink two tablespoons daily for a month. Eat a papaya on an empty stomach daily for a month, and do not eat anything else for about two hours. Drinking roselle tea will relieve hypertension. Watermelon seeds contain 'cucurbocitrin' to aid in lowering blood pressure. Slice immature soursop fruit and soak in water overnight in the fridge and drink daily. Drink a cup of tea made with sour tamarind leaves or mauby bark daily.

BOILS: cashew bark, garlic, guava leaves, onion, parsley, roselle, turmeric

Boils are mainly caused by bacterial staphylococcus germs that enter pores or hair follicles. Cleanse your system thoroughly for treatment of boils. Begin by fasting with a glass of equal portions of orange juice and water for three to four days, or eat only fresh fruits for a week. Onion juice or garlic juice may be applied on boils externally to draw, break, and evacuate pus. Boil parsley in water till it becomes soft and apply it on the boils as a poultice. Cover with a fresh, dry wash cloth. Roast a few roots of turmeric. Then dissolve the ashes in a cupful of water and apply to boils. This solution will cause the boil to burst. Roselle leaves that have been warmed in hot water can be used on the skin to draw out boils. Make a tea of guava and cashew bark and drink a cup three times a day.

CANCERS: brazil nuts, broccoli, cabbage, manioc/cassava, Indian jujube, garlic, ginger, green beans, jackfruit, leeks, lady fingers/ochro/okra, onions, oranges, bok choy, papayas, peanuts, pumpkin, spinach, sugar cane juice, tomato, watermelon Tomatoes reduce the risk of prostate cancer if eaten almost daily. Watermelon contains high concentrations of lycopene that may help reduce the risks of prostate cancer. Eggplant in a poultice may help against skin cancer.

CATARACTS: Almonds are valuable in cataract prevention. To help strengthen the eyes grind seven almonds finely and combine with half a teaspoon black pepper in half a cup of water, and sweeten with honey. Drink daily. An ancient Egyptian remedy for cataracts is to put a few drops of unprocessed pure honey in the eyes. Extract the juice of pumpkin flowers and apply externally on the eyelids twice a day. This is purported to stop further clouding of the eye lens.

CHOLESTEROL (LOWER): almonds, cabbage, onions, garlic, chickpea, dark chocolate, eggplant, lady fingers/ochro/okra, dry orange peel tea, papayas, peanuts, turmeric Cholesterol, a yellowish fatty substance, is one of the essential ingredients of the body. Although it is essential to life, it has a bad reputation, being a major villain in heart disease. Every person with high blood cholesterol is regarded as a potential candidate for heart attack or a stroke. Most of the cholesterol found in the body is produced in the liver. However, about 20-30% generally comes from the food we eat. Cholesterol is measured in milligrams per 100 millimeters of blood. The normal level of cholesterol varies between 150 - 200 mg per 100 ml.

It is also caused by taking rich foods and fried foods; excessive consumption of milk and its products like clarified butter, butter, and cream; white flour, sugar, cakes, pastries, biscuits, cheese, and ice cream; and non-vegetarian foods like meat, fish, and eggs. Other causes of cholesterol increase are smoking and drinking alcohol. Stress has also been found to be a major cause of high cholesterol. Lecithin is a fatty food substance that breaks up cholesterol into small particles easily handled by the system. With sufficient intake of lecithin, cholesterol cannot build up against the walls of the arteries and veins. Lecithin also increases the production of bile acids made from cholesterol, thereby reducing its amount in the blood. Egg yolk, vegetable oils, wholegrain cereals, soybeans, and unpasteurized milk are good sources of lecithin. It can also be taken in powder or capsules. HDL cholesterol, or 'good' cholesterol, appears to scour the walls of blood vessels, cleaning out excess cholesterol. It carries the excess cholesterol - which otherwise might cause coronary artery (heart) disease - back to the liver for processing. Measure a person's HDL cholesterol level, is measuring how their blood vessels are being 'scrubbed' free of cholesterol. HDL levels below 40 mg/dL result in an increased risk of coronary artery disease, even in people whose total cholesterol and LDL cholesterol levels are normal. HDL levels between 40 and 60 mg/dL are considered 'normal'.

However, HDL levels greater than 60 mg/dL may actually protect people from heart disease. LDL cholesterol is called 'bad' cholesterol, because elevated levels of LDL cholester-ol (lipoprotein deposits) increase the risk of coronary heart disease by forming a hard, thick substance called cholesterol plaque. Over time, cholesterol plaque causes thickening of the artery walls and narrowing of the arteries, a process called atherosclerosis. To reduce the risk of heart disease, it is essential to lower the level of LDL and increase the level of HDL. This can be achieved by a change in diet and lifestyle. As a first step, foods rich in cholesterol and saturated fats, which lead to an increase in the LDL level, such as eggs, meats, cheese, butter, bacon, beef, and whole milk, should be eaten minimally. Virtually all foods from animals, and coconut and palm vegetable oils are high in saturated fats. These should be replaced by polyunsaturated fats such as corn, safflower, and soy bean, and sesame oils which can lower the level of LDL. Persons with high blood cholesterol level should drink at least eight to ten glasses of water every day, to eliminate excess cholesterol from the system. Regular physical exercise promotes circulation and helps maintain the blood flow to every part of the body. Gardening, jogging, swimming, and bicycling, are excellent forms of exercise. Onion juice reduces cholesterol and works as a tonic for the nervous system. It cleans blood, helps the digestive system, helps control insomnia, and regulates the heart action. Regular drinking of a decoction of coriander seeds helps lower blood cholesterol. It is a good diuretic and helps stimulate the kidneys. It is prepared by boiling two tablespoons of dry seeds in a glass of water and straining the decoction after cooling. This decoction should be taken twice daily.

COLDS: garlic, ginger, lemon, yellow mustard powder, onion, orange, oregano, pineapple, sapodilla, sugar apple leaves

Lemon is an important home remedy for the common cold. It is beneficial in all types of colds with fever. Vitamin C-rich lemon juice increases body resistance, decreases toxicity, and reduces the duration of the illness. One lemon should be squeezed and diluted in a glass of warm water, and a teaspoon of honey should be added. This should be taken once or twice daily. Garlic soup of four cloves of chopped garlic boiled in a cup of water is an old remedy to reduce the severity of a cold, and should be taken once daily. Garlic contains antiseptic and antispasmodic properties, besides several other medicinal virtues. The oil contained in this vegetable helps to open up the respiratory passages. In soup form, it flushes out all toxins from the system and thus helps bring down fa fever. Another effective treatment for the com-mon cold is five drops of garlic juice combined with a teaspoon of onion juice diluted in a cup of water; drunk two to three times a day. Ginger is another excellent remedy for colds and coughs. A small piece of ginger should be cut into small pieces and boiled in a cup of water, strained, and combined with a half teaspoon of honey or sugar. This decoction should be drunk hot. Ginger tea or a teaspoonful of ginger juice taken with equal quantity of honey brings relief. Lettuce tea made from a half cup of lettuce to a cup of boiling water relieves stress and is a good body tonic fighting cold viruses and asthma. A tea of old, yellowed sapodilla leaves is a reputed remedy for coughs and colds.

CONSTIPATION: bananas, turmeric, coriander, cucumbers, dunks/jujubes, Malay apple. Use spice such as cumin powder, turmeric powder, and coriander while cooking to make food easily digestible.

COUGH: almonds, ginger, lemon, onion, nutmeg, oregano, sage, sapodilla, roselle

Soak almonds overnight. Remove their skin. Make a paste of these almonds with a lit-tle butter and sugar. Very useful for a dry cough. This paste should be taken in the morning and evening. Mix 2 tablespoons of pure aloe vera gel into a glass of apple juice or cranberry juice. Drink this once a day for 5 days, either in the morning or before going to bed. Pour boiling water over two teaspoons of sage leaves. Let mixture steep for ten minutes. Strain, then drink one to two cups per day to relieve symptoms of coughs, tonsillitis, and respiratory infections. Raw onion is should be juiced. One teaspoon of onion juice should be mixed with one teaspoon of honey and be taken twice daily. Onions are also useful in removing phlegm. One medium onion should be crushed and combined with the juice of one lemon, and one cup of boiling water. Add a teaspoon of honey for taste. This remedy should be taken two or three times a day. Turmeric root should be roasted and powdered. This powder should be taken in tablespoon doses twice daily, in the morning and evening. A tea made from the old, yellowed sapodilla leaves is a reputed remedy for coughs and colds. Drinking roselle tea will relieve coughs.

CUTS: cabbage leaf raw in a poultice, coriander, garlic, guava leaves, mint, oranges, Malay apple bark, sage, roselle, thyme, turmeric

A paste of roselle leaves and seeds is used in Egypt as an antibiotic on wounds. Thyme has always been used as a poultice for wounds. Mix turmeric pow-der with cooking oil to make a thick paste. Put on the skin over wounds, bites, or bruises.

DEPRESSION and IMPROVE MOOD: apples, bananas, cardamom, cashew nuts, nutmeg, saffron

The symptoms of depression are a feeling of loss, inexplicable sadness, loss of energy, lack of interest in the surrounding world, and fatigue. A disturbed sleep is a frequent symptom. The diet of a depressed person should completely exclude tea, coffee, alcohol, chocolate, colas, all white flour products, sugar, food colorings, chemical additives, white rice, and strong condiments. Eating apples is a good remedy for mental depression. Substances present iin apples such as vitamin B, phosphorus, and potassium help the synthesis of glutamic acid, which controls the wear and tear of nerve cells. The fruit should be taken with milk and honey. This remedy will act as a very effective nerve tonic and recharge the nerves with new energy and life. Cashew nuts are another valuable remedy for general depression and nervousness. They are rich in vitamins of the B group, especially thiamine, and useful in stimulating the appetite and the nervous system. Cashews are also rich in riboflavin which aids in keeping the body active, cheerful, and energetic. The use of cardamom has proved valuable in depression. The powdered seeds should be boiled in water and a tea prepared in the usual way.

DIABETES: bitter melon juice, cinnamon tea, mango leaves, breadnut leaves, mauby,bark, arambella/pommecythere, Malay apple, annatto/roucou roots, and sage

Diabetes mellitus (commonly referred to as just diabetes) is a blood sugar disease in which the body either does not produce, or does not properly utilize insulin. Insulin is a hor-mone that is needed to convert sugar, starches and other food into energy needed for daily life. Because diabetics have a problem with insulin, their bodies can't use glucose (blood sugar) for energy, which results in elevated blood glucose levels (hyperglycemia) and the eventual urination of sugar out of their bodies. As a result diabetics can literally starve themselves to death. Diabetes is now being found at younger ages and is even being diagnosed among chil-dren and teens. Every cell of our body is surrounded by an oily membrane that separates it from the surrounding extra cellular fluid. This oily membrane is designed to allow nutrients and oxygen to flow in, and carbon dioxide and waste products to be removed. The consumption of fats / oils causes a hardening of the oils / fats in our cells' membrane, preventing hormones and nutrients from passing into them. This is not only a cause of Type 2 Diabetes, but also may also cause many other chronic diseases and health problems. It is caused almost entirely by the replacement of traditional fats and oils by modern fats and oils that harden in the body. Each body cell, for reasons which are becoming clearer, finds itself unable to accept glucose from the bloodstream. The glucose then remains in the bloodstream, or is stored as body fat or as glycogen, or is passed in urine. The pancreas compensates in early stages of insulin resistance, by producing more and more insulin. This stress of excessive insulin production eventually wears out the pancreas. The most common causes of type 2 diabetes are poor diet and/or lack of exercise; both of which can result in insulin resistance, a condition where the cells in our bodies aren't sensitive enough to react to the insulin produced by our pancreas. To fight the incidence of diabetes it is considered best to eliminate almost all vegetable, seed, bean, and nut oils obtained by heat extraction. The only ones that are considered safe state 'cold pressed/ unheated' on the label and most usually are obtained at a health shop. When cooking–frying, roasting, or grilling - the higher the oil temperature, or reusing the oil are more harmful to the body. To regain a soft oily membrane around the billions of cells in our body virgin olive, canola, and sunflower oils may be used. These will gradually dissolve hardened fats and oils and expel them from the body. Iodine as a supplement has many benefits, with just a few drops a day added to a drink. This is a convenient, cost effective way to become healthier and con-tribute to normalizing blood sugar levels. North American ginseng may improve blood sugar control. Magnesium is a mineral found naturally in foods such as green leafy vegetables, nuts, seeds, whole grains, and in nutritional supplements. Magnesium helps regulate blood sugar levels, and is needed for normal muscle and nerve function, heart rhythm, immune function, blood pressure, and for bone health. The mineral zinc plays an important role in the production and storage of insulin. Zinc is found in fresh oysters, ginger root, lamb, split peas, egg yolk, rye, beef liver, lima beans, almonds, walnuts, sardines, chicken, and buckwheat. Some research has discovered cinnamon improves blood glucose control in people with type 2 diabetes. Aloe vera gel, a home remedy for minor burns and other skin conditions, may help people with diabetes. Bitter melon is the best reputed home remedy for diabetes. Eat this vine vegetable or drink at least one tablespoon of the juice daily. Ten fresh mango leaves soaked in a pint of water

Everything that happens to us is a reflection of who we are. ~ Deepak Chopra overnight, squeeze, then filter the water through a cloth, and drink every morning to control early diabetes. Clinical studies have also shown that it can lower blood sugar in cases of diabetes. Boil annatto/roucou roots or mauby bark in water and drink a cup daily for a week.

***Let thy food be thy medicine and thy medicine be thy food*.** ~
Hippocrates (460-377 B.C.

DIARRHEA: West Indian/acerola cherry, allspice, arambella/pommecythere bark, carrot soup, elephant apple/chalta, coriander, Indian jujube/dunks bark, ginger, pomegranate, guava leaves, oregano, sapodilla, lablab/seim beans, soursop leaves, starfruit, yam

Diarrhea occurs due to over indulgence in food and or alcohol, incomplete digestion causing food to rot in the intestines, nervous stress, some antibiotic drugs, and laxatives. Parasites, germs, virus, bacteria, poisons, and allergies can also attribute to this inconstancy of the bowels. Carrot soup is an effective remedy. It supplies water to rehydrate; replenishes sodium, potassium, phosphorus, calcium, sulfur, and magnesium; supplies pectin; and coats the intestine to allay inflammation. It also checks the growth of harmful intestinal bacteria and prevents vomiting. Boil a pound of carrots in a quart of water until soft. Add a tablespoon of salt and give in small amounts to the patient every half an hour. Ginger in a tea, being carminative, aids digestion by stimulating the gastrointestinal tract and will assist with the nausea of diarrhea. Another treatment is combining teaspoons of fresh mint juice, lime juice, and honey: given three times daily. Drinking pomegranate juice will fight diarrhea. Mango seeds should be dried in the shade and powdered, and a dose of about one and a half to two grams with or without honey, should be administered twice daily. Young sapodilla fruits can be boiled and eaten as a remedy for diarrhea. Boiled starfruit will also relieve diarrhea.

DYSENTERY: West Indian/acerola cherry, dunk bark, oregano, Malay rose apple root Dysentery is an inflammatory disorder of the intestine, especially of the co-lon, that results in severe diarrhea. Basically the same remedies as diarrhea apply.

EYE WASH: The liquid squeezed from heated plantain leaves dropped into the eyes, or use cooled green tea water.

FERTILITY: carrots

FEVERS:basil, cardamom, grapefruit, honey and ginger, potato slices, onions, lemon, oregano, sage, starfruit juice, sugar cane juice

Fever is when the body's temperature exceeds 37.5 C, or 98.6 F. Drinking plenty of water helps to lower body temperature. Sleep under a warm blanket. Permit the body to sweat through the night, and in the morning the fever should be gone. A decoction made of about forty holy basil leaves boiled in a pint of water combined with a half cup of milk, one teaspoon of sugar, and a quarter teaspoon of powdered cardamom drank twice daily should lower the temperature. The juice of grapefruit mixed equally with water quenches thirst and removes the burning sensation produced by the fever. Combine three drops of ginger juice with a teaspoonful of honey to help break the fever. Slice potatoes and place the slices on your head, chest, and stomach to help reduce the fever. Drink a tea of lemon juice and honey several times daily. To lower the body temperature tie or tape half onion to the sole of each bare foot as long as possible. Cane juice is beneficial to drink to break fevers.

GAS / FLATULENCE: ginger and lime juice, peppermint, elephant apple/chalta, coriander, cumin, dill, fennel, hot pepper seeds, nutmeg, starfruit, turmeric

Stomach gas can be caused by the presence of excessive bacteria in the intestines, drinking too much beer, eating too much fibrous food like cabbage, broccoli, or cauliflower, or yeasty foods such as breads and cheeses. Carbonated drinks also produce gas. After meals chew fresh ginger slices soaked in lime juice. Chewing peppermint fights flatulence, bloating, and abdominal pain that accompanies gas. A cup of tea daily from hot pepper leaves and seeds helps.

GENERAL RELIEF: Combine 2 TBS aloe vera gel, 1 TBS medicinal charcoal, 1 TS molasses, and 1 egg white, and drink.

HEALTHY HEART: almonds, bananas, carrots, elephant apple/chalta, peanuts, ambarella, saffron, and plenty of exercise.

HEADACHES/ MIGRAINES: apples, basil, bay leaves, grapes, cinnamon, gin-ger, green beans, lemon, marjoram, mint, mustard seeds, starfruit leaves Headaches may be caused by allergies, emotional stress, eye strain, high blood pressure, a hangover, infection, low blood sugar, nutritional deficiency, tension, and or the presence of poisons and toxins in the body. Make a decoction of half a teaspoon mustard seed powder and three teaspoons water and snort into the nostrils to fight migraines. The juice of ripe grapes will fight a mi-graine. Eating apples, peeled and cored, with a little salt every morning on an empty stom-ach for about a week fights all types of headaches. Crush a few cabbage leaves, place in a cloth and apply on the forehead for extended period of time. Use fresh leaves when the compressed leaves dry. Also try bay leaves in a damp wash cloth resting on the eyes. For a sinus headache, eat a jalapeno pepper as soon as possible. Combine a teaspoon of ground cinnamon in a teaspoon water and apply on the forehead. Make a paste of dry ginger with a little water or milk, and apply to the forehead for sinus. Grind a dozen basil leaves with four cloves and a teaspoon of dried ginger into a paste and apply to the forehead for sinus. The juice of three or four slices of lemon in a cup of tea gives immediate relief. Cinnamon should be mixed with water and applied to the temples and forehead to obtain relief. An infusion of the leaves of the herb marjoram drank as a tea is a treatment of a nervous headache.

A man's health can be judged by which he takes two at a time
- pills or stairs. ~ Joan Welsh

HEART DISEASE: corn, peanuts, arambella/pommecythere, spinach, tarragon, turmeric Many diseases are related to the heart, arteriosclerosis and atherosclerosis, angina, heart attack, heart failure, arrhythmias, heart murmurs, rheumatic heart dis-ease, and high blood pressure or hypertension. The heart is a muscle and needs the blood to supply oxygen and nutrients. If coronary arteries become narrowed or clogged, and cannot supply enough blood to the heart, the result is coronary heart disease.

HICCUPS: peanut butter, sugar, elephant apple/chalta, dill Hiccups start from the diaphragm, a dome-shaped muscle in the chest that pulls and pushes air in and out of the lungs. Hiccups are caused when the diaphragm is irritated. Generally hiccups last for a few minutes, yet sometimes they may last for days or weeks. This may be a sign of some other medical problems in the body. There is no special diet recommended for hiccups. However, it is advisable to avoid hot and spicy food because they can irritate the lining of the esophagus. Some natural remedies are to hold your breath, gargle with water, sip ice water quickly, close your eyes and gently press your eye balls, drink a glass of soda water quickly, eat some sugar, or eat one tablespoon of either peanut butter or mustard.

HIGH BLOOD PRESSURE: garlic, grapefruit, lemon, parsley, potatoes, rice, water-melon seeds

Hypertension or high blood pressure is a disease of the modern age. Blood pressure is measured with an instrument called a sphygnomanometer which measures in millimeters of mercury. The highest pressure reached during each heart beat is called systolic pressure, and the lowest between two beats is known as diastolic pressure. Most young adults have blood pressure around 120/80. It increases normally with age, even going to 160/90 The main causes of high blood pressure are stress and a faulty style of living. Hardening of the arteries (athero-sclerosis), obesity, and diabetes cause hypertension. Persons with high blood pressure should always follow a well-balanced routine of a proper diet, exercise, and rest. The pressure can be lowered by eating fruits, and a vegetarian diet consisting of fresh fruits and vegetables. Salt should be avoided. Persons suffering from hypertension must get at least eight hours of good sleep, because proper rest is a vital treatment. Most important of all, the patient must avoid over strain, worries, tension, and anger and create a calm, cheerful attitude, and contented state of mind. Garlic is regarded as an effective means of lowering blood pressure. It slows down the pulse rate and modifies the heart rhythm, besides relieving the symptoms of dizziness, numbness, shortness of breath, and the formation of gas within the digestive tract. It may be taken in the form of two raw cloves daily. Lemon is essential for preventing capillary fragility. Grapefruit is helpful in toning up the arteries. Watermelon seeds, dried, and roasted, should be taken in liberal quantities to lower the blood pressure. Natural brown rice is perfect food for hypertensive people who have been advised to change to a salt-restricted diet because it has low-fat, low-cholesterol, and low-salt content. Calcium in brown rice, in particular, soothes and relaxes the nervous system, and helps relieve the symptoms of high blood pressure. Potatoes are rich in potassium, but not in sodium salts. The magnesium in potatoes has beneficial effects in lowering blood pressure when boiled with their skin. Parsley tea made with a handful of fresh leaves boiled in a liter of water for three minutes, taken several times daily will reduce high blood pressure and help keep the arterial system healthy.

IMMUNE SYSTEM (strengthen: broccoli, cabbage, carrots, cauliflower, greens, jackfruit seeds, long white radish (daikon, red peppers, sweet potatoes, tarragon, yams Emotional stress, lifestyle, and dietary habits can affect the immune system. To strengthen the immune system eat fruits, vegetables, beans, seeds, whole grains, and nuts; especially foods high in carotenes such as yellow and orange squash, dark greens, carrots, yams, sweet potatoes, red peppers, tomatoes, and foods in the cabbage family. Brussel sprouts, cauliflower, broccoli, radish, and turnip can help prevent low immunity. Eat adequate amounts of protein, keep the diet low in fats and refined sugars.

INDIGESTION: allspice, aloe vera, almonds, bananas, cab-bage leaf (boiled), cinnamon, cloves, coriander, lemon, cumin, dill, Indian ju-jube, ginger, mace, mango, marjoram, mint, nutmeg, pineapple, long white radish, rosemary, saffron, lablab beans, soursop leaves, tarragon, thyme, turmeric

Indigestion, also known as dyspepsia, is a stomach problem. Indigestion can cause heartburn due to stomach acid reflux. Following are the causes of indigestion eating without chewing properly, heavy food, excess alcohol, smoking, pregnancy, and stress. Improve eating habits to reduce acidity, heartburn, and gastritis. Avoid drinking liquids during meals, and wait at least 15 minutes to drink after you have eaten. Avoid large meals, take small and frequent meals. Abstain from smoking and alcoholic beverages. Avoid tea, coffee, and other drinks that contain caffeine. Avoid hot spicy and fatty foods. Restrict intake of chocolates. Take at least 30 minute walk daily. Decrease your stress level by relaxing. Regular exercise is good for digestive system. Eat your meals on time and chew them properly. Lemon juice or cider vinegar in a glass of water drank before a meal will help prevent acid indigestion. Take one piece of clove and suck on it slowly for relief from acidity. Eat a cup of vanilla ice cream, or drink a glass of cold milk to get heartburn and acidity relief within minutes. Eat several almonds when heartburn symptoms persist. Cut a lemon into thin strips and dip in salt and eat before meals to prevent heartburn. Mix equal parts of baking soda and water in a glass. Drink as soon as you feel indigestion coming. The gel of aloe vera can also help improve digestion.

INSECT BITES: reduce irritation and swelling: banana skin, elephant apple/chalta, fennel, lemon, lime juice, papaya, plantain, thyme, turmeric

Hundreds of insects that bite or sting, from mosquitoes to spiders, can cause serious reactions in the body. The problem is not the injury, but what the insect leaves behind - the venom. The area becomes swollen, red, extremely painful, or burning. If the symptoms are, hard to breathe or swallow, disorientation, swelling of eyes and mouth seek a doctor as fast as possible because in some cases, especially with allergic reactions, stings can cause unconsciousness or death. Remove the stinger. Do not squeeze it, this will in-ject more of the venom in the patient. Clean the area. Crush plantain leaves extracting the juice and applying it on the injured area. Apply toothpaste on the sting. For mosquito bites apply lime juice diluted with water over the bite, or rub dry soap over the mosquito bite. For wasp or bee stings make a poultice of ripe or green papaya Try a mud pack of fresh dirt (clay if possible) and water. Sapodilla seeds can be crushed into a paste to sooth insect bites. Thyme has always been used as a poultice for insect bites. Mix turmeric powder with cooking oil to make a thick paste. Put on the skin over wounds, bites, bruises.

INSOMNIA: calabash leaves, honey, lettuce, marjoram, nutmeg

Often worrying about falling asleep is enough to keep one awake. The most common cause of sleeplessness is mental tension brought about by anxiety, worries, overwork, suppressed resentment, anger, bitterness, or over excitement. It is also caused by constipation, overeating, and excessive tea/coffee/smoking. A balanced diet with simple modifications in the eating pattern will go a long way in the treatment of insomnia. Avoid white flour products, sugar products, tea, coffee, cola drinks, alcohol, fatty foods, and fried foods. Two teaspoons of honey in a glass of water taken before bedtime induces a sound sleep. Babies generally fall asleep after taking honey. Lettuce soup is also useful in avoiding insomnia. Drink tea made from calabash leaves.

JAUNDICE: beet root, lemons, lime, pigeon peas, long white radish (daikon/morai, sugarcane juice, tomatoes, watermelon, yam

The symptoms of jaundice are extreme weakness, headache, fever, loss of appetite, severe constipation, nausea, and yellow discoloration of the eyes, tongue, skin, and urine, and or a dull pain in the liver region. Obstructive jaundice may be associated with intense itching. Jaundice may be caused by an obstruction of the bile ducts which normally discharge bile salts and pigment into the intestine. The bile gets mixed with blood and this gives a yellow pigmentation to the skin. The obstruction of the bile ducts could be due to gallstones or inflammation of the liver, which is known as hepatitis caused by a virus. The green leaves of the long white radish should be pounded and their juice extracted through cloth. Drink a pint of this juice daily for adults. It induces a healthy appetite and proper evacuation of bowels, and this results in gradual decrease of the trouble within eight or ten days. A glass of fresh tomato juice, mixed with a pinch of salt and pepper, taken early in the morning, is considered an effective remedy. Taking 2 TBS of the juice from the green leaves of pigeon pea daily fights jaundice. One glass of pure sugar cane juice combined with the juice of half a lime taken twice daily will help recovery. Drink a half cup of fresh lemon juice mixed with water several times a day to regenerate damaged liver cells. Sugar cane juice is said to speed up the recovery process after jaundice especially when mixed with lime juice. Watermelon seeds contain 'cucurbocitrin' to improve kidney function.

KIDNEY STONES: apples, basil, celery, grapes, pomegranate, watermelon, chayote leaves, coconut, grapefruit, long white radish (daikon), starfruit

Stones in the kidneys or urinary tract are formed from the chemicals usually found in the urine such as uric acid, phosphorus, calcium, and oxalic acid. Stones grow because a substance in the urine exceeds its solubility. Most kidney stones are composed either of calcium ox-alate or phosphate. A person with kidney stones should avoid foods, which irritate the kidneys, such as alcoholic beverages, condiments and pickles, cucumber, radish, tomato, spinach, onion, beans, cabbage, and cauliflower; meat and gravies; and carbonated waters. A low protein diet is recommended. Try to drink a gallon of water daily to flush the system. A half cup of long white radish/daikon daily will prevent kidney or gall stones. DO NOT EAT COLOCASIA / EDDOES IF YOU HAVE A PROBLEM WITH KIDNEY STONES. One teaspoon basil juice and honey taken daily for six months may expel stones. Eating apples and celery prevents stone formation. Grapes or grape juice is an excellent remedy for kidney stones. two tablespoons cooked mung or udri beans will help dissolve stones. Watermelon is one of the safest and best diuretics which can be used to fight kidney stones because it has plenty of water and potassium. Eating crushed sapodilla seeds are claimed to expel bladder and kidney stones

In order to change we must be sick and tired of being sick and tired. ~ Author Unknown

LEG CRAMPS: bananas

Leg cramps are a painful and extremely discomforting involuntary contraction of a single muscle or a group of muscles, the calf muscle, the hamstring, or the quadriceps in the leg. The duration ranges from less than a minute to several minutes at times. Increase water consumption to stay well hydrated throughout the day. Potassium and calcium-rich foods will help to prevent muscle cramps. Increase your intake of calcium-rich foods to about 1,200 milligrams of calcium a day. Concentrate on fresh yogurts and soy and tofu. Stretch the sore muscle, follow your instinct, your body will automatically guide you in the correct manner. Eat one or two bananas a day, drink plenty of fluids, and stretches help relax muscles.

LIVER AILMENTS: beetroot, carrot juice, Indian jujube / dunks, grapes, grape-fruit, lemon, papaya, lime, chalta, radish, annatto roots and seeds, soursop, sour tamarind

The liver may become damaged or diseased due to bacterial infection or an injury. Gall stones often obstruct the normal low of bile, causing an unnecessary accumulation and infection and/or cholesterol and triglycerides may also begin to accumulate in the liver. This is one of the major causes of infection and liver damage. Chemicals or an over-consumption of minerals may also cause the liver cells to become damaged. Excessive use of alcohol over a long period is the most potent cause of cirrhosis of the liver in adults. The black seeds of papaya have been found beneficial in the treatment of cirrhosis of the liver caused by alcoholism or malnutrition. A tablespoon of juice obtained by grinding the seeds, mixed with ten drops of fresh lime juice, should be given once or twice daily for about a month as treatment while abstaining from alcohol in any form. The liver can be cleansed by fasting, only drinking a juice for seven days. Use red beetroot, lemon, papaya, or grapes. A diet rich in vitamins A and C is the best for keeping your liver healthy. However, if the liver is diseased, a qualified doctor should prepare a diet. Carrot juice can help detoxify the liver on a regular basis. Taken when fasting, soursop juice is believed to relieve liver ailments. Boil the seeds and roots of the annatto tree into a tea and drink three times a day as a liver purifier. Soak sour tamarind in hot water overnight and drink.

MENSTRUAL PROBLEMS: beetroot, carrot, cucumber, fennel, ginger, green beans, guava leaves, papaya, parsley, yam

The two major female sex hormones in the body are estrogen and progesterone. They are produced in a pair of organs in the abdomen, known as the ovaries. Begin with an all-fruit diet for about five days, taking three meals a day of fresh, juicy fruits. Adding a glass of milk to each fruit meal is recommended if this causes weight loss. Parsley is most effective in treatment of menstrual disorders. It increases menstruation and assists in the regularization of the monthly cycle. Cramps are relieved and frequently corrected entirely by the regular use of parsley juice, with beetroot juice; or with beet, carrot, and cucumber juices. The recommended quantity is 75 ml of each of the four juices. A piece of fresh ginger pounded and boiled in a cup of water for a few minutes sweetened with sugar, should be used three times a day after meals. Unripe papaya helps a proper menstrual low. Papaya is especially helpful when menstruation ceases due to stress or fright in young unmarried girls.

MORNING SICKNESS: bananas, cumin, soursop leaves

Morning sickness causes nausea and in many cases vomiting in women who are in the early stages of pregnancy. It usually restricts itself to the first trimester of pregnancy.

Try eating smaller meals more often, so that you are never too hungry or too full at one time. Avoid fatty or fried foods. Keep crackers, bread or toast, cereal, or other bland foods handy. Try eating a few crackers before getting out of bed in the morning. Drink enough fluids, especially if you have been vomiting. Try drinking in between meals rather than with meals. Make a tea from grated root ginger. Steep this in boiled water, leave it to cool and sip it throughout the day. Try sipping peppermint or spearmint.tea.

MOUTH ULCER: basil leaf, coconut milk, turmeric, coriander seeds, tomatoes

Mouth ulcers are small white spots on the inside of the cheek, the tongue, or clustered on the inner side of the lip. They are also known as canker sores. Mouth ulcers may be caused by nutritional deficiencies such as iron, vitamins, especially B12 and C, poor dental hygiene, food allergies, stress, infections - particularly herpes simplex, biting the cheek, or a hormonal imbalance. Apply peppermint oil, or mix coconut milk with honey and massage the gums 3 times a day. Mix a pinch of turmeric powder to one teaspoon glycerin and apply on ulcers. Take one teaspoon of coriander seeds in one cup of water and boil, Gargle with it slightly warm. Gargle 3-4 times a day. Eating raw tomatoes helps mouth ulcers. Also gargle tomato juice four times a day.

MUSCLE ACHES: aloe vera, bay leaf bath, basil, cayenne pepper, peppermint, ginger, guava leaves, mustard oil, turmeric, nutmeg oil Many things can cause muscle soreness. Overdoing it at work or sports can cause small tears in your muscles that can cause extreme discomfort. Dehydration can cause the agonizing pain of a muscle spasm. Muscle cramps can occur from being deficient in certain vitamins. If any of these muscle conditions do not quickly fade away consult a health center. Using a heating pad or hot water bottle may feel good, but it's the worst thing for sore muscles because it dilates blood vessels and increases circulation to the area, which in turn leads to more swelling. Especially if the injury is new – less than 6 weeks - continuous heat can actually increase muscle soreness and stiffness, especially if applied during the first 24 hours after the strenuous activity. Ice is the best remedy for a new muscle ache. It may feel uncomfortable, but the chill numbs the pain as t constricts the blood vessels, relieves the inflammation and limits the bruising. (Cleveland Clinic) As long as it's gentle, a massage can help ease muscle soreness and stiffness. Menthol, a component of peppermint, can be of great bene it when added to a hot, steamy bath. Squeeze the juice of fresh, grated ginger and combine with equal parts of olive or sesame oil to massage into the skin for relief of muscle pain. You can also try bay leaves in a little olive oil to help with pain as well as swelling and sprains. The active compound of cayenne is capsaicin; it can reduce pain at skin level as well as possibly help with deeper pain. It increases blood low and warms the area. Turmeric, a common spice used in India, can be taken internally as a tea and externally in a poultice for pain and inflammation. An aloe vera rub can help the sore area.

NAUSEA: beets, allspice, black pepper, ginger, guava leaves, mace, mango, mint, nutmeg, oregano, soursop leaves Nausea is a symptom of various other conditions. If you are not suffering from upset stomach there are other illnesses that can cause this problem. Nausea is basically a reaction to the presence of some illness in your body. You can treat it by having two capsules of ginger root. It will give you quick relief depending the severity of your nausea. Smaller amounts of food are always good for you diet. Once you start feeling better, have a fiber rich diet such as banana, rice, apple sauce and toast. Mint is a soothing herb which is useful in helping to calm and settle the stomach. Eat half a teaspoon powdered cumin seed to relieve nausea and vomiting. Another natural remedy for nausea is smelling a freshly cut lemon or drink lemon juice in water.

NERVOUSNESS / ANXIETY: almonds, bananas, marjoram, orange, onions, celery, nutmeg, sapodilla, soursop leaves, tarragon Nervousness is a feeling of restlessness, apprehension, and worry. Inhale the fragrance while peeling an orange, or drop an orange in a pot of boiling water and inhale the vapor. Drink the mixture of two tablespoons of honey with one teaspoon nutmeg in a cup of orange juice. Add dried rosemary herb to boiling hot water and let it steep for 15 minutes. Drain the decoction and sip it when it cools. Grate two onions and add two cups celery as a salad to relieve restlessness. Soak ten raw almonds in water overnight. Peel and pound them adding just enough water to form a paste. Add one teaspoon nutmeg, with a pinch of ginger. Eat paste at bedtime. Crushed sapodilla seeds are used throughout the Yucatan as a tranquilizer. Soursop leaves are regarded throughout the West Indies as having sedative or soporific properties. In the Netherlands Antilles, the leaves are put into one's pillowslip or strewn on the bed to promote a good night's sleep. An infusion of the leaves is commonly taken internally for the same purpose. Slice a banana thin, dry in an oven at low heat, and pound into powder. Add two tablespoons of powder to boiling water and drink three times a day.

PARASITES AND WORMS: carrots, carailli/bitter melon seeds, coconut, papaya, pumpkin seeds, starfruit Parasites are organisms that depend on other living organisms (host) for nourishment and protection. There are more than a hundred types of parasitic worms that live in human bodies. They enter the human body through food and water. Eat 2 grated carrots with an empty stomach in the morning every day to remove worms. Fry 10 bitter melon seeds in a little ghee or butter and take 2-3 times in a day. You should see the worms in your stool when they pass. Drinking coconut water daily and chewing coconut continuously for 3 days removes worms. Combine one TBS raw papaya juice with one TS of honey and take on empty stomach in morning followed by a glass full of warm milk with one teaspoon castor oil. This treatment should be done for two to three days. A poultice of crushed starfruit leaves will fight ringworm. Eat raw pumpkin seeds to expel worms.

QUIT SMOKING: bananas, cayenne pepper, ginger, orange juice Every cigarette contains nicotine which is very addictive and causes many problems to the body. Nicotine stimulates different parts of the brain producing a feeling of pleasure, causes adrenaline production to increase, accelerates the heart rate, and increases blood pressure. It also affects the level of some hormones, and the body's temperature. These sudden changes produced by smoking tobacco causes a feeling of pleasure, and is the principal fact that makes quitting smoking so hard. For cravings take powdered cayenne pepper because it desensitizes the respiratory linings to tobacco. It's an antioxidant that stabilizes lung membranes preventing damage. The warm peppery taste reduces cigarette cravings. Ginger prevents nausea and helps quitting, reducing anxiety. Ginger also produces perspiration to flush toxins generated from smoking. Oats reduce or eliminate tobacco cravings, and also reduce the number of cigarettes desired even in those people not trying to quit. Orange juice is acidic in nature and the first step in quitting smoking is to eliminate the nicotine from the body. Drinking orange juice twice a day will be useful.

RESPIRATORY DISEASES / BRONCHITIS: almonds, thyme, cayenne pepper, Indian jujube, fennel, horseradish, garlic, ginger, marjoram, yellow mustard powder, oranges, papayas Signs and symptoms of respiratory disease include shortness of breath after mild physical activity, wheezing and coughing especially at night, blood in cough, drowsiness, loss of appetite, unexplained weight loss, and mild to severe chest pain. Acute bronchitis is most often caused by a virus. Stop smoking and avoid second hand smoke. Get plenty of rest. Avoid any kind of stress. One of the best home remedies for cold and cough is salt water gargles as it can soothe the throat and get rid of mucous. Drink a cup of thyme tea four times a day. For severe bronchitis combine one small cayenne pepper chopped, one clove garlic chopped, one tablespoon crushed horseradish root, one teaspoon yellow mustard powder, one large onion, a pinch of ground ginger, and a pinch of turmeric. Place the cayenne pepper, garlic, ginger, horseradish, mustard, onion, and turmeric in a medium sized saucepan and cover with water. Bring to the boil and then reduce the heat and simmer for twenty minutes. Strain the mixture and allow to cool before drinking.

RHEUMATISM: basil leaf tea, carailli/bitter melon, coffee, green beans, horseradish, marjoram, mint, mustard powder, potato, onions, oranges, papayas, soursop leaves, sugar apple leaves Rheumatism is a painful state in muscles, ligaments, tendons, and joints. It is an acute and chronic illness that affects people of all ages and both sexes. The juice of raw potato is regarded as an excellent remedy for rheumatism. One or two teaspoons of the juice, taken out by pressing mashed raw potatoes, should be taken before meals. This will help to eliminate the toxic condition and relieve rheumatism. The potato skin is exceptionally rich in vital mineral salts, and the water in which the peelings are boiled is one of the best medicines for ailments caused by excess toxic matter in the system. About two big handfuls of potato peelings should be thoroughly washed and boiled in a pint of water until half boils away. Strain this decoction and take four times a day. A cup of bitter melon juice mixed with a teaspoon of honey can be taken daily as treatment for at least three months to provide relief. Take the juice of two or three lem-ons each day. Thoroughly chew six walnuts daily to relieve this pain. A remarkable remedy for rheumatism and gout is said to be as follows: A pint of hot, strong, black coffee, which must be perfectly pure, and seasoned with a teaspoonful of pure black pepper, thoroughly mixed before drinking, and the preparation taken just before retiring. (If you can sleep! Mashed soursop leaves are used as a poultice to alleviate rheumatism. Bathe in sugar apple leaves relieves rheumatism.

SKIN RASHES / DRY SKIN: bay leaves in a bath, beet juice, canistel bark, carrot juice, black pepper, mint, oregano, plantain, Malay apple leaves, soursop leaves, turmeric, yam Dry skin is primarily caused due to a lack of natural oils (sebum being secreted by the pores of the skin. It may also be caused due to an improper or insufficient diet, insufficient intake of water and other fluids. Prolonged exposure to extreme temperatures or harsh chemicals, the sun or prolonged submersion in water, especially sea water, can also dry out your skin. Drink plenty of water and fluids. This is the best thing you can do to keep dry skin at bay. The juice of crushed Malay apple leaves can be used as a skin treatment steeped into baths. Mashed soursop leaves are used as a poultice to alleviate eczema and other skin afflictions. East Indian women with lovely, velvety skin often attribute it to consuming turmeric. Take an aloe vera plant, cut open a leaf and rub the moist insides directly onto the affected skin. This will act as an antiseptic bandage as well as an anti-bacterial, anti-fungal, anti-inflammatory ointment.

SNAKE BITES: plantain, pumpkin One can also use a plantain poultice on a snake or venomous insect bite. **GO TO THE HOSPITAL OR CLINIC IMMEDIATELY!!!!!**

SORE GUMS: basil leaf tea, carrots, basil, coconut, sage Combine sage with your toothpaste. This helps remove plaque and will strengthen bleeding gums.

SORE THROAT: lemon, leeks, sage, sugar cane juice, tamarind For sore throats, try mixing sage tea with apple cider vinegar and salt for gargling. Eating leeks is reputed to keep your voice from becoming hoarse. Tamarind pulp can be a rinse for a sore throat.

STOMACH ULCERS: broccoli, canistel seeds, carrots, basil, coconut, jackfruit

WARTS: banana skin, marigolds, onion

Warts refer to hard growths on the skin. They are common both in children and adults. Warts are capable of spreading, but they are usually harmless. The main cause of warts is a viral infection of the skin. Plantar warts on the soles are usually contracted in swimming pools. Constitutional factors, however, appear to be at the root of the trouble. These factors lead to some defects in the proper development of the skin surface in certain areas. Marigold is anoth-er herb/ lower found as a treatment of warts. The juice of the leaves of this plant can be applied over warts. The sap from the stem has also been found beneficial in the removal of warts. Raw potatoes are a treatment of warts. They should be cut and rubbed on the affected area several times daily, for at least two weeks. This will bring about good results. Onions are also valuable in fighting warts. They are irritating to the skin and stimulate the circulation of the blood. Warts sometimes disappear when rubbed with cut onions.

Let's build wellness rather than treat disease. ~ – Bruce Daggy

Life begins the day you start a garden. – Chinese proverb

THE HAWAIIAN HOME GARDEN GUIDE
GLOSSARY

This glossary defines terms that are used within the previous text. The home rem-edies, or homeopathic/traditional/Ayurvedic treatments mentioned in the following text are meant to be informative and of historical value. The natural remedies mentioned have been derived from various sources. Every remedy **HAS NOT** been tested or verified. Any natural medicine guide does not replace the advice of your doctor. Never use any home remedy or other self-treatment without being advised to do so by a physician. Please be careful when anything is combined with prescription drugs. Always store any herbal preparations away from chil-dren and pets. **NOTE:** Any vitamin or mineral essential to the body's healthy operation has a recommended daily allowance or RDA measured in mg –milligrams or mcg - micrograms.

A **ACHAR**: a pickled article of food or relish as prepared in India.

AMINO ACIDS: There are 22 amino acids, but the human body can only produce 13 naturally. The other 9 amino acids are acquired by eating protein-rich foods. They are termed 'the essential amino acids' because your body needs to acquire them from the foods you eat. They are the building blocks of protein.

ANTI-INFLAMATORY: Anything that helps reduce inflammation, which is the body's reaction to harm, such as infections from germs, damaged cells, or irritants. Inflammation is the body's effort to fight, remove, and heal the tissue. Inflammation does not always mean infection, and is necessary for a wound to heal. Eating a variety of anti-inflammatory foods such as olive oil, walnuts, and pumpkin seeds will reduce inflammation. Avoid refined oils and sugars.

ANTIOXIDANT: In much the same way as oxidation weakens iron and steel with rust causing an eventual breakdown, oxidation inside the body causes a breakdown of cells. An antioxidant helps slow or prevent oxidation of other molecules. Antioxidants are substances that may protect your cells against the effects of free radicals. Within the human body, millions of processes are occurring constantly. These processes require oxygen. Unfortunately, that same life giving oxygen can create harmful side effects, or oxidant substances, which causes cell damage and leads to chronic disease. Free radicals are molecules produced when your body breaks down food, or are caused by stress, alcohol, or by exposures to chemicals, tobacco smoke, and radiation. Free radicals can damage cells, and may play a role in heart disease, cancer, and other diseases. An antioxidant is any substance thought to protect body cells from the damaging effects of oxidation, such as vitamin E, vitamin C, or beta carotene, and beans are some of the best.

AROMATIC OILS: Natural essentialoils used in lotions and perfumes.

ASTRINGENT PROPERTIES: An astringent substance tends to shrink or constrict body tissues, usually locally after topical application. Astringent medicines cause shrinkage of mucous membranes. This can happen with a sore throat, diarrhea, or with peptic ulcers. Externally applied astringents dry, harden, and protect the skin. Acne sufferers use astringents if they have oily skin. Astringents also help heal stretch marks and other scars. Apples, pomegranate, pears, beans, and lentils are very astringent foods. To make a basic astringent to clean skin, slice half a lemon thin, do the same with half an orange (or a lime), and combine with ¾ cup of rubbing alcohol. Strain though a cloth and keep refrigerated. Apply to skin with a cotton ball or gauze pad. Rinse with cool water.

ANTI-SPASMODIC: Prevent spasms of the stomach, intestine, or bladder. Peppermint oil has been traditionally used as an antispasmodic, also yam.

AYURVEDA: This is an alternative natural medicine developed in ancient India and Sri Lanka. The word is a combination of the Sanskrit words *ayur* – meaning *life* and *veda* meaning *knowledge*. Ayurveda, is a system of medicine that relies and is based upon the knowledge of the natural systems of our world inherited from the indigenous peoples. It is believed to have begun in Sri Lanka and India preserved for more than 3,000 years. In Hindu mythology, Dhanvantari was the physician to the gods. Brahma presented him with Ayurveda is attributed to, who received it from Brahma. Its ear-liest ideas were written in the portion of the Vedas more than 5,000 years ago in Sanskrit. The earliest Ayurvedic concepts are written in the Vedas, four sacred texts: the Rig Veda (3000-2500 BCE, Yajur Veda, Sam Veda, and Atharva Veda (1200-1000 BCE. Maharshi Charaka who lived circa 300 BCE, is considered the father of Ayurveda. Ayurveda remains a preferred natural treatment in India, Sri Lanka, and SE Asia. People use this system alone or combined with modern medicine.

B

B COMPLEX VITAMINS: All B vitamins help the body to convert carbohydrates into fuel / glucose, for energy. B vitamins help the body metabolize fats and protein. B complex vitamins promote healthy skin, hair, eyes, and liver and help the nervous system function properly.

BATH PREPARATIONS: For an herbal bath, place a handful of herbs into a teapot or suitable vessel and cover with boiled water. Leave for at least 15 minutes and strain when pouring into bath water. You may add a handful of natural sea salt to the bath. Another method is to place the herbs in a muslin pouch, or tie them in a piece of natural, thin material, and leave to soak in the bath while the hot water is running. The pouch can also be used as a gentle exfoliating rub over the skin after soaking. Use rolled oats to soften the water and soothe irritated skin, particularly if su ering from eczema. A handful of rose petals makes a romantic bathing experience and may ease rheumatic aches and pains. Rosemary soothes aches and pains and stimulates the mind. Soaking with thyme and marjoram will help soothe aches and pains.

Stimulating bath herbs include basil, bay, fennel, lavender lowers, mint, rosemary, sage, savory, and thyme. Soothing bath herbs include catnip, chamomile lowers, comfrey, jasmine lowers, lemon balm, and rose lowers. Bath tonic herbs include comfrey, ginseng root, orange, and raspberry leaves. Bath herbs for muscles and joints include bay, oregano, and sage. Antiseptic bath herbs are eucalyptus and sandalwood. Astringent bath herbs include bay, comfrey, lemongrass, nasturtium lowers, raspberry leaves, rose lowers, rosemary, and witch hazel.

BETA CAROTENE: Beta-carotene is the main source of vitamin A, essential for normal growth and development, immune system function, and vision. Beta-carotene is one of a group of natural chemicals known as carotenes responsible for the red-orange color in many fruits and vegetables as carrots, pumpkins, and sweet potatoes. Beta carotene is converted by the body to vitamin A. It is an antioxidant, like vitamins E and C. Good sources of beta-carotene include carrots, sweet potatoes, squash, spinach, broccoli, romaine lettuce, apricots, and green peppers helps to prevent cancer and heart disease, slow the progression of cataracts, boost immunity, pro-tect the skin against sunburn, asthma, depression, arthritis, and high blood pressure. No RDA.(Recommended Daily Allowance.

BIOTIN: Biotin is vitamin B7 needed to produce fatty acids and glucose for energy. It helps metabolize carbohydrates, fats, and proteins. It is found naturally in food like brewer's yeast, liver, cauliflower, salmon, bananas, carrots, egg yolks, sardines, beans, and mushrooms. Excessive alcohol consumption may increase a person's requirement for biotin A deficiency may cause skin rash, hair loss, cholesterol and heart problems. 300 mcg RDA.

BOTANY: The scientific study of plant life. Scientific efforts to identify edible, medicinal, and poisonous plants make botany one of the oldest sciences. Today botanists study over half a million species of living organisms.

C **CALCIUM:** This is a dietary mineral needed for healthy bones, and proper function of muscles and nerves. It helps to clot blood when you are wounded. Calcium is the most abundant mineral in the body. Milk, yogurt, and cheese are rich sources of calcium, and vegetables, such as bok/pak choy, kale, and broccoli. When blood calcium levels drop too low, the vital mineral is 'borrowed' from the bones. It is returned to the bones from calcium supplied through the diet. The average person loses 400 mg of calcium every day. 2500 mg RDA.

CALORIE: Calories are usually associated with food, but they apply to anything containing energy. For example, a gallon of gasoline contains about 31,000,000 calories. When you hear something contains 100 calories, it's a way of describing how much energy your body could get from eating or drinking it. The body needs calories for energy. Eating too many calories and not burning enough of them off through activity can lead to weight gain. Calories should be represented in a nutrition fact label describing the components of processed food. An easy way to remember calorie content is that every gram of carbohydrates has 4 calories. Every gram of protein has 4 calories while a gram of fat has 9 calories. So, a 10 gram baked potato with 1 gram of butter is 49 calories. Your body needs some calories just to operate, to keep your heart beating and your lungs breathing. A youthful body also needs calories to grow and develop. To lose weight, women should consume 1200 calories per day and men 1600.

DID YOU KNOW? 1 pound of fat burns only 2 calories a day, but 1 pound of muscle burns 50 calories a day. The more muscle you build, your metabolism increases, and the more calories will burn while at rest. When muscle mass decreases, the metabolic rate also drops.

CARBOHYDRATES: When you eat carbs, the body breaks them down into simple sugars, absorbed into the bloodstream. As the sugar level rises in the body, the pancreas release a hormone called insulin. Insulin is needed to move sugar from the blood into the cells, where the sugar can be used as a source of energy. Simple carbohydrates, also called simple sugars, are found in re ined sugars, fruit, and milk. It's better to get your simple sugars from less refined food since they contain vitamins, fiber, and important nutrients like calcium. Complex carbohydrates are also called starches like grain products, bread, biscuits, pasta, and rice. Processing grains removes nutrients and fiber. Unrefined grains contain these valuable food stuffs with vitamins and minerals.

CARMATIVE: A substance that stops the formation of intestinal gas and helps expel gas that has already formed.

CAROTENE: see beta carotene

CATARRH: It consists of inflammation of the mucous membrane of the nose, sinuses, or nasal cavities. It is a common affliction, arising from repeated colds, damp living conditions, wet feet, insufficient clothing, or hot rooms. The symptoms are weariness, pains in the back and limbs, frontal headache, increased discharge from the nose, hoarseness, sore throat, impaired vision, fever, constant hawking, and cough. If the disease continues, partial or complete deafness may result. Constant dripping of the secretions into the throat, the catarrhal inflammation will extend to the mucous membrane of the throat and larynx, causing gastritis, tonsillitis, laryngitis, and bronchitis.

CATHARTIC: a substance which accelerates bowel movement, a laxative.

CHOLESTEROL: This is a yellowish fatty substance and one of the essential ingredients of the body with a bad reputation of being the major cause of heart disease. Every person with high blood cholesterol is regarded as a potential candidate for heart attack or a stroke. Most of the cholesterol found in the body is produced in the liver. LDL cholesterol is called 'bad' cholesterol because it forms a hard, thick substance called cholesterol plaque. HDL cholesterol, or 'good' cholesterol, appears to scour the walls of blood vessels, cleaning out excess cholesterol. However, about twenty to 39% generally comes from the food we eat.

Cholesterol is measured in milligrams per hundred millimeters of blood. The normal level of cholesterol varies between 150 - 200 mg per 100 ml. It is also caused by eating rich foods and fried foods, excessive consumption of milk and milk products like butter, and cheese. It is also raised by consuming white lour, sugar, cakes, pastries, biscuits; and non-vegetarian foods like meat, fish, and eggs. Cholesterol increases with smoking and drinking alcohol. Stress has also been found to be a major cause of high cholesterol. Lecithin is a fatty food substance that breaks up cholesterol into small particles easily handled by the body's system. With sufficient intake of lecithin, cholesterol cannot build up against the walls of the arteries and veins. Lecithin also increases the production of bile acids made from cholesterol, thereby reducing its amount in the blood. Egg yolk, vegetable oils, wholegrain cereals, soybeans, and unpasteurized milk are good sources of lecithin. It can also be taken in powder or capsules. It carries that excess cholesterol, which might cause heart disease back to the liver for processing. Over time, cholesterol plaque causes thickening of the artery walls and narrowing of the arteries, a process called atherosclerosis. To reduce the risk of heart disease, it is essential to lower the level of LDL and increase the level of HDL. This can be achieved by a change in diet and lifestyle. Virtually all foods from animals and coconut and palm vegetable oils, are high in saturated fats These should be replaced by polyunsaturated fats such as corn, safflower, soy bean, and sesame oils which can lower the level of LDL. Persons with high blood cholesterol should drink at least eight to ten glasses of water every day, to eliminate excess cholesterol from the system. Regular physical exercise promotes circulation and helps maintain blood flow to every part of the body. Gardening, jogging, swimming, bicycling are great exercise.

COLLAGEN: A type of fibrous protein, which connects and supports other bodily tissues such as skin, bone, tendons, muscles, and cartilage. It also supports the internal organs and is even in teeth. There are more than 25 types of collagens that naturally occur in the body and makes up about a quarter of the total amount of proteins in the body. Some people refer to collagen as the glue that holds the body together. Without it, the body would, fall apart.

COMPRESS: A compress is a cloth soaked in a water-based herbal infusion, decoction, or diluted tincture which can be held against the skin to relieve swelling, bruising, and pain, or to soothe headaches and cool fevers. Soak a towel in a hot herbal tea and lay it on the affected area. Be careful not to burn yourself when you wring out the towel thoroughly, or the 'patient' when you lay it on the area to be treated. Cover the compress with a dry towel. Leave it in place until it no longer feels warm, and then replace it with another. Keep the area under compresses for up to 30 minutes, depending on the condition and the herb used. Stop the application when the skin becomes uniformly lushed, a tingling sensation begins, or feels better. Stop if the patient feels discomfort. Some herbs are stimulating and warming like cayenne pepper or ginger root. They will increase circulation and energize areas of the body. Other herbs soothe, cool, and reduce fevers or swelling from sprains or bruises.

CONSERVE: A conserve is cooked mixture of fruit, or nuts heated until they become thick and heavily textured. To make a simple conserve put a handful of dried herbs in a small jar and cover the herbs with two inches of apple juice or cider. Add two cinnamon sticks to the jar and steep the mixture in the refrigerator overnight. By morning the herbs should have absorbed the liquid. Remove the cinnamon and add honey to your taste. Spread some on your toast or biscuits, and enjoy right away. Store the remaining conserve in the refrigerator.

CONSTIPATION: Constipation is defined as having a bowel movement fewer than three times per week. With constipation, stools are usually hard, dry, small in size, and difficult to eliminate. Some people who are constipated find it painful to have a bowel movement and often experience straining, bloating, and the sensation of a full bowel.

CURRIED: prepared in or flavored with an Indian-style sauce of hot-tasting spices.

D **DAHL, DAL, DAAL, or DHAL:** a preparation of dried lentils, peas, or beans stripped of their outer hulls and split. It also refers to the thick stew prepared from these.

DECOCTION: A standard quantity to make a 4 dose concoction is add 20 grams of dried or 40 grams fresh herbs to a pint and a half of cold water. Crush, chop, or bruise the herbs and place in a pan. Cover with cold water, bring to a boil and simmer for half an hour, or until the liquid is reduced by about a third. Strain into a clean container, cover, and store in a cool place until required. It is best used fresh that day. A decoction is made by boiling the hard and woody parts of herbs. Be sure to break up the bark or roots into small pieces; the smaller the better. More heat is needed in making decoctions than infusions because these parts of herbs are more difficult to extract active constituents and be absorbed by water.

DIABESITY: It's caused by problems with a chemical in the body (hormone) called insulin. It's often linked to being overweight or inactive, or having a family history of type 2 diabetes. Diabesity is a modern epidemic, which indicates the coexistence of both diabetes and obesity. Linked by various patho-physiological mechanisms, revolving around insulin resistance and hyperinsulinemia, diabesity has important diagnostic and therapeutic implications.

DIETARY FIBER: Dietary iber, or simply iber, refers to plant components not digestible by the body's digestive enzymes. Dietary fiber is classified into soluble dietary fiber and insoluble dietary fiber. Food sources rich in soluble dietary fiber are beans, lentils, Brussels sprouts, cabbage, apples, berries, and oat bran. Insoluble dietary fiber can be found in whole grain breakfast cereals, celery, and carrots. The rate and extent of fermentation of insoluble fiber in the colon is slower than soluble fiber. The body needs an average of about 35 grams a day.

DIURETICS: Diuretics are natural foods or herbs that increase the low of urine, and remove fluids from the body. It may also remove important vitamins and minerals. Foods like salt and sugar cause the body to retain considerable luids, and may cause water retention. Natural foods with high water content such as watermelons and cucumbers help increase urination and better lush out toxins. Cucumbers are rich in sulfur and silicon that stimulate the kidneys into better removal of uric acid. Watercress is a natural diuretic. Asparagus contains asparigine - a chemical alkaloid that boosts kidney performance, thereby improving waste removal from the body. Beets are natural diuretic foods that attack floating body fats and fatty deposits. Oats contain silica, which is a natural diuretic. Eating cabbage is known to break down fatty deposits, especially around the abdominal region. Carrots are a rich source of carotene that speeds the metabolic rate of the body and removes fat deposits and waste. Garlic is a natural diuretic food that aids the breakage of fat. Horseradish, raw onions, and radish speed up your metabolism.

DYSENTERY: A disease that inflames the intestines, especially the colon (large intestine), accompanied by fever, abdominal pains, low volume of diarrhea, and blood in the stool.

E **ECZEMA:** Inflammation of the skin, characterized by redness, itching, and lesions.

EMETIC: Induce vomiting. Controlling nausea and vomiting is anti-emetic Peppermints or peppermint tea, ginger biscuits or ginger beer may help some people.

EXFOLIATE: To remove old dead skin cells from skins surface by chemical or mechanical means.

EXPECTORANT: A medication that helps bring up mucus and other material from the lungs, bronchi, and trachea.

EXTRACT: Extracts and tinctures are more potent than decoctions or infusions, so smaller doses are necessary, in drops, not cups. To make an extract combine 3 ounces of fresh herbs with hundred-proof alcohol, preferably vodka, in an air tight bottle. Shake the extract twice a day to maintain the blend. It takes time for the active ingredients of the herb to be released into the alcohol. This should be good up to six weeks depending on the herb, but can last for over a year because alcohol acts as a preservative. If you prefer not to use alcohol, use vinegar.

F **FAMILY:** Botanical category of groups of similar genus, highest natural grouping.

FAT: This is one of the three nutrients the body uses as energy sources. The energy produced by fats is 9 calories per gram. Proteins and carbohydrates each provide 4 calories per gram. Total fat is the sum of saturated, monounsaturated, and polyunsaturated fats. Intake of monounsaturated and polyunsaturated fats can help reduce blood cholesterol when substituted for saturated fats in the diet.

FREE RADICALS: Free radicals are molecules produced when your body breaks down food, but can be caused by stress, alcohol, or by exposures to chemicals, tobacco smoke, or radiation. Free radicals can damage cells, and may play a role in heart disease, cancer, and other diseases. Substances such as vitamins C and E, and beta carotene, protect body cells from the damaging effects of oxidation caused by free radicals.

FLATUENCE: Excess gas in the intestinal tract.

FLAVONOIDS: Together with carotenes, flavonoids are responsible for the colors of fruits, vegetables, and herbs. They are found in most plant material especially fruits, tea, and soybeans. Green and black tea are a quarter flavonoids. They are known for their antioxidant activity in prevention of cancers and cardiovascular diseases.

FOLIC ACID / FOLATE: Folic acid is a vitamin, B9. It helps the body make healthy new cells. Everyone needs folic acid, but it is really important for pregnant women. Folic acid may reduce the risk of stroke. It can be found in leafy green vegetables, fruits, dried beans, peas, nuts, enriched breads, cereals, and other grain products.

G **GARGLES:** Gargling with herbal preparations is one way to soothe a sore throat, gum pain, and mouth sores such as ulcers. They are inexpensive and easy to prepare. Gargles made from herbal tinctures are one part herbal tincture to five parts clean water used four or more times daily as long as the condition persists. Sage tea is particularly good for mouth health. A pinch of salt provides a soothing element to herbal gargles.

GENUS: The biological classification of living and fossilized organisms. This designation is above species and below family.

H **HERBAL BATH:** Herbal baths include the use of various herbal additives to enhance the natural healing power of the water. There are baths to which plant decoctions or infusions have been added. There are full and partial herbal baths. For a full bath some of the medicinal plant parts should be sewn into a cloth bag and then boiled in a quart of water; the strained mixture is then added to the bath. Sometimes you can put the bag right into the tub for a more thorough extraction of the herbal properties.

HERBAL POWDER: Grind dried plant parts to a powder. The powder can be taken with water, milk, or soup.

HERBAL TEAS: To make a cup of herbal tea, for drinking or as a face wash, boil 10 ounces of water and pour over 1 TBS of dried herbs or a large pinch of fresh herbs. Add honey, cover and steep for 5 minutes. Strain and drink.

HOMEOPATHIC: Homeopathy is a system of natural medicine based on the belief that the body can cure itself. Its basic premise is: similia similibus curentur, or 'like cures like.' Practitioners are termed homeopaths, use tiny amounts of natural plants and minerals to stimulate the healing process. Homeopathy was developed in Germany, circa the late 1700s, and attributed to physician Samuel Hahnemann. It's common in many European countries where treatments, or remedies, are natural substances that are highly diluted.

I **IU:** An international panel of scientists certify that a quantity of each substance produces a certain biological effect for 1 International Unit.

INFUSION or TISANE: To make an infusion, fill a teapot with one quart of boiling water. Then throw in a large handful of fresh herbs, or an ounce or more of dried herbs. Add some honey, if desired, and let the mixture steep for 10-20 minutes. Strain and drink. Another easy way to make an herbal remedy is to bruise one ounce of dried flowers, leaves, or petals of the herb of your choice in a clean cloth. If you are using multiple herbs, the total amount used should equal 1 ounce. Then, pour 3 cups of boiling water over the herb. Cover and let steep for at least half an hour. Strain and drink at room temperature, or cold. Infusions generally will last in the refrigerator for 3 days. Dosage is in cups per day. Herb infused water preparations can be used in a number of ways; as a natural herbal bath infusion, skin rinse, hair rinse, mouthwash and gargle, herbal cleaning infusion, flea wash for cats and dogs, or as an ingredient in a more complex preparation. Use a glass or ceramic, copper, or stainless steel container. Aluminum, iron, tin or other metals will leach into the tea. Most herbalists recommend you use clean glass, ceramic, pottery, or an enameled pot. Use pure water. Fresh spring water or distilled water is best. Strain the finished tea before capping and storing. Refrigerate if kept for more than a few hours.

IRON: This is a dietary mineral needed for transportation of oxygen throughout the body. Iron is essential for the regulation of cell growth and differentiation. Iron deficiency results in fatigue, poor work performance, and low immunity. There are two forms of dietary iron: heme and nonheme. Heme iron is derived from hemoglobin, the protein in red blood cells that delivers oxygen to cells. Heme iron is in animal foods that originally contained hemoglo-bin, such as red meats, fish, and poultry. Iron in plant foods such as lentils and beans is arranged in a chemical structure called nonheme iron. This is the form of iron added to iron-enriched and iron-fortified foods. Heme iron is absorbed better than nonheme iron, but most dietary iron is nonheme iron. Rich sources of dietary iron include red meat, fish, poultry, lentils, beans, leaf vegetables, tofu, chickpeas, black-eyed peas, fortified bread, and breakfast cereals. 8 mg RDA.

L

LYCOPENE: Lycopene is a carotenoid found in red fruits like tomatoes and watermelons. Carotenoids are natural pigments that act as antioxidants for the body. Antioxidants serve to lessen the effects of free radicals, blamed by some in the scientific community for cell damage. Lycopene gets its name from the species classification of the tomato. Numerous studies have shown that eating foods high in lycopene is beneficial in warding off heart disease and several types of cancer such as lung, prostate, cervical, digestive tract, and breast. Good sources of lycopene are pink grapefruit, guava, watermelon, and rosehips. Tomatoes are the best source.

M

MG: milligram measurement abbreviation.

MCG: microgram measurement abbreviation.

MAGNESIUM: a dietary mineral needed for healthy muscle function and other processes in the body. Magnesium is the fourth most abundant mineral in the body. Approximately 50% of total body magnesium is found in bones. The other half is found predominantly inside cells of body tissues and organs. Only 1% of magnesium is found in blood, but the body works very hard to keep blood levels of magnesium constant. It helps maintain nor-mal muscle and nerve function, keeps heart rhythm steady, supports a healthy immune system, and keeps bones strong. Magnesium also helps regulate blood sugar levels, promotes normal blood pressure, and is known to be involved in energy metabolism and protein synthesis. Eating a wide variety of beans, nuts (especially peanut butter), whole grains, bananas, and vegetables will help you meet your daily dietary need for magnesium. 400 mg RDA

MANGANESE: Manganese is a mineral element that is nutritionally essential in the breakdown of amino acids and the production of energy. It is necessary for the metabolism of vitamins B-1 and E. It activates various enzymes important for proper digestion & utilization of foods. Manganese is a catalyst in the breakdown of fats and cholesterol. It helps nourish the nerves and brain, is necessary for normal skeletal development, helps maintain sex hormone production, and to regulate blood sugar levels. Manganese plays an important role in a number of physiological processes as a constituent of some enzymes and an activator of other enzymes. A manganese deficiency may cause joint pain, high blood sugar, bone and spinal disc problems, and or poor memory. Excellent sources of manganese include mustard greens, kale, chard, raspberries, pineapple, romaine lettuce, spinach, collard greens, turnip greens, kale, maple syrup, molasses, garlic, grapes, summer squash, strawberries, oats, green beans, brown rice, chick peas /garbanzo beans, ground cloves, cinnamon, thyme, peppermint, and turmeric. 2.5 to 5 mg RDA.

METABOLISM: This is our bodies' chemical reactions to convert fuel from food into the energy needed to do everything. Speci ic proteins in the body control the chemical reactions of metabolism, and each chemical reaction is coordinated with other body functions. In fact, thousands of metabolic reactions happen at the same time, all regulated by the body, to keep our cells healthy and working. To increase your metabolism, do not skip any meals. Eat six small meals daily. Starving will slow down your metabolism. Exercise daily and build your muscles. Avoid alcohol and sugar. Drink 10 glasses of water daily. Eat only nutritious food. A natural increaser, drink a mixture of 1 ounce of vinegar, honey with 2 crushed garlic cloves 3 times a day.

MINERALS: Dietary minerals are the chemical elements required by living organisms. In nutrition, minerals are elements which the body requires at least 100 mg of per day, and trace minerals are needed in smaller amounts. They are derived from the earth's crust when plants extract the minerals from the soil, and humans eat the plants. There are 7 major minerals. Calcium occurs mainly in the teeth and bones, but a small amount is found in blood plasma and other body fluids where it influences nerve transmission, blood clotting, and muscle contraction. Dairy products and green leafy vegetables are dietary sources of calcium, and an adequate intake of vitamin D is required for calcium absorption. Phosphorus is closely allied to calcium in bone and tooth formation and its association with vitamin D. It is present in every cell and is also found in dairy products. Magnesium is necessary for carbohydrate and protein metabolism, cell reproduction, and smooth muscle action. Food sources include nuts, soy beans, and cocoa. Sodium is in the skeleton and extracellular fluids and is necessary for fluid and acid-base balance, c ell permeability, and muscle function. Its main source is table salt, milk, and spinach. Potassium is found in intra - and extracellular fluid, plays a major role in fluid and electrolyte balance and in heart muscle activity, and is also required for carbohydrate metabolism and protein synthesis. Its sources include beans, whole grains, and bananas. Chlorine helps maintain normal fluid-electrolyte and acid-base balance, and in the stomach it provides the acidic environment necessary for digestion. Table salt is its main dietary source. Sulfur is important to the structure of proteins, is also necessary for energy metabolism, enzyme function, and detoxification. Sulfur is obtained from protein foods, such as meat, eggs, and beans. Some trace minerals are considered 'essential' in human nutrition and include iron, which is a constituent of hemoglobin; iodine, which is necessary for the thyroid gland, and cobalt, which is a component of vitamin B12. Other essential trace minerals are chromium, copper, fluorine, manganese, molybdenum, selenium, and zinc.

MONOUNSATURATED: Monounsaturated fat is considered to be probably the healthiest type of general fat, and may assist in reducing heart disease. Oils high in monounsaturated fats are better oils for cooking. Monounsaturated fat is believed to lower cholesterol and essential fatty acids for healthy skin and the development of body cells.

Monounsaturated fat is also believed to offer protection against certain cancers, like breast and colon cancer. Olive oil is considered the best because it has none of the adverse effects associated with saturated fats, trans-fats, or omega-6 polyunsaturated vegetable oils. Like polyunsaturated fat it remains stable at higher temperatures and does not easily become hydrogenated or saturated. Olive oil, hazelnuts, almonds, Brazil nuts, cashews, avocado, sesame seeds, and pumpkin seeds are the best sources.

N NAICIN: Niacin (nicotinic acid is a B3 vitamin that's used by your body to turn carbohydrates into energy. Niacin also helps to keep your nervous system, digestive system, skin, hair, and eyes healthy. Niacin can be found in chicken, liver, eggs, and milk. It is also available in nuts, seeds, beans, fruits, and vegetables like broccoli. Fish is one of the best sources. 35 mg RDA.

O OINTMENT / SALVE: An ointment, unguent, or salve is made by combining a decoction of an herb with olive oil and simmering it until the water has completely evaporated. A little beeswax is then added to get a irm consistency. An alternate way to make an ointment or salve is to add crushed dried herbs to olive oil and cook at very low heat in the oven for 3-4 hours. Then, strain into melted beeswax and quickly add to containers before the beeswax sets. For ointments, add one to one and a half ounces of melted beeswax (or tallow to any herb oil.

OMEGA: Two important polyunsaturated fatty acids are linoleic acid and alpha-linolenic acid. Linoleic acid is used to build omega-6 fatty acids and alpha-linolenic acid is used to build omega-3 fatty acids. These fatty acids cannot be synthesized in the body and must be supplied by the diet. They are called essential fatty acids. Omega-3 and omega-6 fatty acids are important in the normal functioning of all tissues of the body. You should include good sources of omega-3 and omega-6 each day. Pregnant women have an increased need for omega-3 and omega-6 fatty acids. They are needed for fetal growth, brain development, learning, and behavior. Lactating women should also increase their fatty acids intake, since infants receive their essential fatty acids through breast milk. Omega-6 fatty acid can be found in leafy vegetables, seeds, nuts, grains, and vegetable oils (corn, sa lower, soybean, cottonseed, sesame, and sun lower). Most diets provide adequate amounts of omega-6. Unless you eat a diet that is extremely low in fat, it is very easy to get more than enough omega-6. Supplementation of omega-6 is usually not necessary. It is obtained in oils as canola oil, walnut oil, wheat germ oil, and soybean oil. They're in green leafy vegetables like lettuce, broccoli, kale, and spinach, and in beans such as mung, kidney, navy, pinto, lima, peas, and split peas. They can be found in citrus fruits, melons, and cherries. Omega-3s are damaged by heat, so the oils should not be cooked. They are also damaged by oxidation; that's why you should store the oils in dark bottles in the refrigerator or freezer. The absolute best source of omega-3 are flaxseeds. One tablespoon of ground flaxseed will supply the daily requirement of omega-3. Flaxseeds need to be ground for your body to be able to absorb the omega-3 from them. Once flaxseeds are ground, the shells don't protect them anymore; then store them in the refrigerator or freezer, just like the oils.

P PARASITES: The major groups of parasites include one-cell organisms and parasitic worms such as hookworms. Parasites can get into the intestine through the mouth from uncooked or unwashed food, contaminated water, or by skin contact with larvae from infected soil. When the organisms are swallowed, they move into the intestine, where they can reproduce and cause symptoms. Children are particularly susceptible if they are not thoroughly sterilized after coming into contact with infected soil that is present in environments that they may frequently visit such as sandboxes and school playgrounds. People in developing countries are also at particular risk due to drinking water from sources that may be contaminated with parasites that colonize in the gastrointestinal tract. Extreme temperatures kill parasites and their eggs. Avoid drinking from natural waters without boiling. Always wash and peel fruits carefully. If there are any splits or laws, get rid of the bad parts. Salting might also help because it dries the parasite's body. Educate yourself about common parasites in your area. Avoid known parasite carriers such as mosquitoes.

PHOSPHORUS: Phosphorus is the second most-abundant mineral found in the human body, second only to calcium. It is very involved with bone and teeth formation as well as most metabolic actions in the body, including kidney functioning, cell growth, and the contraction of the heart muscle. It plays an important role in the body's utilization of carbohydrates and fats, and in the synthesis of protein for the growth, maintenance, and repair of cells and tissues. It is also crucial for the production of ATP, a molecule the body uses to store energy. The main food sources are the protein food groups of meat and milk. A meal plan that provides adequate amounts of calcium and protein also provides an adequate amount of phosphorus. Meat, poultry, and fish, as well as eggs, nuts, seeds, milk, carbonated soft drinks, broccoli, almonds, apples, carrots, asparagus, bran, brewer's yeast, and corn contain a good source of phosphorus. 700 mg RDA.

PHYTONUTRIENTS: These are substances derived from plants, such as a pigment, that is beneficial to health, especially one that is neither a vitamin, nor a mineral. These are chemicals in plants apart from vitamins, minerals, and macronutrients that have a beneficial effect on the body. There are hundreds, if not thousands of them, and they have effects such as antioxidants, boosting the immune system, anti-inflammatory, antiviral, an-tibacterial, and cellular repair. Highly colored vegetables and fruits tend to be highest in these chemicals, but tea, chocolate, nuts, and lax seeds are all excellent sources as well.

PLASTER: Make an herb paste as described in 'Poultice,' and place with-in folds of cheesecloth or muslin. Apply to the injured area. Cayenne and mustard powder are best applied as plasters rather than poultices so they don't touch the skin. Other herbs work well as plasters when you want an antiseptic and healing effect on an injury.

POLYUNSATURATED: Polyunsaturated and monounsaturated fats are the two unsaturated fats. Some examples of foods that contain these fats include fish, avocados, olives, walnuts, and liquid vegetable oils such as soybean, corn, safflower, canola, olive, and sun lower. Both polyunsaturated and monounsaturated fats may help lower your blood cholesterol level when you use them in place of saturated and trans-fats. Keep total fat intake to only a quarter of your total calories. They also include essential fats that your body needs, but can't produce as omega-6 and omega-3, which are crucial to brain function and normal growth and development of the body. Polyunsaturated fats - like all fats– contain 9 calories per gram. All fats are equally high in calories.

POTASSIUM: A dietary mineral needed for water balance and healthy muscle function in the body. 3500 mg RDA.

POULTICE: To make a poultice, crush the medicinal parts of the plant to a pulpy mass and heat. Mix with a hot, sticky substance such as moist lour or corn meal. Apply the pasty mixture directly to the skin. Wrap a hot towel around the a ected area and moisten the towel periodically. A poultice will draw impurities from the body. Some poultices require the herbs to be simmered irst for roughly 2 minutes. If you want a paste-like mixture, add lour or oatmeal. Then the excess liquid is squeezed out and the herbs applied to the area, bandaging them in place for up to 3 hours. To prevent the mixture from sticking to the skin apply a little oil such as olive oil to the area before applying the poultice. Another method is to fold crushed herbs in a surgical gauze or muslin to make a pack, place in a dish and pour on just enough boiling water to cover the pack. Soak for 3-5 minutes, drain o the water, allow the poultice to cool to a comfortable temperature and place on the affected area. To make a cold poultice, crush and bruise fresh herbs to make a paste, then spread on a piece of gauze and placed in the freezer for 10 minutes. Remove and place on affected area.

PROTEIN: The protein in our food contributes essential amino acids. Amino acids are used by cells to build new proteins and repair muscles. Protein food is not a high source of energy, however protein is essential in the right amount for proper functioning of our bodies. The amino acids then can be reused to make the proteins your body needs to maintain muscles, bones, blood, and body organs. Protein from animal sources, such as meat and milk, is called complete, because it contains all 9 of the essential amino acids. Most vegetable protein is considered incomplete because it lacks one or more of the essential amino acids. This can be a concern for someone who doesn't eat meat or dairy products. But people who eat a vegetarian diet can still get all their essential amino acids by eating a wide variety of proein-rich vegetable

The greatest medicine of all is teaching people how not to need it. ~ Hippocrates foods. For instance, you can't get all the amino acids you need from pea-nuts alone, but if you have peanut butter on whole-grain bread you're set. Likewise, red beans won't give you everything you need, but red beans and brown - unrefined rice will do the trick. The good news is that you don't have to eat all the essential amino acids in every meal. As long as you have a variety of protein sources throughout the day, your body will grab what it needs from each meal. Good sources of protein are meat, pumpkin seeds, peanuts, cheese, almonds, cashews, eggs, milk, beans, and brown rice. About 50 grams a day RDA.

R **RDA-RECOMMENDED DAILY ALLOWANCE:** (RDA) This is the daily dietary intake level of a nutrient considered su icient by the Food and Nutrition Board to meet the requirements of nearly all (97–98%) healthy individuals in each life-stage and gender group.

RIBOFLAVIN: Also known as vitamin B2, ribo lavin is an easily absorbed micro-nutrient with a key role in maintaining health in humans. It is a member of the water-soluble family of B-complex vitamins required for glucose metabolism so the body can produce energy from carbohydrates, normal red blood cell production, and general body growth. It prevents skin lesions and weight loss, and is necessary for the maintenance of good vision, skin, nails, and hair; alleviates eye fatigue; promotes general health. Exposure to light destroys ribo lavin. Milk, cheese, leafy green vegetables, liver, kidneys, legumes, tomatoes, yeast, mushrooms, and almonds are good sources of vitamin B2. Ribo lavin is yellow or yellow-orange in color and in addition to being used as a food coloring, it is also used to fortify some foods.

S **SAMBAL/SAMBOL:** A spicy condiment used in Indonesia, SE Asia or Sri Lanka that is made with chili peppers and other ingredients such as coconut or sugar.

SASHIMI: This is a delicacy consisting of very fresh raw fish or seafood sliced into thin pieces. Traditionally it's served with only soy sauce, wasabi paste, and ginger slices, but other simple garnishes are also popular.

SATURATED FATS: Saturated fat is the main dietary cause of high blood cholesterol. Saturated fat is found mostly in foods from animals and some plants. Foods from animals include beef, beef fat, veal, lamb, pork, lard, poultry fat, butter, cream, milk, cheeses, and other dairy products made from whole and two percent milk. All of these foods also contain dietary cholesterol. Foods from plants that con-tain saturated fat include coconut, coconut oil, palm oil, palm kernel oil, and cocoa butter.

SELENIUM: Selenium is a trace mineral essential to good health, but should be con-sumed only in small amounts. It makes special proteins, antioxidant enzymes, which play a role in preventing cell damage. Some medical information suggests that selenium may help prevent certain cancers. Selenium may boost fertility, especially among men. Selenium helps in maintaining healthy thyroid function. Brazil nuts are the best source of selenium. Fish, shell ish, red meat, grains, eggs, chicken, liver, and garlic are all good sources of selenium. Too much selenium in the blood can cause loss of hair, nail problems, nausea, irritability, fatigue, and mild nerve damage. 50 IU RDA.

SPECIES: A group of plants or animals having similar appearance. A rank in the classification of organisms, below genus. A group of organisms which resemble each other and can interbreed with each other.

STEAMS AND INHALATIONS: To inhale a facial steam, place a hand-ful of herbs in a wide bowl, pour on freshly boiled water. Then, using a towel draped over the back of your head, sit with your face at a comfortable distance from the water and inhale steam for at least 10 minutes, or as long as is comfortable. At the beginning, do not put your face too close to the water to begin with or the steam may scald you.

STEEP: This method of soaking herbs in water or other liquid helps to extract its flavor or essence. It is important to cover your ot while steeping to prevent the aromatic oils from evaporating into the air. A lid will cause the steam to condense back into the water. Boil the water first, then remove it from the heat and add the herb, or pour over the herb.

SUSHI: Rice with raw fish wrapped in seaweed

SYRUPS: FIRST METHOD: Syrups are made by using equal proportions of herbal infusions or decoctions with honey or unre ined sugar. Herbal infusions or decoctions used in syrups need to be brewed or simmered for longer than normal. Place the infusion or decoction in a saucepan together with the honey or sugar and gently heat, stirring until the honey or sugar has dissolved and the mixture has a thick, syrupy consistency. Remove from the heat and leave to cool. Once cooled, pour into a sterile glass bottle, seal with a cork, and store in a cool, dark place. The cork seal is important because syrups are prone to ferment and may explode if kept in a screw-lid topped bottle. A regular dose for syrups is 1-2 teaspoons taken 3 times daily.

SYRUPS - SECOND METHOD: Honey-based syrups preserve the healing qualities of some herbs and can be used to soothe sore throats and provide re-lief from coughs and colds. To make an herbal syrup, combine 2 ounces of dried herbs with 1 quart of water in a large pot. Boil down until it is reduced to one pint. Add 1-2 ta-blespoons of honey. Store all herbal syrups in the refrigerator for up to one month.

T

TEA: The standard quantity for a cup of herbal 'tea' is 1 teaspoon dried, or 2 teaspoons fresh herbs per cup of freshly boiled water. If you are making your herbal brew in a teapot warm the teapot first with water from the kettle just before it boils, add the ap-propriate quantity of herbs and pour on freshly boiled water. Put the lid on the teapot and leave to infuse for about 5 minutes, then strain into a cup and add honey, lemon, or spic-es to taste as desired. For medicinal brews use twice the standard amount depending on your chosen herb or remedy, and leave to infuse for longer, generally at least 10 minutes.

TBS: tablespoon

TS: teaspoon

THIAMIN: It is also known as vitamin B1 and is an energy building vitamin, which helps you to digest carbohydrates. It also keeps your heart and muscles stable. Assists in production of blood formation, carbohydrate metabolism, and a ects energy levels in the body. 1.2 mg RDA.

TINCTURES: Use 8 ounces dried or 12 ounces of fresh herbs to a quart of high proof alcohol - vodka, whiskey, brandy, or rum. A regular dose is a teaspoon diluted in water or fruit juice, taken 3 times daily. Place the herbs in a clean glass jar, pour in alcohol ensur-ing all the herb is covered put the lid on and shake. Leave in a cool dark place for at least 2 weeks shaking every day. It takes time for the active ingredients of the herb to be released into the alcohol. Strain and pour into clean glass bottles and store in a cool, dark place. Use a small, sterile, leak - proof, air tight bottle or jar. Tinctures can last for over a year. The alcohol acts as a preservative. If you prefer not to use alcohol use apple vinegar instead; or add the tincture when inished as above to one cup of warm water, which will cause most of the alcohol to evaporate. This will also dilute the bitter taste. However, the strength is also changed.

TONICS: Tonic wines are similar to a tincture. Fill a clean jar with the desired herbs then ill with wine. Forti ied port is considered the best to use. The herbs are completely cov-ered. Close securely and leave to mature for at least a month. Regularly top up the jar to ensure the herbs remain covered, replacing with a new batch of herbs as required. This should last for about 4 months. Discard the mixture if any mold occurs. A quicker method is to add the herbs and wine / port to a saucepan roughly 6 ounces of herbs to a quart of liquid. Cover with a lid and heat gently until the wine begins to simmer. Do not allow the mixture to boil unless you wish to eliminate the alcohol content. Remove from the heat and leave covered for a day. Strain and bottle.

TOPICAL: A topical medicinal agent is applied to a certain area of the skin and is intended to affect only the area to which it is applied. This can be applied under a plaster/band aid, or a gauze wrap. A topical's effects are limited to that area depending upon whether the agent stays under a wrap or plaster where it is applied.

U **ULCERS:** An ulcer is an open sore of the skin, eyes, or mucous membranes in the nose or mouth, often caused by an initial abrasion and maintained by an inflammation and/or an infection. Germs can enter an ulcer and make it difficult to heal. The most common ulcers to be sores that develop in the mucous membrane of the stomach, and are more frequent among women. Ulcers in the duodenum, the beginning portion of the small intestine, are more frequent in men. An ulcer develops when the area's ability to resist acids in gastric juice is reduced. It causes a burning ache and hunger-like pain. Ulcers can bleed, perforate the abdominal wall, or block the gastrointestinal tract. Stress and diet were blamed until a specific type of bacteria was discovered. This type of ulcer can be treated with combination drug therapy. Long-term use of aspirin and similar drugs were shown to be the two major causes and the best therapy is stop using those drugs. Cigarette smoking and alcohol consumption slows healing and promotes recurrence.

V **VITAMINS:** A lot of the vitamins in fruits and vegetables are lost between the gar-den and your plate. The longer the foods are stored before you eat them, the more nutrients are lost. Heat, light, and exposure to air all reduce the amount of vitamins, especially vitamin C, B1 - thiamin, and B9 - folic acid. Many people do not eat balanced meals that meet the requirements the body needs by not digesting enough nutrients to sustain the body's health and fuel factors. Research demonstrates almost all varieties of disease can be induced by the deficiency of vitamins, minerals, amino acids, and other nutrients. Vitamins are vital for your skin. The most important factor of nutritional deficiencies is the intense processing and reining of foods like cereals, and over-consumption of sugar. The human body uses food to manufacture all its building blocks as well as to provide fuel. To do this, it performs several thousand different chemical reactions. Water soluble vitamins, B-complex vitamins, and vitamin C are vitamins not stored in the body and must be replaced each day. These vitamins are easily destroyed or washed away during food storage and preparations. Fat soluble vitamins will not be lost by cooking; those the body doesn't need every day are stored in the liver when not used.

VITAMIN A: Vitamin A is referred to as retinol or carotene, and is a vitamin for growth and body repair. It is very vital in the formation of bone and tissues, and also keeps your skin smooth. And if you are night blind, the cure is having more vitamin A. This is a fat soluble vitamin. Vitamin A works together with vitamins D, B, E, zinc, phosphorus, and calcium. It also acts as an antioxidant essential during pregnancy and lactation. It may help protect against can-cer and other diseases. Night blindness occurs if deficient. Good sources of vitamin A are liver, carrots, mangos, papayas, spinach, eggs, dairy products, and sweet potatoes. 800 mcg RDA.

VITAMIN B-1: Also named thiamin, it is a water-soluble vitamin of the B complex, which converts food to energy, promotes healthy nerve function, muscle tone, and growth. It helps digest carbohydrates and keeps your heart and muscles stable. Good sources are baked potatoes, beef kidney and liver, pork, rye and whole grain lour, beans, oranges, oysters, peanuts, peas, raisins, wheat germ, and brown rice. 1.2 mg RDA.

VITAMIN B-2: Also called riboflavin, it is yellow or yellow-orange in color. In addition to being used as a food coloring, it is used to fortify some foods. Aids in formation of red blood cells and antibodies, essential for carbohydrate, protein, and fat metabolism, promotes general health, necessary for the maintenance of good skin, nails, hair and good vision, and maintains cell respiration. If deficient, a painful tongue and fissures occur at the corners of the mouth, and/or chapped lips. Good food sources are liver, eggs, whole grains, soybeans, and green leafy vegetables. 1.3 mg RDA.

VITAMIN B-3: Also referred as niacin, helps the body transform carbohydrates into energy and maintains nervous and digestive systems, skin, hair, and eyes healthy, reduces high

The physician treats, but nature heals.~ Hippocrates

blood pressure, improves circulation, and lowers cholesterol levels while increasing energy. If deficient pellagra - scaly skin sores appear. Good sources are meat, fish, whole grains, beans, nuts, and peas. 15 mg RDA

VITAMIN B-5: Also called pantothenic acid, is an essential nutrient, a water-soluble vitamin required to sustain life, needed to form coenzyme-A. It is critical in the metabolism and synthesis of carbohydrates, proteins, and fats. Aids in the utilization of vitamins, helps in cell building, development of the central nervous system, fights infections. It helps release energy from carbohydrates. If deficient, paresthesia develops, which is abnormal skin sensations as tingling, tickling, itching, or burning usually associated with nerve damage. Good sources are liver, eggs, peanuts, mushrooms, split peas, beans—especially soy beans, and whole grains. 5 IU MG RDA.

VITAMIN B-6: Also called pyridoxine, vitamin B6 is necessary for production of anti-bodies, building blocks of protein, necessary for synthesis and breakdown of amino acids. This vi-tamin promotes healthy skin, reduces muscle spasms and leg cramps, and helps the body maintain a proper balance of phosphorous and sodium. If deficient, anemia and fatigue develop. Good sources are fi tshuna, bananas, poultry, lean meat, whole grains, and potatoes. 1.5 mg RDA.

VITAMIN B-9: Also called folic acid and very important to the growth and reproduction of all body cells, including red blood cells. It is a water-soluble vitamin. Folic acid helps digestion, and the nervous system, improving mental as well as emotional health, and can be used to treat depression and anxiety. Deficiency may cause fatigue, acne, a sore tongue, and cracking at the corners of mouth. A severe deficiency may cause infertility or even sterility. Deficiency during pregnancy is associated with birth defects. Good sources are green leafy vegetables, dried beans, poultry, fortified cereals, whole grains, oranges, and nuts. 400 mcg RDA.

VITAMIN B-12: Also called cobalamin, vitamin B-12 is required for carbohydrate and fat metabolism. This is a water soluble vitamin. This vitamin is a must for children's growth, needed to make red blood cells and DNA, which is the genetic material in cells, and to keep nerve cells healthy. Prevents anemia by helping in formation and regeneration of red blood cells. It's necessary for fat, carbohydrate and protein metabolism, and increases energy. Good food sources are beef, fish, poultry, eggs, and dairy products. 2.4 mcg RDA.

VITAMIN C: Also known as ascorbic acid, vitamin C is water soluble and very essential as it protects your bones, teeth, and gums. It is the ultimate medicine for cur-ing scurvy and fights body infections. Without its support, collagen cannot be synthe-sized in the body. Vitamin C aids in absorption of iron. Good sources are citrus fruits, gua-va, cherries, tomatoes, melons, berries, sweet and hot peppers, and broccoli. 80 mg RDA

VITAMIN D: Vitamin D is also called cholecalciferol, a fat soluble vitamin and very important for children. The common disease seen in kids suffering from malnutri-tion is rickets. Bones cannot grow in a normal way if there is a lack of this vitamin. In adults, osteoporosis is caused due to lack of vitamin D. It is necessary for the reproduction of new skin cells. It is one of only three vitamins that are absorbed by the skin, as are vitamins A and E. It plays a key role in ensuring the absorption of calcium and phosphorus from the intestines. This maintains a stable nervous system. Direct sunlight is a natural source of vitamin D apart from spinach and vegetables. Good sources are egg yolks, fatty fish like herring and mackerel, and milk. This vitamin is also made in the skin when exposed to sunlight. 10 mcg RDA.

VITAMIN E: Also known as alpha-tocopherol, vitamin E is fat soluble, and good to treat cuts and wounds. It helps protect against aging and damage that may lead to cancer by supplying oxygen to the cells. It is an antioxidant vitamin that prevents cell damage by inhib-iting the oxidation of LDL cholesterol. Vitamin E may reduce the risk of heart disease. It is essential to prevent sterility. Prevents and dissolves blood clots. Good sources are vegetable oil, wheat germ, nuts, dark green vegetables, pumpkin, whole grains, and beans. 1000 mg RDA.

VITAMIN H: Also named biotin, vitamin H is a water soluble B complex vitamin essen-tial for metabolic reactions to synthesize fatty acids. The body needs biotin to metabolize carbohydrates, fats, and amino acids. Biotin is recommended to strengthen hair and nails, and it's found in many cosmetic products for hair and skin. Good sources are egg yolk, meat, dairy products, dark green vegetables, nuts, and bananas. It is also formed by microorganisms inside intestinal tract. Deficiency causes hair loss, dry skin, dermatitis, and fungal infections. 30 mcg RDA.

VITAMIN K: This is also referred as phytonadione. Vitamin K is fat soluble and is known as the clotting vitamin, because without it blood would not clot. Some studies indicate that it helps in maintaining strong bones in the elderly. It does this by helping the body transport calcium. Vitamin K plays an important role in the intestines and aids in converting glucose into glycogen for long term energy storage in the liver, while promoting a healthy liver function. Deficiency causes delayed clotting and hemorrhaging, bruising easily, and nosebleeds. Good sources are green leafy vegetables – spinach, avocados, meat –especially liver, cheese, asparagus, coffee, and green tea. Vitamin K is also made by the bacteria that line the gastrointestinal tract. 35 mcg RDA.

VITAMIN P: This is also called bioflavonoids or phytochemicals and Vitamin P is water soluble. It enhances the use of vitamin C by improving absorption and protecting it from oxidation. This vitamin promotes blood vessel health, including improving capillary strength. It prevents the accumulation of cholesterol plaque. Great sources of this vitamin are found in the edible pulp of fruits, green peppers, broccoli, and red wine. There are no daily recommended allowances for this vitamin.

W **WATER:** The most important nutrient for the human body is water. An average per-son can live for about 40-45 days without food, yet only 3-5 days without water. Our body's about 75% water. The average-sized person requires 3 quarts of water a day or 96 ounces. A good rule to follow is divide your body weight in half, and that's the number of ounces of water you should be drinking daily. If you are eating adequate amounts of fresh fruits and vegetables daily, you're already getting about one quart of water. Water is the major ingredient of all fluids in the body including saliva, gastric juice, bile, pancreatic juices, and intestinal secretions. Water is necessary for almost every bodily function. It helps carry essential nu-trients to all our cells, aids in circulation, digestion, and helps remove waste products. Drink at least one 8-ounce glass of water 30 minutes before each meal and then another 8-ounce glass about 2 hours after each meal. Before your largest meal of the day, drink at least 16 ounces of water and you will decrease your appetite and greatly help your digestion. Stay away from drinks with caffeine, such as coffee, tea, and sodas. Caffeine is a diuretic and ac-tually causes you to lose water, so it does more harm than good. Drink filtered water. There are three basic types of filters you can purchase, distillation, carbon filtration, and reverse osmosis. Carbon filters are the most common and least expensive. Change filters regularly.

Z **ZINC:** Zinc is needed for the body's immune system to properly work. It plays a role in cell division, cell growth, healing wounds, and the breakdown of carbohydrates. Zinc is also needed for the senses of smell and taste. Deficiency causes hair loss, poor appetite, and the reduced senses of taste and smell. High-protein foods contain high amounts of zinc. Beef, pork, and lamb contain more zinc than fish. The dark meat of a chicken has more zinc than the light meat. Other good sources of zinc are peanuts, peanut butter, and legumes. Fruits and vegetables are not good sources, because zinc in plant proteins is not as available for use by the body as the zinc from animal proteins. Therefore, low-protein diets and vegetarian diets tend to be low in zinc. 40 mg RDA.

All those vitamins aren't to keep death at bay, they're to keep deterioration at bay. ~ Jeanne Moreau

THE HAWAIIAN HOME GARDEN GUIDE
SOURCES AND REFERENCES

I began composing this work several years ago after realizing there wasn't a practical guidebook on how to create a home garden in a tropical climate. Raised on a farm in the Eastern US, I didn't realize until years later, we'd lived an organic, natural lifestyle. During my time, I've lived and worked in many tropical and sub-tropical areas: the Caribbean, Central America, Hawaii, Sri Lanka, and SE Asia. I always had a home garden for fresh herbs and vegetables, and fruit trees.

Initially, information came from acquaintances, neighbors, and friends. Usual-ly, useful information only came when a garden disaster struck. When luck smiled, with a good crop, I never asked questions. I studied grow box construction and attended every gov-ernment and agri-business course on specific fruit and vegetable courses that I could. I in-quired at libraries, agri-shops, government offices, and the Internet to locate free gardening lessons. I learned new and time-proven, common sense gardening techniques to get the most from garden labor. These courses provided knowledge on using garden chemicals safely.

Most importantly, I never stopped loving the garden. In heat or mud, the process of growing my own food directed me to other questions and more answers. The Internet has been the best information source for most of the crops. If you can correctly phrase the question, somewhere the Internet has the answer. Patience is the main investigative technique working with Mr. Google.

Trying to ascertain the correct name or term for a local fruit, vegetable, or root will always provide work for an earnest gardening detective. Locals have their own vocabulary, and terms may be limited to a family, village, or a section of the country, while the most common name is very different. To avoid confusion, always try to rst ascertain the botanical reference name of the particular fruit or vegetable. This may be difficult and frustrating, but you will eventually succeed. Please remember, if gardens created are at similar latitude, altitude, climate patterns, and reasonable soil, the same crops will grow. Information about one crop is available from many sources in many countries. Purdue University Center For New Crop and Plant Products is my first go-to source for most of my Internet garden info searches. If you can ask the food/crop question with reasonable skill, it's likely this site already has someone organizing the existing research information. Several international universities are accessible for information. Deciding to deal with sources of other countries at the same latitude as the Hawaiian Islands, I found agricultural contacts at the universities, blogs, magazines, and newspapers of Hawaii, Florida, and throughout the Caribbean. There are several international web sites dedicated to tropical plants, fruits, and vegetables. These are too numerous to list. Remember, you'll never know everything about the garden and you'll have to ask someone. Here are some places to look.

Books that were early influences that I recommend:
BACK TO EDEN by Jethro Kloss - Healing Herbs,
Home Remedies, Diet and Health, 1971 USA.

THE FOLK REMEDY ENCYCLOPEDIA:
Olive Oil, Vinegar, Honey and 1,001 Other Home Remedies,
by Frank K. Wood and the Members of FC&A Medical Publishing 2003 USA

INTERNET RESOURCES:
PURDUE UNIVERSITY CENTER FOR NEW CROP AND PLANT
PRODUCTS: *http://www.hort.purdue.edu/newcrop/*

AGRICULTURAL RESOURCES AND CAPABILITIES OF HAWAII BY WM C. STUBBS 1901: https://thishawaiilife.com

AMERICAN SOCIETY for HORTICULTURAL SCIENCE: https://journals.ashs.org/hortsci

ATLAS of FLORIDA PLANTS – INSTITUTE OF SYSTEMATIC BOTANY https://florida.plantatlas.usf.edu/

AYURVEDIC MEDICINAL PLANTS OF SRI LANKA COMPENDIUM Sponsored by Barberyn Ayurvedic Resorts and the University of Ruhuna http://www.instituteofayurveda.org/plants/plants_list.php?s

BIG ISLAND NOW: https://bigislandnow.com/

BOTANICAL GROWER'S NETWORK: http://botanicalgrowersnetwork.net

CENTRE FOR AGRICULTURE AND BIOSCIENCES INTERNATIONAL – CABI:https://www.cabi.org/tag/centre-for-agriculture-and-biosciences-international/

COMMERCIAL VEGETABLE PRODUCTION IN HAWAII IN 1932: https://www.ctahr.hawaii.edu/oc/freepubs/pdf/EB-16.pdf

COMMON FOREST TREES OF HAWAII (NATIVE AND INTRODUCED.) by Elbert L. Little Jr. & Roger G. Skolmen US Dept. of Agriculture Forest Service 1989.PLANTS FOR A FUTURE: https://pfaf.org/

CROP KNOWLEDGE MASTER: http://www.extento.hawaii.edu › kbase › crop › crop

DENGARDEN: https://dengarden.com

EDIBLE HAWAIIAN ISLANDS: https://ediblehi.com/

EVERYTHING WHAT: https://everythingwhat.com/

FLORICULTURE FRUIT & VEGETABLE INDUSTRY IN SRI LANKA: POTENTIALS and OPPORTUNITIES: http://www.bureauleeters.nl/data/105-5NeF8Ldu51dL/horticulture-study-sri-lanka-2016.pdf

FOOD AND AGRICULTURE ORGANIZATION OF THE UNITED NATIONS – FAO: http://www.fao.org/home/en

GARDEN NOTES FROM HAWAII: http://garden-notes-from-hawaii.blogspot.com

GARDENING KNOW HOW: https://www.gardeningknowhow.com/

GARDENING SOLUTIONS UNIVERSITY OF FLORIDA: https://gardeningsolutions.ifas.ufl.edu/

GROW A GARDENER: https://growagardener.blogspot.com/

HAWAII HORTICULTURE: http//hihort.blogspot.com//

HAWAII TRIBUNE-HERALD: https://www.hawaiitribune-herald.com

HAWAIIAN TROPICAL PLANT NURSERY:
http://www.hawaiiantropicalplants.com/

HAWAIISCAPE: https://www.hawaiiscape.com

HOME GARDENING IN HAWAII – CATHR:
https://www.ctahr.hawaii.edu › freepubs

HONOLULU STAR-ADVERTSER: http://www.staradvertiser.com

INDIA BIODIVERSITY PORTAL: https://indiabiodiversity.org/

KEOLA MAGAZINE: https://keolamagazine.com/

KAUAI HAWAII: //www.kauai-hawaii.com/

LEGUMINOUS CROPS FOR HAWAII F. G. KRAUSS, Agronomist. OFFICE OF EXPERIMENT STATIONS, U. S. DEPARTMENT OF AGRICULTUHE.1911
https://scholarspace.manoa.hawaii.edu/bitstream/10125/25507/1/B-23.pdf

MARY'S HEIRLOOM SEEDS: https://www.marysheirloomseeds.com/

MID WEEK OAHU: http://www.midweek.com

NATIONAL IPM DATABASE: https://ipmdata.ipmcenters.org/

NICK SAKOVICH: http://www.gardenguyhawaii.com

PACIFIC ISLANDS ECOSYSTEMS AT RISK (PIER): http://www.hear.org/pier/

PLANTS OF SOUTHEAST ASIA: https://asianplant.**net**

REAEARCHGATE: https://www.researchgate.net

SCIENCE DIRECT:
https://www.sciencedirect.com/topics/agricultural-and-biological-sciences

SINGAPORE INFOPEDIA - National Library Board, Singapore:
https://eresources.nlb.gov.sg/infopedia/

SRI LANKA AGRI DEVELOPMENT - MINISTRY of AGRICULTURE https://www.agrimin.gov.lk/web/index.php/home-1/14-divisions/16-agriculture-development-ta

STAR ADVERTISER: https://www.staradvertiser.com/

STARR ENVIRONMENTAL: http://www.starrenvironmental.com/images/

STATE OF HAWAII DEPARTMENT OF AGRICULTURE:
https://hdoa.hawaii.gov/pi/files

THE GARDENING DAD:
https://thegardeningdad.com

THE MAUI NEWS: https://www.mauinews.com/

THE WEST HAWAII TODAY: https//the westhawaiitoday.com

TROPICAL. INFERNS – USEFUL TROPICAL PLANTS:
https://tropical.theferns.info/

TROPICAL SELF-SUFFICIENCY: https://tropicalselfsufficiency.com/

ULUKAU – THE ELECTRONIC HAWAIAN DICYIONARY: https://ulukau.org/

USDA PLANTS DATA BASE: https://plants.usda.gov › java

VEGETABLE GARDENING IN THE CARIBBEAN AREA:
https://naldc.nal.usda.gov/download/CAT87209120/PD

US NATIONAL LIBRARY OF MEDICINE NATIONAL INSTITUE OF HEALTH: https://www.ncbi.nlm.nih.gov/pmc/articles

UNIVERSITY OF HAWAII at MANOA – COLLEGE OF TROPICAL AGRI CULTURE AND HUMAN RESOURCES: https://cms.ctahr.hawaii.edu/

UNIVERSITY OF FLORIDA - GARDENING SOLUTIONS – INSTITUTE OF FOOD – GARDENING SOLUTIONS: https://gardeningsolutions.ifas.ufl.edu/

SOUTH FLORIDA PLANT GUIDE
: https://www.south-florida-plant-guide.com/florida-gardening-tips.html

WIKIPEDIA: https://www.wikipedia.org

WEBMD.COM: https://www.webmd.com/

YOUR FLORIDA GARDEN: https://upf.com/book.asp?id=9780813008622

Gardening adds years to your life and life to your years.
– Unknown

Agriculture is civilization.

~ E. Emmons

We have neglected the truth that a good farmer is a craftsman of the highest order, a kind of artist.

~ Wendell Barry

Remember that children, marriages, and gardens reflect the kind of care they get.

~ Jackson Brown Jr.

Made in United States
Orlando, FL
04 November 2026

90201561R00321